Molecular Biology of RNA

SECOND EDITION

David Elliott
Institute of Genetic Medicine, Newcastle University

Michael Ladomery
Faculty of Health and Applied Science, University of the West of England, Bristol

OXFORD
UNIVERSITY PRESS

OXFORD
UNIVERSITY PRESS

Great Clarendon Street, Oxford, OX2 6DP,
United Kingdom

Oxford University Press is a department of the University of Oxford.
It furthers the University's objective of excellence in research, scholarship,
and education by publishing worldwide. Oxford is a registered trade mark of
Oxford University Press in the UK and in certain other countries

First edition 2011

Published in the United States of America by Oxford University Press
198 Madison Avenue, New York, NY 10016, United States of America

British Library Cataloguing in Publication Data
Data available

Library of Congress Control Number: 2015938215

ISBN 978-0-19-967139-7

Printed in Great Britain by
Ashford Colour Press Ltd, Gosport, Hampshire

Molecular Biology of RNA

CONTENTS

Acknowledgements ix
Foreword x

1 Introduction to *Molecular Biology of RNA* 1
1.1 Aims of this book 4
1.2 DNA and RNA are composed of slightly different building blocks 4
1.3 Nucleotides are joined together through a phosphodiester backbone to give nucleotide chains 8

2 RNA can form versatile structures 11
2.1 How do RNA molecules form structures? 11
2.2 RNA secondary structure: RNA molecules tend to form a number of shorter helices compared with DNA 13
2.3 RNA and DNA form different kinds of double helix 17
2.4 Five common secondary structure motifs are found within RNA molecules 18
2.5 RNA secondary structures can be worked out experimentally and predicted bioinformatically 19
2.6 The formation of RNA helices is stimulated by positively charged molecules and particularly metal ions 23
2.7 RNA molecules use a set of strategies to build tertiary structures 24
2.8 Summary of how RNAs build structured molecules 27
2.9 RNA structures can be used as thermosensors 27
2.10 RNA structures can be selected which bind to target molecules 28
2.11 Riboswitches are shape-changing RNAs which can flip gene expression patterns on binding specific target molecules 31

3 Catalytic RNAs 36
3.1 Three properties of RNA enable the catalytic function of ribozymes 36
3.2 What kinds of reactions do ribozymes catalyse? 38
3.3 Ribozymes were first discovered through serendipity 40
3.4 Group I introns are spliced through a two-step mechanism which uses metal ions in their active sites 41
3.5 Metal ions play a key role in catalysis by Group I introns 42
3.6 Group II introns are also spliced through a two-step mechanism 43
3.7 RNA is inherently chemically unstable because of its 2′ -OH group 45
3.8 Small ribonucleolytic ribozymes catalyse their own cleavage 46
3.9 The hammerhead ribozyme 48
3.10 The HDV ribozyme 50
3.11 Are ribozymes true catalysts? 52
3.12 The RNA world hypothesis: a time when RNA was used as a genetic material 52
3.13 Experiments have been carried out that might model the early steps that might have occurred during the evolution of life 53

4 The RNA-binding proteins 60
4.1 The RNA recognition motif (RRM) 62
4.2 The K-homology (KH) domain 65
4.3 The cold-shock domain 66
4.4 Double-stranded RNA-binding proteins 69
4.5 The zinc-finger domain 71
4.6 Other RNA-binding domains 73
4.7 Investigating protein-RNA interactions 76

5 Pre-mRNA splicing by the spliceosome 84
5.1 RNA splicing was discovered in a virus 84
5.2 Spliceosomal introns are critical for efficient eukaryotic gene expression 85
5.3 Introns enhance eukaryotic gene expression at several levels 88
5.4 Introns have an important role in evolution 89
5.5 The mechanism of pre-mRNA splicing 90
5.6 Splice sites 91
5.7 The spliceosome 93

5.8 Spliceosomes assemble and disassemble on each intron to be removed in a spliceosome cycle 94
5.9 How the spliceosome works 96
5.10 The spliceosome cycle has been worked out using in vitro extracts 100
5.11 A minor class of eukaryotic spliceosomal introns have different splice sites 101
5.12 Spliceosomes can assemble through intron and exon definition 103
5.13 *Trans*-splicing is common in trypanosome parasites and in the nematode *C. elegans*, where it enables efficient translation 106

6 Regulated alternative splicing 111
6.1 There are several different types of alternative splicing 112
6.2 How frequent is alternative splicing? 112
6.3 How exons are recognized by the splicing machinery 115
6.4 Exon and intron definition control the type of alternative splicing that operates 119
6.5 Three main factors can contribute to alternative splicing regulation 121
6.6 Regulation of RNA splicing is controlled by changes in the concentration of RNA-binding proteins 121
6.7 Signal transduction pathways can regulate alternative splicing by changing the function and location of splicing factors 126
6.8 Transcription elongation speeds can regulate alternative splicing choices 127
6.9 Transcription can also modulate splicing pathways via the recruitment of cofactors 130
6.10 Alternative splicing is critical for normal animal development 130
6.11 RNA splicing regulators play a critical role in nervous system development in animals 134

7 Pre-mRNA splicing defects in development and disease 138
7.1 Mutations affecting the splicing code can be catastrophic for gene function 138
7.2 Mutations in splicing control sequences are very frequent causes of human genetic disease 140
7.3 Genetic mutations create a new splice site in a premature ageing disease 141
7.4 Mutation of an exonic splicing enhancer in a DNA damage control gene leads to breast cancer 144
7.5 How are mutations that cause splicing defects diagnosed? 144
7.6 Diseases caused by mutations affecting components of the spliceosome 147
7.7 Genes encoding important spliceosomal proteins are mutated in patients with the eye disease retinitis pigmentosa (RP) 148
7.8 The genes encoding splicing proteins can become mutated in some kinds of cancer 151
7.9 Splicing changes can change the properties of cancer cells 152
7.10 Diseases caused by mis-expression of levels of splicing factors 155
7.11 Splicing as a target to treat cancer 158
7.12 Manipulating pre-mRNA splicing offers a route to treating muscular dystrophy 158
7.13 Splicing as a route to therapy for infectious diseases like AIDS 162

8 Co-transcriptional pre-mRNA processing 166
8.1 Transcription and the RNA polymerases 166
8.2 Formation of the ends of an mRNA 169
8.3 The C-terminal domain (CTD) of RNA polymerase II 172
8.4 The links between splicing, transcription, and chromatin 173
8.5 The spatial organization of pre-mRNA processing 178
8.6 Histone mRNA 3' end formation 181

9 Nucleocytoplasmic traffic of messenger RNA 186
9.1 Step 1: mRNAs are 'dressed for export' as they are synthesized by the addition of nuclear export adaptors 188
9.2 Step 2: mRNA transcripts reach the nuclear pore by random nuclear diffusion 193
9.3 Step 3: Transit through the nuclear pore requires addition of nuclear export receptors 194
9.4 Step 4: Disassembly of the export competent mRNP 199
9.5 Step 5: Export receptors shuttle between the nucleus and the cytoplasm 200
9.6 mRNA export can be hijacked by some viruses 201
9.7 mRNA export can become defective in human diseases 202

10 Messenger RNA localization 205
10.1 The need for mRNA localization 205
10.2 The machinery of mRNA localization 207
10.3 Classical examples of mRNA localization in development 209

10.4 Localization of mRNA in differentiated somatic cells 212
10.5 Localization of mRNA in algae and plants 217

11 Translation of messenger RNA 222
11.1 What is translation? 222
11.2 The structure and function of the ribosome 222
11.3 Deciphering the genetic code 224
11.4 The three phases of translation 229
11.5 Regulation of mRNA translation 234
11.6 Masked messages 240
11.7 Manipulating translation 244

12 Stability and degradation of mRNA 250
12.1 Messenger RNAs have a half-life 250
12.2 Sites and mechanisms of mRNA degradation 252
12.3 The process of mRNA degradation 253
12.4 Extracellular stimuli influence the stability of mRNA 259
12.5 Nonsense-mediated, non-stop, and no-go mRNA decay 260
12.6 Degradation of mRNA in bacteria and plants 264

13 RNA editing 268
13.1 What is RNA editing and why might it exist? 268
13.2 A→I editing takes place by modification of adenosine through removal of an amino group 269
13.3 The biological consequences of A→I RNA editing: adenosine and inosine form different base pairs in RNA secondary structure 271
13.4 What does A→I mRNA editing do? 272
13.5 A→I editing plays an important role in the function of tRNAs 279
13.6 C→U RNA editing takes place through base deamination (removal of an amino group) of cytidine 281
13.7 C→U RNA editing creates two different forms of the *APOB* mRNA in different tissues, and was the first RNA editing reaction to be discovered in animals 281
13.8 *APOB* mRNAs are editing by an RNA editing complex containing the cytidine deaminase ApoBec1 283
13.9 ApoBec proteins play an important role in innate immunity to retroviruses like HIV and in generating an antibody response 284
13.10 Trypanosome mitochondrial RNA is edited by base insertions and deletions to create ORFs from frameshifted transcripts 287
13.11 RNA editing was discovered in trypanosomes by sequencing cDNAs encoded by mitochondrial genes 289
13.12 Short RNAs called guide RNAs target trypanosome mitochondrial RNA editing 289
13.13 Guide RNAs are used as a template for RNA editing through uridine insertions and deletions 291
13.14 Trypanosome mitochondrial RNA editing requires nuclear-encoded proteins which might be useful therapeutic targets 291

14 The biogenesis and nucleocytoplasmic traffic of non-coding RNAs 295
14.1 The snoRNAs and scaRNAs: multiple roles in RNA biogenesis 296
14.2 Structure and function of the nucleolus 302
14.3 Processing of tRNA and of mitochondrial transcripts 304
14.4 SMN proteins and snRNP assembly 309
14.5 Nucleocytoplasmic traffic of non-coding RNA 314
14.6 Retroviruses have hijacked the RNA export machinery to assist in the export of partially processed mRNAs 328

15 The 'macro' RNAs: long non-coding RNAs and epigenetics 334
15.1 Epigenetic regulation and the epigenetic code 334
15.2 Long ncRNAs are involved in epigenetic gene regulation of gene expression 337
15.3 A long ncRNA called *XIST* epigenetically regulates the inactive X chromosome in female mammals 338
15.4 The X inactivation centre contains a number of non-coding RNAs as well as *XIST* 341
15.5 Non-placental mammals also use a long non-coding RNA to inactivate an X chromosome in females 342
15.6 Fruit flies use a long ncRNA to upregulate expression from a single male X chromosome 343
15.7 The logic of dosage compensation strategies used in flies and mammals 344
15.8 Genetic imprinting uses long non-coding RNAs 345
15.9 Transcription of the *H19* long non-coding RNA acts as a decoy for transcription of the *IGF2* gene 347
15.10 The *AIRN* ncRNA epigenetically represses *IGF2R* gene expression by directing epigenetic chromatin modification 348

15.11 Long ncRNAs play an essential role in establishing animal body plans 349
15.12 Long ncRNAs are involved in transcriptional enhancer function 351
15.13 Antisense RNAs 352

16 The short non-coding RNAs and gene silencing 358
16.1 Key concepts and common pathways 358
16.2 Discovery and mechanism of RNA interference 363
16.3 The uses of RNA interference 366
16.4 Discovery, biogenesis, and developmental roles of microRNAs 370
16.5 Transcriptional silencing by non-coding RNAs in the centromere 376
16.6 RNA-induced transcriptional silencing of transposons 379

17 RNA biology: future perspectives 390
17.1 The emergence of transcriptomics 390
17.2 The growing prominence of non-coding RNAs 393
17.3 RNA-guided genome editing 397
17.4 Concluding remarks 400

Glossary 403
Index 415

ACKNOWLEDGEMENTS

I would especially like to thank Michael Ladomery and Alice Roberts for their hard work, Anne-Marie, Jamie, and Emily for their patience, Jane, Jonathan James, and Aline Elliott, Dorothy Ianzito, and my research group and students present and past. All are much appreciated.

David Elliott
Newcastle, January 2015

Writing a textbook is a significant and exciting undertaking fraught with many challenges. I would like to acknowledge the following people: Len Kelly, my BSc(Hons) supervisor while at the University of Melbourne, who first got me interested in the amazing world of genetics, and John Sommerville at the University of St Andrews, who opened the doors to the wonderful world of RNA biology. I am also deeply indebted to Alice Roberts at Oxford University Press and to my co-author and friend, David Elliott. Their seemingly infinite patience and enthusiasm during this project is something I will always remember. I would also like to thank Cyril Dominguez for some wonderful images of RNA-protein complexes, and Oliver Rackham, Tilman Sanchez-Elsner, and Nicola Gray for help and advice. Lastly I would like to thank friends and family for their continued and much appreciated encouragement.

Michael Ladomery
Bristol, January 2015

FOREWORD

It is a truly remarkable time to be a molecular biologist. Over the last few decades we have witnessed an amazing explosion of knowledge, to the extent that particular areas of molecular biology warrant a dedicated textbook. Several years ago we decided to write a textbook that, for the first time, captured systematically the field of RNA biology. RNA is more ancient than DNA; indeed, according to the *RNA world hypothesis* we are all descended from life forms that relied exclusively on RNA to transmit genetic information. RNA is also structurally versatile, and is able to form three-dimensional structures that even have catalytic properties. It is now abundantly clear that RNA-based processes contribute significantly to the regulation of gene expression through, amongst others, regulated pre-mRNA processing (in particular alternative splicing) and mRNA translation, localization, and stability. A bewildering assortment of non-coding RNAs, which play important regulatory roles in cells, have also been discovered. At the same time, it is clear that aberrations in RNA-based processes are linked to disease.

We are very aware that the field is developing fast. Our intention in the second edition has been to update all chapters with some of the most recent developments. It is of course an impossible task to capture everything; therefore we apologize to any colleagues whose important work has not been mentioned. We have also reorganized the order of the chapters so that they flow better. Each chapter now includes a set of questions that will help students frame their revision. In response to feedback, we have also included some colour illustrations in the middle of the textbook.

The opening chapter now includes a more comprehensive basic introduction for the benefit of newcomers to the field, as well as an overview of the textbook. Specific updates include additional RNA binding domains, more material on the connections between epigenetics and RNA biology, and further examples of non-coding RNAs. We have also added a new chapter at the end of the book ('RNA biology: future perspectives') which covers the challenges of next-generation sequencing and transcriptomics, the growing prominence of non-coding RNAs, and RNA-based genome editing technology. Throughout the textbook we have endeavoured to highlight key discoveries and Nobel Prize winners in the field. We have also underlined the biomedical importance of RNA-based processes, illustrating how they can be relevant in the development of novel therapies. We hope that the second edition will continue to help students of RNA biology grapple with this complex subject, and that they will experience, as we have, the thrill and excitement of this fascinating field.

David Elliott and Michael Ladomery

January 2015

1 Introduction to *Molecular Biology of RNA*

Introduction

The ability to sequence entire genomes has revolutionized molecular biology. Genome sequencing would have been considered science fiction during the 1970s and 1980s, at a time when molecular biology was already advancing by leaps and bounds. Yet, despite this incredible achievement, we still do not fully understand how genes are expressed in a regulated fashion. An appreciation of how genes are regulated is fundamental to understanding not only normal development but also the causes of many important diseases.

There is an interesting problem that has arisen from the Human Genome Sequencing Project. The number of protein-coding genes is relatively small (around 21 000), yet the number of actual proteins (the proteome) is vastly larger (perhaps in excess of a million proteins). How can this be possible? The answer to this question lies to a large extent in the molecular biology of RNA. RNA (ribonucleic acid) is more ancient than DNA (deoxyribonucleic acid). It is widely thought that before DNA evolved, RNA was responsible for carrying genetic information (the **RNA world hypothesis**).[1] According to this hypothesis, we descend from an ancient life form consisting of simple replicating microbes (**ribocytes**) in which fundamental biochemical processes were entirely dependent on RNA. Support for this theory comes from the fact that RNA can form complex three-dimensional structures which include enzymes known as **ribozymes**. Other compelling facts also support this theory. Adenine, one of the basic components of RNA, is a pentamer of the very simple molecule hydrogen cyanide and can form in prebiotic conditions.[2] Ribozymes that self-replicate have been generated in the laboratory through a process called in vitro evolution.[3] Interestingly, RNA still works as a genome for several important viruses, including those that cause AIDS, influenza, and haemorrhagic fever. An example of the latter is Ebola virus, which causes a life-threatening disease characterized by catastrophic bleeding. Figure 1.1 shows a cell infected with Ebola virus.

While the responsibility for carrying genetic information is now fulfilled by DNA, DNA itself still needs to be copied into RNA through the process of transcription for a gene to be expressed. While the regulation of DNA transcription establishes which genes are switched on and off, much of the complexity of the proteome is generated through processes that occur downstream of transcription—as we shall see, at the level of RNA, whereby a single gene can give rise to multiple protein products.

There has been an explosion in RNA research over the last few years. This is illustrated by the emergence of several international RNA meetings dedicated to RNA research (for example, see http://www.rnasociety.org/). Yet despite the exponential growth in RNA research, most textbooks are still catching up. Thus, RNA biology has not been given appropriate coverage (compared with DNA and epigenetic and transcriptional processes). The purpose of this textbook is to redress the balance by providing students with a comprehensive yet concisely written account of the complexity of RNA biology.

We start with an overview of the biochemical properties of RNA, including the versatility of its structure

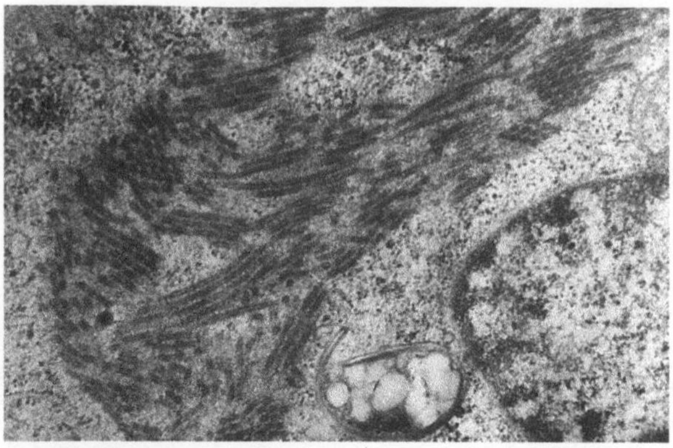

Figure 1.1 Kidney epithelial cells infected with the Ebola virus. The name Ebola is derived from a river in the Democratic Republic of Congo. Ebola virus is a member of a family of viruses called Filoviridae. The name of the family is derived from the Latin *filum*, meaning thread, and refers to the filamentous nature of the virus. Note the stacks of viral particles in this heavily infected cell. The particles contain a viral genome of around 19 000 nucleotides of RNA. Image courtesy of the Center for Disease Control and Prevention/Dr Stan Foster, B. Partin, Department of Health and Human Services, USA.

which enables a multitude of RNA molecules and functions in cells. RNA-binding proteins are at the heart of the processes we describe. Therefore we describe the different types of RNA-binding domains that have evolved. Further, we describe the co-transcriptional processes of capping, and pre-mRNA (pre-messenger RNA) splicing, cleavage, and polyadenylation. We present a detailed discussion of the process of alternative splicing, a major generator of proteomic diversity. We cover the processes of RNA editing, nucleocytoplasmic traffic, mRNA localization, translation, stability, and decay, and rRNA (ribosomal RNA) and tRNA (transfer RNA) processing.

Evidence suggests that up to 85% of the human genome is transcribed into RNA.[4] The collection of all RNA transcripts is known as the **transcriptome**. A surprisingly large proportion of the transcriptome is non-coding (in other words, not translated into amino acid sequences).[5,6] The following facts clearly illustrate the point. In humans the ratio of non-coding to coding sequences is 47:1! By comparison, the ratio is 43:1 in mice, 2.4:1 in *Drosophila melanogaster* (fruit fly), and 1.3:1 in *Caenorhabditis elegans* (nematode). The proportion of non-coding RNA varies across species but it is consistently significant.[3] This abundance of non-coding transcribed RNA is not merely a molecular accident due to transcriptional noise, but rather an additional and important way to regulate gene expression. To illustrate this, towards the end of the book we explore the fascinating and increasingly complex collection of non-coding RNAs and their role in gene expression.

Throughout the text we provide historical perspective, explaining the roles of pioneering scientists in some of the key discoveries in the field. Several RNA biologists have won Nobel Prizes, as shown in Table 1.1. Be aware, however, that even Nobel Prize winners have achieved their successes with the collaboration of hundreds of scientists. The field of RNA biology is where it is now thanks to the hard work of thousands of dedicated scientists.

Table 1.1 Nobel Prize laureates in RNA biology

Nobel Prize winners	Discovery
François Jacob, André Lwoff, Jacques Monod (1965)	'For their discoveries concerning genetic control of enzyme and virus synthesis' (nature and function of messenger RNA)
Robert Holley, Har Gobind Khorana, Marshall Nirenberg (1968)	'For their interpretation of the genetic code and its function in protein synthesis' (first sequenced transfer RNA)
Sidney Altman and Thomas Cech (1989)	'For their discovery of catalytic properties of RNA' (ribozymes)
Richard Roberts and Philip Sharp (1993)	'For their discoveries of split genes' (splicing)
Andrew Fire and Craig Mello (2006)	'For their discovery of RNA interference—gene silencing by double-stranded RNA'
Roger Kornberg (2006)	'For his studies of the molecular basis of eukaryotic transcription'
Venkatraman Ramakrishnan, Thomas Steitz, Ada Yonath (2009)	'For studies of the structure and function of the ribosome'

For details of their achievements and biographical information, visit http://nobelprize.org/

RNA-based processes are also increasingly linked to the aetiology of disease. Therefore we also provide examples of human diseases that arise when post-transcriptional processes have gone awry and of techniques that are useful in studying RNA biology.

This textbook assumes a prior knowledge of the essentials of molecular biology and genetics, and an understanding of the main aspects of gene expression. Basic definitions central to understanding are listed in Table 1.2. The Glossary at the end of the book defines all the key terms and abbreviations that we use.

Table 1.2 The essential terminology of gene expression: basic definitions central to the understanding of how genes are expressed and regulated*

Terminology	Definition
DNA and RNA	DNA (deoxyribonucleic acid) is the most famous biomolecule among the lay audience. In its Watson–Crick double helix conformation, DNA is elegantly and easily replicated. RNA (ribonucleic acid) is perhaps less famous, but equally important. It is less stable but more versatile—to the extent that RNA molecules can even catalyse their own replication. The RNA world hypothesis suggests that RNA preceded DNA as the repository of genetic information
Gene expression	The process whereby genes, encoded by DNA, express their inherited information usually in the form of proteins (but sometimes in the form of non-coding RNAs)
Gene regulation	Genes need to be switched on and off in specific tissues and at specific times in development. Expression levels are also modulated so that a gene can be highly expressed or expressed at very low levels. Traditionally, gene regulation explored how many mRNAs can be transcribed from genes (DNA). A more modern view of gene regulation must now also take into account the multiple ways in which gene expression is regulated at the RNA level
Genome	The genome is essentially the collection of all DNA in a given organism. Genomics is the study of genomes. Comparative genomics looks at the interrelationships between different genomes—their similarities and differences, and what this tells us about the evolutionary relationships between different organisms
Transcriptome	The transcriptome is the collection of all RNAs that are transcribed from the genome. It includes different mRNA isoforms derived from single genes, and also non-coding RNAs that do not encode proteins
Proteome	The proteome is the collection of all of the proteins that are translated from the transcriptome. The proteome also includes isoforms that arise from post-translational modifications such as glycosylation
Chromatin structure	Genomes are organized into separate chromosomes which can incorporate hundreds of millions of bases of DNA. These bases need to be packaged efficiently into a tiny volume—the cell nucleus. DNA packaging is achieved by the histones, and packaged DNA is known as chromatin. Chromatin structure is highly dynamic and, not surprisingly, regulated
Epigenetics	Epigenetics means 'above genetics' and describes changes in gene expression that are not directly due to changes in DNA sequence. The term instead refers to physical changes to the structure of DNA (typically methylation) that prevent specific genes from being switched on
Transcription	Transcription is the process whereby genes are copied into messenger RNA, or other RNA 'transcripts'. Transcription is achieved by the RNA polymerases. Transcription factors are proteins that can bind DNA and help to switch genes on or off by facilitating or preventing the recruitment of RNA polymerases
Pre-mRNA processing	RNA transcripts are processed in the nucleus—not least messenger RNAs, which are modified by the processes of capping, polyadenylation, editing, and splicing
Translation	This is the process whereby the genetic information carried by messenger RNAs is 'read' by ribosomes and transfer RNAs giving rise to a polypeptide chain

*Other terms and abbreviations used in this book are given in the Glossary

1.1 Aims of this book

This book is written for multiple audiences: students of molecular biology and genetics, including both undergraduate and postgraduate students who wish to revise or extend their knowledge of gene regulation, biologists from related fields who wish to learn the basics of RNA biology, and other scientists and clinicians who have encountered RNA-based problems in their biomedical research.

We provide a broad overview of the RNA-based processes that occur in gene expression. However, it is important to realize that what we present in terms of complexity and detail is merely the tip of the iceberg. The best way to study this field is first to gain a basic appreciation of the molecular processes—of their purpose and nature. You can then deepen your understanding of the molecular detail, leading to an appreciation of how the processes are involved in development and disease. It will also become apparent that all molecular processes are closely interlinked (the strict categorization of processes is a convenient human intellectual construct, but there is in fact much overlap in nature).

By the end of the book, you will realize that the field of RNA biology has grown exponentially over the past few years, and you will also realize that there is a lot of work in progress. What we present is a snapshot of where the field is at the moment. However, whereas some of the details of the processes, molecules, and complexes that we describe are likely to change, the key concepts will still apply. In fact, learning the underlying concepts first is the best way to study; if you can master them, it is easier to remember the details. Some of the terms and abbreviations that we use in this chapter might be new to you; if so, please refer to the glossary at the end of the book for their definitions.

Our intention has been to cover all of the important concepts and processes in RNA biology. We hope, therefore, that the keen student of RNA biology will use this textbook as a platform from which to explore the field in greater depth. Each chapter ends with a summary, key questions, and a suggested reading list, so that specific topics can be explored in greater detail. Online resources are also provided, allowing the student to broaden his or her perspective. We hope that you will find, as we have, the RNA field to be both fascinating and intellectually stimulating. We begin the journey into the RNA world by first asking the questions: What is RNA? And how does it differ from DNA?

1.2 DNA and RNA are composed of slightly different building blocks

Although RNA and DNA are made of similar building blocks, the resulting molecules are quite different. This difference is illustrated by Fig. 1.2 which compares side by side an RNA molecule (Fig. 1.2a) and a DNA molecule (Fig. 1.2b). Notice how DNA is a linear double helix, while RNA is much more globular in structure. In fact, Fig. 1.2a contains two RNA molecules—the surface area of a catalytically active RNA called RNAse P is shown, which forms a tight fit with a physically separate tRNA molecule. In terms of the complexity and variety of their three-dimensional structures RNAs are much more like proteins than like DNA. Note that RNA and DNA can adopt very different structures, even though they are built from similar starting materials.

First, we are going to look at the similarities and differences between the building blocks of DNA and RNA before considering how these building blocks are assembled in a hierarchical way to give nucleic acids a primary structure (the nucleic acid sequence). In Chapter 2 we will explore how a secondary structure arises (simple motifs which are based around helices), and then a tertiary structure (how helical and single-stranded regions of RNA fold together to form complex shapes).

Both RNA and DNA are polymers of repeating monomer units called nucleotides (see Box 1.1). Notice in Fig. 1.3 that nucleotides are made from three distinct modules: a single sugar, a single base, and up to three phosphate groups (in Fig. 1.3a a single phosphate group is shown). Next we are going to look at each of these modules in turn.

1.2.1 Sugars: nucleotides contain five-ringed sugars

Figure 1.3 shows the chemical structure of the **ribose sugar used in RNA** and the **deoxyribose sugar used in DNA**. Both ribose and deoxyribose sugars are five-membered ring-shaped molecules containing carbon atoms and a single oxygen atom, with

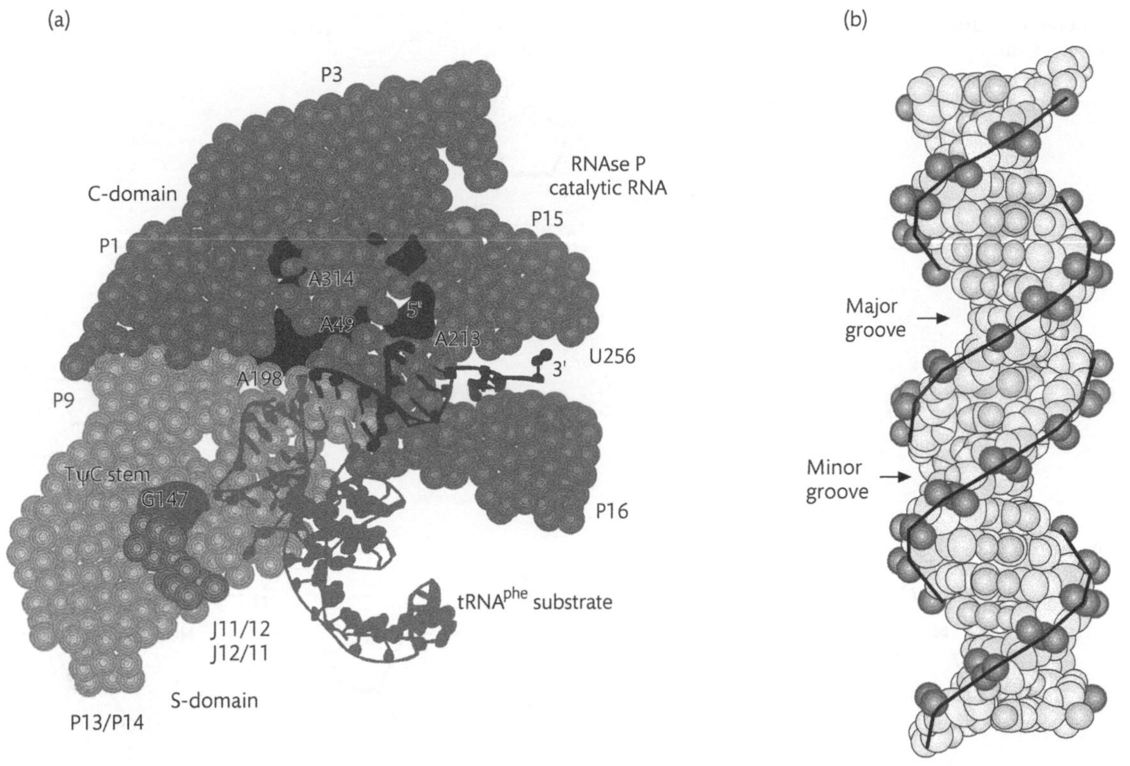

RNAse P attached to substrate tRNA **B-DNA**

Figure 1.2 RNA and DNA have very different structures. (a) Tertiary structure of two physically interacting RNA molecules. The space-filling structure is the RNA component of bacterial RNAse P as determined by X-ray crystallography; notice the globular shape of this RNA structure. RNAse P is a catalytic RNA which is required for the processing of transfer RNAs (tRNAs) in all organisms from bacteria to humans and is discussed in more detail in Chapter 3. The tertiary structure of RNAse P forms a substrate-binding domain into which fits its substrate precursor tRNA. A molecule of $tRNA^{phe}$ is shown as a helical structure which fits neatly into the active site formed by the surface of RNAse P.[7,8] For a full colour version of Fig. 1.2a, see Colour Illustration 1. (Reprinted from Torres-Larios A, et al.[7] with permission from Macmillan Publishers Ltd.) (b) For comparison, a space-filling model of the DNA double helix—notice that this has a more rod-like structure. (Reprinted from Wang et al.[8] with permission from Macmillan Publishers Ltd.)

side groups attached to the carbons. The carbon atoms in both ribose and deoxyribose sugars are numbered from 1′ through to 5′. This nomenclature is important, since it differentiates the carbon atoms in the sugar from those in the base (which are discussed below). The ′ symbol is pronounced 'prime': for example 5′ is referred to as '5 prime'.

Differences between RNA and DNA 1

Note in Fig. 1.4 that DNA and RNA contain different sugars. Look carefully at the ribose and deoxyribose sugars shown in the figure; both have 3′ –OH groups but ribose sugar has an additional 2′ –OH group. The name 'deoxyribose' is used because the five-ringed deoxyribose sugar in DNA is missing a 2′ oxygen atom compared with the ribose sugar.

Box 1.1 How the building blocks of nucleic acids are named

A **nucleotide** is a ribose sugar connected to one or more phosphate groups and a base. A **nucleoside** is a ribose sugar connected only to a base through a β-glycosidic bond.

The additional 2′ –OH group in the ribose sugar has two consequences for the function of RNA compared with DNA.

1 The 2′ –OH group in ribose sugar is polar, making RNA more chemically reactive than DNA. An –OH group carries an asymmetric charge distribution (see Fig. 1.5). The role of the 2′ –OH group

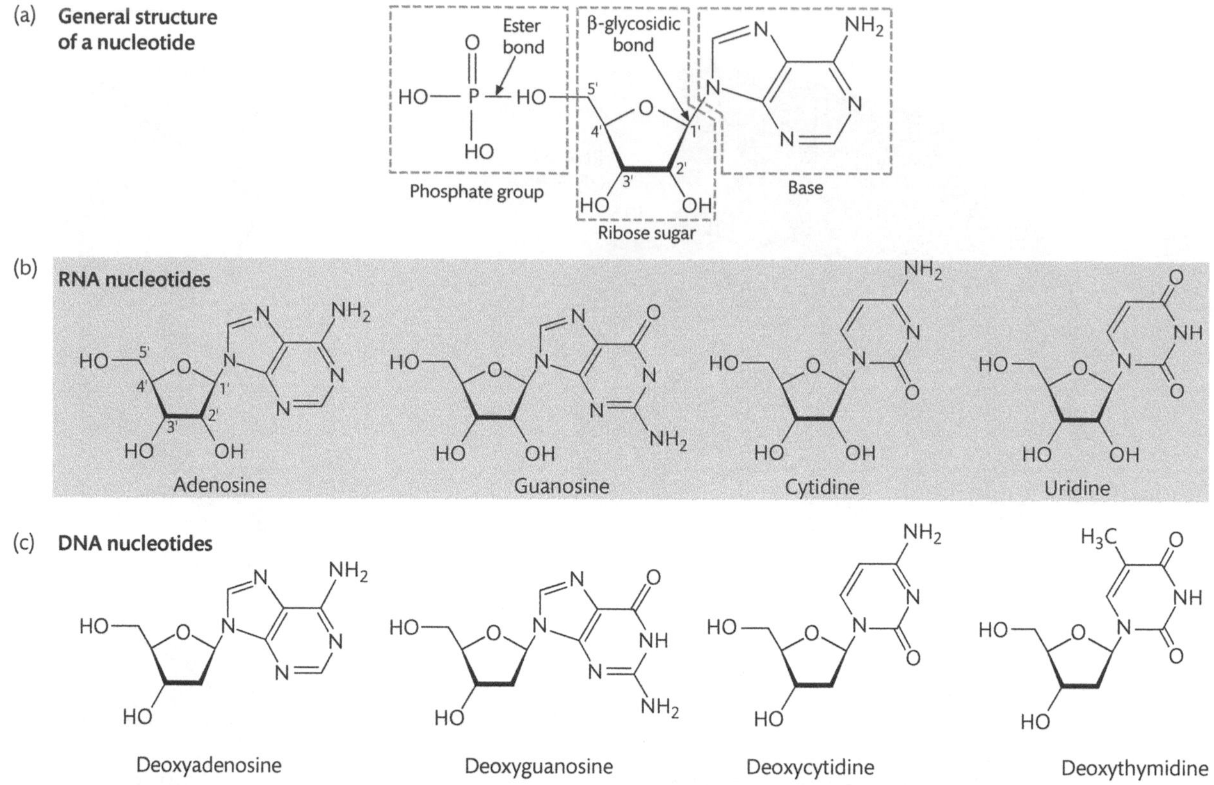

Figure 1.3 Nucleotides are assembled from ribose sugars, bases, and phosphate groups. (a) General structure of a nucleotide showing the important bonds holding these together. Notice that the β-glycosidic bond holds the base to the ribose sugar, and an ester bond holds a phosphorus atom to the ribose sugar. (b) Structure of each of the four nucleotides used in RNA. (c) Structure of each of the four nucleotides used in DNA.

in the chemical reactivity of RNA is discussed in Chapter 3.

2 The ribose sugar molecule has a slightly different shape from the deoxyribose sugar. The ribose sugar ring is slightly twisted to minimize interactions between the polar 2′ -OH group and the other non-bonding atoms attached to the ring. This twisted

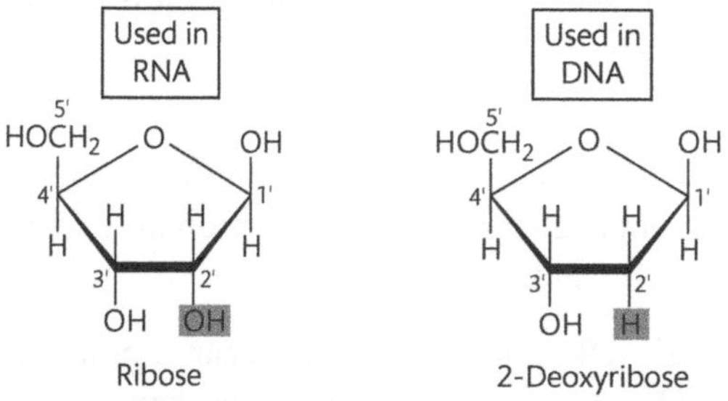

Figure 1.4 Different sugars are used in DNA and RNA. Ribose sugar has a 2′ –OH group, and is used to make RNA nucleotides. DNA is made with a deoxyribose sugar, which lacks this 2′ –OH group.

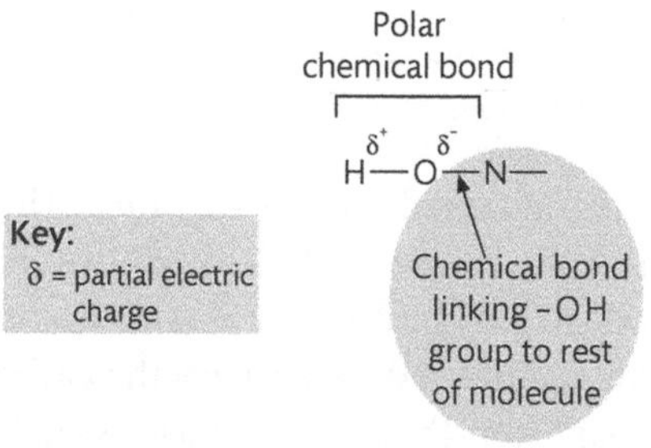

Figure 1.5 –OH groups are polar and can form hydrogen bonds. The hydroxyl groups on ribose and deoxyribose sugars are polar; the electrons joining the hydrogen and oxygen atoms are distributed unequally. This unequal sharing arises because the oxygen atom is highly electronegative and pulls electrons strongly towards itself. By contrast, hydrogen is weakly electronegative and so exerts less of a 'pull' on electrons. The unequal sharing of electrons between the hydrogen and oxygen atoms results in these atoms carrying partial electrical charges when they are joined by a covalent bond; the hydrogen atom carries a partial positive charge (denoted δ^+), and the oxygen atom carries a partial negative charge (denoted δ^-). These partial charges are called dipoles. Because of their charge distribution, these polar groups can form hydrogen bonds.

shape is called a sugar pucker, and has implications for the kinds of secondary structures that RNA forms compared with DNA. (We will discuss these structures in Chapter 2.)

Next we shall compare and contrast the bases used in RNA and DNA.

1.2.2 Bases: RNA contains four different bases

Four different common nucleotides are used to make up RNA molecules, which contain four different bases connected to the ribose sugar. Notice in Fig. 1.6 that the four bases form two chemically distinct groups called *purines* and *pyrimidines*.

Both purine and pyrimidine bases are ring-shaped molecules containing carbon and nitrogen atoms. (Ring-shaped molecules which contain additional atoms to carbon are termed **heterocyclic**.) Purine bases (adenine and guanine, see Table 1.3) contain a double ring, while pyrimidine bases (cytosine and thymine) contain a single ring. Each of the atoms in the rings of bases are numbered; since purines are double ringed there are more carbon atoms (numbered 1–9) than in pyrimidines (numbered 1–6). Although four different bases are commonly present in RNA (two purines and two pyrimidines), in some cases these bases can be chemically modified. Base modification is described in Chapters 13 and 14.

Bases are joined to pentose sugars in nucleotides through a strong chemical bond called a **β-glycosidic bond**. The position of this β-glycosidic bond is shown in Fig. 1.3a; notice that this bond is between carbon atom 1′ of the ribose sugar and a nitrogen atom in the base (pyrimidine nitrogen 1 and purine nitrogen 9). Phosphate groups are attached to the 5′ carbon of the sugar molecule through an ester bond.

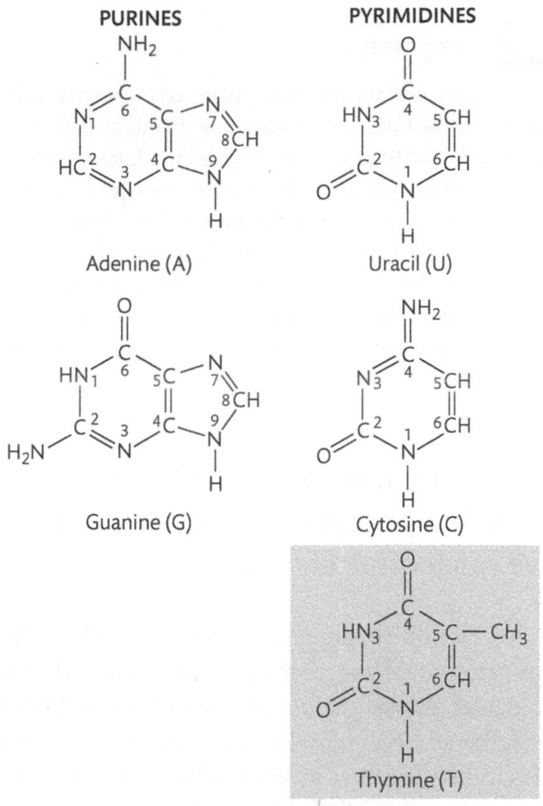

Figure 1.6 Overlapping groups of four bases are used in RNA and DNA. In DNA, thymine is used as a base instead of uracil.

Differences between RNA and DNA

Note that RNA and DNA use a different but overlapping set of bases. Although both RNA and DNA contain nucleotides with four different bases, a clear difference between RNA and DNA is that RNA uses uracil as a base whereas DNA uses thymine. The reason for this important difference is related to the chemical stability of nucleotides and the repair of nucleic acid damage in the cell (DNA has evolved to use thymine as a base instead of uracil as a mechanism to protect it from spontaneous damage; see Fig. 1.7).

Table 1.3 The five different bases which are commonly used in DNA and RNA

Base	Nucleoside	Abbreviation	Special properties	Type
Adenine	Adenosine	A		Purine
Cytosine	Cytidine	C		Pyrimidine
Guanine	Guanosine	G		Purine
Thymine	Thymidine	T	Only used in DNA	Pyrimidine
Uracil	Uridine	U	Only used in RNA	Pyrimidine

Key points

RNA molecules are polymers made of monomer units called nucleotides. The nucleotide building blocks of DNA and RNA have the same overall modular organization and are made up of three subcomponents: pentose sugars, bases, and phosphate groups. The ribose sugar used to make RNA has an extra hydroxyl group compared with the deoxyribose sugar used in DNA, which makes RNA more chemically reactive than DNA. DNA and RNA also contain a different but overlapping set of bases.

1.3 Nucleotides are joined together through a phosphodiester backbone to give nucleotide chains

In this section we are going to look at how nucleotides are joined together to make RNA and DNA molecules. Nucleic acid molecules (RNA and DNA) are polymers composed of long chains of nucleotides joined together in linear chains. Short nucleic acid chains are called **oligonucleotides**, and long nucleic acid chains are called **polynucleotides**. Since both RNA and DNA contain four different nucleotides (ACGT in DNA, and ACGU in RNA) that can be joined together in different combinations, an almost infinite variety of RNA and DNA molecules can be made. A good example of this is to look at the possible number of sequence variations from even a very short DNA or RNA molecule: for a 10-nucleotide sequence there would be 4^{10} possible different sequence versions (a grand total of 1 048 576 variants).

Nucleotides in both DNA and RNA molecules are joined together in chains by the same type of chemical bond, which is called a **phosphodiester bond**. Phosphodiester bonds are identical in DNA and RNA.

As its name suggests, a phosphodiester bond is composed of two individual ester bonds (see Fig. 1.8). One ester bond links a phosphorus atom to the oxygen atom on the 3′ carbon of the upstream ribose sugar, and the other ester bond links the same phosphorus atom to the oxygen atom attached to the 5′ carbon atom of the downstream ribose sugar. Looking at Fig. 1.8 you will notice that the central

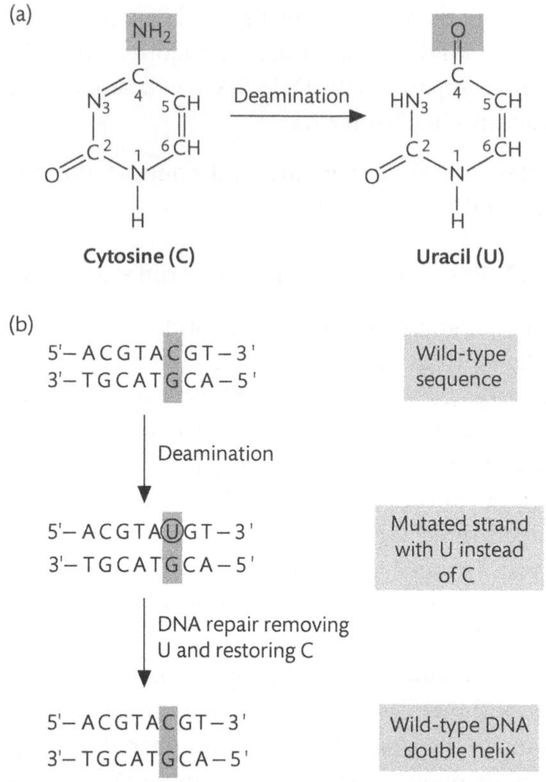

Figure 1.7 The use of thymine as an alternative to uracil as a base in DNA is linked to the DNA repair machinery. (a) Cytosine is an unstable base because the amino group ($-NH_2$) on carbon atom number 4 can spontaneously be replaced with an oxygen (=O) atom. (This process is called **deamination**.) If you look carefully at the chemical structures of the bases you will see that deamination of cytosine chemically converts it into a uracil molecule. Both thymine and uracil have similar base-pairing properties: both base pair with adenine. Thymine and uracil are distinguished by an extra methyl group on carbon-5 in thymine. (b) The DNA repair machinery monitors the spontaneous appearance of uracil in DNA, which naturally occurs through base deamination, and changes any uracils thus formed back to cytosine. By acting as a monitor for detecting DNA damage, the use of thymine as a base in DNA reduces the occurrence of point mutations caused by spontaneous cytosine deamination. RNA does not have such a mechanism to safeguard it. Since RNA normally uses uracil as a base, the cell cannot monitor RNA cytosine deamination. Although cytosines can also become deaminated in RNA and remain undetected, this is less of a problem; it is DNA which is used for long-term storage of genetic information rather than RNA (except in some viruses). Compared with DNA, RNA has a short half-life in the cell, so deterioration through base deamination poses less of a genetic danger.

phosphorus atom in the phosphodiester bond is surrounded by four oxygen atoms, two of which are non-bridging oxygen atoms not involved in the ester

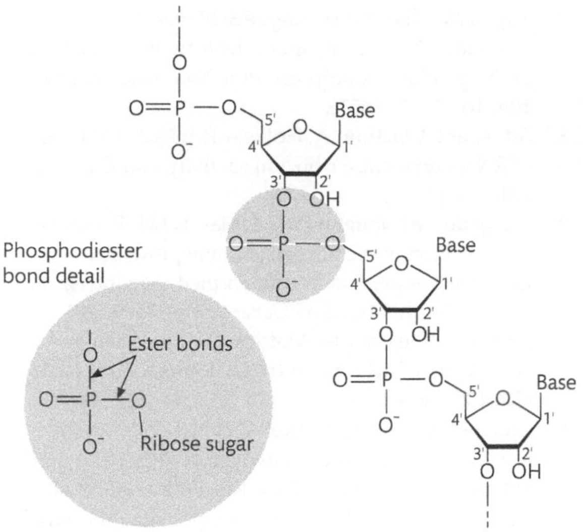

Figure 1.8 Individual nucleotides in an RNA molecule are joined together by phosphodiester bonds. A phosphodiester bond is composed of a single phosphorus atom joined to two oxygen atoms through two ester bonds. Long linear chains of nucleotides are joined through sequential phosphodiester bonds: the alternating phosphodiester bonds and sugars in chains of DNA and RNA are called the phosphodiester backbone. The bases project from this backbone.

Key points

Nucleic acids are made from four different nucleotides joined together in linear chains. Since four different nucleotides can be joined together in any combination in an RNA chain, an enormous number of primary sequence variants are possible. The length of an RNA chain depends on how many nucleotides it contains. Short chains are called oligonucleotides, and long chains polynucleotides. Within these chains, nucleotides are held together by phosphodiester bonds between the ribose sugars to form a **phosphodiester backbone**. The bases are held through their attachment to ribose sugars and project from the backbone.

bonds. This high density of oxygen atoms around the linking phosphorus atoms in the phosphodiester bonds within RNA molecules is also important for the chemical reactivity of RNA, as we shall see in Chapter 3. The bases are not themselves directly involved in the phosphodiester bonds, and instead project from the phosphodiester backbone of the molecule.

Because of the phosphodiester bonds which join nucleotides together in a polynucleotide chain, every RNA molecule starts with a 5′ phosphate group and finishes with a 3′ -OH group. This gives RNA strands a directional polarity, denoted 5′–3′. The phosphodiester bond plays a critical role in RNA biology, so we shall now look at this bond in more detail. Cleavage of phosphodiester bonds in both RNA and DNA is an important reaction in cells. We discuss some RNA cleavage reactions in Chapter 3.

Although they do not contribute to the phosphodiester backbone of RNA, bases do play a critical role in interactions between nucleotide chains. Hydrogen bonding between the bases of RNA creates the double helices of RNA secondary structure—these are the topic of the next chapter.

Summary

Molecular biology has grown spectacularly over the last several decades. RNA has always been at the heart of this new field. Pioneering research in the mid-twentieth century showed how messenger RNA, transfer RNA, and ribosomal RNA work together to translate genetic information into proteins. It then became apparent that additional RNA-based processes provide additional layers of complexity to the regulation of gene expression—in particular, the discovery of pre-mRNA splicing and the realization that the genome transcribes a multitude of non-coding RNAs that have critical regulatory roles. Given the complexity of RNA-based processes, a distinct subfield has emerged, namely RNA biology. RNA differs from DNA in a number of ways. Both are polymeric chains of nucleotides, joined together through a phosphodiester backbone. Whereas RNA uses the five-ringed sugar ribose, DNA uses the less chemically reactive deoxyribose. As in DNA, there are four heterocyclic bases in RNA, but RNA uses the base uracil instead of thymine. According to the RNA world hypothesis, RNA was the original repository of genetic information. In time, this role was taken over by DNA, but RNA retains critically important roles in gene expression and regulation.

Questions

1.1 Can you name the Nobel Prize winners who won recognition for their pioneering work in RNA biology?
1.2 What is the 'RNA world hypothesis' and what is the evidence to support it?
1.3 What proportion of the human genome is transcribed?
1.4 What proportion of the human transcriptome is coding and what proportion is non-coding?
1.5 Can you explain the key differences between RNA and DNA?

References

1. **Yarus M.** *Life from an RNA World: The Ancestor Within.* Cambridge, MA: Harvard University Press (2010).
2. **Roy D, Najafian K, von Ragué Schleyer P.** Chemical evolution: the mechanism of the formation of adenine under prebiotic conditions. *Proc Natl Acad Sci USA* **104**, 10 727–7 (2007).
3. **Attwater J, Wochner A, Holliger P.** In-ice evolution of RNA polymerase ribozyme activity. *Nat Chem* **5**, 1011–18 (2013).
4. **Hangauer MJ, Vaughn IW, McManus MT.** Pervasive transcription of the human genome produces thousands of previously unidentified long intergenic non-coding RNAs. *PLoS Genet* **9**, e1003569 (2013).
5. **Frith MC, Pheasant M, Mattick JS.** The amazing complexity of the human transcriptome. *Eur J Hum Gen* **13**, 894–7 (2005).
6. **Djebali S, Davis CA, Merkel A, et al.** Landscape of transcription in human cells. *Nature* **489**, 101–9 (2012).
7. **Torres-Larios A, Swinger KK, Krasilnikov AS, et al.** Crystal structure of the RNA component of bacterial ribonuclease P. *Nature* **437**, 584–7 (2005).
8. **Wang AH, Quigley GJ, Kolpak FJ, et al.** Molecular structure of a left-handed double helical DNA fragment at atomic resolution. *Nature* **282**, 680–6 (1979).

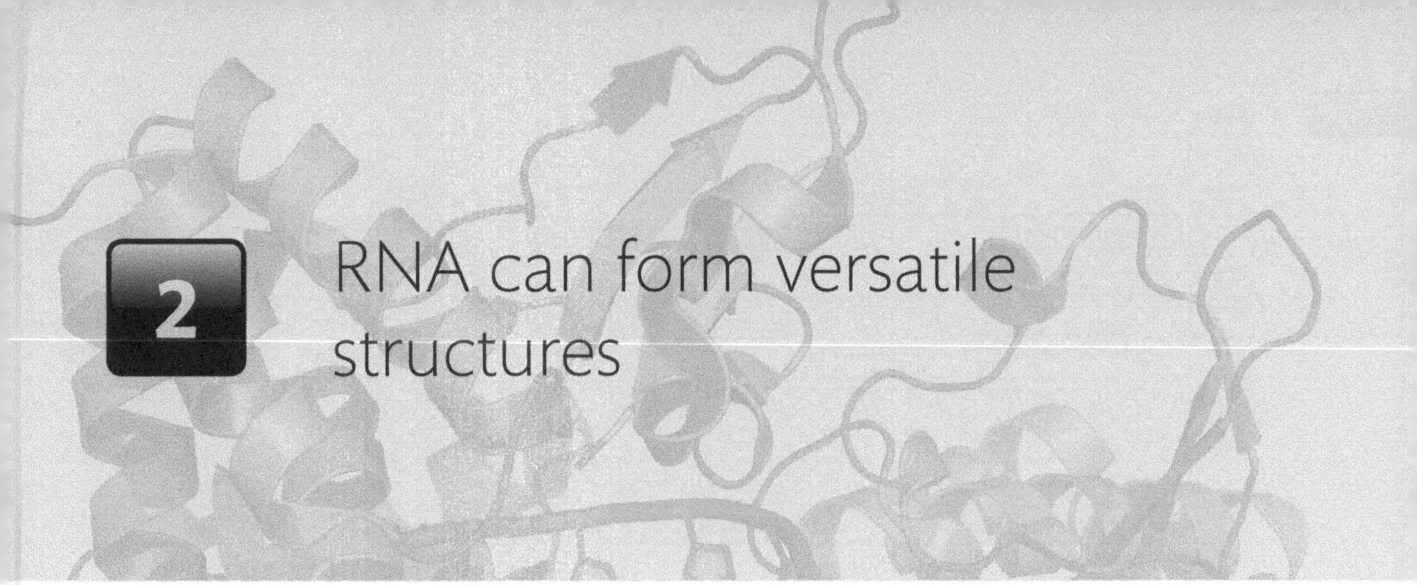

2 RNA can form versatile structures

Introduction

Within chromosomes DNA exists as a continuous double helix. For example, each nuclear chromosome in human cells is a long linear double helix, and each mitochondrial genome is a circular double helix. Bacterial chromosomes are usually circular double helices. The double-helix structure of DNA was established by James Watson and Francis Crick in the 1950s based on X-ray diffraction images of DNA taken by Rosalind Franklin and Maurice Wilkins. (The gripping story of the discovery of the DNA double helix is described by James Watson in his book *The Double Helix*.) Once the structure of DNA had been found to be a double helix the next challenge for crystallographers was to work out if there was a similar molecular structure for RNA.

Double helices are also important for RNA secondary structure. However, while DNA has a fairly constant and predictable secondary structure enabling its crystallization, RNA can adopt a range of different secondary and tertiary structures. This structural complexity has made the structure of RNA more difficult to analyse, but is also very important for its biological functions.

This ability of RNA to fold into diverse structures enables it to be involved in a number of biological processes, and for RNA to have more roles in the cell than DNA. DNA's principal role is as a store of genetic information. In contrast, and as we shall see throughout this book, RNA is much more versatile with diverse cellular roles (see Fig. 2.1 for some functionally diverse RNAs, and where they are discussed in this book). The structural complexity of RNA has even enabled it to develop catalytic functions (described in more detail in Chapter 3).

The same strategies are used by many RNA molecules to form three-dimensional structures, and these are the topic of the first part of this chapter (Sections 2.1–2.7). Then, in the remainder of the chapter, we discuss some of the biological functions of these RNA structures, including their use as thermosensors in pathogenic bacteria and RNA structures that strongly bind target molecules.

2.1 How do RNA molecules form structures?

RNA molecules use three hierarchical levels of structural organization (see Fig. 2.2).

1 Primary structure. This is the linear sequence of nucleotides in a nucleic acid, or *nucleotide sequence*. The RNA nucleotide sequence is encoded by its template DNA, and copied into RNA by transcription.

2 Secondary structure. This is composed of helices which form through base pairing. RNA helices can form both within single molecules of RNA (intramolecular base pairing) and between different RNA molecules (intermolecular base pairing).

3 Tertiary structure. This is the highest level of organization, in which RNA molecules with secondary structure fold up into very compact and highly organized structures.

While both DNA and RNA have primary structures composed of nucleotide sequences, they typically form different kinds of secondary structure in

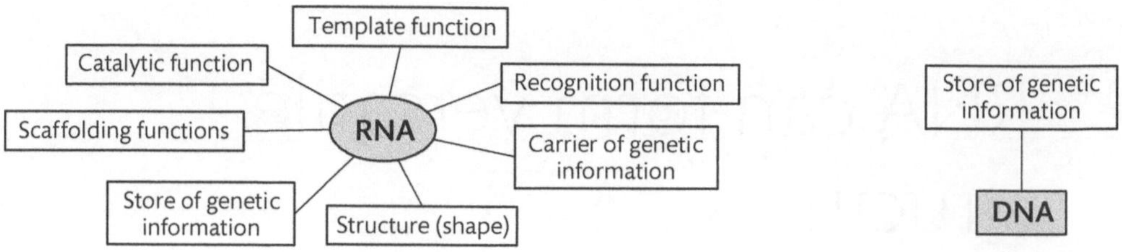

Function of RNA	Example	Chapter
A store of genetic information	Some viruses have RNA rather than DNA genomes RNA is thought to have been used as the main genetic material earlier in evolution	 **See Chapter 3**
Recognition function: using base pairing RNA can specifically recognize other RNA or DNA molecules	**Codon-anticodon interaction** between mRNA and tRNA and **16S rRNA-Shine–Dalgarno sequence recognition** **Splice site recognition** **snoRNA and substrate rRNA** **MicroRNAs and target mRNAs**	**See Chapter 11** **See Chapter 5** **See Chapter 14** **See Chapter 16**
Template function:	**Telomerase RNA** acts as a template for making new DNA ends of chromosomes	**See Chapter 14**
Catalytic function:	Ribozymes can catalyse chemical reactions, including the peptidyltransferase site of rRNA, Group I and II introns etc.	**See Chapter 3**
Scaffolding functions to build large RNP complexes:	Signal recognition particle and the ribosome The spliceosome	**See Chapter 11** **See Chapter 5**
Structure (shape). RNA molecules often have to have specific shapes so they can dock with either other RNAs or proteins	The interaction between the two subunits of the ribosome The interaction between RNA aptamers and their ligands	**See Chapter 11** **This chapter**
Carrier of genetic information	**mRNA**	

Function of DNA	Example
Store of genetic information	DNA is the repository of genetic information in eukaryotic, prokaryotic and organellar chromosomes

Figure 2.1 Different functional repertoires for RNA and DNA in the cell. Note that there are several different functional classes of RNA in the cell. In comparison, DNA is solely involved in information storage.

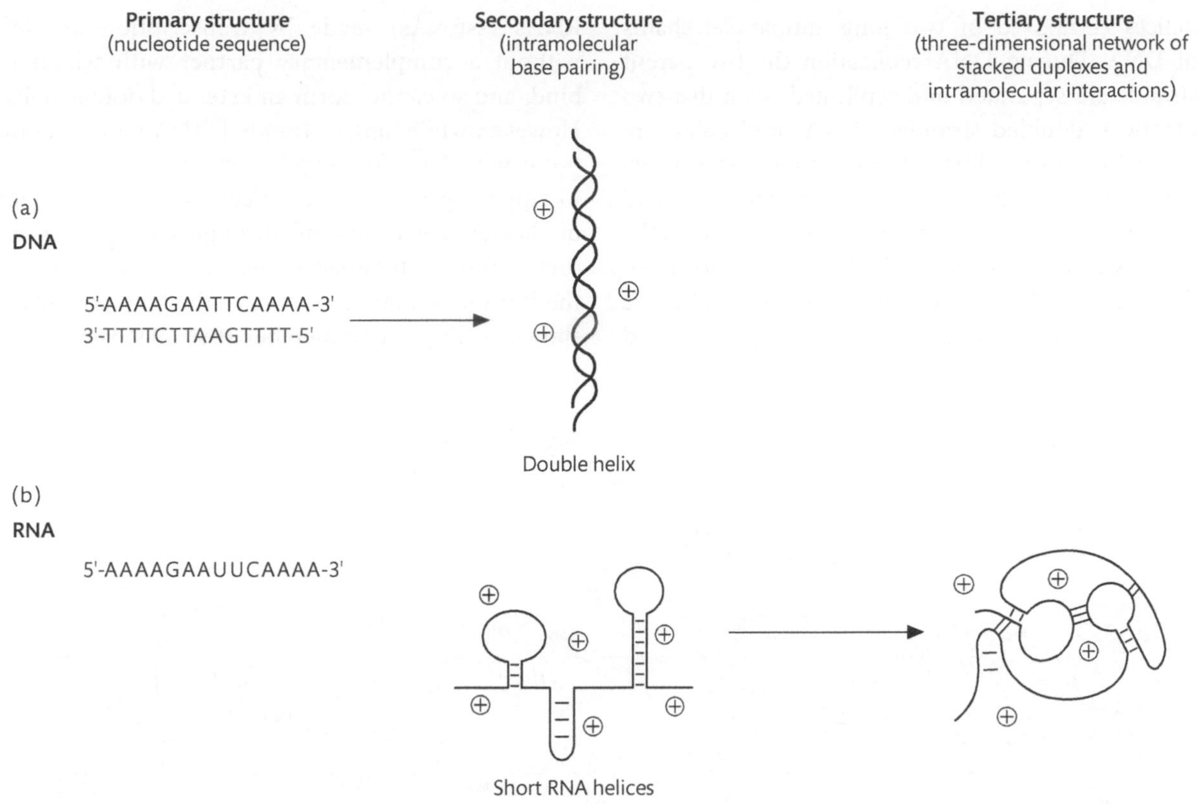

Figure 2.2 **Three different levels of structural organization in nucleic acids.** (a) DNA. The primary structure of DNA is also the sequence of the nucleotides. The secondary structure of DNA is the extended double helix, which forms between two complementary DNA strands over their full length. (b) RNA. The primary RNA structure is its nucleotide sequence. Secondary RNA structures result from the formation of double-stranded RNA helices. (Double-stranded regions of RNA are also called RNA duplexes.) Formation of RNA helices is helped by positively charged molecules (shown as + signs) in the environment, which balance the negative charges on RNA and so make it easier to bring RNA strands together. RNA tertiary structure occurs due to folding and packing of RNA helices into more compact globular structures. RNA tertiary structures can rival proteins in complexity.

cells (see Fig. 2.2). DNA molecules are found in long double helices which extend along their full lengths (the classical double helix), while RNA molecules form shorter double helices between single-stranded regions. RNA molecules can form extensive tertiary structures through further interactions between regions of secondary structure.

2.2 RNA secondary structure: RNA molecules tend to form a number of shorter helices compared with DNA

Why are there these differences in structural organization between RNA and DNA molecules? The answer lies in how RNA and DNA molecules are made in the cell (see Fig. 2.3). Chromosomes are double

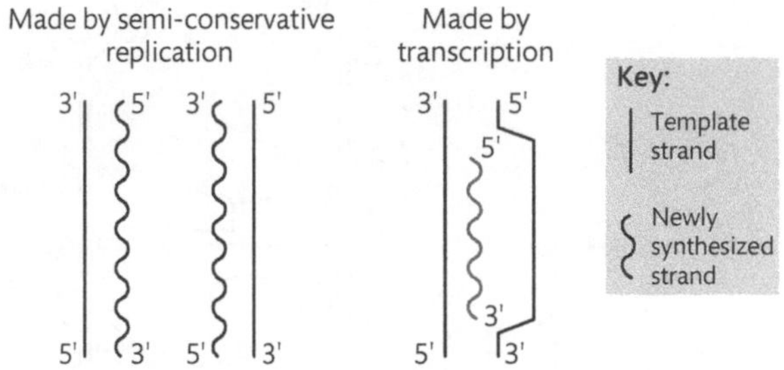

Figure 2.3 **Why do DNA molecules form long double-helix structures and RNA molecules form much shorter helices?** Differences in the way RNA and DNA molecules are made in the cell mean that DNA molecules have ready-made full-length complementary partner strands with which to form double helices, while RNAs usually do not. DNA is replicated by semi-conservative replication, which produces two daughter helices, each containing a parental strand and a newly replicated strand of DNA. RNA is made by transcription, in which just a single strand of DNA is copied into RNA.

helices composed of two long antiparallel chains of DNA. During DNA replication the two parent strands are separated and replicated, such that two identical doubled-stranded DNA molecules are made. In contrast, during transcription, the two parent strands of DNA are only transiently separated and only one strand is used as a template for RNA synthesis. Newly made RNA is then separated from the template DNA, and the two strands of DNA come together immediately. Hence the single-stranded RNAs (ssRNAs) made by transcription are left without a complementary partner with which to bind, and so cannot form an extended double helix. However, while 'single-stranded' RNA molecules do not usually have full-length partner strands of RNA to zipper up with, RNA molecules form a number of shorter helices instead through base pairing either within or between molecules. As an example, the RNA secondary structure of the HIV genome is shown in Fig. 2.4 (remember that HIV is an RNA

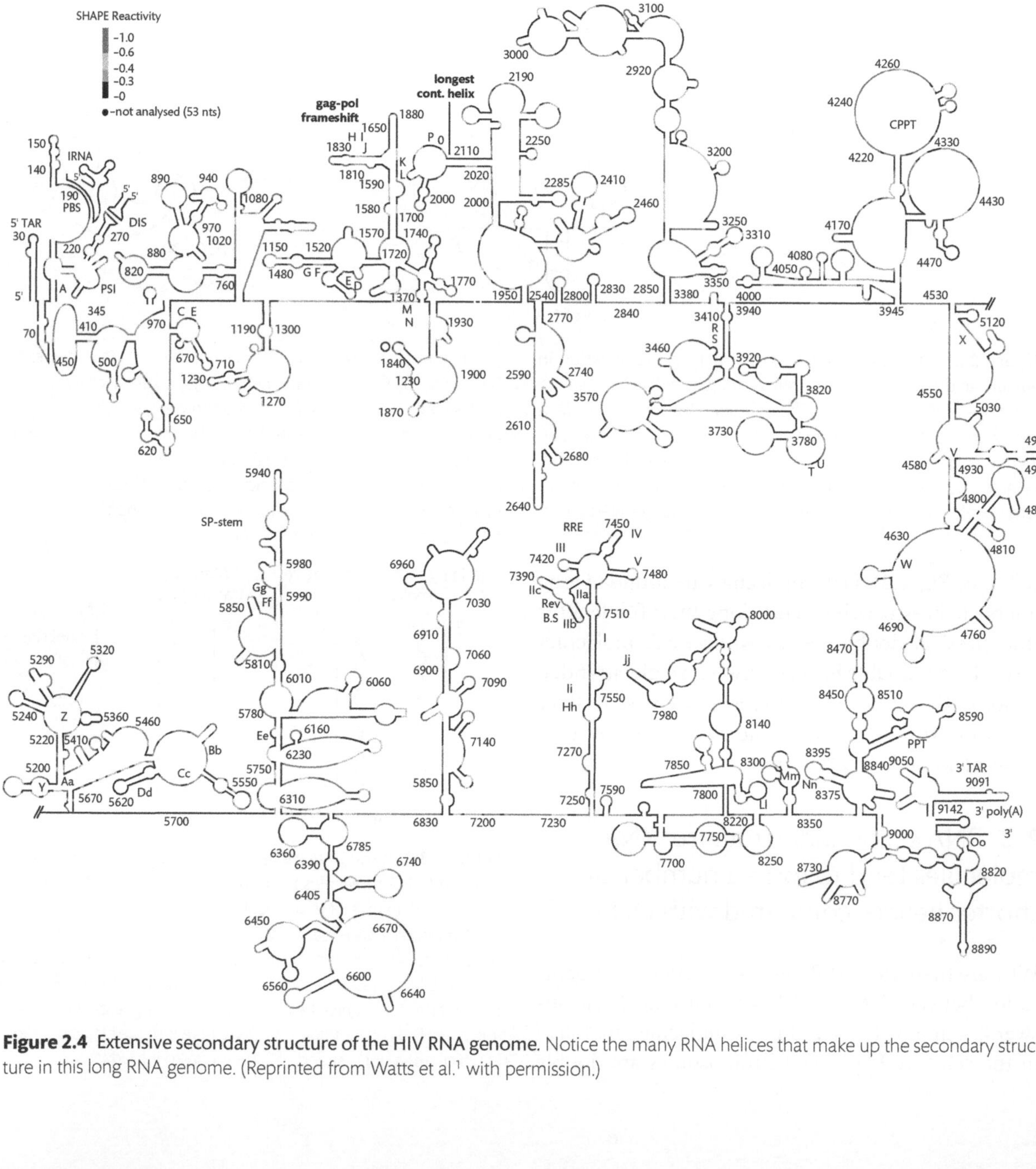

Figure 2.4 Extensive secondary structure of the HIV RNA genome. Notice the many RNA helices that make up the secondary structure in this long RNA genome. (Reprinted from Watts et al.[1] with permission.)

virus, and is converted into DNA after infection via the enzyme reverse transcriptase).[1]

Four common principles apply to both RNA and DNA double helices.

1 Phosphodiester bonds are on the outside of the helix. In a double helix the phosphodiester backbones of the nucleotide chains are on the outside and the bases are on the inside.

2 Nucleic acid strands are antiparallel in a helix. The two phosphodiester backbones in a nucleic acid helix are antiparallel. We discussed the 5′–3′ directional polarity of phosphodiester backbones in Section 1.3. In double helices the strand arrangement of the two chains is antiparallel; this means that one phosphodiester backbone is in a 5′–3′ orientation and the phosphodiester backbone to which it is paired is in the opposite (3′–5′) orientation.

3 Hydrogen bonding holds helices together. Inside the helix, the bases projecting from the two phosphodiester backbones interact through hydrogen bonding. RNA and DNA double helices are held together by Watson–Crick base pairing. Hydrogen bonds form between bases as a result of the asymmetric charge distributions which occur when hydrogen atoms are covalently bonded to electronegative atoms such as nitrogen or oxygen (see Box 2.1). These partial charges are called dipoles, and oppositely charged dipoles (hydrogen bond donors and acceptors) will attract to form a hydrogen bond. Watson–Crick hydrogen bonding takes place between hydrogen bond donors and acceptors which are brought together when the inner faces of A and U bases and of G and C bases are aligned (see Fig. 2.6). A–U and G–C base pairs form since they have the same kind of overall structure, and so will not distort the helical structure of double-stranded RNA. Other base-pair combinations have different widths and so would distort the helix.

4 Hydrogen bonds are individually weak but gain strength in numbers. Since it is held together by three separate hydrogen bonds the G–C base pair is stronger than the A–U pair. Longer double-stranded nucleic acids with higher G–C content are held together by more hydrogen bonds and so are more stable.

5 Helices are stabilized by base stacking. Stacking means that the bases project into the centre of the helix, and are literally stacked on top of each other. Stacking enables the electrons in the aromatic heterocyclic rings of the bases to interact with each other, and hydrophobic groups of the bases are positioned on the inside of the helix away from any aqueous solution.

Box 2.1 Hydrogen bonds

Hydrogen bonds (usually indicated in figures by a broken line) are non-covalent bonds that form between a hydrogen atom and an adjacent electronegative atom (usually either oxygen or nitrogen). To participate in a hydrogen bond, the hydrogen atom must itself be covalently bound (indicated by an unbroken line) to a strongly electronegative atom like oxygen or nitrogen that affects the distribution of charges across the bond such that there are slight positive (δ^+) and negative charges. The characteristic properties of hydrogen bonds are as follows.

1. They are relatively weak non-covalent interactions.
2. They are longer than covalent interactions. Positively charged hydrogen atoms will behave as hydrogen bond donors, which can interact electrostatically with closely located and partially negatively (δ^-) charged atoms (called hydrogen bond acceptors).
3. Straight hydrogen bonds are strongest, and so they tend to form between molecules with complementary shapes and opposing hydrogen bond donors and acceptors.

The helical structures formed by DNA and RNA have many important similarities (Fig. 2.5). These

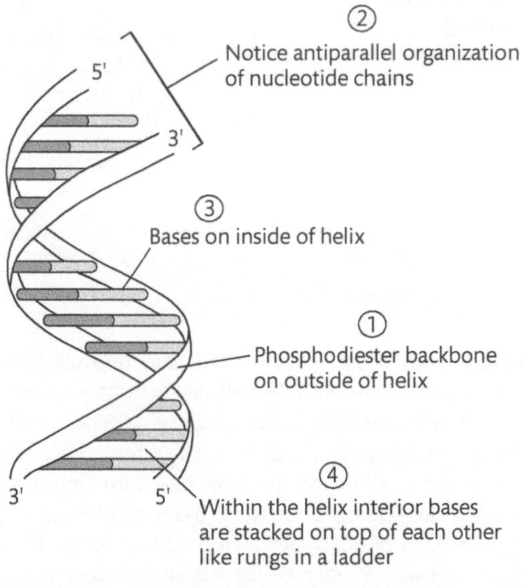
②
Notice antiparallel organization of nucleotide chains
5'
3'
③
Bases on inside of helix
①
Phosphodiester backbone on outside of helix
④
3'
5'
Within the helix interior bases are stacked on top of each other like rungs in a ladder

Figure 2.5 Four important properties of nucleic acid helices.

Table 2.1 Summary of differences between RNA and DNA double helices

Nucleic acid	Type of helix	Base pairs allowed	Lengths of helices
RNA	A-helix	A–U, G–C, and non-Watson–Crick base pairs such as G–U	Often helices are shorter than the full length of RNA molecules, and form as a result of short regions of complementarity
DNA	B-helix	A–T, G–C	Long double helix forms between complementary DNA strands

similarities include that RNA double helices are held together by hydrogen bonding between the bases of nucleotides, including Watson–Crick base pairing (see Fig. 2.6a). However, RNA and DNA double helices have a number of differences (summarized in Table 2.1). One important difference is that, unlike in DNA helices, non-Watson–Crick base pair combinations can sometimes hold RNA helices together.[3] Investigations into the atomic structure of RNA molecules have indicated that, in fact, most nucleotides can base pair with each other within RNAs. Because the width of the paired bases can vary as a result of these unusual pairings, the phosphodiester backbone distorts to accommodate non-Watson–Crick base pairs (Fig. 2.6 (b) and (c) shows some common non-Watson–Crick base pairing).

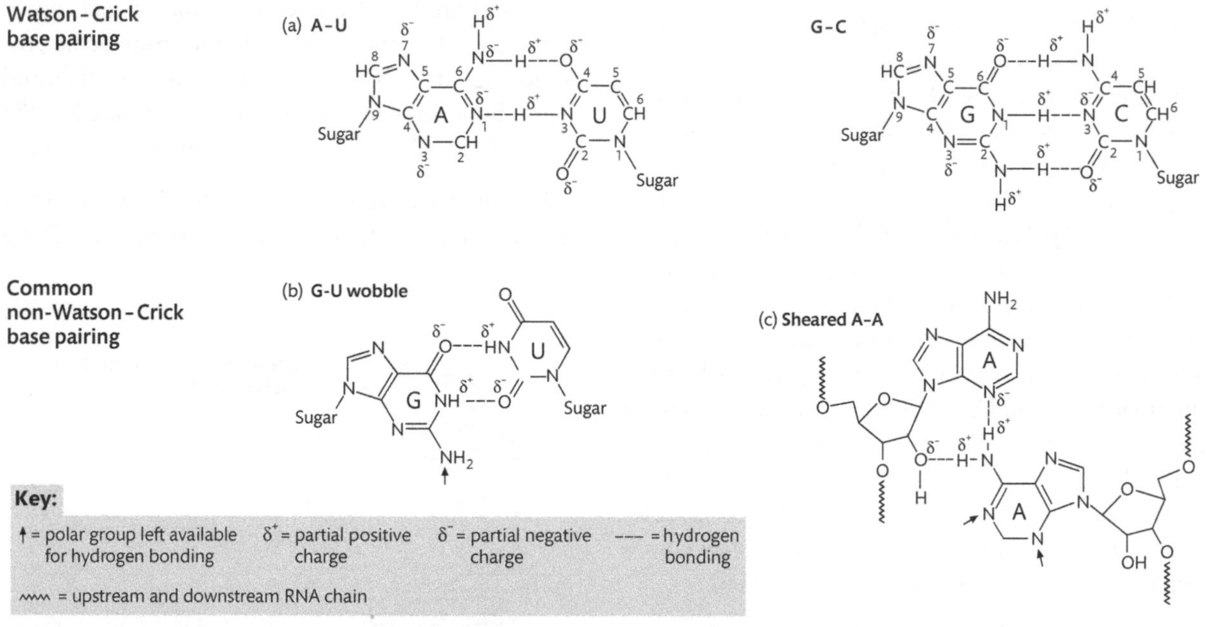

Figure 2.6 **Both usual Watson-Crick and unusual kinds of base pairing are found in RNA helices.** (a) Watson-Crick base pairing in RNA double helices takes place between A–U and G–C (according to rules worked out for DNA double helices these combinations do not distort the DNA helix). (b) Non-Watson-Crick pairing can also occur between bases, most frequently including hydrogen bonding between G–U base pairs. The G–U base pair has two hydrogen bonds like an A–U base pair but a slightly different shape, so it introduces a minor but allowable distortion to an RNA helix. The G–U base pair is called a wobble base pair. (c) *Ribose sugars* can also form hydrogen bonds between nucleotides. In one of these bonds called the sheared *A–A base pair*, the 2′ –OH on the ribose sugar of one of the adenosines is used as a hydrogen bond donor. This base pair leaves the Watson-Crick face of one of the nucleotides visible in the shallow groove of the helix (shown by arrows) and free to base pair with another nucleotide. (Parts (b) and (c) reprinted from Strobel and Doudna[2] with permission from Elsevier.)

2.3 RNA and DNA form different kinds of double helix

Although RNA and DNA double helices are similar in overall structure, double-stranded RNAs and DNAs form slightly different types of double helix called A-form and B-form double helices, respectively.

1 The B-form DNA double helix (see Fig. 2.7a). Within cells complementary DNA strands usually assemble into a B-form helix. The bases project at right angles to the axis of the B-form helix. Gaps called grooves lie between the phosphodiester backbone, and spiral around the outside of the helix. The B-form helix has a deep major groove and a shallow minor groove. These grooves are important for protein–DNA interactions. DNA double helices are typically long, symmetric, very stable, and held together by Watson–Crick hydrogen bonding between A–T and G–C pairs.

2 The A-form RNA double helix (see Fig. 2.7b). A-form helices are slightly more compact than the B-form helices formed by DNA, and the bases on the inside of the helix are tilted rather than being at right angles to the axis of the helix. This difference in shape is because the electrically charged 2′ -OH group in the ribose sugar prevents RNA from forming the less compact DNA B-form helix.[4] The 2′ -OH group of ribose affects sugar pucker which, in turn, affects the inter-phosphate distance, giving rise to the different

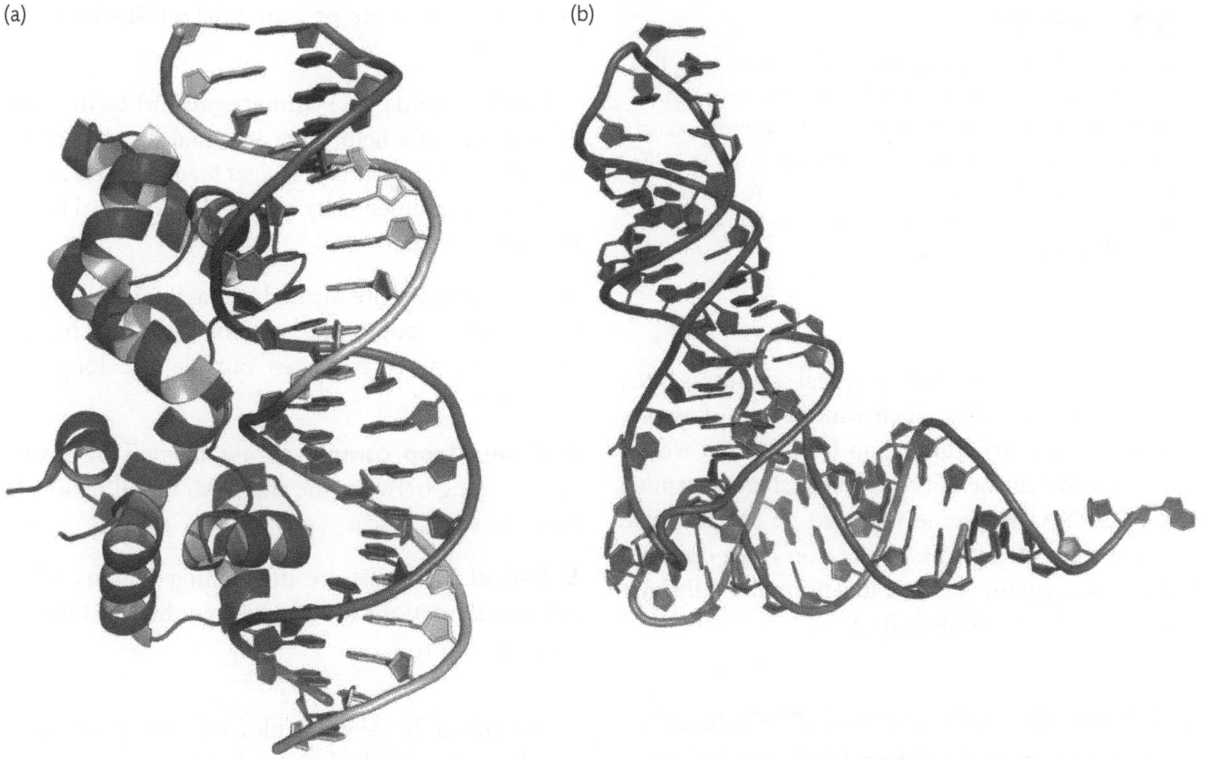

Figure 2.7 Double-stranded RNA forms an A-type helix while double-stranded DNA usually forms a B-type helix in cells. This figure shows (a) a B-form helix DNA molecule next to (b) an A-form helix tRNA molecule. Notice there are two distinctive features that distinguish the A- and B-form double helices: (1) the A-form helix is more compact (there are 11 nucleotides/turn of helix in A-form DNA compared with 10 nucleotides/turn of helix in B-form DNA); (2) in B-form DNA the bases are at right angles to the phosphodiester backbone of the DNA, while in A-form RNA the bases are tilted relative to the phosphodiester backbone. In the RNA helix the major groove is very deep and narrow compared with the major groove of DNA. This means that proteins project into the major groove of DNA but not RNA. Here α-helices of the C2 repressor protein are shown projecting into the DNA major groove containing the binding site sequences for this protein. (RNA and DNA structures redrawn from the RCSB Protein Data Bank by Jonathan Crowe, Oxford University Press.)

helical structure. For the same reason (namely, the presence of the 2′ -OH group) a DNA–RNA hybrid will also form an A-form helix. Unlike the B-form helix, in the A-form helix the major groove is deep and narrow, making the bases fairly inaccessible for protein interactions.

The formation of RNA helices can be controlled. There is evidence that the bacterium *Escherichia coli* RNA polymerase can stall during transcription to enable newly synthesized RNA to fold properly before continuation of polymerization.[5] There are also a set of enzymes called **RNA helicases** which unwind RNA helices and so dynamically reorganize RNA structures (RNA helicase enzymes can also release proteins bound to RNA).[6]

Key points

Double helices form as a result of hydrogen bonding between bases which project from the phosphodiester backbone. Bases contain a number of chemical groups that act as possible hydrogen bond donors and acceptors. In addition to Watson–Crick base pairing between A–U and G–C in double helices, a number of other hydrogen bonds can form in RNA helices.

Because they form shorter helical structures RNA molecules can adopt a much wider range of associated secondary structures than DNA, which we are going to consider next. (Note that a single-stranded DNA (ssDNA) molecule will also be able to form many of these structures. The reason DNA does not form these structures is that it is normally fully base paired in a DNA double helix.)

2.4 Five common secondary structure motifs are found within RNA molecules

By forming a number of shorter helical regions connected by single-stranded regions, RNA can form a much greater diversity of secondary structures than the extended double helix of DNA. Five frequent forms of RNA secondary structure motifs[7] are shown in Fig. 2.8.

1 Helices are the basic RNA secondary structure. Remember RNA helices are formed through base pairing between antiparallel complementary nucleotide sequences—this can occur either within an RNA molecule, or between RNA molecules. This antiparallel organization is indicated by opposing arrows in Fig. 2.8.

2 Loops are single-stranded regions within helices. A **hairpin stemloop**[8] forms when nearby regions of complementary nucleotides form a short 'hairpin' helix, separated by a sequence which forms the loop. The loop sequences are themselves important; they allow the RNA strands to bend back on themselves to allow the helices to form. Adjacent complementary regions of RNA cannot fold back to form helices without some sort of internal loop. The shortest possible intervening loop is likely to be three or four nucleotides long (see Fig. 2.8b).

Internal loops are symmetrical, and form where two strands of a helix have an equal number of unpaired bases; by contrast, *bulges* form where a region of non-complementarity on one RNA strand bulges out of the helix.

3 Pseudoknots[9] are formed by base pairing of RNA sequences in loops (such as those found in hairpin stemloops) and sequences outside the loop (see Fig. 2.8c).

4 Kissing loop complexes are formed by hydrogen bonding between the single-stranded regions of loops (see Fig. 2.8d).

5 Helical junctions are the joining regions which act as intersections to link different helices together (see Fig. 2.8e).

Individual RNA molecules will often contain a number of individual secondary structure motifs. Figure 2.9 shows an example of secondary structures in the RNAse P RNA which is involved in processing tRNA,[10] and which we also encountered in the last chapter (Fig. 1.2). Notice that several secondary structure motifs are present in this RNAse P molecule, in addition to non-Watson–Crick G–U base pairing within helices.

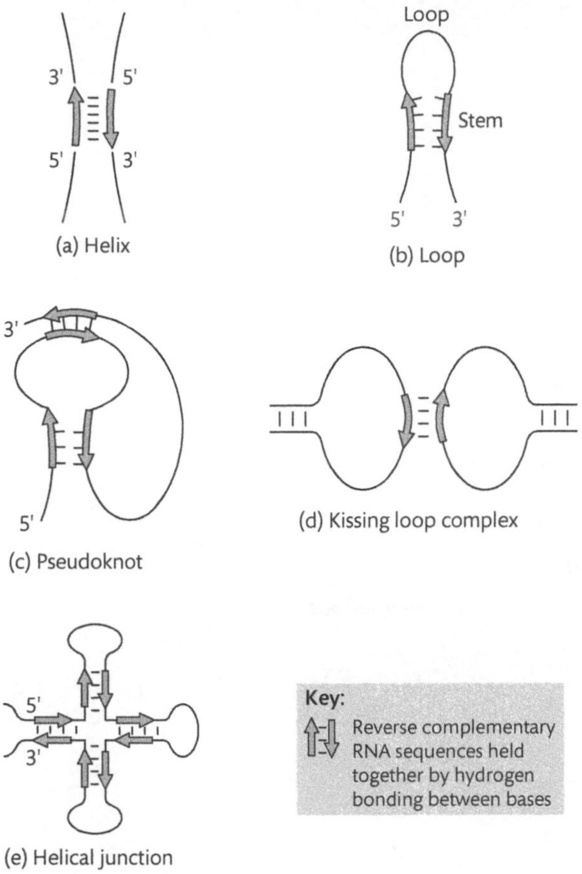

Type of RNA secondary structure	Description
Helix	A-form helix formed by hydrogen bonding between bases.
Loop	Region within a helix which is not base paired.
Pseudoknot	A pseudoknot forms by base pairing between a single-stranded region of RNA and a loop.
Kissing loop complex	Base pairing interaction between two loop sequences.
Helical junction	Junction where helices meet. A three-way junction is formed between three helices etc.

Figure 2.8 Base pairing between nucleotides can produce different kinds of RNA secondary structure.

Key points

RNAs in the cell can either be single stranded or form secondary structures held together by hydrogen bonds which form between complementary bases on antiparallel strands. There are five common secondary structure motifs in RNA.

2.5 RNA secondary structures can be worked out experimentally and predicted bioinformatically

The secondary structures of RNA molecules have been worked out using experimental techniques like SHAPE (see Fig. 2.10).[11,12] SHAPE is based on a chemical reaction that modifies only the single-stranded nucleotides within RNA molecules. After this reaction, modified nucleotides can no longer be copied by reverse transcriptase, causing a stop in cDNA synthesis. Mapping these stops in cDNA synthesis shows where the single-stranded regions were in the starting RNA molecule. Notice this protocol uses a reverse transcriptase step, and this enzyme is blocked by the modified nucleotides.

RNA secondary structures have been worked out using these SHAPE approaches for the whole RNA genome of the HIV pathogen that causes AIDS (see Fig. 2.4), and also for mammalian transcriptomes. Such experiments showed that different parts of individual mRNAs have a different propensity to form secondary structure, with the 5′ and 3′ **UTRs** (untranslated regions) having much more

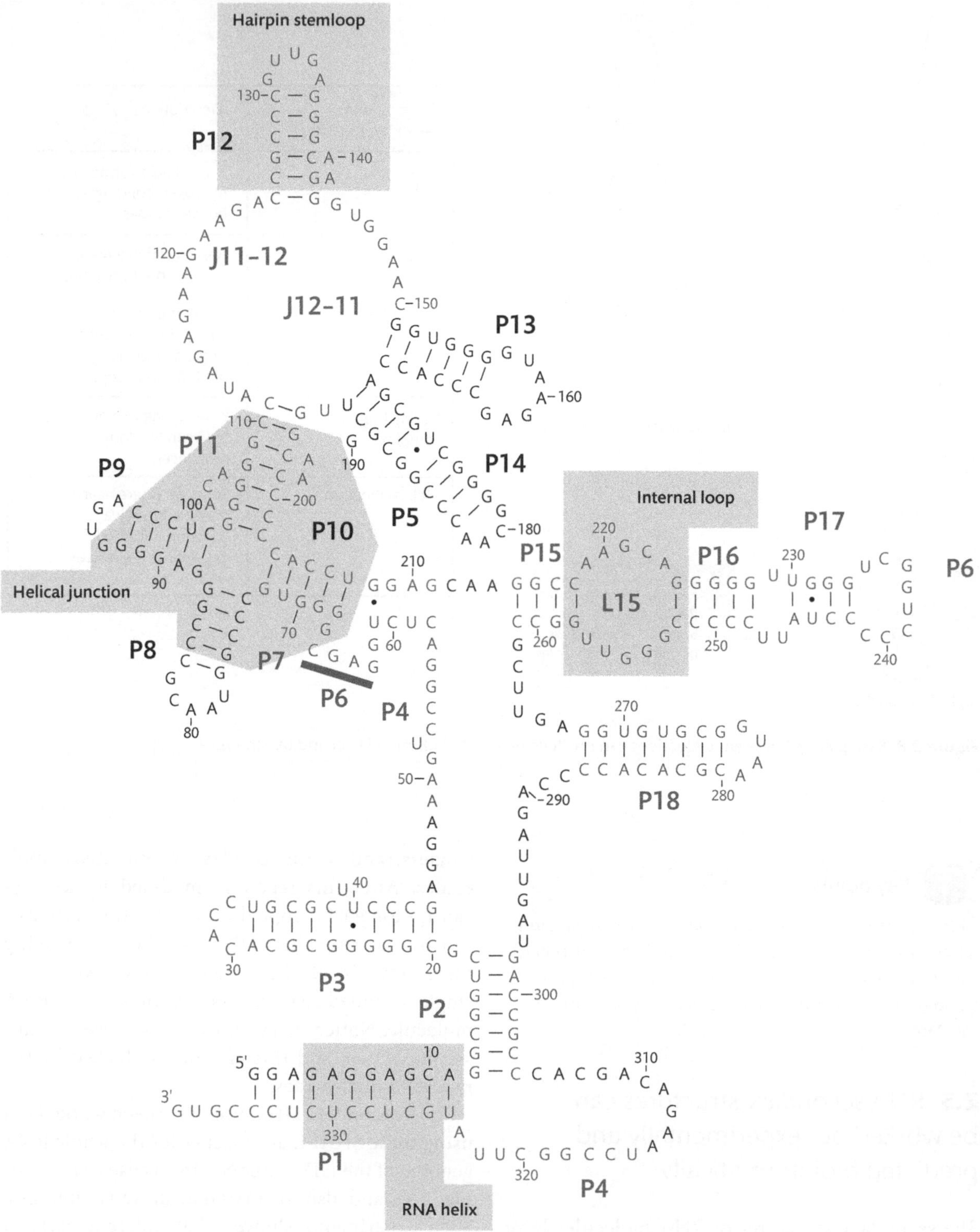

Figure 2.9 Long folded RNAs like RNAse P contain a number of different secondary structure motifs. Non-Watson–Crick base pairing is shown as circles between bases. (Reprinted from Torres-Larios et al.[10] with permission from Macmillan Publishers Ltd.)

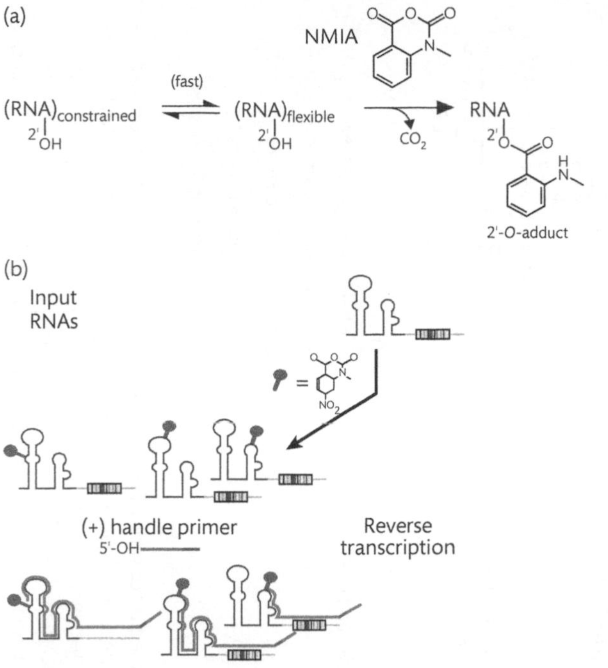

Figure 2.10 **RNA structures can be experimentally determined using selective 2′-hydroxyl acylation analysed by primer extension (SHAPE).** The chemical N-methylisatoic anhydride (NMIA) reacts with single-stranded nucleotides only at the 2′-OH group of the ribose sugar. This produces a chemical modification in the RNA molecule, that prevents passage of the enzyme reverse transcriptase during cDNA synthesis. Part (a) shows the chemical reaction, and part (b) the extension of a primer on the RNA using reverse transcriptase to make cDNA molecule (the primer and cDNA copy is the thick grey line: notice this stops at the chemical modification of the RNA template). By mapping where these 'stops' occur in an cDNA synthesis compared to a sequencing ladder as a marker, it is possible to find where the single-stranded nucleotides are in the original RNA molecule. (Adapted from Lucks et al.[12] with permission.)

secondary structure than the open reading frames (see Fig. 2.11).

Why might this difference be important? As we discuss in more detail in Chapter 16, UTRs are targeted for binding by groups of short RNAs called microRNAs. Thus UTR secondary structure might be important to prevent inappropriate binding of miRNAs to mRNAs (this existing base pairing would prevent subsequent miRNA base pairing). Another observation is that regions of secondary structure can be found within mRNAs at key points between the coding information for protein domains. Such regions of secondary structure might be important for slowing the ribosome and stalling translation. It is thought that these regions of secondary structure may allow nascent protein sequences to fold before new protein sequences are translated. In contrast, start codons and stop codons tend to be free from secondary structure to enable the ribosome easy access.

RNA folding patterns can also be predicted by comparing the sequences of RNA molecules between species. These so-called phylogenetic comparisons (see Fig. 2.12) give clues about important RNA secondary structures. Because of the rules of base pairing, for nucleotides involved in base-paired helices, any changes in RNA sequence between species need to take place in pairs to maintain hydrogen bonding. For example, if a G is mutated to a U in one strand of an RNA helix, the original complementary C nucleotide on the other strand of the helix will have to change to A to maintain hydrogen bonding. By maintaining paired changes

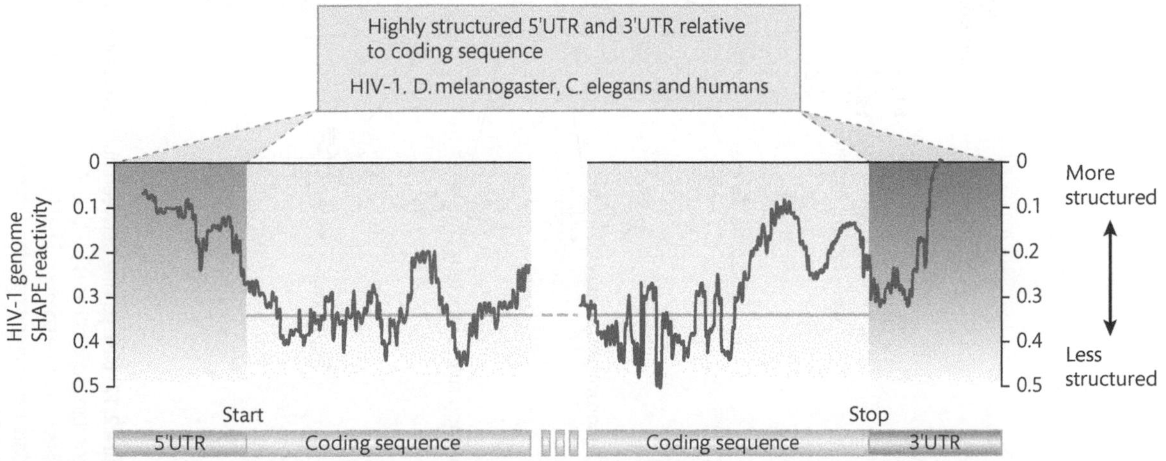

Figure 2.11 **There is more RNA secondary structure in the untranslated regions of mRNAs.** (Reproduced from Mortimer et al.[11] with permission.)

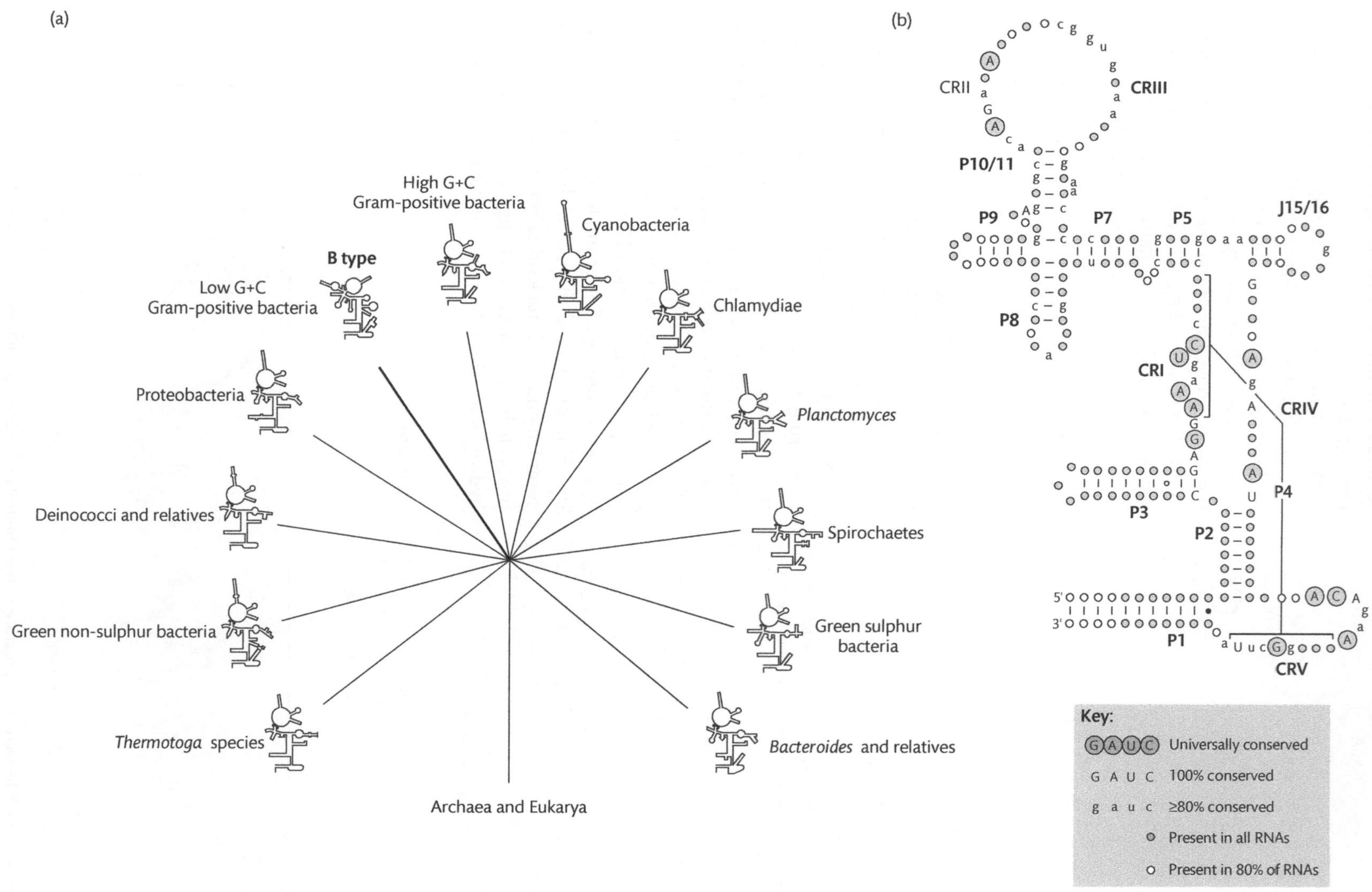

Figure 2.12 **Base pairing between nucleotides is often phylogenetically conserved even if the actual RNA sequence varies.** Base pairing most frequently occurs between particular combinations of bases (A–U, G–C, and G–U). If a part of an RNA molecule is involved in base pairing to form an important helix, nucleotide changes in that part of RNA between species should be compensated by secondary changes in its partner RNA strand to preserve base pairing. Changes in RNA sequence between species would occur in pairs to accommodate base pairing patterns. From such phylogenetic comparisons of RNA sequences it is possible to predict the helices of RNA secondary structure. (a) Comparison of shapes of RNAse P from different bacterial species. (b) Consensus structure—notice that base pairing is conserved even though actual nucleotides are not necessarily. (Reprinted from Kazantsev and Pace[14] with permission from Macmillan Publishers Ltd.)

in the nucleotide sequence to preserve hydrogen bonding, structured RNAs can maintain their secondary structure in different species even if the nucleotide sequences of the RNAs show variation. Figure 2.12 shows how RNAse P in different bacterial species has a similar shape.[14] Note that RNAse P is a non-coding RNA, so there is no requirement to maintain nucleotide sequence for coding information.

A given RNA sequence can usually fold into a number of different combinations of helices. Bioinformatic programs can calculate which RNA structures are energetically most likely to form for a given input sequence.[14] The most likely structures will have the lowest free energy. One of these programs, the M-fold program, is frequently used for analysing RNA secondary structure in this way and is available on the World Wide Web (http://mfold.rna.albany.edu/?q=mfold).

Key points

RNA secondary structures can be experimentally determined, and predicted from phylogenetic conservation. Detection of such covariation enables theoretical secondary structures of RNA molecules to be deduced. Some computer programs calculate the most likely RNA structure to form.

2.6 The formation of RNA helices is stimulated by positively charged molecules and particularly metal ions

The formation of RNA secondary structures depends to some extent on the environment. The reason for this is that efficient formation of RNA double helices faces a problem: the oxygens in the phosphate groups of the phosphodiester backbone of RNA each have a strong negative charge, which makes RNA molecules negatively charged overall. In principle, negatively charged RNA molecules will have a tendency to repel one another rather than undergo base pairing and form helical secondary structures.

This is where environmental effects come in. Although RNA strands will show a natural tendency to repel each other, stable RNA secondary structures are promoted when RNAs are in solution with positively charged molecules.[16–18] Notice in Fig. 2.1 that these positively charged molecules are shown as + signs. Positively charged molecules have two roles.

1. They help RNA strands to come together in helices by balancing the strong negative charges of the RNA phosphodiester backbone which would otherwise repel each other.
2. They stabilize the burying of hydrophobic regions of bases within the interior of double-helix structures, with the charged phosphodiester backbone on the outside.

Metal ions provide important positive charges for helping RNA secondary structures to form, particularly magnesium ions because of their high charge density. Figure 2.13 illustrates how positively charged

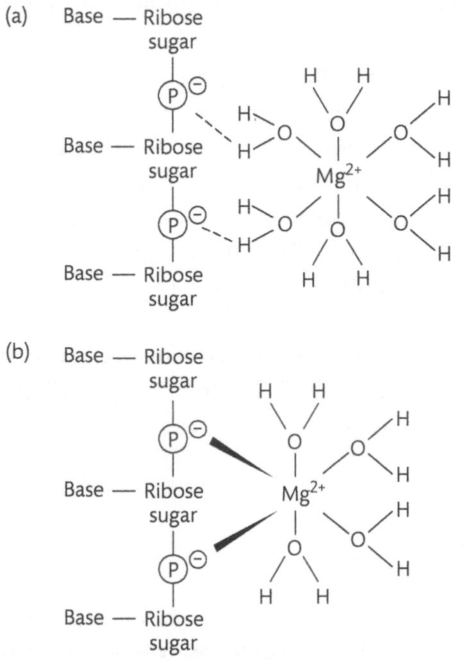

Figure 2.13 **Metal ions bind to RNA and help to stabilize the formation of RNA helices.** Magnesium ions in water are usually surrounded by six water molecules to give magnesium hexahydrate. (a) Hydrogen atoms in the water molecules in magnesium hexahydrate can bond to oxygen molecules in the phosphate groups of RNA. Although not shown here, magnesium ions can interact with other charged groups in RNA as well as the phosphates, particularly to polar groups which are exposed at the edges of the bases and the ribose 2′ -OH groups. (b) Although magnesium hexahydrate is coordinated with water, the water molecules can be replaced by phosphate ions to enable direct coordination with RNA.

magnesium ions interact with an RNA phosphodiester backbone. Notice that there are two possibilities.

1 Indirect interactions. Magnesium ions can interact indirectly with RNA through water molecule intermediates. In solution, magnesium ions normally interact with six water molecules (to give magnesium hexahydrate). Water molecules complexed with Mg^{2+} ions can hydrogen bond with the negatively charged oxygens in the phosphate groups of RNA. Notice that the hydrogen bonds in Fig. 2.13 are shown as broken lines.

2 Direct interactions. Positively charged magnesium ions can also directly interact with negatively charged oxygens in the phosphodiester bond. In this case, electrostatic interactions with oxygen atoms in some of the water molecules which surround magnesium hexahydrate are directly replaced by similar interactions with the oxygen atoms in the phosphate groups of RNA.

As well as stabilizing the formation of RNA double helices, some metal ions bound by RNA can also be important in catalysis. We discuss the role of metal ions in catalysis in Chapter 3. Metal ions also play an important role in establishing the tertiary structures of RNA molecules, where they act as a glue to hold RNA helices together. We explore RNA tertiary structure further in the next section where we look at how RNA secondary structure structures are linked together to give RNAs with complex shapes.

Key points

The overall negative charge of RNA molecules—which results from the electronegative oxygens drawing electrons into the phosphate groups of the phosphodiester backbone—can be electrically balanced by positively charged metal ions, particularly sodium (Na^+), potassium (K^+), and, most importantly, magnesium (Mg^{2+}). Magnesium ions are particularly effective in assisting RNA folding because of their small size and high charge density. The binding of metal ions to RNA balances negative charges in the phosphodiester backbone and enables RNA strands to associate with each other to form duplexes. Without this stabilization, negatively charged RNA strands would tend to repel each other.

2.7 RNA molecules use a set of strategies to build tertiary structures

RNA molecules fold into highly complex tertiary structures which can rival protein structures in terms of their sophistication.[10,14] Structures of some large RNA molecules have been solved using techniques such as nuclear magnetic resonance (NMR) and X-ray crystallography. For example, in the large subunit rRNA (ribosomal RNA), over 100 distinct RNA helices are held together in a stable tertiary structure.

How are these tertiary structures formed, particularly given the challenge of bringing together negatively charged molecules in close proximity?[2,19] RNA molecules use three main strategies to pack together negatively charged helices to form compact overall shapes. Each of these strategies has been used in several different RNA tertiary structures, indicating that they are general tools used to construct RNA shapes. To illustrate these strategies we are going to look at part of the catalytic RNA of the *Tetrahymena* Group I intron.[20] (We revisit this RNA in Chapter 3.)

The Group I intron structure is formed from a number of RNA helices and loop secondary structures, and these fold into a compact tertiary RNA structure (see Fig. 2.14). The three strategies used for assembling RNA tertiary structures in the Group I intron are as follows (see Table 2.2).

1 Coaxial stacking. The shorter helices formed by RNA molecules are often stacked on top of each other to give longer helical structures that are more stable. These are called coaxially stacked helices; notice the arrangement of helices on top of each other in Fig. 2.14. Another example of coaxial stacking of RNA helices is found in tRNAs—see Fig. 11.5, and notice that the acceptor stem stacks on top of the T loop, and the D loop stacks on top of the anticodon loop in the tertiary structure. Stacked helices are more stable since they help bury the hydrophobic groups of the bases from the aqueous environment.

2 Hydrogen bonding between nucleotides helps to hold RNA tertiary structures together. While Watson–Crick and other hydrogen bond

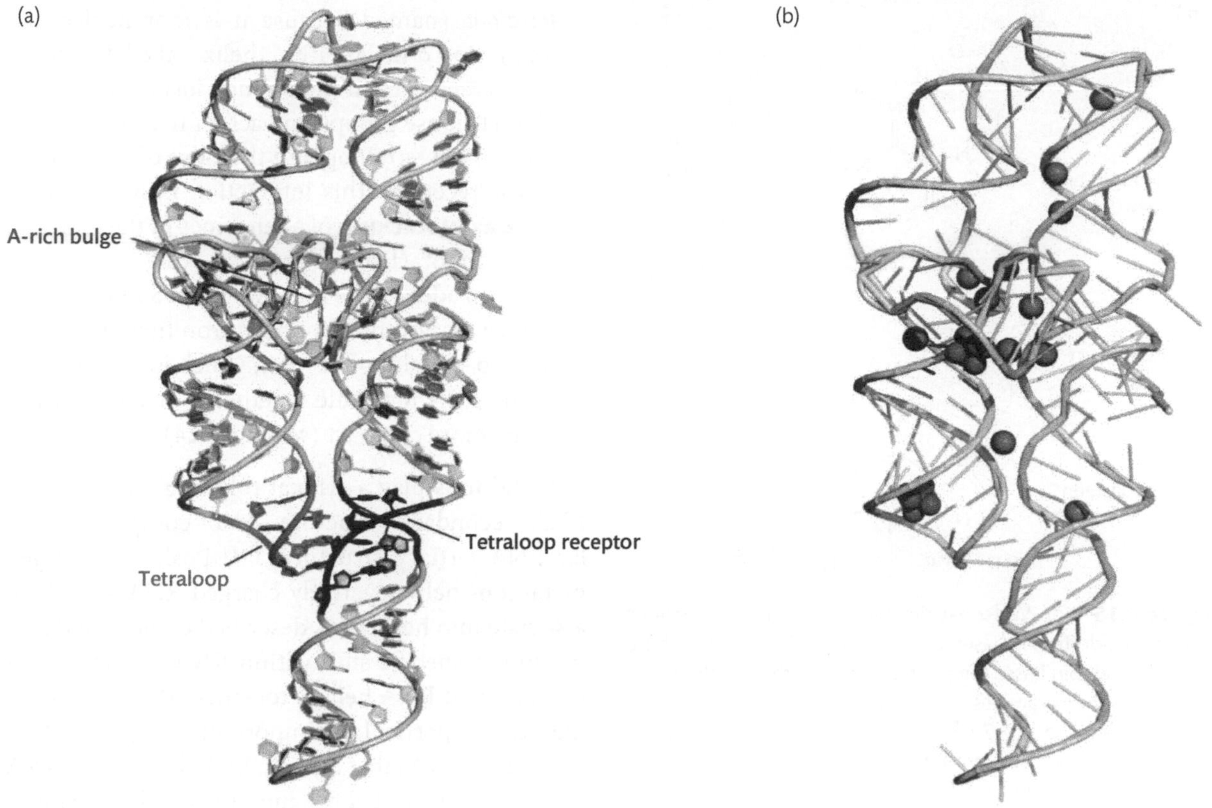

Figure 2.14 Some important tertiary structures contributing to the shape of the Group 1 intron. The Group I intron structure is extremely well characterized and has established many of the principles through which much longer RNAs, including those found in the ribosome, are folded into globular structures. Two important features are (a) a tetraloop–tetraloop acceptor interaction that makes the shape of the molecule more compact, and (b) metal ions that help stabilize negatively charged RNA strands (blue spheres, cobalt hexamine; red spheres, magnesium ions). For a full colour version of this figure, see Colour Illustration 2. (Redrawn from the RCSB Protein Data Bank 1GID by Alice Mumford, Oxford University Press.)

interactions hold RNA double helices together to create RNA secondary structures, other types of hydrogen bond interactions link single-stranded regions of RNA and helices together in RNA tertiary structures.

a Base triples. As well as the hydrogen bond donors and acceptors used for Watson–Crick base pairing, other chemical groups in bases can be used for hydrogen bonding interactions. Figure 2.15 shows hydrogen bonding interactions between an A nucleotide and an A–U base pair, and between a G nucleotide and a G–C base pair. Notice that in both these cases a normal Watson–Crick base pair forms but additional hydrogen bonds are used to

Table 2.2 RNA tertiary elements holding together the Group I intron RNA structure

RNA tertiary element	Role in tertiary structure
Coaxial stacking	RNA helices stack on top of each other in tertiary structures
Ribose zippers	Binds two parts of the RNA molecule together
Metal ions	Positively charged ions enable RNA strands and helices to come close together

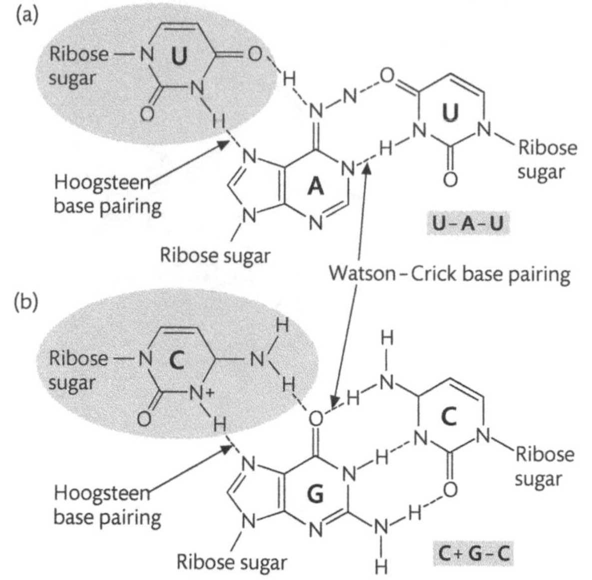

Figure 2.15 Non-Watson-Crick hydrogen bond interactions can hold additional bases onto existing Watson-Crick base pairs, and so join bases together in groups of three (called base triples). The two examples here are a U bound to a U-A base pair, and a C bound to a C-G base pair. Notice that the adenine and guanidine use two different faces to enable both Watson-Crick pairing and a second kind of hydrogen bond called Hoogsteen pairing to occur at the same time. (Reprinted from Strobel and Doudna[2] with permission from Elsevier.)

attach the third base. These extra hydrogen bonds are called *Hoogsteen bonds*, and can be used simultaneously to build up groups of interacting nucleotides in RNA tertiary structure.

b Ribose zippers are an important group interaction holding together RNA tertiary structures.[21] They involve the participation of ribose 2′ -OH groups in hydrogen bonding. This helps RNA strands to interact since the 2′ -OH groups on the ribose sugars of RNA molecules project to the outside of RNA helices and so are available for hydrogen bonding with adjacent RNA strands, specifically with charged groups on the sides of bases which are not involved in Watson–Crick base pairing. In RNA tertiary structures, ribose zipper interactions join together RNA strands like a zipper on a piece of clothing.

An important group of ribose zippers are called tetraloop–tetraloop receptor interactions. In these, a single-stranded RNA sequence called the *tetraloop* (named because it is four nucleotides long) zips onto an RNA helix called the *tetraloop receptor*. Hydrogen bonds form between the 2′–OH ribose groups on each of the four nucleotides in the tetraloop and the sides of nucleotides in the receptor. This interaction is strengthened by **coaxial stacking** of adenosines in the loop and the receptor. (Each of these stacking interactions between adenosines is shown as an S in Fig. 2.16.) Notice that in the Group I intron there is one tetraloop–tetraloop receptor, which joins two parts of the RNA molecule together to make the RNA shape more compact (see Fig. 2.14).

3 Metal ions play a critical role in gluing together RNA secondary structures into compact globular RNA tertiary structures.[16–18] Positively charged metal ions help negatively charged RNA strands to associate into helices (as described earlier), and also to bind to specific sites within RNA tertiary structures to hold RNA helices together. Metal ion binding sites are particularly important in regions within structured RNAs that contain high densities of RNA strands (such as helical junctions). The positions of known metal ions in the Group I intron tertiary structure are shown in Fig. 2.14 as spheres. Notice that both these metal binding sites have a high concentration of tightly packed RNA strands.

Magnesium ions are very important for helping RNA tertiary structures to form, followed by sodium ions. There can be many metal ions within complex

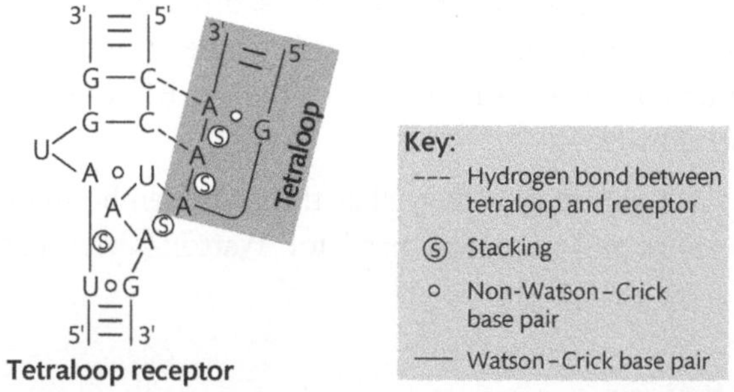

Figure 2.16 The tetraloop-tetraloop receptor interaction holds together a stemloop structure and an RNA helix. Interactions take place through hydrogen bonding between the 2′ -OH groups on ribose sugars and by base stacking. The name tetraloop refers to the four GAAA nucleotides in the loop, although the G forms a non-Watson-Crick base pair with the last A residue in the loop (indicated by a circle).

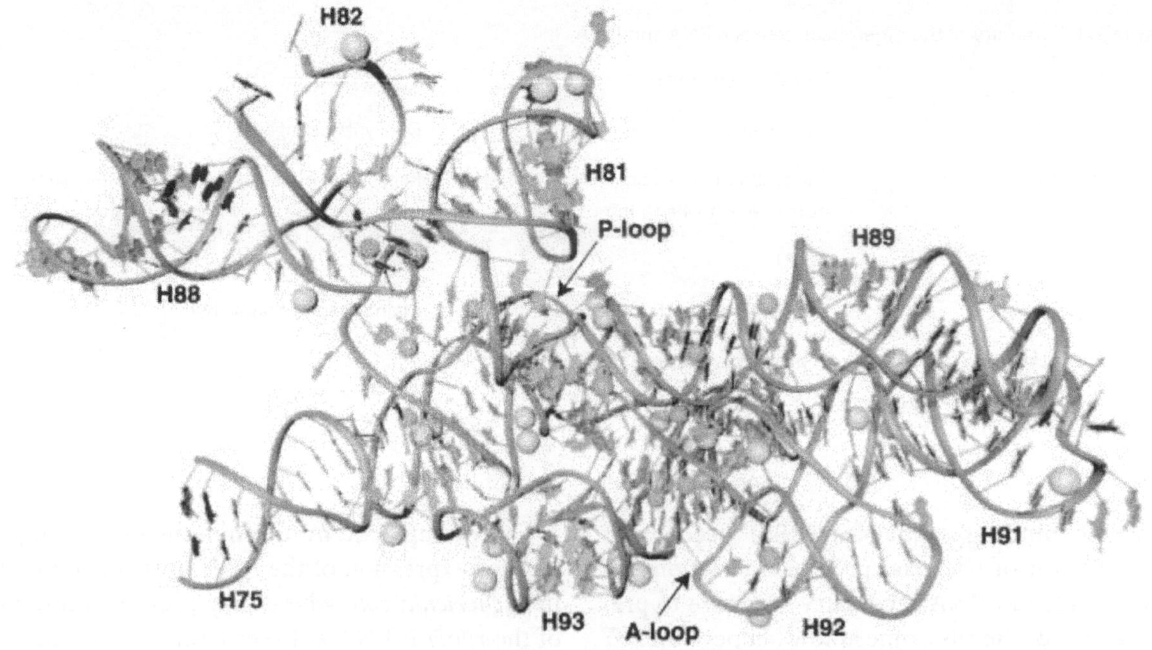

Figure 2.17 **RNA tertiary structures are frequently stabilized by binding of metal ions.** Positively charged metal ions help to stabilize RNA tertiary structures by binding to specific RNA sequences within and around helices. This RNA tertiary structure is that of the peptidyl transferase centre of the ribosome, determined by X-ray crystallography. Notice that the metal ions are shown as spheres. Metal ions in RNA structure function to nucleate RNA tertiary structure, in a manner analogous to the hydrophobic cores of protein tertiary structures. (Reprinted from Klein et al.[22] with permission from Cold Spring Harbor Laboratory Press.)

RNA structures. The RNA tertiary structure of the peptidyl transferase centre of the ribosome which helps catalyse peptide bond formation is shown in Fig. 2.17.[22] Notice that multiple metal ions, including magnesium and sodium ions, are embedded within this structure. These positively charged metal ions help to assemble RNA helices and pack them together in the tertiary structure of the ribosome. In this way metal ions help to hold different parts of the structured RNA together. In using metal ions in this way, complex RNA tertiary structures like the ribosome can fold around a core of metal ions, in a manner similar to the way hydrophobic regions of proteins become embedded in the interior of the protein tertiary structure.

Key points

RNAs can fold into complex tertiary structures which rival those of proteins in terms of their intricacy. Tertiary RNA structures are held together by interactions between distant regions of the RNA, which knit the RNA shape together. Metal binding sites bind metal ions, which form a positively charged core around which RNA molecules can fold.

2.8 Summary of how RNAs build structured molecules

At the start of this chapter we noted that RNA and DNA molecules have quite different repertoires of secondary and tertiary structures despite being formed from somewhat similar chemical subunits (Table 2.3 shows a summary of these differences). The end result is that while DNA forms extended double helices, RNA molecules are able to form distinct structures with an array of functionally important roles in the cell.

In the next sections we shall discuss examples of biologically important RNA structures. First, we consider how RNA secondary structures enable bacteria to sense changes in their environment.

2.9 RNA structures can be used as thermosensors

Some bacteria have developed methods that use RNA secondary structure to control translation including in response to ambient temperature. The

Table 2.3 Summary of the differences between RNA and DNA

Property	Version used in RNA	Version used in DNA
Sugars	Ribose sugar	Deoxyribose sugar
Mode of synthesis	Transcription makes single strands of RNA from one template strand	Replication makes two complementary strands which can base pair with one another
Bases	ACGU	ACGT
Secondary structure	Short regions of A-form of double helix	Mainly B-form double helix
Tertiary structure	Sometimes complex folding patterns produce compact globular molecules	

ribosome binding sequences (RBS) needed for the initiation of bacterial protein translation can be embedded in RNA secondary structure to prevent access of the ribosome at low temperature.[11,23] Increases in temperature can then melt these secondary structures, enabling ribosome binding to the RBS and translation initiation. Such temperature dependent elements are called thermosensors (see Fig. 2.18).

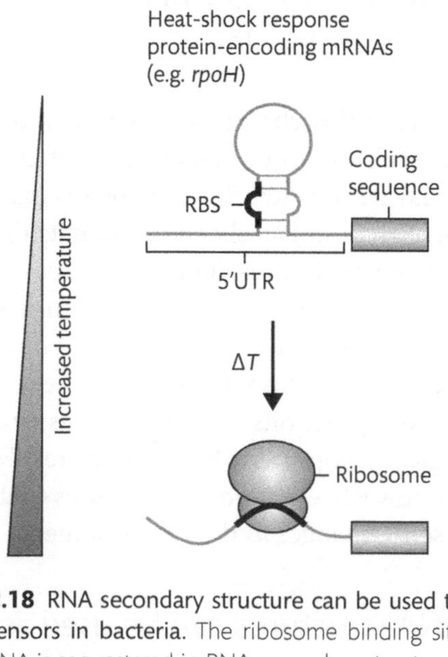

Figure 2.18 RNA secondary structure can be used to create thermosensors in bacteria. The ribosome binding site of the *rpoH* mRNA is sequestered in RNA secondary structure, but this melts at increased temperature, allowing access of the ribosome and subsequent translation. Translation of the *rpoH* mRNA produces a sigma factor for RNA polymerase which coordinates the heat shock response of *E. coli*. (Reproduced from Mortimer et al.[11] with permission.)

For example, RNA thermosensors are important for expression of the *rpoH* heat shock protein in *Escherichia coli*, where they prevent translation of the *rpoH* mRNA at lower temperatures, but enable translation after heat shock. The *rpoH* mRNA in turn encodes a component of RNA polymerase II (called a sigma factor) needed for transcription of downstream genes involved in the heat-shock response. By forming different RNA secondary structures at different temperatures, other thermosensors can activate gene expression only at lower temperatures and enable translation of cold-shock proteins.

RNA secondary structure thermosensors also play an important role in some bacterial pathogens including *Yersinia pestis* (the causative agent of the Black Death) and *Listeria monocytogenes* (causes listeriosis, a form of meningitis in newborns transmitted through soft cheeses eaten by mothers).[23] In these cases, the increase in temperature that takes place when these pathogens enter a mammal is monitored by the thermosensor RNA secondary structure, and this leads to expression of virulence genes only in the host.

2.10 RNA structures can be selected which bind to target molecules

RNA can form structures which can bind to target molecules like a protein. RNA molecules can also be copied and amplified, which has enabled highly structured target-binding RNA molecules called

aptamers to be purified. Aptamers are analogous to protein antibodies in their affinity and specificity of interaction with their targets.

RNA structures that bind to specific target sequences have been isolated experimentally using an in vitro selection technique called **SELEX** (Systematic Evolution of Ligands by EXponential enrichment). The purpose of SELEX is to select from a randomized pool structured RNAs which bind to the molecule of interest. This 'bind and catch' selection procedure is shown as a net in Fig. 2.19; the net acts to 'catch' RNAs with particular binding properties (the aptamers).

For example, RNA aptamers which bind to the molecule theophylline have been isolated.[24] Theophylline is used in medicine as a bronchial dilator, and is chemically similar to caffeine found in coffee.

- **Step 1: Selection.** A library of random RNAs were passed through a column containing theophylline (as in Fig. 2.19). Random RNAs unable to bind to theophylline passed straight through the column and were discarded, while some specific RNAs had a structure which enabled them to become bound to the column. These theophylline-binding RNAs were then eluted from the column by running a solution of theophylline through it. This 'bound and eluted' group comprised two groups of RNAs: some RNAs that could specifically bind theophylline, and some contaminants. Bound and eluted RNAs were converted back into DNA by reverse transcriptase, amplified by PCR, and then transcribed back into a new pool of RNAs, completing a single round of SELEX. This procedure was carried out a total of five times to give a purer population of binding RNA molecules.
- **Step 2: Counter-selection.** Three further rounds of SELEX were carried out. In these further rounds a step was introduced to remove RNA

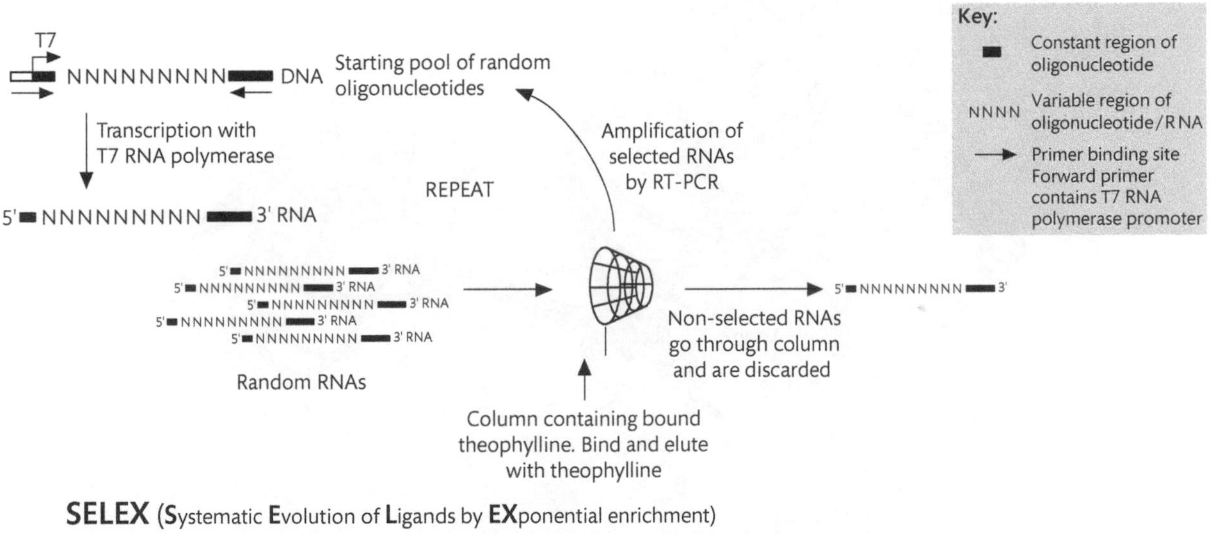

Figure 2.19 SELEX has enabled RNAs to be selected which specifically bind to ligands such as caffeine. A SELEX experiment starts with a pool of oligonucleotides, all of which contain a T7 promoter and so can be transcribed by RNA polymerase, and a randomized core sequence. This randomized core sequence should contain every possible DNA sequence combination. For example, a two-nucleotide randomized sequence would contain AA, AC, AG, AT, CA, CC, CG, CT ... to give 4^2 different combinations. A 25-nucleotide random sequence would contain 4^{25} variants (equal to 1.2×10^{15} different individual DNA sequences). Since there are a large number of possible DNA sequences even for a relatively short randomized 25-mer, each of these individual 25-mer sequences of DNA would only be found at a very low concentration in the pool. Transcription of this random pool of random oligonucleotides using T7 RNA polymerase produces in turn a random pool of RNA molecules. SELEX enables RNAs which bind to particular molecules to be pulled out of the random pool and selectively amplified. This selection is shown as a net catching RNAs. After selection, recovered RNAs can be converted back into DNA by reverse transcriptase and amplified by PCR. Amplified cDNAs can then be converted back into RNA for another round of SELEX.

> aptamers which could bind to caffeine, a closely related molecule. RNA aptamers were first eluted with caffeine (to remove cross-reacting aptamers) and then theophylline (to give specific anti-theophylline aptamers).

Such experiments enabled the purification of RNA aptamers that could very specifically bind to theophylline but not to the closely related caffeine molecule. The ability to select RNAs that can bind to specific target molecules has applications in medicine (where aptamers might substitute for antibodies in binding and detecting specific proteins). For example, aptamers have been isolated which can bind to molecules found in important pathogens, including components of HIV, which causes AIDS.[25] Some such aptamers are very effective in reducing the production of viruses from infected cells. For example, expression of aptamers specific to the reverse transcriptase protein of HIV reduced virus particle formation in infected cells by more than 95%.

Structurally, RNA aptamers require a specific shape and charge distribution to attach to their target molecule. (This is analogous to the way that an antigen is only recognized by an antibody whose epitope possesses a shape that is complementary to the antigen.) Some aptamer structures bound to their target sequences have been resolved at atomic resolution—this includes the aptamer which binds to and inhibits HIV reverse transcriptase (the RNA aptamer is shown as a thread in Fig. 2.20). The aptamer sequence recognizing reverse transcriptase contains

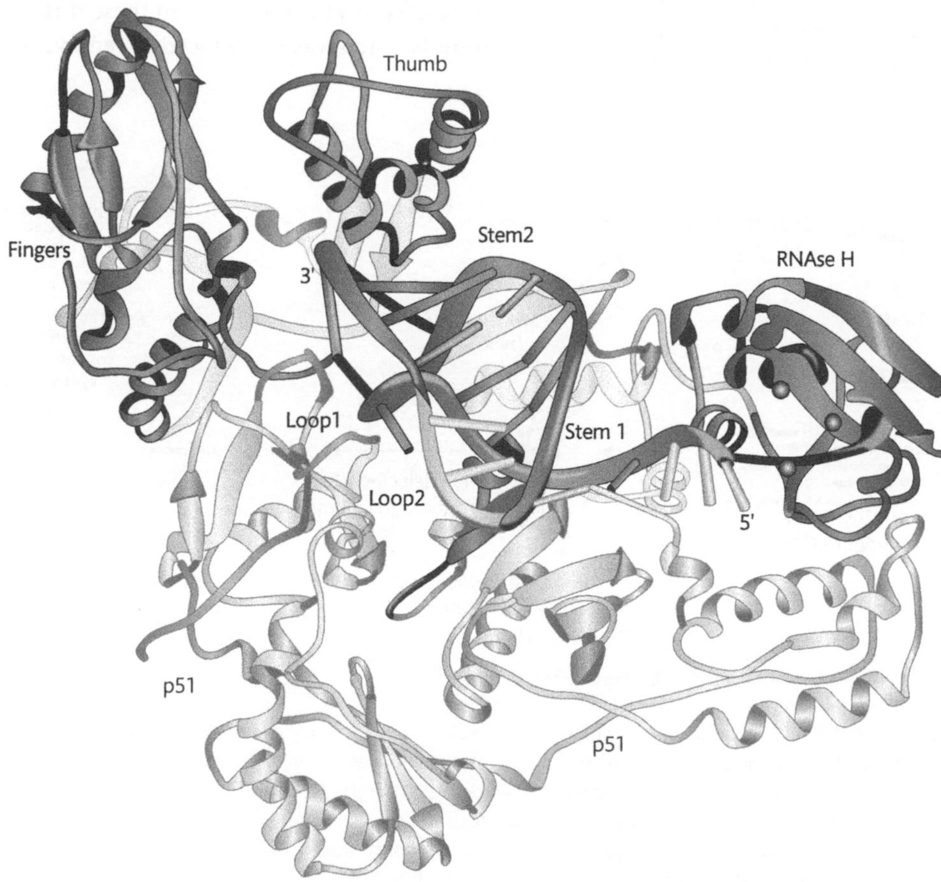

Figure 2.20 The structure of an RNA aptamer bound to the protein reverse transcriptase from HIV has been solved. The HIV reverse transcriptase protein heterodimer is shown as a standard protein ribbon diagram indicating pleated sheets (arrows) and α-helices (helices) with the different domains indicated (the fingers, thumb, and RNAse H domain of the p66 subunit, and the p51 subunit). The RNA aptamer has a complementary shape and so fits into this protein structure. For a full colour version of this figure, see Colour Illustration 3. (Reprinted from Jaeger et al.[25] with permission from Macmillan Publishers Ltd.)

a pseudoknot which binds to the substrate-binding site of reverse transcriptase protein. Pseudoknots are frequently found in aptamer structures because they have a rather flexible structure and so can produce different binding surfaces by switching hydrogen bonding partners. In addition, most protein ligands bound by RNA aptamers contain amino acids with aromatic groups. These aromatic groups behave like bases to stack with the RNA bases present in the RNA aptamer.

Key points

RNA molecules can fold into tertiary shapes which can bind to other molecules very tightly and specifically. The specific binding of one molecule to another by virtue of shape is a theme seen throughout biology, including the binding of enzymes to their substrates and the binding of antibodies to their antigens.

2.11 Riboswitches are shape-changing RNAs which can flip gene expression patterns on binding specific target molecules

In the SELEX experiments described in the previous section, aptamers are selected experimentally. However, there are also some biologically occurring examples of RNA structures which bind to specific targets and have been shaped by natural selection over millions of years. The particular group of RNA structures that we are going to look at are called **riboswitches**. Riboswitches are named since they change shape in response to the binding of targets, and switch patterns of gene expression in bacteria.[26–28]

The identification by SELEX of RNA aptamers that can bind to molecular targets suggested to some scientists, including Ronald Breaker of Yale University, that RNA aptamers could exist not just in the laboratory but also in nature. If such naturally occurring aptamers existed, Ronald Breaker and his colleagues had a hunch where they might be found. First they predicted that aptamers would be located in the UTRs (untranslated regions) of mRNAs. (Remember that most mRNAs contain a protein-coding **open reading frame** (ORF) and upstream and downstream UTRs which control expression of this ORF.) Secondly, aptamers were predicted to be found in bacterial genes known to be regulated by metabolites, but where no other protein regulator had yet been identified.

These speculations, together with shrewd laboratory detective work, led to the discovery of an important class of gene-regulatory RNA elements called riboswitches, which fulfilled both these predictions. Riboswitches are naturally occurring RNA aptamers which can bind target molecules but can also switch between different conformations depending on whether they are bound to their ligand or not. The ligands to which riboswitches bind are basic metabolites, and include bases, amino acids, and coenzymes. Binding of the ligand to the riboswitch regulates expression of the connected RNA by switching the RNA folding patterns.

The first riboswitches were discovered in bacterial mRNAs which encode proteins that control the cellular levels of the vitamin coenzyme B_{12}. The *btuB* mRNA encodes a protein required to import coenzyme B_{12} into cells of the bacterium *E. coli*. Expression of *btuB* mRNA was known to be regulated by concentrations of coenzyme B_{12} within the cell; production of the btuB protein is prevented in conditions of high concentrations of coenzyme B_{12}. The *btuB* mRNA has a very long 5′ UTR of 240 nucleotides, but no protein regulator had ever been identified which bound to this 5′ UTR.

The discovery of riboswitches came when the structure of the *btuB* mRNA was monitored, and structural changes in the 5′ UTR were identified when coenzyme B_{12} was added, as illustrated in Fig. 2.21. The conclusion drawn from these experiments was that the 5′ UTR region of the *btuB* mRNA contains structured RNA (a riboswitch), whose change in conformation in response to binding coenzyme B_{12} then regulated the translation of *btuB* mRNA into protein. Riboswitches often enable this useful kind of feedback control; riboswitches bound by a particular metabolite are found in mRNAs which are involved in the synthesis or metabolism of that metabolite.

In principle, riboswitches are similar to allosteric proteins, whose structures also change when they bind ligands. This similarity in operation between

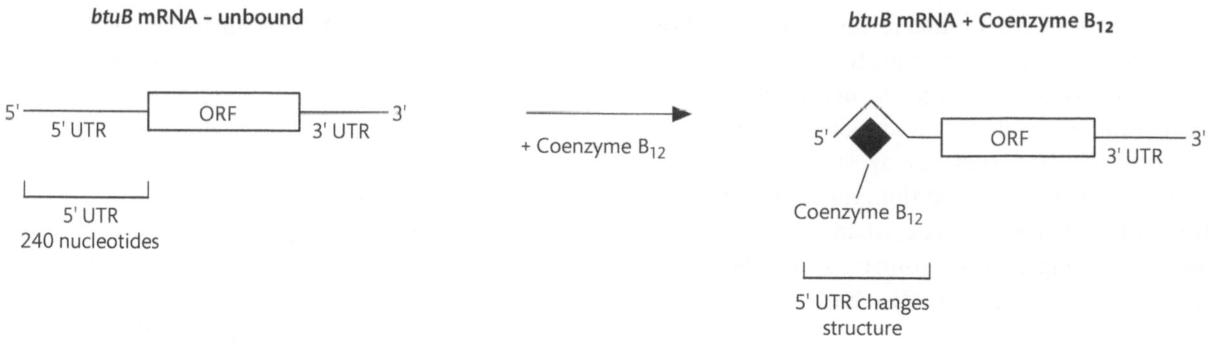

Figure 2.21 The 5′ untranslated region (UTR) of the *E. coli btuB* mRNA is unusually long and contains an RNA sequence which is able to bind and change structure in response to coenzyme B_{12}. Abbreviations: UTR, untranslated region; ORF, open reading frame.

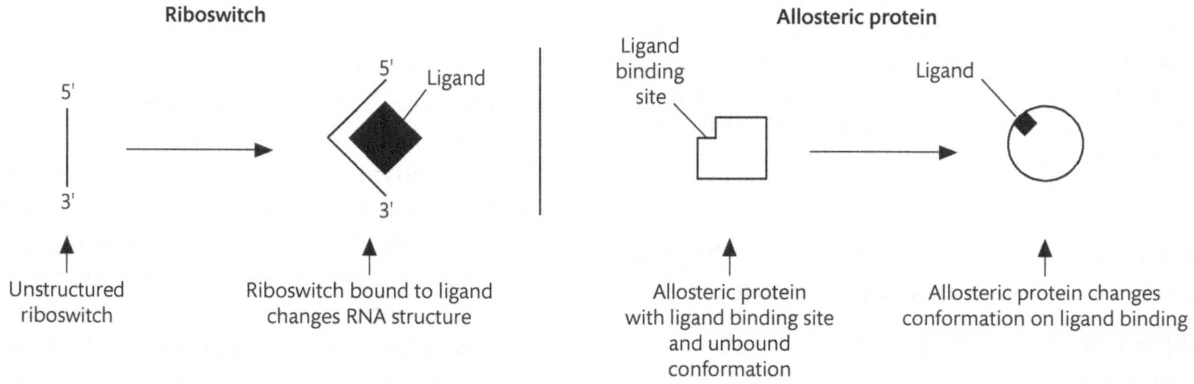

Figure 2.22 Riboswitches are conceptually similar to allosteric proteins. Riboswitches are RNAs which bind to metabolites and change structure. As a comparison, allosteric enzymes are proteins which bind to metabolites and change protein structure and function.

these two classes of molecule is shown in Fig. 2.22. Remember, however, that ribozymes are RNAs, while allosteric proteins are made up of amino acids.

How do changes in the RNA structure of riboswitches regulate gene expression? Riboswitches are composed of two RNA sequence elements: the RNA aptamer itself which binds to a target molecule, and an expression platform which controls gene expression in response to this binding (see Fig. 2.23). Structural changes in the aptamer portion of a riboswitch are induced upon binding of the target molecule, and these in turn affect the expression platform, causing changes in translational initiation and termination of the bound mRNA (Fig. 2.24).

Riboswitches are often found in the 5′ UTR of mRNAs. The 5′ UTR is a key strategic location for controlling gene expression at the level of transcriptional and translation initiation. Similar riboswitch aptamers are often tagged up to different gene expression platforms in different mRNAs so that they bind the same molecule but have

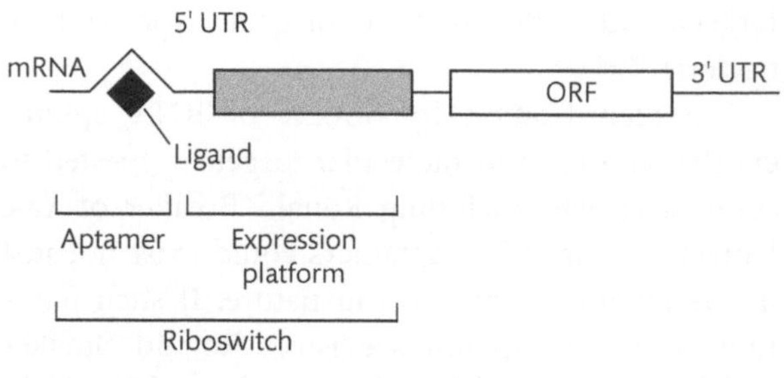

Figure 2.23 Riboswitches are composed of an aptamer domain which binds a ligand and an expression platform which regulates gene expression. Abbreviations: UTR, untranslated region; ORF, open reading frame.

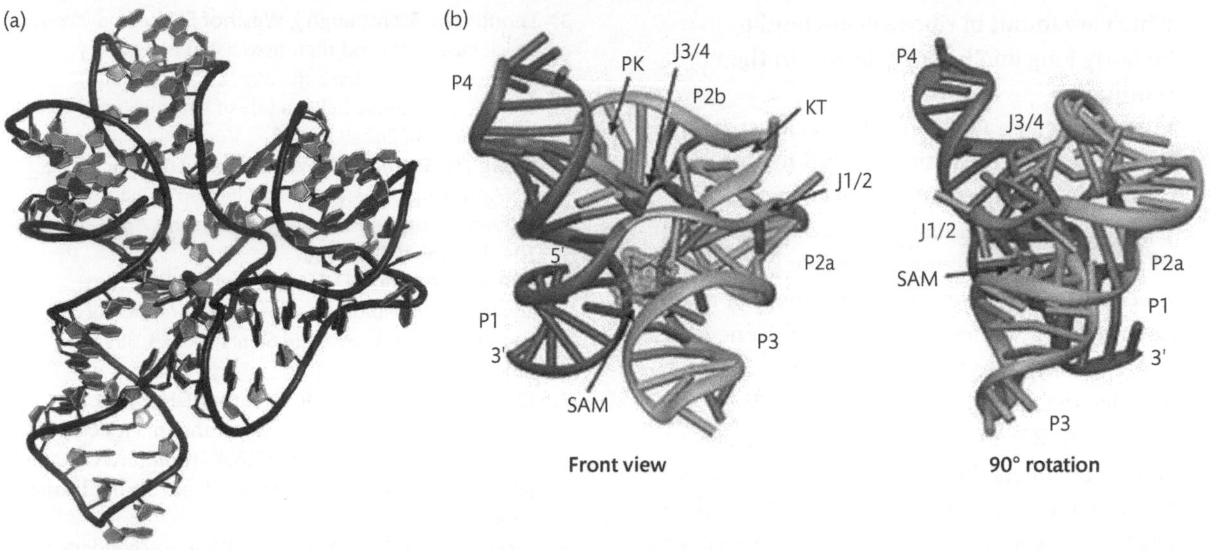

Figure 2.24 RNA riboswitch structures have been characterized in three dimensions. (a) The molecular structure of the *S*-adenosylmethionine riboswitch determined by X-ray crystallography. (Redrawn from the RCSB Protein Data Bank entry 2GIS by Jonathan Crowe, Oxford University Press.) (b) This particular riboswitch can bind to *S*-adenosylmethionine (SAM). Riboswitches change their structure to bind and hold their target molecules—notice that the SAM molecule is embedded within the RNA structure of the riboswitch. Abbreviations: P, P-base paired regions; J, single-stranded joining regions; KT, kink turn in the RNA strand; PK, pseudoknot. (Reprinted from Montange and Batey[29] with permission from Macmillan Publishers Ltd.)

different effects on gene expression. For example, riboswitches binding to coenzyme B_{12} can regulate either transcription or translation depending on the mRNA in which they are embedded.

Key points

Both naturally occurring aptamers and aptamers selected in vitro by SELEX bind to their targets selectively and with high affinity. This makes aptamers conceptually similar to protein antibodies. RNA aptamers have been fashioned by millions of years of evolution where they have competed with proteins which might have been able to replace them by doing the same job.

Summary

- RNA molecules can form extensive secondary structures through intermolecular and intramolecular base pairing. Formation of secondary structure is helped by positively charged metal ions which prevent the negative charges on the phosphodiester bonds from repelling each other.
- RNA molecules can form complex tertiary structures, which frequently involve precise packing together of helical structures and binding of metal ions.
- RNA folding can create shaped RNAs which are able to efficiently and selectively bind to ligands. Ligand-binding RNAs, called aptamers, can be selected by a continual in vitro evolution method called SELEX.
- A class of naturally occurring aptamers play an important role in controlling gene expression in bacteria. These naturally occurring aptamers exist in riboswitches, and are found in mRNAs encoding proteins involved in synthesizing or importing important cellular metabolites. Binding of metabolites to these riboswitches causes a switch in RNA folding to repress either translation or transcription of the respective mRNA.
- Riboswitches have been conserved over evolution. Naturally occurring aptamers

which are found in riboswitches tend to be fairly long and highly selective in their binding.

- One well-characterized bacterial riboswitch blocks further translation of *btuB* mRNA in high cellular concentrations of coenzyme B_{12}. By blocking *btuB* translation, the concentration of coenzyme B_{12} provides a feedback control mechanism to prevent further import of more coenzyme B_{12} (since the *btuB* mRNA encodes a protein component of the import machinery). This feedback control is mediated by the riboswitch RNA structure; a change in the structure affects the accessibility of the mRNA to ribosomes. Hence, although a full-length *btuB* mRNA may be transcribed by the bacterial cell, if its riboswitch structure adopts the 'off' conformation, it is unable to be translated into protein in conditions of high coenzyme B_{12} concentrations.

Questions

2.1 Name four properties of an RNA helix.

2.2 Why do RNAs typically form multiple short helices rather than the long double-helix structure of chromosomal DNA?

2.3 Why does RNA form an A-type helix rather than the B-form helix typical of DNA?

2.4 Draw diagrams illustrating five commonly found RNA secondary structures.

2.5 How is the RNA secondary structure monitored?

2.6 What contributions do metal ions make to RNA structure?

2.7 What is an aptamer and how would such an RNA molecule be selected in the laboratory?

References

1. **Watts JM, Dang KK, Gorelick RJ, et al.** Architecture and secondary structure of an entire HIV-1 RNA genome. *Nature* 460, 711–16 (2009).
2. **Strobel SA, Doudna JA.** RNA seeing double: close-packing of helices in RNA tertiary structure. *Trends Biochem Sci* 22, 262–6 (1997).
3. **Leontis NB, Stombaugh J, Westhof E.** The non-Watson-Crick base pairs and their associated isostericity matrices. *Nucleic Acids Res* 30, 3497–531 (2002).
4. **Rich A.** The double helix: a tale of two puckers. *Nat Struct Biol* 10, 247–9 (2003).
5. **Wong TN, Sosnick TR, Pan T.** Folding of noncoding RNAs during transcription facilitated by pausing-induced nonnative structures. *Proc Natl Acad Sci USA* 104, 17 995–18 000 (2007).
6. **Chu VB, Herschlag D.** Unwinding RNA's secrets: advances in the biology, physics, and modeling of complex RNAs. *Curr Opin Struct Biol* 18, 305–14 (2008).
7. **Mathews DH, Schroeder SJ, Turner DH, Zuker M.** Predicting RNA secondary structure. In: Gesteland R, Cech T, Atkins JF (eds), *The RNA World* (3rd edn), pp. 631–57. Cold Spring Harbor, NY: Cold Spring Harbor Laboratory Press (2006).
8. **Svoboda P, Di Cara, A.** Hairpin RNA: a secondary structure of primary importance. *Cell Mol Life Sci* 63, 901–8 (2006).
9. **Brierley I, Pennell S, Gilbert RJ.** Viral RNA pseudoknots: versatile motifs in gene expression and replication. *Nat Rev Microbiol* 5, 598–610 (2007).
10. **Torres-Larios A, Swinger KK, Krasilnikov AS, et al.** Crystal structure of the RNA component of bacterial ribonuclease P. *Nature* 437, 584–7 (2005).
11. **Mortimer SA, Kidwell MA, Doudna JA.** Insights into RNA structure and function from genome-wide studies. *Nat Rev Genet* 15, 469–79 (2014).
12. **Lucks JB, Mortimer SA, Trapnell C, et al.** Multiplexed RNA structure characterization with selective 2′-hydroxyl acylation analyzed by primer extension sequencing (SHAPE-Seq). *Proc Natl Acad Sci USA* 108, 11 063–8 (2011).
13. **Wilkinson KA, Gorelick RJ, Vasa SM, et al.** High-throughput SHAPE analysis reveals structures in HIV-1 genomic RNA strongly conserved across distinct biological states. *PLoS Biol* 6, e96 (2008).
14. **Kazantsev AV, Pace NR.** Bacterial RNase P: a new view of an ancient enzyme. *Nat Rev Microbiol* 4, 729–40 (2006).
15. **Zuker, M.** Mfold web server for nucleic acid folding and hybridization prediction. *Nucleic Acids Res* 31, 3406–15 (2003).
16. **Cate JH, Hanna RL, Doudna JA.** A magnesium ion core at the heart of a ribozyme domain. *Nat Struct Biol* 4, 553–8 (1997).
17. **Pyle AM.** Metal ions in the structure and function of RNA. *J Biol Inorg Chem* 7, 679–90 (2002).
18. **Reichow S, Varani G.** Structural biology: RNA switches function. *Nature* 441, 1054–5 (2006).
19. **Batey RT, Rambo RP, Doudna, JA.** Tertiary motifs in RNA structure and folding. *Angew Chem Int Ed Engl* 38, 2326–43 (1999).

20. **Hendrix DK, Brenner SE, Holbrook SR.** RNA structural motifs: building blocks of a modular biomolecule. *Q Rev Biophys* 38, 221–43 (2005).
21. **Tamura M, Holbrook SR.** Sequence and structural conservation in RNA ribose zippers. *J Mol Biol* 320, 455–74 (2002).
22. **Klein DJ, Moore PB, Steitz TA.** The contribution of metal ions to the structural stability of the large ribosomal subunit. *RNA* 10, 1366–79 (2004).
23. **Kortmann J, Narberhaus F.** Bacterial RNA thermometers: molecular zippers and switches. *Nat Rev Microbiol* 10, 255–65 (2012).
24. **Jenison RD, Gill SC, Pardi A, Polisky B.** High-resolution molecular discrimination by RNA. *Science* 263, 1425–9 (1994).
25. **Jaeger J, Restle T, Steitz TA.** The structure of HIV-1 reverse transcriptase complexed with an RNA pseudoknot inhibitor. *EMBO J* 17, 4535–42 (1998).
26. **Mandal M, Breaker RR.** Gene regulation by riboswitches. *Nat Rev Mol Cell Biol* 5, 451–63 (2004).
27. **Tucker BJ, Breaker RR.** Riboswitches as versatile gene control elements. *Curr Opin Struct Biol* 15, 342–8 (2005).
28. **Breaker RR.** Riboswitches and the RNA world. *Cold Spring Harb Perspect Biol* 4 (2012).
29. **Montange RK, Batey RT.** Structure of the *S*-adenosylmethionine riboswitch regulatory mRNA element. *Nature* 441, 1172–5 (2006).

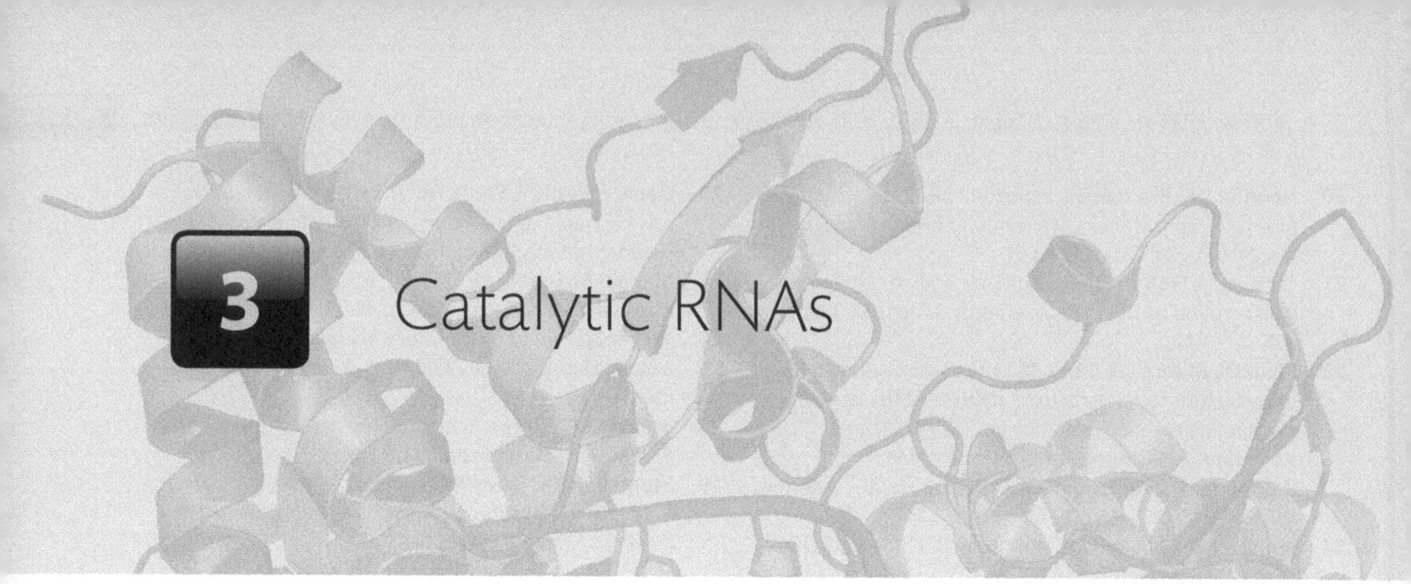

3 Catalytic RNAs

Introduction

In Chapter 2 we looked at how RNA molecules can form structures in cells which rival those formed by proteins. Furthermore, these RNA structures can bind to target molecules and change shape in response to them. In this chapter we are going to raise the bar for RNA function, and look at how RNA molecules catalyse chemical reactions, a domain which was previously thought to be reserved only for proteins.

Proteins make good enzymes because of the chemical diversity of the 20 different amino acid side groups within them. In contrast, RNA has a more limited set of functional groups for building catalysts, confined to just four different nucleotides (A, C, G, and U). However, seminal discoveries made in the 1980s by Sidney Altman at Yale University (discovery in bacteria of a catalytic RNA called RNAse P, which can cleave the 5′ end of pre-tRNA) and Tom Cech at the University of Colorado at Boulder (discovery of a catalytic RNA that can remove an internal RNA intron sequence called a Group I intron) led to the identification of enzymatically active RNAs. These enzymatically active RNAs are now called **ribozymes** to reflect their ribonucleic acid chemical identity. For these discoveries Sidney Altman and Tom Cech shared the 1989 Nobel Prize in Chemistry.

It is now known that RNA catalysis plays a central role in life itself. For example, one of the central reactions of life—peptide bond formation, through which proteins are assembled from amino acids—is catalysed by an RNA active site. The important role of RNA catalysts in critical chemical reactions in the cell supports the hypothesis that early in the evolution of life RNA was used as both the genetic material and as a catalyst for its replication. At the end of this chapter we shall look at some of the evidence for this **RNA world hypothesis**, which also postulates that some of the catalytic and non-coding structural RNAs in living cells may be 'living fossils' remaining from this early time.

Explanations of the important terminology used in this chapter to describe the molecular biology of ribozymes are given in the Glossary. The first time these terms are used in the text they are given in **bold**.

3.1 Three properties of RNA enable the catalytic function of ribozymes

Ribozymes speed up chemical reactions as effectively as protein enzymes. For example, the ribosome speeds up peptide bond formation by a factor of the order of 10^7.

Four general molecular properties of RNA provide catalytic activity in ribozymes.[1–6]

3.1.1 RNA molecules can fold into tertiary structures

All enzymes (both protein and RNA) use active sites to bring substrates close together, and to stabilize the transition states of reactions that occur as substrates are converted into products. The ability of RNA molecules to form structures also enables assembly of RNA active sites (as one example, the structure of the peptidyl transferase centre of the ribosome is shown in Fig. 2.17). Ribozymes can base pair with RNA substrates and hold them in place during chemical reactions.

3.1.2 RNA binds metal ions that function in catalysis

Metal ions have several important functions in ribozymes. First, metal ions help RNA structures form (see Chapter 2), and balance the strong negative charge in ribozyme active sites that result from high densities of RNA strands (see Chapter 2).

Secondly, metal ions also have specific catalytic roles in ribozymes.

1 In ribozyme active sites metal ions function as hydrogen atom donors and acceptors. This is called **general acid/base catalysis**. Metal ions coordinate with water molecules to function in this way. A metal ion coordinated with a water molecule (M–OH_2) can act as a hydrogen donor (**general acid**), and a metal ion coordinated to a hydroxyl group from water (M–OH) can act as a hydrogen atom acceptor (**general base**). For example, an -OH group in a ribozyme substrate can be activated into a potent $-O^-$ nucleophile, by M–OH accepting its hydrogen atom (becoming M–OH_2). Similarly an O^- leaving group can be stabilized by addition of a hydrogen atom to give OH (in this case the general acid M–OH_2 would become M–OH).

2 Metal ions can bind to unpaired electrons to stabilize substrates, products, and reaction intermediates.

3 Metal ions can function as **electrophilic catalysts** which help reactions proceed by balancing charges on reaction intermediates.

Because of their high charge density, magnesium ions are frequently used metal ions in ribozymes.

3.1.3 The bases in RNA can accept and donate protons

The purine and pyrimidine bases of RNA have NH groups (see Fig. 3.1) that can potentially act as hydrogen donors or acceptors in acid–base catalysis. However, the actual pK_a values of these bases mean they will

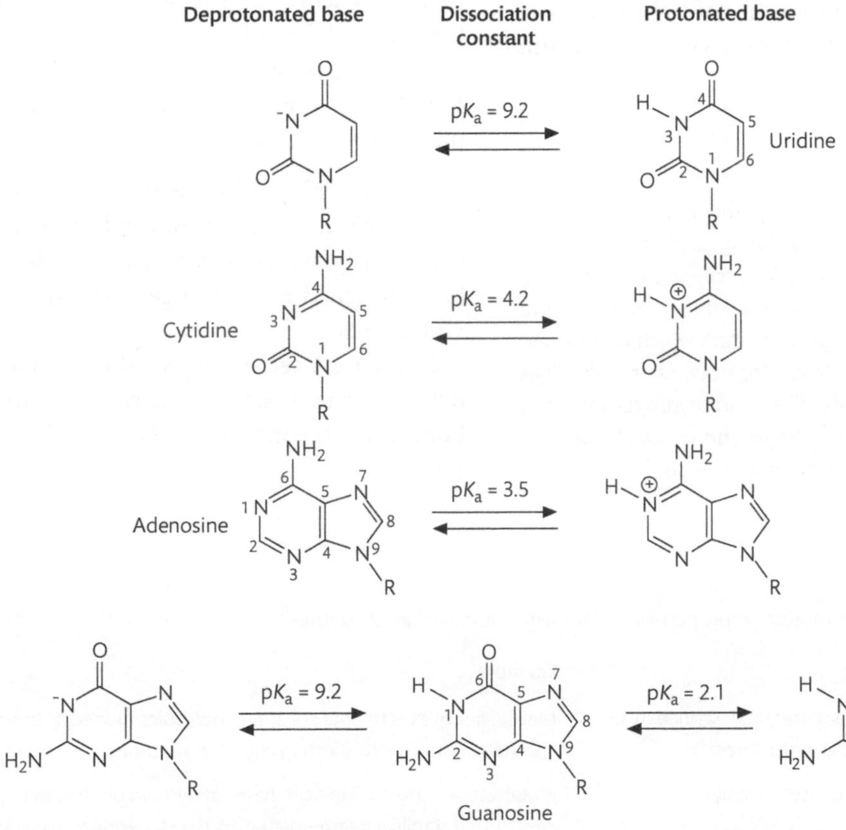

Figure 3.1 The bases of RNA can accept and release protons for acid-base catalysis. pK_a is the dissociation constant of an acid, which is the pH at which half the protons have dissociated. Physiological pH is 7—notice that each of the bases used in RNA has a pKa which indicates that protons are released or accepted at non-physiological pHs. (Redrawn from Fedor and Williamson[2] with permission from Macmillan Publishers Ltd.)

actually function only very weakly as acids or bases at physiological pH. For example, cytidine has a pK_a of 4.2 making it an acid, but at physiological pH only one out of every 1000 molecules will be protonated, so cytidine is not a good source of hydrogen atoms.[5] However, while RNA bases do not easily give up or accept hydrogen ions at physiological pH, some ribozymes still use this as a catalytic strategy as these exchanges can happen more easily within the highly charged environment of catalytic RNA molecules. Examples of ribozymes that use this strategy are the nucleolytic ribozymes discussed in Sections 3.8–3.11.

3.1.4 The 2′ -OH groups on the ribose sugars of RNA can be a source of hydrogen ions

The 2′ -OH groups on the ribose sugars of RNA can also be a source of hydrogen ions during catalysis (e.g. in catalysis of peptide bond formation in the ribosome—see Chapter 11), and for mediating interactions with substrate molecules.

These four general properties of RNA that enable the catalytic properties of ribozymes are summarized in Table 3.1.

Key points

RNA tertiary structures are very important for positioning reactive chemical groups during catalysis. Metal ions bound to RNA are important for accepting and donating protons during catalytic reactions, and for balancing the high concentration of negative charges which occur when RNA strands are folded closely together. Chemically, RNA is an acid at physiological pH (releasing hydrogen ions), but its bases can also accept hydrogen atoms. Catalytically, ribozymes frequently use acid-base catalysis.

3.2 What kinds of reactions do ribozymes catalyse?

Most of the naturally occurring ribozymes catalyse reactions which break ester bonds in the phosphodiester backbone of RNA, either by hydrolysis (reaction with water) or by *trans*-esterification (breaking one ester bond and re-forming a new one). The exception is the ribosome, a ribozyme which catalyses peptide bond formation (how the ribosome works is discussed in more detail in Chapter 11).

Several ribozymes have now been identified in nature, and fall into one of two major groups based on their size (Fig. 3.2).

1 The large ribozymes have sequences longer than 200 nucleotides. Historically these ribozymes were discovered first.

2 The small ribozymes contain RNA sequences up to 200 nucleotides in length. Each of the small ribozymes catalyses its own cleavage at specific sites. These short ribozymes are used by RNA viruses as tools to cut viral RNA multimers into genome-sized pieces after replication.

Some members of these two groups of ribozymes are shown in Table 3.2. Beyond these naturally occurring ribozymes, some artificial ribozymes have been designed for use in gene therapy and as antiviral agents.

In the next section we shall look in a little more detail at the reactions carried out by these ribozymes, starting with the *Tetrahymena* rRNA intron.

Table 3.1 Three general molecular properties of RNA which are used in ribozymes

Molecular ability of RNA	Example
1. RNA molecules use the strategies described in Chapter 2 to fold into tertiary structures	Making active sites to hold substrate molecules in specific orientations and stabilize transition states in chemical reactions
2. RNA binds metal ions to use in catalytic mechanisms	Metal ions are involved in acid-base catalysis, in binding to unpaired electrons to stabilize intermediates, and as electrophilic catalysts
3. The bases in RNA can accept and release protons	Purine and pyrimidine molecules are used in some ribozymes to bind and release protons during acid-base catalysis

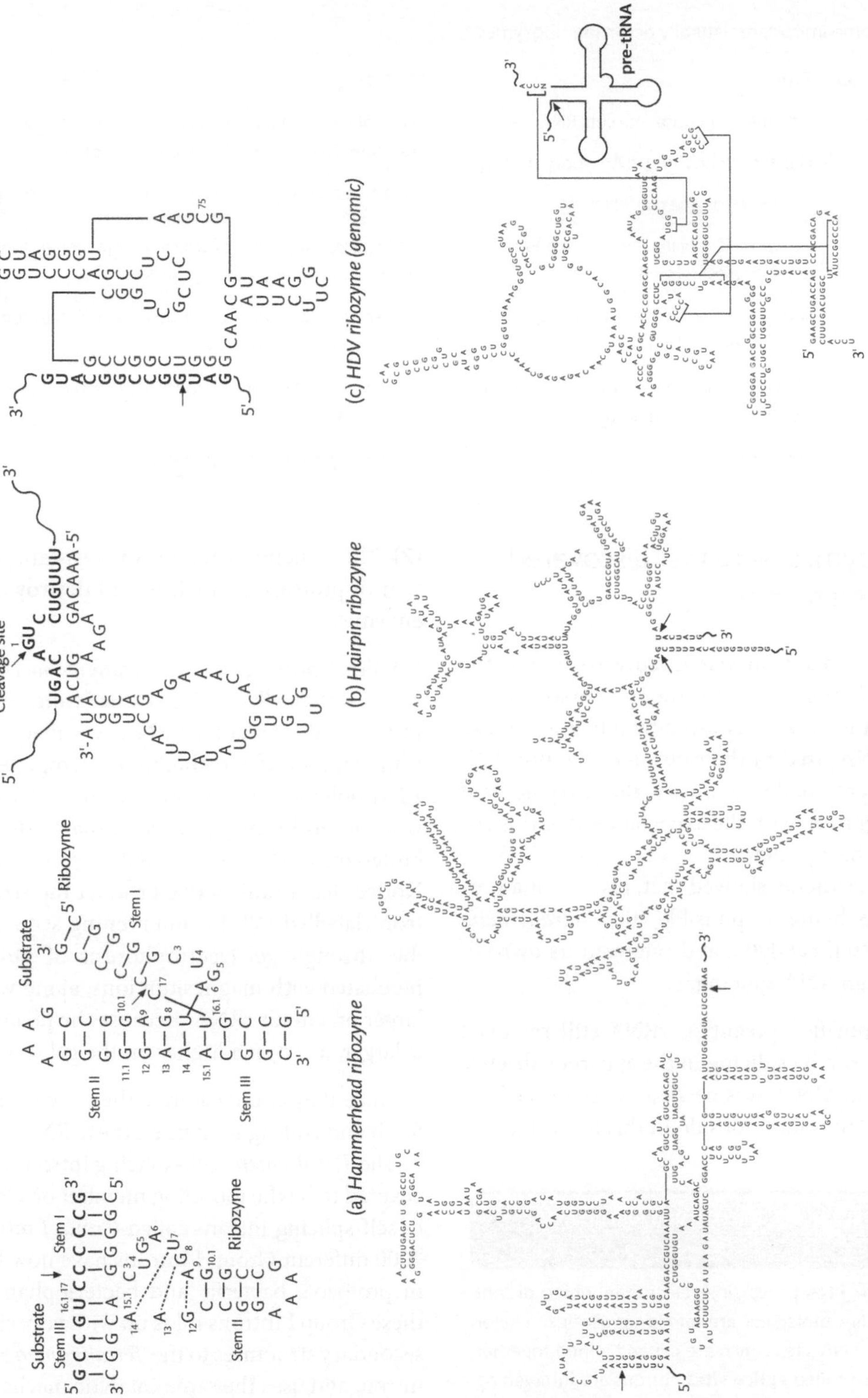

Figure 3.2 Ribozymes fall into large and small groups. Notice the increase in size from the small hammerhead ribozyme, hairpin ribozyme, and hepatitis delta virus (HDV) ribozyme (top line) to the large Group I ribozyme, Group II ribozyme, and RNAse P ribozyme on the bottom line. (Redrawn from Takagi et al.[6] with permission from Oxford University Press.)

Table 3.2 Some important naturally occurring ribozymes

Ribozyme group	Type	Notes
Small ribozymes	Hammerhead ribozyme cuts RNA	Each of these ribozymes cuts itself via a *trans*-esterification reaction, generating a 2′, 3′ cyclic phosphate
	Hepatitis delta virus (HDV) ribozyme cuts RNA	
	Hairpin ribozyme cuts RNA	As a group these are called ribonucleolytic ribozymes
Large ribozymes	Group I self-splicing introns cut and re-join phosphodiester bonds	Self-splices via a *trans*-esterification reaction to remove an intron and join two exons
	Group II self-splicing introns cut and re-join phosphodiester bonds	Self-splices via a *trans*-esterification reaction to remove an intron and join two exons
	RNAse P cleaves pre-tRNA in all species to give mature tRNA molecules	Processes pre-tRNA to remove 5′ leader through a hydrolysis reaction
	Ribosome	Catalyses peptide bond formation

3.3 Ribozymes were first discovered through serendipity

In the 1980s an intron was discovered within the protozoan *Tetrahymena* in the precursor of an rRNA. This intron was removed by splicing to make the final rRNA used in the ribosome (see Box 3.1). Attempts were made to purify the enzyme that spliced this intron, with the expectation that this enzyme would be a protein.

These experiments showed that, instead of a protein enzyme being responsible, the *Tetrahymena* intron was itself catalytic, and catalysed its own removal through RNA splicing.[7]

(1) Highly purified precursor rRNA still removed the rRNA intron by splicing in the apparent absence of any protein. $MgCl_2$ was needed–we shall see later what these magnesium ions do in this reaction.

Box 3.1 Splicing reactions cut and join up RNA molecules

Splicing is an RNA processing reaction in which distant sections of RNA molecules are joined up either in *cis*- or in *trans*-. The RNA sites which are cut and joined together in this way are called **splice sites**. Introns are intervening sequences (IVSs) that are encoded by the gene and transcribed, but removed from the primary transcript. The exon sequences of this primary transcript are spliced together to generate the final RNA product.

(2) This splicing reaction was resistant to the addition of proteases, which would destroy any protein enzymes.

(3) Final proof that the *Tetrahymena* rRNA intron was catalytically active came from experiments in which the precursor RNA was made by in vitro transcription of a recombinant template and *E.coli* RNA polymerase. This highly pure rRNA precursor was spliced in the presence of magnesium ions and buffer only. This experiment is shown in Fig. 3.3. Notice that a radioactive product the size of the intron (labelled IVS for intervening sequence) is visible through gel electrophoresis of purified RNA incubated with magnesium ions, along with a much larger precursor rRNA. Because the precursor rRNA is larger, it is much higher on the gel.

Since they could catalyse their own splicing, Tom Cech and colleagues named these RNAs *ribozymes*.

The *Tetrahymena* self-splicing intron was later discovered to be the founding member of a larger group of self-splicing introns called *Group I introns*. About 2000 different Group I introns have now been found in protozoa, bacteria, and bacteriophages. Each of these Group I introns folds up with very similar RNA secondary structures to the *Tetrahymena* self-splicing intron, and uses the same catalytic mechanism.

What do Group I introns do? The answer is apparently not much for their host cell. They are parasitic genetic elements with no obvious benefit to their host, and move about and propagate in their host

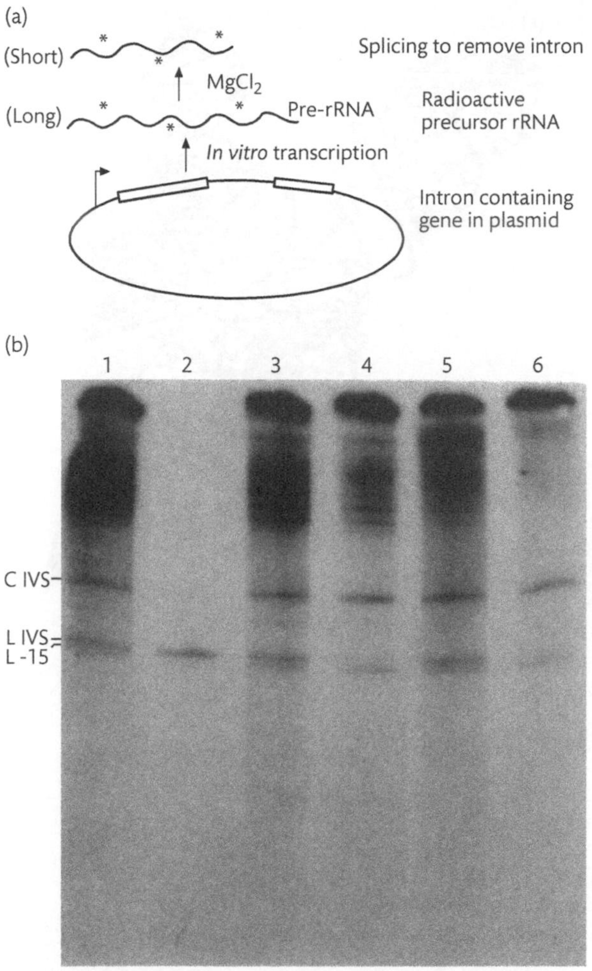

Figure 3.3 The discovery of ribozymes. Seminal experiments in the 1980s showed that splicing of *Tetrahymena* rRNA introns can take place in the absence of any protein. (a) Radioactive rRNAs were transcribed from a plasmid containing the rRNA intron and incubated with magnesium ions. (b) After purification, RNAs were analysed by gel electrophoresis. This revealed long precursor RNAs at the top of the gel, and shorter RNAs the size of the intron (IVS, linear (L-IVS), and circular (C-IVS) forms of the intron were detectable). (From Kruger et al.[7] Reproduced by kind permission of *Cell* 1982 Nov, 31(1), 147–57.)

genomes. Once transcribed into RNA, Group I introns splice themselves out—since they are removed by splicing their host transcript remains functional. Hence, because of splicing, these introns do not affect the function of the gene in which they are embedded, and so do not harm the viability of the host. Integration of Group I introns into the genome is determined by a DNA-cutting endonuclease.

Ribozymes were also discovered independently in another laboratory. In the same timeframe that the Cech laboratory identified the Group I intron ribozymes, Sidney Altman and colleagues at Yale University discovered another ribozyme.[8,9] This was the bacterial RNAse P enzyme that processes tRNA molecules (see Box 14.5). RNAse P is discussed more fully in Chapter 14.

3.4 Group I introns are spliced through a two-step mechanism which uses metal ions in their active sites

3.4.1 The Group I intron splicing pathway

The reaction pathway through which Group I introns are spliced is shown in Fig. 3.4. The precursor is made up of an upstream 5′ exon and a downstream 3′ exon that become spliced together, and an intron

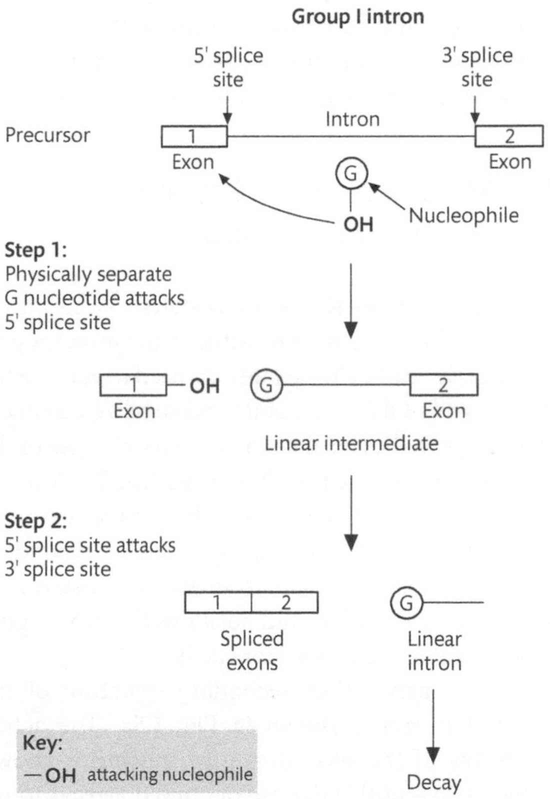

Figure 3.4 Splicing of Group I introns takes place through a two-step mechanism. In the first step of splicing, the hydroxyl group of a physically separate G nucleotide (G-OH) attacks the 5′ splice site. In the second step of splicing the 3′ -OH group on exon 1 attacks the 3′ splice site to join exon 1 to exon 2 and release the intron.

between the exons that is removed by the splicing reaction. A separate GTP (guanosine 5′ triphosphate) nucleotide provides the attacking 3′ –OH group. The conserved secondary structure of Group I introns is important for holding this attacking nucleophile in the first reaction step.[10] Base pairing between the intron and the 5′ splice site positions the 5′ splice site ready for the first step of the splicing reaction.

- In **step 1** of the splicing reaction, the 3′ –OH group of the GTP molecule attacks the phosphodiester bond linking the intron with exon 1 of the pre-rRNA. As a result of this chemical attack the phosphodiester backbone is broken and the GTP becomes attached to the 5′ end of the intron by a new 3′–5′ phosphodiester bond.
- In **step 2** of the splicing reaction, the 3′ –OH group on exon 1 chemically attacks the phosphodiester bond linking the intron with the 3′ exon of the pre-rRNA. This joins the upstream and downstream exons with a new phosphodiester bond, releasing the intron with the original GTP nucleophile still attached.

3.5 Metal ions play a key role in catalysis by Group I introns

Group I introns are RNA-based catalysts—this means that they will splice in vitro without any proteins present. Catalytically, Group I introns need to activate the attacking 3′ –OH nucleophile and stabilize leaving 5′ O^- ions, as well as balance the negative charges of the oxygen atoms in the transition state. The RNA-based Group I intron satisfies these catalytic requirements by creating an active site through its RNA secondary structure. This active site arranges the substrates in the correct orientation, and positions the two magnesium ions that are needed for catalysis.[11–13]

The extensive RNA secondary structure of the Group I intron is shown in Fig. 3.5a. The actual chemistry of the second step of splicing is shown in Fig. 3.5b—notice the use of curved arrows to indicate movement of electrons during the chemical reaction. Two positively charged magnesium ions (Mg^{2+}) in the active site have critical catalytic roles:

1 Two magnesium ions provide a positive charge in the Group I intron active site. This positive charge is important because the active site of the Group I intron contains a high density of folded RNA, and the transition state contains five negatively

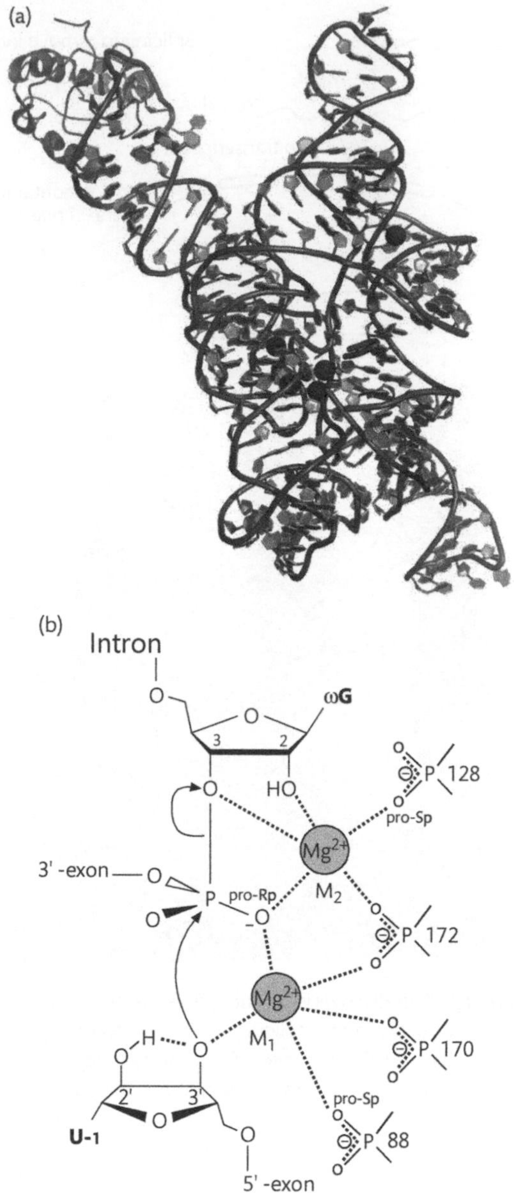

Figure 3.5 **Metal ions play a central role in catalysis by Group I introns.** (a) The Group I intron is highly structured through multiple secondary and tertiary interactions. (Redrawn from the RCSB Protein Data Bank (NDB ID PRO170) by Jonathan Crowe, Oxford University Press.) (b) The active site of the Group I ribozyme contains two magnesium ions that are involved in catalysis. Here the second catalytic step is shown, where the 5′ splice site is attacking the 3′ splice site. The last nucleotide in the intron is a G, and is identified as ωG. (From Stahley and Strobel 2005.[12])

charged oxygen atoms around the phosphorus atom of the phosphodiester bond. The positive charge of the metal ions in the active site balances the high negative charge of the reaction centre so that it carries no overall charge and is consequently more stable.

2 Two magnesium ions in the active site play an active catalytic role in Group I introns. In the second step of the splicing reaction, the 5′ exon is joined to the 3′ exon, releasing the intron. Notice that magnesium ion M_1 coordinates with the attacking 3′ $-O^-$ ion, and magnesium ion M_2 coordinates with the leaving 3′ $-O^-$ ion of exon 1. Magnesium ion coordination stabilizes these attacking and leaving groups and so helps the reaction to proceed.

In the Group I intron, the catalytic magnesium ions are held in the active site by coordinating with phosphate groups in the phosphodiester backbone (notice the coordinating phosphates shown in Fig. 3.5 as P88, P170, P172, and P128). Amazingly, this two-metal-ion catalytic mechanism of Group I introns is very similar to catalytic strategies used by protein enzymes to carry out similar reactions. Some protein nucleases which cut phosphodiester bonds similarly use two metal ions in their active site that are held 3.8–4Å apart in the active sites but by key amino acids in the protein sequence.[11,13] Examples include the 3′–5′ exonucleolytic activity of *Ecoli* DNA polymerase I, alkaline phosphatase, and RNAse H. Hence, through convergent evolution, RNA-based ribozymes have adopted similar catalytic strategies to protein enzymes to cleave phosphodiester bonds.

Key points

Group I introns actually catalyse their own removal from their host transcripts. Splicing of Group I introns takes place through a two-step reaction pathway using an -OH group as a nucleophile to initiate the splicing reaction that is provided by a separate GTP nucleotide held in place by the intron secondary structure. Metal ions in the active site of the Group I introns are key for stabilizing negative charges in attacking nucleophiles and leaving groups.

3.6 Group II introns are also spliced through a two-step mechanism

Another important group of self-splicing intronic ribozymes are the *Group II self-splicing introns.*[14] Group II introns have extensive RNA tertiary structure (see Fig. 3.6), and this creates the active site which catalyses the reaction.

Group II introns use the 2′ –OH group of an adenosine nucleotide that is encoded within the sequence of the intron itself as a nucleophile to initiate the splicing reaction.[15]

The pathway of Group II intron splicing is shown in Fig. 3.7.

- In the first step of the reaction the 2′ –OH of an adenosine residue actually encoded within the intron attacks the phosphodiester bond at the 5′ splice site (linking exon 1 with the intron). This breaks the old 3′–5′ phosphodiester bond to release exon 1, and creates a new 2′–5′ phosphodiester bond. The intron is looped but still connected to exon 2. Because of the loop, this is called a **lariat intermediate**.

Figure 3.6 The tertiary structure of a Group II intron from the bacterium *Oceanobacillus iheyensis.* This RNA structure is composed of short helices stacked on top of each other and held together by ribose zippers (see Chapter 2). (Redrawn from the RCSB Protein Data Bank entry NDB ID NA0107 by Jonathan Crowe, Oxford University Press.)

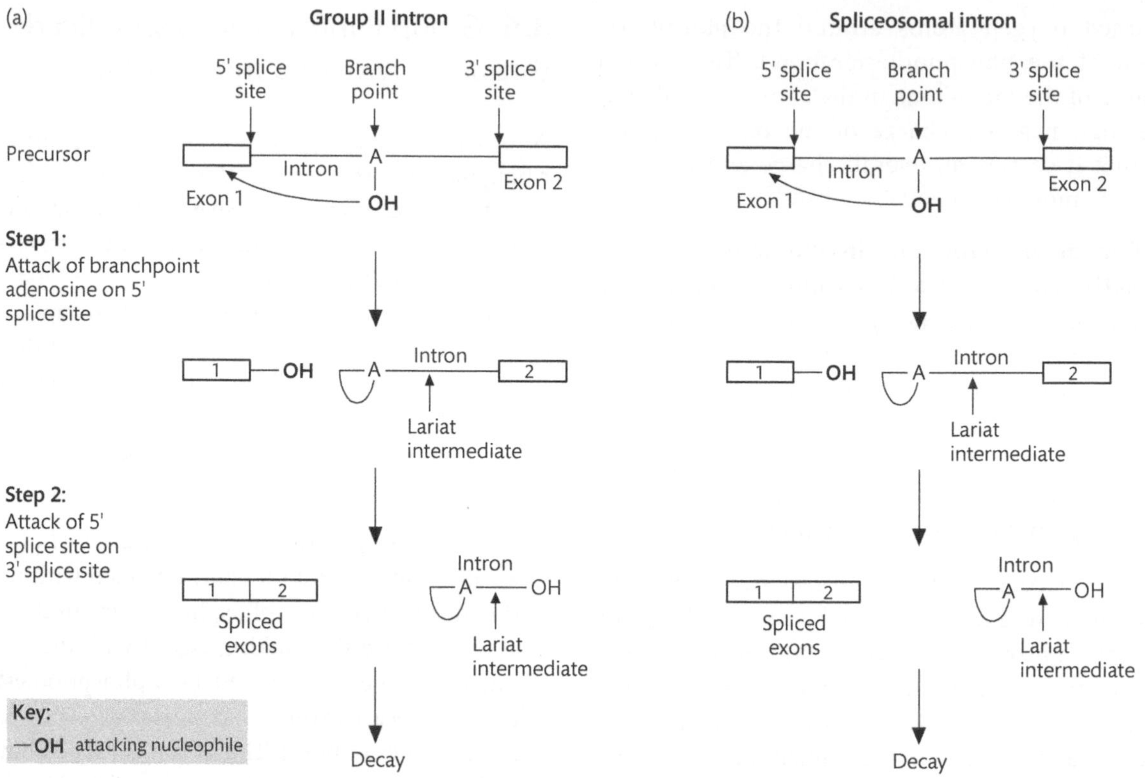

Figure 3.7 Group II introns are spliced through a two-step mechanism. (a) Group II introns are spliced through a two-step reaction pathway. (b) Spliceosomal introns follow a similar splicing pathway to Group II introns, and this is shown here in parallel. (A full description of the pathway of spliceosomal intron removal is given in Chapter 5.)

- In the second step of the reaction the 3′ –OH group at the end of exon 1 chemically attacks the phosphodiester bond linking the intron with the 3′ splice site. This breaks the old phosphodiester bond to release the lariat, splicing together exons 1 and 2 via a new 3′–5′ phosphodiester bond.

Magnesium ions play a central role in catalysis by Group II introns at both step 1 and step 2 splicing reactions. This common role of magnesium ions at both steps of the splicing reaction is shown in Fig. 3.8. The active site magnesium ions activate the nucleophile –OH group before it attacks the phosphodiester bond (creating a more potent O^- attacking group), and stabilize the leaving 3′ O^- group after phosphodiester bond cleavage.

Like Group I introns, Group II introns are selfish genetic elements. Excised Group II intron RNA can re-insert back into the genome (via reverse transcription back into DNA, primed by nicked genomic DNA), and in this way can move around the genome.

The splicing pathway followed by Group II introns is of particular interest since it is strikingly similar to the spliceosomal pre-mRNAs which we will discuss in more detail in Chapter 5 (for comparison, the spliceosomal pathway is shown side by side with the Group II splicing pathway in Fig. 3.7). Both Group II introns and pre-mRNAs use an adenosine 2′ –OH at the branchpoint as a nucleophile to initiate the splicing reaction by attacking the 5′ splice site, and the 3′ –OH of the upstream exon as the nucleophile in the second step of splicing.

We discuss the pathway followed by spliceosomal introns in more detail in Chapter 5. However, it is worth mentioning here that metal ions play a critical and similar role in the active sites of both the spliceosome and Group II introns.[16,17] In Group II introns these metal ions are held in a conserved part of the intron called domain V which binds the catalytic magnesium ions (see Fig. 3.9). In the spliceosome, a similar RNA secondary structure in a small nuclear RNA called U6 snRNA is responsible for holding the

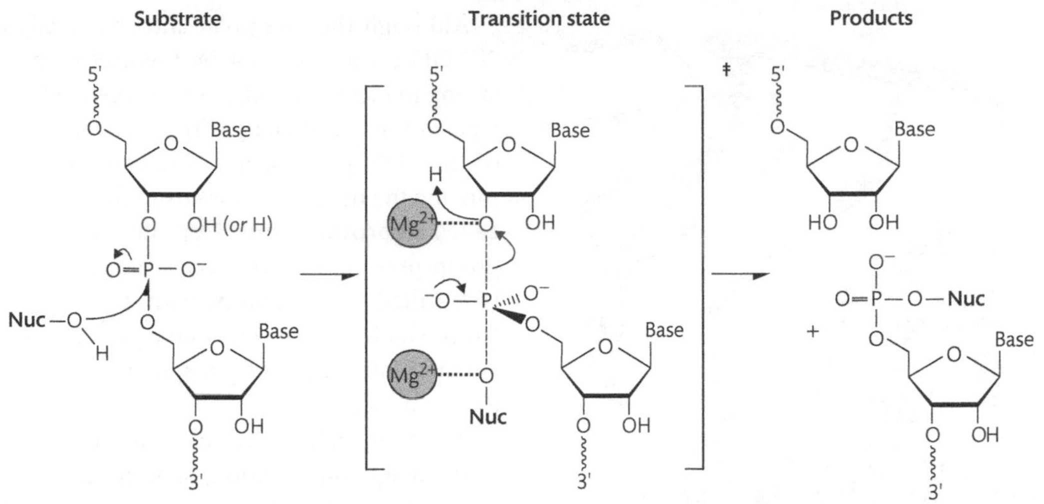

Figure 3.8 Two metal ions play a key role in catalysis by **Group II introns.** Both step 1 and step 2 of the splicing mechanism of Group II introns occur similarly through attack by a nucleophile on a phosphodiester bond to be cleaved. In the first step of splicing the nucleophile is the 2′ -OH of the branchpoint adenosine. In the second step the nucleophile is the 3′ -OH of the upstream exon. There are two magnesium ions in the active site of the Group II intron. One of these coordinates with the nucleophile (attacking O^- group), and the other coordinates with the leaving O^- group. (Reproduced from Pyle[15].)

catalytic magnesium ions in place. Because of these similarities, it is thought that Group II introns probably provided the evolutionary precursor to spliceosomal introns.

Key points

Group II introns are catalytic RNAs that self-splice. Both step 1 and step 2 of the Group II intron splicing pathway involve a chemical attack by a nucleophilic -OH group on a phosphodiester bond. In the first step of splicing the 2′ -OH of the branchpoint adenosine attacks the phosphodiester bond at the 5′ splice site. In the second step of splicing, the 3′ -OH of the upstream exon attacks the phosphodiester bond at the 3′ splice site. Group II introns have strong similarities to spliceosomal introns and may have been their evolutionary precursor.

3.7 RNA is inherently chemically unstable because of its 2′ -OH group

The Group I and Group II introns discussed in the previous section belong to the large ribozymes group. Next we are going to discuss the other group of small ribozymes. Many of these small ribozymes catalyse RNA cutting reactions, in which the phosphodiester backbone of the RNA is cleaved. This RNA cleavage reaction can also occur spontaneously. As a chemical, RNA is more reactive than DNA, and will spontaneously cleave in solution without the need for any catalysis. Before we discuss the small ribozymes in more detail, we will look first at why and how this RNA cutting reaction happens spontaneously.

RNA is inherently more chemically reactive than DNA because of the 2′ –OH group in the ribose sugar of RNA (remember, this 2′ –OH is not found in the deoxyribose sugar of DNA). The –OH is a polar group: electrons are drawn towards the oxygen and away from the hydrogen. This makes the oxygen in the 2′ –OH group a **nucleophile** that can chemically attack the phosphorus atom of the adjacent phosphodiester bond. Because it can launch a chemical attack on its own phosphodiester backbone, the 2′ –OH group on the ribose sugar of RNA has been described by Professor David Lilley of Dundee University as 'a dagger pointing to its own heart'.[3]

The spontaneous nucleophilic attack mediated by the 2′ –OH group on the phosphodiester backbone of RNA is shown in Fig. 3.10. Note that this reaction occurs in the absence of any catalysis. Nucleophilic attack leads to a **transition state**, in which the phosphorus atom of the phosphodiester bond forms transient bonds with five oxygen atoms. To complete the RNA cutting reaction and release the downstream RNA chain, the ester bond between the phosphorus atom

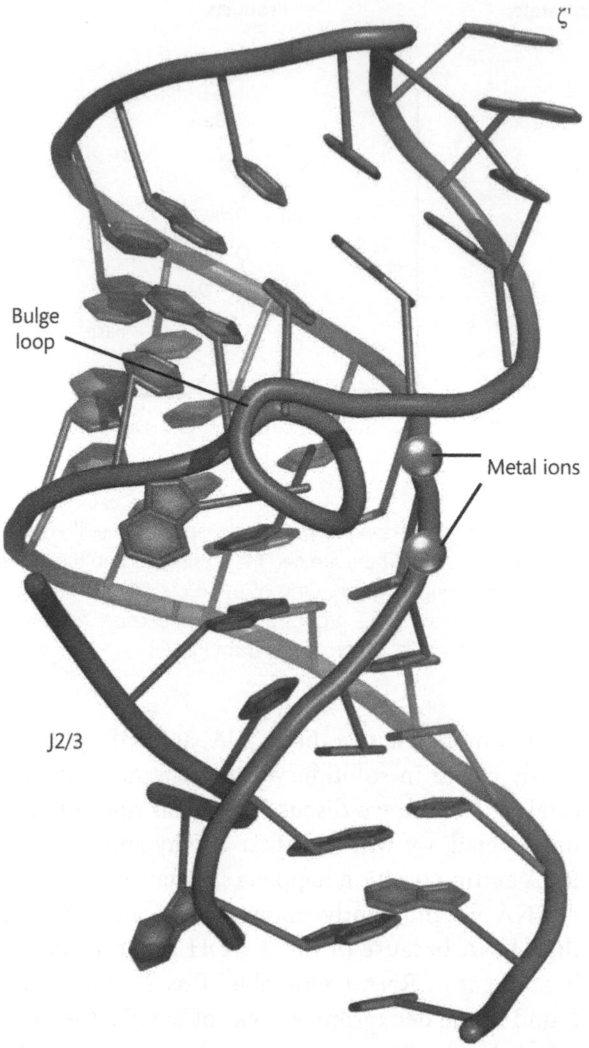

Figure 3.9 **A region of RNA structure called domain V holds the catalytic magnesium ions in place within the active sites of Group II introns.** The two metal ions are shown as small spheres held in place by the RNA structure of the intron. For a full colour version of this figure, see Colour Illustration 4. (Redrawn from the RCSB Protein Data Bank (PDB ID 3IGI) by Alice Mumford, Oxford University Press.)

and the 5′ oxygen breaks, and a new ester bond simultaneously forms between the 2′ oxygen and the phosphorus atom. As a result, after cleavage of the RNA chain the upstream RNA chain terminates in a 2′3′ cyclic phosphate, and the downstream chain initiates in a 5′ -OH.

Within the transition state three particular atoms line up—these are the attacking oxygen atom, the phosphorus atom under attack, and the leaving oxygen atom. This is called an **in-line configuration** (notice that these three atoms are joined by a broken line in the transition state in Fig. 3.10).

Although they are polar and chemically reactive, 2′ -OH groups are not by themselves particularly strong nucleophiles since the oxygen atom only has a *partial* negative charge. The nucleophilic strength of the -OH group can be increased by *entirely* removing the hydrogen atom from the -OH (a process called **deprotonation**). Deprotonation of an -OH group produces a negatively charged oxygen atom (O^-, called an **oxyanion**) with a free pair of electrons which can efficiently take part in a subsequent chemical attack on the phosphorus of the phosphodiester bond.

Although RNA cleavage can occur spontaneously in aqueous solution, deprotonation of 2′ -OH groups to produce 2′ O^- oxyanions takes place at a higher frequency in basic solutions (containing OH^- ions. These OH^- ions accept protons from the 2′ -OH group to produce H_2O and an 2′ $-O^-$ ion). The increase in spontaneous cleavage of phosphodiester bonds through 2′ -OH deprotonation makes RNA molecules unstable in solutions that contain OH^- ions—this is called **base catalysis**. For example, look at the experiment shown in Fig. 3.11. In this experiment, the same RNA was analysed by gel electrophoresis after incubation under different conditions. Notice the strong ladder of RNA cleavage products in the RNA which has been incubated in the solution with the OH^- ions (-OH lane). In contrast, the RNA incubated without OH^- ions remained largely intact (a very few smaller cleavage products, which occur due to spontaneous cleavage, are visible in the NR lane in Fig. 3.11b).

Key points

The extra 2′ -OH on the ribose sugar in RNA makes it more chemically reactive than DNA. These reactions depend on the polar hydroxyl groups of RNA chemically attacking the phosphodiester backbone, and is speeded up in basic solutions.

3.8 Small ribonucleolytic ribozymes catalyse their own cleavage

Many small ribozymes catalyse reactions in which the phosphodiester bond of RNA is cut using the same RNA cutting reaction discussed in the previous section, which occurs during spontaneous cleavage

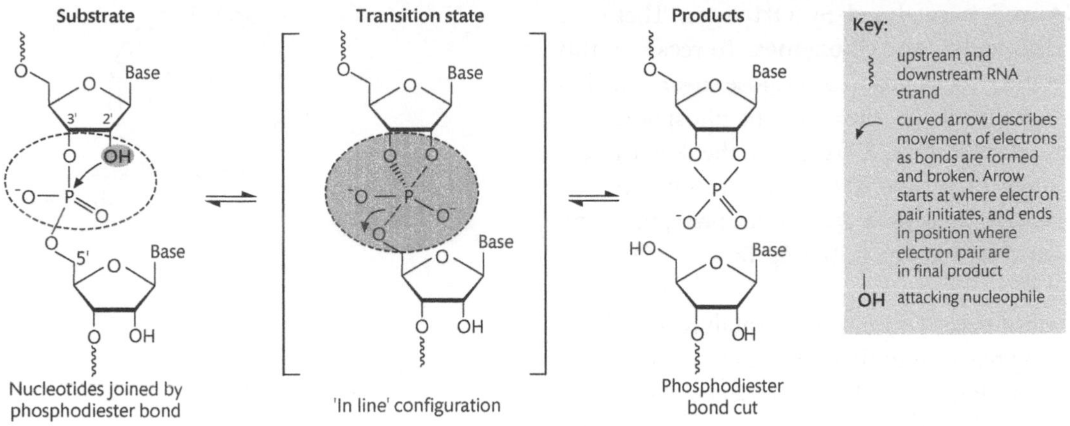

Figure 3.10 The phosphodiester backbone of RNA is chemically unstable. The key reacting parts of the RNA molecule are highlighted in grey. The oxygen of the 2′ -OH group on the ribose sugar can launch a nucleophilic attack on the phosphorus atom in the adjacent phosphodiester bond in the RNA molecule (notice the use of curved arrows to indicate movement of electrons). The attacking 2′ oxygen atom, phosphorus atom, and leaving oxygen atom need to adopt a linear configuration in the transition state—this configuration is described as being 'in-line': the three atoms which have to line up for cleavage to occur are connected by a broken line in the transition state. When the in-line configuration of atoms happens, the 2′ oxygen atom can form a new ester bond with the phosphorus atom, which simultaneously releases the linking 5′ oxygen of the downstream nucleotide. The new phosphodiester bond is formed at the same time as the leaving oxygen group on the old phosphodiester bond is displaced. The chemical term for this kind of reaction is an SN_2 reaction. (Reprinted from Mandal and Breaker[18] with permission from Macmillan Publishers Ltd.)

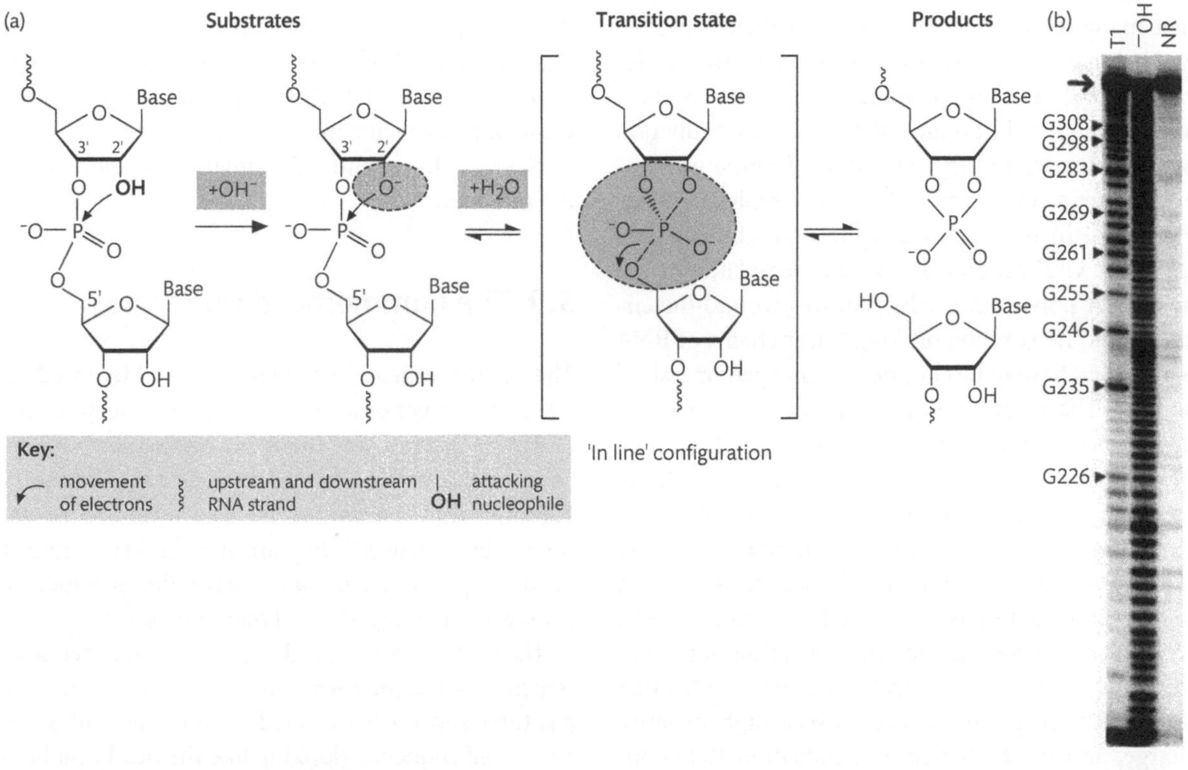

Figure 3.11 Spontaneous RNA cleavage is catalysed in a basic solution containing -OH groups. (a) Outline of the reaction. Notice that the -OH group activates the 2′ -OH group of the ribose sugar (circled by dashed line). (Adapted from Mandal and Breaker[18] with permission from Macmillan Publishers Ltd.) (b) In this experiment, radioactively labelled RNA molecules have been analysed on a polyacrylamide gel after incubation in solutions with or without -OH groups. Notice that lane NR is RNA that has not been pretreated, while lane -OH corresponds to RNA that has been incubated in an alkaline solution. RNA incubated in an alkaline solution shows multiple bands because the RNA has been cleaved at multiple points. The T1 lane represents RNA that is cut with a protein nuclease. (Redrawn from Nahvi *et al.*[19] with permission from Oxford University Press.)

of RNA and is catalysed by OH^- ions. These are called **ribonucleolytic ribozymes**. To recap, in this reaction the 2′ -OH group of a ribose sugar launches a chemical attack on the adjacent phosphodiester bond and displaces the 5′ oxygen of the downstream ribose sugar to cut the RNA chain into two pieces, one terminating with a 2′3′ cyclic phosphate and the downstream chain initiating in a 5′ -OH (see Fig. 3.11a).

The ribonucleolytic enzymes catalyse this RNA self-cutting reaction so that it takes place around 10^6 times faster than it would in the absence of catalysis. However, unlike general base catalysis by OH^- ions, the sites of the cleavage produced by this ribozyme-catalysed reaction are extremely specific.

Why do the small ribozymes manage to be smaller than the Group I intron ribozymes? One reason is that this ribonucleolytic reaction is a bit simpler—the small ribozymes do not need to hold a nucleophile in place (like GTP for the Group I introns), so they do not need to be so elaborate. Another reason for the small size of ribonucleolytic ribozymes is that many are used by viruses to cut their genomes, and being in viral genomes have probably been under very strong selective pressure to maintain a short length. The small ribonucleolytic ribozymes function in the processing of the viral RNA genome multimers that are produced by rolling-circle replication (see Fig. 3.12). In rolling-circle replication, circular copies of the viral genome are used as templates for the host RNA polymerase, which moves around the circular genome running off long linear chains of RNA which then have to be chopped up into genome-sized lengths. The small size and self-cleaving activity of these ribonucleolytic ribozymes enables these long RNA chains to be cut into single genomes, without devoting much genome space to this function.

Each of the small ribonucleolytic ribozymes uses acid–base catalysis to enhance the spontaneous RNA self-cutting reaction we discussed in Section 3.7. The catalytic challenges include the activation of the nucleophile to initiate chemical reactions, stabilization of the leaving groups which have a high negative charge, and the existence of a transition state containing a high concentration of negatively charged oxygen atoms. Some protein enzymes catalyse this same reaction (e.g. RNAse A). Compared with protein RNAses that carry out similar reactions, one key

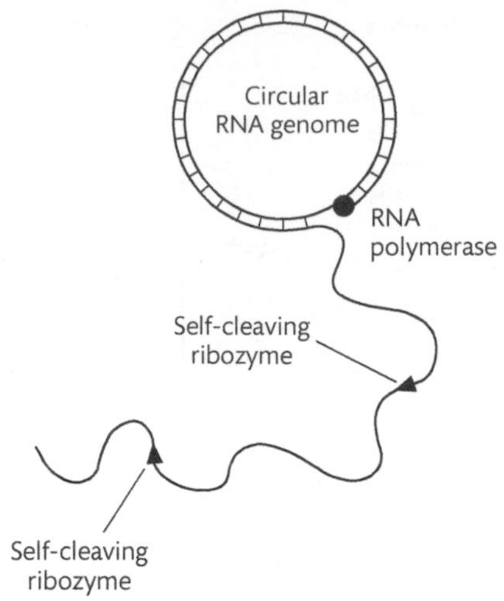

Figure 3.12 Rolling circle replication is used by some viral RNAs. Note that linear chains of RNA corresponding to multiple copies of the genome are run off a circular copy of the genome by RNA polymerase. Self-cleaving ribozyme sites at the ends of each linear genome cut these linear genomes up into single-genome-length RNA copies.

advantage of ribozymes is that through base pairing they can provide high specificity in target-site selection for RNA cleavage (e.g. the protein enzyme RNAse A enzyme which catalyses this same reaction is much less specific in target site).

Next we discuss how the small ribozymes work at the molecular level.

3.9 The hammerhead ribozyme

The hammerhead ribozyme was identified in 1986 in the genome of the tobacco ringspot virus (TRSV),[4,20,21] and was the third ribozyme to be discovered (after the Group I introns and RNAse P). Hammerhead ribozymes are also found in several other plant viruses. They are the shortest naturally occurring ribozymes, and cleave the products of rolling-circle replication of plant viruses.

The name 'hammerhead' comes from the secondary structure of hammerhead ribozymes which comprises just three short helices joined at a junction with a very conserved sequence (looking like the head of a hammerhead shark). This hammerhead secondary structure is shown in Fig. 3.13. Two of the helices in the full-length hammerhead ribozyme contain loops. An interaction between these loops enhances the catalytic

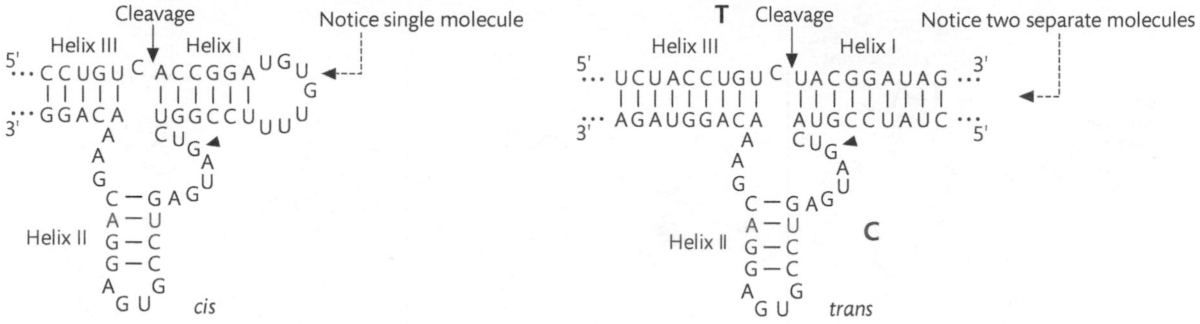

Figure 3.13 Hammerhead ribozymes can cut in *cis*- or in *trans*-. In a *cis*- configuration the hammerhead ribozyme is formed within a single RNA molecule (notice the continuous RNA chain running from 5′ to 3′). In the *trans*- configuration the hammerhead ribozyme is formed between two separate RNA molecules. The target and catalytic portions of the hammerhead are marked T and C, respectively, and a G residue essential for catalysis is arrowed. (Reproduced with permission from Samarsky et al.[22] ©1999 National Academy of Sciences of the USA.)

activity of the ribozyme. The shape of the hammerhead RNA molecule is critical for ensuring that the reacting molecules are brought sufficiently close together for catalysis to take place. The hammerhead ribozyme also folds into a tertiary structure as shown in Fig. 3.14—notice how compact this structure is.

Hammerhead ribozymes are very good catalysts, increasing the rate of RNA cleavage 10^9-fold over the uncatalysed RNA cleavage reaction. The reaction pathway followed by the hammerhead ribozyme is shown in Fig. 3.15. The cytidine residue at position 17 of the ribozyme (abbreviated C17) is the site at which the RNA is cut in the hammerhead ribozyme. The 2′ -OH of nucleotide C17 acts as a nucleophile to initiate the RNA cleavage reaction, a transition state forms, and then this resolves in RNA cleavage.

How does the hammerhead ribozyme catalyse this RNA cleavage reaction?[4,13] Hammerhead ribozymes use acid–base catalysis (see Fig. 3.15) via the bases of two guanosine residues in the RNA sequence of the hammerhead ribozyme. These two guanosine residues are called G12 and G8.

1. **G12 is involved in starting the reaction:** nitrogen 1 of the base of G12 removes a proton from the 2′ -OH group of C17 to activate the C17 2′ -OH group as a nucleophile.
2. **G8 carries out the second catalytic step:** the 2′ -OH group of G8 then donates a proton to the 3′ oxygen of the leaving group after the cleavage reaction.

Hammerhead ribozymes have also been of interest since they can be constructed as artificial or synthetic

Figure 3.14 Three-dimensional structure of the hammerhead ribozyme prior to RNA cleavage. The RNA tertiary structure is held together by base pairing to create helical stems and substantial tertiary interactions within the RNA molecule. (Redrawn from Protein Database ID URO121 by Jonathan Crowe, Oxford University Press.)

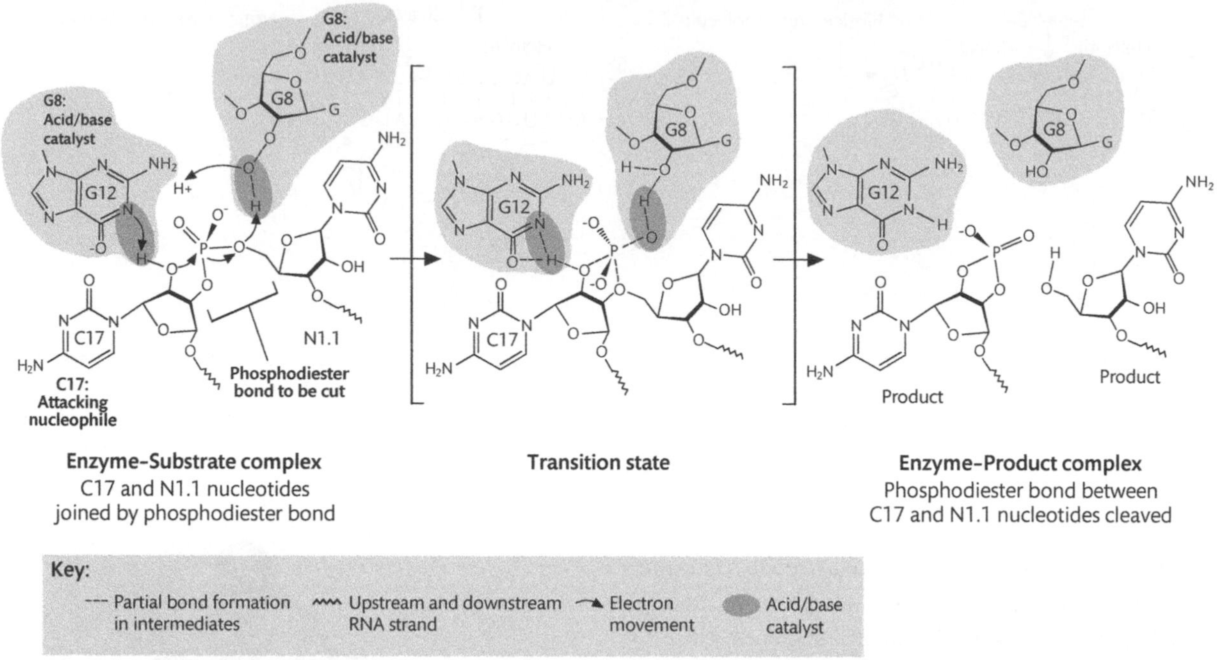

Figure 3.15 Hammerhead ribozymes use a nucleotide base to catalyse RNA cleavage. Electron movement is shown as curved arrows, and partial bonds are shown as broken lines. (Redrawn from Chi et al.[23])

hammerhead ribozymes to cut RNA. Synthetic hammerhead ribozymes have also been designed which are made up of two RNA molecules, one of which is catalytic and the other of which becomes cleaved by the reaction. Synthetic hammerhead ribozymes have been expressed in cells, and targeted through base pairing to cut specific RNA targets.[22,24] A comparison of a single-molecule hammerhead and a hammerhead formed by the association of two separate RNA molecules is shown in Fig. 3.13.

Several viruses have adopted small ribonucleolytic ribozymes to process their genomes. This includes a small nucleolytic ribozyme in the hepatitis delta virus (HDV). The HDV ribozyme also uses one of its bases for catalysis but, unlike the hammerhead ribozyme, also uses a metal ion for catalysis. We will look at this HDV ribozyme next.

3.10 The HDV ribozyme

Hepatitis B virus (HBV) is a pathogen that infects and causes disease in human liver cells. HDV is a small RNA virus which can superinfect cells already infected with HBV, worsening the symptoms and often leading to liver cirrhosis. HDV replicates by a rolling-circle mechanism which produces strings of multiple RNA genomes which then need to be separated into single genomes. The cutting of these long chains of genomes into single HDV genomes is carried out by a short ribozyme sequence encoded at the end of the HDV genome, called the HDV ribozyme, which folds up into the tertiary structure shown in Fig. 3.16.

How does the HDV ribozyme work? The HDV ribozyme uses a metal ion to activate the 2′ –OH nucleophile that initiates the nucleophilic attack on its adjacent phosphodiester bond.[2,25] Then one of the cytidine bases in the HDV ribozyme acts as a proton donor to stabilize the leaving group. Figure 3.17 illustrates this reaction in more detail:

- **Substrate.** The initial step of HDV ribozyme catalysis uses a metal ion for base catalysis. A magnesium ion in the ribozyme activates the 2′ –OH group of the ribose sugar immediately upstream of the phosphodiester bond to be cut. This makes a more potent O^- nucleophile. This activated 2′ O^- then launches a nucleophilic attack on the phosphorus atom of the phosphodiester bond. (The movements of electrons in this nucleophilic attack are shown as curly arrows.)

Figure 3.16 Tertiary structure of the hepatitis delta virus (HDV) ribozyme. The crystal structure shown is of the HDV ribozyme in its pre-cleaved state and attached to the RNA-binding domain U1A to assist crystallization (shown as lighter-shade ribbon at the top). (Redrawn from Protein Database ID PRO122 by Jonathan Crowe, Oxford University Press.)

- **Intermediate.** As a result of this nucleophilic attack the 2′ oxygen of the ribose sugar forms a new bond with the phosphorus atom in the scissile phosphate group. This breaks the phosphodiester backbone of the RNA.
- **Products.** The HDV ribozyme makes a second catalytic contribution. Negatively charged oxygen atoms are not good leaving groups after a chemical reaction. The HDV ribozyme uses an NH group on cytidine base C75 as a general acid to protonate the oxygen leaving group, resulting in the formation of a 5′ -OH group. This means that the cleaved RNA molecules can efficiently separate from each other.

Key points

Small ribozymes catalyse RNA cleavage at very specific sites within RNA chains. RNA cutting is achieved by acid-base catalysis. The hammerhead ribozyme uses bases within the RNA molecule for catalysis, while the HDV ribozyme uses a combination of its bases and metal ions. The small size of these ribonucleolytic ribozymes means that they are efficient tools for carrying out this reaction. This is a practical advantage, since they fit compactly into the genomes of these viruses, which are size-limited. Ribonucleolytic ribozymes catalyse RNA strand cutting at very specific sites (unlike the equivalent reaction catalysed by OH^- ions or the RNAse A protein enzyme).

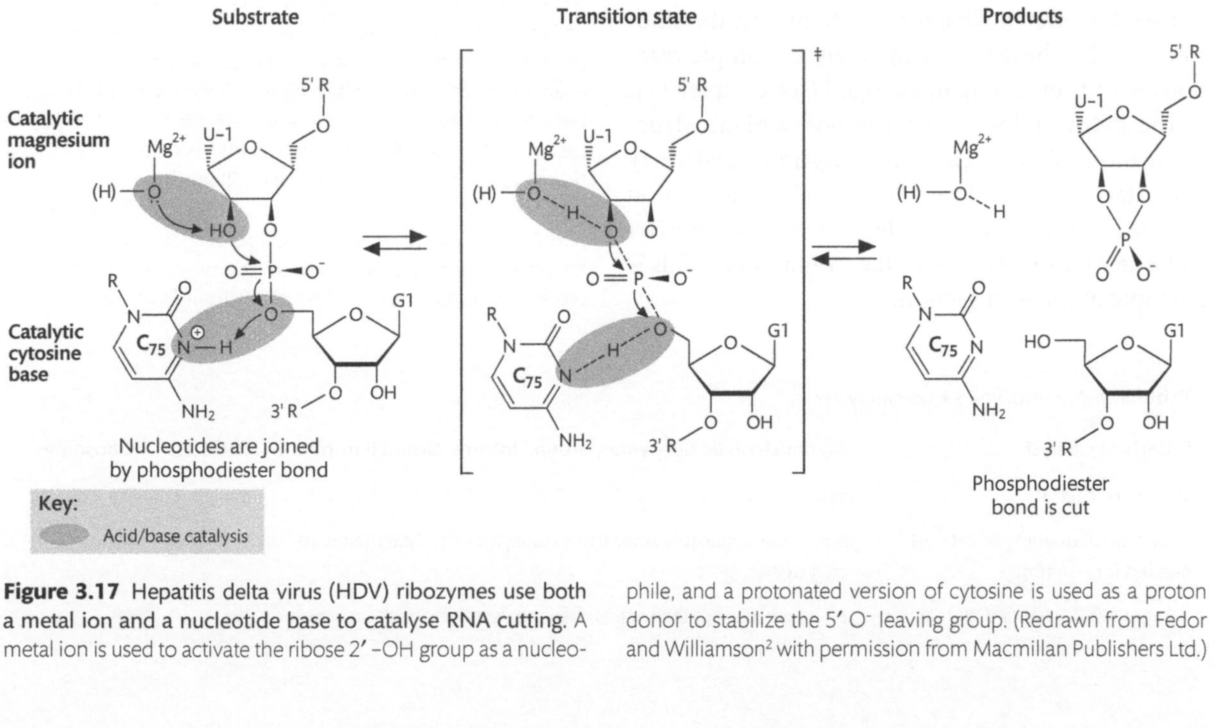

Figure 3.17 Hepatitis delta virus (HDV) ribozymes use both a metal ion and a nucleotide base to catalyse RNA cutting. A metal ion is used to activate the ribose 2′ -OH group as a nucleophile, and a protonated version of cytosine is used as a proton donor to stabilize the 5′ O⁻ leaving group. (Redrawn from Fedor and Williamson[2] with permission from Macmillan Publishers Ltd.)

3.11 Are ribozymes true catalysts?

So far in this chapter we have been discussing ribozymes as catalysts. In fact, catalysts have a somewhat precise definition based on the three criteria shown in Table 3.3. Tom Cech's group gave the archetypal catalytically active Group I introns the name 'ribozymes'. This name was controversial since Group I introns were actually removed by the splicing reaction, thus destroying the ribozyme. In contrast, true catalysts are unchanged by the reactions that they catalyse and so are able to carry out multiple catalytic events.

According to this formal definition of a catalyst, the Group I introns do not quite qualify as fully fledged catalysts. Although Group I introns speed up splicing around 10^{10}-fold, they are changed by this reaction and can only catalyse their own excision once. Although the Group I introns can only remove themselves once by splicing, subsequent experiments showed that engineered introns that were prevented from splicing themselves by the removal of their own splice sites *could* splice multiple other RNA molecules that did have splice sites if they were able to base pair with them, thus satisfying the formal definition of a catalyst.

Unlike the Group I introns, RNAse P satisfies the true criteria of a catalyst in that it can process the 5′ ends of multiple tRNA molecules, and yet remain unchanged by these reactions (we come back to how RNAse P works in Chapter 14). Similarly, the ribosome is RNA based and can catalyse multiple reactions whilst remaining unchanged (see Chapter 11).

The ability of RNA to work as both a highly structured molecule that can bind substrates and carry out catalysis, as well as carrying out information storage in some viruses, has led to the idea that RNA had a central role to play in the origin of life. This is the topic of the next section.

3.12 The RNA world hypothesis: a time when RNA was used as a genetic material

The **RNA world hypothesis** suggests that, at early stages of evolution, RNA instead of DNA was used as a replicating molecule (this period is called the **RNA era** in Fig. 3.18).[26–31]

Why should RNA come first before DNA in the history of life? The advantages of RNA as an early

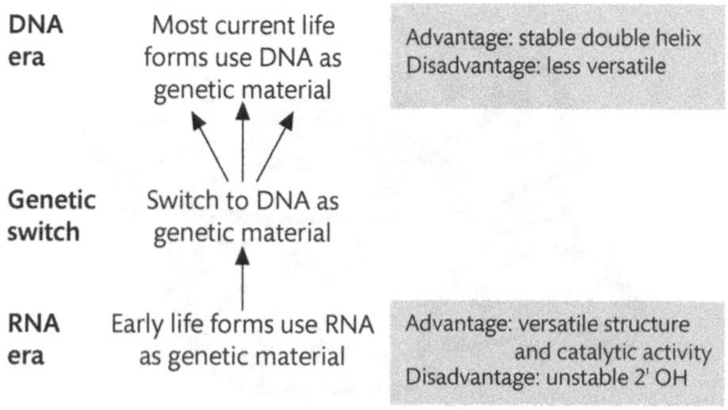

Figure 3.18 **RNA might have been used by very early replicating organisms.** The evolutionary tree leading to modern organisms might have included a period in which self-replicating RNAs were used as both genetic material and catalysts (the RNA era). RNA molecules replicate themselves and so are used as genetic material. By promoting their own replication, variant RNAs which were self-replicating would have become more numerous and had a selective advantage over RNAs unable to self-replicate. Variant RNAs which were more efficient replicators would be preferentially copied and so would become more numerous. To enable their efficient multiplication, self-replicating RNAs would need some mechanism to prevent them from replicating competitor RNAs. This might have led to the evolution of cell membranes to contain the self-replicating RNA within its own compartment. Over the course of time self-replicating RNAs would have been replaced by DNA replicators encoding RNA templates for proteins. Proteins now take on the major role as structural proteins and catalysts in the cell (the current DNA era).

Table 3.3 Are ribozymes true catalysts?

Criteria of catalysis	Ribonucleolytic ribozymes, Group I introns, Group II introns	RNAse P	Ribosome
Speed up reaction	Yes	Yes	Yes
Only a small quantity of catalyst needed for reaction	Not really applicable since these ribozymes catalyse their own cleavage	Yes	Yes
Catalyst left unchanged by reaction	No: since they cut themselves, they are destroyed by the reaction	Yes	Yes

replicator are that as well as fulfilling a genetic role, RNA could fulfil additional structural and catalytic roles. These would include the evolution of RNA molecules which could behave as catalysts to assist their own replication. Once primitive RNA catalysts capable of self-replication had developed, they could replicate themselves and then be fine-tuned by natural selection. Better RNA replicators would in turn become more numerous as a result of replication.

The RNA world theory also provides a rationale for why early cells might have evolved. The development by RNAs of catalytic functions to facilitate RNA replication of RNAs would drive the development of cell membranes, as these would provide a physical barrier preventing successful RNA replicators from copying other RNAs in the environment. Membrane-bound compartments which contained more efficiently self-replicating RNAs would have a selective advantage and become more numerous.

Although RNA has many advantages as a functional molecule, at some point in evolution RNA replicators must have been replaced by DNA replicators which are copied into RNA transcripts encoding largely protein-based catalysts. The switch to DNA might have occurred because RNA is less stable as a molecule because of its 2′ -OH group.

Experimental evidence supports the existence of an early RNA world. Catalytic RNA molecules have been evolved in vitro using **SELEX** approaches which can replicate RNA molecules and so might mimic early RNA-based replicating molecules. These important experiments are described in more detail in the following sections.

3.13 Experiments have been carried out that might model the early steps that might have occurred during the evolution of life

In modern cells RNA molecules are made by transcription, which copies a DNA template using highly evolved protein enzymes called DNA-dependent RNA polymerases. These RNA polymerases are DNA dependent because they copy a DNA template by catalysing the addition of nucleotide triphosphates (NTPs) to the growing RNA strand. Modern protein RNA polymerases have several important properties (see Fig. 3.19).

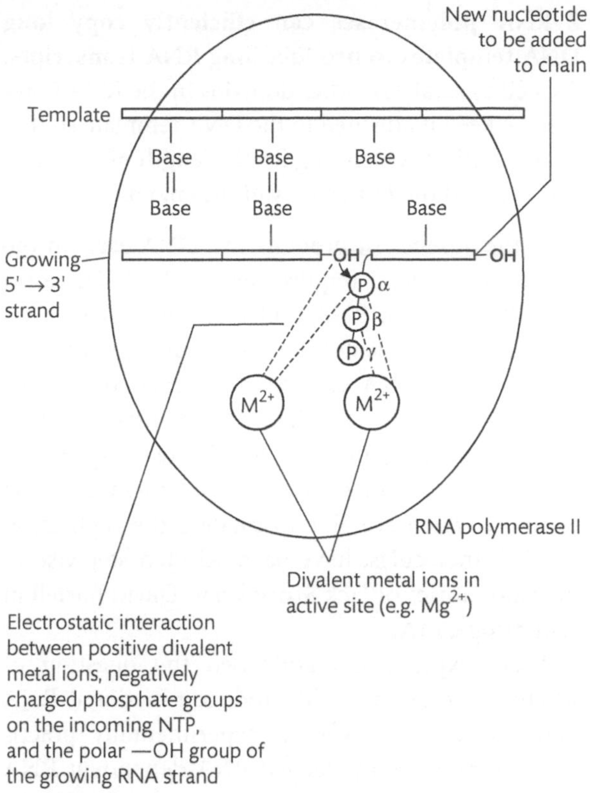

Figure 3.19 Protein RNA polymerases carry out transcription in modern cells. Protein side chains in the RNA polymerase position two positively charged divalent metal ions in the active site of the RNA polymerase where they activate the 3′ -OH group for catalysis, interact with the negatively charged phosphate groups in the incoming NTP (nucleotide triphosphate), and stabilize the reaction intermediate. Abbreviation: M, metal.

1 RNA polymerases catalyse the formation of new phosphodiester bonds. RNA polymerases catalyse the 3′ -OH group on the last nucleotide in the growing RNA chain to attack the α-phosphate of the incoming NTP and create a new phosphodiester bond. RNA polymerases are accurate. To ensure that only the correct NTP is added to the growing RNA strand, the catalytically functional active site of the RNA polymerase is generated only if the template nucleotide and the incoming NTP are correctly base paired: this is called an *induced fit mechanism*. Although it has to hold itself in position during catalysis, the RNA polymerase needs to release itself after catalysis has occurred and move forwards along the template to add the next nucleotide.

2 RNA polymerases can efficiently copy long DNA templates to produce long RNA transcripts. As well as catalysis, other domains in the RNA polymerase keep it attached to the DNA template so that it can continue extending for the length of the gene: this is called the *processivity* of the enzyme.

In the absence of proteins and DNA early in the evolution of life, a requirement of the RNA world hypothesis is that RNA molecules must have been able to copy (replicate) themselves using RNA templates. Could RNA catalysts have carried out this copying role early in evolution? The answer is likely to be yes. Using in vitro SELEX techniques like those discussed in Chapter 2, polymerases made of RNA, which are able to catalyse the replication of RNA molecules, have been selected stepwise in the laboratories of Jack Szostak and David Bartell in Cambridge, MA.

These experiments addressed the question of whether evolution could produce functional RNA polymerases, but over an experimentally practical timeframe. Fully fledged RNA-dependent RNA polymerase ribozymes were not experimentally selected in the laboratory in a single step. Instead, the first experimental step was the laboratory isolation of a group of ribozyme catalysts called **RNA ligases** that carry out a similar reaction (Fig. 3.20). Then, following this, more sophisticated RNA polymerases were made. Making a catalytic RNA ligase is a bit simpler experimentally. As well as being the first step in making an RNA polymerase ribozyme in the laboratory, catalytic RNA ligases made of RNA might have had an early evolutionary role in replicating primordial RNA molecules by stitching together short RNA strands before fully fledged RNA polymerases evolved.

Let us now look at how RNA polymerase ribozymes were experimentally evolved in the laboratory.

3.13.1 Stage 0: the uncatalysed reaction between two RNA chains

The initial experimental design was to use two RNA chains held by Watson–Crick hydrogen bonding. Using two RNA strands in this way, rather than trying to add a single nucleotide, had a major advantage. It juxtaposed the two molecules so that the 5′ triphosphate group of the longer RNA was directly adjacent to the 3′ -OH group of the shorter RNA. These two groups were then poised for the first chemical reaction (see Fig. 3.21, uncatalysed RNA ligation).

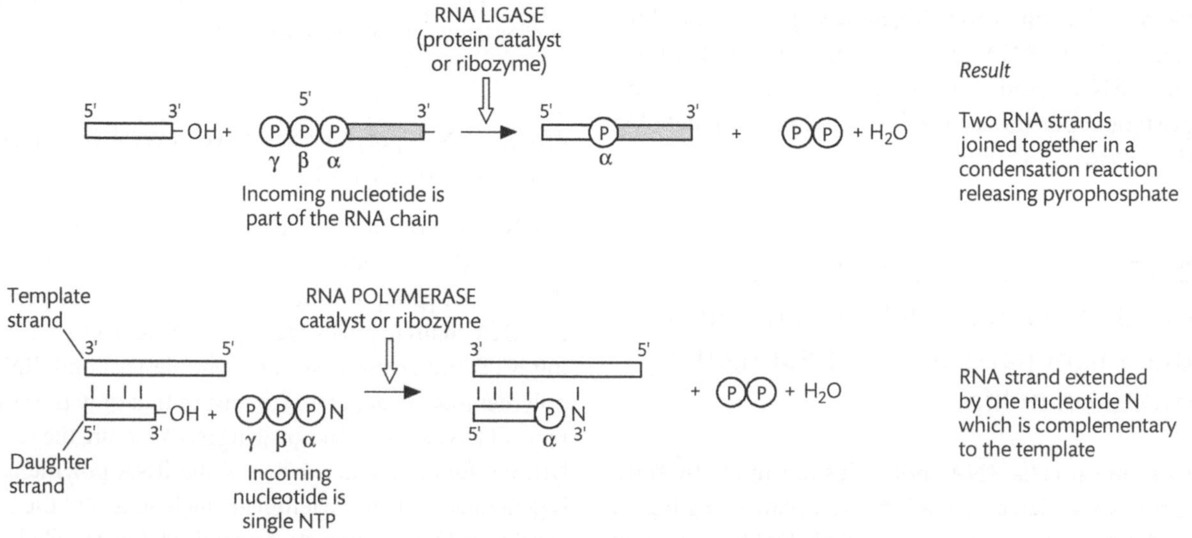

Figure 3.20 **RNA ligases and RNA polymerases catalyse the same chemical reaction.** Both catalyse the chemical attack of a 3′ -OH group on the α-phosphate of an incoming nucleotide to create a new phosphodiester bond. This reaction releases H_2O and pyrophosphate (PP). In the case of RNA ligases the incoming nucleotide is on an RNA chain, while for polymerases the incoming nucleotide is a single NTP (nucleotide triphosphate). Notice that the α-phosphate of the incoming nucleotide is retained in the elongated nucleotide chain as the linking phosphate of the new phosphodiester bond.

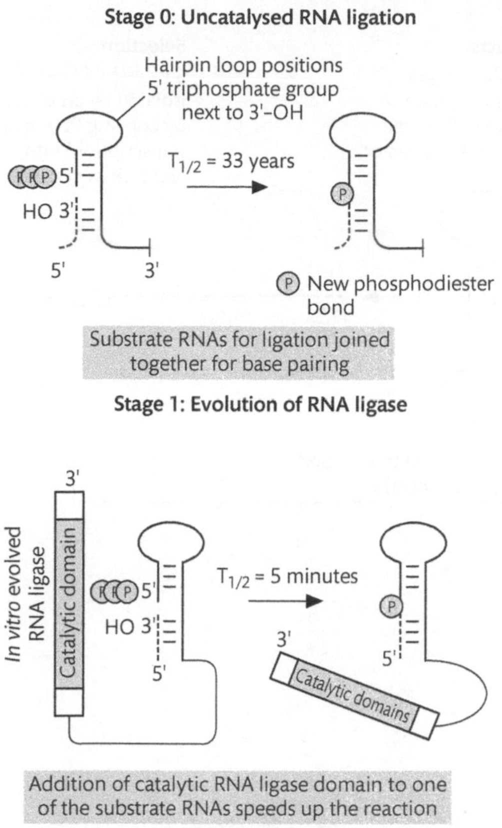

Figure 3.21 Laboratory-based evolution of an RNA ligase from a random pool of sequences. **Stage 0** shows the uncatalysed RNA ligation reaction. **Stage 1** shows that the addition of an RNA catalytic domain to the longer RNA speeds up this reaction so that it takes place with a half-life ($T_{1/2}$) of 5 minutes.

In the absence of any catalyst, the ligation reaction between the 3′ -OH group and the 5′ triphosphate would only take place with a half-life of 33 years (i.e. after 33 years, half the molecules would have been converted into product). This reaction would clearly be too slow to be of any use, and was improved by next evolving a catalytic ribozyme.

3.13.2 Selection of catalytic RNA ligases

To try and speed up the reaction, a randomized 220-nucleotide insert was attached to the longer RNA molecule (see Fig. 3.21, evolution of an RNA ligase). This randomized insert contained a pool of 10^{15} different variant RNA sequences, some of which by chance might be catalytically active. In the event of a successful ligation reaction, the catalytic RNA on the longer RNA chain would become attached to the short RNA chain through a new phosphodiester bond and so be amplifiable by RT-PCR across the join.

The random pool of RNAs was incubated in the presence of magnesium ions and buffer to enable any chemical reactions between the 3′ -OH and 5′ triphosphate group to take place (since it was known that magnesium ions play an important role in modern RNA polymerases). To select catalytically active ribozymes, ligated RNAs were then amplified using reverse transcription and then PCR between primers which bound to sequences present in the two initially separate RNA molecules; only ligated RNA molecules would be amplified by this step.

In order to fine-tune the RNA ligase ribozymes, the sequences of the selected RNAs were then mutated, and the SELEX process carried out again (see Fig. 3.22). In a series of 10 steps of mutagenesis–amplification, a set of RNA ligases were identified that could carry out the RNA ligation reaction with a half-life of about 5 minutes, which was a *7-million-fold improvement* over the uncatalysed reaction. The structure of an in vitro selected ligase has been worked out using atomic resolution and is shown in Fig. 3.22(b)—notice that this catalytic RNA incorporates significant secondary and tertiary structure, giving a globular molecule. This structure is likely to be important for the RNA to work as a catalyst, similar to the ribozymes we discussed earlier in this chapter.

3.13.3 Selection of catalytic RNA polymerases

Could this RNA-based ligase also function as an RNA polymerase and add nucleotides? The answer was yes. In further experiments, the RNA ligase isolated as described in the previous section was further improved. A first improvement was attachment to a template strand by base pairing. Once attached to a template strand, the RNA ligase catalyst could extend the 3′ end of an existing RNA strand just like an RNA polymerase.

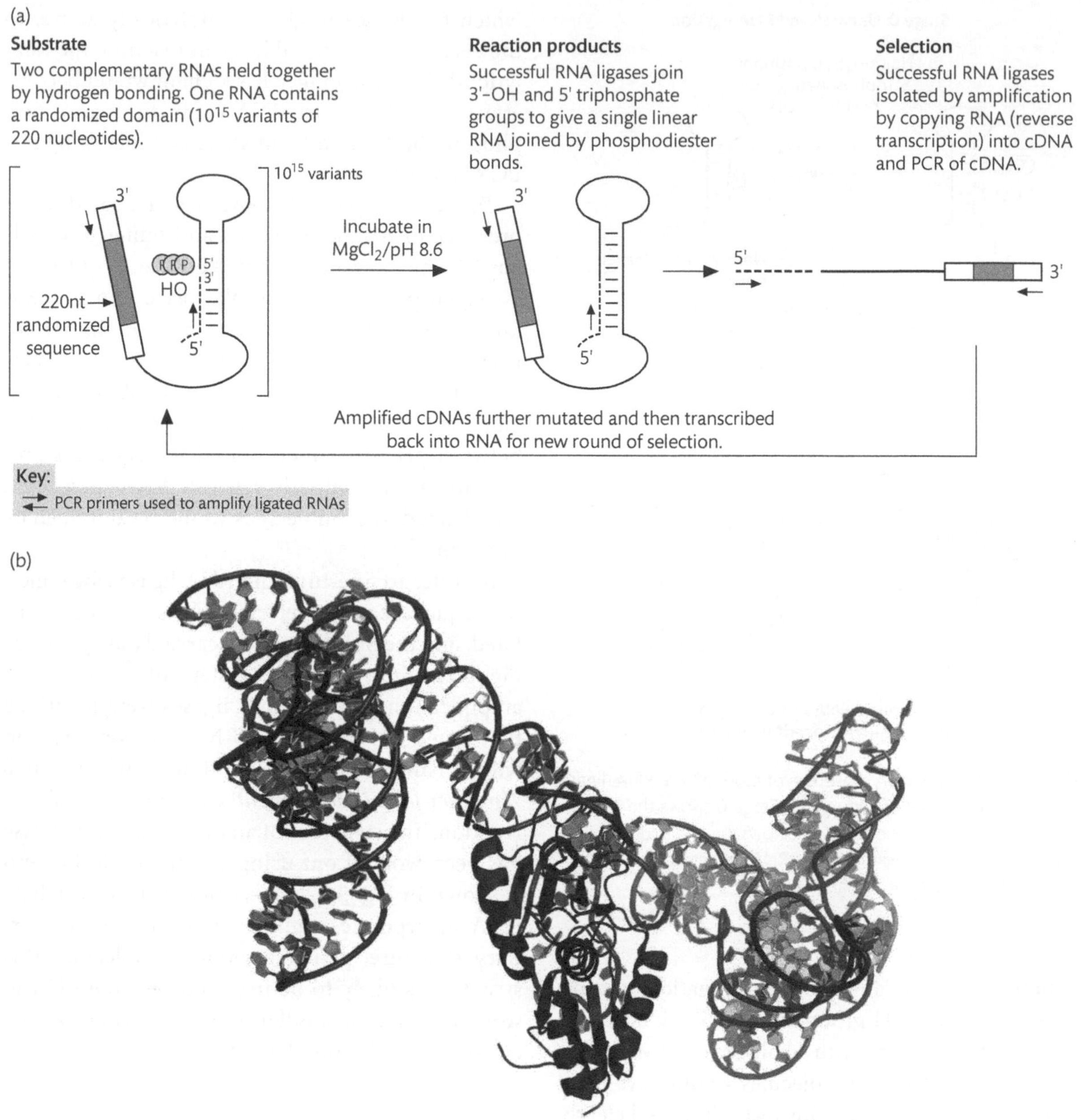
(a)
Substrate
Two complementary RNAs held together by hydrogen bonding. One RNA contains a randomized domain (10^15 variants of 220 nucleotides).
10^15 variants
3'
5'
3'
HO
5'
220nt
randomized sequence
Incubate in MgCl2/pH 8.6
Reaction products
Successful RNA ligases join 3'-OH and 5' triphosphate groups to give a single linear RNA joined by phosphodiester bonds.
3'
5'
Selection
Successful RNA ligases isolated by amplification by copying RNA (reverse transcription) into cDNA and PCR of cDNA.
5'
3'
Amplified cDNAs further mutated and then transcribed back into RNA for new round of selection.
Key:
PCR primers used to amplify ligated RNAs
(b)

Figure 3.22 In vitro selection of RNA ligase ribozymes using SELEX. (a) How the experiment was designed to select ligases. The longer RNA has a 3′ extension containing a 220-nucleotide (nt) randomized sequence with 10^{15} different sequence variants. The expectation was that some of these sequence variants might have catalytic activity to speed up the RNA ligation reaction. To select catalytic RNA sequences, the pool of sequence variants was incubated in solution with magnesium ions. Notice that the shorter RNA chain is shown as a broken line, the longer catalytic RNA as a solid line, and the randomized regions as rectangles. Successful ligases joined the gap between the short and longer RNA molecules. Ligated RNAs could be amplified with RT-PCR using primers (shown as arrows) complementary to the previously separate RNA molecules. Notice that this step will also amplify the randomized sequence providing the catalytic activity, but only ligated RNA molecules can be amplified. The successful ribozyme catalysts were then further mutated and transcribed back into RNA for another 10 rounds of this selection procedure. (b) Crystal structure of an in vitro selected RNA ligase ribozyme including a U1A RNA-binding sequence and coupled to the RNA-binding domain of U1A protein. (Redrawn from the RCSB Protein Data Bank (NDB ID PRO381) by Jonathan Crowe, Oxford University Press.)

3.13.4 What did these experiments mean?

These in vitro evolution experiments were successful in that they enabled RNA-dependent RNA polymerases to be selected in the laboratory. The implication of these experiments is that similar evolutionary steps could have occurred early in the history of life. Naturally evolved RNA polymerases would have developed in the wild over hugely longer timeframes, over which time they would have perfected their abilities. The laboratory-developed RNA polymerases still had some limitations—they could only add 14 new NTPs onto a 3′ -OH group. To continue with polymerization, an efficient RNA polymerase needs to release its hold on the template/elongating RNA strand and move forward. The in vitro evolved RNA polymerases had a tendency to totally release after catalysis, preventing further NTPs from being added. Also, the high magnesium/slightly basic pH required for catalytic reactions in the laboratory caused the input RNAs to become rather unstable and degrade.

Although the RNA era is hypothesized to have occurred early in the history of life, a number of RNA molecules still perform important catalytic roles in modern cells. According to the RNA world hypothesis, these catalytic RNA molecules might be molecular fossils from early stages of life on Earth.

Key points

Scientists have speculated that RNA might have been used as a replicating molecule in the very earliest organisms before DNA was adopted. The advantages of RNA over DNA as an early replicator are that RNA can not only store genetic information but can also form more complex structures than the double helix and catalyse chemical reactions. In fact, the very earliest RNA replicators might have catalysed their own replication. Catalytically active RNAs might have evolved cell membranes to prevent them replicating competing RNAs in the environment ('freeloaders'). Consistent with this model for the early steps of life, SELEX experiments have been used to isolate first RNA ligase ribozymes which can join RNA chains and then RNA polymerase enzymes which can copy RNA templates.

Summary

- Catalysts speed up reactions while themselves remaining unaltered. Enzymes are catalysts used by living organisms to speed up chemical reactions needed by the cell so that they can occur rapidly under physiological conditions. A group of RNA-based catalysts play important roles in cells, including in translation, splicing, and tRNA processing. Some ribozymes are 'true catalysts' which remain unaltered after catalysis. Other ribozymes, like the Group I introns and the ribonucleolytic ribozymes, are more formally semi-catalytic in that they speed up reactions but are themselves altered.
- The ability to form tertiary structures (giving rise to active sites with precisely positioned reacting groups) is an important feature of catalytic RNA. The 2′ -OH group is important for some but not all ribozymes.
- Many ribozymes use acid–base catalysis in their reactions. In these reactions, removal of protons converts polar -OH groups into more highly reactive O^- oxyanions. In the reverse of this, the addition of a proton converts a released O^- ion into an -OH group which can more easily leave the active site of a ribozyme.
- The ability to bind to metal ions is an important property of RNA which equips it to form ribozymes. Metal ions can stabilize the build-up of negative charges in the closely adjacent RNA strands which make up ribozyme catalytic sites, can stabilize charged reaction intermediates, and can bind and release protons within the active site of the ribozyme.
- Some ribozymes use base protonation/deprotonation as a mechanism to catalyse reactions.
- Two groups of ribozymes are found in nature. The 'large' class of ribozymes include the Group I introns of *Tetrahymena* rRNA. The group of smaller ribozymes include small RNA modules which process viral RNAs into genome-sized packages.
- RNA might have been used as an early replicating molecule before the use of DNA evolved. In vitro experiments have been carried

out which can select self-replicating RNA molecules from pools of random molecules, mimicking steps which might have occurred early in the evolution of life on Earth.

Questions

3.1 What are the chemical properties of RNA that enable it to function as a catalyst?
3.2 How were ribozymes first discovered?
3.3 What reaction pathway do Group I introns follow?
3.4 What mechanism do Group I introns use to catalyse this reaction pathway?
3.5 What reaction pathway do Group II introns follow?
3.6 What mechanism do Group II introns use to catalyse this reaction pathway?
3.7 The reaction pathway followed by Group II introns is similar to another RNA processing pathway. What is this other pathway, and what does this similarity in mechanism suggest?
3.8 Why is RNA unstable in basic solutions?
3.9 For what purposes have viruses evolved ribozymes?
3.10 How does the hammerhead ribozyme catalyse RNA cleavage?
3.11 What is the RNA world hypothesis and what does it seek to explain?
3.12 How have laboratory experiments tested some of the predictions of the RNA world hypothesis?

References

1. **Fedor MJ.** The role of metal ions in RNA catalysis. *Curr Opin Struct Biol* **12**, 289–95 (2002).
2. **Fedor MJ, Williamson JR.** The catalytic diversity of RNAs. *Nat Rev Mol Cell Biol* **6**, 399–412 (2005).
3. **Lilley DM.** The origins of RNA catalysis in ribozymes. *Trends Biochem Sci* **28**, 495–501 (2003).
4. **Lilley DM.** Structure, folding and mechanisms of ribozymes. *Curr Opin Struct Biol* **15**, 313–23 (2005).
5. **Lilley DM.** Mechanisms of RNA catalysis. *Philos Trans R Soc Lond B Biol Sci* **366**, 2910–17 (2011).
6. **Takagi Y, Warashina M, Stec WJ, et al.** Recent advances in the elucidation of the mechanisms of action of ribozymes. *Nucleic Acids Res* **29**, 1815–34 (2001).
7. **Kruger K, Grabowski PJ, Zaug AJ, et al.** Self-splicing RNA: autoexcision and autocyclization of the ribosomal RNA intervening sequence of *Tetrahymena*. *Cell* **31**, 147–57 (1982).
8. **Altman S.** Nobel lecture. Enzymatic cleavage of RNA by RNA. *Biosci Rep* **10**, 317–37 (1990).
9. **Guerrier-Takada C, Gardiner K, Marsh T, et al.** The RNA moiety of ribonuclease P is the catalytic subunit of the enzyme. *Cell* **35**, 849–57 (1983).
10. **Bass BL, Cech TR.** Specific interaction between the self-splicing RNA of *Tetrahymena* and its guanosine substrate: implications for biological catalysis by RNA. *Nature* **308**, 820–6 (1984).
11. **Steitz TA, Steitz JA.** A general two-metal-ion mechanism for catalytic RNA. *Proc Natl Acad Sci USA* **90**, 6498–502 (1993).
12. **Stahley MR, Strobel SA.** Structural evidence for a two-metal-ion mechanism of group I intron splicing. *Science* **309**, 1587–90 (2005).
13. **Strobel SA, Cochrane JC.** RNA catalysis: ribozymes, ribosomes, and riboswitches. *Curr Opin Chem Biol* **11**, 636–43 (2007).
14. **Michel F, Jacquier A, Dujon B.** Comparison of fungal mitochondrial introns reveals extensive homologies in RNA secondary structure. *Biochimie* **64**, 867–81 (1982).
15. **Pyle AM.** The tertiary structure of group II introns: implications for biological function and evolution. *Crit Rev Biochem Mol Biol* **45**, 215–32 (2010).
16. **Fica SM, Mefford MA, Piccirilli JA, Staley JP.** Evidence for a group II intron-like catalytic triplex in the spliceosome. *Nat Struct Mol Biol* **21**, 464–71 (2014).
17. **Fica SM, Tuttle N, Novak T, et al.** RNA catalyses nuclear pre-mRNA splicing. *Nature* **503**, 229–34 (2013).
18. **Mandal M, Breaker RR.** Gene regulation by riboswitches. *Nat Rev Mol Cell Biol* **5**, 451–63 (2004).
19. **Nahvi A, Barrick JE, Breaker RR.** Coenzyme B12 riboswitches are widespread genetic control elements in prokaryotes. *Nucleic Acids Res* **32**, 143–50 (2004).
20. **Prody GA, Bakos JT, Buzayan JM, et al.** Autolytic processing of dimeric plant virus satellite RNA. *Science* **231**, 1577–80 (1986).
21. **Scott WG, Horan LH, Martick M.** The hammerhead ribozyme: structure, catalysis, and gene regulation. *Prog Mol Biol Transl Sci* **120**, 1–23 (2013).
22. **Samarsky DA, Ferbeyre G, Bertrand E, et al.** A small nucleolar RNA:ribozyme hybrid cleaves a nucleolar RNA target in vivo with near-perfect efficiency. *Proc Natl Acad Sci USA* **96**, 6609–14 (1999).
23. **Chi YI, Martick M, Lares M, et al.** Capturing hammerhead ribozyme structures in action by modulating general base catalysis. *PLoS Biol* **6**, e234 (2008).
24. **Dower K, Kuperwasser N, Merrikh H, Rosbash M.** A synthetic A tail rescues yeast nuclear accumulation of a ribozyme-terminated transcript. *RNA* **10**, 1888–99 (2004).

25. **Cochrane JC, Strobel SA.** Catalytic strategies of self-cleaving ribozymes. *Acc Chem Res* **41**, 1027–35 (2008).
26. **Joyce GF, Orgel LE.** Progress towards understanding the RNA world. In Gesteland R, Cech TR, Atkins JF (eds), *The RNA world* (3rd edn) pp. 23–56. Cold Spring Harbor, NY: Cold Spring Harbor Laboratory Press (2006).
27. **Szostak JW, Bartel DP, Luisi PL.** Synthesizing life. *Nature* **409**, 387–90 (2001).
28. **Benner SA.** Catalysis: design versus selection. *Science* **261**, 1402–3 (1993).
29. **Ekland EH, Szostak JW, Bartel DP.** Structurally complex and highly active RNA ligases derived from random RNA sequences. *Science* **269**, 364–70 (1995).
30. **Levy M, Ellington AD.** The descent of polymerization. *Nat Struct Biol* **8**, 580–2 (2001).
31. **McGinness KE, Joyce GF.** In search of an RNA replicase ribozyme. *Chem Biol* **10**, 5–14 (2003).

4 The RNA-binding proteins

Introduction

RNA-binding proteins are at the heart of all the co-transcriptional and post-transcriptional processes described in this textbook.[1-4] Therefore it is important to appreciate the structure of RNA-binding proteins and the RNA-binding domains that enable them to bind RNA.[5-11] RNA-binding domains have several important properties and are very versatile. They can bind single-stranded RNA (ssRNA) or double-stranded RNA (dsRNA). They can recognize RNA sequences (the primary structure of the RNA) or alternatively they can recognize RNA structures at a three-dimensional level. RNA-binding proteins can work by combining several RNA-binding domains, often assisted by **auxiliary RNA-binding domains** (parts of the protein that can help by providing additional RNA-binding activity or the ability to promote protein–protein interactions).[8] Contacts between the RNA and the protein can be made with the RNA bases, the ribose sugar and phosphate groups. In this chapter you will gain an appreciation of the best-studied RNA-binding domains, which are listed in Table 4.1. It is worth noting that some organisms have become highly dependent on RNA-binding proteins—mostly notably the trypanosomes (Box 4.1).[12,13]

The study of RNA-binding proteins began with the **hnRNP proteins** (hn stands for 'heterogeneous nuclear' and RNP stands for 'ribonucleoprotein').[7,14] In the 1960s and 1970s, when these proteins were first being studied, the function of hnRNA was not fully understood. Nowadays hnRNA is generally referred to as **pre-mRNA**. The hnRNP proteins bind to hnRNA (pre-mRNA); however, the term hnRNP protein has persisted. They were first described as chromatin-associated RNA-binding proteins, but it then became apparent that they bind to nascent transcripts (pre-mRNA). Bear in mind that all nascent transcripts, whether coding (mRNA translated into protein) or non-coding (e.g. microRNAs, snRNAs), are packaged by RNA-binding proteins. The need for the packaging of transcripts is twofold: it protects the delicate chains of RNA from degradation, and it facilitates pre-mRNA processing events.

In early experiments that used **gradient centrifugation**, the hnRNP proteins were observed to package pre-mRNA into large complexes that could be broken down into particles of an apparently uniform size—the core 30S particles (the S refers to Svedberg units of sedimentation rate). hnRNP proteins were further broken down into major and minor hnRNP proteins. The major hnRNP proteins are amongst the most abundant RNA-binding proteins in the nucleus, and include the following: the well-studied RNA-binding protein hnRNP A1, involved in alternative splicing; hnRNP A2/B1 (involved in splicing and mRNA trafficking); hnRNP C1/C2 (involved in pre-mRNA packaging, alternative splicing, and nuclear retention); hnRNP F (alternative splicing); and the multifunctional hnRNP K (alternative splicing, translation regulation, mRNA stability, and even transcription). The less abundant minor hnRNP proteins include HuR, involved in the regulation of mRNA stability and transport.

Several hnRNP proteins shuttle between the nucleus and cytoplasm and can be detected in the

Box 4.1 Trypanosomes rely on RNA-binding proteins to regulate the expression of their genes

Trypanosomes are unicellular parasitic protozoa. *Trypanosoma brucei* is the cause of the disease 'sleeping sickness' (trypanosomiasis), which is fatal if left untreated. Trypanosomes are transmitted by invertebrates that feed on blood (such as the infamous tsetse fly, genus *Glossina*). The infection begins with a bite that inoculates subcutaneous tissue; it then moves into the lymphatic system and bloodstream, and eventually enters the central nervous system. Once the infection reaches the brain the consequence is extreme lethargy and eventually death. Trypanosomes control the expression of their genes in an unconventional way. Primary transcripts often include multiple open reading frames that are cleaved into individual mRNAs making it more difficult to regulate gene expression through transcription. Instead, gene expression is mostly regulated through post-transcriptional mechanisms, particularly by altering mRNA decay rates. Less stable mRNAs produce less stable proteins, whereas highly stable mRNAs are more productive. The regulation of mRNA stability is achieved through RNA-binding proteins.

To illustrate how post-transcriptional control works, consider the adaptation to heat shock. Trypanosomes need to adapt to different temperature environments depending on their location (in the insect vector or in various tissues of the mammalian host). The *T. brucei* zinc-finger protein ZC3H11 is essential for the survival of the parasite in the blood. ZC3H11 binds and stabilizes mRNAs induced by heat shock that encode chaperone proteins.

cytoplasm (e.g. hnRNP A1), whereas others are restricted to the nucleus (e.g. hnRNP C1). A common theme with many hnRNP proteins is that they are able to keep RNA single-stranded, facilitating pre-mRNA splicing and other processing events (3′ end formation). The picture that has now emerged is that the hnRNP proteins are not only involved in pre-mRNA packaging, but are multifunctional and are involved in several post-transcriptional processes. Note that the phenomenon of multifunctionality is widespread among RNA-binding proteins—we shall see this time and time again.[6,7] This suggests that RNA-binding proteins are able to coordinate several steps in gene expression.

Key points

RNA-binding proteins can package RNA, protect it, organize it, and prepare it for post-transcriptional processes. They contain different kinds of RNA-binding and auxiliary domains, which enable them to bind RNA in a versatile way. They can recognize both primary sequences of RNA and three-dimensional structures.

The first RNA-binding proteins to be studied in some detail were the hnRNP proteins. The term hnRNA (heterogeneous nuclear RNA) is no longer used; instead, they are known as 'pre-mRNA'. The hnRNP proteins package pre-mRNA and are involved in multiple post-transcriptional processes. Some, but not all, hnRNP proteins remain bound to messenger RNA in the cytoplasm in mRNP particles.

Table 4.1 The known RNA-binding domains

RNA-binding domain	Structure of the RNA-binding domain	Type of nucleic acid recognized	Examples of RNA-binding proteins with the domain
RRM (RNA recognition motif)	βαββαβ barrel-like; with RNP1 and RNP2 motifs	ssRNA, ssDNA	hnRNP A1, PTB, SRSF1, HuD, Sxl, RBM3
KH (K-homology)	Three-stranded β-sheet packed against three α-helices	ssRNA, ssDNA	hnRNP K, FMR1, Nova
CSD (cold-shock domain)	β-barrel structure (βββββ) similar to OB-fold	ssRNA, ssDNA	CspB, YB-1, UNR
dsRBD	αβββα fold	dsRNA	PKR, Staufen, Dicer, RNAse III, ADAR1
Zinc finger	ββα structure	dsRNA, ssRNA	TFIIIA, WT1, Tra-1, ZRANB2
Arginine-rich	Alone or in glycine/aromatic-rich domains	ssRNA, dsRNA	Tat, YB-1, hnRNP A1

... continued

RNA helicase	Domain 1 and domain 2 with conserved motifs	dsRNA substrate	eIF4A, DHH1
PAZ	β-barrel reminiscent of OB-fold	dsRNA substrate	Dicer
PIWI	RNAse H core; five-stranded β-sheet surrounded by α-helices	dsRNA substrate	Argonaute (Ago)
PUF	Eight repeats of a three α-helix bundle of 36 amino acids	ssRNA	Pumilio
Pentatricopeptide repeats	2–30 repeats of two antiparallel α-helices	ssRNA	PPR10
Homeodomain	Helix-turn-helix	dsRNA	Jerky, bicoid

*The table lists the names of the best-known RNA-binding domains, their structure, the nature of the RNA that they bind, and examples of RNA-binding proteins that contain the domains. Note that several RNA-binding proteins contain multiple RNA-binding domains, and that some RNA-binding proteins also bind DNA.

Abbreviations: ssRNA/DNA, single-stranded RNA/DNA; dsRNA/DNA, double-stranded RNA/DNA; OB-fold, oligosaccharide/oligonucleotide-binding fold.

4.1 The RNA recognition motif (RRM)

We begin our discussion of RNA-binding domains with the best-studied and most prevalent example, the RNA recognition motif (referred to as the **RRM**, and occasionally as the RBD or RNP motif).[9,10,15–18] Note the distinction between a motif and a domain. A motif is a sequence of amino acids or a specific arrangement of secondary structure, whereas a domain is a part of a protein that has a distinct function. For example, the leucine zipper motif is part of a dimerization domain in some transcription factors, and the RNP1 and RNP2 motifs are part of the RRM. Confusingly, although RRM stands for RNA recognition motif, it is in fact an RNA-binding domain. The RRM consists of 80–90 amino acids which form a four-stranded antiparallel β-sheet with two additional α-helices arranged in the order βαββαβ, which gives rise to a barrel-like topology. Within the RRM there are two highly conserved motifs, called RNP1 (present in β-sheet 1) and RNP2 (in β-sheet 3); both are located in the central β-strands on the same face of the barrel. Their amino acid consensus sequences are K/R–G–F/Y–G/A–F/Y–V/I/L–X–F/Y for the RNP1 motif and I/V/L–F/Y–I/V/L–X–N–L for the RNP2 motif. Amino acids in the RNP1 and RNP2 motifs are involved in RNA recognition. A single RRM can recognize between four and eight specific nucleotides in a single-stranded conformation.

4.1.1 Key features of the RRM

The main feature of the RRM is its ability to recognize specific RNA sequences. When the RRM and target RNA interact, the ssRNA target sequence lies across the surface of the β-sheet in the RRM, and amino acids in the RNP1 and RNP2 motifs provide the basis for base stacking and ionic interactions with the RNA target sequence. The use of multiple RRMs in a single binding protein, with varying lengths of linker sequences between them, can provide additional versatility in target sequence recognition and the kinetics of binding.

The co-crystal structure (in other words, the structure of the RNA-binding domain complexed to its target RNA) of PABP (poly(A) binding protein) is shown in Fig. 4.1. The co-crystal does not contain the entire RNA-binding protein, but rather two N-terminal RRMs that are connected by a short linker sequence. Together, the two RRMs form a continuous RNA-binding trough. The polyadenylate sequence that they bind to adopts an extended conformation that runs the full length of the trough. Contacts between protein and RNA are made by amino acids in the conserved RNP1 and RNP2 motifs in each RRM.

It has been estimated that up to 1% of all genes encode proteins that include one or more RRMs (including over 500 in humans). RRMs are also found in prokaryotes, where they tend to occur as single domains in small proteins, typically around 100

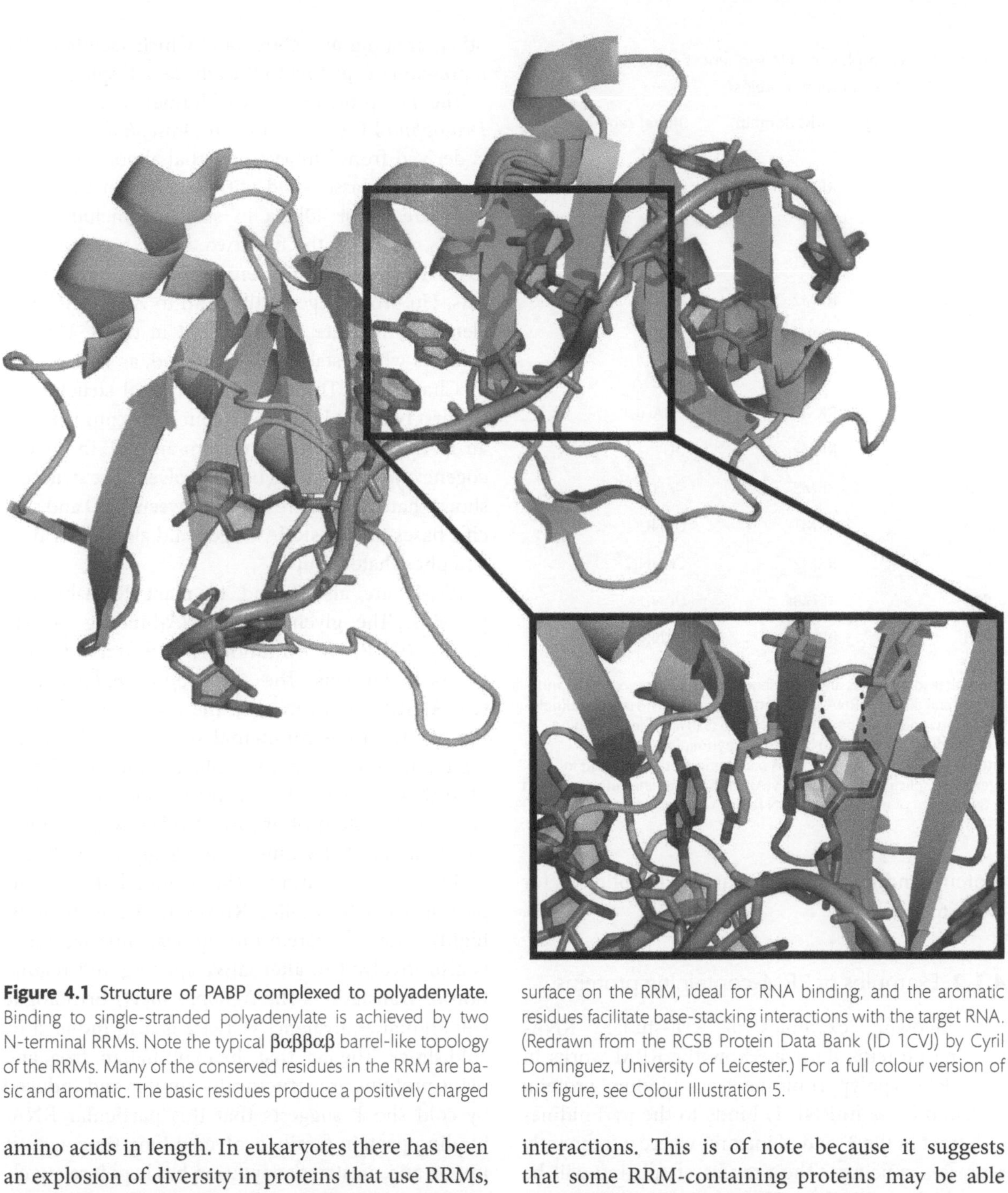

Figure 4.1 Structure of PABP complexed to polyadenylate. Binding to single-stranded polyadenylate is achieved by two N-terminal RRMs. Note the typical βαββαβ barrel-like topology of the RRMs. Many of the conserved residues in the RRM are basic and aromatic. The basic residues produce a positively charged surface on the RRM, ideal for RNA binding, and the aromatic residues facilitate base-stacking interactions with the target RNA. (Redrawn from the RCSB Protein Data Bank (ID 1CVJ) by Cyril Dominguez, University of Leicester.) For a full colour version of this figure, see Colour Illustration 5.

amino acids in length. In eukaryotes there has been an explosion of diversity in proteins that use RRMs, with many proteins containing multiple RRMs (up to six) that are often associated with additional auxiliary RNA-binding domains. RRMs are able to recognize very different sequences. Table 4.2 lists examples of RRMs and the diverse range of ssRNA RNA targets that they recognize.

It should be noted that some RRMs are also able to bind ssDNA and to mediate protein–protein interactions. This is of note because it suggests that some RRM-containing proteins may be able to influence and coordinate multiple steps in gene expression. The ability of an RNA-binding domain such as the RRM to perform multiple tasks is not unusual; as we shall see, the same applies to the K-homology domain and zinc-finger motifs. Note that some RRMs are also involved in protein binding. This type of RRM is now referred to as a UHM (U2AF homology motif, named after a

Table 4.2 Examples of RNA sequences recognized by several RRM-containing proteins*

RRM-containing protein	RRM domain	Target sequence
Sex-lethal (Sxl)	RRM1	UUUUUUU
Sex-lethal (Sxl)	RRM2	UGU
hnRNP A1	RRM1	UAGG
hnRNP A1	RRM2	UUAGG
PABP	RRM1	AAAA
PABP	RRM2	AAAA
HuD	RRM1	UUAUUU
HuD	RRM2	UU
HuD	RRM3	UAU
PTB	RRM1	UCNU
PTB	RRM2	CNUNN
PTB	RRM3	UCNU
PTB	RRM4	UCNN

*Note that some RRMs are more sequence specific than others, and that several different RNA targets are recognized. The use of multiple RRMs can modulate specificity. Sex lethal (Sxl) and the pyrimidine tract-binding protein (PTB) bind to the pyrimidine-rich tract of introns. hnRNP A1 is a pre-mRNA packaging protein involved in alternative splicing. PABP is a poly(A) binding protein. HuD is involved in the control of mRNA stability. N refers to any nucleotide.

protein binding RRM present in the splice factor U2AF65).[17]

4.1.2 Examples of RRM-containing proteins

The following examples of well-studied RNA-binding proteins illustrate how the RRM works in vivo. PTB (polypyrimidine tract binding protein; also known as hnRNP I) binds to the pyrimidine-rich tract in the 3′ end of introns, where it frequently represses splicing. (The process of splicing will be described in detail in Chapter 5.) This is because PTB does not contain an SR (serine–arginine rich) domain (unlike the splice factor U2AF65, which also binds the pyrimidine tract). **SR domains** promote splicing by facilitating protein–protein interactions with other splice factors. PTB contains four RRMs, each of which binds RNA with a different specificity.[18] RRM3 and RRM4 can also interact with each other, creating an 'RNA loop' which facilitates the repression of splicing to the adjacent 3′ splice site.

The Hu proteins are the human equivalent of *Drosophila* ELAV proteins. In *Drosophila*, the name is derived from 'embryonic lethal abnormal vision', a phenotype associated with a defect in this gene. They are about 40kDa in size and include three RRMs, of which the first two are juxtaposed (next to each other) and sufficient for binding to RNA targets. Hu proteins generally bind to AREs (AU-rich elements) that are often present in the 3′ UTR of mRNAs whose stability is regulated, as we shall see in Chapter 12. The three-dimensional structure of the first two RRMs of the protein HuD complexed to an 11-nucleotide fragment from an ARE in the oncogene c-*fos* mRNA has been resolved. The structure shows that contacts are made between HuD and specific bases in the ssRNA target, and also with ribose and phosphate groups.

RRMs are also found in plant RNA-binding proteins. The glycine-rich RNA-binding protein family (GRP) is evolutionarily conserved from plants to humans. The *Arabidopsis thaliana* protein AtGRP7 is involved in the response of flowering plants to environmental stresses, probably by regulating the alternative splicing and translation of mRNAs associated with stress response. AtGRP7 is a small protein of around 15kDa, with a single RRM in the N-terminus, and a glycine-rich domain in the C-terminus. The mammalian counterpart of AtGRP7 is called RBM3. RBM3 is the most highly induced protein during bear hibernation; it is also involved in alternative splicing and regulation of translation. The *Xenopus laevis* orthologue (functional equivalent) xCIRP2 is also induced by cold shock. The fact that the mammalian, frog, and plant proteins are structurally similar and induced by cold shock suggests that this particular RNA-binding protein family evolved a long time ago. Its importance in stress adaptation has kept it relatively unaltered for millions of years.

4.1.3 The quasi-RRM

The hnRNP proteins, abundant proteins that coat pre-mRNA, comprise several subclasses which are distinguished by the ribohomopolymers that they preferentially bind to in vitro (e.g. poly(rG) in the case of hnRNP F and H). They belong to a subfamily of proteins which includes, in humans, hnRNP H, hnRNP H′, and hnRNP F. All three contain three repeats of a domain that has a loose similarity to the RRM. There are notable discrepancies with the canonical RNP1 and RNP2 motifs of the RRM. These domains are termed **quasi-RRMs** (qRRMs).[19]

The hnRNP H family proteins have since emerged as important regulators of alternative splicing. The structure of the qRRMs of hnRNP F is now well understood. HnRNP F binds to a G-rich tract of the *Bcl*-X pre-mRNA. *Bcl-X* belongs to the Bcl-2 family of genes which regulate apoptosis (programmed cell death). Alternative splicing produces Bcl-X isoforms with antagonistic functions: Bcl-X_L is anti-apoptotic whereas Bcl-X_S is pro-apoptotic. HnRNP H and F bind an RNA element made of three consecutive G-rich tracts which promotes Bcl-X_S expression. The three qRRMs in hnRNP F do adopt the canonical βαββαβ fold,[20] and both qRRM1 and qRRM2 are required for recognition of the G-rich tracts.[20] However the amino acids required for binding to the G-rich tracts are not on the surface of the β-sheet but are instead presented to the RNA target by a short β-hairpin and two adjacent loops.

SRSF1 (previously known as ASF/SF2) is a well-studied splice factor, best known for its involvement in the regulation of alternative splicing. SRSF1 also shuttles to the cytoplasm where it is thought to be involved in the regulation of translation. Interestingly, SRSF1 also has the properties of a proto-oncogene because it promotes tumorigenesis when overexpressed in transgenic mice. It contains an RRM followed by a quasi-RRM and an SR domain in the C-terminus. The quasi-RRM of SRSF1 is very unusual as it binds RNA through an α-helix and not through the surface of a β-sheet.[21]

Key points

The RNA recognition motif (RRM) is present in up to 1% of all proteins. Single or multiple RRMs are present in a wide range of proteins, many of which are evolutionarily conserved. The RRM forms a barrel-like structure which primarily recognizes ssRNA sequences, but it also occasionally binds ssDNA and can be involved in protein interactions. Some proteins use a related structure known as a quasi-RRM to bind target RNAs.

4.2 The K-homology (KH) domain

The K-homology (or **KH domain**) was first identified in the protein hnRNP K, which gave the domain its name.[22–24] The KH domain can recognize both ssRNA and ssDNA, and is involved in a wide range of processes which include translation, splicing, transcription, and even chromatin remodelling. The KH domain is not restricted to eukaryotes—it is also found in eubacteria and **Archaea**. The KH domain is about 70 amino acids long and includes the signature sequence (I/L/V)IGXXGXX(I/L/V) in the middle of the domain. Its structure consists of a three-stranded β-sheet packed against three α-helices. The KH domain is subdivided into two subfamilies: type I (βααββα) is mostly found in eukaryotic proteins, and type II (αββααβ) is found in prokaryotic proteins. Nucleic acid recognition by the KH domain is achieved through hydrogen bonding, electrostatic interactions, and shape complementarity. Both type I and type II KH domains bind a target of four nucleotides by means of a cleft formed by the GXXG loop within the signature sequence and flanking α-helices and β-strands. Like RRMs, KH domains can bind both ssRNA and ssDNA.

Human hnRNP K includes three KH domains denoted KH1, KH2, and KH3. The KI domain (K-protein interactive region) is sandwiched between KH2 and KH3. A series of proline-rich docking sites in the KI domain (P-P-X-P) are responsible for the ability of hnRNP K to interact with the SH3 domains of the oncogenic Src family of tyrosine protein kinases. SH3 stands for the Src homology 3 domain and consists of a 60 amino acid β-barrel similar to the OB-fold proteins described later, in Section 4.3. SH3 domains are found in several proteins involved in signal transduction pathways that regulate cytoskeletal architecture or cellular proliferation. Thus hnRNP K illustrates the potential for RNA-binding proteins to interact directly with the signal transduction machinery—which means that environmental cues can rapidly affect gene expression at the post-transcriptional level.

HnRNP K is one of many multifunctional proteins which can affect gene expression at several levels. For example, hnRNP K interacts with the Polycomb group protein (PcG), a chromatin remodelling factor; and with DNA methyltransferases (both chromatin remodelling and DNA methylation directly affect gene expression). Its ability to bind ssDNA as well as ssRNA means that it can also be involved in transcription; it binds to a CT-rich element in the c-*myc* promoter and helps to activate transcription. Figure 4.2 shows the KH3 domain of hnRNP K complexed to a CT-rich ssDNA sequence. In terms of its post-transcriptional functions, hnRNP K is involved in the regulation of alternative splicing, translation, and mRNA stability. In the context of alternative splicing, hnRNP K interacts with the splice factor SRSF3 (SRp20) and has been shown to be a component of an intronic enhancer complex that binds to *β-tropomyosin* pre-mRNA. hnRNP K can repress the translation of *LOX* (15-lipoxygenase) mRNA—it binds to a CU-rich element in the 3′ UTR and blocks the recruitment of the large ribosomal subunit. On the other hand, hnRNP K stimulates translation of c-*myc* mRNA by promoting the initiation of translation. hnRNP K also binds to the 3′ UTR of *renin* mRNA—but in this case it regulates mRNA stability as opposed to translation. With respect to c-*myc*, hnRNP K promotes c-*myc* expression both transcriptionally, and at the level of mRNA translation. So hnRNP K illustrates very powerfully how multifunctional RNA-binding proteins are involved in diverse steps in gene expression through the use of a versatile RNA-binding domain.

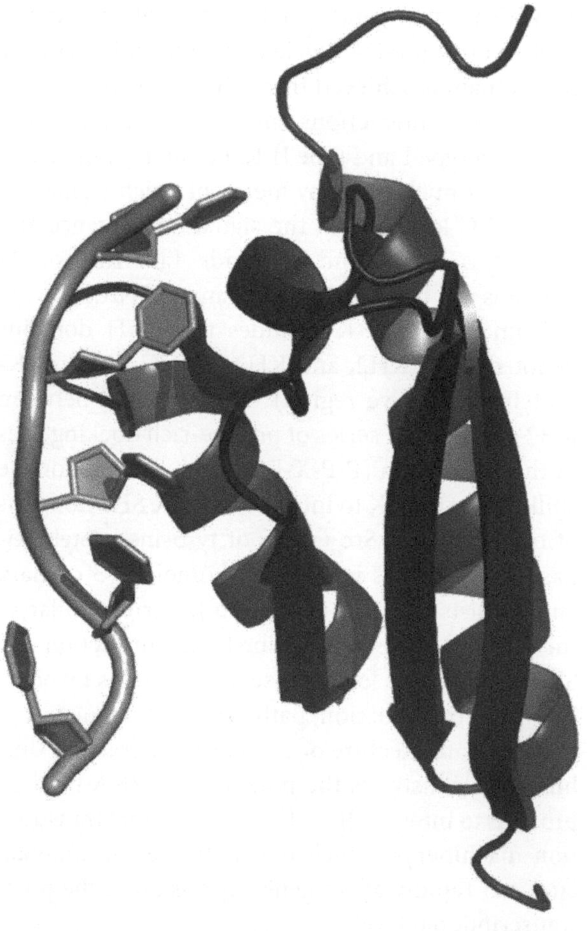

Figure 4.2 **Structure of the third KH domain of hnRNP K complexed to ssDNA.** Like the RRM, the KH domain is also mainly responsible for interactions with ssRNA or ssDNA. In this image the third KH domain (KH3) of hnRNP K is shown complexed to CT-rich ssDNA. (Redrawn from the RCSB Protein Data Bank (ID 1ZZI) by Jonathan Crowe, Oxford University Press.)

Several other proteins contain KH domains, most of which are involved in post-transcriptional processes; a notable example is FMRP, encoded by the *FMR1* gene. The protein FMRP contains two KH domains and is involved in translation regulation. A failure to express FMRP is associated with fragile X syndrome, a neurological disorder characterized by several behavioural and developmental phenotypes. An aggravated phenotype in fragile X syndrome is associated with a specific amino acid substitution of a highly conserved hydrophobic residue in the second α-helix of the KH domain, disrupting its ability to recognize RNA. This illustrates the fact that point mutations can dramatically affect the RNA-binding protein domains, leading to severe phenotypic consequences.

Key points

The K-homology (KH) motif is present in a wide range of often multifunctional proteins. The hnRNP K protein was the first protein found to contain this domain, and gave the domain its name. KH domains recognize both ssRNA and ssDNA.

4.3 The cold-shock domain

Bacteria, like all organisms, need to respond rapidly to sudden changes in environmental conditions, including drastic changes in temperature. They do this by expressing a set of cold-shock proteins, the most abundant of which are RNA-binding proteins. Similar cold-shock proteins are found in a wide range of bacteria, both Gram-positive and Gram-negative, suggesting that they are ancient.[25,26] In *E. coli*, CspA is a 7.4kDa protein whose expression is induced at 10°C. The structure of CspA has been named the *cold-shock domain* (CSD).

4.3.1 Key features of the cold-shock domain

When analysed by X-ray crystallography the cold-shock domain is seen to consist of a 'β-barrel structure'. In this structure, five antiparallel β-strands form two β-sheets, and highly conserved aromatic and basic side chains protrude from the solvent face of β-sheet 1. The structure of a typical CSD is shown in Fig. 4.3. The arrangement of positive charges creates an attractive potential for nucleic acids, and the aromatic rings have the potential to stack with bases in ssRNA or even DNA. These features are also present in the RRM, and are particularly reminiscent of the RNP1 and RNP2 motifs of the RRM. However, note that the RRM consists of an arrangement of β-strands and α-helices (βαββαβ) as opposed to the five β-strands of the CSD (βββββ).

What then is the function of these small conserved proteins in the cold-shock response? The prevailing theory is that they work as **RNA chaperones** (described in Chapter 11). In this role they help to couple transcription and translation.[27-30] In bacteria, translation occurs soon after or during transcription—after all, bacteria lack some of the complexities of RNA processing in eukaryotes (pre-mRNA splicing, editing, and so forth). In the presence of a cold shock, the lower temperature favours the formation of secondary structure in RNA which impedes translation. Binding of cold-shock proteins in these conditions tends to promote ssRNA, thereby enabling translation to progress.

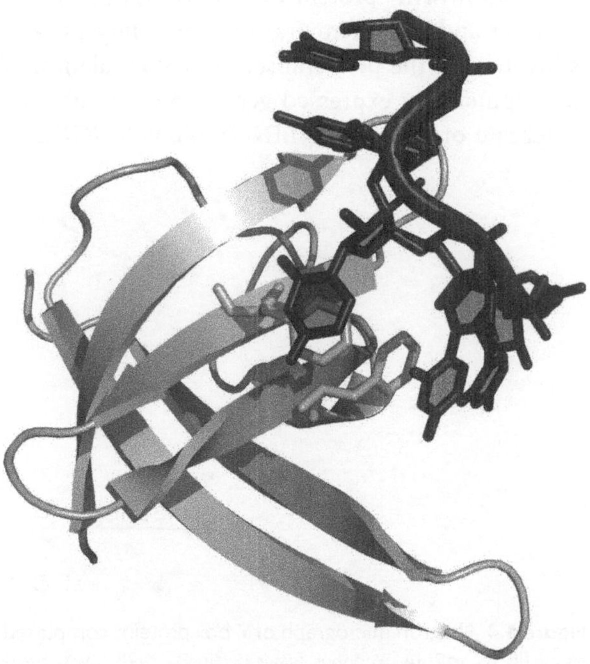

Figure 4.3 β-barrel structure of CspB from *Bacillus subtilis*. The structure of the cold-shock domain is reminiscent of the RRM. The RNP1-like motif (KGFGFIEV) forms the β2 strand and the RNP2-like motif (VFVHF) is located within the β3 strand. Contacts with ssRNA are made through exposed aromatic and basic residues. (Redrawn from the RCSB Protein Data Bank (ID 3PF5) by Cyril Dominguez, University of Leicester.)

The β-barrel structure of the CSD of the cold-shock proteins is also reminiscent of the **OB-fold** (oligosaccharide/oligonucleotide-binding fold) proteins. Examples of OB-fold proteins include staphylococcal nuclease, the anticodon-binding domain of yeast asp-tRNA synthetase, and the yeast **RNA helicase** PRP22. Like the CSD, the β-barrel structure of the OB-fold proteins is formed from two antiparallel β-sheets derived from five β-strands. Despite sharing these structural properties with the CSD and the RRM, there is little direct amino acid sequence conservation across classes of OB-fold proteins. This indicates the convergent evolution of a common RNA-binding surface. In general, the OB-fold proteins are involved in the regulation of RNA translation and in RNA turnover, processes which require the ability of proteins to recognize ssRNA targets.

4.3.2 The cold-shock domain in eukaryotic proteins

Nature likes to maximize the use of successful domains; the CSD is a good example.[28-31] This point is illustrated by the eukaryotic Y-box proteins, a multifunctional family of proteins originally discovered in the context of the transcriptional regulation of the MHC (major histocompatibility) class II cluster genes. The expression of the MHC class II genes in the immune system is regulated transcriptionally by a series of *cis*-acting elements which include the TATA box, the octamer motif, and the W, X, and Y boxes. The Y-box motif, to which these transcription factors bind, gives the Y-box proteins their name.

As might be expected with typical transcription factors, several transcriptional targets were defined such that the Y-box proteins both activate and repress genes associated with, for example, cell proliferation and spermatogenesis. However, as is the case with many nucleic-acid-binding proteins, the Y-box proteins are multifunctional. It came as a major surprise at the time to find that they are amongst the most abundant proteins associated with stored translationally repressed mRNP particles in *Xenopus* oocytes and mouse spermatocytes; these will be described in Chapter 11. Y-box proteins are now well-established regulators of mRNA translation in both the germline and somatic cells; they are even implicated in alternative splicing and have been detected in P-bodies, the latter being areas of concentration of translationally repressed mRNAs. Their multifunctional nature is described in Box 4.2.

Box 4.2 Y-box proteins are involved in transcriptional and post-transcriptional regulation

Y-box proteins were first discovered as transcription factors, but in fact they are also involved in post-transcriptional processes. How might the transcriptional activities of Y-box proteins, then, be coupled to their post-transcriptional activities? An answer to this question came through experiments described by Bouvet and Wolffe in 1994. mRNA encoding histone H1, a maternally expressed mRNA, was injected directly into oocytes, but this was not sufficient to achieve translational repression by Y-box proteins. Instead, *de novo* transcription of mRNAs was required to direct them to translational repression. The implication was that Y-box proteins (known as FRGY2 in *Xenopus* oocytes) bind to mRNA co-transcriptionally and mark them for future translational repression.

Xenopus oocytes are also well known for their 'lampbrush' chromosomes. Extensively studied by the eminent scientist Harold Callan at the University of St Andrews in the 1960s and 1970s (but first observed in the late nineteenth century), lampbrush chromosomes are highly transcriptionally active chromosomes. They form brush-like loops along the main axis of the chromosome from which a thick assortment of nascent transcripts can be observed. The Y-box protein FRGY2 is clearly detected on nascent transcripts along the *Xenopus* lampbrush chromosomes. Such a generalized staining pattern is more reminiscent of an abundant splice factor or hnRNP protein as opposed to a transcription factor.

Although the Y-box proteins were originally discovered in their role as transcription factors, they could equally have been first discovered as mRNA packaging proteins or regulators of translation or splicing.

The most remarkable feature shared by all Y-box proteins is a highly conserved CSD consisting of about 70 amino acids which are 43% identical to the *E. coli* cold-shock protein CspA. Proteins that contain the CSD are found in all eukaryotes. In the Y-box protein family, a single CSD is followed by a series of alternating basic and acidic charged islands, thought to participate in RNA binding and multimerization. The basic islands are rich in arginine, proline, tyrosine/phenylalanine, and glutamine/asparagine; these are reminiscent of the RNA-binding regions of the HIV-1 (human immunodeficiency virus) RNA-binding proteins Tat and Rev. The acidic islands are rich in aspartate/glutamate, adopt an α-helical conformation, and contain potential casein kinase II phosphorylation sites (consensus target SXXE/D). When phosphorylated, Y-box proteins are more efficient at binding mRNA and repressing translation. Figure 4.4 illustrates an electron micrograph of purified Y-box proteins complexed to *cyclin B1* mRNA.

Evolution has also coupled the CSD with different auxiliary domains. The *Caenorhabditis elegans* (a nematode worm) protein LIN-28 comprises a single CSD and two retroviral-type zinc fingers, and is involved in the post-transcriptional regulation of developmentally expressed genes. Interestingly, the vertebrate orthologue of LIN-28 binds to *IGF-2* (a

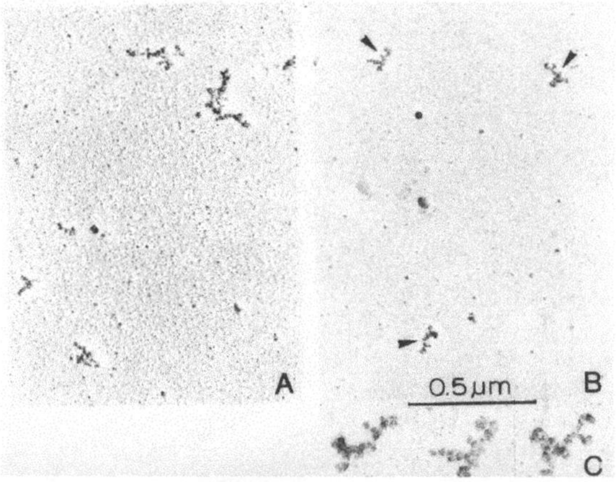

Figure 4.4 Electron micrograph of Y-box proteins complexed to cyclin B1 mRNA. *Xenopus laevis* cyclin B1 mRNA was transcribed in vitro and mixed with purified Y-box protein (FRGY2). The ensuing complexes were visualized by electron microscopy (panel B). Note the remarkably regular structure of three independent cyclin B1:FRGY2 complexes (panel C). In contrast, native mRNPs are heterogeneous in size (panel A) because of the varying shape and size of mRNAs complexed by several different RNA-binding proteins. ©Michael Ladomery.

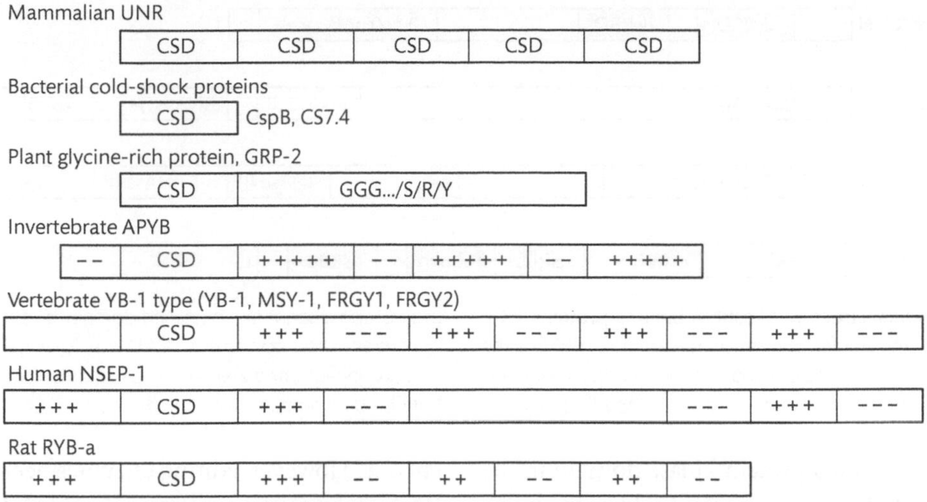

Figure 4.5 Examples of RNA-binding proteins with cold-shock domains. The CSD is present in a single copy in the small bacterial cold-shock proteins such as *B.subtilis* CspB or *E.coli* CS7.4. In eukaryotes, five CSDs are present in the protein UNR. Several examples of a single CSD coupled with highly basic (+ + +) or acidic (−) auxiliary domains are present in plants, vertebrates, and invertebrates. ©Michael Ladomery.

key growth factor) mRNA and is involved in skeletal myogenesis. The protein UNR is unusual, in that it contains five copies of the CSD with no obvious auxiliary domains. It is thought to be involved in the process of mRNA turnover; for example, of the proto-oncogene c-*fos* mRNA.[28] A diagram of RNA-binding proteins that have CSDs is shown in Fig. 4.5.

Key points

The cold-shock domain (CSD) is another ancient ssRNA-binding domain. Bacterial cold-shock proteins, thought to work as RNA chaperones, are essentially a single CSD. The CSD adopts a β-barrel structure with sequence motifs reminiscent of the RRM. Exposed basic and aromatic residues mediate contacts with single-stranded nucleic acid. In evolution the CSD has associated with several auxiliary domains and is repeated up to five times in the protein UNR.

4.4 Double-stranded RNA-binding proteins

RNA molecules fold dynamically into conformationally favourable structures. A proportion of RNA, such as the stem part of RNA hairpins, is double-stranded at any given time. Proteins have evolved the ability to bind dsRNA, with binding through the **dsRBD** domain (also known as the DBRD) being the most prevalent.[32-34] The dsRBD is 70 amino acids long and folds into an αβββα structure. What is particularly interesting about the dsRBD is the fact that it does not only recognize nucleotide sequences. Instead, it also contacts RNA through 2′-OH groups in the ribose moiety as well as the phosphate backbone. In other words, dsRBDs recognize both the sequence and the overall shape of a dsRNA target.

Double-stranded RNA generally adopts the A-form of the Watson–Crick double helix, in which the minor groove is shallower and wider compared with DNA which generally adopts the B-form. How then do dsRBD domains achieve sequence specificity? One potential mechanism is through cooperation with additional RNA-binding domains, either within the same protein or in an associated protein. An alternative mechanism is one in which multiple dsRBDs work together to recognize subtle variations in RNA double helices, which reflect differences in the primary structure of the RNA. The length of dsRNA sequence recognized by dsRBDs can be as short as 11 base pairs. The number of individual dsRBDs in proteins ranges from one to five, as illustrated in Fig. 4.6. The following examples of proteins with dsRBDs illustrate how they work in vivo.

One of the first proteins with dsRBDs to be discovered was PKR, the dsRNA-dependent serine/threonine protein kinase. PKR has two dsRBDs in

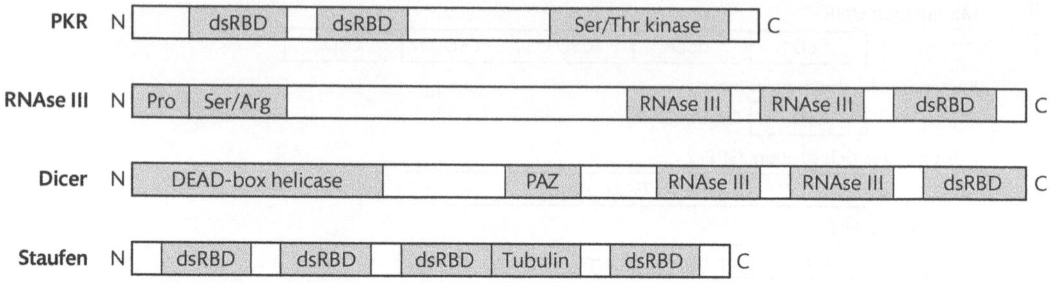

Figure 4.6 Modular structure of dsRNA-binding proteins. PKR, RNAse III, Dicer, and Staufen are shown as examples. dsRBD is the dsRNA-binding domain, present in one or more copies. PKR also contains a serine/threonine kinase domain. RNAse III contains a proline and a serine/arginine-rich domain, and the RNAse III domain itself. Dicer contains a DEAD box RNA helicase and a PAZ (Piwi, Argonaute, and Zwille) domain. Staufen contains a microtubule-binding domain (Tubulin). (Based on Saunders and Barber.[33])

the N-terminus, and a protein kinase domain in the C-terminus. PKR has an important role in the host's defence against viruses. This role stems from the way viruses tend to contain, in their life cycles, a large proportion of dsRNA which can be sensed by the cell and interpreted as an infection. PKR recognizes dsRNA in viruses via its dsRBD, becomes activated, and then inhibits translation, generally by phosphorylating the translation initiation factor eIF2α. Viruses have also evolved the ability to defend themselves against PKR by encoding a dsRBD-containing decoy protein which competes for dsRNA.

The Ebola virus causes haemorrhagic fever. It encodes the dsRNA-binding protein VP35, one of eight Ebola viral proteins. The function of VP35 is thought to be that of masking dsRNA targets which would otherwise be recognized by the cell as a sign of infection. The structure of VP35 bound to dsRNA is shown in Fig. 4.7.

Staufen is a *Drosophila melanogaster* protein involved in oocyte development. It is a large protein, comprising in excess of 1000 amino acids, and is specifically involved in mRNA localization and translational regulation of developmentally important mRNAs such as *bicoid* and *oskar*. There are four dsRBDs in *Drosophila* Staufen, with a tubulin-binding domain between the third and the fourth. The human orthologue of Staufen also interacts with dsRNA and tubulin, and is associated with the rough endoplasmic reticulum and polyribosomes, suggesting a similar involvement in the regulation of mRNA translation.

RNA editing, which is covered in Chapter 13, is mediated by enzymes that can modify RNA bases directly, altering their base-pairing properties and, potentially, the sequences of the polypeptides they encode. However, editing enzymes need to find their RNA targets very precisely. The ADARs (adenosine deaminases acting on RNA) are involved in adenine to inosine editing. Human ADAR1 contains three dsRBDs, a nuclear localization signal, and a deaminase domain in the C-terminus. One of the best-known substrates of ADARs is mRNA that encodes a glutamate-gated ion channel (GluR) where an editing event changes the biochemical properties of the channel. Mouse knockouts of *ADAR2* display early onset epilepsy and premature death as a result of a lack of appropriate editing of *GluR* pre-mRNA.

The RNAse III family of ribonucleases is found widely in prokaryotes and eukaryotes. RNAse III targets dsRNA, creating a nick in one of the RNA strands, and even double-stranded breaks. The enzyme contains an N-terminal endonuclease and a C-terminal single dsRBD; the best-known function of RNAse III is in the processing of pre-rRNA and tRNA, which we will describe in Chapter 14. Furthermore, enzymes that are involved in miRNA biogenesis (such as Drosha) and RNA interference (Dicer) belong to the RNAse III family and also contain dsRBDs. In their case, the dsRBDs enable the enzyme to find its correct substrate.

Key points

The dsRBD enables proteins to recognize dsRNA. The dsRBD is 70 amino acids long and folds into an αβββα structure. It makes contact with RNA through 2′-OH ribose groups and the phosphate backbone. dsRBDs are present in single or multiple copies and are found in proteins that need to dock with dsRNA in order to perform their functions. Notable examples include RNAse III, ADAR1, PKR, Dicer, and Staufen.

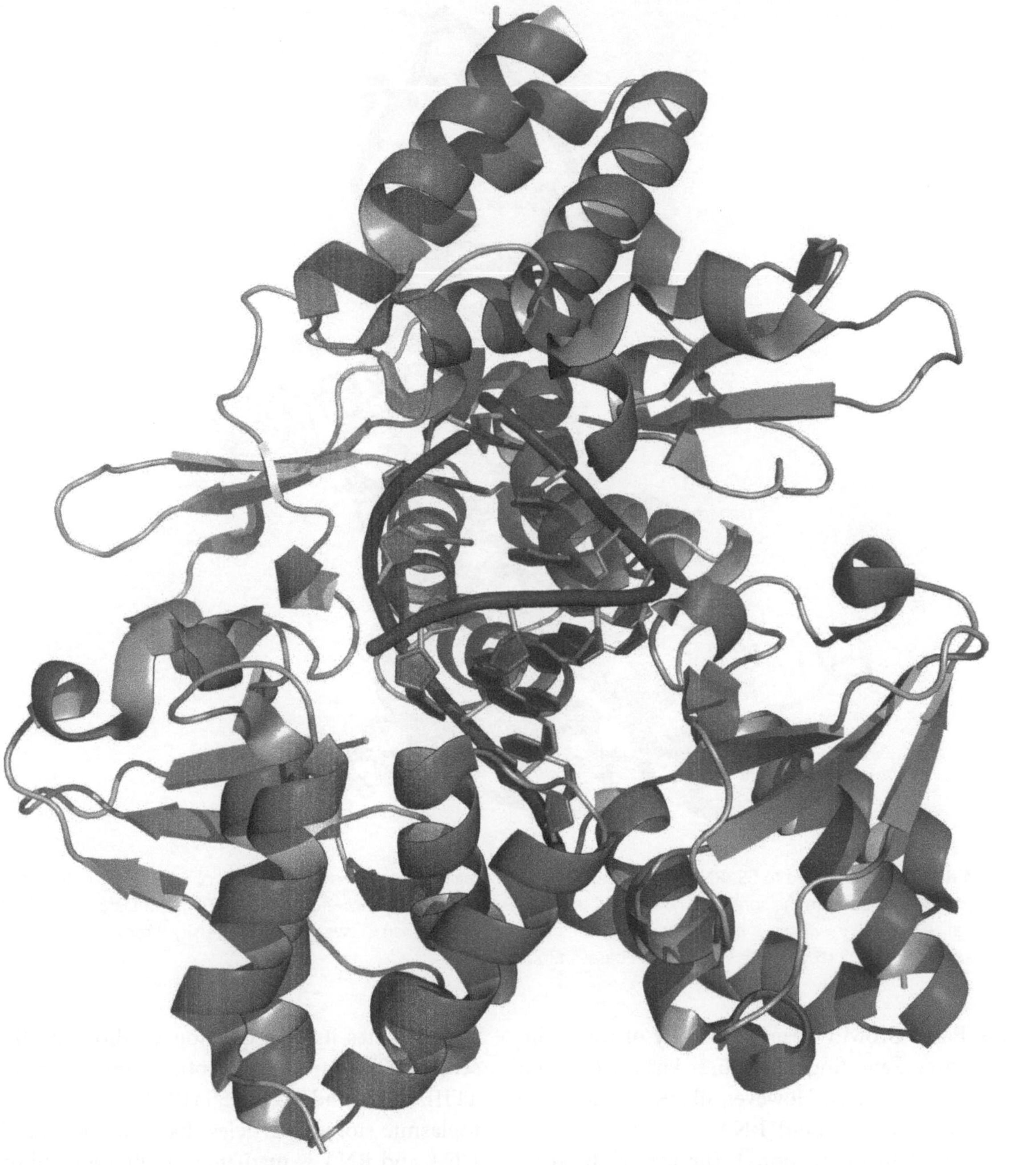

Figure 4.7 Zaire Ebola virus protein VP35 bound to dsRNA. VP35 is shown bound to eight base pairs of RNA. Note how the dsRBD of VP35 envelopes and masks the dsRNA target. For a full colour version of this figure, see Colour Illustration 6. (Redrawn from the RCSB Protein Data Bank (ID 3L25) by Jonathan Crowe, Oxford University Press.)

4.5 The zinc-finger domain

Zinc fingers are among the most prevalent nucleic acid-binding motifs in eukaryotes. Zinc-finger domains are around 30 amino acids long and consist of two antiparallel β-strands followed by an α-helix (ββα arrangement). A zinc ion is an integral part of the zinc-finger structure. The zinc ion coordinates a set of four cysteines or a combination of cysteine and histidine residues. The steroid receptor superfamily zinc fingers are characterized by four zinc fingers, whereas there are two cysteines and two histidines

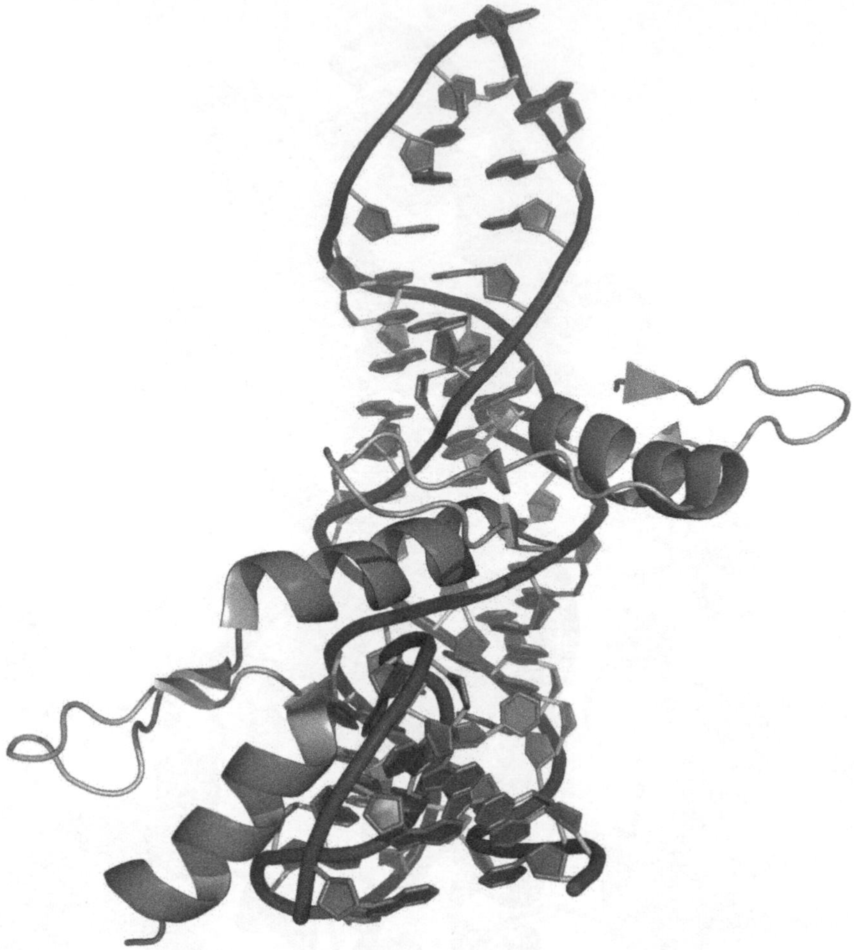

Figure 4.8 TFIIIA complexed to 5S rRNA. Zinc fingers 4–6 are shown bound to 55 bases in the core of 5S rRNA. Its zinc fingers can recognize 5S rRNA in two ways: by interacting with the phosphate backbone and by recognizing specific RNA bases in loop regions of the 5S rRNA. For a full colour version of this figure, see Plate 7. (Redrawn from the RCSB Protein Data Bank (ID 2HGH) by Jonathan Crowe, Oxford University Press.)

in the Early Growth Response family of transcription factors. Zinc fingers are best known for their ability to bind DNA. However, like RRMs and KH domains, they also bind RNA and even mediate protein–protein interactions.[6] The use of the term 'finger' arose because of the way the domains are presented in two-dimensional diagrams; however, their true three-dimensional structure does not resemble a finger!

The transcription factor TFIIIA was one of the first zinc-finger proteins to be characterized. It is unusual as it consists of nine zinc fingers and is best known for its ability to switch on the expression of the gene that encodes 5S rRNA. TFIIIA binds to the internal control region (ICR) of the 5S rRNA gene and activates its transcription by directing the assembly of an active transcription complex. However, TFIIIA also binds to the 5S rRNA product in 7S cytoplasmic storage particles. Its ability to bind both DNA and RNA is mediated by different subsets of its zinc fingers.[35,36] Zinc fingers 1–3 are particularly important in DNA binding and recognize the major groove of DNA, while zinc fingers 4–7 are involved in RNA recognition. Figure 4.8 shows the interaction between TFIIIA zinc fingers and 5S rRNA.

There is an important distinction to be made between recognition of DNA and RNA by TFIIIA; whereas its DNA recognition is dependent on the DNA sequence, RNA recognition is more dependent on structure (i.e. TFIIIA recognizes

Box 4.3 Generation of artificial RNA-binding zinc fingers

Using as a template the fourth Cys2His2 zinc finger of TFIIIA, Friesen and Darby[37] used the technique known as *phage display* to evolve zinc fingers that bind specifically to HIV-1 RNA. Amino acids required for the formation of the basic structure of the zinc finger were retained, but selected amino acids were randomized, particularly in the α-helix. The library was displayed on the surface of bacteriophages and phages that bound to specific RNA ligands, 5S rRNA, or a high-affinity site in HIV-1 that binds to the HIV-1 RNA-binding protein Rev.

The result was the isolation of several artificial zinc fingers that bound the HIV-1 RNA targets with high efficiency, comparable to the binding of native TFIIIA or Rev. RNA specificity was found to be mostly determined by amino acids in the α-helix and just upstream, in the tip of the zinc finger. Artificial zinc fingers could be used as experimental tools or in a therapeutic setting, given their ability to bind to specific viral RNA sequences.

three-dimensional structure; the same applies to the dsRBD domain). Thus the importance of the three-dimensional structure of RNA, as opposed to the primary sequence, is a shared feature in the recognition of RNA sequences by several RNA-binding proteins. It is also possible to genetically engineer zinc fingers so that they can recognize specific sequences, as illustrated in Box 4.3.[37]

There are additional examples of zinc-finger proteins that bind RNA. The Wilms tumour suppressor gene *WT1* expressed in vertebrates encodes a protein with four Cys2His2 zinc fingers (with two cysteines and two histidines) in its C-terminus and is a member of the Early Growth Response family of transcription factors. WT1 can work as a transcriptional activator or repressor and does so typically by recognizing GC-rich sequences in the promoters of its target genes. However, WT1 also binds to RNA through its zinc fingers, and may regulate alternative splicing and translation. Experiments suggest that WT1 uses its zinc fingers differentially to recognize DNA or RNA structures; zinc finger 1 is particularly important in RNA but not DNA recognition. The ability of different WT1 zinc fingers to recognize DNA or RNA is reminiscent of the functional versatility of TFIIIA's nine zinc fingers.[6]

Tra-1 is another example of a Cys2His2 zinc-finger protein, a member of the GLI family of transcription factors. In *C. elegans*, Tra-1 binds to the 3′ UTR of *tra-2* mRNA and is required for its nuclear export. The implication is that several other zinc-finger proteins, previously discovered and characterized as transcription factors, may also bind to RNA and thus coordinate transcription with post-transcriptional regulatory events.

RNA-binding zinc fingers are also present in Cys4 zinc-finger proteins. An example is the ZRANB2 family of proteins. This unusual family of proteins consists of two Cys4 zinc fingers in the N-terminus separated by a flexible 24-residue linker, followed by a highly acidic domain in the middle of the protein and an SR-like (serine–arginine rich) domain in the C-terminus; the latter is characteristic of splice factors that contain RRMs. ZRANB2 interacts with the splice factors U170K and U2AF35 and can alter the alternative splicing of *Tra2β1* pre-mRNA.[38] The zinc fingers of ZRANB2 each bind to a single-stranded target, combining to recognize the sequence AGGUAA(Nx)AGGUAA. The recognition of the target RNA sequence involves hydrogen bonds with the RNA bases and the stacking of a conserved tryptophan between two guanines.[39]

Key point

The zinc-finger domain is a widespread and multipurpose domain that can bind DNA and RNA and mediate protein-protein interactions. Combinations of cysteines and histidine residues are coordinated by a zinc atom in a structure that comprises two β-strands followed by an α-helix. Sequences in the α-helix are critical for nucleic acid recognition.

4.6 Other RNA-binding domains

Additional RNA-binding domains exist in nature. Most are well known but others may remain to be discovered, particularly in less abundant or less studied proteins. We now briefly discuss additional RNA-binding domains: PAZ and PIWI, arginine-rich domains, the RNA helicases, PUF domains, pentatricopeptide repeats, and the homeodomain.

4.6.1 PAZ and PIWI domains

The PAZ domain is present in the enzyme Dicer, a key player in the process known as RNA interference (see Chapter 16). Note in Fig. 4.6 that Dicer also contains a dsRBD. PAZ is a 110 amino acid domain which forms a β-barrel structure reminiscent of the OB-fold proteins; it also features an additional αβ clamp structure that facilitates the recognition of two-nucleotide overhangs in Dicer's substrates.

The PIWI domain (the name is derived from ***P-element induced wimpy testis*** in *Drosophila*) is present in several RNA-binding proteins that bind and cleave RNA; one of the best studied examples are the **Argonaute** proteins. The PIWI domain core has a tertiary structure related to the RNase H family of enzymes. RNAse H degrades RNA present in RNA–DNA duplexes. The PIWI domain core has a five-stranded β-sheet surrounded by α-helices. Argonaute proteins are part of the RNA-induced silencing complex (RISC) in RNA interference; they are involved in siRNA (small interfering RNA) mediated cleavage of target mRNAs. (See Chapter 16 for a more detailed discussion of the role of Dicer and Argonaute proteins in RNA interference.)

4.6.2 The RNA helicases

We now turn our attention to the RNA helicases. These are a special class of RNA-binding proteins which combine the ability to bind RNA with a catalytic activity—namely, the ability to unwind RNA. As we saw in Chapter 2, RNA adopts a complex secondary structure thanks to its structural versatility; this versatility facilitates its diverse cellular functions. However, in all processes that involve RNA there is a time when base pairing needs to be reversed in order to generate a stretch of ssRNA, for example to enable ribosomes to translate mRNA into protein or the splicing machinery to locate splice sites. This activity is catalysed by RNA helicases. RNA helicases are also known as RNP helicases, because their substrate is not usually naked RNA (RNA alone), but rather RNA in complex with proteins (RNP, or ribonucleoprotein).

RNA helicases belong to an abundant family of proteins with structural features that have been conserved from bacteria to humans. The first RNA helicase to be described was the eukaryotic translation initiation factor eIF4A, which is involved in cap-dependent translation initiation. RNA helicases are subdivided into five superfamilies, SF1–SF5. The most numerous are the SF2 RNA helicases which are further subdivided based on conserved motifs: the DEAD box, DEAH, and DExH subfamilies. Conserved motifs in RNA helicases are arranged into two distinct domains as shown in Fig. 4.9. Domain 1 contains motifs involved in NTP binding and hydrolysis (required for the catalytic activity of RNA helicases) and also substrate binding. Domain 2 contains motifs involved in substrate binding and in coupling NTP hydrolysis with helicase activity. Surrounding these conserved domains, additional domains confer functional and RNA-binding specificity.

4.6.3 Arginine-rich domains

Many of the RNA-binding domains that we have described are structurally complex. However, there are

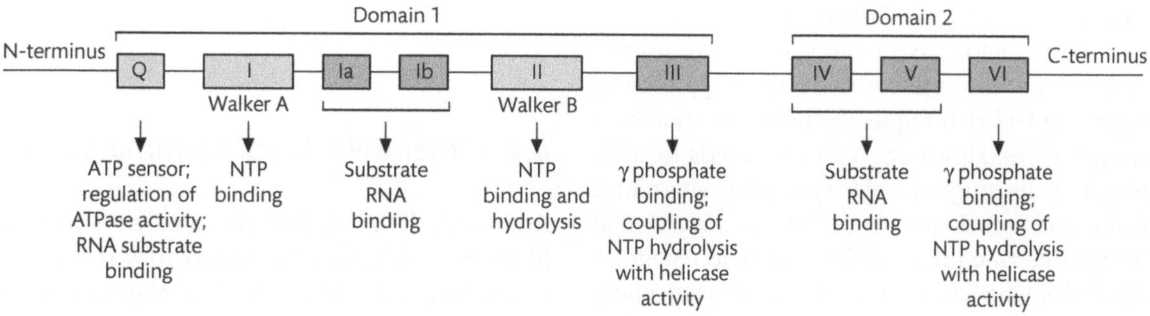

Figure 4.9 Conserved motifs in the core RNA helicase domains 1 and 2. NTP binding and hydrolysis requires motifs I and II (also known as Walker A and B) in domain 1. Substrate binding is achieved by motifs Ia and Ib in domain 1, together with motifs IV and V in domain 2. NTP hydrolysis is coupled to unwinding activity through motifs in both domains. (From Bleichert F and Baserga SJ,[40] with permission from Elsevier.)

other simpler types of domain, many of which are essentially areas rich in arginine. Arginine is unique in that its side chain contains two NH_2 groups that are free to interact with RNA. The HIV-1 RNA-binding protein Tat binds to a bulged RNA element in HIV-1 called TAR RNA. Only a single arginine is responsible for the recognition of TAR RNA, and it interacts through hydrogen bonding between both its free NH_2 groups and adjacent phosphates in TAR RNA. This interaction has been described as the **arginine fork**. In general, arginine-rich peptides are flexible and relatively unstructured in solution, but they adopt different conformations depending on the structure of the bound RNA: an α-helix in the case of the HIV-1 protein Rev binding to the Rev-Response Element (RRE), and a β-hairpin in the case of Tat binding to TAR.

Arginine is also found in association with glycine in **RGG boxes**.[41] RGG boxes are present in the 'auxiliary domain' of hnRNP A1 (along with its two RRMs). Auxiliary domains (including the RGG box) are believed to aid RNA binding, conferring stronger or more specific binding. However, they are also involved in protein interactions. The importance of the RGG box is powerfully illustrated in the context of fragile X syndrome, the most common form of inherited intellectual disability. The syndrome is caused by the inactivation of the *FMR1* gene that encodes FMRP (previously discussed in Section 4.2). FMRP is an RNA-binding protein with two KH domains and a C-terminal RGG box. The RGG box of FMRP is essential for the recognition of an RNA target known as the 'G quartet', consensus $(DWGG)_4$ where D is any nucleotide except C, and W is U or A. The G quartets are present in physiologically relevant target mRNAs that are important for neuronal function.[42]

The Y-box proteins, described in Section 4.3, contain alternating arrangements of basic and acidic islands. These islands are also thought to promote multimerization and aid RNA binding.[43] The basic islands in the Y-box proteins also contain aromatic residues, which may further contribute to arginine-mediated RNA binding through base stacking with RNA bases. Arginine is also prevalent in the SR domains that give the SR proteins their name. In this case, arginine is associated with serine; the SR domain is believed to promote protein–protein interactions and to contribute to RNA binding.

There is an added twist to the versatility of arginine. Nitrogen atoms in the guanidine groups can be methylated by PRMT proteins (protein arginine methyl transferases).[44] This modification increases hydrophobicity while retaining the positive charge potential of the nitrogen. The consequence of arginine methylation is the modulation of protein–protein interactions and RNA binding. The RGG box in FMRP is methylated on four arginine residues, inhibiting RNA binding. Thus the methylation of arginine is a way of regulating the affinity of RNA-binding proteins for their RNA targets.

4.6.4 Pentatricopeptide repeats, PUF domains, and homeodomains

Additional RNA-binding domains have been described. We end our survey of RNA-binding domains by describing three more, starting with the pentatricopeptide repeats (PPRs). PPR proteins are a large family of eukaryotic RNA-binding proteins which are particularly abundant in mitochondria and chloroplasts of terrestrial plants. They contain a 35 amino acid long motif that forms two antiparallel α-helices. PPR10 is a maize chloroplast protein with 19 repeats of the motif. In the absence of RNA the 19 repeats of PPR10 form a superhelical spiral. Upon binding to target RNA the repeats undergo a rather complex conformational change, with six repeats interacting with six nucleotides of the target RNA, mainly through hydrogen bonds.[45]

PUF (**Pu**milio and **F**BF homology) proteins are eukaryotic RNA-binding proteins generally involved in development and differentiation. A well-known example is the *Drosophila* protein Pumilio, involved in germline development, gonadogenesis, and embryogenesis. PUF proteins regulate the translation of target mRNAs. The PUF domain consists of eight repeats of a three α-helix bundle of 36 amino acids. Amino acids at positions 12 and 16 of the repeat interact with RNA bases through hydrogen bonding and van der Waals contacts, and the amino acid at position 13 stacks with an RNA base.[46]

The homeodomain is a well-known 60 amino acid long helix–turn–helix domain present in several transcription factors, best known for its ability to recognize specific DNA sequences in target genes. Bicoid (bcd) is a *Drosophila* homeodomain protein that forms a concentration gradient in an anterior to posterior direction in embryos, activating genes involved in segmentation. There is another homeodomain protein, caudal (cad), which forms a concentration gradient in the opposite direction. The cad gradient fails to form in *bcd* mutants. It turns out that bicoid binds to *cad* mRNA through its homeodomain and regulates its translation.[47] In mice there is a protein called Jerky—its loss causes recurrent seizures. The N-terminus of Jerky contains two tandemly arranged helix–turn–helix motifs that are very reminiscent of the homeodomain. Jerky is detected in neuronal mRNP (messenger RNA—protein) complexes. It binds to mRNA through the homeodomain-like motifs.[48] Lastly, the ribosomal protein L11 also contacts rRNA (ribosomal RNA) through a structure that is strikingly reminiscent of the homeodomain.[49]

Key points

Additional RNA-binding domains include the PAZ and PIWI domains, the RNA helicase domain, arginine-rich sequences, pentatricopeptide repeats, PUF domains, and homeodomains. PAZ and PIWI domains are found in proteins involved in RNA interference. The RNA helicase domain contacts a dsRNA substrate, facilitating its unwinding in response to NTP hydrolysis. Arginine-rich domains are present in several auxiliary domains of RNA-binding proteins.

4.7 Investigating protein–RNA interactions

There are several approaches to the study of protein–RNA interactions. When a protein or an RNA-binding domain is known to bind a specific RNA target, it is possible to co-crystallize the complex in preparation for X-ray crystallography. X-ray crystallography provides the precise three-dimensional details of the interaction and the residues involved. Despite its power as a technique, X-ray crystallography only provides a static picture of three-dimensional structure. Understanding binding kinetics is important because RNA binding is a dynamic process. The efficiency with which an RNA-binding protein can lock onto its target, remain bound to it, and then dissociate when the time comes are all important considerations. Nuclear magnetic resonance (NMR) spectroscopy is another useful method for studying protein–ssRNA complexes.[50] There are a number of reasons for this: RNA-binding domains, such as the RRM or KH domain, are relatively small (around 100 amino acids) and so suitable for NMR, and single-stranded RNA is highly flexible which is not compatible with crystal formation. In this section we shall discuss traditional methods for studying protein–RNA interaction in vitro, followed by an overview of techniques used to study protein–RNA interactions in living cells.

4.7.1 In vitro RNA-binding assays

It is useful to consider the more traditional approaches to the study of protein–RNA interactions as they can be very informative. The study of the composition of *Xenopus* oocyte messenger RNP particles can be used as an example. *Xenopus* oocyte messenger RNP complexes are isolated by oligo(dT) chromatography by virtue of the poly(A) tail of the RNA molecule hybridizing to a column in the presence of salt. Protein–RNA interactions can then be destabilized in 8M urea or high salt, facilitating separation of the protein and RNA, and the RNP reconstituted by dialysing the urea or salts out. Radiolabelled RNA (a **riboprobe**) can be added to the mixture and the binding of mRNP proteins (or specific purified proteins) to the riboprobe studied in several ways. Figure 4.10 outlines a range of ways of studying these protein–RNA complexes in vitro.

One approach is to use the **gel retardation assay**, also known as the electrophoretic mobility shift assay (EMSA). If free riboprobe is loaded into an agarose or acrylamide gel it will run further than when bound to proteins, when its movement will be slowed down. The gel can be dried down and the proportion of free riboprobe determined by taking an autoradiograph. The amount of protein added to the riboprobe can be titrated carefully. What this potentially tells you is the number of RNA molecules that can be bound by a particular protein. It can also be used to determine what sequences can compete for binding and what the best binding conditions are.

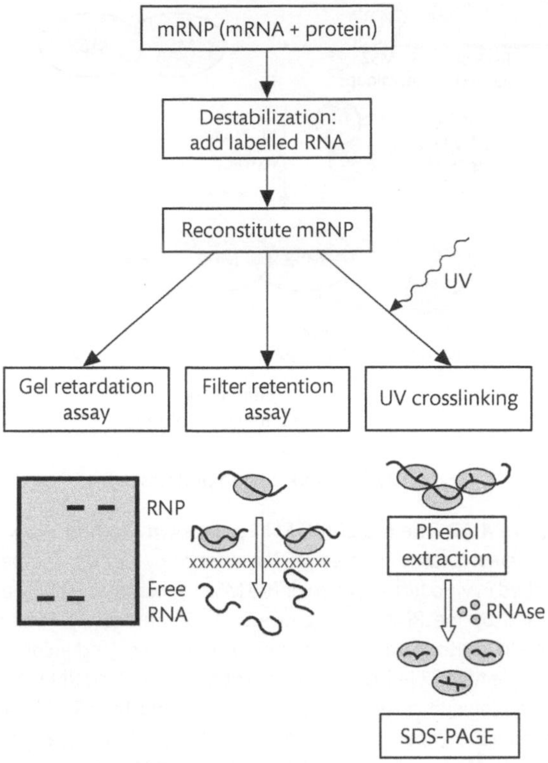

Figure 4.10 Classical RNA-binding assays. Messenger RNP complexes can be destabilized and radiolabelled RNA added to the mix. Alternatively, specific purified proteins can be incubated with specific in vitro transcribed radiolabelled RNAs. The complexes are formed or reconstituted in an appropriate binding buffer, and then the binding is analysed in three ways: a gel retardation assay (EMSA), a filter retention assay, or UV crosslinking followed by RNAse treatment and SDS–PAGE. ©Michael Ladomery.

Figure 4.11 UV crosslinking assay with radiolabelled RNA. The *Xenopus* oocyte mRNP fraction was incubated with a radiolabelled cyclin B1 mRNA before UV crosslinking. After RNAse digestion, the Y-box proteins FRGY2a and FRGY2b have become radiolabelled, showing that they can be crosslinked to RNA. ©Michael Ladomery.

The filter retention assay is a variation of this procedure. In this approach, free riboprobe can pass through a filter. The radioactive signal left on the filter is a measure of the proportion of riboprobe bound by protein. The binding can then be modified by altering ion concentrations in the binding buffer or by adding competitor molecules. This technique achieves essentially the same objectives as the gel retardation assay.

Another approach is to crosslink proteins to RNA using ultraviolet (UV) light with a wavelength of 254nm.[51,52] The crosslinks produce stable covalent bonds between proteins and RNA within the region of RNA that interacts with the protein. Complexes can then be treated with RNAse (ribonuclease), so that only the crosslinked RNA remains bound. 'Free' RNA, which is not crosslinked to protein, is digested by the RNAse. This process effectively labels the RNA-binding proteins with fragments of RNA. It is then possible to run them out on sodium dodecyl sulphate–polyacrylamide gel electrophoresis (SDS–PAGE). An example of an ultraviolet (UV) crosslinking assay is shown in Fig. 4.11. This method can be used to identify novel RNA-binding proteins and even specific RNA-binding domains using a radiolabelled riboprobe. It is also possible to digest the protein enzymatically or cleave it chemically after the crosslinking step. In this way, only the parts of the protein that bind to the riboprobe are radiolabelled.[51]

There is also a variation of the standard immunoblot or Western blot technique, where proteins are run on SDS–PAGE, transferred to polyvinylidine fluoride (PVDF) or nitrocellulose, and then incubated with a riboprobe—this is called a **Northwestern blot**. The advantage of this technique is that it can be applied to a complex mixture of proteins, perhaps where the identity of the RNA-binding proteins is not known. Because of the two-dimensional separation achieved by the SDS–PAGE, the radioactive signal (corresponding to riboprobe retained on the filter) can be associated with proteins of a particular molecular weight.

Box 4.4 The MS2:MBP RNA pull-down method

The MS2:MBP method was established in the laboratory of Robin Reed at Harvard University. The purpose was to devise a strategy to efficiently pull down specific protein–RNA complexes out of a complex mixture. The method is based on a very strong and specific affinity between the bacteriophage MS2 coat protein and its target, the MS2 RNA stemloop. The MS2 protein is fused to maltose-binding protein (MBP) which can bind to an amylose column. The MS2 RNA stemloops can in turn be fused to any RNA sequence of interest. A complex that includes the MS2:MBP, the MS2:RNA fusion, and either a purified RNA-binding protein or extract is assembled in vitro and purified with an amylose column as shown in Fig. 4.12. This method has been used successfully to isolate the active spliceosome.

To further study protein–RNA interactions, the target RNA sequence can be modified by introducing mutations, insertions, or deletions to define precise RNA target sequences. As applies to any method, however, the MS2:MBP system has its limitations, so additional assays are always needed to confirm that in vitro interactions occur in vivo. These can include knockdown or overexpression of the RNA-binding proteins (then assessing the consequences on the expression of the target RNA); or the **CLIP assay** to confirm that the interaction between an RNA-binding protein and a given RNA sequence occurs in vivo.

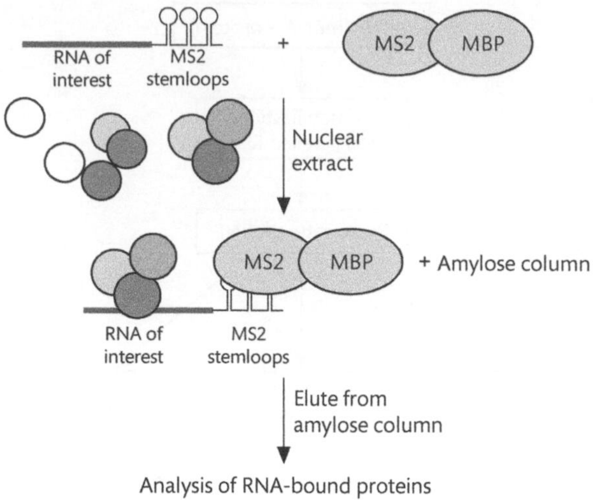

Figure 4.12 The MS2:MBP RNA pull-down method. An RNA of interest is fused to three MS2 stemloops by cloning and transcribed in vitro from a plasmid. The MS2 stemloops are efficiently bound by the RNA-binding protein MS2. An MS2:MBP fusion protein is also expressed and purified from *E. coli* and incubated with the MS2:RNA fusion. Nuclear extract (or purified RNA-binding proteins) is mixed with the MS2:MBP and MS2:RNA fusion forming a complex. The complex is then isolated using an amylose column (which binds the MBP moiety). ©Michael Ladomery.

We now briefly discuss two other approaches that can be used to study protein–RNA interactions in vitro. One is the MS2:MBP pull-down method, described in Box 4.4. In this method, an RNA sequence of interest is fused to another RNA (the bacteriophage MS2 stemloop).[53] Because the MS2 stemloop is very efficiently bound by the MS2 protein, it is possible to exploit this interaction to isolate specific protein–RNA complexes, as illustrated in Fig. 4.12. The second approach is called SELEX (**s**ystematic **e**volution of **l**igands by **ex**ponential enrichment).[54] In SELEX a pool of RNA ligands of random sequence, also described as **aptamers**, are first incubated with a protein of choice. Unbound aptamers are removed and bound aptamers are then eluted, converted into cDNA, and cloned into a suitable vector so that they can be transcribed again. The new pool of aptamers is mixed with the protein again and the cycle is repeated several times, thus selecting for the strongest binders. The strongest binders are then sequenced and a consensus binding sequence can then be deduced.

4.7.2 Examining protein–RNA interactions in vivo

Studying protein–RNA interactions in vitro is relatively straightforward. However, the key challenge in the study of RNA-binding proteins is the identification of their exact in vivo target-binding sequences. One in vivo approach is the yeast three-hybrid technique,[55] a method that is based on the principles of the yeast two-hybrid technique. The yeast two-hybrid technique is used to study protein–protein interactions; the interaction between proteins results in the activation of transcription of a reporter gene. In the three-hybrid variation the activation of the reporter gene requires a third element, a hybrid RNA molecule. The way this works is illustrated in Figure 4.13.

Another important method of studying in vivo protein–RNA interactions is the CLIP assay (**c**ross**l**inking and **i**mmuno**p**recipitation). It combines crosslinking of RNA-binding proteins and their target RNA sequences within the cell using UV radiation, followed by co-immunoprecipitation of proteins

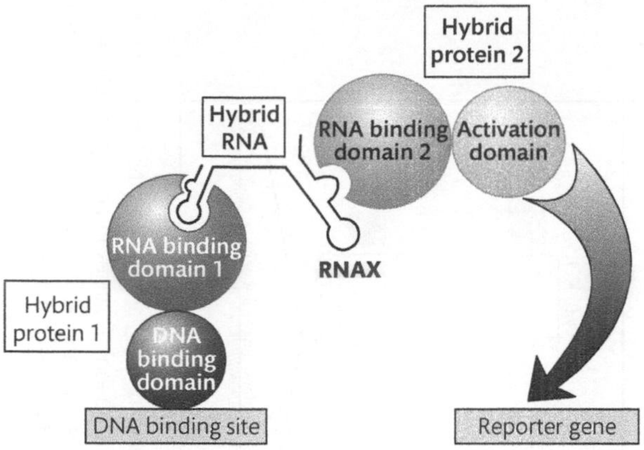

Figure 4.13 The yeast three-hybrid approach. There are three components. The first hybrid protein consists of a DNA-binding domain fused to an RNA-binding domain. The DNA-binding domain binds to a promoter in a reporter gene. The RNA-binding domain binds to a specific RNA (e.g. an MS2 stemloop) that is fused to an RNA of interest (labelled RNA X). RNA X could be a library of RNAs, or a specific RNA of interest. The second hybrid protein consists of a second RNA-binding domain fused to a transcription activation domain. When the second hybrid protein interacts with RNA X, the reporter gene is activated. (Reproduced with permission from Bernstein et al.[55])

with their target RNAs. UV radiation only links together nucleic acids and protein which are very close together in the cell (typically those separated by about 1Å). Notice that ribonucleoprotein complexes in cells are first 'frozen' by crosslinking, and then the RNA-binding protein is purified along with any attached RNA, which is then amplified by RT-PCR and sequencing.

The first step in the procedure is to introduce crosslinks which chemically bond the RNA-binding proteins with their target RNAs. This is generally carried out by UV irradiation of *living cells*. Because whole cells are irradiated, crosslinks only form between proteins and RNAs which are in direct contact within the cell. Importantly all these crosslinks are formed in intact cells, and represent contacts which are 'frozen in time'. Proteins are not themselves crosslinked together by UV irradiation (unless an excessively strong dose is given). Protein–RNA complexes are then purified using a very stringent procedure based on two steps: the isolation of protein–RNA complexes, and then their purification based on size.

The first step in purification uses an antibody which is highly specific to the RNA-binding protein of interest to precipitate it from an extract made from the UV-irradiated cells. This technique is called immunoprecipitation. This not only precipitates the RNA-binding protein itself, but will also co-precipitate any RNA that was bound to it initially, together with any interacting proteins. The bound RNA is partially digested using RNAse (which cuts RNA non-specifically) so that only very short lengths remain associated with the RNA-binding protein (typically 40–60 nucleotides) and is then radiolabelled.

In the next step in purification, the immunoprecipitated material sample is boiled in the presence of a strong detergent (sodium dodecyl sulphate (SDS)) to detach any non-crosslinked protein or RNAs. The mixture is then separated out using polyacrylamide gel electrophoresis (PAGE), followed by exposure to photographic film. Because of the radioactive label on the RNA, any protein attached to RNA will appear as a dark band on the photographic film. As the RNA-binding protein is of known size, this labelled protein–RNA complex can be purified from the gel. These RNAs correspond to the RNAs originally bound in the cell by the RNA-binding protein.

In the final stage of the purification, the RNAs purified from the polyacrylamide gel are converted into DNA copies called CLIP tags. This is done by first attaching short fragments of DNA, called linkers, to the RNAs. The protein is removed by the enzyme proteinase K to release the short free RNAs, and then the CLIP tags are made followed by reverse transcription primed by the DNA linkers. CLIP tags are then amplified by PCR and sequenced with high-throughput sequencing approaches and the data are analysed bioinformatically to define the RNA consensus sequence recognized by a particular RNA-binding protein.

The use of CLIP has made it possible to generate maps of the binding sites of specific RNA-binding proteins. These maps are useful for deciphering the function of these RNA-binding proteins in the cell, because the RNA targets may be present in mRNAs that correspond to a family of genes or a biochemical pathway. An example of a particularly comprehensive RNA-binding protein map for the NOVA protein is shown in Fig. 4.14. This figure shows the distribution of tags identified for NOVA along the length of mouse chromosome 2 relative to CLIP tags identified from the brains of two individual mice.

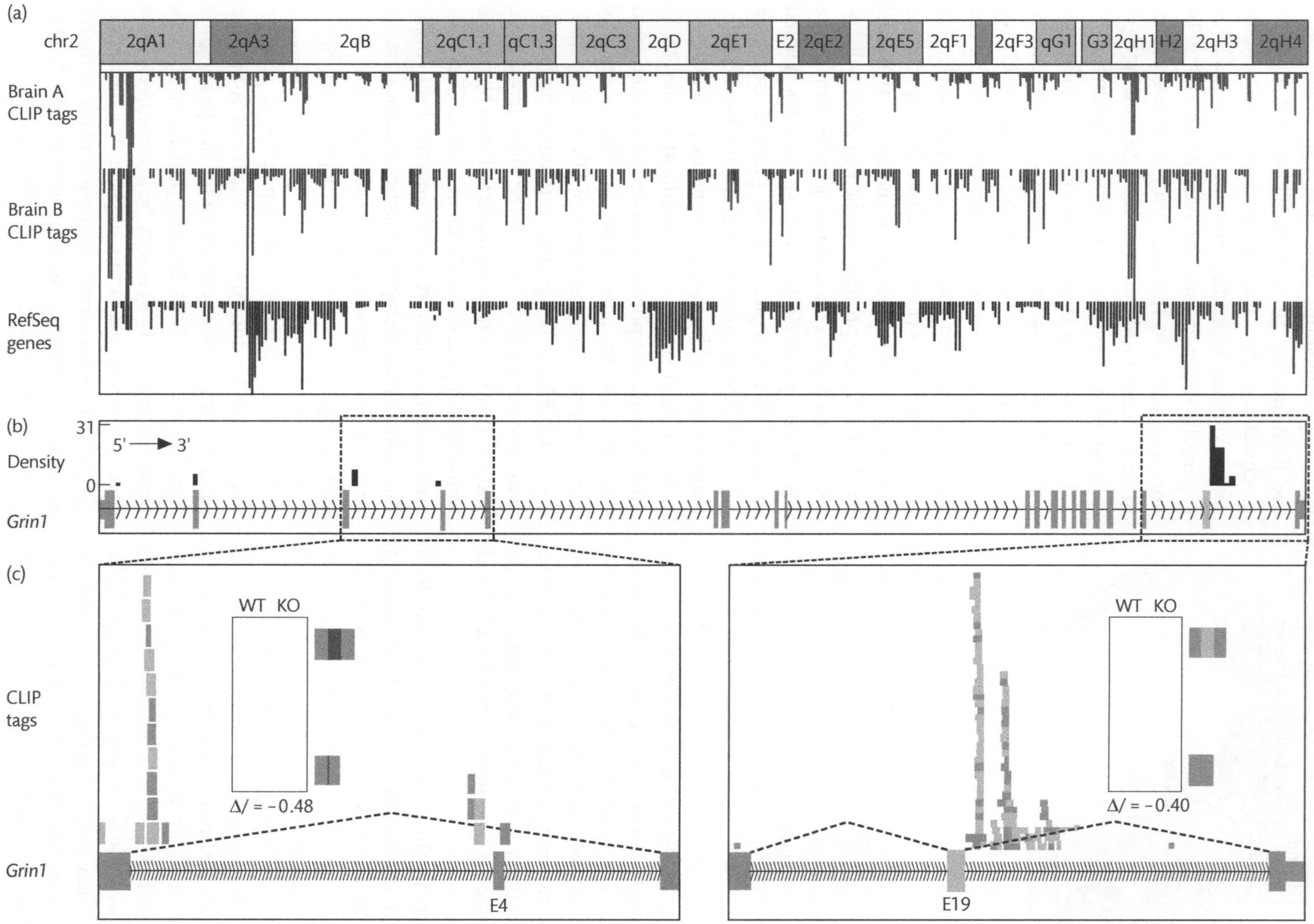

Figure 4.14 **CLIP genome-wide map of Nova RNA-binding sites.** (a) A low-resolution map of chromosome 2, with the position of CLIP tags indicated along with the position of genes (the height of the columns is an indication of the number of CLIP tags or genes). (b) The density of CLIP tags within the ***Grin1*** gene. (c) The position of these CLIP tags (at higher resolution) relative to exons (boxes) and introns (lines between the boxes, with the arrowheads indicating the direction of transcription). Changes in alternative splicing of *Grin1* are apparent when comparing wild-type (WT) and Nova knockout (KO) mice. (From Licatalosi et al.[56] with permission from Macmillan Publishers Ltd.)

At the bottom of this figure is a higher-resolution picture of CLIP tags on one particular gene called *Grin1*, showing that tags are clustered around particular exons (shown as boxes). The inset shows the results from an RT-PCR when the NOVA protein is removed from the mouse brain by mutagenesis. Notice that the splicing pattern of these pre-mRNAs changes as a result.

Technical modifications have further improved the CLIP procedure.[57] The problem with the initial CLIP approach is that there is a requirement for the reverse transcriptase to proceed from the 3′ ligated adapter (attached to the 3′ end of the precipitated RNA) all the way to the 5′ adapter (attached to the 5′ end). This is because the presence of both adapters is required for PCR amplification. However, the reverse transcriptase stalls (stops working) at the site of the protein crosslink (residual amino acids remain crosslinked). This problem has been overcome by using a clever technical trick—intramolecular cDNA circularization. The technique is called *iCLIP* (individual-nucleotide resolution CLIP), which can provide information about the position of the crosslink site at nucleotide resolution, providing further insights into how proteins bind to their target RNAs.

Key points

There are several ways to study protein–RNA interactions. Traditional methods include gel retardation and filtration assays, UV crosslinking followed by RNAse treatment and SDS-PAGE, and Northwestern blots. More sophisticated methods include co-crystallization, detailed binding kinetics, and the MS2 pull-down assay. The CLIP method (in vivo crosslinking followed by immunoprecipitation) is very popular because it enables the identification of in vivo RNA targets.

Summary

Evolution has generated a wide range of RNA-binding domains in proteins. Given the prevailing theory that RNA preceded DNA in evolution, it is reasonable to assume that RNA-binding domains are more ancient than DNA-binding domains. RNA-binding proteins are at the heart of post-transcriptional processes, and play an integral and vital role in gene expression.

- The first RNA-binding proteins to be described in detail were the hnRNP proteins, abundant proteins that are involved in several post-transcriptional processes. Some are exclusively nuclear while others shuttle between nucleus and cytoplasm.
- The RNA recognition motif (RRM) is among the most frequently occurring RNA-binding domains in nature. It generally binds to specific ssRNA targets.
- The K-homology (KH) domain also binds to ssRNA targets, but it can also bind ssDNA.
- The dsRNA-binding domain (dsRBD) allows RNA-binding proteins to make contact with dsRNA. The dsRBD generally appears to recognize RNA structure as opposed to RNA sequence.
- Zinc-finger motifs are also able to recognize specific RNA sequences. Zinc fingers can be engineered to recognize specific RNA targets.
- Additional RNA-binding domains exist, including the PAZ and PIWI domains, arginine-rich motifs, the core RNA helicase domain, pentatricopeptide repeats, PUF domains, and the homeodomain.
- Several RNA-binding domains are multifunctional; in this way RNA-binding proteins are involved in multiple processes, even transcription.
- Several methods are available to analyse protein–RNA binding, including X-ray crystallography, NMR spectroscopy, gel retardation (EMSA), filter retention, UV crosslinking, CLIP, and the MS2 pull-down system.

Questions

4.1 How many different types of RNA-binding domains are known?

4.2 What strategies are used by RNA-binding proteins to make molecular contacts with RNA?

4.3 Do RNA-binding domains recognize base sequences alone, or the sugar phosphate backbone and three-dimensional structure?

4.4 Why is arginine such an important amino acid in RNA-binding proteins?
4.5 Can proteins bind both RNA and DNA?
4.6 What techniques can be used to study protein–RNA interactions in vitro?
4.7 What techniques can be used to study protein–RNA interactions in vivo?

References

1. **Castello A, Fischer B, Hentze MW, Preiss T.** RNA-binding proteins in Mendelian disease. *Trends Gen* **29**, 318–27 (2013).
2. **Glisovic T, Bachorik JL, Yonge J, Dreyfuss G.** RNA-binding proteins and post-transcriptional gene regulation. *FEBS Lett* **582**, 1977–86 (2008).
3. **Dreyfuss G, Kim VN, Kataoka N.** Messenger-RNA-binding proteins and the messages they carry. *Nat Rev Mol Cell Biol* **3**, 195–205 (2002).
4. **Baltz AG, Munschauer M, Schwannhäuser B, et al.** The mRNA-bound proteome and its global occupancy profile on protein-coding transcripts. *Mol Cell* **46**, 674–90 (2012).
5. **Maris C, Dominguez C, Allain FH.** The RNA recognition motif, a plastic RNA-binding platform to regulate post-transcriptional gene expression. *FEBS J* **272**, 2118–31 (2005).
6. **Ladomery M, Dellaire G.** Multifunctional zinc-finger proteins in development and disease. *Ann Hum Genet* **66**, 331–42 (2002).
7. **Chaudhury A, Chander P, Howe PH.** Heterogenous nuclear ribonucleoproteins (hnRNPs) in cellular processes: focus on hnRNP E1's multifunctional regulatory roles. *RNA* **16**, 1449–62 (2010).
8. **Biamonti G, Riva S.** New insights into the auxiliary domains of eukaryotic RNA binding proteins. *FEBS Lett* **340**, 1–8 (1994).
9. **Antson A.** Single stranded RNA binding proteins. *Curr Opin Struct Biol* **10**, 87–94 (2000).
10. **Lunde B, Moore C, Varani B.** RNA-binding proteins: modular design for efficient function. *Nat Rev Mol Cell Biol* **8**, 479–90 (2007).
11. **Chen Y, Varani G.** Engineering RNA-binding proteins for biology. *FEBS J* **280**, 3734–54 (2013).
12. **Clayton C.** The regulation of trypanosome gene expression by RNA-binding proteins. *PLoS Pathogens* **9**, e1003680 (2013).
13. **Droll D, Minia I, Fadda A.** Post-transcriptional regulation of the trypanosome heat shock response by a zinc-finger protein. *PLoS Pathogens* **9**, e1003286 (2013).
14. **Krecic A, Swanson M.** hnRNP complexes: composition, structure, and function. *Curr Opin Cell Biol* **11**, 363–71 (1999).
15. **Handa N, Nureki O, Kurimoto, K.** Structural basis for recognition of the tra mRNA precursor by the Sex-lethal protein. *Nature* **398**, 579–85 (1999).
16. **Wang X, Tanaka Hall TM.** Structural basis for recognition of AU-rich element RNA by the HuD protein. *Nat Struct Biol* **8** 141–5 (2001).
17. **Zhang Y., Madl T, Bagdiul I, et al.** Structure, phosphorylation and U2AF65 binding of the N-terminal domain of splicing factor 1 during 3′-splice site recognition. *Nucleic Acids Res* **41**, 1343–54 (2013).
18. **Oberstrass FC, Auweter FD, Erat M, et al.** Structure of PTB bound to RNA: specific binding and implications for splicing regulation. *Science* **309**, 2054–7 (2005).
19. **Honoré B, Rasmussen HH, Vorum H, et al.** Heterogenous nuclear ribonucleoproteins H, H′ and F are members of a ubiquitously expressed subfamily of related but distinct proteins encoded by genes mapping to different chromosomes. *J Biol Chem* **270**, 28 780–9 (1995).
20. **Dominguez C, Allain F.** NMR structure of the three quasi RNA recognition motifs (qRRMs) of human hnRNP F and interaction studies with Bcl-x G-tract RNA: a novel mode of RNA recognition. *Nucleic Acids Res* **34**, 3634–45 (2006).
21. **Cléry A, Sinha R, Anczuków O, et al.** Isolated pseudo-RNA-recognition motifs of SR proteins can regulate splicing using a noncanonical mode of RNA recognition. *Proc Natl Acad Sci USA* **110**, E2802–11 (2013).
22. **Musco G, Stier G, Joseph C, et al.** Three-dimensional structure and stability of the KH-domain: molecular insights into the fragile X syndrome. *Cell* **85**, 237–54 (1996).
23. **Bomsztyk K, Denisenko O, Ostrowski J.** hnRNP K: one protein multiple processes. *BioEssays* **26**, 629–38 (2004).
24. **Valverde R, Edwards L, Regan L.** Structure and function of KH domains. *FEBS J* **275**, 2712–26 (2008).
25. **Graumann P, Marahiel MA.** A superfamily of proteins that contain the cold-shock domain. *Trends Biochem Sci* **23**, 286–90 (1998).
26. **Sachs R, Max KE, Heinemann U, Balbach J.** RNA single strands bind to a conserved surface of the major cold shock protein in crystals and solution. *RNA* **18**, 65–76 (2012).
27. **Sommerville J.** Activities of cold-shock domain proteins in translation control. *BioEssays* **21**, 319–25 (1999).
28. **Sommerville J, Ladomery M.** Masking of messenger RNA by Y-box proteins. *FASEB J* **10**, 435–43 (1996).
29. **Sommerville J, Ladomery M.** Transcription and masking of messenger RNA in germ cells-involvement of Y-box proteins. *Chromosoma* **104**, 469–78 (1996).
30. **Kohno K, Izumi H, Uchiumi T, et al.** The pleiotropic functions of the Y-box-binding protein, YB-1. *BioEssays* **25**, 691–8 (2003).
31. **Chang TC, Yamashita A, Chen CY, et al.** UNR, a new partner of poly(A)-binding protein, plays a key role in translationally coupled mRNA turnover mediated by the c-fos major coding-region determinant. *Genes Dev* **18**, 2010–23 (2004).

32. **Fierro-Monti I, Mathews M.** Proteins binding to duplexed RNA: one motif, multiple functions. *Trends Biochem Sci* **25**, 241–6 (2000).
33. **Saunders L, Barber G.** The dsRNA binding protein family: critical roles, diverse cellular functions. *FASEB J* **17**, 961–83 (2003).
34. **Stefl R, Oberstrass FC, Hood JL, et al.** The solution structure of the ADAR2 dsRBM-RNA complex reveals a sequence-specific readout of the minor groove. *Cell* **143**, 225–37 (2010).
35. **Pieler T, Theunissen O.** TFIIIA: nine fingers–three hands? *Trends Biochem Sci* **18**, 226–30 (1993).
36. **Searles M, Lu D, Klug A.** The role of the central zinc fingers of transcription factor IIIA in binding to 5 S RNA. *J Mol Biol* **301**, 47–60 (2000).
37. **Friesen WJ, Darby MK.** Specific RNA binding proteins constructed from zinc fingers. *Nat Struct Biol* **5**, 543–6 (1998).
38. **Mangs H, Morris B.** ZRAN B2: structural and functional insights into a novel splicing protein. *Int J Biochem Cell Biol* **40**, 2353–7 (2008).
39. **Loughlin F, Mansfield RE, Vaz PM, et al.** The zinc fingers of the SR-like protein ZRANB2 are single-stranded RNA-binding domains that recognize 5′ splice site-like sequences. *Proc Natl Acad Sci USA* **106**, 5581-6 (2009).
40. **Bleichert F, Baserga SJ.** The long unwinding road of RNA helicases. *Mol Cell* **27**, 339–52 (2007).
41. **Thandapani P, O'Connor TR, Bailey TL, Richard S.** Defining the RGG/RG motif. *Mol Cell* **50**, 613–23 (2013).
42. **Darnell JC, Jensen KB, Jin P, et al.** Fragile X mental retardation protein targets G quartet mRNAs important for neuronal function. *Cell* **107**, 489–99 (2001).
43. **Ladomery M, Sommerville J.** Binding of Y-box proteins to RNA: involvement of different protein domains. *Nucleic Acids Res* **22**, 5582–9 (1994).
44. **Blackwell E, Ceman S.** Arginine methylation of RNA-binding proteins regulates cell function and differentiation. *Mol Rep & Dev* **79**, 163–75 (2012).
45. **Yin P, Li Q, Yan C, et al.** Structural basis for the modular recognition of single-stranded RNA by PPR proteins. *Nature* **504**, 168–71 (2013).
46. **Filipovska A, Razif MF, Nygard KK, Rackham O.** A universal code for RNA recognition by PUF proteins. *Nat Chem Biol* **7**, 425–7 (2011).
47. **Rivera-Pomar, Niesing D, Schmidt-Ott U, et al.** RNA binding and translational suppression by bicoid. *Nature* **379**, 746–9 (1996).
48. **Liu W, Seto J, Sibille E, Toth M.** The RNA binding domain of jerky consists of tandemly arranged helox-turn-helix/homeodomain-like motifs and binds specific sets of mRNAs. *Mol Cell Biol* **23**, 4083–93 (2003).
49. **Draper D, Reynaldo L.** RNA binding strategies of ribosomal proteins. *Nuc Acids Res* **27**, 381–8 (1999).
50. **Foot JN, Feracci M, Dominguez C.** Screening protein—single stranded RNA complexes by NMR spectroscopy for structure determination. *Methods* **64**, 288–301 (2013).
51. **Ladomery M, Sommerville J.** Binding of Y-box proteins to RNA: involvement of different protein domains. *Nucleic Acids Res.* **22**: 5582–9 (1994).
52. **Hartley R, Le Meuth-Metzinger V, Osborne HB.** Screening for sequence-specific RNA-BPs by comprehensive UV crosslinking. *BMC Mol Biol* **3**, 1–8 (2002).
53. **Zhou A, Reed R.** Purification of functional RNA-protein complexes using MS2-MBP. *Curr Protoc Mol Biol* Chapter 27:Unit 27.3 (2003).
54. **Stoltenburg R, Reinemann C, Strehlitz B.** SELEX—a (r) evolutionary method to generate high-affinity nucleic acid ligands. *Biomol Eng* **24**, 381–403 (2007).
55. **Bernstein DS, Buter N, Stumpf C, Wickens M.** Analyzing mRNA-protein complexes using a yeast three-hybrid system. *Methods* **26**, 123–41 (2006).
56. **Licatalosi DD, Mele A, Fak JJ, et al.** HITS-CLIP yields genome-wide insights into brain alternative RNA processing. *Nature* **456**, 464–9 (2008).
57. **Huppertz I, Attiq J, D'Ambrogio A, et al.** iCLIP: protein-RNA interactions at nucleotide resolution. *Methods* **65**, 274–87 (2014).

5 Pre-mRNA splicing by the spliceosome

Introduction

This chapter will focus on an important development in eukaryotic RNA processing: pre-mRNA splicing catalysed by the **spliceosome**. Spliceosomal splicing is needed since a large proportion of eukaryotic genes are split between segments called **introns** and **exons** (Fig. 5.1). Splicing occurs in the nucleus, often during transcription (see Chapter 8). Both the introns and exons of split genes are transcribed within the nucleus into long precursor mRNAs (pre-mRNAs). However, only the exons are spliced together to give mRNA molecules which are then exported into the cytoplasm. The name exon originates from the fact that these exon sequences are EXpressed, while introns are removed by splicing and so are not expressed in the mature mRNA. Although introns were initially considered 'junk', more recent work has shown that they have important functions in gene expression and can act as reservoirs for other genes (both protein coding and ncRNA genes). In this chapter we will consider the biology of splicing. **Alternatively spliced exons**, which are optionally included in some mRNAs, are themselves an important area of RNA biology, and are covered in Chapter 6. Terms which may be unfamiliar are defined in the Glossary and highlighted in **bold**.

5.1 RNA splicing was discovered in a virus

This first section will describe how splicing was discovered. In the 1970s, the intriguing observation that in eukaryotes nuclear transcripts seemed much larger than cytoplasmic transcripts puzzled molecular biologists. These large nuclear RNAs were called heterogeneous nuclear RNAs (**hnRNAs**). The explanation for this conundrum came in 1977 from experiments on adenovirus carried out by two teams, one led by Phil Sharp at the Massachusetts Institute of Technology (MIT) and the other led by Richard Roberts at Cold Spring Harbor Laboratory, which led to the discovery of introns.

The discovery of introns was enabled by a technique called R-looping (see Fig. 5.2). Both DNA and RNA form duplexes by hydrogen bonding between bases (see Chapter 2). Hydrogen bonds are rather weak and break at high temperatures or in chemicals like formamide. The key to R-looping is

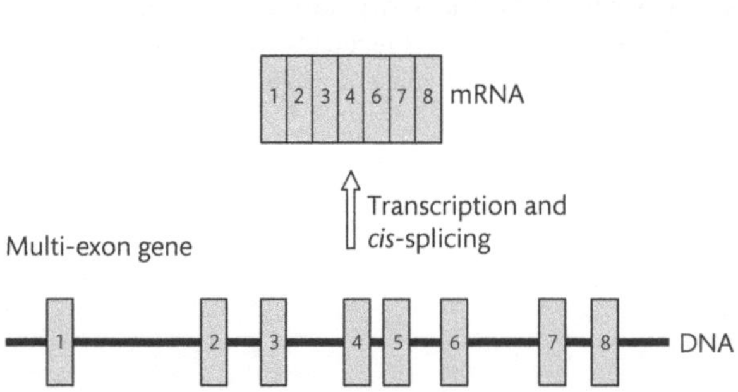

Figure 5.1 *Cis*-splicing is needed to decode split eukaryotic genes. In most eukaryotes spliceosomal splicing occurs in *cis*-, in which exons are joined from within a single pre-mRNA transcript. Exons are shown as boxes, and introns as straight lines. In *cis*-splicing, the preliminary transcript or pre-mRNA from a multi-exon gene is much longer than the final spliced mRNA. Exons in the pre-mRNA are recognized and joined together by the spliceosome.

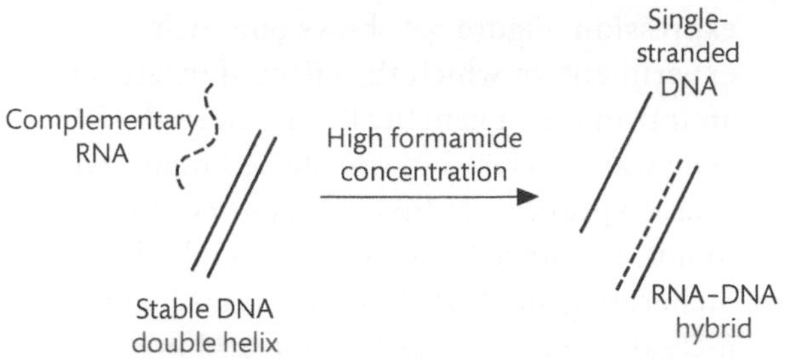

Figure 5.2 R-looping can be used to visualize RNA-DNA helices. Regions of double- and single-stranded nucleic acids in RNA-DNA hybrids can be visualized under the electron microscope. The DNA strands are shown as unbroken lines and the complementary RNA is shown as a broken line.

that in high concentrations of formamide, DNA-RNA hybrids are more stable than DNA-DNA duplexes. Because of this, double-stranded DNAs (dsDNAs) will be replaced by complementary DNA-RNA hybrids which can be visualized by electron microscopy—any non-complementary regions appear as loops.

Adenoviruses were used as a model system for investigating eukaryotic RNA synthesis since they infected human cells grown in culture and could be purified in large amounts from these cells. By 1977 the 35kb adenovirus genome had already been isolated and characterized by restriction mapping. To map adenoviral transcripts onto this genome, the RNA transcript encoding the major viral coat protein (called the hexon protein) from the late stages of adenovirus-infected cells was purified from translating ribosomes. Complementarity of these late transcripts with the adenovirus genome was tested by R-looping with restriction fragments of adenovirus DNA. The expected result was that the late transcripts would hybridize perfectly with the region of the adenovirus genome which encoded them. These RNA-DNA duplexes would be visible under the electron microscope, and so map the hexon transcripts onto the adenoviral genome.

The adenoviral hexon RNA did indeed hybridize with a fragment of the adenoviral genome containing part of the hexon gene, but not over its entire length (see Fig. 5.3). The conclusion of these experiments was that the hexon mRNA was transcribed as a long precursor hnRNA comprising exons (which are spliced into the final mRNA) connected by

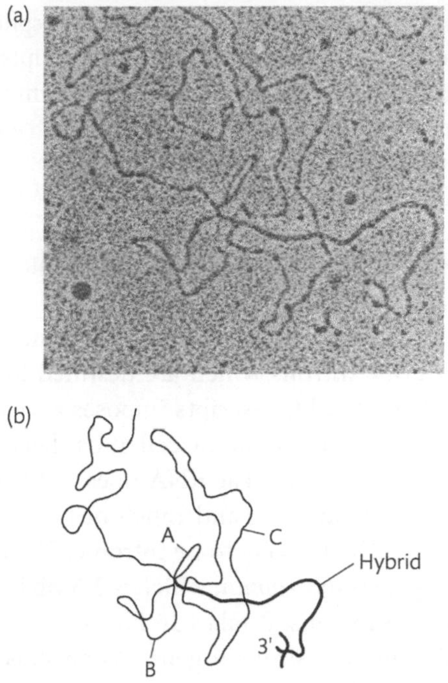

Figure 5.3 Split genes were first identified in a viral genome. (a) R-looping pattern obtained by hybridization of a single-stranded DNA probe with the hexon mRNA from adenovirus. (b) Interpretation of this pattern. Notice that an RNA-DNA hybrid forms where the mRNA is complementary to the gene (bold line), but there are also loops and single-stranded regions where the mRNA is not complementary to the gene probe. Regions of non-complementarity between the mRNA and gene are due to the presence of introns in the gene and not the mRNA, and the addition of a poly(A) tail to the mRNA which is not encoded by the gene. (From Berget *et al.*[1] ©PNAS, 2010.)

separating intervening sequences. The discovery of introns and splicing in these classic R-looping experiments provided new and exciting insights into the structure of eukaryotic genes, and led to the award of a Nobel Prize to Phil Sharp and Richard Roberts in 1990. For background to this section see references 1–4.

5.2 Spliceosomal introns are critical for efficient eukaryotic gene expression

Splicing is not restricted to adenoviruses. Far from it. Viral genes often use gene expression pathways provided by their hosts, and all current eukaryotes have at least some split nuclear genes containing introns, such as the adenovirus hexon gene. These split genes are transcribed into long primary transcripts called **pre-mRNAs**, and then the introns are removed by a large molecular machine in the nucleus called the **spliceosome**.

What are the functions of introns? At first glance, intron-containing genes are longer than required for their open reading frames (ORFs), and do not seem to represent very streamlined vehicles for gene expression. Taking the time and energy to transcribe introns which are destined to be removed from final transcripts imposes a seemingly wasteful requirement on the cell. It has been calculated that 95% of nuclear RNA does not leave the nuclei of human cells, and much of this nuclear-restricted RNA is likely to be intronic. The longest human gene is *dystrophin*, which is 2.5Mb long and contains 78 introns. It takes 16 hours to transcribe the full-length *dystrophin* gene. As an illustration of this, if an RNA polymerase started transcribing the *dystrophin* gene at breakfast (8a.m.), it would not finish until midnight of the same day. Since 99.4% of the full length of the *dystrophin* gene comprises intron sequences and so is not used for protein coding, this makes a *dystrophin* mRNA just 14kb long.

As well as the time taken for transcription, another penalty of having extremely large genes is that, given a random distribution of mutations, it is much more likely that a longer gene like *dystrophin* will pick up a mutation somewhere along its length than will a shorter gene. Mutations in *dystrophin* cause muscular dystrophy; we will come back to the impact of splicing in disease in Chapter 8. Even intronic mutations can have catastrophic effects on gene expression by disrupting the *cis*-acting elements required for pre-mRNA splicing.

Contrary to these perceived shortcomings, experiments performed in mammals, fruit flies, and yeast have shown that introns do make an important positive contribution to efficient gene expression despite their early exit from the transcript. Two reciprocal kinds of experiment have been carried out.

- **Intron insertion.** Insertion of introns into intron-less genes leads to increases in gene expression. Figure 5.4 shows one such experiment, in which the effect of inserting an intron into a genetically engineered *TP1* gene was measured in transfected mammalian cells. Expression of this *TP1* gene could be monitored since it contained an embedded luciferase gene. Note that intron insertion increased luciferase expression and mRNA production from this gene (particularly if this intron was inserted into a 5′ location in the gene).
- **Intron removal.** The converse experiments have also been carried out, in which introns have been *removed* from genes. Removing introns from the *alcohol dehydrogenase* gene in fruit flies reduced both the expression of the *alcohol dehydrogenase* transcript and the amount of alcohol dehydrogenase protein made in flies carrying the deletion.

Both these experiments show that the presence of introns increases eukaryotic gene expression. A similar picture has come from global analyses of gene expression in the yeast *Saccharomyces cerevisiae*. Very few genes in this yeast contain introns—only around 4% of the total gene number of 6000 genes (left-hand pie chart in Fig. 5.5). However, over a third of the total cellular mRNA transcripts are derived from this 4% of intron-containing genes (right-hand pie chart in Fig. 5.5).

Intron-containing mRNAs comprise more than 10 000 of the nearly 38 000 mRNA molecules made per hour by a single yeast cell. Why are these intron-containing yeast mRNAs needed at such high levels? The answer is that almost half of the intron-containing genes in yeast encode ribosomal proteins (and 102 out of 139 genes encoding yeast ribosome proteins contain introns), which are required at particularly high levels. Removal of introns from some of these ribosomal protein genes has been directly shown to reduce the rate of yeast cell division under conditions in which nutrients are limiting. Hence introns are needed for efficient expression of these ribosomal protein genes. For further information about this section see references 5–10.

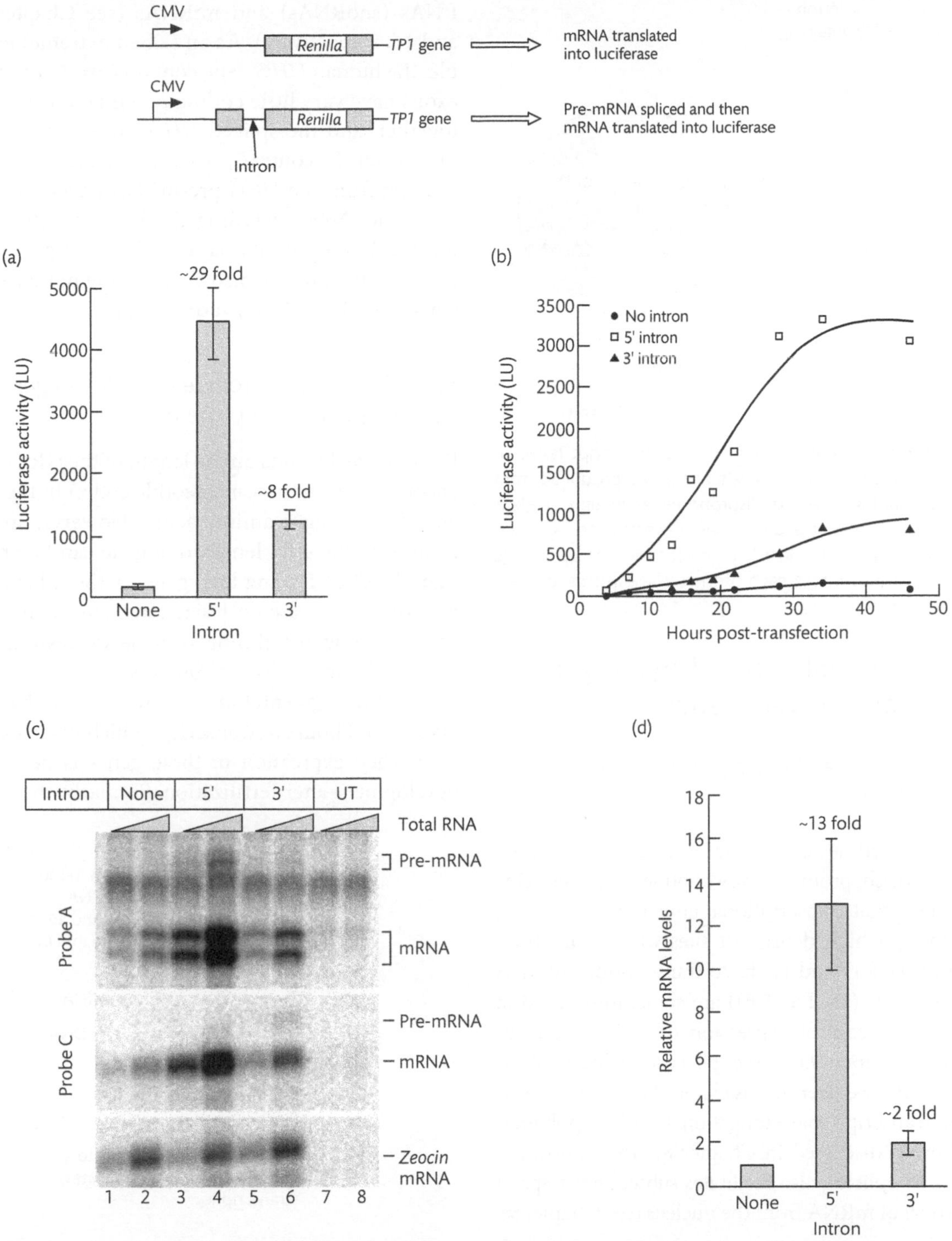

Figure 5.4 **Introns increase eukaryotic gene expression levels.** Introns have been added to intron-less genes and the effect on gene expression tested. In this case an intron was added to a genetically engineered *TP1* gene which contained an embedded *Renilla* luciferase gene. (a), (b) The levels of luciferase produced by the intron-containing and intron-less genes were compared in cultured cells. Notice that the level of luciferase is much higher in the version of the gene which contains the 5′ or 3′ inserted intron. These are introns in the upstream or downstream parts of the *TP1* reading frame, respectively. LU = light units, which is a measure of luciferase expression. (c), (d) Levels of mRNA produced from a gene are much higher when the *TP1* gene includes an intron. Levels of mRNA were quantified using Northern blotting—notice there are much higher relative levels of the engineered *TP1* mRNA relative to the *zeocin* mRNA loading control when an intron is included. (From Nott *et al.*,[5] with permission from Cold Spring Harbor Laboratory Press.)

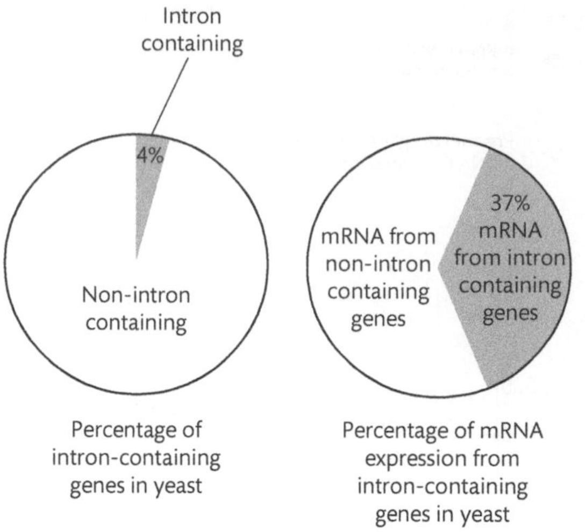

Figure 5.5 **Microarray quantification of transcript levels in yeast have shown that the levels of mRNAs produced from intron-containing genes are disproportionately high.** The left-hand pie chart shows the percentages of yeast genes with and without introns, and the right-hand chart shows the percentage of transcripts made from each of these classes of gene. (Data from Ares *et al.*[6])

5.3 Introns enhance eukaryotic gene expression at several levels

5.3.1 Introns can have a positive impact on transcription

Introns sometimes contain some of the *cis*-acting elements which promote transcription of genes. One example is that transcriptional enhancers (sequences containing a high density of sites which bind transcription factors and facilitate transcription by RNA polymerase II (see Fig. 5.6)) are sometimes found in introns, and control expression of their host gene. Secondly, components of the spliceosome (particularly U1 snRNP—see later in this chapter) contribute to efficient transcriptional elongation by RNA polymerase (how is discussed in Chapter 8). The removal of introns by splicing also facilitates subsequent steps in the export of mRNA from the nucleus (see Chapter 9).

5.3.2 Introns can themselves contain genes

In some cases distinct genes can be embedded in the introns of other genes. These intron-embedded genes can be either protein coding or encode non-protein-coding RNAs including small nucleolar RNAs (snoRNAs) and miRNAs (see Chapters 14 and 16, respectively). As an (albeit extreme) example, the human *UHG* gene contains exons, but these exons have very little coding potential once spliced together, and the spliced *UHG* mRNA is unstable in the cell. In contrast, the introns excised during splicing from the *UHG* pre-mRNA encode a set of eight snoRNAs, including the U22 snoRNA (giving the *UHG* gene its name—U22 host gene) (see Fig. 5.7). Introns can also encode long ncRNAs (see Chapters 15 and 16 for more details).

5.3.3 Introns can provide flexibility in gene size independent of protein size

Protein size depends on the length of the ORF which encodes it (each three-nucleotide codon in the ORF encodes a single amino acid). However, introns mean that the total length of a gene can be greater than the ORF. Having long genes without having to necessarily increase ORF size can be useful in terms of controlling the timing of gene expression during development. The *Drosophila* homeotic genes involved in segmental identity are so long that they take several hours to transcribe, which fits in exactly with when expression of these genes is needed in development after fertilization. Similarly, the length

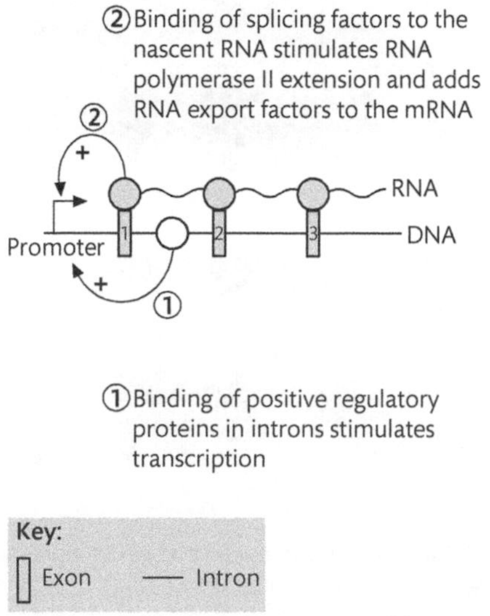

Figure 5.6 **Introns can positively affect transcription.** Introns provide *cis*-acting sequences promoting transcriptional initiation, such as enhancers. Components of the assembling spliceosome also assist RNA polymerase elongation.

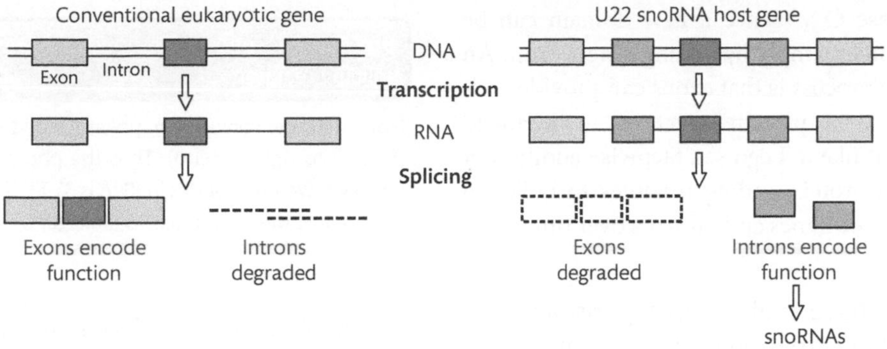

Figure 5.7 Introns can themselves contain genes. The introns of the human *UHG* gene encode multiple small nucleolar RNAs (snoRNAs), and in this case the exons do not encode protein. (Reprinted from Moore *et al.*[11], with permission from Macmillan Publishers Ltd.)

of time required to transcribe the muscle-specific *dystrophin* gene (16 hours) might prevent the expression of dystrophin protein in muscle precursor cells which have a cell division time shorter than 16 hours, but not in differentiated muscle cells which no longer divide.

For further information about this section, see references 11–14.

Key points

Introns have an important positive role in gene expression in eukaryotes. Introns contribute to transcriptional control and coordination of RNA-processing steps in the nucleus leading to efficient RNA export from the nucleus, act as a reservoir of non-coding RNA genes, and contribute to timing of gene expression during development.

5.4 Introns have an important role in evolution

As well as having a role at the centre of eukaryotic gene expression, introns have made important evolutionary contributions to eukaryotic organisms. This is the topic of this section.

5.4.1 Introns can enable the evolution of new proteins through the addition and subtraction of exons which encode extra amino acids to their cognate genes

Introns provide a route to the efficient evolution of new proteins. New exons can be inserted into genes along with flanking introns (e.g. by genetic duplication, or insertion of mobile genetic elements). After transcription, these new exons will be recognized by the spliceosome and spliced into mRNAs so the intron–exon structure ensures a clean insertion. The resulting longer mRNAs will be translated into protein, with a new peptide insert provided by the new exon.

Whole protein domains can be encoded by exons, so addition of a new exon to a gene can add a new functional domain to the encoded protein (Fig. 5.8 shows a hypothetical example). For new amino acids to be added in this way, any new exons need to be added in frame with the existing reading frame so that their splicing inclusion does not cause a frameshift. There is some evidence that very old exons might indeed start their reading frames from the initial nucleotide

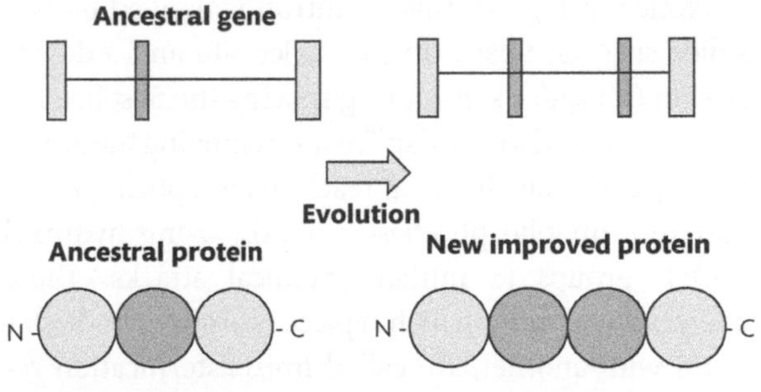

Figure 5.8 Introns facilitate evolution of new proteins. In this hypothetical example the ancestral gene has three exons. Over the course of evolution, a new exon becomes inserted into the intron—this could be through a region of local genome duplication, or through insertion of a transposable element. Since it has splice sites, this new exon will be spliced into the mRNA transcribed from the new gene, inserting new coding information into the resulting protein. This might result in the evolution of a new and improved protein.

(so-called 'phase O'). Hence a new domain can be added to a protein simply by adding a new exon. An analogy to this process is that exons can provide coding information for protein structures to be added to mRNAs a bit like a 'Lego set'. Stepwise addition of exons can add protein-coding modules to build up more sophisticated genes and proteins over time.

5.4.2 Introns have enabled an expansion in the coding information of the genome by enabling alternative splicing

In alternative splicing different exon combinations can be included in transcripts from the same gene. This helps to increase the information content of the genome, and has been extensively used in more complex organisms (we return to alternative splicing and its advantages in detail in Chapter 6).

For further information about this section, see reference 15.

5.5 The mechanism of pre-mRNA splicing

In this section we are going to discuss how splicing happens; in other words, what its mechanism is. During splicing, pre-mRNAs need to be cut at their exon–intron boundaries and the exons joined together. The sites of cutting and joining are called **splice sites**.

Notice in Fig. 5.9 that an intron is flanked by two splice sites, an upstream (5′) splice site and a downstream (3′) splice site. During splicing the 5′ splice site becomes joined to the 3′ splice site, removing the intron in between. The chemical reactions in splicing break and re-form phosphodiester bonds, using hydroxyl (–OH) groups to initiate chemical attacks. These chemical reactions, which replace one phosphodiester bond with another, are called *trans*-esterification reactions. See Box 5.1 for the correct nomenclature of phosphodiester bonds in splicing reactions.

5.5.1 Splicing follows a two-step reaction pathway

Splicing takes place by the two-step reaction shown in Fig. 5.9.

Box 5.1 Phosphodiester bond nomenclature in splicing reactions

Note that by convention polynucleotides go from left (5′ end) to right (3′ end). Thus the phosphodiester bond between two nucleotides in RNA is 3′–5′. Similarly, the linkage holding the lariat is a 2′–5′ linkage.

In the first step the 2′ –OH of the branchpoint (BP) adenosine (within the intron) attacks the phosphodiester bond at the 5′ splice site (which links the upstream exon with the intron), breaking this and creating a new 2′–5′ linkage. The resulting branched structure is called a lariat intermediate; notice that at this point the intron is still attached to the downstream exon but is no longer attached to the upstream exon (giving a loop or lariat).

In the second step, the upstream exon is spliced to the downstream exon, releasing the intron as a lariat

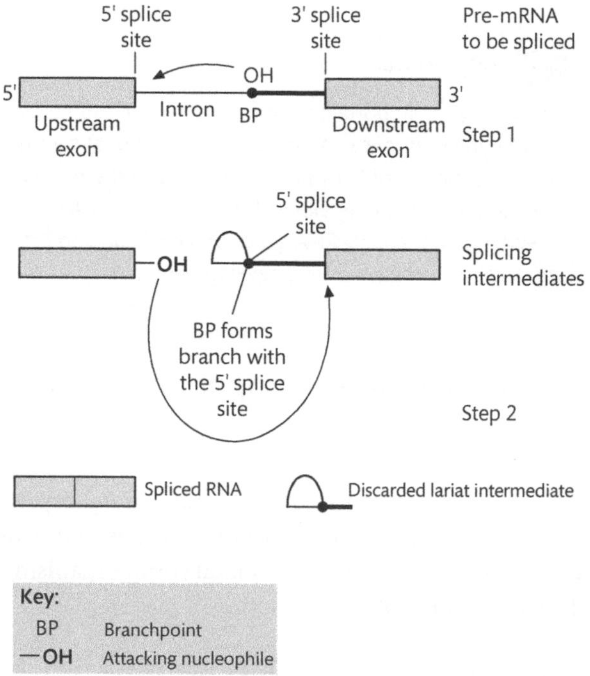

Figure 5.9 Splicing takes place in a two-step chemical reaction between an upstream splice site called the 5′ splice site (also called the splice donor) and a downstream 3′ splice site (also called the splice acceptor). The 3′ splice site region comprises the 3′ splice site itself, and an internal site called the branchpoint sequence (BP). These important positions on the pre-mRNA play a chemical role in the splicing reaction.

product (in which the branchpoint nucleotide is still connected to the 5′ splice site).

To summarize these two steps:

- **Step 1**. The 2′ -OH of the branchpoint adenosine chemically attacks the 3′–5′ phosphodiester bond connecting exon 1 to the intron. This breaks this 3′–5′ phosphodiester bond and creates a new 2′–5′ phosphodiester bond. Because the 2′–5′ bond is between the start of the intron and the branchpoint adenosine, a looped reaction intermediate called a **lariat intermediate** is formed. The upstream exon, terminating in a 3′ -OH group, is released by this reaction.
- **Step 2**. The 3′ -OH on the end of the upstream exon attacks the 3′–5′ phosphodiester bond joining the intron and the 3′ splice site. The result is a new 3′–5′ phosphodiester bond joining the two exons, releasing the intron as a lariat intermediate. The lariat intermediate is then debranched in the nucleus, and either used as an ncRNA or degraded.

Key points

Exons are spliced together via a two-step mechanism. In the first step a 2′ -OH group on the branchpoint adenosine nucleotide attacks the phosphodiester bond linking the upstream exon with the intron. The products of this reaction are a free upstream exon and a lariat intermediate joined to the downstream exon. In the second step of splicing the 3′ -OH group on the upstream exon forms a new phosphodiester bond splicing it to the downstream exon, releasing the intron as a lariat product.

5.6 Splice sites

5.6.1 Pre-mRNAs are punctuated by splice sites at intron–exon junctions

Splice sites are where exons are joined together. In the early 1980s it was noticed that the nucleotide sequences found around splice sites are very similar to each other. The consensus sequences found in human splice sites are shown in Fig. 5.10; notice that the conserved nucleotides are mainly within the intron, but also partly within the exon across the intron–exon junction. These conserved splice site sequences are essential for splicing. Experiments have shown that mutation of the six most 5′ bases and the 24 most 3′ bases in an intron (corresponding to the 5′ and 3′ splice sites, respectively) prevent splicing of β-globin pre-mRNAs.

5.6.2 The 5′ splice site sequence

In humans the consensus 5′ splice site sequence is CAG↓GURAGU. The downward pointing arrow (↓) represents the exon–intron boundary which is the actual splice site (also called the splice donor site).

5.6.3 The 3′ splice site sequence

The 3′ splice site sequence comprises three distinct parts.

1 **The 3′ splice site itself** usually has an AG ↓ dinucleotide, where the ↓ represents the boundary of the intron with the downstream exon.

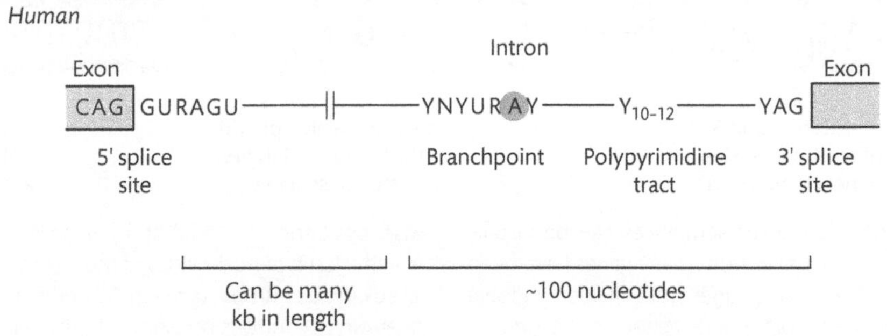

Figure 5.10 Conserved splice site sequences are found at exon–intron junctions. Conserved sequences at splice sites were first identified in 1981 after the alignment of a small number of genes with their mature transcripts. Splice sites were not all identical but fitted into a consensus sequence. Splice site consensus sequences for a human gene are shown here. Nucleotides which are the same in every splice site are abbreviated (according to the nucleotide A, C, G, or U). In some positions of the consensus nucleotides can be any pyrimidine (abbreviated Y) or any nucleotide (abbreviated N). An invariant A residue is found at the branchpoint (circled), and provides the 2′ -OH group used in the first catalytic step of splicing.

2 **A polypyrimidine tract** is found immediately upstream of the 3′ splice site; the polypyrimidine tract is usually a run of T residues, but can contain C residues.
3 **The branchpoint sequence**: in humans the branchpoint sequence has the consensus sequence YNYURAC (Y = pyrimidine, R = purine).

5.6.4 Degenerate splice site sequences can be visualized using pictograms

Although similar enough to fit into a consensus sequence, 5′ and 3′ splice site sequences still show variation. Although certain positions within splice sites are almost always the same nucleotide, others can be more variable. Another way of saying this is that splice site sequences are degenerate.

Because of this degeneracy, instead of writing an actual nucleotide sequence for a splice site it is sometimes more informative to illustrate each nucleotide combination that can occur at each site. This is done using a **pictogram** which shows the relative frequency of bases at each position. Pictograms for splice sites of several species are shown in Fig. 5.11. Notice that between species the pictograms contain slightly different sequences. For example, the branchpoint sequence for the yeast *S. cerevisiae* is much more conserved than that in humans (*H. sapiens*). The yeast branchpoint sequence follows a TACTAAC sequence consensus (phonetically pronounced WAC-WAC in pre-RNA, as the T is transcribed into a U in RNA to give the RNA sequence UACUAC). The polypyrimidine tract upstream of the 3′ splice site is strongly conserved in humans but is not found in *S. cerevisiae*.

For further background reading for this section, see references 16 and 17.

Key points

Eukaryotic genes are composed of introns and exons. As a result of splicing, exons are joined together at splice sites after they are transcribed into RNA: 5′ splice sites are at the 5′ end of introns, and 3′ splice sites are at the 3′ end of introns. The nucleotide sequences of splice sites show some variability but follow a consensus sequence. These conserved splice site sequences are essential for splicing; deletion of either the 5′ or 3′ splice site consensus sequence prevents splicing from happening.

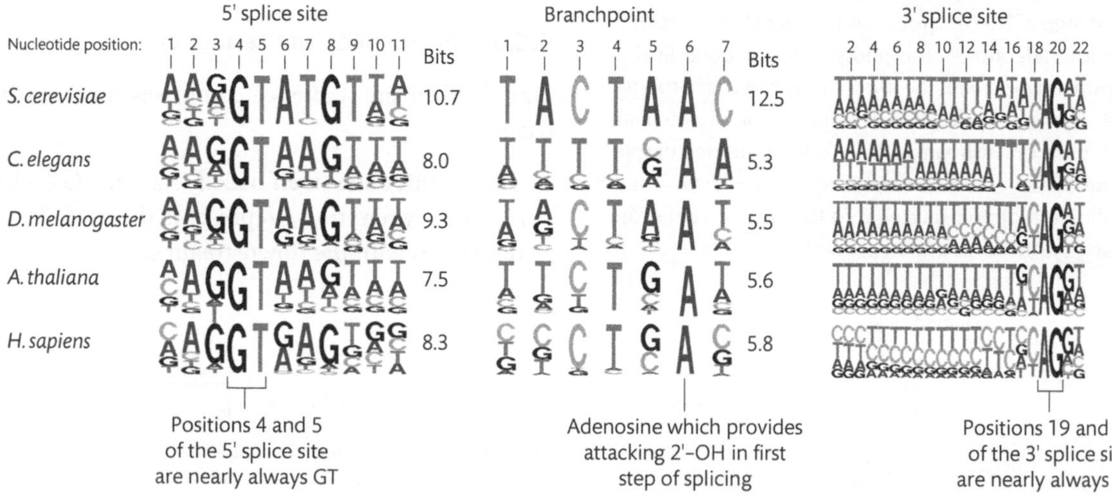

Figure 5.11 **Splice site consensus sequences can be visualized using pictograms.** Sequences from many genes have been used to compare splice sites on a large scale, including global comparison of splice sites from entire genomes. Pictograms show conserved sequences at exon–intron junctions and which nucleotides can vary at each position. The relative sizes of the nucleotides at each point in the pictogram provide an indication of how often these nucleotides appear at these positions. Positions 4 and 5 of the 5′ splice site consensus in the pictogram are almost always GT—the first two nucleotides of the intron. Likewise, positions 19 and 20 of the 3′ splice site are almost always AG—the last two nucleotides of the intron. This is because almost all eukaryotic introns start with GT and finish in AG (a minor class of eukaryotic introns not shown in these pictograms do not follow this GT–AG rule, and are discussed later in the chapter). On the other hand, the nucleotide immediately upstream of GT in the 5′ splice site is usually G, but can also more occasionally be A, T, or C. Hence for this position all four nucleotide possibilities are shown. (Reproduced with permission from Lim and Burge.[16] ©2001, National Academy of Sciences USA.)

5.7 The spliceosome

In this section we are going to start to introduce the large macromolecular complex where pre-mRNA splicing takes place—this is called the **spliceosome**. Spliceosomes assemble stepwise on pre-mRNAs to remove each intron. For example, a typical human gene contains eight exons, and so seven independent spliceosome complexes need to assemble to remove all the introns and join the exons together. After splicing is complete, each spliceosome disassembles, mRNA is released, and splicing components are recycled. Together, the full sequence of events in spliceosome assembly and subsequent recycling of its components is called the **spliceosome cycle**.

5.7.1 Small nuclear ribonucleoproteins (snRNPs) are essential components of the spliceosome

The reason why splice sites in pre-mRNA are so conserved is that the sequences at the exon–intron junctions are recognized by components of the assembling spliceosome (see Fig. 5.12). These include members of an important group of small nuclear RNAs (**snRNAs**) called U1 and U2 which base pair with the 5′ splice site and branchpoint respectively. Proteins are also important in early spliceosome assembly. In particular, proteins called U2AF65 and U2AF35 bind to the 3′ splice site, and help to stabilize U2 snRNA interactions (U2AF is the acronym for U2 Auxiliary Factor).

The full spliceosome also contains three other snRNAs called U4, U5, and U6. The U prefix is based on the high levels of uridine nucleotides within the sequence of these snRNAs. The number refers to how abundant these are in the cell, with U1 snRNA being the most abundant, U2 the next most abundant, and so on. Base-pairing interactions involving snRNAs are critical for assembling the spliceosome (see Box 5.2), and we shall

Box 5.2 RNA–RNA base pairing in the spliceosome is important for mediating interactions between its components

RNA–RNA interactions form through Watson–Crick base pairing (see Chapter 2), and play a key role in the spliceosome. Complementary RNA sequences in the spliceosome can efficiently hybridize together, even when they are diluted by much higher concentrations of competing non-complementary RNA sequences. This means that snRNAs can efficiently search complementary pairing partners within the nucleus. However, here are many sequence-specific nucleic-acid-binding proteins (e.g. bacterial restriction enzymes) which can cut nucleic acids, and other enzymes (ligases) which can join them together again. Exactly why the key role for RNA interactions has been conserved in the spliceosome will be explained later in this chapter.

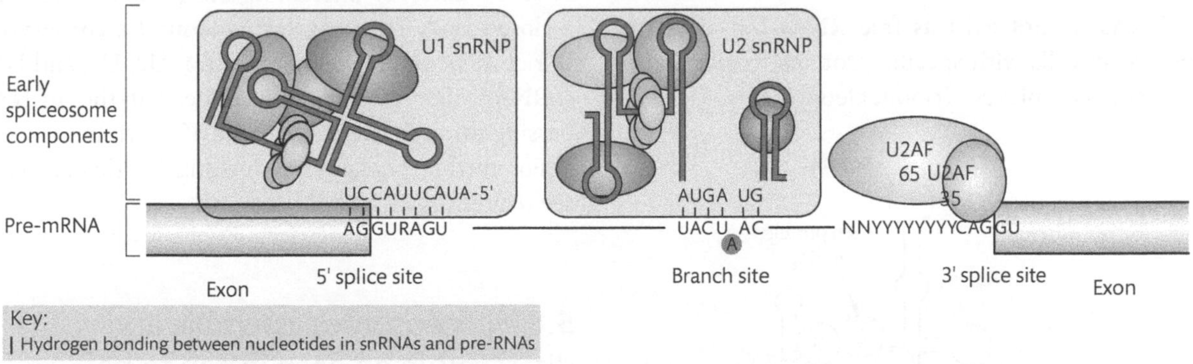

Figure 5.12 Consensus sequences at the 5′ splice site and branchpoint base pair with U1 and U2 snRNAs. U1 snRNA interacts with the 5′ splice site and U2 snRNA interacts with the branchpoint sequence of the pre-mRNA, and are important in early spliceosome assembly. These base-pairing interactions explain why the pre-mRNA 5′ splice site and branchpoint sequences are conserved—to maintain base pair complementarity with U1 and U2 snRNAs, respectively. The degree of base pair complementarity between splice sites and snRNAs also forms the basis of strong and weak splice sites (see Chapter 6). Two proteins also associate with the 3′ splice site—U2AF65 and U2AF35. (Adapted from Wang and Cooper[18] with permission from Macmillan Publishers Ltd.)

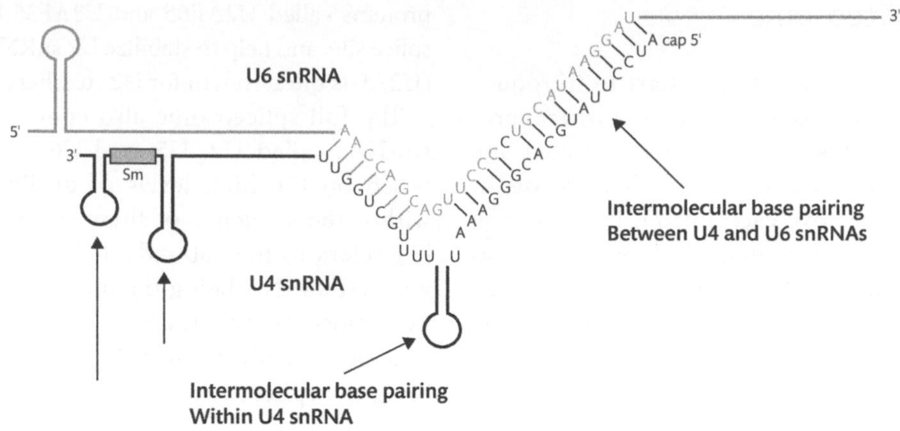

Figure 5.13 SnRNAs have both secondary stem and loop structures, and base pair with other molecules. For example we show the U4/U6 snRNAs which can interact by base pairing. (From Patel and Steitz[19], with permission from Macmillan Publishers Ltd.)

discuss this in the next section. snRNAs can base pair within themselves (to build up snRNA secondary structures), and between themselves (U4 is extensively base paired to U6 snRNA early in spliceosome assembly—Fig. 5.13 shows this intermolecular interaction as well as some intramolecular base pairing within U4 and U6 snRNAs). Notice that each of the pre-mRNA–snRNA and snRNA–snRNA contacts are short (only extending for part of the snRNA or pre-mRNA). This partial base pairing enables an snRNA to have more than one partner and so build up a network of interactions.

5.7.2 snRNAs are associated with proteins in cells to give snRNPs

snRNAs do not exist as free RNAs but are complexed in cells with specific proteins to give RNA–protein complexes (ribonucleoproteins (RNPs)) called snRNPs (small nuclear ribonucleoproteins, pronounced 'snurps').[20] An important group of snRNP proteins are the Sm proteins, which were first discovered in patients with autoimmune diseases. In these diseases patients develop antibodies against their own Sm proteins. The Sm group of proteins were so named as they were first recognized in auto-antisera from a patient called Stephanie Smith who had the autoimmune disease SLE (systemic lupus erythematosus). Rather than recognizing a single protein, Sm antisera recognize a number of nuclear snRNP proteins which each share a similar motif, called the Sm motif.

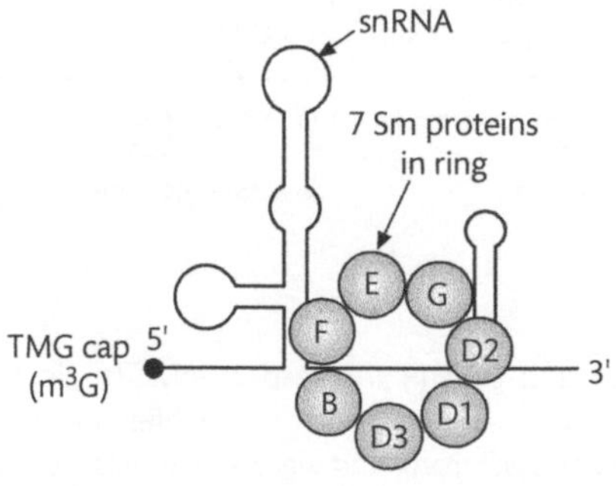

Figure 5.14 snRNAs are associated with proteins to give snRNPs. One group of these proteins is the Sm proteins, which form a closed loop around snRNAs. Notice also that snRNAs have a 5′ m^3G cap.

The arrangement of these Sm proteins on an snRNA molecule is shown in Fig. 5.14; notice a set of seven different interacting Sm proteins forming a closed ring. This ring forms around a conserved U-rich RNA sequence motif on U1, U2, U4, and U5 snRNAs called the Sm-binding site, with the snRNA passing through the central hole of the ring.

For further background reading for this section see references 21 and 22.

5.8 Spliceosomes assemble and disassemble on each intron to be removed in a spliceosome cycle

snRNPs make up the major building blocks of the spliceosome.[21,23–30] Since the nucleotide sequences of snRNAs are partially complementary to the conserved splice site sequences found within pre-mRNA and also sequences in other snRNAs, they are able to associate by base pairing to generate a framework of interacting molecules. These base-pairing interactions help the spliceosome to assemble, a bit like a jigsaw puzzle (see Box 5.2). With this information in mind, let us now look at the key steps in the spliceosome cycle (see Fig. 5.15).

5.8.1 Step 1: First association of proteins with newly transcribed RNA

Immediately after RNA is transcribed it associates with a group of RNA-binding proteins called the **hnRNP proteins** (see Chapter 6) to make an RNA–protein complex called the **H complex**. While some protein–RNA interactions in the H complex together may be non-specific, some key splicing regulators also bind to the transcript at this early point (see Chapter 6).

5.8.2 Step 2: Formation of the E (early) complex

The next step in spliceosome assembly is the formation of the **E complex** through snRNPs and proteins binding to the 5′ and 3′ splice sites.

- **The 5′ splice site: U1 snRNP base pairs with the 5′ splice site.** This is stabilized by a protein component of U1 snRNP called U1C. This is a general theme in spliceosome assembly: RNA–RNA interactions are often stabilized by proteins.
- **The 3′ splice site is bound by three proteins, SF1, U2AF65, and U2AF35, which interact with the branchpoint, polypyrimidine tract, and 3′ splice site, respectively.** These three proteins act to stabilize each other on the 3′ splice site: U2AF65 interacts with both SF1 and U2AF35.

The E complex is the earliest real step in spliceosome assembly (the **H complex** forms on any nuclear RNA). This makes the E complex biologically very significant, since choice of splice sites is established at this very early stage.

Key molecular event in creating the E complex

Addition of U1 snRNP to the 5′ splice site and RNA-binding proteins to the 3′ splice site

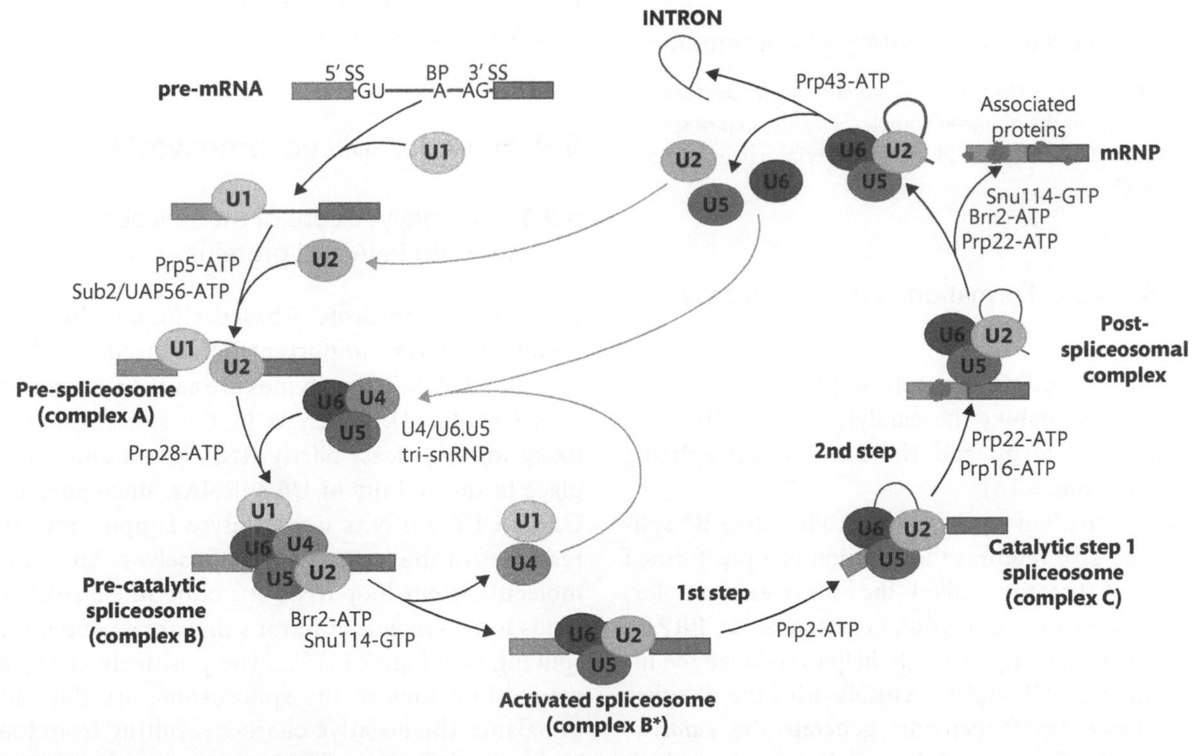

Figure 5.15 The spliceosome cycle. (From Wahl et al. [21])

5.8.3 Step 3: Formation of the A complex

To make the next stage in spliceosome assembly, the **A complex**, SF1 is displaced and replaced by U2 snRNP through base pairing with the **branchpoint sequence** in the pre-mRNA. Base pairing between U2 snRNP and the branchpoint is not perfect and causes the branchpoint adenosine to bulge out. The 2′ -OH group of this bulged branchpoint adenosine is involved in initiating the first catalytic step of splicing, so this bulge prepares it for its role in splicing. Binding of U2 snRNP to the branchpoint is stabilized by interactions with U2AF65 bound to the polypyrimidine tract.

Key molecular event in creating the A complex

Replacement of SF1 with U2 snRNP

5.8.4 Step 4: Formation of the pre-catalytic B complex

In the next step of spliceosome assembly, the **B complex** is formed by the addition of the U5•U4/U6 tri-snRNP.

Key molecular event in forming the B complex

Addition of the U4/U6•U5 tri-snRNP and its associated proteins, and the nineteen complex. The first step of splicing catalysis takes place in the catalytically active spliceosome B*

5.8.5 Step 5: Formation of the catalytically active spliceosome B*

U1 and U4 snRNPs are released from the spliceosome. This enables the catalytic core of the spliceosome to form, and the first step of splicing catalysis (Fig. 5.16).

The transition to the catalytically active B* spliceosome also requires the addition of a pre-formed group of proteins called the 'nineteen complex', named after one of its protein components, PRP19. The nineteen complex might help to stabilize the interactions of U5 and U6 snRNP with the 5′ splice site. These rearrangements generate the *catalytic core* for the first step of the splicing reaction which occurs in the **B* complex**. In the first catalytic step the 2′ -OH group of the branchpoint adenosine attacks the phosphodiester bond at the 5′ splice site to form a new 2′–5′ phosphodiester bond.

5.8.6 Step 6: Splicing catalysis

The second step of splicing catalysis takes place in the next complex in the spliceosome cycle, called the **C complex**, which contains the three remaining snRNPs (U2, U5, and U6) and in excess of 100 different proteins. The C complex lacks some of the proteins involved in early spliceosome formation, including U2AF35, U2AF65, and U1 snRNP components which are discarded earlier in the spliceosome cycle.

Key molecular event in forming the C complex

The second step of splicing catalysis

5.8.7 Step 7: Post-splicing disassembly

After splicing is complete, the mRNA is released by the spliceosome and exported into the cytoplasm, the lariat product is degraded, and the snRNPs are recycled to be used again.

5.9 How the spliceosome works

5.9.1 The catalytic core of the spliceosome contains both RNA and proteins

Most enzymes are protein-based, although there are a number of very important RNA-based enzymes as well (RNA-based enzymes are called **ribozymes** –see Chapter 3). Catalysis in the spliceosome is likely to be at least partly RNA based and takes place in the vicinity of U6 snRNAs, since purified U2 and U6 snRNAs can catalyse simple splicing reactions in the test tube by themselves. An intramolecular stem-loop RNA structure in U6 snRNA binds to two magnesium ions that are essential for splicing (see Fig. 5.16).[31,32] The positively charged magnesium ions in the spliceosome are thought to balance the negative charges resulting from the high concentration of RNA in the catalytic core,

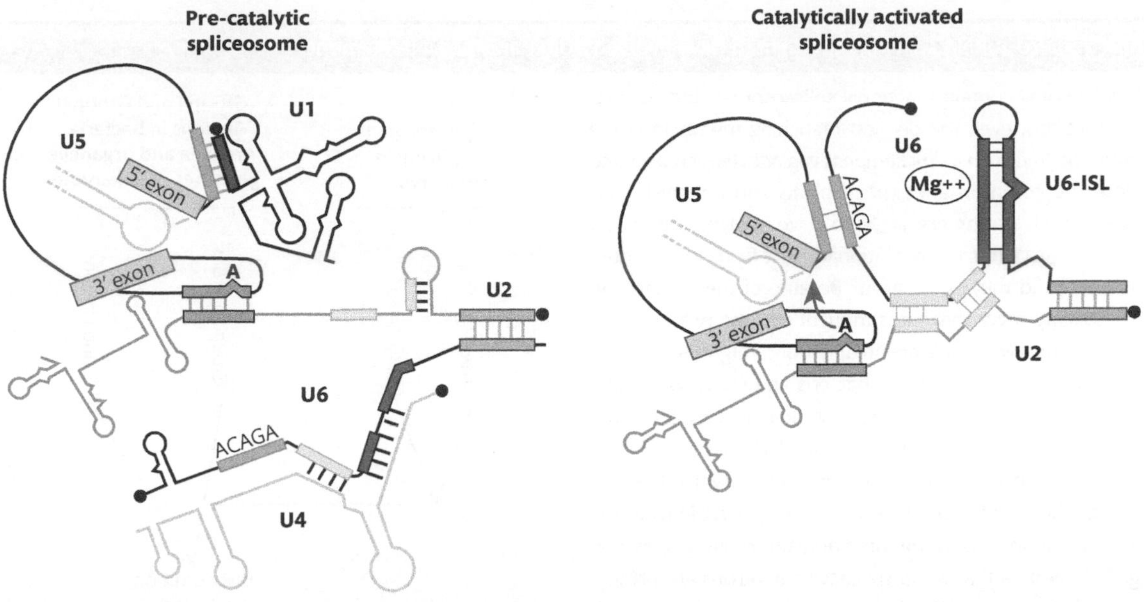

Figure 5.16 A switch in the snRNA content in the catalytically active spliceosome. Pre-catalytic spliceosome: The pre-catalytic spliceosome contains U1, U2, U4/U6 (notice that these are held together by base pairing independent of the pre-mRNA), and U5. U5 snRNP base pairs with exon 1 and exon 2, and holds these in the catalytic core using non-canonical base pairs. Hence U5 can base pair with different exon sequences and equally hold them in position. **Catalytically activated spliceosome**: A transition takes place to make the catalytic core of the spliceosome, which involves loss of U1 and U4. U6 snRNP base pairs with both the 5′ splice site of the pre-mRNA (which has been vacated by the exit of U1 snRNP) and U2 snRNP. Further base pairing of the pre-mRNA with U2 snRNP causes the branchpoint adenosine in the pre-mRNA to bulge out. At physiological pH the phosphodiester bonds of RNA are negatively charged (see Chapter 2), so would normally repel any additional attacking negatively charged oxygen groups. U6 snRNA binds to magnesium ions which balance this negative charge and stabilizes the leaving oxygen ion on the first exon. (Reproduced from Sontheimer et al.[27] with permission from Macmillan Publishers Ltd.)

and also to stabilize the negatively charged oxygen atom which is released after the *trans*-esterification reactions.

A group of self-splicing introns called **Group II introns** also follow the same two-step reaction pathway as spliceosomal introns (see Fig. 3.7 where these reactions are shown side by side). Group II intron splicing is known to be RNA-based since it can take place without proteins, with the RNA secondary structure provided by the introns themselves holding magnesium ions (and other divalent cations) in key catalytic positions (see Chapter 3). However, Group II introns fold themselves into an RNA-based catalytic core, which then catalyses their own excision (see Chapter 3). Modern spliceosomal introns do not themselves fold up into active catalytic sites, but instead their splicing excision uses folded RNAs provided by separate snRNA molecules.

Spliceosomal introns have protein components in their catalytic core, notably a protein called PRP8. RNA-binding domains in PRP8 physically contact the 5′ splice site, 3′ splice site, and branchpoint sequences. Although Group II introns can splice without accessory proteins in vitro, they are assisted by maturase proteins in vivo. Recent atomic resolution of the structure of PRP8 protein revealed protein domains (including reverse transcriptase and RNAse H domains) which are shared with the maturase proteins of Group II introns. These recent data further emphasize the links between Group II introns and spliceosomal pre-mRNA intron. In fact, Group II introns might have been the evolutionary precursors of modern-day pre-mRNA introns (see Box 5.3).

Box 5.3 Pre-mRNA splicing is thought to have evolved from parasitic DNA elements[15,33]

All eukaryotes contain functional spliceosomes and spliceosomal introns, even the deepest-branching members of the eukaryotic tree such as *Trichomonas vaginalis* (which diverged before the common ancestor of plants and animals). Thus spliceosomal introns are likely to have evolved either just before or just after the evolutionary split from prokaryotes, but where did they come from? Because of the similarity in their splicing mechanism it is thought that the precursors to modern eukaryotic spliceosomal introns might have been the Group II introns found in bacteria and some subcellular organelles. These are discussed more fully in Chapter 3, but briefly Group II introns are parasitic DNA elements which catalyse their own excision from the RNAs in which they are located. The distribution of Group II self-splicing introns and spliceosomal introns in the three domains of life is shown in Fig. 5.17. Notice that while spliceosomal introns are present only in eukaryotes, self-splicing introns are present in both bacteria and the Archaea as well as in subcellular organelles such as mitochondria and chloroplasts, which themselves are thought to have had a bacterial origin. Entry of Group II introns into the first eukaryotic cells might have occurred through endosymbiosis of bacteria in the evolution of eukaryotes; the acquisition of bacterial endosymbionts which contained Group II introns would have initiated intron spread through the very earliest eukaryotes, as well as the development of mitochondria and chloroplasts. Group II introns would have subsequently evolved into spliceosomal introns. Although they might first have evolved from parasitic DNA elements, spliceosomal introns are now very numerous in many eukaryotic genes. While translation in bacteria occurs at the same time as transcription, the key defining feature of eukaryotes is the presence of a nuclear membrane. The appearance of spliceosomal introns might have been one of the driving forces in the development of the nuclear membrane in eukaryotes, to prevent unspliced mRNAs from being translated since splicing is comparatively slow compared with translation. Introduction of spliceosomal introns may have driven the development of the nuclear membrane to provide distinct cellular compartments to enable, first, transcription and splicing of mRNAs in the nucleus and then translation in the cytoplasm.

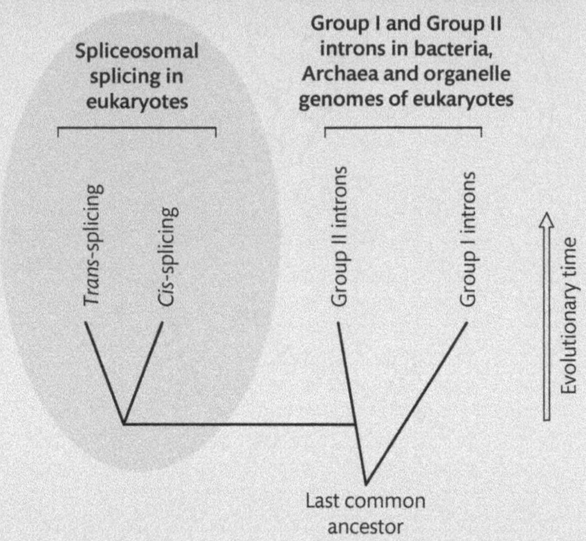

Figure 5.17 Evolutionary distribution of spliceosomal introns and group II introns.

Table 5.1 Stepwise addition of proteins drives assembly of the spliceosome

Motor	Function	Source	
PRP5*	Remodels U2 to bind BP		Early spliceosome
SUB2/UAP56*	Displaces SF1		
PRP28*	Releases U1 so U6 can bind 5′ splice site	U5 snRNP	Brought in by U5 to reorganize the spliceosome
Brr2*	Releases U4	U5 snRNP	
Snu114-GTPase	Releases U4, regulates Brr2	U5 snRNP	
PRP8	Part of catalytic domain, regulates Brr2	U5 snRNP	
PRP2*	Remodels B complex into catalytic B^{ACT}		Modelling active spliceosome
PRP16*	Remodels spliceosome into complex C		
PRP43*	Release of snRNPs from excised lariat		

*DEAD box helicases.

5.9.2 Stepwise addition of proteins drives assembly of the spliceosome

What drives the spliceosome cycle forward[21,34]? Although the snRNAs act as a framework to hold the spliceosome together through base pairing, and form much of the catalytic core of the spliceosome, spliceosome assembly is driven forward by the proteins which are added at each stage. These protein components drive the RNA and protein rearrangements, and are added to the spliceosome at specific times, ensuring that things happen in the correct order. A particularly important role is played by a group of helicases called *DEAD box helicases*. These helicases were so named because they contain a motif comprising Asp–Glu–Ala–Asp, or DEAD in the single-letter code for amino acids. DEAD box helicases use energy from ATP hydrolysis to rearrange inter- or intramolecular RNA structures or dissociate RNA–protein complexes.

The important proteins which drive the reorganization and assembly steps of the spliceosome assembly and where they work are shown in Fig. 5.15 (the DEAD box RNA helicases are asterisked in Table 5.1). Notice that each of the major steps after complex E requires the input of energy and the function of RNA helicases, so splicing is a fairly energy-consuming process. The energy consumption is shown in Fig. 5.15 as ATP or GTP, depending on which nucleotide provides the energy source. The individual jobs that these proteins do in reassembling the spliceosome are also summarized in Table 5.1. Notice that each of the DEAD box RNA helicases has the job of unwinding RNA helices, or RNA–protein interactions, to enable the next stage of spliceosome assembly to take place. While DEAD box helicase proteins play important roles over the whole of spliceosome assembly, to illustrate how they work we are going to look at the switch between complex A and complex B, in which U4/U6•U5 tri-snRNP is added into the assembling spliceosome.

U4/U6•U5 tri-snRNP addition adds three very important proteins which remodel the spliceosome to generate the *catalytically active spliceosome* B* which carries out the first step of splicing. These critical new proteins are called PRP28 and Brr2 (both DEAD box RNA helicases) and snu114 (a GTPase which modulates the activity of Brr2).

- **PRP28 releases U1 snRNP from the spliceosome** (this frees the 5′ splice site so that it can be bound by U6 snRNA in the catalytic core of the spliceosome).
- **Brr2 unwinds the base pairing holding U4/U6 snRNAs together.** This results in U4 snRNP being released from the spliceosome, while U6 becomes a key component of the spliceosome catalytic core. Snu114 protein controls the activity of Brr2, so the unwinding of U4/U6 only occurs at the correct time in spliceosome assembly.

In addition to the important role played by RNA helicases, other proteins also contribute to the assembly of the spliceosome, including proteins which increase the stability of different interactions. Many of the RNA–RNA interactions in the spliceosome are weak and need to be stabilized by proteins. For example, the interaction of U1 snRNP with the 5′ splice site occurs through hydrogen bonding, but is stabilized by a U1 snRNP protein called U1C, and another protein called U2AF stabilizes the association of U2 snRNP with the branchpoint. A further group of important proteins, called SR proteins, also stabilize interactions during spliceosome assembly, and we shall discuss these proteins further in the next chapter.

For further information for this section see references 35–37.

Key points

Splicing takes place in a large macromolecular machine called the spliceosome. Spliceosome assembly and disassembly follow a cyclical pathway. Catalytically active spliceosomes assemble on pre-mRNAs, splicing takes place, and after splicing the snRNPs and other spliceosomal components are recycled for the next step in splicing. Base pairing interactions within the spliceosome are also critical during splicing, and a group of proteins called RNA helicases dynamically regulate the RNA pairing partners during splicing. The very early steps of spliceosome assembly involve binding of U1 snRNP to the 5′ splice site and a group of proteins to the 3′ splice site. Next, U2 snRNP binds to the branchpoint, causing the branchpoint adenosine to bulge out so that it is ready to chemically attack the 5′ splice site. Addition of the U4/U6/U5 tri-snRNP to the spliceosome brings in a number of RNA helicases which rearrange the spliceosome to generate a catalytic site.

5.10 The spliceosome cycle has been worked out using in vitro extracts

The process of spliceosome assembly is understood in a great deal of detail. In this section we are going to look at some of the experiments which have provided this knowledge. An important contribution to the detailed understanding of the spliceosome has been made by analysis of the splicing of radiolabelled pre-mRNAs in cell extracts (Box 5.4). By monitoring the radiolabel, splicing reactions have been analysed to follow both the pre-mRNA (on denaturing gels), and assembly of spliceosome complexes (on native gels). In these experiments, the radioactive signal from spliced mRNA starts to appear after about 30 minutes of incubation in splicing extracts, although the lariat intermediate can be seen after about 20 minutes of incubation (Box 5.4). Each of the complexes which we discussed in the previous section appear in sequence in these experiments. First, the H complex appears (immediately after the radioactive pre-mRNA and cell extract are mixed). Then the E complex (in the gel shown this migrates with the H complex), the A complex, the B complex, and the C complex form progressively. The catalytically active C complex starts to appear 15–30 minutes after incubation of the radioactive pre-mRNA in splicing extract.

Proteomics have recently been used to identify many of the component proteins in human spliceosomes. Individual complexes have also been resolved by electron microscopy. The biochemistry of splicing is a conserved process in all eukaryotes. Spliceosomal exons are spliced together in single-celled yeasts in exactly the same way as they are in humans, and yeast genetics has been used to dissect spliceosome assembly. Together, all these experimental approaches have shown that the spliceosome is a huge complex in the cell comprising a number of different snRNAs and over 200 different proteins.

For further information about this section see references 38–46.

Box 5.4 In vitro analysis of pre-mRNA splicing has revealed the major steps in spliceosome assembly and reaction intermediates

A key tool is to use a radiolabelled pre-mRNA. To make the radiolabelled pre-mRNA, an intron-containing gene is first cloned into a plasmid containing a promoter (usually recognized by an RNA polymerase made by a bacteriophage promoter, either SP6 or T7 RNA polymerase). The plasmid is then incubated in a 'transcription buffer' containing nucleotides (NTPs) and either SP6 or T7 RNA polymerase to make a pre-mRNA. The pre-mRNA is labelled radioactively by including a radioactive NTP in the transcription mix. This radioactive pre-mRNA is then used to monitor the splicing reaction in a cell extract which contains all the components needed for splicing.

- **Splicing of the radiolabelled pre-mRNAs is visualized on denaturing polyacrylamide gels (Fig. 5.18).** These denaturing gels disrupt RNA-protein complexes, so each of the RNA substrates, intermediates, and splicing products discussed in Section 5.8 can be visualized. Different RNA molecules migrate to a different position on the gel according to their size and shape (the splicing intermediates and products containing lariats migrate unusually because of their loop structure; notice that they are located towards the top of the denaturing gel, even though they are smaller

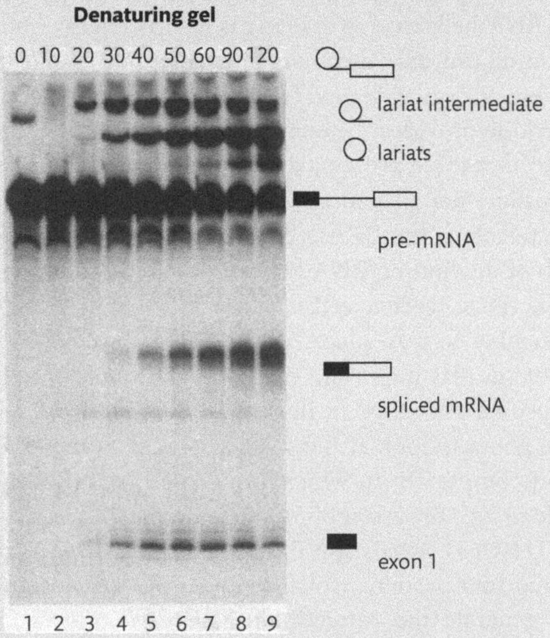

Figure 5.18 Visualization on denaturing polyacrylamide gel. (From Moore and Sharp[47].)

than the pre-mRNA). At time zero—before the reaction—only the pre-mRNA is visible on the film (lane 1 on the denaturing gel). After increasing incubation times in nuclear extract, a number of new radiolabelled RNAs are seen on the gel—these correspond to each of the splicing reaction products and intermediates expected from the splicing pathway (these RNA products are identified next to the gel).

- **Spliceosome complexes can also be visualized on native gels (Fig. 5.19).** Native gels do not disrupt stable protein-RNA contacts, so the different RNA-protein complexes associated with each stage of splicing can be seen. The first complex (complex H) forms immediately when RNA and proteins are mixed—notice that the H complex is present at the earliest time point taken (time zero, immediately after the RNA and proteins are mixed). Complex H is the smallest of the spliceosome complexes, and migrates towards the bottom of the native gel, but after a longer incubation in splicing extracts larger RNA-protein complexes called E, A, B, and C appear on the native gel, corresponding to more advanced stages in spliceosome assembly. The use of native gels to analyse RNA-protein complexes in splicing is useful since it enables the size of the complexes to be measured (by how far they migrate in the gel). Native gels can be further analysed by Northern blotting (to detect which complexes contain particular snRNAs) and by Western blotting (to detect which complexes contain specific proteins).

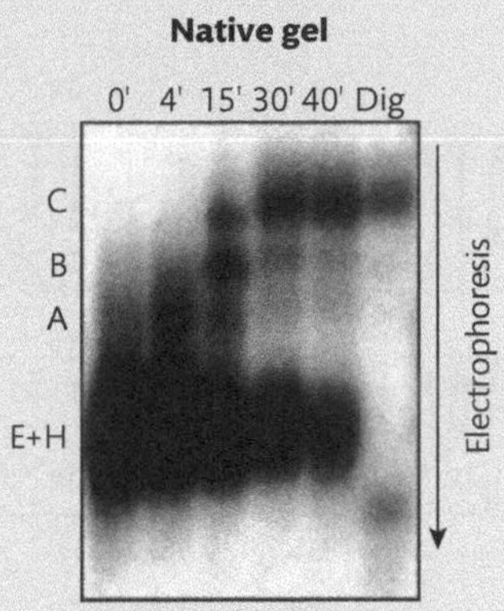

Figure 5.19 Visualization on native gel. (From Jurica and Moore.[40])

Key points

A considerable amount of analysis has been carried out on the splicing of simple exon-intron-exon pre-mRNA substrates. In these cases, spliceosomes assemble stepwise as complexes on pre-mRNAs.

5.11 A minor class of eukaryotic spliceosomal introns have different splice sites

Most eukaryotic spliceosomal introns—like the ones discussed in the previous sections—follow a 'GT–AG' rule, meaning that they start with the nucleotide sequence GT and finish with AG. However, in the 1990s a minor class of introns following a different 'AT–AC' rule (frequently starting with AT and finishing with AC) were discovered in metazoans and plants.[19] Because of their unusual splice site sequences, these rare introns were first given the catchy name 'AT–AC introns'. However, they have since been renamed *U12-dependent introns*, because while not all the introns in this class contain terminal AT–AC nucleotides, all are spliced by an alternative spliceosome which contains U12 snRNP. A second major difference of the U12-dependent introns is that they do not have a polypyrimidine tract upstream of the 3′ branch site. Pictograms comparing the splice site sequences of U2 snRNP-dependent and U12 snRNP-dependent introns are shown in Fig. 5.20.

U12-dependent introns are found in animals and plants, but not in the yeasts *S. cerevisiae* and *Schizosaccharomyces pombe* or in protists. In comparison, U2-dependent introns are found in all eukaryotic organisms. This means that the alternative spliceosome is ancient, and must have been conserved for at least a billion years of eukaryotic evolution. For conservation over such a long period of time, this second spliceosome must have a critical function. Consistent with a conserved and ancient function, U12 introns have in some cases been found in the same gene from very different species. One important parameter which makes U12 introns different is that some of them seem to be spliced more slowly, so their presence might be important for regulating gene expression.

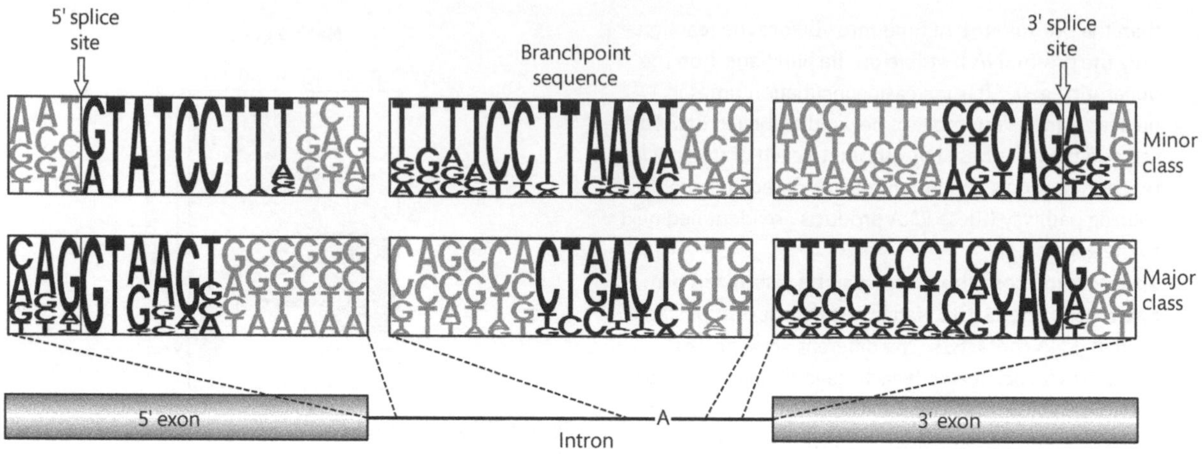

Figure 5.20 U2- and U12-dependent introns have slightly different splice sites. (From Patel and Steitz[19] with permission from Macmillan Publishers Ltd.)

U12-dependent introns are relatively rare (present at 0.15–0.34% of the level of U2-dependent introns). Although rare as a class, U12-dependent introns have been particularly found in genes which encode proteins involved in gene expression. As a rule, genes do not contain exclusively U12 introns or U2 introns; when they do occur, U12 introns are found in the same genes as the major group of U2-dependent introns.

5.11.1 Major and minor spliceosomes coexist in most eukaryotes

U12-dependent exons are spliced in a two-step splicing pathway biochemically identical to the U2-dependent exon pathway but catalysed by a different spliceosome called the U12 spliceosome or minor spliceosome which uses a different collection of snRNPs (see Tables 5.2 and 5.3). The minor spliceosome uses U11 snRNA for recognizing the 5′ splice site (so U11 is functionally equivalent to U1 snRNA) and U12 snRNA for recognizing the 3′ splice site (U12 binds to the branchpoint sequence to bulge out the branchpoint adenosine, so functionally is equivalent to U2 snRNA).

The minor spliceosome also has its own versions of the U4 and U6 snRNAs called U4atac and U6atac, and these play a similar role to U4 and U6 snRNA in the major U2-dependent spliceosome. Recent data show that $U6_{ATAC}$ snRNA stability is controlled by cellular signalling pathways to control inclusion of U12-dependent exons.[48]

The minor spliceosome is essential: fruit flies lacking the gene encoding one of the essential components of the minor spliceosome, U12 snRNA, fail to complete embryonic development. Although the snRNAs used in the minor spliceosome are distinct, they both use U5 snRNP, and are associated with many of the same protein components as the major spliceosome including PRP8.

Table 5.2 Major and minor spliceosomes

Major spliceosome	Minor spliceosome
In human cells the major splicing U snRNAs are fairly abundant (e.g. there are > 1 × 10^6 U1 snRNPs within the nucleus of a HeLa cell, which is about one-tenth of the cellular quantity of ribosomes). The cellular level of snRNPs within a species correlates with the level of splicing: snRNPs are much less abundant in *S. cerevisiae*, corresponding to the extremely low frequency of introns in this yeast.	U12-dependent introns are much rarer than U2-dependent introns, and each of the minor spliceosomal snRNAs are found at a much lower concentration of about 10^4 copies per cell.

Table 5.3 U2- and U12-dependent spliceosomes follow parallel assembly pathways, and have both parallel and shared components

Splicing complex	U2-dependent spliceosome snRNP component	U12-dependent spliceosome snRNP component
H complex	Very early complex which contains hnRNPs but no snRNPs.	
E complex	U1 snRNP bound to 5′ splice site.	
A complex	U1 snRNP bound to 5′ splice site and U2 snRNP to branchpoint.	U11 bound to 5′ splice site, U12 bound to branchpoint.
B complex	Entry of U4/U6•U5 snRNP	Entry of U6atac/U4atac•U5 snRNP
B* complex	Release of U1 and U4 snRNPs, U6 snRNA binding to the 5′ splice site and to U2 snRNA, bringing together the 5′ splice site and the branchpoint.	Release of U11atac and U4atac. U6atac snRNA binding to the 5′ splice site and to U12 snRNA, bringing together the 5′ splice site and the branchpoint.
C complex	Contains U2, U6, and U5 snRNPs and enables the free 2′ -OH group of the branchpoint adenosine to attack the 5′ splice site. Finally the two exons are bound and kept in alignment by U5 snRNA in the second catalytic step of splicing, in which the 3′ -OH group of exon 1 attacks the 3′ splice site.	Contains U12, U6atac, and U5 snRNPs and enables the free 2′ -OH group of the branchpoint adenosine to attack the 5′ splice site. Finally the two exons are bound and kept in alignment by U5 snRNA in the second catalytic step of splicing, in which the 3′ -OH group of exon 1 attacks the 3′ splice site.
Post-splicing complex	Recycles snRNPs	Recycles snRNPs

Key points

Most metazoan exons follow a GT-AG rule, although a second type of intron with different splice sites is recognized and spliced by a second spliceosome which contains a different but functionally overlapping set of snRNPs. Because this second spliceosome contains U12 snRNP instead of U2 snRNP it is called the U12 spliceosome, and the introns it splices are called U12-dependent introns. Splicing of U12-dependent introns follows the same two-step biochemical pathway as for U2-dependent introns. U5 snRNP is found in both the U12-dependent spliceosome and the U2-dependent spliceosome. U12-dependent introns are spliced more slowly than U2-dependent introns.

5.12 Spliceosomes can assemble through intron and exon definition

In the previous sections we have discussed how major and minor spliceosomes assemble across introns to splice together exons. This model of splicing is called intron definition—this means that the spliceosome first forms across the intron that is going to be spliced. However, in higher eukaryotes the spliceosome first recognizes exons within pre-mRNAs. This is called exon definition. These mechanisms of early spliceosome assembly are the topic of this section.

Exon and intron definition are important for gene expression. The in vitro splicing experiments discussed in the previous sections usually use a single short intron flanked by exons. In contrast, particularly in multicellular animals, genes typically have many small exons separated by huge stretches of intron. However, despite the vast amount of background intron sequence, spliceosomes still manage to make functional mRNAs from long pre-mRNAs by splicing together authentic exons. Spliceosomes achieve this feat by **exon definition.**

- In **exon definition** (Fig. 5.21, left-hand side) the exons are recognized first by binding of early spliceosome factors. In binding to the pre-mRNAs in this way, these early splicing components act like 'punctuation marks', telling the spliceosome where the exons are. Particularly important 'punctuation marks' are binding of the 3′ splice site by U2AF and binding of the 5′ splice site by U1 snRNP. Exon definition complexes are stable because of molecular interactions across the exon: U1 snRNP bound to the 5′ splice site on one side of the exon is stabilized by U2AF binding to the 3′ splice site on the other side of the exon, and vice versa. These molecular interactions are shown as double-headed arrows in Fig. 5.21, and are cooperative or synergistic interactions—they stabilize each other. Early exon definition is followed by assembly of fully fledged spliceosomes across introns. Although early stages of spliceosome assembly can take place through exon definition, the later stages of spliceosome assembly always involve interactions across the intron to be removed as splicing proceeds.

- In **intron definition** (see Fig. 5.21, right-hand side), early spliceosome components first assemble across the intron to be removed, followed by full spliceosome assembly and splicing. Notice in intron definition that interactions between 5′ and 3′ splice sites to be joined together take place straight away across the intron to be removed. In effect, the spliceosome defines the intron which is to be removed.

What are the advantages of recognizing exons through exon definition first, compared with the spliceosome recognizing introns first? The answer is that since exons are quite small they are 'easier' to identify as discrete units compared with most (considerably longer) vertebrate introns. Exon definition also means that short sequences in the transcriptome which resemble either 5′ or 3′ splice sites alone are not always selected by the spliceosome.

Exon definition also helps the spliceosome recognize bona fide exons. As we have seen, splice site

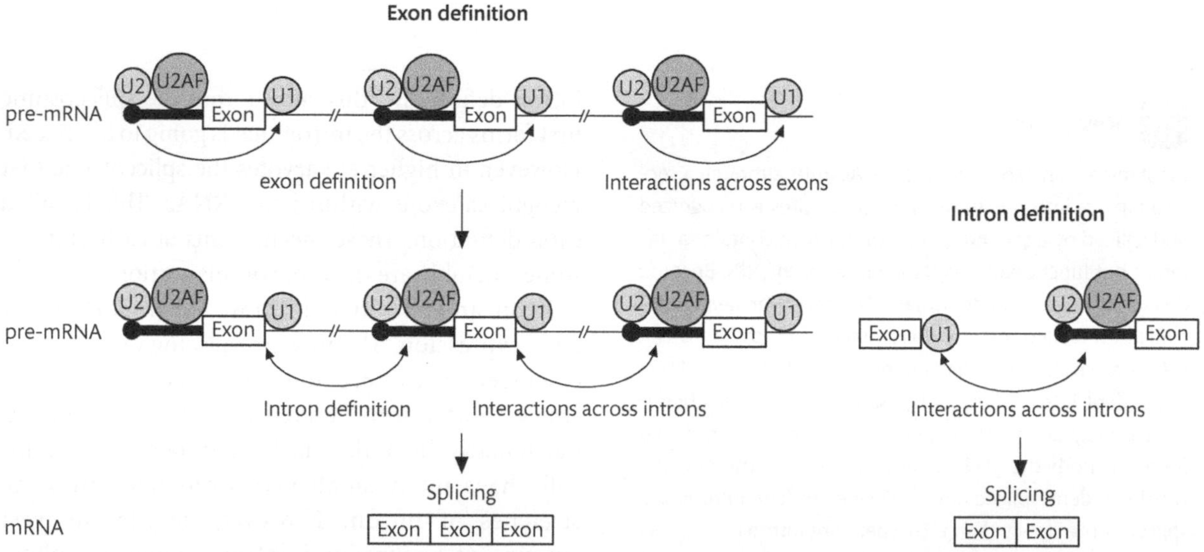

Figure 5.21 Exon and intron definition take place at the very earliest stages of spliceosome assembly. In intron definition (right-hand side), early spliceosome components interact straight away across the intron to be removed, followed by full spliceosome assembly and splicing. Exon definition (left-hand side) works differently in that the exons are recognized first. To achieve this, factors that associate with pre-mRNA early in spliceosome assembly recognize and bind to exons in pre-mRNA. Once bound, these factors act like 'punctuation marks' in the pre-mRNA, telling the spliceosome where the exons are. Particularly important 'punctuation marks' are binding of the 3′ splice site by U2AF and binding of the 5′ splice site by U1 snRNP. Although early stages of spliceosome assembly can take place through exon definition, the later stages always involve interactions across the intron to be removed as splicing proceeds. Note that the molecular interactions taking place early in spliceosome assembly are shown as double arrows.

sequences are short, and in fact sequences resembling splice sites can occur in genome sequences, yet these are not usually recognized by the spliceosome. Exon definition means to be recognized as exon sequences have to be flanked by both 5′ and 3′ splice sites. Single sequences in the transcriptome which resemble either 5′ or 3′ splice sites, but do not have a closely adjacent partner, are not selected by the spliceosome. Real exons are flanked by both 5′ and 3′ splice sites.

For background reading about the minor spliceosome see reference 19. For background reading about intron and exon definition see references 49 and 50.

5.12.1 The rules of exon versus intron definition

Sometimes the spliceosome uses intron and sometimes exon definition. What are the rules governing this?

Whether spliceosomes use intron or exon definition depends on whether genes have long or short introns. As a general rule, if introns are less than about 200–250 nucleotides long the spliceosome will use intron definition. If introns are longer than 250 nucleotides the spliceosome will use exon definition. With short exons it is easiest to set up cooperative interactions across the exon leading to stable exon definition complexes. With short introns it is easier for the interactions between components of the assembling spliceosome to occur straight away across the intron to be removed. Note from Fig. 5.21 that even when early spliceosome assembly takes place via exon definition, later molecular interactions in spliceosome assembly still occur across the intron to be spliced out.

Although exon definition is frequently used in humans and other animals, intron definition predominates in organisms which have very short introns such as the yeast *S. cerevisiae*. In this yeast, the first interactions in the assembling spliceosome take place across the intron to be spliced out. There are no prior interactions across the exon.

Thus understanding exon and intron definition is important for understanding how split eukaryotic genes are decoded into spliced mRNAs in different organisms. Exon definition is also particularly important for two other aspects of RNA biology which we will discuss later in the book. These aspects are flagged up in Table 5.4, and are where alternative exons are selected (Chapter 6) and in diseases where pre-mRNAs cannot be properly decoded (Chapter 7).

The **splicing code** also plays an important role in helping the spliceosome to differentiate between exons and introns. The splicing code is covered in more detail in the next chapter when we talk about alternative splicing.

Key points

The genes of most multicellular animals (metazoans) frequently contain short exons separated by very long introns, and very early spliceosome components interact across exons. Once exons have been 'tagged', fully fledged spliceosomes start to assemble across introns and exons to be spliced together. In species like the yeast *S. cerevisiae* with short introns, spliceosomes assemble straight away across the intron which is to be removed. As a general rule, irrespective of the species, if introns are shorter than about 200–250 nucleotides the spliceosome will use intron definition. If introns are longer than 250 nucleotides the spliceosome will use exon definition.

Table 5.4 Important impacts of exon definition on RNA biology

Process	Impact of exon definition
Alternative splicing (see Chapter 6)	Exon definition in multicellular animals means that alternative splicing occurs most frequently by the inclusion or exclusion of whole exons
Genetic disease (see Chapter 7)	Exon definition means that mutations in splicing signals often cause whole exons to be missed out of transcripts

Box 5.5 Biological functions of *trans*-splicing

The protozoan parasite *Trypanosoma brucei* causes the disease sleeping sickness in men and cattle, and is spread by the tsetse fly (see Fig. 5.22). All mRNAs in trypanosomes (encoded by ~9500 protein coding genes) are *trans*-spliced, but only two are known to be *cis*-spliced. *Trans*-splicing has two biological functions.

1. ***Trans*-splicing separates polycistronic mRNAs into individual translatable mRNAs.** Trypanosomes organize their genes in long polycistronic transcription units containing as many as 100 individual ORFs. Since translation normally initiates at the 5′ end of a eukaryotic transcript, and then stops once the first translational stop codon is reached, internal genes within a polycistronic transcription unit would not be well translated. *Trans*-splicing solves this by converting polycistronic mRNAs into separate mRNAs, each containing single ORFs and thus enables each to be translated. *Trans*-splicing also has this important function in the nematode worm *C. elegans*, in which a quarter of the genes are transcribed as long polycistronic transcription units.

2. ***Trans*-splicing adds a cap structure needed for translation.** Another idiosyncrasy of trypanosomes is that their mRNAs are transcribed by RNA polymerase I. In most eukaryotes mRNAs are transcribed by RNA polymerase II, and the 5′ cap structure of mRNA transcripts made by RNA polymerase II is important for translational initiation (see Chapter 11). The SL RNA is transcribed by RNA polymerase II, so addition of the SL mini-exon by *trans*-splicing provides trypanosome mRNAs with the 5′ cap structure needed for efficient translation.

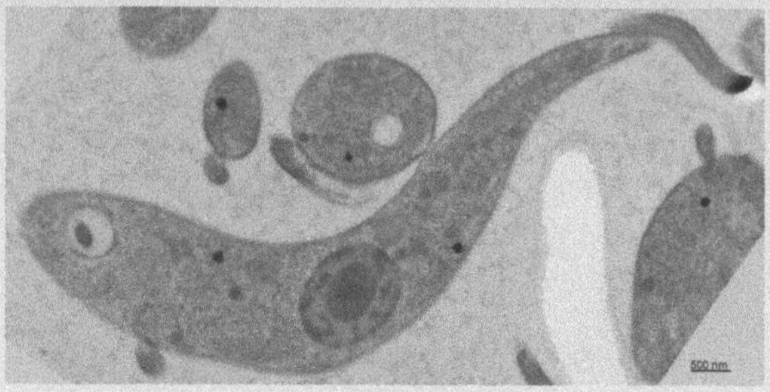

Figure 5.22 Electron micrograph of a trypanosome. Image taken from http://blog.wellcome.ac.uk/2010/10/27/ice-cold-trypanosome courtesy of Johanna Höög.

5.13 *Trans*-splicing is common in trypanosome parasites and in the nematode *C. elegans*, where it enables efficient translation

The spliceosomal-based splicing that we have discussed so far in this chapter joins exons together which were originally contained within the same pre-mRNA. This is called *cis*-splicing (see Fig. 5.1). There is another kind of splicing called *trans*-splicing, named since it actually splices together pre-mRNAs which are transcribed from separate genes. (The prefix *trans*- means separate molecules, while *cis*- means within the same molecule.)

We shall look at *trans*-splicing in this section. *Trans*-splicing has been found in eukaryotes from different lineages, suggesting that it has a similar ancient ancestry to *cis*-spliceosomal splicing. The frequency of *trans*-splicing versus *cis*-splicing is species dependent. In the parasitic protozoan *Trypanosoma brucei*, all genes are *trans*-spliced (so it is very frequent). In trypanosomes *trans*-splicing of an RNA sequence called the **spliced leader** (SL) is important for preparing mRNAs to be ready for translation (see Box 5.5). In humans and other metazoans, *trans*-splicing is rarer, and does not include a spliced leader RNA.

5.13.1 The chemistry of *trans*-splicing follows a similar two-step pathway to *cis*-splicing

The spliced leader RNA is transcribed from around 20 distinct genes by RNA polymerase II. The SL RNA contains a 39-nucleotide mini-exon and 5′ splice site, along with some downstream 'intron' (see Fig. 5.23).

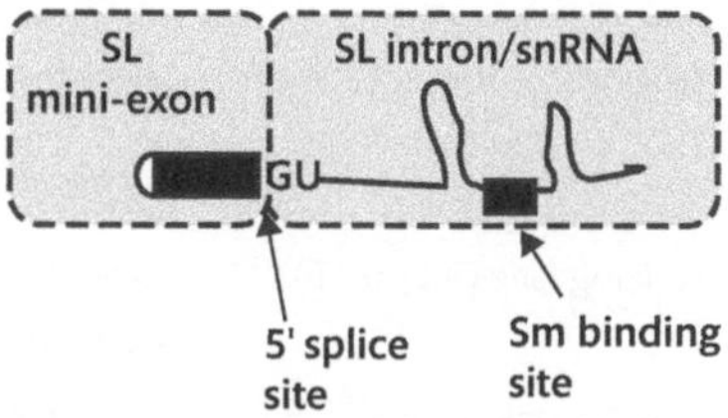

Figure 5.23 Structure of the SL mRNA used in trypanosome *trans*-splicing.

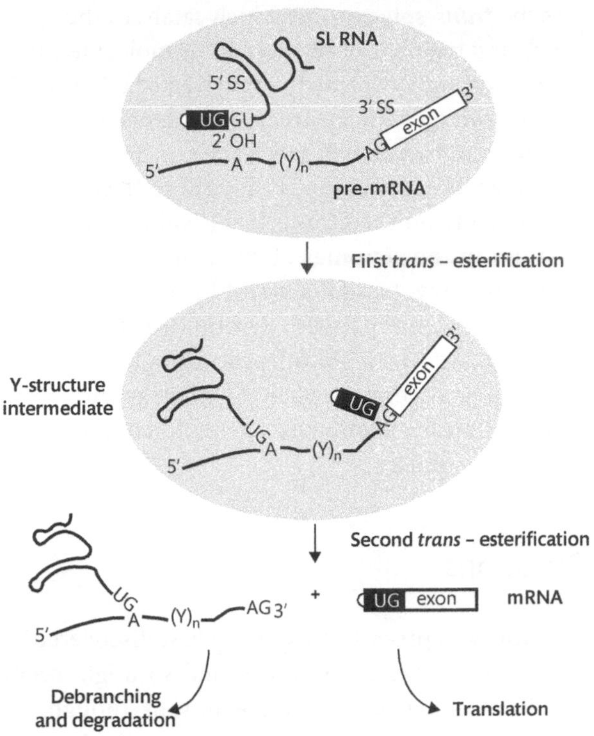

Figure 5.24. *Trans*-splicing follows a similar two-step chemical pathway to *cis*-splicing. *Trans*-splicing also requires a conserved branchpoint adenosine, a polypyrimidine tract, and 5′ and 3′ splice site. (This figure is taken from Preusser et al.[51])

This 39-nucleotide mini-exon becomes joined by *trans*-splicing to each of the ORFs encoded by the trypanosome genome in a two-step pathway (shown in Fig. 5.24).

- **The first step of splicing takes place between two physically separate RNA molecules.** The 2′ –OH group of the branchpoint sequence adenosine residue in one RNA molecule chemically attacks the 5′ splice site of the physically separate SL RNA. This releases the mini-exon of the SL, and results in a branched Y-shaped molecule which is equivalent to the lariat intermediate in *cis*-splicing.
- **In the second step of splicing** the 3′ –OH group at the end of the SL mini-exon attacks the 3′ splice site in exon 2, splicing together the SL and exon 2, and releasing the branched intermediate.

Notice the similarity of *trans*-splicing to *cis*-splicing. Both involve two steps and sequential *trans*-esterification reactions.

Trans-splicing is catalysed by a similar spliceosome to *cis*-splicing, including most of the same spliceosomal snRNAs. An interesting difference is that the *trans*-spliceosome does not contain U1 snRNA. Instead the U1 snRNA function is provided by the SL RNA itself. The function of U1 snRNP (initial recognition of the 5′ splice site) is carried out by the spliced leader RNA itself, which also binds Sm proteins like an snRNA (see Fig. 5.23). Trypanosomes still have a U1 snRNP, but it is needed only for splicing of the two genes which have conventional *cis*-exons, and is expressed at a much lower level than the spliced leader RNA.

For background reading about *trans*-splicing see references 51 and 52.

Key points

Trans-splicing joins together two initially separate RNA molecules. The 5′ splice site is in one RNA molecule and the 3′ splice site is in another. *Trans*-splicing takes place through a similar two-step mechanism to *cis*-splicing, and is catalysed by a spliceosome. The *trans*-spliceosome is very similar to the conventional *cis*-spliceosome, but lacks U1 snRNA. Instead, the function of U1 snRNA is provided by the upstream spliced leader RNA itself. *Trans*-splicing is very common in some parasitic protozoa and in the nematode worm *C. elegans*, where it plays an important role in making mRNAs which can be easily translated.

Summary

- Historically, introns were first discovered in the genome of adenovirus using R-looping experiments. These showed that adenoviral mRNAs were encoded in pieces called exons which were then stitched together. The intervening sequences were called introns.
- The majority of eukaryotic genes are split into exons and introns. Exons are spliced together in pre-mRNA by a large nuclear machine called the spliceosome.
- Introns mean that most eukaryotic genes are much longer than their final cytoplasmic mRNAs. The average human gene is 25kb long with eight exons, while its final spliced mRNA is only 1.5kb long. The longest human gene is

dystrophin, which is 2.5Mb long and it takes around 16 hours to transcribe.
- Introns are not junk. They contain transcription regulatory elements and harbour genes. The process of intron removal by splicing is important for RNA polymerase transcriptional elongation and the export of mature mRNA from the nucleus.
- The chemical reactions of splicing occur between nucleotides at exon–intron junctions in the pre-mRNA (the 5′ and 3′ splice sites) and the branchpoint inside the intron. These chemically important pre-mRNA positions contain conserved sequences recognized by complementary snRNAs which are components of the spliceosome.
- Splicing occurs through two successive *trans*-esterification reactions, which break and re-form the phosphodiester bonds which hold the RNA backbone together. These splicing reactions join exons together, and also produce a branched structure called a lariat product from the intron.
- Catalytically active spliceosomes assemble stepwise on pre-mRNA. U1 snRNP recognizes the 5′ splice site, and U2 snRNP and RNA-binding proteins recognize the 3′ splice site. Addition of U5 snRNP to the spliceosome brings in a number of kinases and helicases which remodel the spliceosome, in particular releasing U1 and U4 snRNPs. U5 snRNP also contributes the PRP8 protein to the spliceosome. The catalytic core of the spliceosome is largely RNA based, and contains U2 and U6 snRNAs as well as pre-mRNA and the PRP8 protein. The positively charged magnesium ion (Mg^{2+}) is important for stabilizing the charge on the negatively charged oxygen which forms the remnant of the broken phosphodiester bond. After splicing, spliceosomes disassemble and their components are recycled. This series of events is called the spliceosome cycle.
- There are three variant spliceosomes. The major spliceosome recognizes introns which start with GT and finish with AG, and contains the U1, U2, U4/U6, and U5 snRNPs. The minor spliceosome recognizes a variant set of introns, often starting with AT and finishing with AC. The minor spliceosome contains a different set of snRNPs with the exception of U5 snRNP which is shared between both. The third kind of spliceosome is the *trans*-spliceosome which catalyses the splicing together of separate RNA molecules. In the *trans*-spliceosome the function of U1 snRNP is provided by the upstream RNA molecule which contains the 5′ splice site.
- Spliceosomal splicing is thought to have evolved from the Group II self-splicing introns found in the genomes of some mitochondria and prokaryotes. This might have involved the invasion of parasitic Group II introns from newly installed mitochondria into the nuclear genome once the nuclear and mitochondrial genomes started to collaborate and consolidate.

Questions

5.1 How was pre-mRNA splicing first discovered?

5.2 Introns make genes longer and so might seem disadvantageous. Why are introns thought to have a positive effect on gene expression?

5.3 Describe how splicing takes place in two sequential *trans*-esterification reactions.

5.4 Describe how snRNPs associate with the pre-mRNA over the spliceosome cycle.

5.5 What are the functions of the proteins within the spliceosome in driving the spliceosome cycle forward?

5.6 How did the spliceosome evolve?

5.7 Distinguish between the major and minor spliceosomes.

5.8 Distinguish between exon and intron definition. Why is exon definition important in higher eukaryotes?

5.9 What is *trans*-splicing, and what does it achieve?

References

1. **Berget SM, Moore C, Sharp PA.** Spliced segments at the 5′ terminus of adenovirus 2 late mRNA. *Proc Natl Acad Sci USA* **74**, 3171–5 (1977).
2. **Sharp PA.** Split genes and RNA splicing. *Cell* **77**, 805–15 (1994).
3. **Sharp PA.** The discovery of split genes and RNA splicing. *Trends Biochem Sci* **30**, 279–81 (2005).
4. **Chow LT, Gelinas RE, Broker TR, Roberts RJ.** An amazing sequence arrangement at the 5′ ends of adenovirus 2 messenger RNA. *Cell* **12**, 1–8 (1977).

5. **Nott A, Meislin SH, Moore MJ.** A quantitative analysis of intron effects on mammalian gene expression. *RNA* **9**, 607–17 (2003).
6. **Ares M, Jr, Grate L, Pauling MH.** A handful of intron-containing genes produces the lion's share of yeast mRNA. *RNA* **5**, 1138–9 (1999).
7. **Jackson DA, Pombo, A, Iborra F.** The balance sheet for transcription: an analysis of nuclear RNA metabolism in mammalian cells. *FASEB J* **14**, 242–54 (2000).
8. **Juneau K, Miranda M, Hillenmeyer ME, et al.** Introns regulate RNA and protein abundance in yeast. *Genetics* **174**, 511–18 (2006).
9. **Le Hir H, Nott A, Moore MJ.** How introns influence and enhance eukaryotic gene expression. *Trends Biochem Sci* **28**, 215–20 (2003).
10. **McKenzie RW, Brennan MD.** The two small introns of the *Drosophila affinidisjuncta* Adh gene are required for normal transcription. *Nucleic Acids Res* **24**, 3635–42 (1996).
11. **Moore MJ.** Gene expression. When the junk isn't junk. *Nature* **379**, 402–3 (1996).
12. **Swinburne IA, Miguez DG, Landgraf D, Silver PA.** Intron length increases oscillatory periods of gene expression in animal cells. *Genes Dev* **22**, 2342–6 (2008).
13. **Swinburnel A, Silver PA.** Intron delays and transcriptional timing during development. *Dev Cell* **14**, 324–30 (2008).
14. **Tennyson CN, Klamut HJ, Worton RG.** The human *dystrophin* gene requires 16 hours to be transcribed and is cotranscriptionally spliced. *Nat Genet* **9**, 184–90 (1995).
15. **Roy SW, Gilbert W.** The evolution of spliceosomal introns: patterns, puzzles and progress. *Nat Rev Genet* **7**, 211–21 (2006).
16. **Lim LP, Burge CB.** A computational analysis of sequence features involved in recognition of short introns. *Proc Natl Acad Sci USA* **98**, 11 193–8 (2001).
17. **Mount SM.** A catalogue of splice junction sequences. *Nucleic Acids Res* **10**, 459–72 (1982).
18. **Wang GS, Cooper TA.** Splicing in disease: disruption of the splicing code and the decoding machinery. *Nat Rev Genet* **8**, 749–61 (2007).
19. **Patel AA, Steitz JA.** Splicing double: insights from the second spliceosome. *Nat Rev Mol Cell Biol* **4**, 960–70 (2003).
20. **Khusial P, Plaag R, Zieve GW.** LSm proteins form heptameric rings that bind to RNA via repeating motifs. *Trends Biochem Sci* **30**, 522–8 (2005).
21. **Wahl MC, Will CL, Luhrmann R.** The spliceosome: design principles of a dynamic RNP machine. *Cell* **136**, 701–18 (2009).
22. **Tycowski KT, Kolev NG, Conrad NK, et al.** *The Ever Growing World of Small Ribonucleoproteins.* Cold Spring Harbor, NY: Cold Spring Harbor Laboratory (2006).
23. **Abelson J.** Is the spliceosome a ribonucleoprotein enzyme? *Nat Struct Mol Biol* **15**, 1235–7 (2008).
24. **Gordon PM, Sontheimer EJ, Piccirilli JA.** Metal ion catalysis during the exon-ligation step of nuclear pre-mRNA splicing: extending the parallels between the spliceosome and group II introns. *RNA* **6**, 199–205 (2000).
25. **Selenko P, Gregorovic G, Sprangers R, et al.** Structural basis for the molecular recognition between human splicing factors U2AF65 and SF1/mBBP. *Mol Cell* **11**, 965–76 (2003).
26. **Sontheimer EJ.** The spliceosome shows its metal. *Nat Struct Biol* **8**, 11–13 (2001).
27. **Sontheimer EJ, Sun S, Piccirilli JA.** Metal ion catalysis during splicing of premessenger RNA. *Nature* **388**, 801–5 (1997).
28. **Turner IA, Norman CM, Churcher MJ, Newman AJ.** Roles of the U5 snRNP in spliceosome dynamics and catalysis. *Biochem Soc Trans* **32**, 928–31 (2004).
29. **Valadkhan S.** The spliceosome: a ribozyme at heart? *Biol Chem* **388**, 693–7 (2007).
30. **Valadkhan S, Manley JL.** Intrinsic metal binding by a spliceosomal RNA. *Nat Struct Biol* **9**, 498–9 (2002).
31. **Strobel SA.** Biochemistry: metal ghosts in the splicing machine. *Nature* **503**, 201–2 (2013).
32. **Fica, SM, Tuttle N, Novak T, et al.** RNA catalyses nuclear pre-mRNA splicing. *Nature* **503**, 229–34 (2013).
33. **Martin W, Koonin EV.** Introns and the origin of nucleus-cytosol compartmentalization. *Nature* **440**, 41–5 (2006).
34. **Brow DA.** Allosteric cascade of spliceosome activation. *Annu Rev Genet* **36**, 333–60 (2002).
35. **Bonnal S, Valcárcel J.** RNAtomy of the Spliceosome's heart. *EMBO J* **32**, 2785–7 (2013).
36. **Galej WP, Oubridge C, Newman AJ, Nagai K.** Crystal structure of Prp8 reveals active site cavity of the spliceosome. *Nature* **493**, 638–43 (2013).
37. **Valadkhan S, Mohammadi A, Jaladat Y, Geisler S.** Protein-free small nuclear RNAs catalyze a two-step splicing reaction. *Proc Natl Acad Sci USA* **106**, 11 901–6 (2009).
38. **Grabowski PJ, Seiler SR, Sharp PA.** A multicomponent complex is involved in the splicing of messenger RNA precursors. *Cell* **42**, 345–53 (1985).
39. **Hartmuth K, Urlaub H, Vornlocher HP, et al.** Protein composition of human prespliceosomes isolated by a tobramycin affinity-selection method. *Proc Natl Acad Sci USA* **99**, 16 719–24 (2002).
40. **Jurica MS, Moore MJ.** Capturing splicing complexes to study structure and mechanism. *Methods* **28**, 336–45 (2002).
41. **Konarska MM, Sharp PA.** Electrophoretic separation of complexes involved in the splicing of precursors to mRNAs. *Cell* **46**, 845–55 (1986).
42. **Krainer AR, Maniatis T, Ruskin B, Green MR.** Normal and mutant human beta-globin pre-mRNAs are faithfully and efficiently spliced in vitro. *Cell* **36**, 993–1005 (1984).

43. **Lamond AI, Konarska MM, Grabowski PJ, Sharp PA.** Spliceosome assembly involves the binding and release of U4 small nuclear ribonucleoprotein. *Proc Natl Acad Sci USA* **85**, 411–15 (1988).
44. **Neubauer G, King A, Rappsilber J, et al.** Mass spectrometry and EST-database searching allows characterization of the multi-protein spliceosome complex. *Nat Genet* **20**, 46–50 (1998).
45. **Rappsilber J, Ryder U, Lamond AI, Mann M.** Large-scale proteomic analysis of the human spliceosome. *Genome Res* **12**, 1231–45 (2002).
46. **Ruskin B, Krainer AR, Maniatis T, Green MR.** Excision of an intact intron as a novel lariat structure during pre-mRNA splicing in vitro. *Cell* **38**, 317–31 (1984).
47. **Moore MJ, Sharp PA.** Site-specific modification of pre-mRNA: the 2′-hydroxyl groups at the splice sites. *Science* **256**, 992–7 (1992).
48. **Younis I, Dittmar K, Wang W, et al.** Minor introns are embedded molecular switches regulated by highly unstable U6atac snRNA. *Elife* **2**, e00780 (2013).
49. **Berget SM.** Exon recognition in vertebrate splicing. *J Biol Chem* **270**, 2411–14 (1995).
50. **Black DL.** Finding splice sites within a wilderness of RNA. *RNA* **1**, 763–71 (1995).
51. **Preusser C, Jae N, Bindereif A.** mRNA splicing in trypanosomes. *Int J Med Microbiol* **302**, 221–4 (2012).
52. **Hastings KE.** SL *trans*-splicing: easy come or easy go? *Trends Genet* **21**, 240–7 (2005).

6 Regulated alternative splicing

Introduction

Alternative splicing increases the coding capacity of the genome by challenging the 'one gene–one protein' rule. While there are of the order of 21 000 genes in the human genome at least 90 000 proteins are made by a typical human cell. Therefore each human gene must encode more than one protein.[1,2] This is partly achieved by alternative splicing (see Fig. 6.1), and has started to challenge the very definition of a gene.[3,4]

This topic has received a lot of attention. In 2005, the journal *Science* published a list of the 125 most important questions then outstanding for science. Question number 25 was 'Why do humans have so few genes?'[6] Total protein coding gene numbers do not easily correlate with biological complexity. Even though humans are seemingly more complex organisms than the small plant *Arabidopsis thaliana*, both species have a similar number of genes (~21 000). The fruit fly *Drosophila melanogaster* has fewer protein-coding genes (14 000) than the nematode *Caenorhabditis elegans* (19 000), although the fly has a more complex physiology. Each of these gene numbers in multicellular organisms are not orders of magnitude different from the 6000 identified protein-coding genes in the single-celled yeast *Saccharomyces cerevisiae*. This discrepancy between gene copy number and apparent complexity has been called the *gene number paradox*. Organisms use alternative splicing to

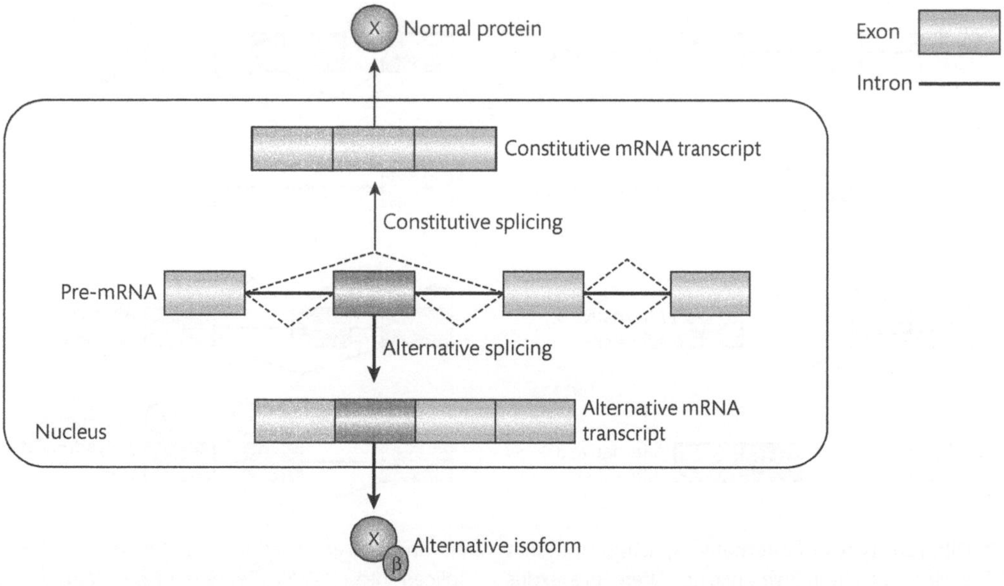

Figure 6.1 Alternative splicing pathways mean that more than one protein can be made from a gene. (From Rajan et al.[5] with permission.)

expand the information content of their genomes by enabling multiple mRNAs and proteins to be made from each gene.

6.1 There are several different types of alternative splicing

What is alternative splicing? Some exons are variably included into mRNAs—these are called **alternative exons**,[7–9] and this occurs through alternative splicing. There are seven distinct types of alternative splicing (see Fig. 6.2):

1 Alternative exons. This is the simplest situation and the most commonly found in humans. Whole exons are either spliced into mRNA or 'skipped' (this means left out).

2 Alternative 5′ splice sites. In this case a choice is made between possible 5′ splice sites for an exon (two alternative 5′ splice sites are shown in Fig. 6.2). Note that by selecting alternative 5′ splice sites a longer or shorter version of the same exon is spliced into mRNA.

3 Alternative 3′ splice sites. In this case a choice is made between two 3′ splice sites for an exon. Note that by selecting alternative 3′ splice sites a longer or shorter version of the same exon is spliced into mRNA.

4 Mutually exclusive exons. In this case a choice is made between exons. Either one or the other is included.

5 Intron retention. Although normally removed by splicing, whole introns can sometimes be retained in mRNAs (i.e. remain unspliced).

Alternative exons can also be included into mRNAs by:

6 use of alternative promoters

7 use of alternative poly(A) sites.

Although they make alternative mRNA isoforms, alternative promoters and polyadenylation sites are not strictly 'alternative splicing mechanisms' since they do not involve the spliceosome.

6.2 How frequent is alternative splicing?

Only a subset of exons are alternatively spliced—these are the **alternative exons**. Other exons are always included in mRNAs—these are called **constitutive exons**.

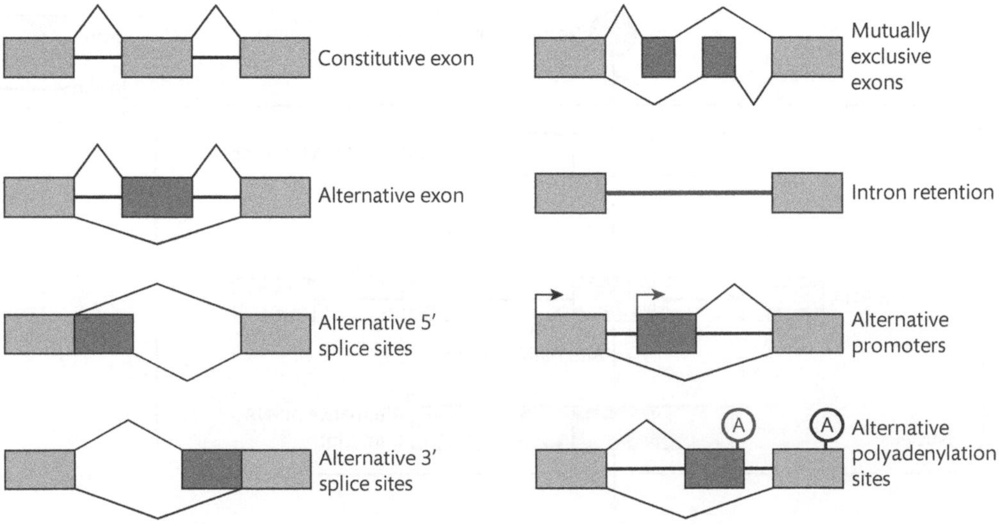

Figure 6.2 Different types of alternative splicing. Note that there are five kinds of alternative splicing. Alternative exons can also be introduced through the use of alternative promoters and polyadenylation sites. Constitutive exons are always spliced into mRNAs. (Reprinted from Ward and Cooper[7] with permission.)

6.2.1 Alternative splicing is frequent in animals and plants

Most human genes contain both constitutive and alternative exons, and so most single genes will produce several different mRNA transcripts through alternative splicing. Splice isoforms can be displayed on genome browsers, particularly for species with well-annotated genomes like human and mouse. As an example, the human *NASP* gene and its known mRNA isoforms on the human genome browser are shown in Fig. 6.3. Notice that this single *NASP* gene encodes three alternative mRNA isoforms through different alternative events.

Alternatively spliced mRNA isoforms made from the same gene will often differ from each other in both sequence and length as a result of different exon structures. Alternative splicing can be analysed by an experimental procedure known as **RT-PCR** (abbreviation of reverse transcription–polymerase chain reaction). The principles behind an RT-PCR experiment are shown on the left-hand side of Fig. 6.4. Primers are selected to hybridize to exon sequences flanking the alternative event. The downstream primer is used to prime reverse transcription to make cDNA which is amplified using the downstream primer and a second primer sequence located in an upstream exon. The resulting products are resolved by agarose gel electrophoresis. Actual data from an RT-PCR experiment technique to detect splicing inclusion of a cassette exon are shown on the right-hand side of Fig. 6.4—notice that this exon has different levels of splicing inclusion in different mouse tissues (different levels of the upper and lower products).

6.2.2 Alternative splicing patterns are revealed by '-omic' approaches

Recent technologies can profile the splicing patterns not just of single exons but of all exons in the transcriptome at the same time.[12–16] 'Transcriptomic' profiling of mRNA populations using approaches like microarrays and RNAseq (see Fig. 6.5) have shown that in the order of 95% of human genes encode alternatively spliced mRNAs (the transcriptome is the complete set of RNAs made from the genome). Alternative splicing is also frequent in plants, where it plays an important role in stress responses (including to drought and infection). Alternative splicing is less common in the yeast *Saccharomyces cerevisiae* (partly because of the lower frequency of introns in this species of yeast—see Chapter 5), but still plays an important biological role in this species. We shall come back to this later when we talk about splicing mechanisms.

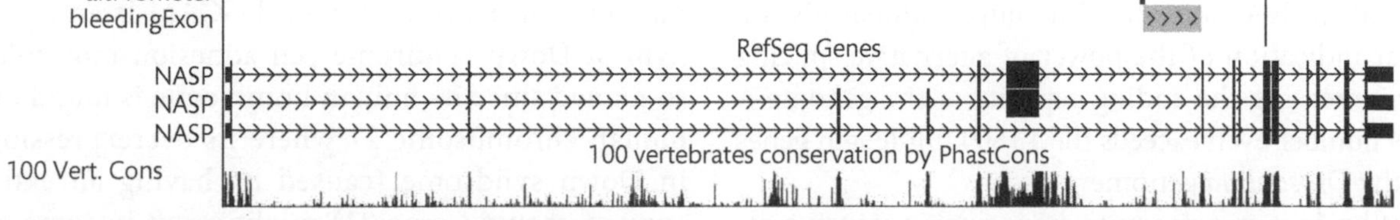

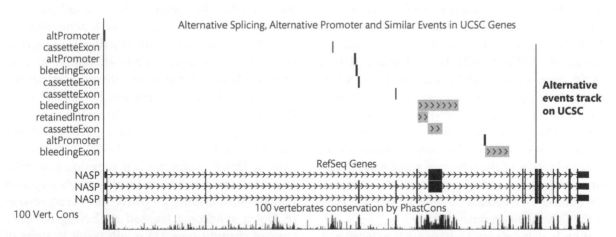

Figure 6.3 Information about different mRNA isoforms encoded by genes can be displayed on genome browsers.[10] As an example here we show the human *NASP* gene. Notice that this single *NASP* gene encodes three distinct mRNAs containing different exon structures (these are Refseq annotations—this means known human protein-coding and non-protein-coding genes taken from the NCBI RNA reference sequence collection). The exons are shown as vertical bars, with coding exons shown as thicker vertical bars than non-coding exon sequences. Introns are shown as lines. The sequences of mRNAs encoded by this single genetic locus are shown as 'RefSeq Genes'. Alternative events are separately annotated as 'Alternative Splicing, Alternative Promoter and Similar Events in UCSC Genes'. (This screenshot is downloaded from the UCSC genome browser.[10])

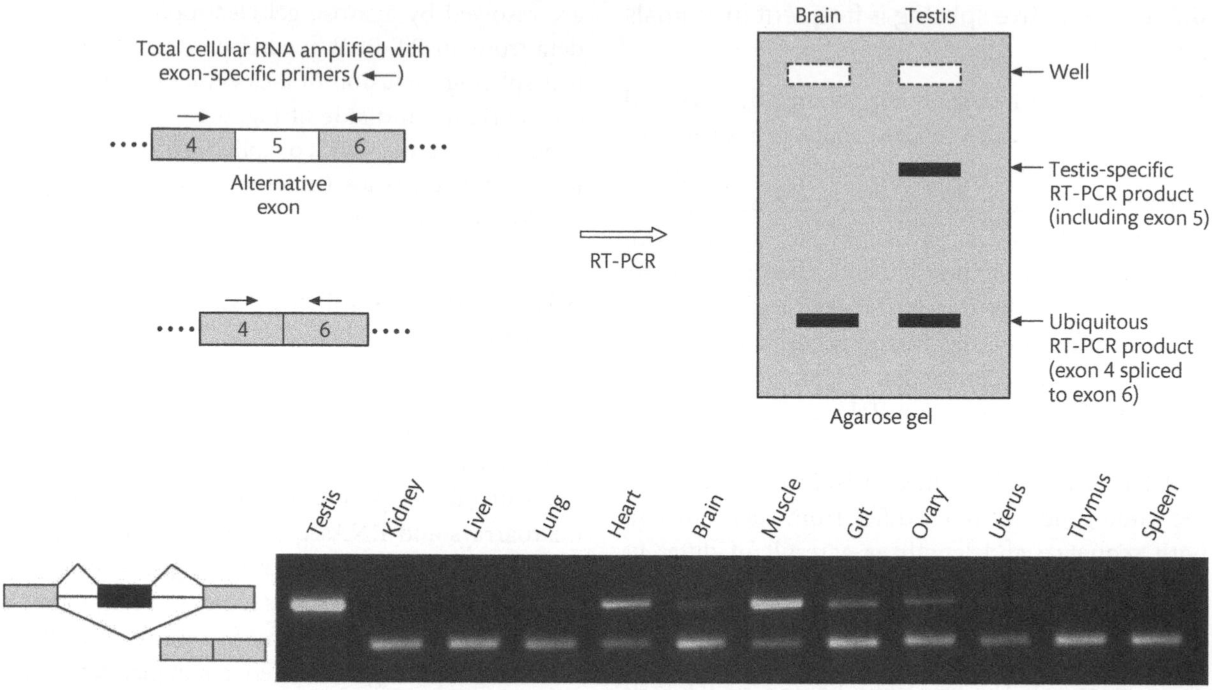

Figure 6.4 Alternative splicing patterns can be detected by RT-PCR. (From Elliott and Grellscheid.[11] ©2006 Society for Reproduction and Fertility.)

6.2.3 Alternative splicing of the fruit fly *DSCAM* gene produces more mRNA variants than there are genes in the fruit fly genome

Some genes encode huge numbers of mRNA isoforms. An extreme example is the fruit fly *DSCAM* gene.[17] Sequencing of individual mRNAs made from the *DSCAM* gene in the year 2000 showed that, although all *DSCAM* mRNAs were encoded by a single gene, these mRNAs were very extensively alternatively spliced. Following the specific rules about exon combinations shown in Fig. 6.6, *DSCAM* pre-mRNAs can be potentially spliced into over 38 000 mRNA isoforms. This huge number gives a clear indication of the power of alternative splicing for increasing the coding capacity of the genome—this number even exceeds the 13 600 full-length genes in the *Drosophila* genome.

Why is it necessary to have such extensive alternative splicing of the *DSCAM* transcript? *DSCAM* encodes a set of proteins important for axon guidance during development of the nervous system (see Fig. 6.6). (Axons are the long extensions of nerve cells that carry electrical impulses and follow elaborate pathways in the nervous system guided to their proper destination by a 'growth cone'. Once at their destination, axons form complex wiring connections with other nerve cells.) Axon growth is guided by proteins like the DSCAM protein, which is embedded in the cell membrane of the growth cone where it recognizes guidance signals and directs changes in cell growth. The high levels of alternative splicing of *DSCAM* contribute to the diversity in neuron connections that are important for normal brain development in the fruit fly. Alternative splicing is also very important in generating diversity in the mammalian nervous system. DSCAM is an acronym of Down syndrome cell adhesion molecule, so named since its human homologue is found on human chromosome 21 where its overexpression in Down syndrome (caused by having an extra copy of chromosome 21) might result in some of the resulting problems in brain function. As well as the developing nervous system, another region of the body that requires highly variable proteins is the immune system, and recent data also implicate DSCAM proteins with a role in cell-mediated immunity.[18]

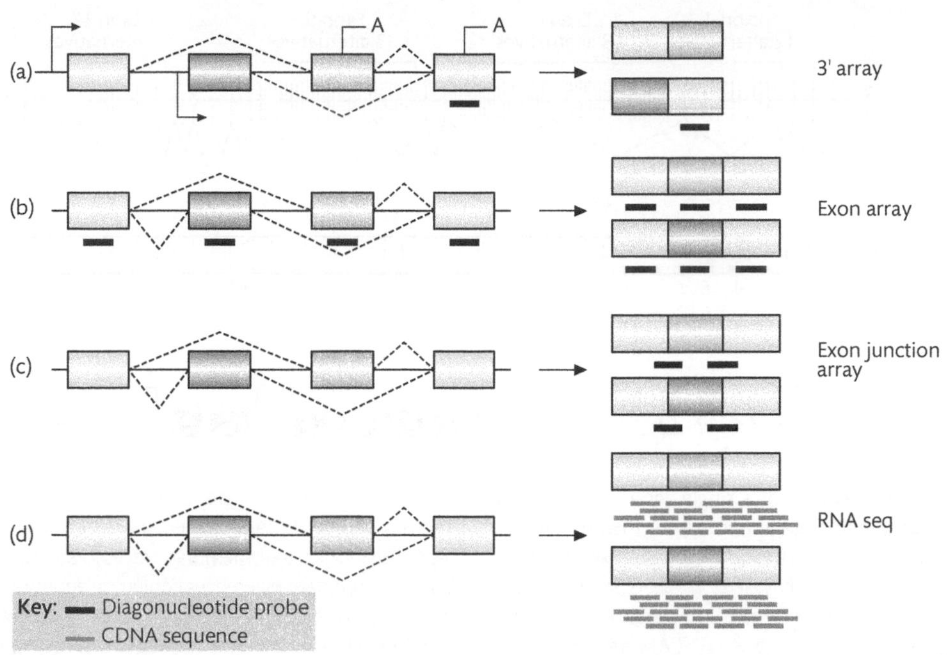

Figure 6.5 High-throughput '-omic' approaches developed to study RNA structure. Transcriptomic approaches analyse mRNA expression from large fractions of genes in parallel. One technology uses microarrays, which are glass slides containing thousands of single-stranded oligonucleotides that are complementary to specific RNAs. Microarrays sample RNA expression by hybridizing to fluorescently labelled RNA samples. (a) A common type of microarray is the 3′ UTR array, frequently used to compare transcript abundance in different samples using probes complementary to just one position on the target RNA, usually within the 3′ UTR. Other microarray platforms have been generated which can detect alternative mRNA splice isoforms. (b) Exon arrays: the second type of splicing array contains probes corresponding to all the exons in the human or mouse genomes (typically four probes per exon are used, and an average signal taken). (c) Exon junction arrays contain a series of oligonucleotides which are complementary to spliced exon junctions. These exon junction oligonucleotides will only stably hybridize to their RNA targets when they have spliced together in particular combinations of exon junctions to create a complementary sequence to the oligonucleotide probe. Usually exon junction arrays also contain additional internal exon probes which can detect the overall frequency with which each exon is spliced. (d) An alternative method for massive parallel analysis of transcriptomes is high-density 'deep sequencing' of transcriptomes, also called RNAseq. RNA is first converted to cDNA, and then this cDNA is sequenced at near-complete coverage of all individual molecules to give the complete sequences of all the transcribed RNAs (including alternative splice forms) in the starting RNA mixture. (Reprinted from Rajan et al.[5] with permission from Macmillan Publishers Ltd.)

Key points

Alternative splicing is a mechanism for producing different mRNA isoforms from the same gene, and helps the genome encode many more proteins than it contains genes. In some species different alternative mRNA isoforms are annotated on genome browsers. By enabling more than one mRNA to be made from each gene, alternative splicing also vastly extends the coding capacity of the human genome. This enables many more proteins to be made than the approximately 21 000 full-length protein-coding genes in the genome. An extreme example of alternative splicing is found in the fruit fly gene *DSCAM* which can produce more alternative transcripts from a single gene than there are actual genes encoded in the fruit fly genome.

6.3 How exons are recognized by the splicing machinery

Before we look in more detail at how alternative splicing is regulated, we first need to take a step back and discuss more about how the spliceosome distinguishes between introns and exons. We shall look at this in the next two sections.

Distinguishing between exons and introns represents a challenge for the spliceosome. Typical human genes have short exons (often around 150 nucleotides long); while introns can be thousands of nucleotides in length (see Fig. 6.7). Moreover, introns can contain sequences called pseudoexons which

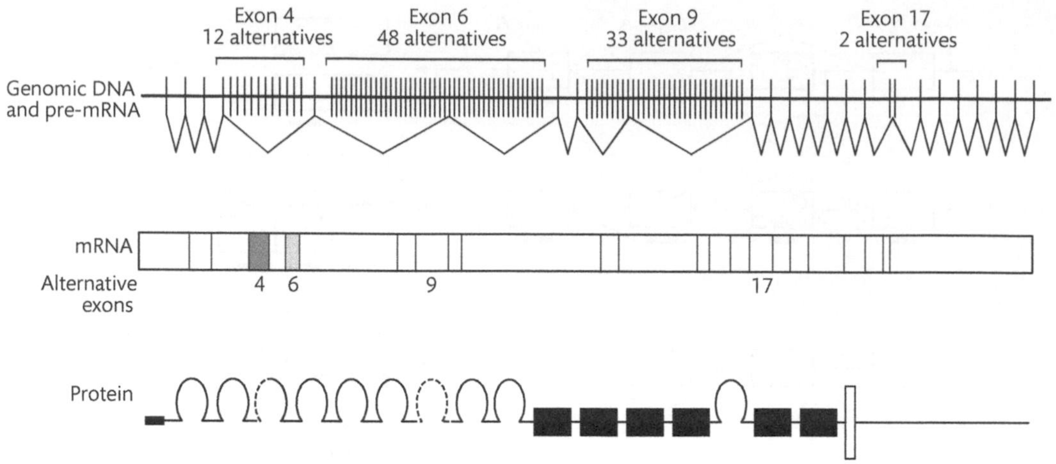

Figure 6.6 Alternative splicing of a single gene in the fruit fly can actually exceed the total gene number for this species. The *DSCAM* gene is expressed from a multiple-exon gene. After splicing, the mature *DSCAM* mRNA contains 24 exons, but is first transcribed as a much more complicated pre-mRNA. Alternative splicing results in multiple different versions of the *DSCAM* transcript being made, each of which contains the same total number of exons but different combinations of exons 4, 6, 9, and 17. The spliceosome includes only one of 12 different versions of exon 4, 48 different versions of exon 6, 33 different versions of exon 9, and two different versions of exon 17. Alternatively spliced *DSCAM* mRNA isoforms are translated into transmembrane receptors similar in overall shape but with important differences conferring enormous variability into the wiring pattern of the *Drosophila* brain. Three of the alternative exons (exons 4, 6, and 9) are in regions encoding immunoglobulin domains important for protein interactions and therefore might affect the way that axons grow and make connections with other cells. (Reprinted from Black[17] with permission from Elsevier.)

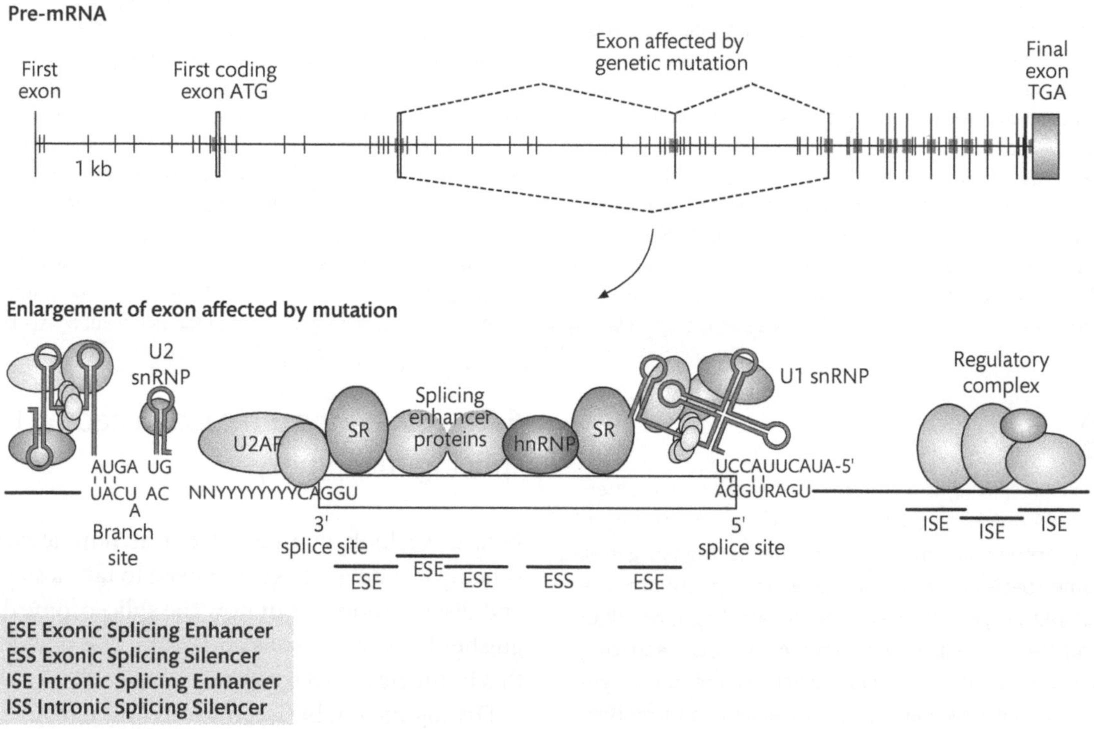

Figure 6.7 The splicing code comprises binding sites for splicing activator proteins (splicing enhancers) and splicing repressor proteins (splicing silencers). A multi-exon gene is shown at the top, with exons as vertical lines and introns as horizontal lines. Pseudo-exons are shown as short vertical lines. The lower diagram focuses in detail on the components binding to the pre-mRNA in and around an alternative exon. Notice that U1 snRNP binds to the 5′ splice site, and U2 snRNP binds to the 3′ splice site. Other components bind to splicing enhancers or silencers to activate or repress exon splicing, respectively. (From Wang and Cooper[20] with permission.)

have splice site sequences but are not recognized as exons by the spliceosome.

6.3.1 The splicing code comprises binding sites for nuclear RNA-binding proteins embedded in transcribed sequences

The process through which spliceosomes distinguish between exons and introns is assisted by a code embedded in the nucleotide sequence of the pre-mRNA itself. This **splicing code** is the topic of this section.[1,17,19,20]

It was originally thought that the only sequences needed for splicing were at the splice sites (splice site sequences were described in Chapter 5). However, groundbreaking experiments in the 1980s showed that sequences deep within the exon were also needed for exon splicing (Box 6.1). It is now known that important nucleotide sequences within both exons and introns control splicing of most exons, in addition to splice sites. Together these intron and exon sequences are called the splicing code, which is critically needed for gene expression (analogous to the genetic code that reads triplets of nucleotides to encode amino acids).

Box 6.1 Discovery of the splicing code

The first indication of a splicing code came from experiments performed in the 1980s by Francisco Baralle and colleagues (now at the ICGEB, Trieste) on an exon within the fibronectin gene called the ED exon, and by Tom Maniatis and Robin Reed at Harvard University working on β-globin. The fibronectin ED exon had consensus 5′ and 3′ splice sites, so should be strongly recognized by the spliceosome. The expectation at the time the experiment was done was that just the splice sites and not the internal portion of the ED exon would be needed for splicing. Experiments proved this expectation wrong. When the central portion of the exon was cut out, splicing of the fibronectin exon was prevented. The reason that this deletion prevented exon splicing was that the internal region of the fibronectin ED exon contains binding sequences for specific nuclear RNA-binding proteins; binding of these proteins is needed to enable spliceosome assembly and splicing of the ED exon to take place. Such RNA-protein-binding sites are not limited to the fibronectin ED exon. Most exons are now known to contain binding sites for RNA-binding proteins needed for normal splicing, collectively known as the splicing code.

The splicing code is made up of enhancers and silencers. The relative positions of these sequences are shown in Fig. 6.7. *Splicing enhancers* activate the splicing of their associated exons, and are either present within exons (exonic splicing enhancers, abbreviated **ESEs**) or within introns (intronic splicing enhancers, abbreviated **ISEs**). *Splicing silencers* in contrast repress exon splicing, and are either within exons (exonic splicing silencers, abbreviated **ESSs**) or within introns (intronic splicing silencers, abbreviated **ISSs**). Splicing enhancers and splicing silencers play an important role in helping the spliceosome distinguish between exons and introns in pre-mRNA. The splicing code also helps the spliceosome to decide between real exons and pseudoexons. Pseudoexons are not spliced by the spliceosome partly because they do not contain splicing enhancers, even if they do have splice sites.

Both alternative and constitutive exons depend on the splicing code as well as their consensus splice site sequences for recognition by the spliceosome. Since they are needed for gene expression, the nucleotide sequences of splicing enhancers and repressors are often highly conserved across evolution. Exonic splicing enhancers usually have to contain codons encoding amino acids as well. Even though they do not encode amino acids, the intron sequences flanking alternative exons are often highly conserved by up to 200 nucleotides into the intron.

6.3.2 The splicing code is deciphered by RNA-binding proteins

Splicing enhancers and silencers function by binding proteins that activate and repress exon splicing, respectively. Some examples of important splicing activator and repressor proteins are shown in Table 6.1. Splicing silencers bind proteins that repress splicing, and splicing enhancers bind proteins that activate splicing. Amongst splicing activators, a group of proteins called the **SR proteins** frequently bind to ESEs and help stabilize the association of spliceosome components with the exon. Other splicing activators (e.g. the TIA1 protein in Table 6.1) often bind just downstream of 5′ splice sites within introns, and stabilize the interaction of U1 snRNP with upstream exons.

Table 6.1 Types and examples of splicing regulatory protein*

Class	Function	Examples†
SR and SR-related proteins	Typically activate splicing by recruiting components of the splicing machinery	nSR100 (SRRM4), SC35, SF2 (ASF), SRM160 (SRRM1), SRp30c, SRp38, SRp40, SRp55, SRp75, TRA2α, TRA2β
hnRNPs	Typically repress splicing by a variety of poorly understood mechanisms	hnRNP A1, hnRNP A2/B1, hnRNP C, hnRNP F, hnRNP G (RBMX), hnRNP H, hnRNP L, nPTB (PTBP2), PTB (PTB1)
Other RNA-binding proteins	Activate or repress splicing U1snRNP is essential for constitutive splicing but can also repress splicing	CELF4 (BRUNOL4), CUGBP, ESRP1, ESRP2, FOX1 (A2BP1), FOX2 (A2BP2), HuD, MBNL1, NOVA1, NOVA2, PSF (SPFQ), quaking, SAM68 (KHDRBS1), SLM2 (KHDRBS3), SPF45 (RBM17), TIA1, TIAR (TIAL1), U1 snRNP

*Some of the important proteins that activate or repress exon selection in pre-mRNAs are shown here. As well as the RNA-binding proteins that decipher the splicing code, the most recent data indicate that changes in the concentration of core spliceosomal proteins can also result in changes in splicing patterns.

†Synonyms are given in parentheses.

From Nilsen and Gravely[2] with permission.

Splicing repressor proteins include the group of proteins called **hnRNPs** (heterogeneous ribonucleoproteins). Splicing repressors typically block recognition of exon sequences by the spliceosome, either by sequestering exons in looped RNA structures or by creating zones of silencing by coating RNA with repressor molecules. In both cases this is thought to prevent other proteins from binding to the RNA, or protein–protein interactions across them that would be needed for the spliceosome to assemble. SR proteins and hnRNPs are discussed in more detail in Chapter 4.

The nuclear concentrations of RNA splicing regulators help maintain proper splicing patterns in the transcriptome. Many splicing regulator proteins, including SR and hnRNP proteins, actually control their own pre-mRNA splicing patterns (see Fig. 6.8). In the case of SR proteins, increases in SR protein concentration activate non-coding exons called poison exons.[22] These poison exons are so called because when they are spliced into the mRNA, they introduce premature translation termination codons (PTCs). These PTCs prevent translation of the mRNA encoding the SR protein, and also destabilize it (by targeting the mRNA down a decay pathway called NMD (see Chapter 12)). Poison exons also contain many ESE sequences which bind to the SR protein encoded by their gene. If SR protein gene expression becomes too high, the increased cellular concentration of SR protein activates splicing inclusion of the poison exons. This adds stop codons to the mRNA, thereby blocking the production of more SR protein.

A slightly different mechanism (although with the same result) operates for the hnRNP proteins.[23] In this case, since hnRNPs are usually splicing repressors an increase in their concentration blocks splicing of specific exons into their mRNAs (see Fig. 6.8). Loss of these important exons by splicing repression similarly causes these mRNAs to contain frameshifted reading frames, introducing premature stop codons and causing mRNA decay. This block in splicing leads to the same effect—a decrease in the concentration of hnRNP proteins.

These mechanisms of splicing factor autoregulation are probably very important. An indication of this is given by looking at the sequences in and around the regulated poison exons in SR protein encoding genes, and in the skipped exons in the genes encoding hnRNPs. Very high levels of nucleotide conservation are seen. Note that these examples of splicing autoregulation by RNA-binding proteins also show that not all alternative exons need to encode proteins—in this case the effect of including poison exons actually stops translation.

6.3.3 The splicing code can be predicted online

Given an input sequence it is possible to predict which nucleotides will be involved in controlling splicing using online sites. These predictions are based on the experimentally determined binding sites for RNA-binding proteins like SR proteins, and short-sequence motifs enriched in exons (predictive of splicing

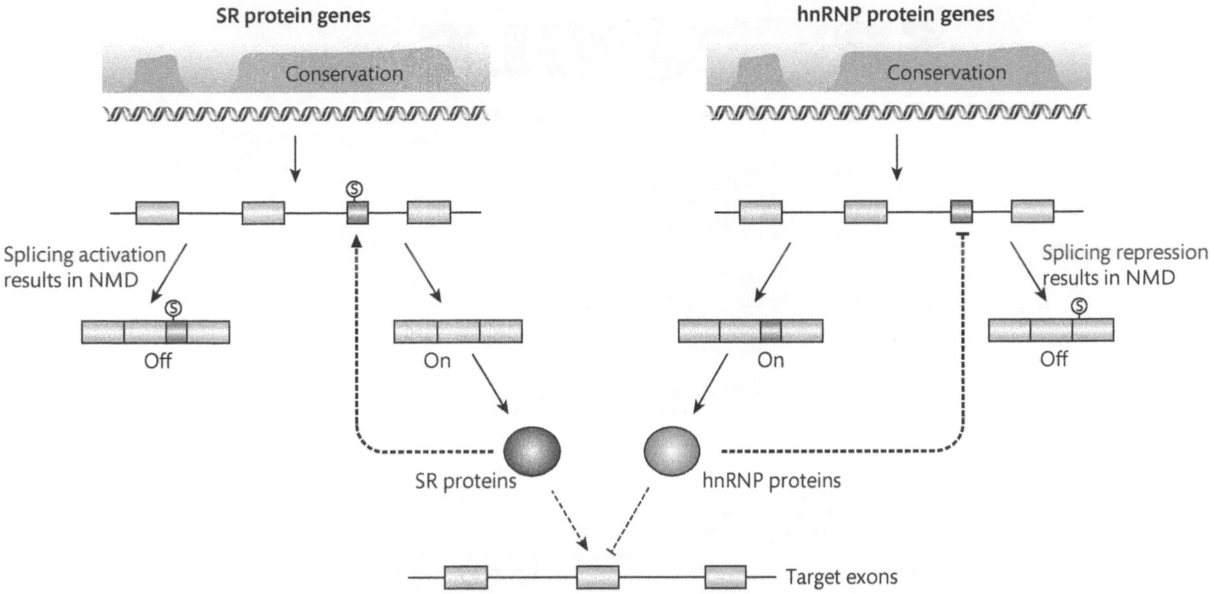

Figure 6.8 Alternative splicing regulators often regulate their own levels through splicing of poison exons. (From Kalsotra and Cooper[21] with permission.)

enhancers) and sequence motifs enriched in introns (predictive of splicing silencers). A screenshot from one of these websites is shown in Fig. 6.9. Strong exons typically have a strong splicing code—splice sites that conform to the expected consensus sequences we discussed in Chapter 5—and contain binding sites for splicing enhancers rather than splicing repressors.

Key points

The splicing code helps the spliceosome to identify whether pre-mRNA sequences are within exons or within introns. The splicing code contributes to exon splicing patterns in all metazoans and is 'decoded' by RNA-binding proteins which are present in the nucleus and which then bind to their target sequences, marking them as being in exons or in introns. The splicing code includes binding sites for activators and repressors of splicing. Exons have a high concentration of binding sites for splicing activators and introns are often bound by splicing repressors. Alternative splicing can both introduce new coding information into reading frames (in which case the number of alternatively spliced nucleotides must be a multiple of three) or can change reading frames (when the number of nucleotides added to the mRNA is not a multiple of three). In the case of poison exons, premature stop codons, which prevent translation and cause decay of cytoplasmic mRNA, are spliced into the mRNA. Poison exons are particularly important for maintaining proper expression levels of splicing regulator proteins through splicing autoregulation.

6.4 Exon and intron definition control the type of alternative splicing that operates

An important mechanism used by humans and most vertebrates to recognize exons from introns is called exon definition.[25-27] We introduced exon and intron definition in Chapter 5. As a reminder, in exon definition early spliceosome components bind to and interact across the exons to be spliced together (see Fig. 5.21). Thus, exon definition marks which of the pre-mRNA sequences are exons. Following exon definition, splicing proceeds by molecular interactions across the introns to be removed, involving assembly of fully fledged spliceosomes across introns. In intron definition, the spliceosome assembles directly through cross-intron interactions.

Exon and intron definition are very important for establishing which kind of alternative splicing takes place. In humans, where exon definition is primarily used, the most frequent form of alternative splicing is where whole exons are spliced into the mRNA or ignored by the spliceosome (exon skipping—see Section 6.1). This makes sense, since the whole exon is recognized by exon definition, and so the whole exon is left out in alternative splicing. In yeast, where intron definition is used, the most frequent form of alternative

(a)

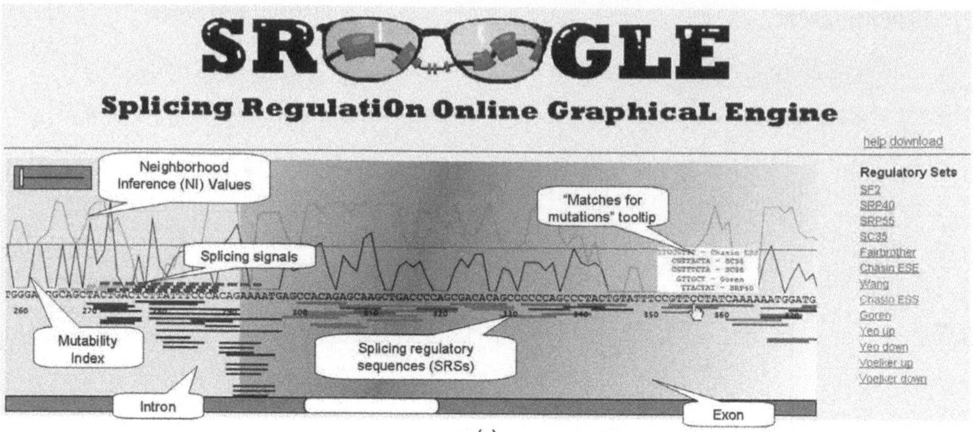

(b)

⊟ **Splice Site Scores**

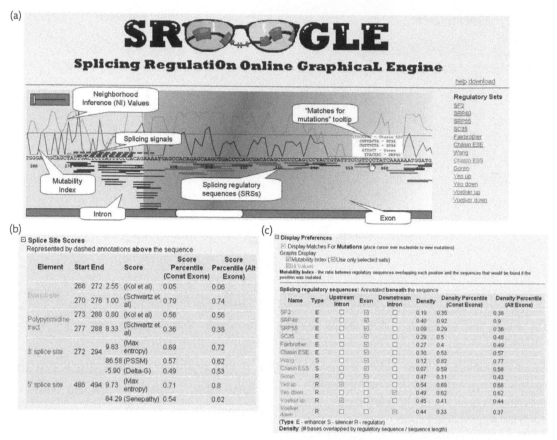

Represented by dashed annotations **above** the sequence

Element	Start	End	Score		Score Percentile (Const Exons)	Score Percentile (Alt Exons)
Branch site	266	272	2.55	(Kol et al)	0.05	0.06
	270	276	1.00	(Schwartz et al)	0.79	0.74
Polypyrimidine tract	273	288	0.80	(Kol et al)	0.56	0.56
	277	288	8.33	(Schwartz et al)	0.36	0.38
3' splice site	272	294	9.83	(Max entropy)	0.69	0.72
			86.58	(PSSM)	0.57	0.62
			-5.90	(Delta-G)	0.49	0.53
5' splice site	486	494	9.73	(Max entropy)	0.71	0.8
			84.29	(Senepathy)	0.54	0.62

(c)

⊟ **Display Preferences**

☑ Display Matches For **Mutations** (place cursor over nucleotide to view mutations)

Graphs Display

☑ Mutability Index (☑ Use only selected sets)

☑ NI Values

Mutability Index - the ratio between regulatory sequences overlapping each position and the sequences that would be found if the position was mutated

Splicing regulatory sequences: Annotated **beneath** the sequence

Name	Type	Upstream Intron	Exon	Downstream Intron	Density	Density Percentile (Const Exons)	Density Percentile (Alt Exons)
SF2	E	☐	☑	☐	0.19	0.35	0.38
SRP40	E	☐	☑	☐	0.40	0.92	0.9
SRP55	E	☐	☑	☐	0.09	0.29	0.36
SC35	E	☐	☑	☐	0.29	0.5	0.48
Fairbrother	E	☐	☑	☐	0.27	0.4	0.49
Chasin ESE	E	☐	☑	☐	0.30	0.53	0.57
Wang	S	☐	☑	☐	0.12	0.82	0.77
Chasin ESS	S	☐	☑	☐	0.07	0.59	0.58
Goren	R	☐	☑	☐	0.47	0.31	0.43
Yeo up	R	☑	☐	☐	0.54	0.69	0.68
Yeo down	R	☐	☐	☑	0.49	0.62	0.62
Voelker up	R	☑	☐	☐	0.45	0.41	0.44
Voelker down	R	☐	☐	☑	0.44	0.33	0.37

(**Type**: E - enhancer S - silencer R - regulator)

Density: (# bases overlapped by regulatory sequence / sequence length)

Figure 6.9 How strongly an exon will be recognized by the spliceosome can be assessed online using computer programs. Bioinformatic analysis is very important for predicting splicing regulator binding sites in RNA and the properties of exons. A useful site for monitoring splice site strength is called SROOGLE (http://sroogle.tau.ac.il/). This site calculates how closely a splice site will match the optimum possible splice site sequences. For example, the 5′ splice site is recognized by U1 snRNP, and a perfect 5′ splice site will be complementary to U1 snRNA over the maximum possible stretch of nine nucleotides, while a poorer 5′ splice site will have a shorter region of complementarity. Sroogle also identifies binding sites for particular RNA-binding proteins, and possible ESE and ESS sequences. (From Schwartz et al.[24] with permission from Oxford University Press.)

splicing is intron retention. This makes sense because in yeast whole introns are recognized by the spliceosome, so whole introns are left in in alternative splicing.

Splicing activators like SR proteins, which bind to ESE sequences within the exon, help exon definition to take place by also binding and stabilizing association of spliceosome components at the 5′ and 3′ splice sites. This stabilization is shown in Fig. 6.10. In this figure, SR proteins bound to ESE sequences within the exon stabilize both U1 snRNP at the 5′ splice site and the U2AF proteins at the 3′ splice site. This cross-exon stabilization helps the early spliceosome to recognize and start assembling around the exons to be spliced. SR proteins might also help to bridge interactions across introns later in spliceosome assemble (cross-intron interactions).

Key points

In most multicellular animals (metazoans) genes frequently contain short exons separated by very long introns. In metazoans very early spliceosome components bind to pre-mRNA and interact across exons. Once exons have been 'tagged' in this way, fully fledged spliceosomes start to assemble across introns, enabling exons to be spliced together. This means that alternative splicing frequently occurs through entire exons being included or left out. In some other species such as the yeast *S. cerevisiae*, where introns are fairly short, spliceosomes assemble straight away across the intron which is to be removed. In this yeast, alternative splicing occurs primarily through intron retention.

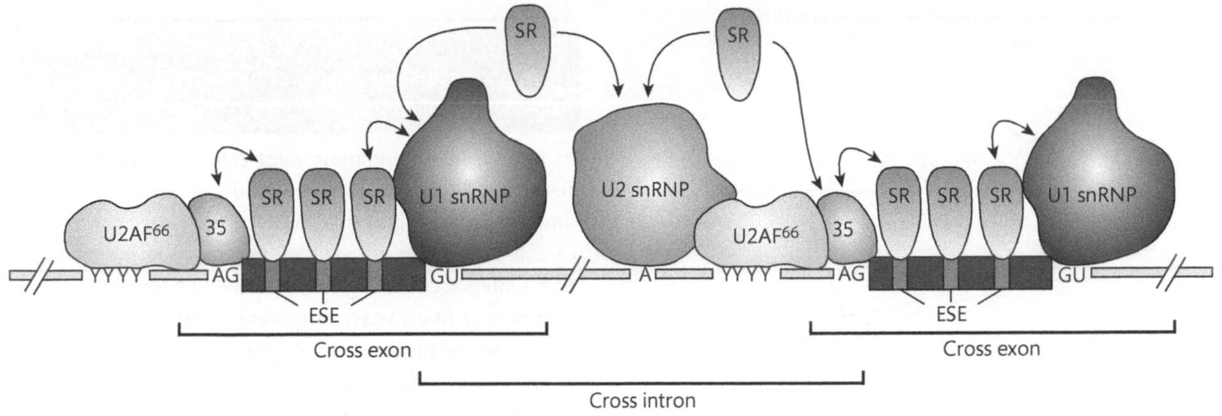

Figure 6.10 Components of the splicing code help to stabilize exon definition complexes. (From Maniatis and Tasic[1].)

6.5 Three main factors can contribute to alternative splicing regulation

From the chapter so far it should be clear that eukaryotic genes contain a set of instructions within them to control the splicing of their pre-mRNAs. This set of instructions is called the splicing code and it is read by a set of RNA-binding proteins in the cell which act as a 'code book' able to decipher this information. To be spliced into mRNA, an exon has to be recognized by the spliceosome. Splice junctions that are efficiently recognized by the spliceosome are always spliced into mRNAs, while splice junctions that are more weakly recognized are less efficiently handled by the splicing machinery. The strength (efficiency of recognition) of exons is affected by the strength of their splice sites and how many associated enhancers and silencers they have.[28]

In the following sections we are going to discuss the mechanisms controlling alternative splicing in a bit more detail.

Three main factors contribute to alternative splicing regulation.

1 **Changes in the expression of RNA-binding proteins that bind to pre-mRNAs in and near exons** (these proteins recognize the splicing code embedded in pre-mRNAs).
2 **Changes in upstream signalling pathways.** These signalling pathways may initiate outside the cell but control pre-mRNA splicing patterns in the nucleus.
3 **Changes in rates of transcription.** One of the key factors controlling alternative splicing decisions is in fact transcription.

6.6 Regulation of RNA splicing is controlled by changes in the concentration of RNA-binding proteins

In this section we are going to discuss some examples of where RNA-binding proteins control alternative splicing decisions.

6.6.1 Negative regulation of alternative splicing regulates ribosome protein levels in yeast

As the first example we are going to look at regulated splicing repression of an exon in the RPL30 mRNA from the yeast *S. cerevisiae*. Although alternative splicing is rarer in yeast than in metazoans, the example of RPL30 demonstrates an important principle of splicing regulation. This mechanism involves a nuclear RNA-binding protein which binds to pre-mRNA and blocks splicing from taking place (see Fig. 6.11).

Splicing regulation of RPL30 mRNA was the first example of regulated alternative splicing discovered in *S. cerevisiae*.[29] Like many other yeast intron-containing genes (see Section 5.2), *RPL30* encodes a ribosomal protein. Expression levels of RPL30 protein are important for normal ribosome assembly and so are tightly controlled. The RPL30 protein is an RNA-binding protein which

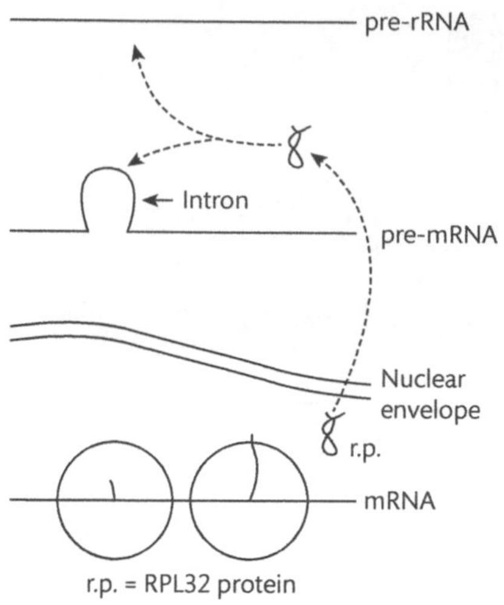

Figure 6.11 **The ribosomal protein RPL32 regulates its own splicing and translation through a feedback loop.** The *RPL32* gene contains a single intron. The RPL32 gene is transcribed and the pre-mRNA spliced, exported, and translated. RPL32 protein is then imported into the nucleus where it binds to an RNA sequence in the large subunit rRNA to become a component of the ribosome. Once the rRNA-binding sites in the cell are saturated, any excess RPL32 protein in the cell binds to a similar RNA target sequence in the *RPL32* pre-mRNA, and in doing so blocks the first step of *RPL32* pre-mRNA splicing. As well as inhibiting splicing of its own pre-mRNA, RPL32 protein can also bind to the spliced *RPL32* mRNA in the cytoplasm blocking its translation (not shown here). (From Dabeva et al.[30] with permission.)

binds to an RNA target sequence found in both the large rRNA subunit and its own pre-mRNA. Most RPL30 protein will bind to ribosomes, but any excess RPL30 protein will bind to its own transcript and block spliceosome assembly on the *RPL30* pre-mRNA.

Splicing of the pre-mRNA is important for expression of RPL30 protein. In unspliced *RPL30* pre-mRNA, the intron between exons 1 and 2 blocks the ORF, so the unspliced pre-mRNA cannot be translated and instead is degraded by nonsense-mediated decay (Chapter 12). By inhibition of splicing, increasing concentrations of RPL30 protein in the nucleus switch off production of spliced *RPL30* mRNA. This prevents any excess RPL30 protein being made which might then interfere with normal ribosome assembly (see Fig. 6.11).

Box 6.2 Only a small subset of genes in baker's yeast contain introns but several of these are regulated by alternative splicing[21,31,32]

In nutrient-rich conditions yeast cells proliferate via mitosis (this is called vegetative growth). When nutrients are limiting, yeast cells switch to dividing by meiosis, a form of cell division which reduces chromosome numbers by half. Only 5% of yeast genes (290/6000) have introns (see Section 5.2). Many yeast introns are found in genes encoding ribosomal proteins (these genes need to be expressed at high levels during vegetative growth—see Chapter 5 on how introns help this). An additional 13 introns are found in genes that are specifically expressed in meiosis. These meiotic introns are important in the yeast life cycle since their splicing is regulated to control meiotic timing. Two mechanisms control this meiosis splicing. First, during meiosis there is expression of specific splicing regulators like Mer1. Secondly, the intron-containing ribosomal protein genes are not expressed in yeast meiosis, meaning that there are more spliceosomes available to splice other pre-mRNAs. This leads to more efficient splicing of other regulated introns during meiosis. Remember that yeast uses intron definition and so alternative splicing takes place by intron retention.

6.6.2 A mechanism of activated splicing inclusion regulates developmental timing in yeast meiosis

Our second example of an RNA-binding protein controlling splicing patterns also comes from yeast. Changes in the nuclear concentration of RNA-binding proteins can also activate splicing. We are going to illustrate this by looking at the yeast *Mer2* gene [21,31,32,34,35] which is regulated during meiosis.

Meiosis-specific splicing in yeast genes was discovered in 1991 in the transcript from the *Mer2* gene, which encodes a protein important for meiotic recombination.[33] *Mer2* is transcribed throughout the cell cycle, but the 5′ splice site of the *Mer2* pre-mRNA is normally inefficiently recognized by U1 snRNP as it has a slightly different sequence from the consensus 5′ splice site sequence (consensus splice sites are discussed in Chapter 5). As a result of its unusual splice site, the intron in the *Mer2* pre-mRNA is normally very inefficiently removed, except during meiosis.

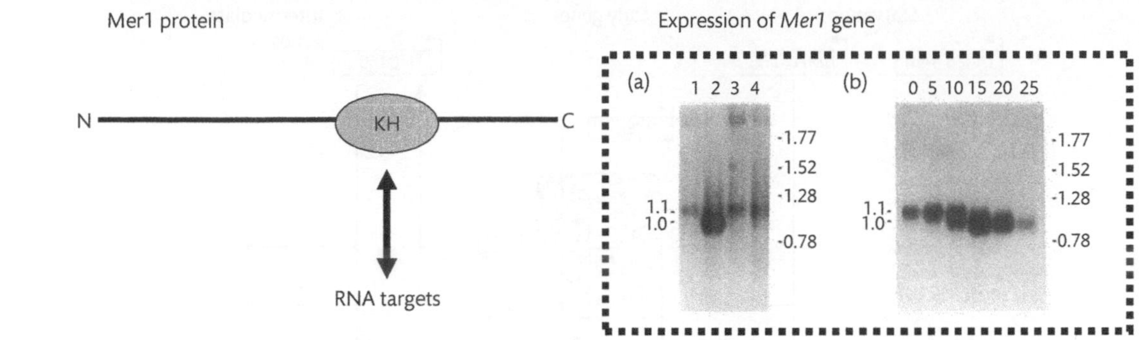

Figure 6.12 **The yeast Mer1 protein is a meiosis-specific splicing factor.** The *S. cerevisiae* Mer1 protein contains a KH domain which binds to RNA. **Inset:** Expression of the *Mer1* gene analysed by Northern blotting. *Mer1* RNA is specifically expressed during meiosis. (a) Expression of Mer1 RNA in wild-type (lanes 1 and 2) and Mer1 knockout cells (lanes 3 and 4). Lanes 1 and 3 are mitotically grown yeast cells. Lanes 2 and 4 are meiotic yeast cells. (b) Expression of Mer1 protein at different stages of meiosis. (Inset reprinted from Engebrecht et al.[33] with permission.)

During meiosis, splicing of the *Mer2* pre-mRNA is controlled by a meiosis-specific RNA-binding protein called Mer1. The Mer1 RNA-binding protein is only expressed during meiosis (Fig. 6.12 shows this expression pattern). During meiosis Mer1 protein binds to an intronic splicing enhancer (ISE) RNA sequence near to the inefficient 5′ splice site of the *Mer2* pre-mRNA. Once bound, Mer1 protein helps U1 snRNP to bind to the weak 5′ splice site of *Mer2* pre-mRNA and the intron to be spliced out (see Fig. 6.13).

Individual RNA-binding proteins can regulate multiple downstream alternative splicing targets. By binding to different target RNAs which contain the same binding site sequence, the meiotic Mer1 protein additionally regulates the meiotic splicing of additional pre-mRNAs as well as *Mer2* pre-mRNA (Box 6.2). The splicing regulation of these multiple mRNA targets is critical for timing in yeast meiosis (shown in more detail in Fig. 6.14).

6.6.3 Exonic splicing enhancers in the *Doublesex* gene are critical for sex determination in fruit flies

The examples of alternative splicing control mechanisms given in the previous section have been taken from yeast. However, similar mechanisms of splicing

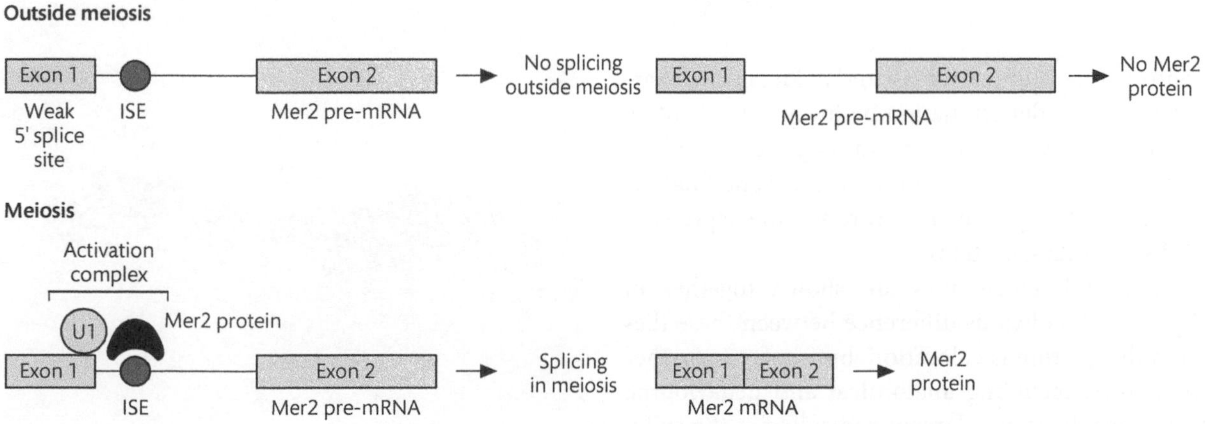

Figure 6.13 **Mer1 protein positively regulates the meiosis-specific splicing of *Mer2* pre-mRNA by binding within the intron and stabilizing U1 snRNP association with the 5′ splice site.** Mer1 protein binds to an intronic splicing enhancer (ISE) element—these elements are discussed in Chapter 7. *Mer2* pre-mRNA splicing regulation illustrates a general principle: a regulated splice can be introduced by weakening a splice site, meaning that it is less easily seen by the spliceosome, and then introducing a protein which strengthens recognition by the spliceosome.

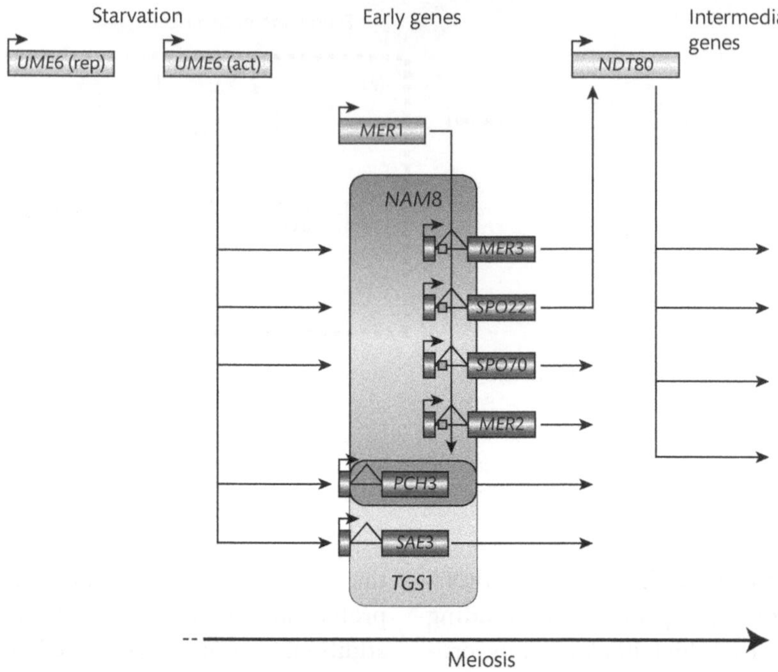

Figure 6.14 **Mer1-regulated alternative splicing controls multiple mRNAs in yeast meiosis.** Meiotic gene expression in yeast is regulated at the levels of both transcription and alternative splicing. Yeast genes are transcribed in early, mid, or late meiosis depending on when their protein products are needed during meiosis. Initiating meiosis in response to starvation, a transcription factor called Ume6 switches from a transcriptional repressor to a transcriptional activator. Ume6 then activates a wave of transcription of early meiosis genes, including the Mer1 splicing regulator, and four other genes whose splicing is regulated by Mer1. These Mer1-regulated genes are called *MER2, MER3, SPO22*, and *SPO70*. The time needed for Mer1 to be transcribed, spliced, and translated delays the expression of these four target pre-mRNAs. Even though they are transcribed at the same time, they need to wait for *MER1* mRNA to be spliced and translated before they themselves can be spliced into productive mRNAs. This lag in MER1-regulated splicing is important for timing of gene expression stages in meiosis. MER2 protein is needed for the formation of double-stranded DNA breaks needed for recombination, SPO70 protein helps control the meiosis cell cycle, and MER3 protein is a helicase needed to resolve double-stranded breaks. MER3 and SPO22 proteins are needed for transcriptional activation for expression of the transcription factor NDT80 which controls the genes expressed during the intermediate wave of meiosis transcription. (From Kalsotra and Cooper[21] with permission.)

control also occur in animals and in fact in all eukaryotes. The sex determination pathway in the fruit fly *Drosophila melanogaster* provides a good example of this, and this is where we find our third and final example of splicing regulation through the expression of RNA-binding proteins.

Male and female flies are shown together in Fig. 6.15. An obvious difference between these flies from this picture is coloration, but there are further differences including anatomical and behavioural. Sex determination is largely controlled at the splicing level in fruit flies, and through a cascade of splicing choices[36] (summarized in Box 6.3). Each of these different splicing choices is controlled by RNA-binding proteins present in female cells, and

Figure 6.15 **A male fruit fly approaching a female fruit fly.** Notice the differences in patterning of the abdomen of male and female flies. Sexual dimorphism in the fruit fly is controlled by alternative splicing. (Image from http://www.sciencedirect.com/science/article/pii/S0092867407004680.)

Box 6.3 An overview of the splicing regulatory cascade in *Drosophila* sex determination

This shows the pathway in female fruit flies only since this is the regulated pathway. Female sex determination starts with production of the **Sex-lethal** RNA-binding protein (abbreviated Sxl)—males do not make this initial protein. Female Sex-lethal protein production initiates a regulatory cascade of alternative splicing choices which results in the production of a female fly (Fig. 6.16). Female Sex-lethal protein regulates three key RNA splicing choices in the sex determination pathway.

1. Sex-lethal binds to its own pre-mRNA to positively control its own splicing. Male flies express Sex-lethal pre-mRNA, but because they never express Sex-lethal protein they can never splice a productive Sex-lethal mRNA.
2. In females Sex-lethal protein also binds and blocks translation of the MSL-2 mRNA, which encodes a protein involved in dosage compensation. Dosage compensation is not needed in female flies, so MSL-2 is not needed.
3. Finally Sex-lethal protein controls splicing of the mRNA encoding a critical protein called Tra, meaning that Tra protein is only made in females (male flies express the Tra gene, but make a splice isoform of the Tra mRNA which contains stop codons and so cannot be translated).

Once produced in female flies, Tra protein forms part of the complex that controls splicing of the mRNAs in the next part of the cascade. Tra protein expression controls splice events in pre-mRNAs encoding the transcription factors Doublesex and Fruitless. Because of Tra regulation in females, distinct male and female spliced mRNA isoforms of Doublesex and Fruitless are made, and these encode transcription factors that control male and female fly development.

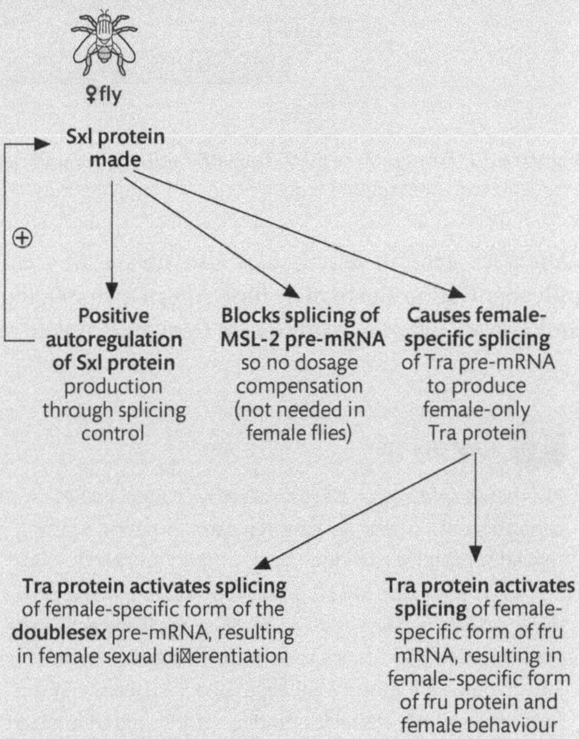

Figure 6.16 Splicing regulatory cascade in *Drosophila* sex determination.

following these eventually ends up in a female fly. We are not going to discuss all of these splicing choices in detail, but at the bottom of this cascade is a gene called *Doublesex*. Although transcribed both in males and females, *Doublesex* pre-mRNA is spliced differently in males and females to produce distinct male and female transcription factors. In turn these transcription factors select different promoters, and are responsible for male and female differentiation.

Exon 4 of the *Doublesex* gene is only spliced into mRNA in female flies. Only female flies express the complete set of proteins which bind to and activate exon 4 splicing. Within the nucleotide sequence of *Doublesex* exon 4 are six individual ESEs (see Fig. 6.17), each of which binds a group of three proteins in females called TRA, TRA2, and RBP1. In female flies assembly of these three proteins forms a series of six complexes on Doublesex exon 4 which together stabilize the association of the spliceosome with its weak 3′ splice site. Interestingly, although both TRA2 and RBP1 bind to RNA, the TRA protein does not directly bind RNA itself, but interacts with the TRA2 and RBP1 proteins bound to the exon.

This example of *Doublesex* splicing illustrates how splicing regulator proteins often assemble splicing complexes. In this case, TRA, TRA2, and RBP1 proteins are each required to make a stable splicing complex in female flies. Even though male flies express RBP1 and TRA2 proteins, they do not express the TRA protein. As a result male flies skip *Doublesex* exon 4—therefore exon 3 is directly spliced onto exon 5. Another thing to notice about Fig. 6.17 is that as well as following a different splicing pathway, the

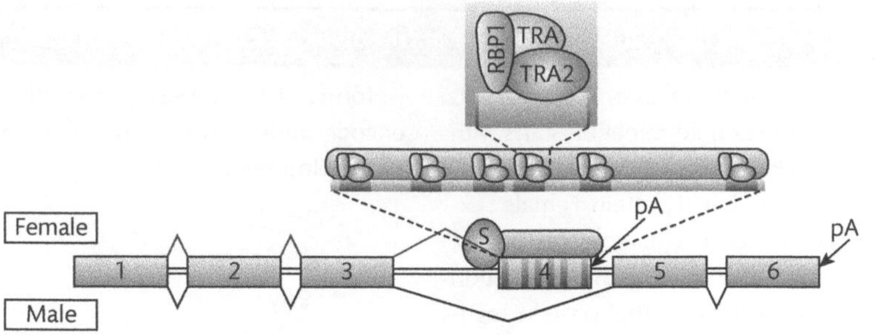

Figure 6.17 Formation of a Tra–Tra2–RBP splicing complex promotes splicing of *Doublesex* exon 4 in female flies. (From Maniatis and Tasic[1] with permission.)

Doublesex gene in female flies also uses a different polyadenylation site to male flies. Alternative splicing and polyadenylation patterns are frequently linked.

Key points

Alternative splicing can be controlled by changes in the concentration of nuclear splicing regulator proteins. Splicing regulator proteins can bind to pre-mRNAs and block splicing, as in the yeast *RPL30* pre-mRNA. RNA-binding proteins can also activate splicing. In yeast a splicing activator protein called Mer1 binds to an intronic splicing enhancer just downstream of the weak regulated 5′ splice site in the *Mer2* pre-mRNA, stabilizes binding of U1 snRNP, and so activates splicing. Mer1p is only expressed in meiosis, so only activates *Mer2* splicing at this time in the cell cycle. Exonic splicing enhancers play a key role in controlling sex determination in fruit flies. Splicing regulator complexes assemble onto these ESEs and control sex-specific splicing by binding a protein complex. While Tra protein is female specific, the other proteins which contribute to the splicing regulator complex on *Doublesex* pre-mRNA exon 4 are not sex specific. A stable splicing regulation complex on *Doublesex* pre-mRNA exon 4 cannot form in the absence of Tra. Thus Tra acts as the switch controlling the pattern of *Doublesex* pre-mRNA splicing in female flies.

6.7 Signal transduction pathways can regulate alternative splicing by changing the function and location of splicing factors

This section discusses the second parameter affecting alternative splicing—the role of cellular signalling pathways which integrate splicing control and the environment. Alternative splicing is frequently a target for signal transduction pathways.[37–39] Changes in the cellular environment can trigger dynamic changes in the concentration and activity of nuclear RNA-binding proteins, to which specific transcripts with appropriate *cis*-acting sequences respond. The effects of signal transduction on pre-mRNA splicing are often mediated through addition or removal of phosphate groups from nuclear RNA-binding proteins (see Box 6.4).

By changing the charge distributions and hydrogen bonding properties of amino acid side groups, phosphorylation can affect molecular interaction partners and the spatial location of splicing factors in the nucleus. This in turn can lead to changes in alternative splicing pathways. In this section we are going to look at an important example where a signalling pathway directly modulates pre-mRNA splicing choices in the nucleus through the addition of phosphate groups to splicing regulator proteins.

Some RNA-binding proteins, including the splicing repressor protein hnRNP A1, are dynamically mobile inside the cell. If you take a snapshot of a cell, at any one time most of the hnRNP A1 will be concentrated in the nucleus. However, hnRNP A1 only appears to be an exclusively nuclear protein since it spends most of its time in the nucleus. In fact, hnRNP A1 is very rapidly shuttling between the nucleus and the cytoplasm.[39,40] Once in the cytoplasm hnRNP A1 is usually rapidly imported back into the nucleus by the nuclear protein import machinery. Transport into and out of the nucleus is controlled by a sequence in hnRNP A1 called the M9 peptide which acts as both a nuclear import and an export signal (see Chapter 11).

Cells respond to stress by activating signalling pathways leading to cascades of protein phosphorylation. Both osmotic shock and ultraviolet irradiation

Box 6.4 Protein phosphorylation can change protein function

The amino acids serine, threonine, and tyrosine can be modified by phosphate addition. These three amino acids have a polar hydroxyl group (-OH) into which the oxygen draws electrons, giving the hydrogen atom a slight positive charge (see also Chapter 2). The addition of a phosphate group to each of these polar amino acids in place of the -OH group creates a strong negative charge. In this way phosphorylation can convert polar regions of proteins into negatively charged regions. Addition of a phosphate group also increases by three the potential number of hydrogen bonds that can be made by the amino acid side group. There are three kinds of kinase: (i) serine and threonine have chemically similar side groups and are phosphorylated by serine/threonine kinases; (ii) tyrosines are phosphorylated by tyrosine kinases; (iii) some kinases have dual specificity and can phosphorylate side groups on both serine/threonine and tyrosines.

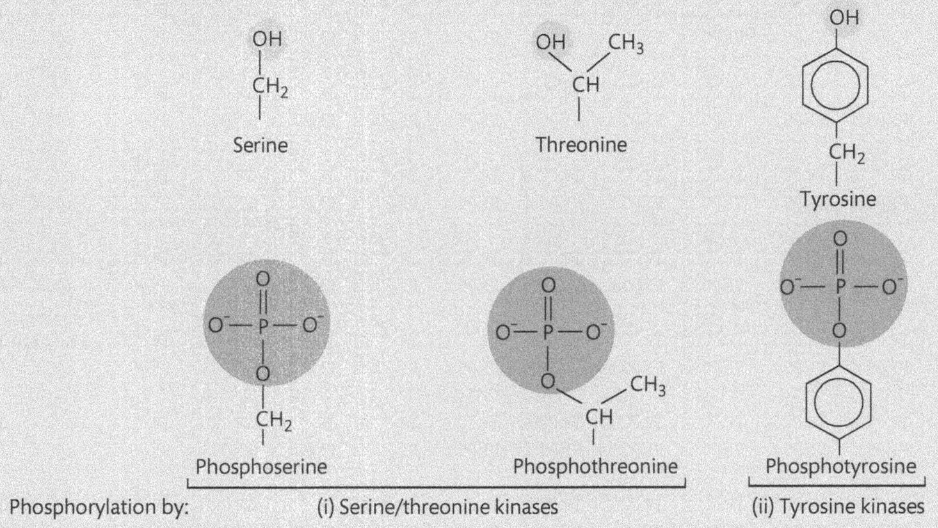

Figure 6.18 Protein phosphorylation

lead to activation of the stress response signalling pathway, which includes the MAP kinase kinase p38 (see Fig. 6.19). Activation of the stress signalling pathway results in the phosphorylation of hnRNP A1 at a site immediately adjacent to the M9 peptide, preventing its recognition by the nuclear import machinery. As a result hnRNP A1 continues to be exported from the nucleus but the phosphorylation blocks its re-import to the nucleus and leads to accumulation of hnRNP A1 in the cytoplasm. hnRNP A1 is an important splicing repressor so its depletion in the nucleus leads to changes in splicing patterns.

6.8 Transcription elongation speeds can regulate alternative splicing choices

In this section we are going to discuss the final major parameter which can regulate whether exons are alternatively spliced or not. Surprisingly, this is not directly related to the spliceosome at all, but instead involves the speed of transcription.

How can transcription affect alternative pre-mRNA splicing? Pre-mRNA splicing often occurs on nascent pre-mRNAs while transcription is still taking place (see Chapter 8, and in particular Fig. 8.8). Two models have been proposed to explain the connection between transcription and splicing—one involves the speed of transcription (this section) and the other involves the recruitment of cofactors via transcription (Section 6.9).

How fast RNA polymerase II moves through genes can have an important effect on the alternative splicing of exons—this is the **kinetic coupling model**. In fact, cultured cells which express mutant 'slower' versions of RNA polymerase II show increased splicing of normally weakly included alternative exons (see Fig. 6.20, where it should be noted that the levels of inclusion of the alternatively spliced fibronectin *EDI* exon are much higher in the cells expressing

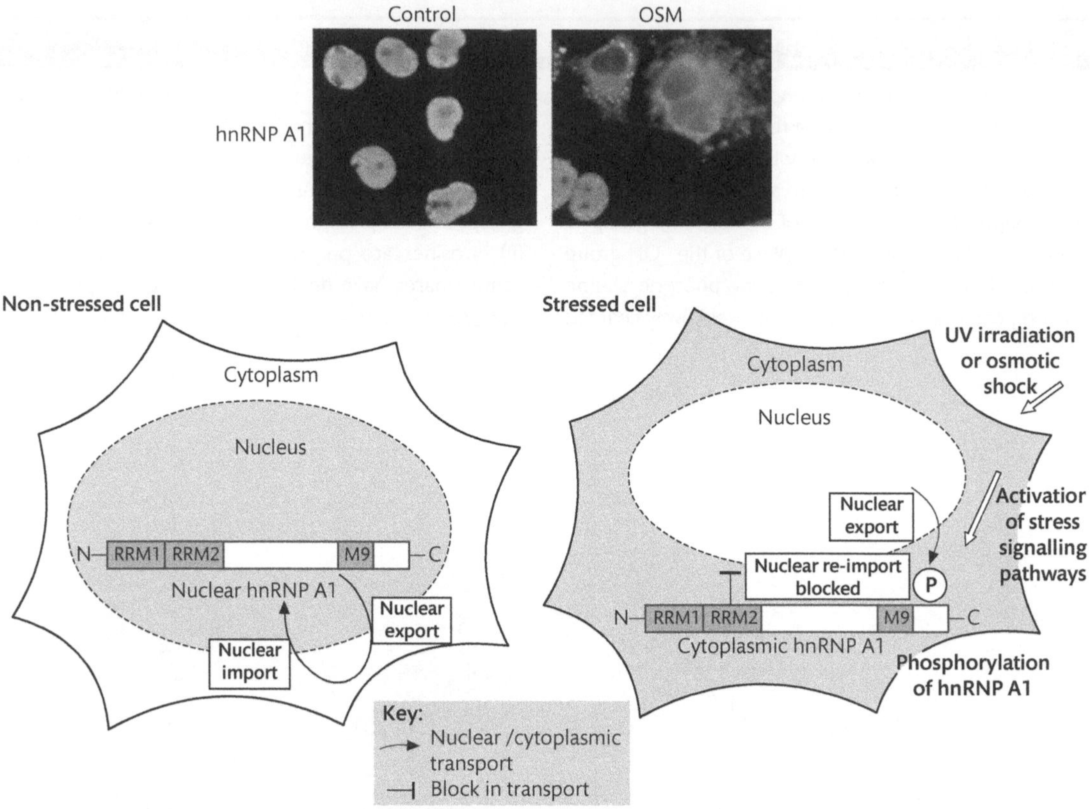

Figure 6.19 **The splicing repressor hnRNP A1 relocalizes to the cytoplasm in response to cellular stress.** Stressing mammalian cells with either osmotic shock or UV irradiation changes the steady state cellular location of hnRNP A1 from the nucleus to the cytoplasm. Notice that hnRNPA1 is nuclear in the control cells. After osmotic shock hnRNPA1 has moved to the cytoplasm—notice the holes in staining where the nuclei are located. A cartoon showing the mechanism is shown below the fluorescent localization images. (Images from van der Houven van Oordt et al.[40], with permission from Rockefeller University Press.)

the mutant slow form of RNA polymerase hC4 compared with the wild-type RNA polymerase).

The reason why the speed of RNA polymerase II is important for splicing is because of competition between exons to be spliced. Faster movement of RNA polymerase through a gene means more rapid appearance of exons in the pre-mRNA (see Fig. 6.21). If exons are supplied faster than splicing can take place, the spliceosome will have a choice of exons in the same pre-mRNA to splice onto the growing mRNA. Generally, the stronger exon will be chosen (i.e. with the strongest splice sites), and weaker exons will be left out. Conversely, slow transcription favours the completion of splicing of each exon in turn before the next exon in the gene is transcribed and so is available for splicing. Slow RNA polymerases will favour the inclusion of weak exons, since competing exons will be made more slowly (this explains

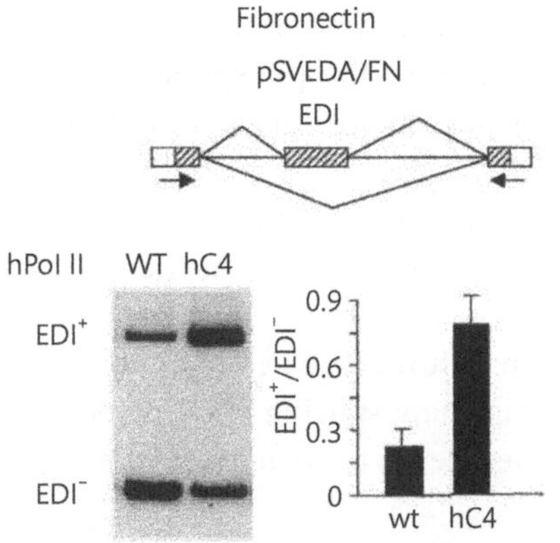

Figure 6.20 **Cells expressing slowly elongating forms of RNA polymerase II show different splicing patterns.** Levels of inclusion of the alternative EDI exon from fibronectin were measured by RT-PCR from cells expressing a wild-type RNA polymerase (wt) or a mutant slow form of RNA polymerase (hC4). Notice that there is more EDI $^{+}$ product with hC4. (Reprinted from de la Mata et al.[41] with permission from Elsevier.)

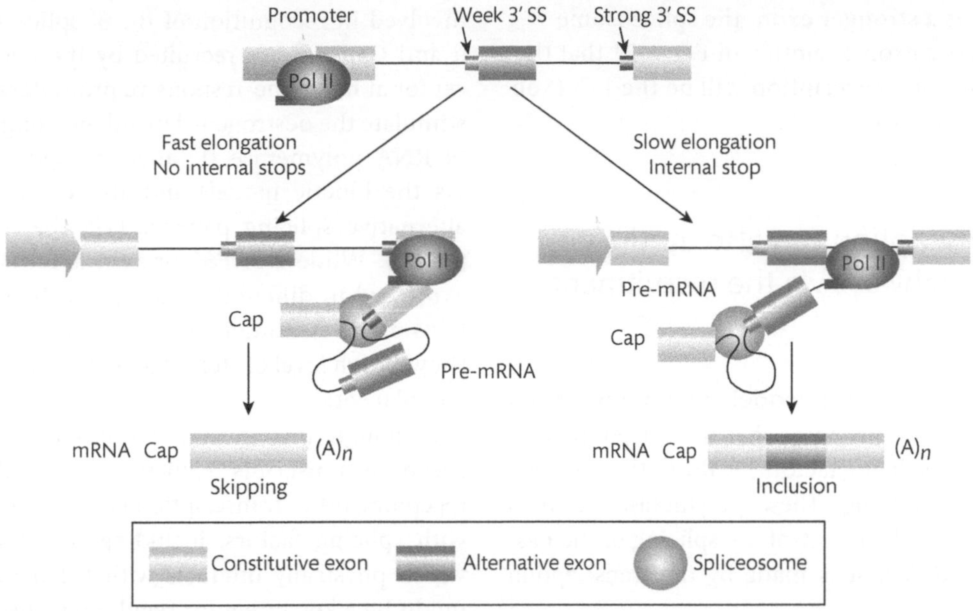

Figure 6.21 **Elongation times of RNA polymerase II can affect alternative splicing of variable exons.** In the example shown, exon 3 is transcribed very shortly after exon 2. Hence, instead of the spliceosome being presented with a pre-mRNA from only one exon at a time, there is a choice between splicing either exon 2 or exon 3 onto exon 1. (Adapted from Kornblihtt [42] with permission from Macmillan Publishers Ltd.)

the inclusion of weak variable exons in cultured cells with slow RNA polymerases, shown in Fig. 6.20).

The rates of RNA polymerase extension within a gene can vary through internal pause sites which cause RNA polymerase II to slow down. We come back to these important effects later in the book, in Chapter 8, Section 8.4.

Another important variable connecting transcription and splicing is gene structure. If exons are separated by long introns there will be a time lag in exon synthesis during which time transcription of the intervening intron takes place. This time lag enables exons to be spliced onto the mRNA before competing downstream exons are transcribed. In contrast, short introns means that there is no significant time delay between the production of adjacent exons, and the spliceosome will be presented with a choice of exons to be spliced into the growing mRNA.

Looking at Figure 6.21, exons 1 and 2 are always spliced into mRNA, but the middle exon (exon 2) has a weaker 3′ splice site that is inefficiently recognized by the splicing machinery. Exons 2 and 3 are separated by a short intron and so are transcribed in reasonably quick succession. There are two different splicing patterns: the exons can be spliced together either 1–2–3 or 1–3 in the final mRNA. The choice between these two splicing patterns will depend on the transcription time across the second intron.

- Pattern 1: slow elongation leads to inclusion of exon 2. In the case of slow transcription, exon 2 will be spliced onto the growing mRNA before exon 3 is transcribed. Afterwards, exon 3 will be transcribed and spliced onto the mRNA. The result of this splicing pattern is that all three exons will be included into the final mRNA. Notice in Fig. 6.21 that the result of slow transcription will be the 1–2–3 exon splicing pattern.
- Pattern 2: fast elongation will lead to skipping of exon 2. If the transcriptional elongation rate is fast, exon 3 is transcribed before exon 2 has been spliced onto the mRNA. Hence the spliceosome now has a choice of splicing either exon 2 or exon 3 onto exon 1.

Since it is a stronger exon, the spliceosome will choose exon 3. Notice in Fig. 6.21 that the result of fast transcription will be the 1–3 exon splicing pattern.

6.9 Transcription can also modulate splicing pathways via the recruitment of cofactors

A second **recruitment model** links transcription and splicing, and proposes that transcription complexes recruit splicing factors locally to the gene that they are transcribing. These polymerase-recruited splicing factors then control the splicing of the nascent pre-mRNA that is made by the transcription complex.

A group of splicing co-regulators are recruited to elongating RNA polymerases by nuclear steroid receptors called the Caper proteins (see Fig. 6.22). Caper proteins look like splicing factors—in particular, they are very similar in sequence and structure to the splicing protein U2AF65 which is involved in recognition of the 3′ splice site. Caper α and Caper β are recruited by the oestrogen receptor at hormone-responsive promoters, and both stimulate the oestrogen-dependent elongation rates of RNA polymerase II (affecting splicing choices via the kinetic model) and also directly regulate alternative splicing patterns (via the recruitment model). While U2AF65 protein is fairly generally expressed in different tissue types, the Caper proteins are preferentially expressed in tissues which show a high level of steroid activity, such as the liver and placenta.

Although the recruitment model was devised based on transcripts regulated by steroid hormone receptors, other transcription factors also have links with splicing factors, including the WT1 protein which physically interacts with U2AF65. U2AF65 might be a key target for regulation, since it is part of a very early complex in spliceosome assembly (see Chapter 5).

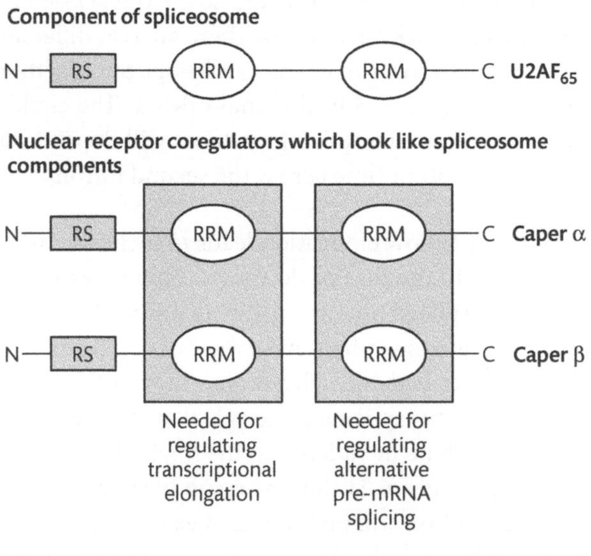

Figure 6.22 The steroid co-regulator proteins Caper α and Caper β have similar structures to the splicing factor U2AF65. Both Caper proteins contain two RNA recognition motifs (RRMs): the first RRM of the Caper proteins is important in controlling transcription elongation rates, and the second RRM is important in controlling alternative splicing. Although similar in protein structure, U2AF65 only functions in pre-mRNA splicing and not transcription.

Key points

The coupling of transcription and splicing adds an element of competition to the selection of different splice sites which can be influenced by transcriptional speeds and intron lengths. Different splicing patterns have been observed for otherwise identical pre-mRNAs when they are made from genes which are tagged up to different promoters by genetic engineering. The classic example of this is control of transcriptional elongation and splicing patterns by steroid hormones. Slower transcriptional elongation speeds enable alternative exons with poor splicing signals to be spliced into mRNAs before more strongly recognized competing downstream exons are transcribed. As well as affecting the speed of transcriptional elongation, splicing factor analogues are recruited as steroid hormone co-regulators at promoters. These splicing factor analogues affect both splicing choices and transcriptional elongation times.

6.10 Alternative splicing is critical for normal animal development

In this last part of this chapter we are going to move from the mechanism of alternative splicing to discuss the biological importance of alternative splicing. Numerous examples of functional alternative splicing are known.[8] Alternative splicing can affect protein domains (by inserting new peptide coding information), protein modifications (e.g. by including or excluding peptide sequences that undergo phosphorylation), and protein–protein interactions (by inserting peptide coding information for protein interaction domains). Sometimes alternative splicing can produce dramatically different proteins. Examples of this are the male and female isoforms of the doublesex transcription factors that control sexual differentiation in fruit flies that we discussed earlier.

In this final section we are going to discuss two further examples of how alternative splicing can be important in development. These examples are also medically very important, and so provide a good introduction to the next chapter (Chapter 7). The first example comes from mammalian development and relates to alternative splicing changes occurring between the embryo and adult stages of development and the disease myotonic dystrophy.

6.10.1 Changes in splicing patterns occur during development between the embryo and adult

Extremely important alternative splicing changes occur between embryos and adults during human development.[21,43-45] These splicing changes help to provide protein isoforms of different specifications needed by embryos and adults, and are controlled by two families of splicing regulator proteins called the muscleblind (MBLN) and CUG-binding (CUGBP) proteins (see Table 6.1). Muscleblind and CUG-binding proteins bind to target RNAs containing CUG-rich motifs, although not to exactly the same sequences (also the mode of RNA binding is different: muscleblind proteins use zinc fingers to bind to RNA, and CUG-binding proteins use RRMs (see Chapter 4)).

Alternative splicing changes between the embryo and adult are caused by expression shifts in muscleblind and CUG-binding proteins. Muscleblind proteins predominate in normal adult cells, while CUG-binding proteins predominate in embryos. These switches cause switches in mRNA splicing patterns. Muscleblind and CELF proteins often regulate the same mRNAs during development. Particularly important exons regulated by muscleblind and CUG-binding proteins are in the insulin receptor, the chloride channel ClC, and cardiac troponin T pre-mRNAs. In each of these cases the embryo needs a different protein to the adult.

6.10.2 Embryo→adult splicing switches are disrupted in the disease myotonic dystrophy

The embryo→adult alternative splicing switches controlled by the muscleblind and CUG-binding proteins are so important that a developmental disease called **myotonic dystrophy** is caused when they do not take place properly (see Box 6.5). Mutations causing both forms of myotonic dystrophy have been mapped by geneticists. Surprisingly, these disease mutations do not directly affect protein coding. Instead, myotonic dystrophy is caused by expansions of short repeat elements within the untranslated regions from two separate genes. The form of myotonic dystrophy called DM1 is caused by expansion of a CTG nucleotide triplet within the 3′ UTR of the *dystrophonia myotonica protein kinase* (*DMPK*) gene. Normally there are fewer than 5–37 copies of this CTG repeat, but in individuals affected by myotonic dystrophy this increases to 50–4000 repeats (see Fig. 6.23). The other form of myotonic dystrophy, called DM2, is caused by from 75 up to 11 000 repeats of CCTG within an intron of the *zinc-finger protein 9* (*ZNF9*) gene.

The CUG repeat expansions found in patients with myotonic dystrophy are in untranslated parts of the *DMPK* and *ZNF9* mRNAs, so they do not change the amino acid sequence of the encoded proteins. Instead, the mutations cause a pathogenic RNA to be made. Note that a CTG triplet at the gene

Box 6.5 Myotonic dystrophy

Myotonic dystrophy is the most common form of muscular dystrophy in adults, and is a multisystem disease, i.e. several organs are affected including myotonia (the inability to relax a muscle after contraction), heart problems (cardiac arrhythmias), cataracts, testicular failure, and a tendency towards diabetes. There are two forms of the disease, DM1 and DM2, with different symptoms and frequency.

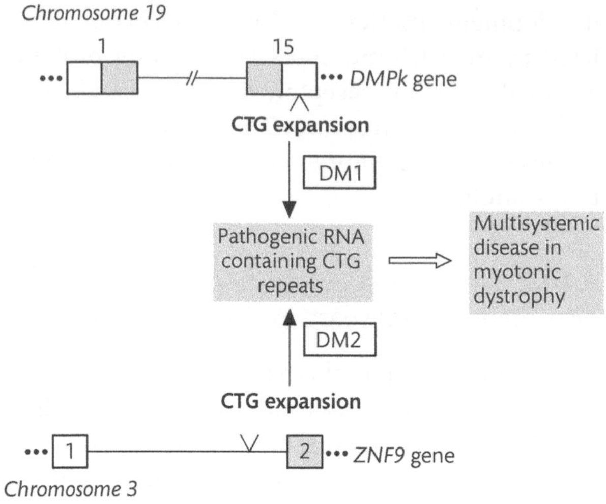

Figure 6.23 Nucleotide expansions within the non-coding regions of DMPK and ZNF9 pre-mRNAs produce the dominant negative pathogenic RNA transcripts responsible for myotonic dystrophy.

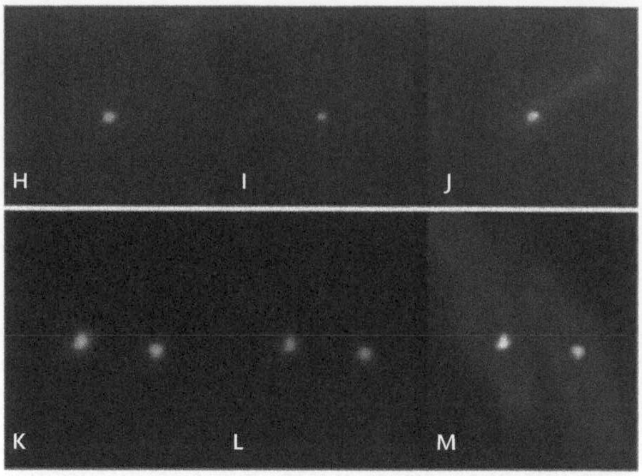

Figure 6.25 Abnormal nuclear foci of muscleblind proteins form in cells from patients with myotonic dystrophy DM1 (upper panel) and DM2 (lower panel). In this experiment, the mRNA containing expanded CUG repeats is localized by in situ hybridization (left), muscleblind protein is localized by immunofluorescence in the same cell (middle), and both are visualized with DNA by DAPI staining (right). (From Mankodi et al.[47])

level becomes a CUG in RNA. The pathogenic RNAs then affect two groups of proteins—muscleblind and CUGBPs.

1 In myotonic dystrophy cells, the CUG repeats form double-stranded RNA structures through Watson–Crick base pairing (see Fig. 6.24). These double-stranded RNAs spatially localize in dots in the nucleus, called nuclear foci, that can be visualized under the microscope (see Fig. 6.25). **Muscleblind protein binds to these double-stranded CUG RNAs.**

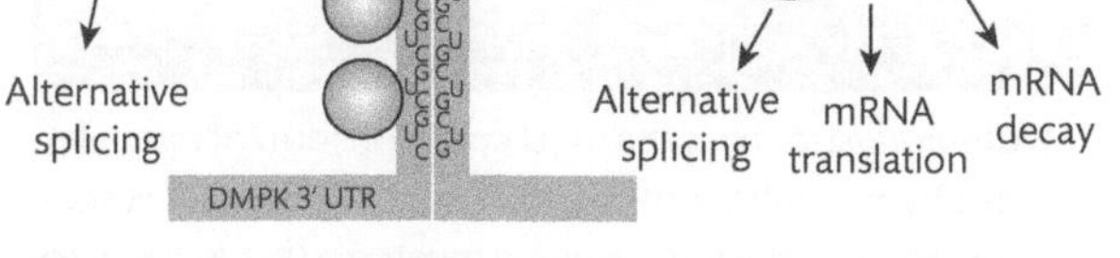

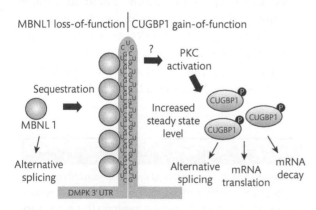

Figure 6.24 CUG repeats directly bind to muscleblind proteins in the nucleus, and indirectly change the expression of CUGBP1. This in turn leads to splicing changes in pre-mRNAs normally controlled by these RNA-binding proteins. (From Lee and Cooper[46].)

As a result, in myotonic dystrophy cells muscleblind proteins concentrate in nuclear foci and are no longer functional.

2 The other group of proteins that are affected by these repeats, CUG-binding proteins, normally bind single-stranded CUG sequences but their nuclear concentration is affected by the CUG expansion. CUG-binding proteins do not themselves directly bind to the expanded pathogenic CUG repeats in myotonic dystrophy, so do not localize in the nuclear foci containing these repeats. Instead, the pathogenic expanded CUG repeats increase the nuclear concentration of CUG-binding proteins through activating signalling pathways involving protein kinase C (PKC). Hence the CUG RNA expansions in myotonic dystrophy result in a reduced concentration of available nuclear muscleblind protein, but an increased concentration of nuclear CUG-binding proteins.

6.10.3 Changes in splicing patterns are observed in myotonic dystrophy

At least 20 pre-mRNAs are mis-spliced in myotonic dystrophy. The changes in the nuclear concentration of muscleblind and CUG-binding proteins cause these splicing defects in myotonic dystrophy patients. There are three particularly important examples that result in patient symptoms.

1 Defects in splicing of the insulin receptor pre-mRNA (IR in Fig. 6.26). Insulin is a hormone that is important for maintaining blood sugar levels, and it circulates around the bloodstream after a meal to control blood sugar. Insulin works by binding to receptors in skeletal muscle, causing cells to take up glucose and store it as glycogen. The splicing defects in the insulin receptor in myotonic dystrophy patients prevent them from absorbing glucose normally from their bloodstream leading to glucose intolerance. Increased CUGBP causes an adult exon to be left out in adult myotonic dystrophy patients.

2 Defects in splicing control of a pre-mRNA encoding a chloride channel. A second splicing defect affects muscle contraction in myotonic dystrophy patients. Muscle relaxation involves a chloride channel called ClC. Myotonic dystrophy patients cannot relax their muscles after contraction (this is called myotonia) because they splice poison exons to create non-functional *ClC* mRNAs (poison exons contain stop codons and so prevent protein translation—see Section 6.3.2). Increased CUGBP and decreased muscleblind causes an adult exon in ClC to be left out in myotonic dystrophy patients.

3 Defects in alternative splicing of the pre-mRNA encoding cardiac troponin T. Defects in alternative splicing also affect heart functions in myotonic dystrophy patients. Normally exon 5 of the *cardiac troponin T* mRNA is not spliced into mRNA in the adult heart. The higher nuclear concentrations of CUG-binding proteins in myotonic dystrophy patients lead to exon 5 being spliced and result in heart defects.

Each of these three splicing defects in myotonic dystrophy are actually embryonic patterns happening in the adult.

There is some hope that therapies to help patients with muscular dystrophy could be developed in the future.[44,45] In mouse models, muscleblind gene knockouts and upregulation of CUG-binding proteins mimic the symptoms of myotonic dystrophy. Mouse models that express extended CUG RNA repeats have also been made. Recently, some drugs have been found that bind to the pathogenic double-stranded CUG repeats that are expressed in myotonic dystrophy and prevent muscleblind protein from associating with them. The hope is that in the future such drugs might be useful for therapy.

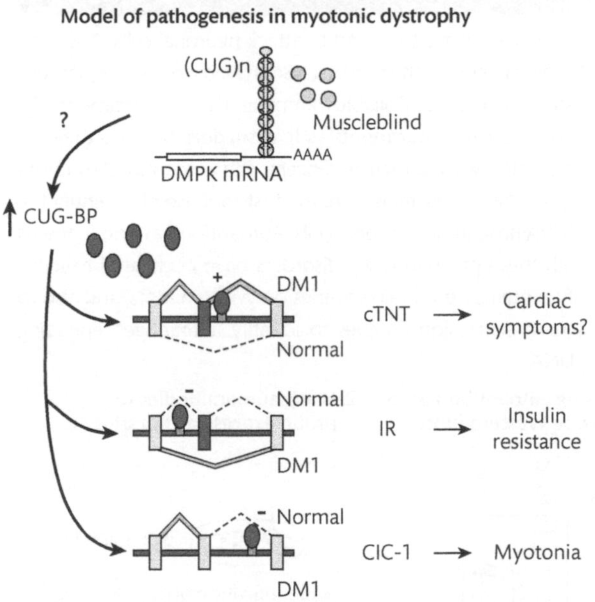

Figure 6.26 The pathogenic RNA repeats in myotonic dystrophy create an embryonic splicing environment in adult DM cells. Normal adult cells contain low concentrations of CUG-binding proteins and high concentrations of muscleblind protein (MBLN1). Normal embryonic cells have high nuclear levels of CUG-binding protein and low nuclear levels of MBLN1. In myotonic dystrophy there is a drop in nuclear MBLN and upregulation of nuclear CUG binding protein, resulting in embryonic splicing patterns of the Insulin Receptor (IR), Chloride Channel 1 (ClC-1), and cardiac Troponin T (cTNT) pre-mRNAs. (From Lee and Cooper[46] with permission.)

Key points

Changes in splicing patterns normally happen between the embryo and adult so as to make different proteins appropriate for each stage. These splicing pattern changes are disrupted in a disease called myotonic dystrophy. In myotonic dystrophy patients, pathogenic expanded CUG RNA repeat sequences change the cellular distribution of MBLN and CUG-binding protein splicing regulator proteins. The overall effect of the pathogenic RNA repeats is that adult DM cells have an embryonic splicing environment, and so embryonic patterns of splicing are continued, resulting in the multi-systemic disease characteristic of myotonic dystrophy.

6.11 RNA splicing regulators play a critical role in nervous system development in animals

From our discussion of myotonic dystrophy we can see that alternative splicing has a key role in the functional development of the heart, the insulin receptor, and muscle. Another tissue in which alternative splicing has been shown to be very important is the nervous system.

Several RNA splicing regulators regulate splicing patterns in the nervous system. These include an RNA-binding protein called **NOVA.**[48,49] Humans have two similar NOVA proteins called NOVA1 and NOVA2 (how these proteins were first discovered relates to human disease pathology, and is described in Box 6.6).

Knockout experiments have removed the *NOVA* genes from mice. The results are dramatic: mice without either *NOVA1* or *NOVA2* die shortly after birth as a result of apoptotic cell death of motor neurons. Mice lacking the NOVA protein are significantly smaller than their wild-type littermates, and also cannot stand up (a side-by-side comparison of these mice is shown in Fig. 6.28). Although NOVA1 and NOVA2 seem to be similar proteins, they are expressed in different parts of the brain, which is probably why they cannot complement for the loss of each other in these knockout experiments.

What do NOVA proteins do that is so important for survival in mice? To find out a group based at the Rockefeller University in New York analysed splicing differences between wild-type and NOVA

Box 6.6 NOVA proteins were first identified through their roles in suppressing cancers in humans with apparent neurological disorders

Along with their major role in controlling splicing patterns in nervous system development, NOVA proteins have a connection with cancer. NOVA1 and NOVA2 proteins are normally expressed in the brain and central nervous system, but become highly expressed in some tumours, where they can be ectopically expressed on the cancer surface. People suffering from these NOVA-expressing cancers raise antibodies in their bloodstream against NOVA proteins (see Fig. 6.27). This overexpression is initially helpful. This immune response to NOVA protein on the cancer cells causes the tumours to shrink. Later on these same autoantibodies cause neurological difficulties in the patients since they start to attack neuronal cells. Antibody binding to cells in the nervous system leads to neurological defects such as uncontrollable trembling. These autoimmune diseases are called **paraneoplastic disorders** because they are connected with particular cancers. The autoantibodies raised against NOVA proteins were the first tool used to identify the NOVA proteins in neuronal cells. Autoantibodies from patients with these paraneoplastic disorders have been used to identify proteins from brain samples on Western blots, and also to screen expression libraries to identify autoantigen-encoding cDNAs.

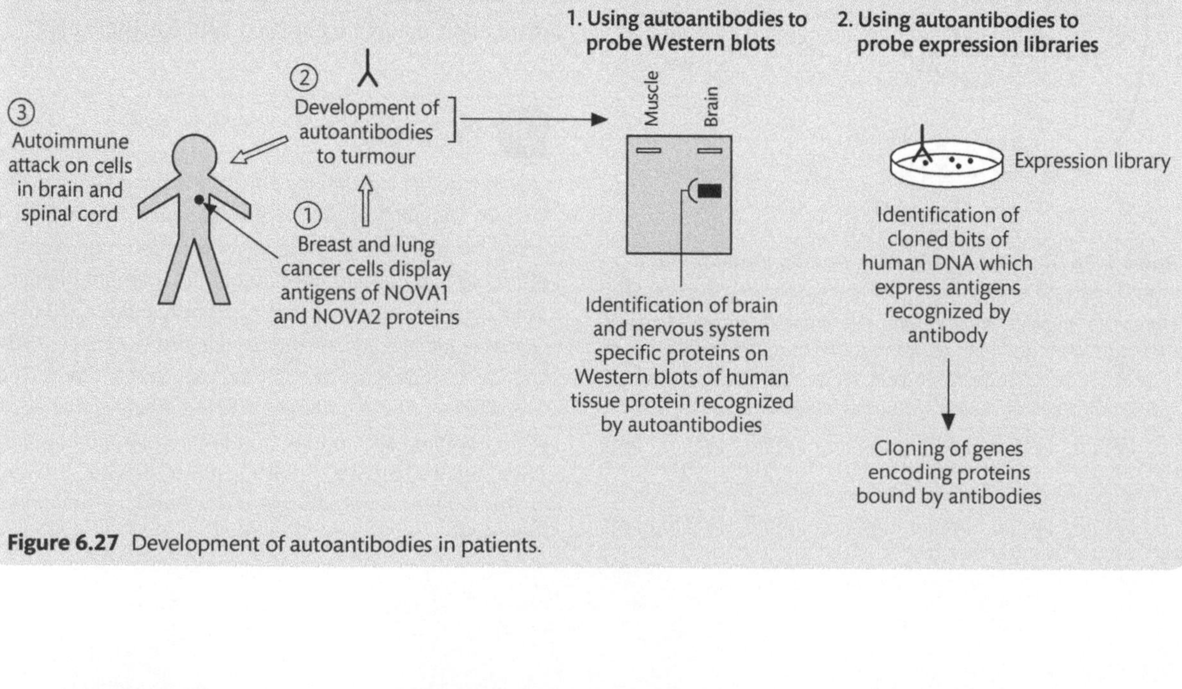

Figure 6.27 Development of autoantibodies in patients.

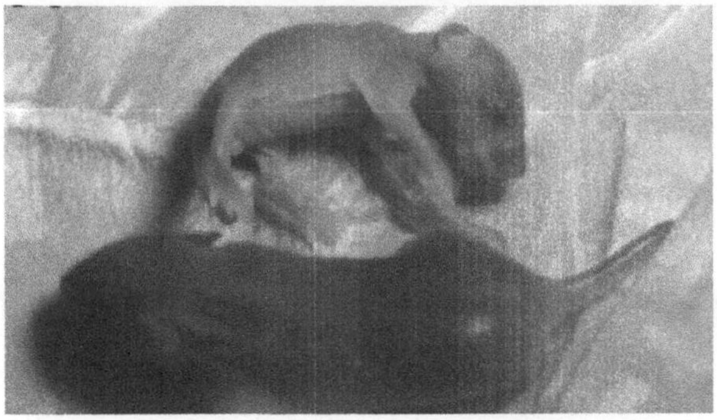

Figure 6.28 **The RNA-binding protein NOVA1 is critical for mouse development.** Mice containing a homozygous knockout of NOVA1 (top) were found to be significantly smaller than their wild-type littermates (bottom), and could not stand up. (From Jensen et al.[50])

knockout mice. They found that around 50 exons were affected after NOVA protein expression was eliminated in the mouse brain after genetic knockout. Many splicing targets for NOVA protein were within exons encoding proteins involved in synapse function. This shows that NOVA proteins are particularly important for shaping the synapse.

One of the exons regulated by NOVA is in the GlyRα2 pre-mRNA, which encodes a subunit of the glycine receptor in the brain. This exon is shown as an example in Fig. 6.29. In the absence of NOVA protein this GlyRα2 and many other exons are misregulated, cumulatively leading to the phenotype of the knockout mice shown in Fig. 6.28. NOVA is just one of the proteins important for neural splicing patters; notice in Fig. 6.29 that an adjacent exon is under the control of another splicing regulator protein expressed in the brain called brPTB.

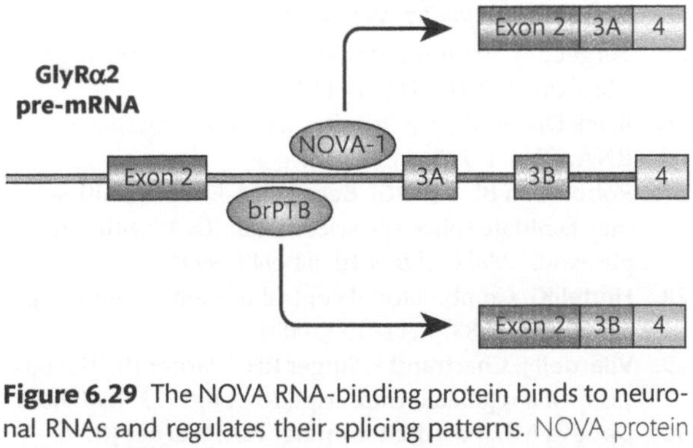

Figure 6.29 **The NOVA RNA-binding protein binds to neuronal RNAs and regulates their splicing patterns.** NOVA protein binds to the GlyRα2 pre-mRNA and regulates its splicing by promoting inclusion of exon 3A. A second neuronal specific splicing factor called brPTB represses exon 3A, and as a result exon 3B is included instead. (From Dredge et al.[50] with permission.)

Summary

In this chapter we have discussed the important role of alternative pre-mRNA splicing in expanding the coding capacity of the genome and in gene regulation, both in response to the environment and in development.

- Alternative splicing produces different mRNA isoforms from the same gene and helps to generate proteomic diversity from a finite number of genes.
- Different kinds of alternative splicing occur depending on whether whole exons are alternatively spliced or bits of exons are left out.
- Splicing can be positively regulated (by splicing activators) or repressed (by splicing repressors). The level of splicing of exons of intermediate strength will depend on how much the splicing machinery is helped to recognize them (positive regulation) or whether recognition is blocked (negative regulation).
- Different species have different kinds of alternative splicing. In multicellular animals most alternative splicing is through exon skipping, while in yeast most alternative splicing is through intron retention.
- Alternative splicing is particularly prevalent in metazoans, but plays a role in all eukaryotes.
- The major factors which affect alternative splicing are the concentration of nuclear RNA-binding proteins, the speed of transcription, and the activity of cellular signalling pathways.
- Alternative splicing can control complex patterns of development. Key examples are sexual differentiation in the fruit fly, embryo–adult switches in splicing in development, and patterns of splicing in the nervous system.

Questions

6.1 Name the different kinds of alternative splicing that occur.

6.2 Which form of alternative splicing is most frequent in animal cells, and which in yeast? Why is there this difference?

6.3 How is alternative splicing detected at a single gene and at global level?
6.4 What is meant by the splicing code?
6.5 How do proteins binding to exons assist the spliceosome to assemble?
6.6 How does the RPL32 protein control its own expression levels in yeast?
6.7 How is sex-specific splicing of the Doublesex pre-mRNA regulated in fruit flies?
6.8 How does a sex-specific difference in Doublesex splicing cause phenotypic differences in male and female flies?
6.9 How can changes in transcription influence splicing patterns?
6.10 What is the mechanism behind the disease myotonic dystrophy? How is this connected with alternative splicing?
6.11 How was the alternative splicing regulator NOVA discovered?

References

1. **Maniatis T, Tasic B.** Alternative pre-mRNA splicing and proteome expansion in metazoans. *Nature* **418**, 236–43 (2002).
2. **Nilsen TW, Graveley BR.** Expansion of the eukaryotic proteome by alternative splicing. *Nature* **463**, 457–63 (2010).
3. **Gerstein MB, Bruce C, Rozowsky JS, et al.** What is a gene, post-ENCODE? History and updated definition. *Genome Res* **17**, 669–81 (2007).
4. **Mudge JM, Frankish A, Harrow J.** Functional transcriptomics in the post-ENCODE era. *Genome Res* **23**, 1961–73 (2013).
5. **Rajan P, Elliott DJ, Robson CN, Leung HY.** Alternative splicing and biological heterogeneity in prostate cancer. *Nat Rev Urol* **6**, 454–60 (2009).
6. **Pennisi E.** Why do humans have so few genes? *Science* **309**, 80 (2005).
7. **Ward AJ, Cooper TA.** The pathobiology of splicing. *J Pathol* **220**, 152–63 (2010).
8. **Kelemen O., Convertini P, Zhang Z, et al.** Function of alternative splicing. *Gene* **514**, 1–30 (2013).
9. **Matlin AJ, Clark F, Smith CW.** Understanding alternative splicing: towards a cellular code. *Nat Rev Mol Cell Biol* **6**, 386–98 (2005).
10. **Karolchik D, Barber GP, Casper J, et al.** The UCSC Genome Browser database: 2014 update. *Nucleic Acids Res* **42**, D764–70 (2014).
11. **Elliott DJ, Grellscheid SN.** Alternative RNA splicing regulation in the testis. *Reproduction* **132**, 811–19 (2006).
12. **Blencowe BJ, Ahmad S, Lee LJ.** Current-generation high-throughput sequencing: deepening insights into mammalian transcriptomes. *Genes Dev* **23**, 1379–86 (2009).
13. **Djebali S, Davis CA, Merkel A, et al.** Landscape of transcription in human cells. *Nature* **489**, 101–8 (2012).
14. **Ellis JD, Barrios-Rodiles M, Colak R, et al.** Tissue-specific alternative splicing remodels protein-protein interaction networks. *Mol Cell* **46**, 884–92 (2012).
15. **Pan Q, Shai O, Lee LJ, et al.** Deep surveying of alternative splicing complexity in the human transcriptome by high-throughput sequencing. *Nat Genet* **40**, 1413–15 (2008).
16. **Syed NH, Kalyna M, Marquez Y, et al.** Alternative splicing in plants—coming of age. *Trends Plant Sci* **17**, 616–23 (2012).
17. **Black DL.** Protein diversity from alternative splicing: a challenge for bioinformatics and post-genome biology. *Cell* **103**, 367–70 (2000).
18. **Watson FL, Püttman-Holgado R, Thomas F, et al.** Extensive diversity of Ig-superfamily proteins in the immune system of insects. *Science* **309**, 1874–8 (2005).
19. **Barash Y, Calacro AJ, Gao W, et al.** Deciphering the splicing code. *Nature* **465**, 53–9 (2010).
20. **Wang GS, Cooper TA.** Splicing in disease: disruption of the splicing code and the decoding machinery. *Nat Rev Genet* **8**, 749–61 (2007).
21. **Kalsotra A, Cooper TA.** Functional consequences of developmentally regulated alternative splicing. *Nat Rev Genet* **12**, 715–29 (2011).
22. **Lareau LF, Inada M, Green RE, et al.** Unproductive splicing of SR genes associated with highly conserved and ultraconserved DNA elements. *Nature* **446**, 926–9 (2007).
23. **Ni JZ, Grate L, Donohue JP, et al.** Ultraconserved elements are associated with homeostatic control of splicing regulators by alternative splicing and nonsense-mediated decay. *Genes Dev* **21**, 708–18 (2007).
24. **Schwartz S, Hall E, Ast G.** SROOGLE: webserver for integrative, user-friendly visualization of splicing signals. *Nucleic Acids Res* **37**, W189–92 (2009).
25. **Berget SM.** Exon recognition in vertebrate splicing. *J Biol Chem* **270**, 2411–14 (1995).
26. **Black DL.** Finding splice sites within a wilderness of RNA. *RNA* **1**, 763–71 (1995).
27. **Robberson BL, Cote GJ, Berget SM.** Exon definition may facilitate splice site selection in RNAs with multiple exons. *Mol Cell Biol* **10**, 84–94 (1990).
28. **Hertel KJ.** Combinatorial control of exon recognition. *J Biol Chem* **283**, 1211–15 (2008).
29. **Vilardell J, Chartrand P, Singer RH, Warner JR.** The odyssey of a regulated transcript. *RNA* **6**, 1773–80 (2000).
30. **Dabeva MD, Post-Beittenmiller MA, Warner JR.** Autogenous regulation of splicing of the transcript of a yeast ribosomal protein gene. *Proc Natl Acad Sci USA* **83**, 5854–7 (1986).

31. **Munding EM, Igel AH, Shiue L, et al.** Integration of a splicing regulatory network within the meiotic gene expression program of *Saccharomyces cerevisiae*. *Genes Dev* **24**, 2693–704 (2010).
32. **Munding EM, Shiue L, Katzman S, et al.** Competition between pre-mRNAs for the splicing machinery drives global regulation of splicing. *Mol Cell* **51**, 338–48 (2013).
33. **Engebrecht JA, Voelkel-Meiman K, Roeder GS.** Meiosis-specific RNA splicing in yeast. *Cell* **66**, 1257–68 (1991).
34. **Spingola M, Armisen J, Ares M, Jr.** Mer1p is a modular splicing factor whose function depends on the conserved U2 snRNP protein Snu17p. *Nucleic Acids Res* **32**, 1242–50 (2004).
35. **Spingola M, Ares M, Jr.** A yeast intronic splicing enhancer and Nam8p are required for Mer1p-activated splicing. *Mol Cell* **6**, 329–38 (2000).
36. **Forch P, Valcarcel J.** Splicing regulation in *Drosophila* sex determination. *Prog Mol Subcell Biol* **31**, 127–51 (2003).
37. **Stamm S.** Regulation of alternative splicing by reversible protein phosphorylation. *J Biol Chem* **283**, 1223–7 (2008).
38. **Shin C, Manley JL.** Cell signalling and the control of pre-mRNA splicing. *Nat Rev Mol Cell Biol* **5**, 727–38 (2004).
39. **Shomron N, Alberstein M, Reznik M, Ast G.** Stress alters the subcellular distribution of hSlu7 and thus modulates alternative splicing. *Journal of Cell Science* **118**, 1151–9 (2005).
40. **van der Houven van Oordt W, Diaz Meco MT, Lozano J, et al.** The MKK(3/6)-p38-signaling cascade alters the subcellular distribution of hnRNP A1 and modulates alternative splicing regulation. *J Cell Biol* **149**, 307–16 (2000).
41. **de la Mata M, Alonso CR, Kadener S, et al.** A slow RNA polymerase II affects alternative splicing in vivo. *Mol Cell* **12**, 525–32 (2003).
42. **Kornblihtt AR.** Chromatin, transcript elongation and alternative splicing. *Nat Struct Mol Biol* **13**, 5–7 (2006).
43. **Kanadia RN, Johnstone KA, Mankodi A, et al.** A muscleblind knockout model for myotonic dystrophy. *Science* **302**, 1978–80 (2003).
44. **Ward AJ, Rimer M, Killian JM, et al.** CUGBP1 overexpression in mouse skeletal muscle reproduces features of myotonic dystrophy type 1. *Hum Mol Genet* **19**, 3614–22 (2010).
45. **Warf MB, Nakamori M, Matthys CM, et al.** Pentamidine reverses the splicing defects associated with myotonic dystrophy. *Proc Natl Acad Sci USA* **106**, 18 551–6 (2009).
46. **Lee JE, Cooper TA.** Pathogenic mechanisms of myotonic dystrophy. *Biochem Soc Trans* **37**, 1281–6 (2009).
47. **Mankodi A, Urbinati CR, Yuan QP, et al.** Muscleblind localizes to nuclear foci of aberrant RNA in myotonic dystrophy types 1 and 2. *Hum Mol Genet* **10**, 2165–70 (2001).
48. **Licatalosi DD, Mel A, Fak JJ, et al.** HITS-CLIP yields genome-wide insights into brain alternative RNA processing. *Nature* **456**, 464–9 (2008).
49. **Ule J, Ule A, Spencer J, et al.** Nova regulates brain-specific splicing to shape the synapse. *Nat Genet* **37**, 844–52 (2005).
50. **Jensen KB, Dredge BK, Stefani G, et al.** Nova-1 regulates neuron-specific alternative splicing and is essential for neuronal viability. *Neuron* **25**, 359–71 (2000).
51. **Dredge BK, Polydorides AD, Darnell RB.** The splice of life: alternative splicing and neurological disease. *Nat Rev Neurosci* **2**, 43–50 (2001).

7 Pre-mRNA splicing defects in development and disease

Introduction

This chapter is going to consider the important contribution that RNA splicing defects make to disease. We have touched upon this topic already in Chapter 6, when we discussed myotonic dystrophy, which is a disease of alternative splicing; and the NOVA proteins that are important in cancer autoimmunity as well as in splicing regulation in the nervous system. This current chapter goes into the topic of splicing and disease in much more depth.

7.1 Mutations affecting the splicing code can be catastrophic for gene function

Much of the information upon which this chapter is based comes from human genetics (the study of inheritance in humans). The defective genes in many inherited diseases, and many of the actual mutations (see Box 7.1) within these genes, have been pinpointed by sequencing. Some of these mutations affect the *genetic code*, by introducing codons for new amino acids and even stop codons which prematurely end open reading frames. However, other mutations can occur in splicing signals, and these latter mutations can prevent the intron–exon structure of pre-mRNAs from being properly decoded by the spliceosome.[1] These latter mutations are in the *splicing code* (see Chapter 6).

Mutations that affect splicing signals can be particularly severe since they cause changes in the structure of mRNAs. One frequently observed effect of mutations in splicing signals is exons being left out, or skipped, by the spliceosome. The reason why mutations in splicing signals cause exon skipping is that the intron–exon structure of most pre-mRNAs is recognized by a process called *exon definition* (see Section 5.12). To illustrate this, let us look at the hypothetical gene shown in Fig. 7.1. In the absence of any mutations to splicing signals, the normal splicing pattern of the pre-mRNA made from this gene will be for exons 1 to 3 to be spliced together in order in the mRNA, using the correct splice sites and removing the internal introns (this is the pattern in Fig. 7.1a). However, because of exon definition, if a mutation inactivates the 3′ splice site of exon 2, this will stop this whole second exon being properly recognized by the spliceosome. As a result exon 2 will be skipped, so exon 3 will be directly

Box 7.1 Mutations in DNA sequences are a source of inherited disease in humans

Point mutations cause single nucleotide changes to DNA sequences. Point mutations in exons can result in changes to protein coding information—these are effects on the genetic code. Point mutations in splicing control sequences can affect whether an exon becomes spliced into an mRNA or not—these are effects on the splicing code.

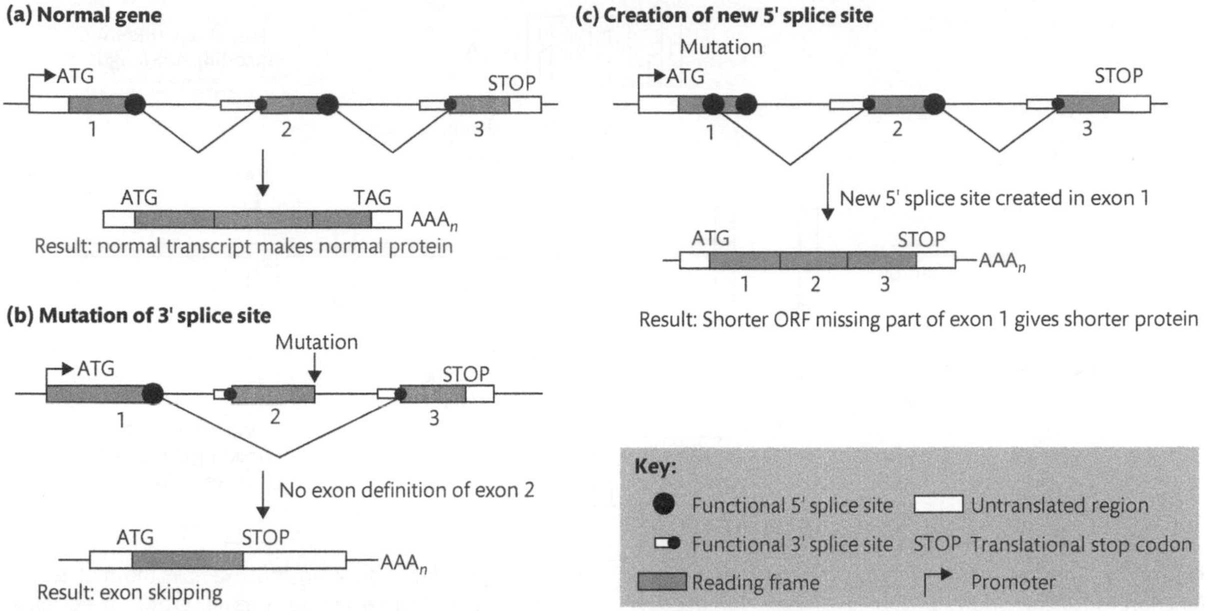

Figure 7.1 Mutations in splicing control sequences frequently result in whole exons being left out of the spliceosome in humans. To illustrate the effects of splicing co-mutations, this example shows a three-exon gene. (a) The wild-type gene is transcribed, and then all three exons are joined together in the order 1-2-3 by pre-mRNA splicing. (b) If there are mutations in the 3′ splice site of exon 2, they prevent this exon being recognized by the spliceosome (they prevent exon 2 definition). As a result exon 2 is skipped by the spliceosome, and the order of exons in the mRNA is 1-2. This kind of exon skipping is the most frequent outcome of mutations that affect splicing signals. (c) Mutations can also create new splice sites in internal exons.

spliced onto exon 1 in the mRNA (this is the pattern in Fig. 7.1b).

Exon skipping is the most frequent outcome of mutations affecting splicing signals in humans. Less frequent splicing consequences downstream of mutations affecting the splicing code include the creation of new splice sites within the pre-mRNA (e.g. Fig. 1c). Later in this chapter we shall discuss a mutation which causes a premature ageing syndrome through the creation of a new 5′ splice site (see the discussion of Hutchinson–Gilford progeria syndrome (HGPS) in Section 7.3). In some other cases whole introns can be left in mRNAs; this is called **intron retention**.

Since they affect the exon structure of mRNAs, mutations in splicing signals can have particularly devastating effects on gene function. To illustrate this, let us look at another hypothetical example in which a gene undergoes a single point mutation. This mutation is shown as a lightning bolt in Fig. 7.2, and causes a single nucleotide change in the gene sequence shown as an asterisk. Let us compare what happens to gene function in two different scenarios:

Scenario 1 Point mutations that change the genetic code. This kind of mutation is shown in the upper pathway of Fig. 7.2. In this case the point mutation changes a nucleotide, and this in turn changes one codon in the open reading frame, and as a result changes one amino acid in the encoded protein. (Note that an exception to this milder effect would be point mutations that create a stop codon that leads to premature termination of transcription.)

Scenario 2 Point mutations which affect RNA splicing signals. This kind of mutation is seen in the lower pathway in Fig. 7.2. In this case, the mutation is in a splice signal, and because of exon definition this mutation leads to the whole of exon 2 being missed (skipped) by the splicing machinery. Hence this splicing defect can amplify a single nucleotide change in a gene into a hugely altered mRNA transcript made from the gene.

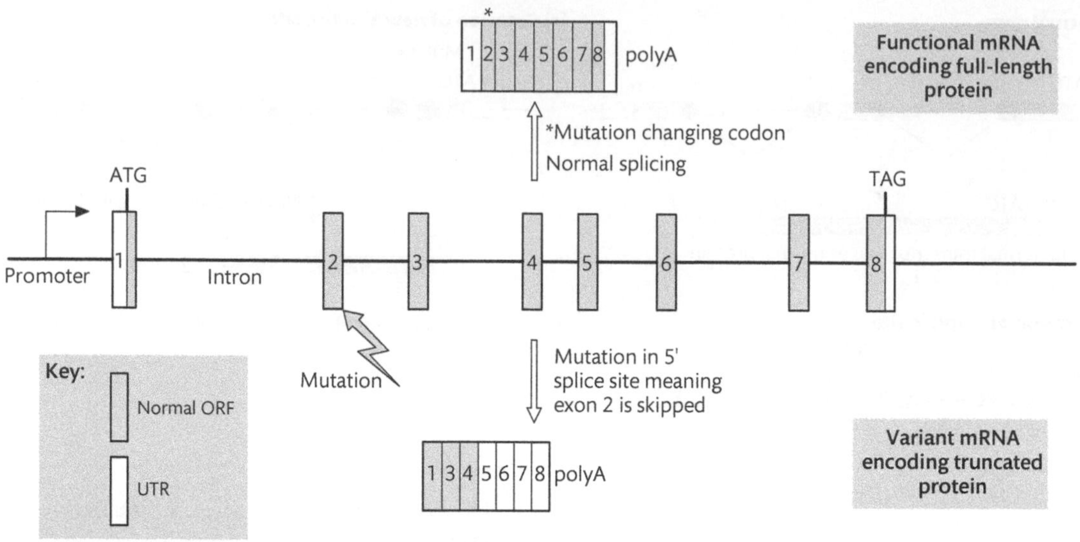

Figure 7.2 Splicing mutations can be catastrophic for gene function. This figure shows a hypothetical gene that becomes mutated (mutation shown as a lightning bolt) in a single nucleotide position (this is called a point mutation) leading to a genetic disease. In the upper pathway the mutation changes a codon within exon 2 of the gene (the mutated codon is indicated with an asterisk) but does not affect splicing. In this case a protein is still made, but with a slightly different sequence. In the lower pathway the point mutation affects an important splicing signal for exon 2—in this example the mutated sequence is the 5′ splice site. As a result, exon 2 is no longer recognized by the spliceosome and so is skipped (not spliced into the mRNA). Since exon 2 is not a multiple of three nucleotides in length, missing this exon out leads to a different downstream reading frame, either encoding a very different protein product or leading to the introduction of early stop codon into the mRNA (resulting in short truncated protein products and unstable mRNA).

Hence, point mutations that change the genetic code can be less severe than those that change the splicing code, since in scenario 1 the mRNA still has a full-length open reading frame and so a (variant) protein is still made.

7.2 Mutations in splicing control sequences are very frequent causes of human genetic disease

Disease-causing mutations that affect splicing patterns are very frequent. An early survey published in 1992 reported that ~15% of the total number of disease-causing point mutations were within either the 5′ or 3′ splice sites.[3] More focused studies have analysed the molecular effect of mutations in the *NF1* and *ATM* genes which result in neurofibromatosis and ataxia telangiectasia, respectively.[4,5] In these studies ~50% of point mutations leading to disease caused defects in mRNA splicing. For a group of patients with splicing mutations within the *ATM* gene no full-length ATM protein was made at all, indicating the severe effect of these splicing defects on *ATM* gene expression.

As we saw in Fig. 7.2, mutations in splice site sequences (see Chapter 5 for more details of splice site sequences) can have dramatic effects on how exons are recognized by the splicing machinery. However, changes to splicing control sequences outside splice sites, including changes to splicing enhancers and splicing silencers, can also affect splicing (remember from Chapter 6 that the complete set of instructions in the pre-mRNA that control how an exon is spliced is called the splicing code). The locations of the different mutations that affect the splicing code are shown in Fig. 7.3. In the next sections we are going to describe two specific examples where the splicing code is compromised—one where a splice site is affected, and one where an ESE is affected.

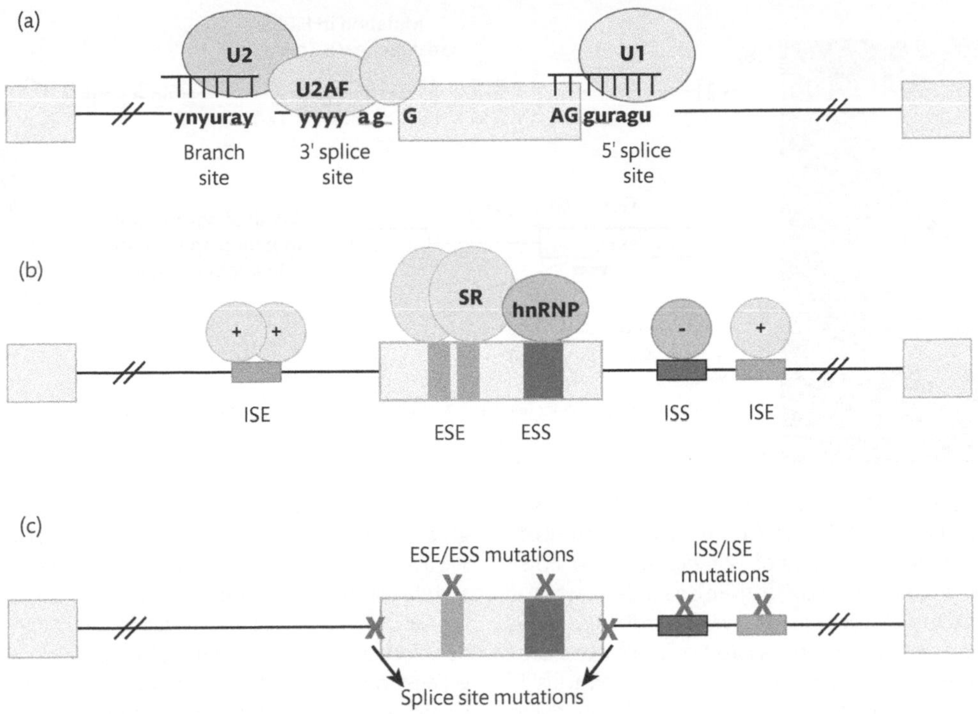

Figure 7.3 Mutations affecting splicing control sequences are found in different locations in the pre-mRNA. The *cis*-elements which are important for controlling pre-mRNA splicing include (a) the 5′ and 3′ splice sites, and (b) enhancer sequences which promote splicing (ESEs and ISEs in exons and introns, respectively) and silencer elements that repress exon splicing (ISEs and ISSs in exons and introns, respectively). (c) Mutations on each of these sequence elements can change splicing patterns in disease. (Reprinted from Ward and Cooper[2] with permission.)

7.3 Genetic mutations create a new splice site in a premature ageing disease

An example of a mutation involving a splice site is found in the devastating disease *Hutchinson–Gilford progeria syndrome* (HGPS). In HGPS, instead of a mutation destroying a splice site, what happens is that a new 5′ splice site is created within an exon.[6,7]

HGPS is a disease of accelerated ageing. People who suffer from HGPS have a lifespan of only about 13 years, and as children develop the physical features of old people, including wizened skin, osteoporosis, baldness, and clogged arteries. All these problems in HGPS patients are secondary to a primary defect in the structure of the nucleus. HGPS is caused by mutations within the *LMNA* gene which encodes a protein called prelamin A, a precursor to lamin A. Lamin A is a component of the nuclear lamina, which is a network of intermediate filaments just under the inner nuclear membrane that provides structural support for the nucleus. Lamin A protein is also found throughout the nucleoplasm.

Complete deletion of the *LMNA* gene causes a form of *muscular dystrophy*, a severe disease affecting muscles. However, in 2003 point mutations causing HGPS were also identified in the *LMNA* gene.[6,7] The point mutations in the *LMNA* gene are different from those that cause muscular dystrophy in that they are much more subtle, and affect splicing of the *LMNA* pre-mRNA by creating a new 5′ splice site within *LMNA* exon 11. Exon 11 is the last but one of the exons in the *LMNA* gene (there are 12 exons in the *LMNA* gene—see Fig. 7.4). This newly created 5′ splice site in HGPS patients means that part of exon 11 is missed out by the spliceosome. Use of the upstream mutant 5′ splice site in HGPS patients removes 150 nucleotides from the encoded *LMNA* mRNA. This results in the splicing pattern shown in Fig. 7.4, where both a shorter mRNA transcript missing 150 nucleotides and the wild-type transcript can be detected in the RNA of patients with HGPS (note that HGPS patients contain

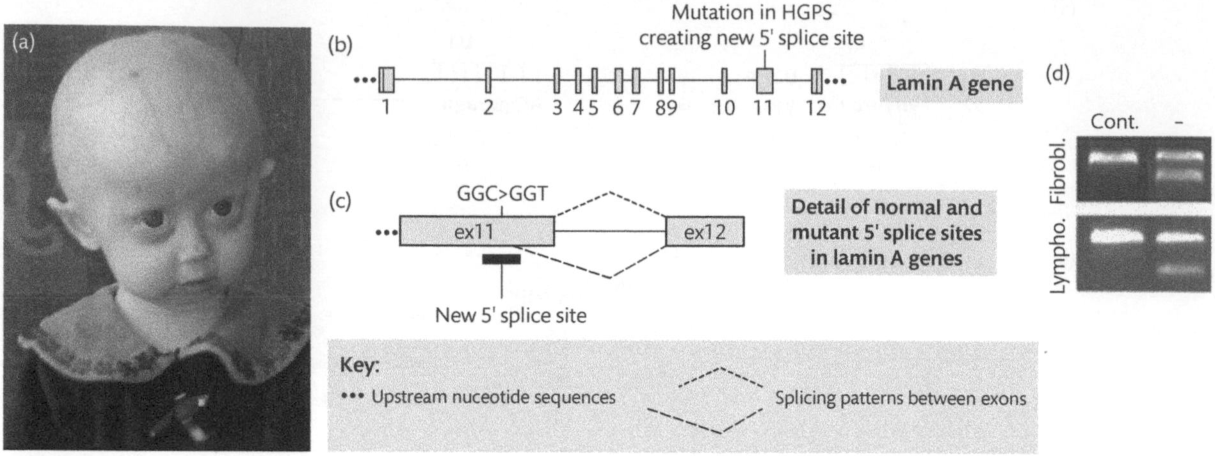

Figure 7.4 **A point mutation which creates a new 5′ splice site in *LMNA* exon 11 causes HGPS.** (a) Picture of a child with HGPS. (b) The *LMNA* gene is composed of 12 exons, shown as boxes 1–12. (c) Internal to exon 11 of *LMNA* there is a sequence which is almost identical to a 5′ splice site, but is not normally recognized by the spliceosome. The point mutations that cause HGPS convert this 5′ splice-site-like sequence within exon 11 to an actual 5′ splice site that is recognized by the spliceosome. Selection of this internal 5′ splice site by the spliceosome means that part of exon 11 is spliced out of *LMNA* mRNA (the part downstream of the internal 5′ splice site becomes an intron sequence), giving a shorter transcript. (d) HGPS patients have one normal copy of the *LMNA* gene and one exon 11 mutated copy; one copy of the *LMNA* gene still uses the normal exon 11 5′ splice site to make *LNMA* mRNA, and the other copy uses the alternative 5′ splice site. Both the shorter *LMNA* transcript and the full-length *LMNA* transcript can be detected using RT-PCR from patient RNA, followed by agarose gel electrophoresis. In fibroblasts and lymphoblasts from normal individuals who do not have HGPS a single product is seen that includes the whole of exon 11. (Part (a) is reproduced from Scaffidi et al.[8]; parts (c) and (d) are reproduced from Scaffidi and Misteli[9] with permission from Macmillan Publishers Ltd.)

only one mutated copy of the *LMNA* gene—the other copy on the other chromosome is normal).

The code mutation that causes HGPS is dominant. In other words, only one copy of the *LMNA* gene is mutated in HGPS patients, while the copy on the other chromosome remains normal. Why a change in the *LMNA* 5′ splice site in one gene copy results in HGPS is understood in some detail.[7] Removal of 150 nucleotides from the *LMNA* mRNA changes the C-terminus of the encoded LMNA protein. This change is very important. In the processing pathway normally followed, lamin A protein is first translated as a precursor protein called prelamin A which is modified by addition of a farnesyl group. This farnesyl group acts as a signal that moves the prelamin A protein to the nuclear membrane. Once at the nuclear membrane, the prelamin A protein is cut by a protease called ZMPSTE24 to release mature lamin A protein. Because of the splicing change in *LMNA* in HGPS patients, the coding sequences for the ZMPSTE24 binding site become removed from the encoded prelamin A protein in HGPS patients (Fig. 7.5). This prevents the proper processing of prelamin A protein.

The mutant prelamin A protein made in HGPS patients is called **progerin** (named after progeria, the medical term for diseases that cause early ageing). This progerin protein is toxic to the cell, since it can associate with the nuclear membrane but cannot be released from the nuclear membrane by ZMPSTE24 protease cleavage. It is this accumulation of progerin at the nuclear membrane that causes membrane instability and premature ageing effects of HGPS.

The defects on cell morphology from an HGPS patient are very obvious if cells are examined under the microscope. Antibodies that recognize lamin A protein normally stain a ring-like structure that surrounds the nucleus—this is the nuclear lamina. This nuclear lamina pattern changes in HGPS patients to become more irregular because progerin has not correctly localized.

HGPS is a disease caused by a mutation creating a new 5′ site that is then used to make lots of progerin protein. However, lower levels of *LMNA* splicing defects may also contribute to the normal ageing process.[11] In fact, the sequence within exon 11 of the *LMNA* pre-mRNA that is altered in HGPS already somewhat resembles a 5′ splice site

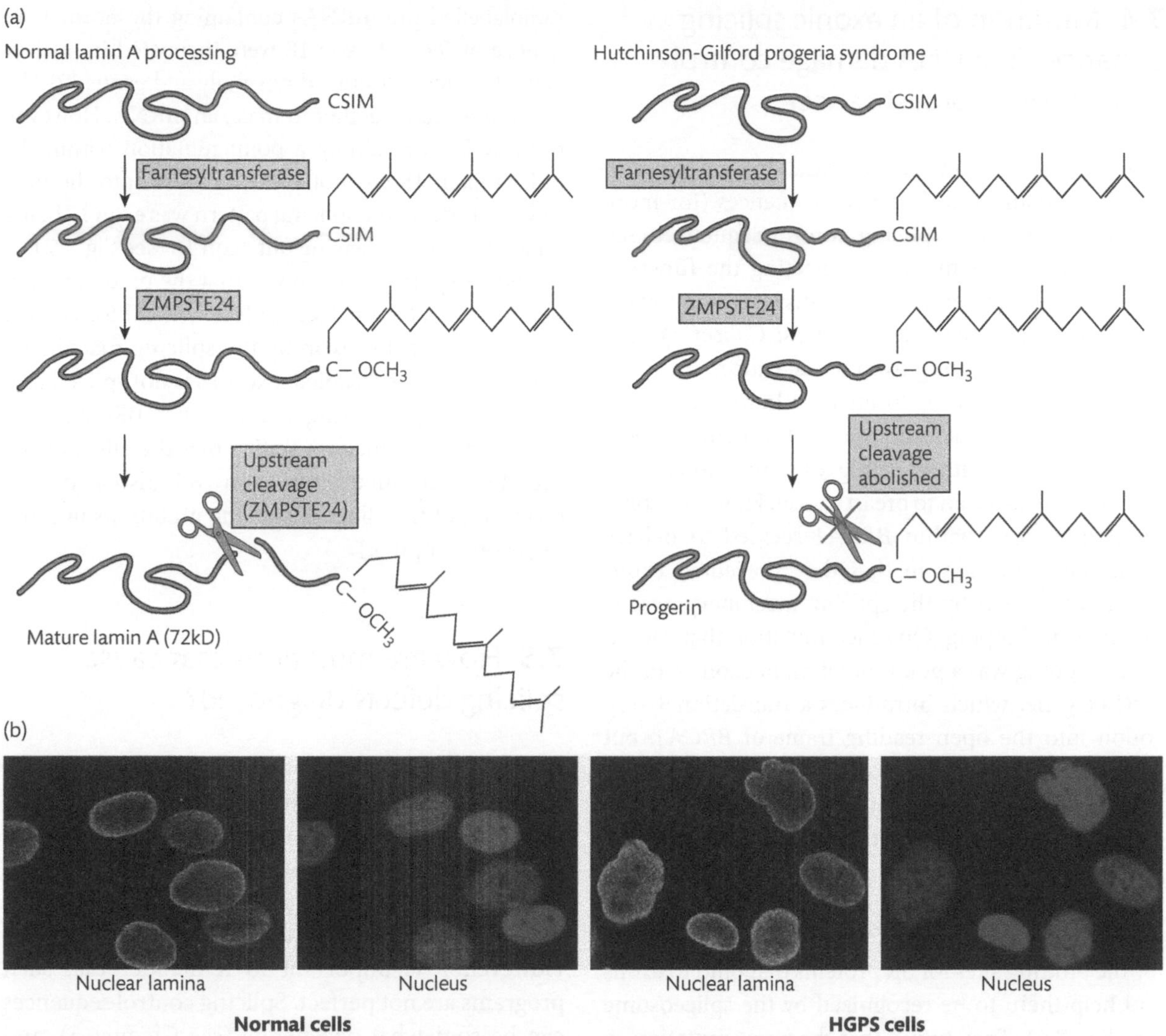

Figure 7.5 **The production pathways used for lamin A protein are different in HGPS patients.** (a) Production of lamin A protein involves first the production of a precursor protein called prelamin A, which is modified by the addition of a farnesyl group and then moves to the nuclear membrane and is cleaved to release mature lamin A protein. This final cleavage is blocked in HGPS patients. (From Gordon et al.[10] with permission from *Journal of Cell Biology*.) (b) Fibroblast cells stained for nuclear lamina (using an antibody to lamin A protein) or nuclear DNA (using a dye called DAPI). The cells are either from a patient with HGPS or an unaffected parent. Note that the nuclear lamina has an irregular structure in the HGPS patient. For a full colour version of Fig. 7.5b, see Colour Illustration 8. (From Eriksson et al.[7] with permission of Nature Publishing Group.)

even without any point mutations, and is used by the spliceosome at low frequency even in normal cells (this 5′ splice site is called a *cryptic*, meaning hidden, splice site). Splicing using this upstream 5′ cryptic splice site sequence leads to the production of low levels of progerin in normal ageing cells.

What do these low levels of progerin do in normal ageing cells? Over time, the gradual accumulation of progerin may result in some of the cellular defects associated with normal ageing. Experiments have been performed in which use of the *LMNA* exon 11 cryptic 5′ splice site is blocked using complementary oligonucleotides.[9] This procedure is effective at stopping progerin mRNA from being made, and also rescues some of the cellular defects observed in cells from older individuals. It is possible that in the future this kind of approach might provide the basis of an anti-ageing treatment.

7.4 Mutation of an exonic splicing enhancer in a DNA damage control gene leads to breast cancer

Genetic diseases can also be caused by mutations in splicing enhancer and silencer sequences (for more discussion of these splicing control sequences, see Chapter 6). Point mutations affecting the function of exonic splicing enhancers (ESEs) were first identified in 2001 in the *BRCA1* (*Breast Cancer 1*) gene from patients suffering from breast cancer.[12,13]

The *BRCA1* gene is located on human chromosome 17 and encodes an important protein involved in DNA repair. As its name suggests, mutations in the *BRCA1* gene can lead to breast cancer. However, some point mutations within *BRCA1* seemed to behave somewhat unusually in that they prevented exons being recognized by the splicing machinery, resulting in exon skipping. One such mutation that caused exon skipping was a point mutation in exon 18 of the *BRCA1* gene, which introduces a translational stop codon into the open reading frame of *BRCA1*, but which in patients caused exon 18 to be skipped.

How was this point mutation in *BRCA1* exon 18 causing exon skipping? Adrian Krainer and his colleagues at Cold Spring Harbor Laboratory screened *BRCA1* exon 18 using an online computer program they had developed called *ESEfinder*[14] which can look for the binding sites for SR proteins that bind to exons and help them to be recognized by the spliceosome (see Fig. 7.6). They found that the point mutation in exon 18 of the *BRCA1* gene which caused exon skipping corresponded exactly to the site of a high-affinity binding site for the splicing factor SRSF1.

Based on this observation, Krainer and colleagues tested if this ESE-associated point mutation could affect splicing of exon 18. Part of the *BRCA1* gene containing exons 17, 18, and 19 and their surrounding introns was cloned into a plasmid downstream of a T7 promoter, and this plasmid was then transcribed in vitro with radioactive nucleotides to generate a radiolabelled pre-mRNA (see Fig. 7.7).

Radiolabelled pre-mRNAs made from normal and mutated *BRCA1* were then incubated in a nuclear extract, and this enabled their splicing patterns to be analysed (further details about this kind of analysis, called *in vitro splicing*, is given in Chapter 5). When radiolabelled pre-mRNAs containing the normal sequence of *BRCA1* exon 18 were incubated in nuclear extract, splicing occurred normally and exons 17–18–19 were spliced together to make an mRNA. However, pre-mRNAs containing a point mutation within the ESE of exon 18 did not splice exon 18 into the final mRNA. Instead, the splicing pattern was exon 17 being joined to exon 19, leaving out exon 18 (see Fig. 7.7).

These experiments showed that the predicted ESE within *BRCA1* exon 18 is indeed critical for correct recognition of this exon by the splicing machinery. When the ESE is disabled, exon 18 is not spliced into the mRNA, preventing a functional BRCA1 protein from being made. (Notice that the effect of the *BRCA1* point mutation in exon 18 is also a good illustration of the effect of splicing mutations in causing exon skipping.)

7.5 How are mutations that cause splicing defects diagnosed?

How are mutations that affect splicing detected? The first task is to identify the location of the disease-causing mutation by DNA sequencing. Next, the nucleotide sequences can be analysed with computer programs that are designed to recognize important splicing sequences (e.g. ESEfinder in Fig. 7.6). However, it is important to remember that such programs are not perfect. Splicing control sequences can be somewhat degenerate (see Chapter 5), and so the effects that mutations in the DNA sequence have on pre-mRNA splicing often need to be tested experimentally.[15]

One experimental approach used to assess the effect of mutations on splicing is to directly compare patterns of exons spliced into mRNAs purified from diseased and unaffected individuals, ideally by examining changes within affected tissues. However, these comparisons are not always possible, since tissues affected in genetic diseases can sometimes be difficult to biopsy without harming the patient (e.g. the heart in myotonic dystrophy or the brain in neurodegenerative diseases).

A useful alternative to using RNA purified from tissue biopsies has been to use cell lines developed from a patient's blood. Figure 7.8 shows an example

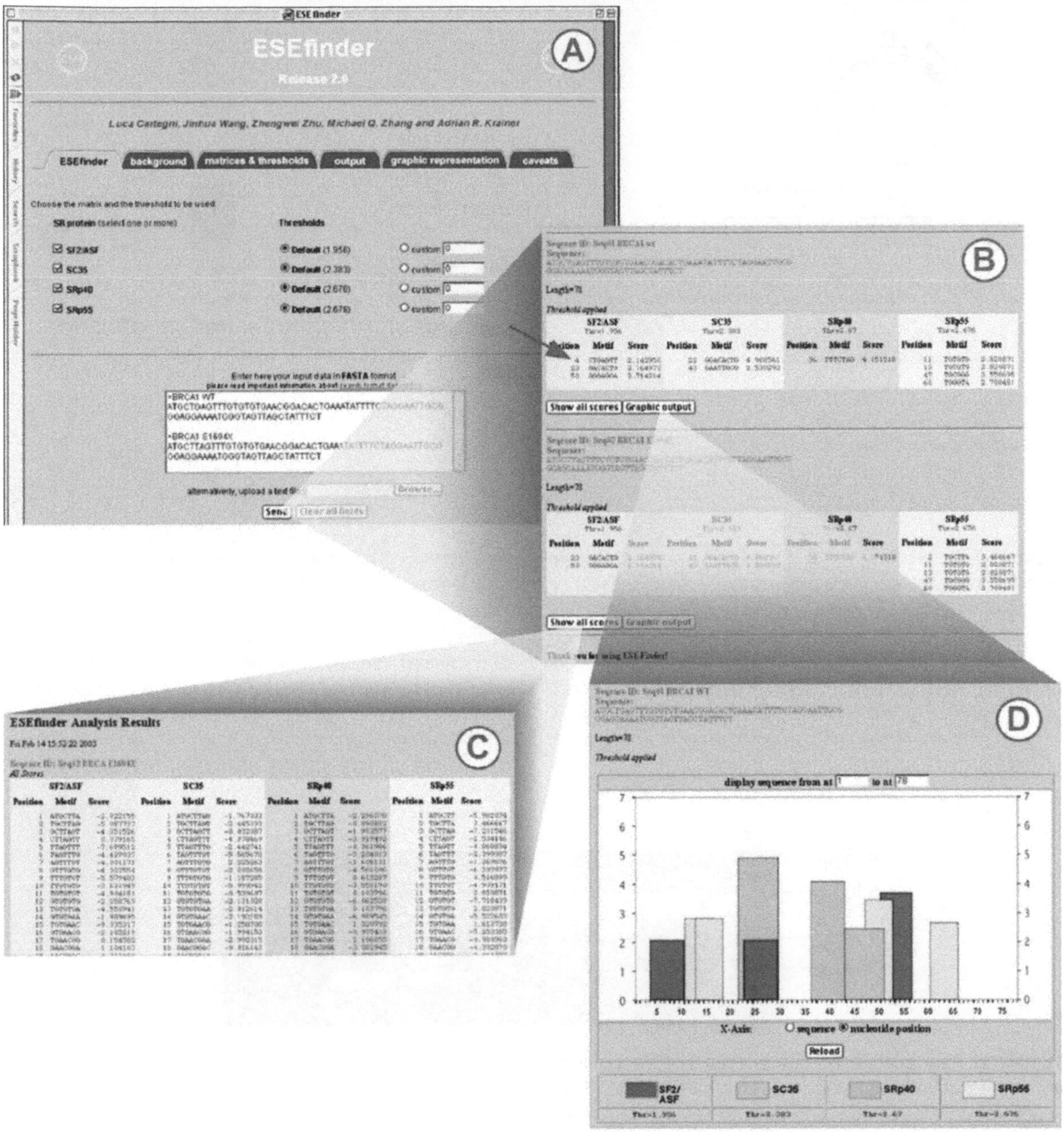

Figure 7.6 The online program ESEfinder (http://exon.cshl.edu/ESE/) can be used to search input sequences for high-affinity binding sites for the SR proteins SC35, SRp40, ASF/SF2, and SRp55. (A) An input sequence can be cut and pasted into ESEfinder. Although ESEfinder searches for RNA sequence elements, the input sequence is DNA (i.e. using ACGT nucleotides). ESEfinder searches for high-affinity binding sites for four SR proteins. ESEfinder tabulates (B) identified sites which have a strength above a threshold value and (C) all possible sites in the input sequence. (D) SR protein-binding sites are also shown as a graph, with the sequence along the *x*-axis and the *y*-axis representing the strength of the binding site. The graph shown here is an analysis of exon 18 of the *BRCA1* gene. This *BRCA1* exon was the first important ESE identified as a target for mutation in human genetic disease. (Reproduced from Cartegni *et al.*[14] with permission from Oxford University Press.)

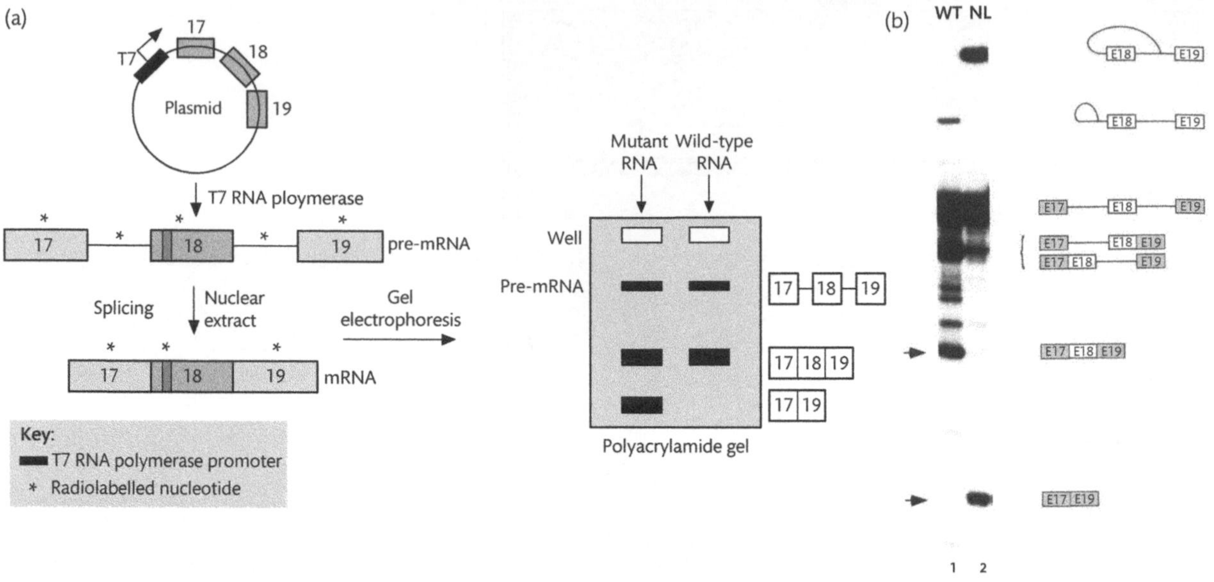

Figure 7.7 **Defects in splicing of the *BRCA1* pre-mRNA were detected in in vitro splicing experiments.** To test this, the splicing pattern of an in vitro transcript was monitored by incubation in nuclear extract and then electrophoresing the reaction products on a polyacrylamide splicing gel. (a) Exons 17–19 of *BRCA1* were cloned into a plasmid with a T7 promoter, and then transcribed in vitro with a radiolabelled nucleotide. This resulted in a radiolabelled pre-mRNA (the radiolabelled nucleotides are indicated with asterisks). This radiolabelled pre-mRNA was then incubated in nuclear extracts where it assembled spliceosomes and became spliced into an mRNA. Gel electrophoresis was used to separate out the pre-mRNA and reaction products (17–18–19 mRNAs and 17–19 spliced mRNAs, along with lariat intermediates from the splicing reaction. (b) The actual gel from the original experiment. (Part (b) reproduced from Liu et al.[13] with permission from Macmillan Publishers Ltd.)

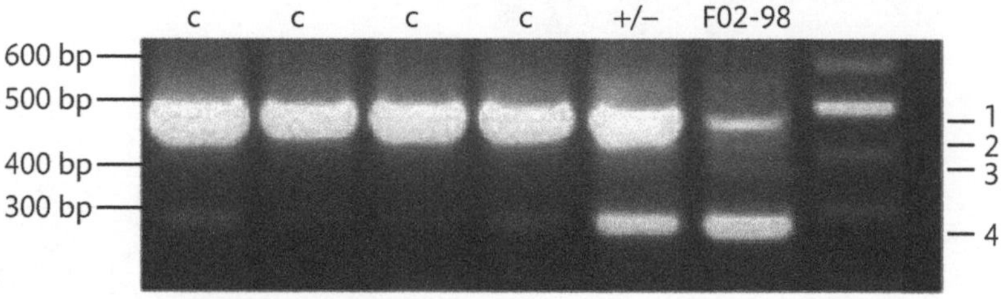

Figure 7.8 **Abnormal mRNA isoforms resulting from splicing mutations can be identified by RT-PCR of mRNAs made from lymphoblastoid cell lines made from patients.** In this case a mutation affected splice site choices in exon 9 of the *ATR* gene which encodes a protein involved in the DNA damage response. Notice that a different splicing pattern is detected in wild-type (c) and mutant (f02–98) mRNA and some in unaffected parent (+/–). (Reprinted from O'Driscoll *et al.*[16] with permission from Macmillan Publishers Ltd.)

where this approach was used. RNA was purified from a cell line derived from a patient, and RT-PCR was used to test the effect of a mutation within the *ATR* gene on pre-mRNA processing. An abnormal splicing product was detected by gel electrophoresis, with a new RT-PCR product appearing in the patient compared with RNA purified from a cell line without the mutation. In this case, a point mutation in the *ATR* gene led to the spliceosome selecting an abnormal splice site within exon 9 in the patient, leading to genetic disease (in this case Seckel syndrome).

Sometimes the effects mutations have on the splicing patterns of mRNAs in patients have been analysed using minigenes made from a patient's genomic DNA. These minigenes are transfected into cultured cells, and again RT-PCR is used to analyse splicing patterns. Information from minigenes can then be used to predict the splicing patterns used in the original patient.

> **Key points**
>
> Mutations in splicing signals are a frequent cause of genetic diseases. Mutations in splicing signals can result in severe effects on gene function, since they often result in segments of the mRNA (and often whole exons) being missed out by the spliceosome. We discussed two specific examples of diseases caused by mutations in splicing control sequences. The premature ageing syndrome HGPS is caused by a mutation which creates a new 5′ splice site within an exon of the *LMNA* gene, and results in the production of a toxic protein. Point mutations disabling an ESE required for efficient splicing of an exon in the *BRCA1* pre-mRNA cause exon skipping, leading to breast cancer. Effects on splicing patterns in disease need to be tested at the RNA level, but can sometimes be predicted using bioinformatic tools.

7.6 Diseases caused by mutations affecting components of the spliceosome

From the preceding part of the chapter, it should be clear that mutations that change important splicing signals in pre-mRNAs, whether they be splice sites or enhancer sequences, can lead to genetic diseases through loss or damage to single proteins. Since splicing is needed to express almost every gene in the human body, it might be expected that mutations occurring in the genes encoding spliceosome components would cause death very early in development rather than disease later on in life. Surprisingly, however, mutations in components of the spliceosome have been reported. These mutations occur in both the genes encoding **snRNAs** and **spliceosomal proteins**. Figure 7.9 shows a summary of mutations found in splicing components that cause human disease, organized

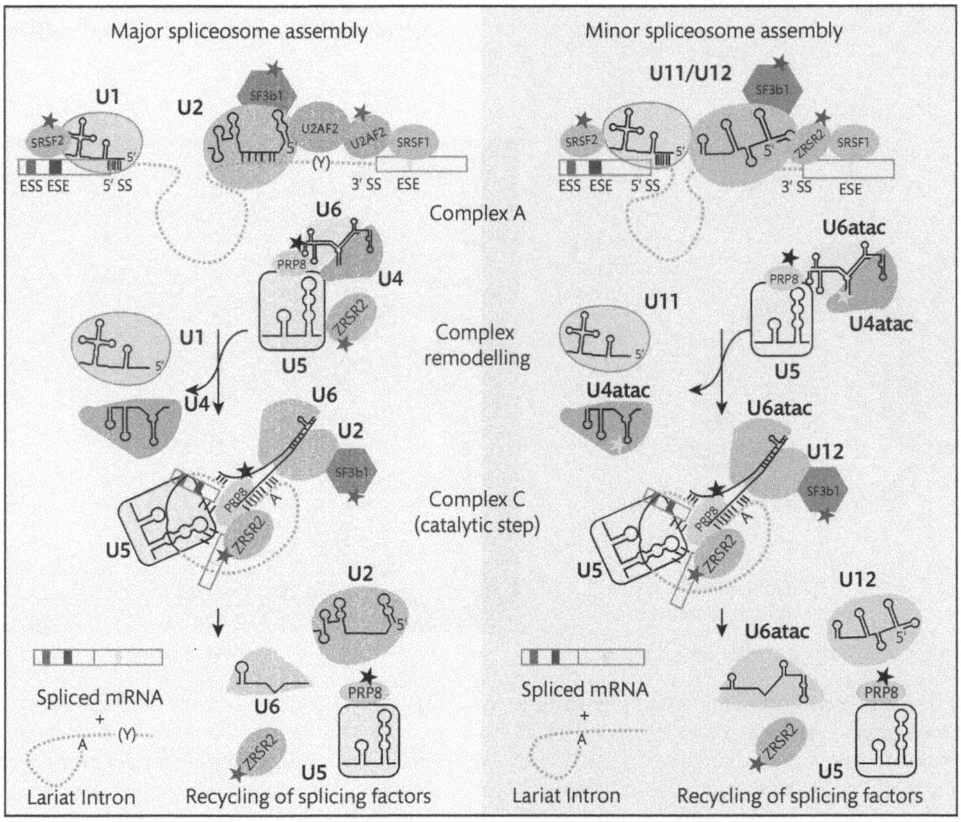

Figure 7.9 Mutations in the splicing machinery resulting in genetic disease. Notice that defects in components of both the major and minor spliceosome can cause disease (disease-causing mutations are indicated by stars). Notice also that disease-causing mutations can occur in proteins involved at each step of spliceosome assembly. (From Singh and Cooper 2012[17] with kind permission.)

according to where the corresponding proteins and snRNAs fit into the spliceosome assembly pathway. Proteins with identified mutations are indicated by stars.

Disease-causing mutations within spliceosomal snRNA genes

A very devastating mutation has been identified in the gene encoding the $U4_{ATAC}$ snRNA component of the spliceosome.[18,19] You will recall from Chapter 5 that $U4_{ATAC}$ is the U4 snRNA found in the minor spliceosome (for more details of the minor spliceosome see Chapter 5). A single nucleotide change in the gene for $U4_{ATAC}$ causes a form of dwarfism called *microcephalic osteodysplastic primordial dwarfism*. This is a multisystem disease, including growth defects and brain abnormalities, and death occurs by three years of age. In this form of dwarfism splicing changes, caused by the change in $U4_{ATAC}$ sequence, occur in exons recognized by the minor spliceosome, leading to developmental abnormalities in the patients.

7.7 Genes encoding important spliceosomal proteins are mutated in patients with the eye disease retinitis pigmentosa (RP)

Splicing mutations also play a key role in the pathology of a disease called *retinitis pigmentosa* (RP).[20,21] In this case the mutations affect the protein components of the spliceosome.

The retina is located at the back of the eye, and contains the photoreceptor cells needed for vision. Patients with RP begin life with normal vision, but this gradually deteriorates with age (the structure of the human eye and observable defects in the retina of an RP patient are shown in Fig. 7.10). In RP, night vision is first affected as a result of defects within the rod cells of the retina responsible for black-and-white vision. This is followed by restricted 'tunnel vision' during daylight, as the rod cells around the periphery of the retina further deteriorate, resulting in vision only in the centre of the retina where the cone cells are located. Eventually the cone cells within the eye

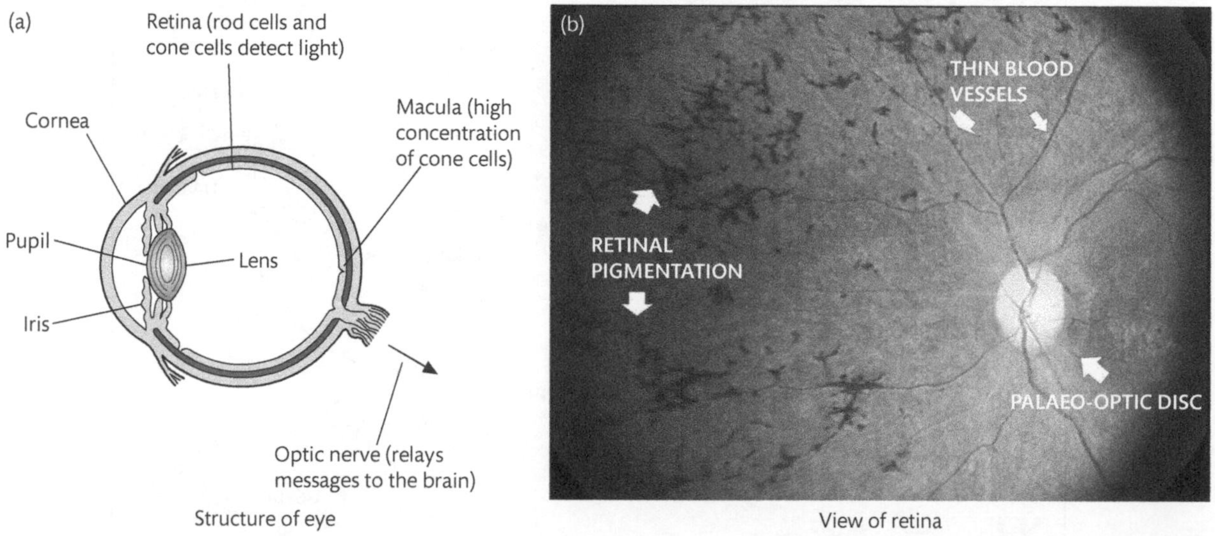

Figure 7.10 The retina is affected in retinitis pigmentosa. (a) Structure of the human eye. Light comes into the pupil at the front of the eye and is focused onto the retina at the back of the eye by the lens. The retina is composed of light-sensitive photoreceptors (rod and cone cells—the neural retina) and pigmented epithelium, and detects light. The cone cells are sensitive to colour; the rods are more numerous but are not colour-sensitive. The macula is the region of the retina with a high concentration of cone cells. Retinitis pigmentosa (RP) affects the rod cells in the retina. The symptoms of RP include poor vision in low light and poor night vision. (b) View of the retina of someone with retinitis pigmentosa, showing key pathological defects (thin optic disc, pigmentation, and blood vessels). (Part (b) is reproduced courtesy of Professor Michel Michaelides, UCL Institute of Ophthalmology and Moorfields Eye Hospital, with kind permission.)

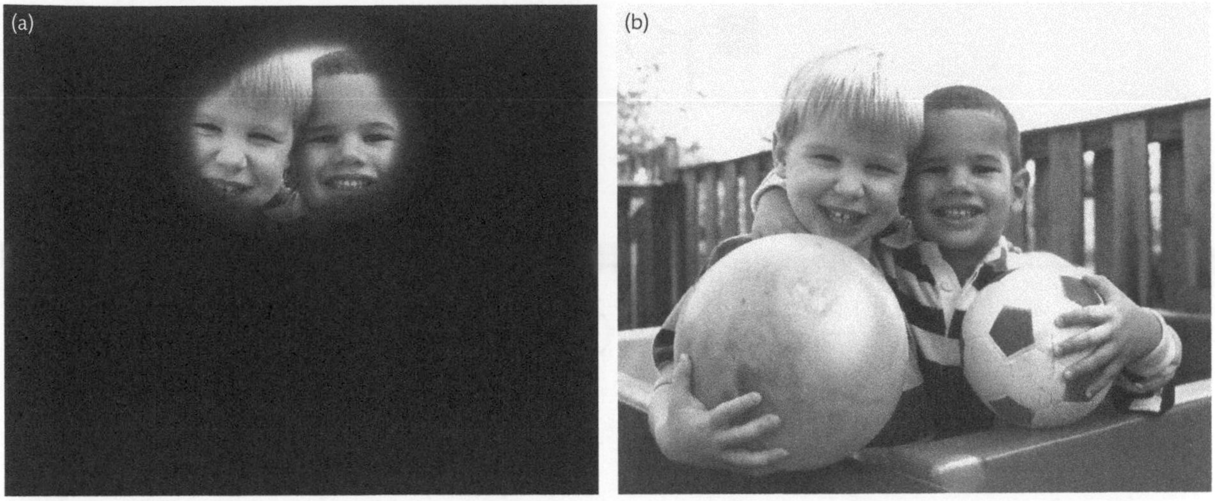

Figure 7.11 'Tunnel vision' in a patient with RP. Notice that (a) clear vision is restricted to the centre of the field of view compared with (b) the normal field of vision.

also deteriorate, affecting colour vision and leading to eventual loss of sight (see Fig. 7.11).

RP is a genetic disease, and mutations have been identified which cause RP. Not surprisingly, some of these mutations are in genes encoding proteins with critical roles in photoreceptor cells, including phototransduction proteins such as rhodopsin and structural proteins needed for the integrity of the photoreceptor cells themselves. Also within the list of affected genes in RP are some photoreceptor-specific transcription factors. Because of their photoreceptor-specific expression, mutation of each of these genes would be expected to give a disease phenotype within the eye.

However, a big surprise came in 2001 when a group of scientists in Leeds, UK, used genetic mapping in families of RP patients to identify autosomal mutations that cause RP within the gene for a spliceosome protein (these mutations were autosomal dominant, which means that even a single copy of the mutant gene is enough to cause disease—remember that human nuclear genes come in pairs with one copy on each chromosome). Some data from these early genetic studies are shown in Fig. 7.12; notice the correspondence between a clinical diagnosis of RP and a genetic mutation within the *PRP8* gene.

The reason why discovery of genetic defects in a spliceosomal protein was such a surprise was that, prior to this, PRP8 was expected to play an absolutely critical role in spliceosome assembly—one that is conserved through evolution (both yeast and human PRP8 proteins are essential in the spliceosome). Box 7.2 provides a reminder of why PRP8 is important in the spliceosome. PRP8 is not the only part of the spliceosome to be affected in RP. Following the initial discovery of mutations in the *PRP8* gene in RP, mutations in other genes encoding protein components of the spliceosome have been identified in other RP patients (Table 7.1).[21,22]

Why should mutations inactivating ubiquitous splicing proteins with apparently critical roles in

Box 7.2 The PRP8 protein has a critical function in the spliceosome

As discussed in more detail in Chapter 5, the PRP8 protein has a very important role in splicing. As a reminder, PRP8 is part of the critical U5/U4/U6 tri-snRNP that is added to the assembling spliceosome, and PRP8 protein might be very close to or in the active catalytic site of the spliceosome.

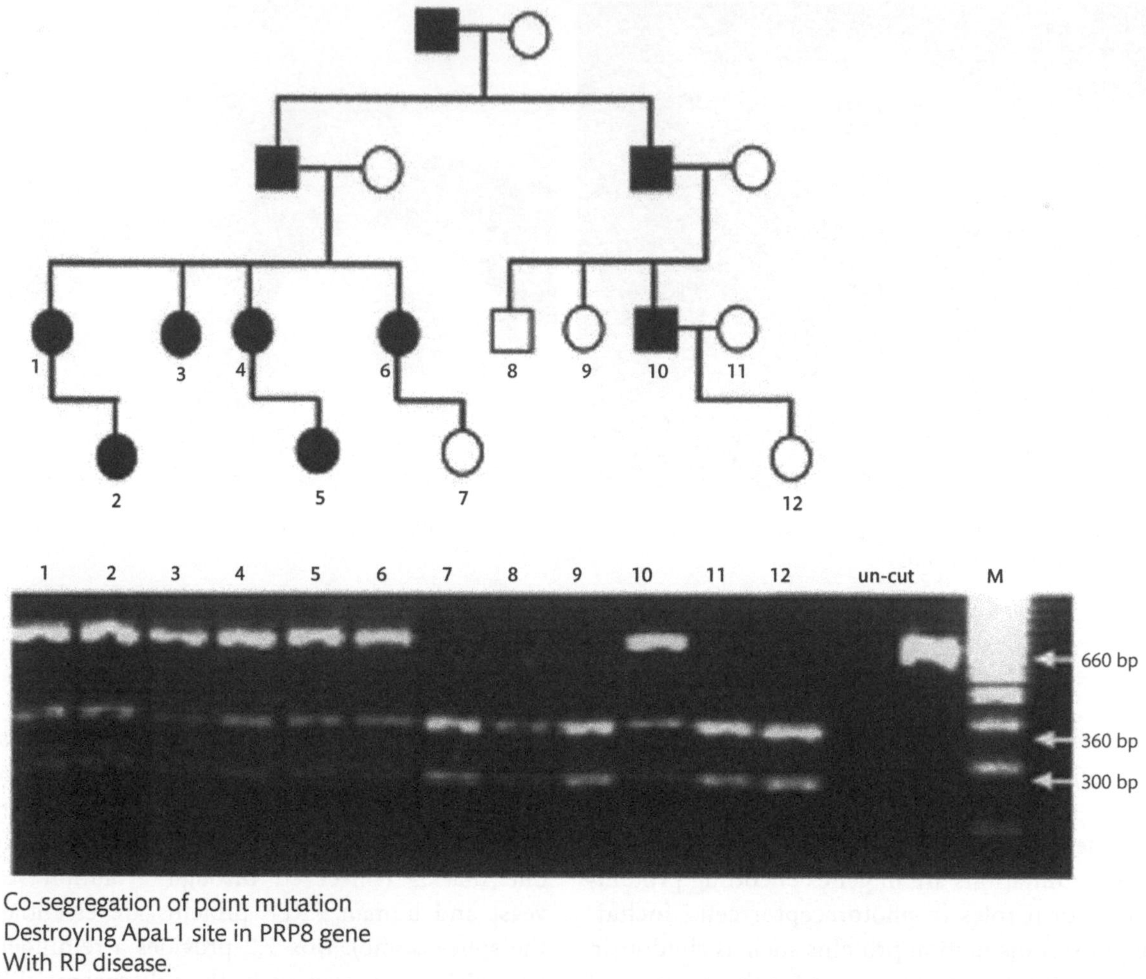

Figure 7.12 A mutation in the gene encoding the splicing factor PRP8 causes retinitis pigmentosa (RP). A family tree affected by RP is shown here with the mutation in diseased individuals. The squares represent males and the circles females, with affected individuals shaded in black and the unaffected individuals in white. Mutation screening in this family and others identified a mutation within exon 42 of the *PRP8* gene. This mutation could be detected easily since it destroyed a restriction site recognized by the enzyme *Apa*L1. To test for the occurrence of this mutation, exon 42 from each of the patients was PCR-amplified from genomic DNA, and then cut with the restriction enzyme *Apa*L1. In the unaffected individuals (without the disease), *Apa*L1 cut the PCR product into two fragments of length around 300bp and 360bp. In contrast, each of the RP-affected individuals also had a larger 660 nucleotide product after *Apa*L1 digestion (this 660bp product results from the PCR product not being cut by *Apa*L1 as a result of the point mutation within *PRP8*). Notice that RP patients still have 300bp and 360bp restriction fragments as well as the uncut 660bp PCR product. This is because RP is a dominant mutation, and each of the affected individuals has a wild-type *PRP8* gene as well as a mutated allele. (Reproduced from McKie et al.[20] with permission from Oxford University Press.)

spliceosome assembly lead to a defect restricted to photoreceptors rather than being lethal by preventing all splicing? Part of the answer explaining why these splicing protein mutations cause RP rather than death is that they are dominant mutations, and patients with RP still contain a functional copy of the gene encoded by their other non-mutated chromosome which can still provide some activity in the spliceosome. In RP patients, this lower level of spliceosome activity might be sufficient in the general body but not the retina, which is why the disease just manifests in the eye.[23]

Table 7.1 Components of the spliceosome that are mutated in retinitis pigmentosa

Protein affected by genetic mutations in RP	Known protein interaction partners	Role of protein in spliceosome activity	Defect in RP
PRP8	RNA helicase Brr2 required to unwind U5/U6 helicase	Important for formation of U4/U5/U6 tri-snRNPs and in active site of the spliceosome	Nuclear assembly of PRP8–Brr2 complex blocked
PRP3	PRP4, snRNAs, and PAP	Required to form stable U4/U6 snRNPs and U4/U6/U5 tri-snRNPs May unwind U4/U6 snRNP to give catalytic centre of spliceosome	Mutant proteins form aggregates in photoreceptor cells
PRP31		Required to form stable U4/U6 snRNPs and U4/U6/U5 tri-snRNPs May unwind U4/U6 snRNP to give catalytic centre of spliceosome	
PAP	PRP3, another protein implicated in RP		

Notice that three proteins are known to be part of the U4/U5/U6 tri-snRNP that we discussed in more detail in Chapter 5.

7.8 The genes encoding splicing proteins can become mutated in some kinds of cancer

Retinitis pigmentosa is an inherited disease involving mutations in genes encoding the core splicing machinery (see also Box 7.3). RP is not the only disease in which genes encoding spliceosomal components become mutated. The genes that encode spliceosome components can also become mutated in some cancers (see Box 7.4). This was discovered with the advent of next generation sequencing, which can produce huge amounts of sequence data from patients. Genomic DNA purified from isolated tumours was sequenced in this way, with the intention of trying to identify causative mutations behind the cancer.

Such experiments revealed splicing factor mutations in a syndrome called *myelodysplastic syndrome*, a condition which can develop into the blood cancer *leukaemia*.[24] Genomic sequencing data showed that in 45–85% of patients with myelodysplastic syndrome the genes encoding proteins involved in the early stages

Box 7.3 Assembly of spliceosome components is affected by mutations in the genetic disease spinal muscular atrophy (SMA)

SMA affects a key protein required for assembly of the snRNP components which then go on to make the spliceosome. SMA is discussed in more detail in Chapter 14 because of its close connection to the maturation of ncRNAs.

Box 7.4 What is cancer?

Cancer is a group of diseases that arise from inappropriate cell division sometimes followed by metastasis around the body. This is an important group of diseases. About half of all adults in the developed world will suffer some form of cancer during their lives. Cancer is not a single disease. Cancers can arise in many different parts of the body, making it a heterogeneous disease, and even cancers arising from the same tissue can show differences in disease initiation and progression. To prevent cancers forming, cell division within the body is normally under tight control.

of spliceosome assembly, including U2AF35, SRSF2, and SF3B1, had become mutated. Each of these three proteins is normally involved in selection of 3′ splice sites by the spliceosome (see Chapter 5). Mutations of SF3B1 also frequently occur in leukaemia patients.[25]

In lung cancer, similar sequencing of cancer cells from tumours showed that there were frequent mutations in the genes encoding the splicing regulator proteins RBM5, RBM6, and RBM10. Following mutation, changes in these proteins in turn lead to changes in the splicing patterns of pre-mRNAs important to lung cells.[26,27] One particularly important splicing target that is affected encodes a repressor of the Notch signalling pathway called Numb. The result of this change in Numb splicing is that the Notch pathway becomes hyperactivated, leading to increased cell proliferation in lung cancer.

Note that in the above mutations affecting spliceosome components that lead to cancer, a difference from the mutations that cause RP is that in the case of cancer these mutations are not inherited from a parent, but occur within cancer cells or their direct precursors, perhaps giving them a proliferative advantage over other cells and helping the cancer to develop.

7.9 Splicing changes can change the properties of cancer cells

Changes in splicing patterns can cause different mRNAs to be made from the same gene, sometimes with quite different effects on the cell. Sometimes inappropriate splice forms can contribute to the properties of cancer cells.

'Hallmark properties' that distinguish cancer cells from normal cells in the body were catalogued in 2011 by Robert Weinberg at MIT and Doug Hanahan at the Swiss Institute for Cancer Research. These two scientists picked out 10 hallmarks of cancer cells[29]—features frequently associated with cancer cells as opposed to normal body cells. It is now known that most of these hallmarks of cancer can be affected by alterations in splicing patterns in cancer cells (see Fig. 7.13).[28,30]

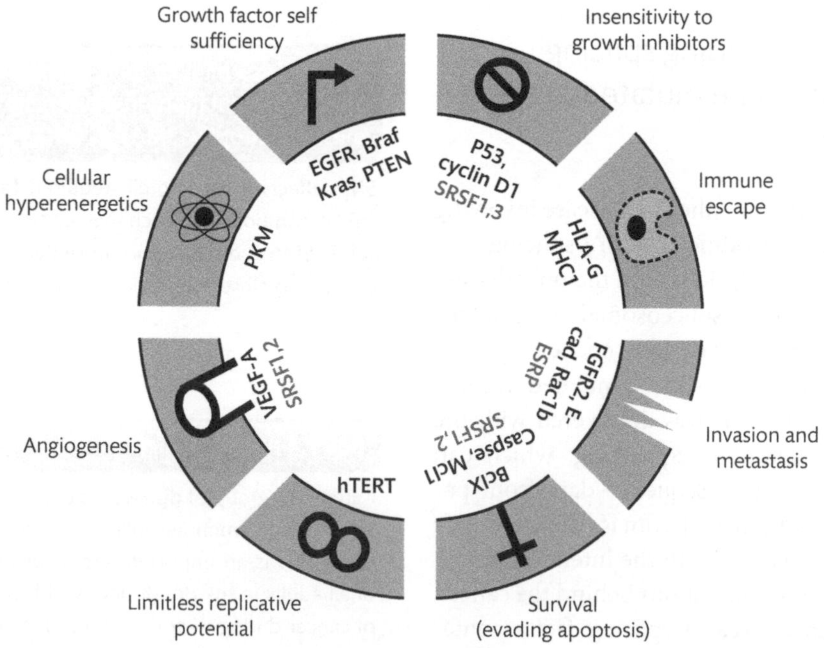

Figure 7.13 Hallmarks of cancer and their connection with splicing. Eight hallmarks of cancer cells are under known splicing control. These eight hallmarks (cellular hyperenergetics etc.) are shown outside of the circle, and some of the genes involved (*PKM* etc.) are shown inside the circle. Notice that several splicing patterns that alter in cancer are known to be regulated by the splicing factor SRSF1 that is itself upregulated in cancer (see text), as well as other splicing factors including ESRP and SRSF2. (Modified from Oltean and Bates[28] with permission from Nature Publishing Group.)

What are these splicing changes in cancer cells? One important example of a splicing change in cancer relates to how cancer cells generate energy. Cancer cells use a different way of generating energy than normal cells, switching from oxidative metabolism involving mitochondrial respiration to breaking down sugars in the cytoplasm by glycolysis. This effect was first observed in the 1950s and was called the *Warburg effect* after its discoverer, the German biochemist Otto Warburg.

The Warburg effect originates in a change in splicing pattern. Research carried out in 2008 showed there is a splicing switch in cancer cells compared with normal cells in a key metabolic enzyme called *pyruvate kinase M*.[31,33] Pyruvate kinase M is part of the glycolytic pathway, where it catalyses the conversion of phosphoenolpyruvate into pyruvate. Normal cells in the body make pyruvate kinase M mRNA by including splicing of exon 10, and this mRNA is translated into an enzyme that efficiently feeds pyruvate into the citric acid cycle (see Figs 7.14 and 7.15). This exon 10 splicing pattern is an important difference between normal and cancer cells. Instead of splicing exon 10, cancer cells (and embryos) splice exon 9 into pyruvate kinase M. As a result of using exon 9 instead of exon 10, the pyruvate kinase M isoform translated in cancer cells works more slowly. This means that in cancer cells there is an accumulation of upstream components in the glycolytic pathway; these components are used by cancer cells to make other things they need. The isoform of pyruvate kinase M made in cancer cells also feeds pyruvate into the production of lactic acid and the glycolytic pathway rather than into mitochondria-based citric acid cycle, which in turn results in the Warburg effect.

A second important splicing difference between tumour cells and normal cells helps developing tumours to 'organize' blood vessels so that they can receive oxygen and nutrients (see Fig. 7.15). Cancer cells express a growth factor called VEGF-A which controls blood vessel growth. There is a single *VEGF-A* gene, but through alternative splicing two types of mRNA isoform can also be made from this single *VEGF-A* gene that have the opposite effect of promoting or inhibiting blood vessel growth: these isoforms are called pro-angiogenic and anti-angiogenic isoforms.[34,35]

In the case of *VEGF-A*, the important alternative splicing event that distinguishes cancer cells and normal cells is alternative selection of the 3′ splice site in the last exon of the *VEGF-A* pre-mRNA (the regulated exon is exon 8 in the *VEGF-A* gene (see Fig. 7.15)). Pro-angiogenic splice isoforms of VEGF-A are called VEGF-Axxx isoforms. These splice isoforms are expressed in cancer cells (and also in some other diseases, and in some normal tissues and stages of development that require angiogenesis).

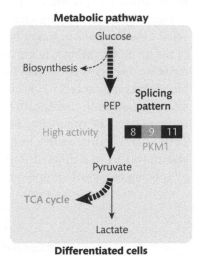

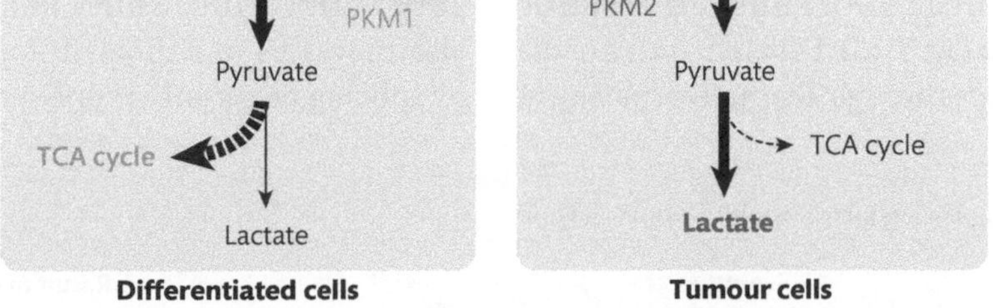

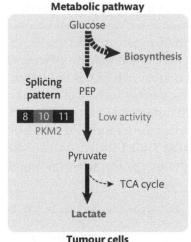

Figure 7.14 **Different isoforms of the pyruvate kinase M enzyme are found in normal cells and cancer cells.** Normal differentiated cells in the body splice exon 9 into the protein kinase M mRNA to encode the high activity PKM1 protein isoform. PKM1 protein efficiently converts PEP into pyruvate. This pyruvate is then fed into the TCA cycle to make energy. However, tumour cells splice exon 10 into the protein kinase M mRNA, to encode a less efficient protein kinase M isoform called PKM2. Instead of mainly feeding the TCA cycle, PKM2 protein makes pyruvate that goes into lactate production for aerobic glycolysis. (Adapted from Chen et al.[31]).

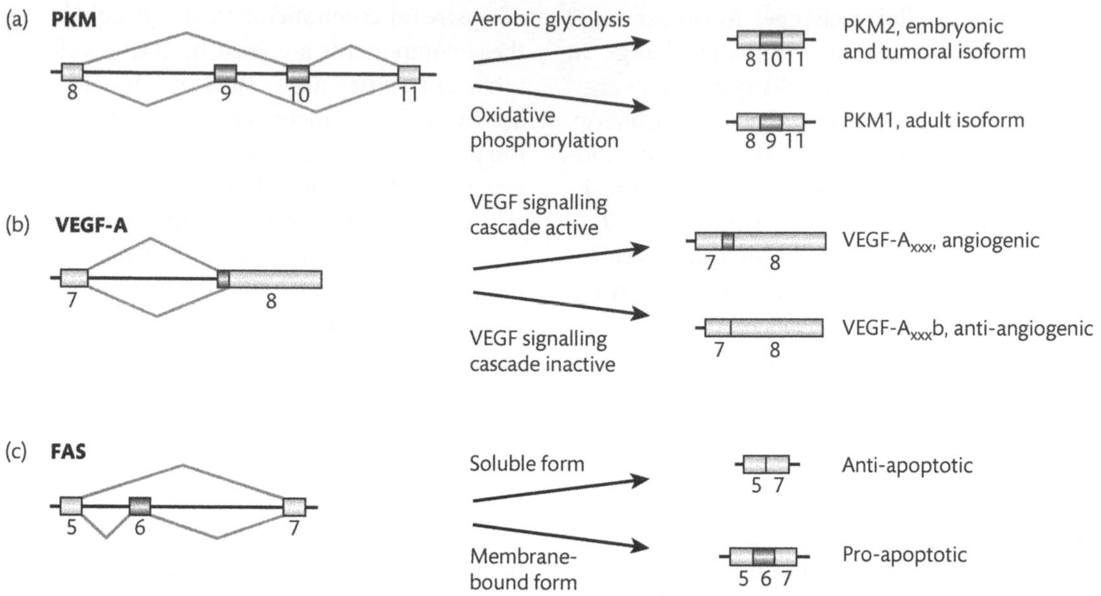

Figure 7.15 **Three important splicing changes occurring in cancer cells.** Splicing switches in (a) pyruvate kinase M, (b) VEGF-A, and (c) FAS. (From Bonal et al.[32] with kind permission of *Nature*.)

The anti-angiogenic isoforms are called VEGF-Axxxb, and are generally expressed in normal adult tissues that already have blood supplies (so they do not need to stimulate growth of new blood vessels). When translated into protein, the pro-angiogenic and anti-angiogenic isoforms of VEGF protein differ by just six amino acids at their C termini. These six amino acids cause differences in the way that the VEGF protein (which is a growth factor) interacts with its receptor, to transmit a signal either encouraging or inhibiting proliferation of blood vessels.

A third example of important splicing differences between cancer and normal cells relates to survival; cancer cells need to escape mechanisms of cell death called *apoptosis*. A cell surface receptor called the FAS cell receptor is important for transmitting apoptotic signals into cells (see Fig. 7.15). Different forms of the FAS receptor are made through alternative splicing of the *FAS* pre-mRNA[36] to make membrane-bound FAS receptor (this enables apoptosis) or a secreted form of the receptor (which prevents cell death). The differences between these two isoforms of FAS protein are controlled by whether exon 6 is spliced into the FAS mRNA or not. Exon 6 encodes the transmembrane domain of the FAS protein, so if this exon is skipped the resulting FAS protein will be the soluble form.

How do these splicing changes arise in cancer cells? One mechanism might be through mutations in the genes encoding splicing components (such as those we discussed above that have been detected in blood and lung cancers). However, another source of splicing changes in cancer cells originates in changes in expression of otherwise 'healthy proteins' (see Table 7.2). We discuss this in the next section. This also moves us to a different topic—diseases caused by splicing factor mis-expression.

Table 7.2 Changes in splicing factor levels in cancer cells

Change in cancer cell	Examples	Result of change
Mutation in protein involved in spliceosome	SF3B1 mutated in blood cancer and lung cancer RBM5, RBM6 and RBM10 mutated in lung cancer	Changes in splicing patterns in cancer cells
Change in expression of splicing factor	The expression level of SRSF1 changes in many cancers	Changes in splicing patterns in cancer cells

7.10 Diseases caused by mis-expression of levels of splicing factors

As well as their mutation, some diseases can also be caused by changes in the nuclear concentration of splicing factors. Diseases caused by changes in splicing factor concentration are conceptually different from those which are caused by the mutation of splicing factors such as occurs in RP. In the former, the mis-expressed splicing factors are not themselves mutated or changed in any way other than that their level of expression is changed. In Chapter 6 we have already come across one disease, **myotonic dystrophy**, caused by mis-expression of splicing components (see also Box 7.5). In this section we are going to discuss how splicing components can become misregulated in cancer.

Some splicing factors can be oncogenic

Recently work has found that the splicing regulator protein SRSF1 (this protein was previously known as ASF/SF2,[37] so this name is used in some of the older literature) is upregulated in a number of cancers. The mechanisms upregulating SRSF1 expression in cancer can include amplification of the *SRSF1* gene (on a region on human chromosome 17 which is known to become amplified in breast cancer which would increase *SRSF1* gene copy number), and sometimes cancer-associated changes in transcription factors like Myc that have binding sites in the *SRSF1* promoter (increases in Myc in cancer cells would then increase transcription of the *SRSF1* gene) (see Fig. 7.16).

How was it established that changes in SRSF1 expression can be an important driver in cancer? First, changes in *SRSF1* gene expression increase expression of the SRSF1 protein[40] detected in experiments like those shown in Fig. 7.17. Notice that SRSF1 is expressed at higher levels in tumours than in normal tissues from the same organs. Secondly, two linked assays used to monitor tumorigenicity indicated that these changes in SRSF1 protein expression in tumours are likely to be very important for establishing cancer (these assays are shown in Fig. 7.18):

1 The NIH 3T3 cell assay. The first of these assays uses a mouse cell line called NIH 3T3 that is grown in culture. NIH 3T3 cells are immortal (this means that they can be grown indefinitely in culture) but they are not transformed (meaning that they are not quite cancer cells). Unlike cancer cells, NIH 3T3 cells normally stop growing when they contact each other on the tissue culture plate so they form a flat lawn of cells (this is called *contact inhibition*). However, when NIH 3T3 cells are transformed into cancer cells through expression of an oncogene, contact inhibition will break down and the cells form colonies rather than growing one layer deep on a plate. Looking for such colonies is used as an assay for detecting oncogenes. This result was observed with

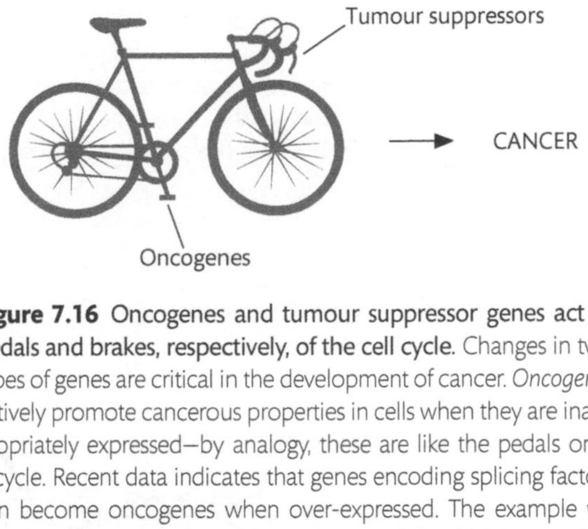

Figure 7.16 Oncogenes and tumour suppressor genes act as pedals and brakes, respectively, of the cell cycle. Changes in two types of genes are critical in the development of cancer. *Oncogenes* actively promote cancerous properties in cells when they are inappropriately expressed—by analogy, these are like the pedals on a bicycle. Recent data indicates that genes encoding splicing factors can become oncogenes when over-expressed. The example we discuss in the text is the over-expression of the *SRSF1* gene. The SRSF1 protein is needed by normal cells, but becomes an oncogene when over-expressed. On the other hand, *tumour suppressor genes* are needed to keep cell division in check, and are often inactivated in cancer cells. By analogy, tumour suppressors are like the brakes on a bicycle. A protein associated with splicing that operates as a tumour suppressor is Wilms' tumour protein (WT1). WT1 is encoded by a tumour suppressor gene on human chromosome 11 which is inactivated in a paediatric cancer called Wilms' tumour. WT1 is a zinc-finger protein which can act as a transcriptional regulator or a splicing regulator depending on its own splice isoforms.[38]

Box 7.5 Changes in splicing factor protein concentration also take place in the disease myotonic dystrophy

Cancer is just one of the diseases caused by changes in splicing factor concentrations. Another well-known disease caused by splicing factor mis-expression is **myotonic dystrophy**, in which changes in splicing factor concentrations lead to embryonic patterns of splicing in the adult, and as a result multi-system disease. We discuss myotonic dystrophy in more detail in Chapter 6 because of its close connection with alternative splicing.

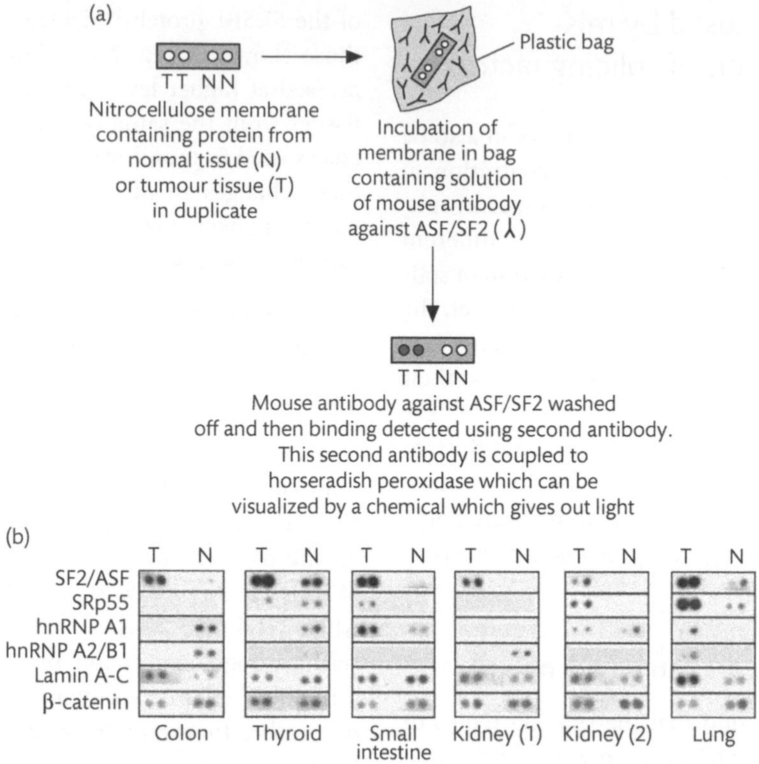
(a)
TT NN
Nitrocellulose membrane containing protein from normal tissue (N) or tumour tissue (T) in duplicate
Plastic bag
Incubation of membrane in bag containing solution of mouse antibody against ASF/SF2
TT NN
Mouse antibody against ASF/SF2 washed off and then binding detected using second antibody. This second antibody is coupled to horseradish peroxidase which can be visualized by a chemical which gives out light
(b)
T N T N T N T N T N T N
SF2/ASF
SRp55
hnRNP A1
hnRNP A2/B1
Lamin A-C
β-catenin
Colon
Thyroid
Small intestine
Kidney (1)
Kidney (2)
Lung

Figure 7.17 Upregulation of the splicing regulator ASF/SF2 can be detected in several tumours compared with adjacent non-cancerous tissue. (a) Protein from tumours (T) was spotted onto a nitrocellulose membrane next to protein from adjacent normal tissue (N). This membrane was then incubated in a solution containing a mouse monoclonal antibody specific to the splicing factor ASF/SF2. Antibody not bound to the membrane was washed off, and then bound α-ASF/SF2 antibody was detected using a secondary α-mouse antibody coupled to the enzyme horseradish peroxidase (HRP). Finally the membrane was incubated in a solution with a substrate for HRP which gives out light when it is cleaved (this method is called enhanced chemiluminescence (ECL)). Using this technique, high levels of ASF/SF2 were detected within the tumour tissue but not within the surrounding non-cancerous tissue. (b) Results from such an experiment. (Part (b) is from Karni et al.[39] with permission from Macmillan Publishers Ltd.)

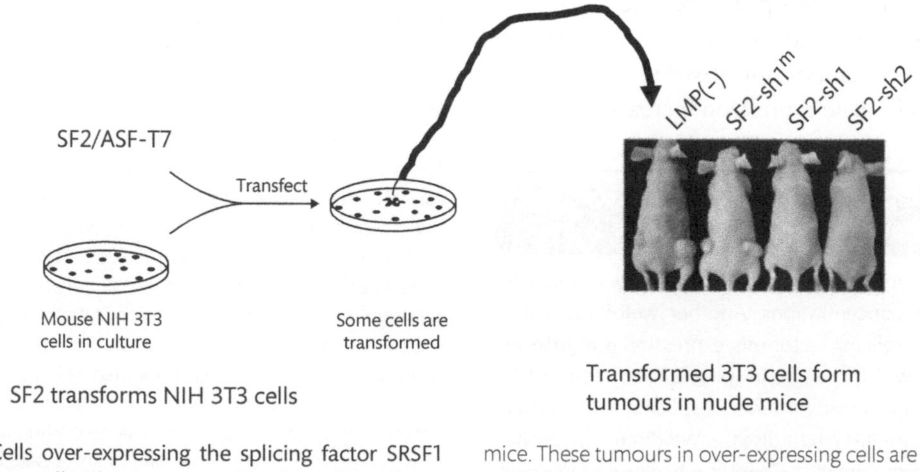
SF2/ASF-T7
Transfect
Mouse NIH 3T3 cells in culture
Some cells are transformed
SF2 transforms NIH 3T3 cells
LMP(-)
SF2-sh1m
SF2-sh1
SF2-sh2
Transformed 3T3 cells form tumours in nude mice

Figure 7.18 Cells over-expressing the splicing factor SRSF1 behave as cancer cells. The gene encoding the splicing factor SRSF1 can transform NIH 3T3 cells and cause tumours in nude mice. These tumours in over-expressing cells are cured by down-regulation of SRSF1 using shRNAs 1 and 2. (From Karni *et al.*[39] with permission from Macmillan Publishers Ltd.)

SRSF1. As shown in Fig. 7.18, just a twofold increase in SRSF1 expression was enough to transform NIH 3T3 cells into colonies of transformed cancer cells.

2 Forming tumours in nude mice. The second assay uses the same transformed NIH 3T3 cells, but injected into nude mice (these are mice with compromised immune systems that prevent them rejecting the NIH 3T3 cells). NIH 3T3 cells transformed by expression of SRSF1 are able to grow into tumours within animals, confirming them as cancer cells.

How does increased SRSF1 protein expression modify the properties of normal cells to make them become cancerous? An RNA-binding protein that can bind to many RNAs in the cell SRSF1 probably affects multiple targets in cancer cells. However, one important change driven by upregulated levels of SRSF1 is the migratory behaviour of cells. Cells with different migratory properties are found in the body. *Epithelial cells* grow in sheets of cells held together by intercellular junctions called tight junctions and gap junctions. Cells can move about within these epithelial sheets, but they do not leave them. In contrast, *mesenchymal cells* are more mobile and are held together by much weaker connections. In cancer cells, a switch occurs to convert epithelial cells (which are anchored in the body) to mesenchymal cells (which are able to metastasize—migrate around the body and form new tumours). This change in cell properties is called an **epithelial to mesenchymal transition** (EMT).

How SRSF1 causes this change in cell mobility is understood and involves a cell surface receptor tyrosine kinase called **RON**. The RON receptor tyrosine kinase is composed of an α and a β chain, and is normally activated by binding **MSF** (macrophage stimulating factor to induce an EMT (see Fig. 7.19). SRSF1 controls splicing of the RON pre-mRNA. The splicing pattern of the critical exons of the *RON* pre-mRNA is shown in Fig. 7.20. Notice that there are two alternatively spliced isoforms: an isoform containing exon 11 and an alternatively spliced isoform called Δ*RON* which lacks exon 11. Loss of exon 11 deletes 49 amino acids from the β chain of the receptor to make a constitutively active form of the RON receptor which induces an EMT (this form of the RON receptor is always active, even when it has not bound MSF). Δ*RON* is normally expressed at low levels in most tissues, but in cancer its splicing is indirectly upregulated by elevated SRSF1. In cancer cells SRSF1 increases the Δ*RON* splice isoform by binding exon 12 of *RON* pre-mRNA—this strengthens recognition of *RON* exon 12 relative to exon 11 (exon 11 and exon 12 are separated by just a short intron so are both transcribed within a short timeframe, meaning that the spliceosome will have a choice between them). Since it binds to elevated levels of SRSF1 protein, and also has stronger splice sites, in cancer cells exon 12 will be preferentially selected by the spliceosome compared with exon 11, creating the ΔRON splice isoform.

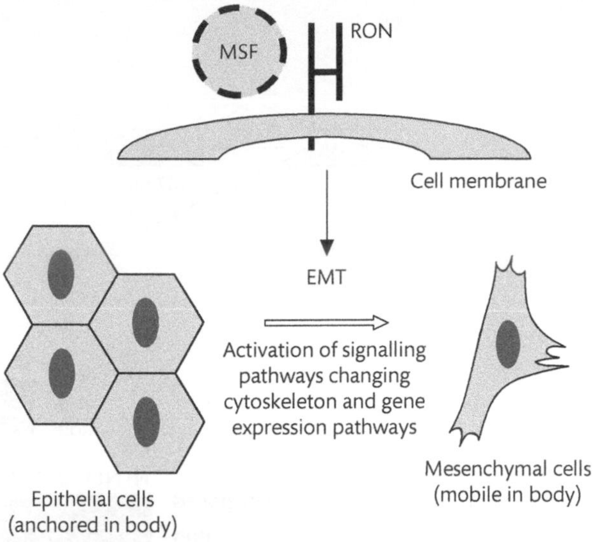

Figure 7.19 Activation of the RON receptor tyrosine kinase changes the properties of epithelial cells so that they become more mobile. Epithelial cells grow together in sheets while mesenchymal cells tend to occur singly and are much more mobile. The transition between sheets of cells and mobile cells capable of metastasis is an important step in cancer (the epithelial to mesenchymal transition (EMT)). One of the molecules which controls this transition is the RON receptor tyrosine kinase. RON protein is composed of an α chain and a β chain. The RON receptor is embedded within the cell membrane, and is activated by binding of the hormone macrophage stimulating factor (MSF) to the extracellular portion. This results in autophosphorylation of the RON receptor and downstream activation of signalling pathways which promote cell migration. ΔRON is a constitutively active form of the RON receptor which lacks 49 amino acids in the β chain because of exon skipping of exon 11 in the *RON* pre-mRNA, and which continually activates signalling pathways even in the absence of MSF. *RON* is a proto-oncogene. A proto-oncogene is a gene which is normally required by the cell but which becomes an oncogene when it is either over-expressed or mutated.

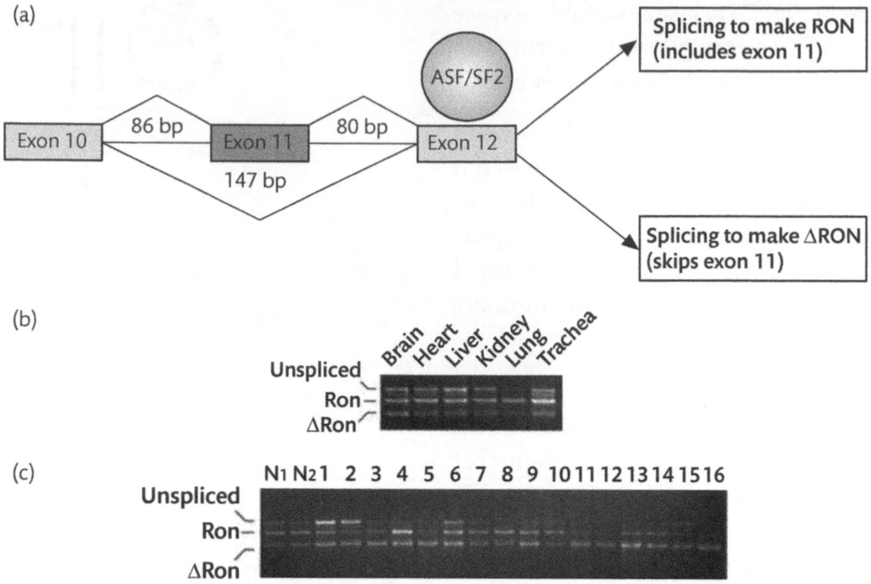

Figure 7.20 Aberrant splicing of the RON transcript promotes metastasis in cancer. (a) Exon 11 of *RON* is flanked by short introns. (b) Exon 11 of *RON* is variably included in different tissues to give a RON and ΔRON mRNA. (c) ΔRON mRNA is produced at higher levels in different cancers (numbered 1–16) compared with normal tissue (N1 and N2). (Reprinted from Ghigna et al.[41] with permission from Elsevier.)

7.11 Splicing as a target to treat cancer

As well as having important roles in the cause of disease, splicing has provided a route to cure disease. Because of the important role of splicing in disease discussed in the previous sections, therapeutically targeting the spliceosome might be a possible route for treating cancer. Some drugs that are active against the spliceosome are being developed as cancer treatments[32] (one of these drugs, *spliceostatin*, is shown in Fig. 7.21). In this last part of this chapter, we discuss two other examples of *splicing therapy*: first the use of oligonucleotides to block splicing of disease causing exons in the muscle disease muscular dystrophy, and secondly the use of small molecules to block the spliceosome in infectious disease.

7.12 Manipulating pre-mRNA splicing offers a route to treating muscular dystrophy

As we discussed at the start of this chapter, many genetic diseases are caused by point mutations in exons which lead to changes in translated proteins (different amino acids, or stop codons to generate truncated proteins). What if these mutation containing exons could be left out of the mRNA? The exciting idea of using exon skipping as a therapeutic tool to miss out stop codons within mRNAs made from mutated disease genes has particularly focused on two muscle diseases—**Becker muscular dystrophy** and **Duchenne muscular dystrophy** (see Fig. 7.22). Both these diseases are caused by different kinds of mutation in the *dystrophin* gene, one of the longest genes found in humans.

Muscular dystrophies are a group of diseases in which muscles become progressively weaker; they are caused by defects in muscle proteins. There are about 30 separate muscular dystrophies, the most common being Duchenne muscular dystrophy (DMD), with an incidence of 1 in 3500 boys, and Becker muscular dystrophy (BMD), which is five times rarer. Both these diseases are caused by mutations in the same gene, *dystrophin*, located on the X chromosome. The *dystrophin* gene is enormous, encompassing 2.5Mb of genomic DNA (2.5×10^6 nucleotides). Mutations causing muscular dystrophy are autosomal recessive. Boys have only one copy of the X chromosome, and so are more at risk of developing muscular dystrophy than girls (who have two copies of the X chromosome). Girls who have a mutation in one copy of the *dystrophin* gene will still

Synthetic analogues

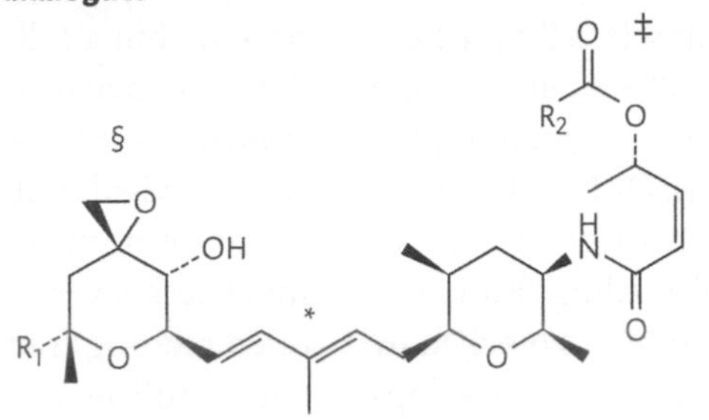

Spliceostatin A and meayamycins

Molecule	Formula	R_1	R_2
Spliceostatin A	$C_{28}H_{43}NO_8$	OCH_3	CH_3
Meayamycin	$C_{28}H_{43}NO_7$	CH_3	CH_3
Meayamycin B	$C_{30}H_{46}N_2O_8$	CH_3	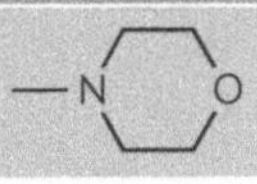

Figure 7.21 Drugs that target components of the spliceosome could be novel therapies for treating cancer. These anti-cancer drugs include spliceostatin which is shown here. Spliceostatin is a synthetic derivative of an anti-tumour agent first identified in *Streptomyces* bacteria. Spliceostatin binds to the spliceosome component protein SF3B, thereby affecting splicing patterns in cancer cells. Interestingly, the gene encoding this same SF3B protein can become mutated in cancer, particularly in leukaemias and myelodysplastic syndrome which can precede development of leukaemia (see Section 7.9). (The chemical structure of spliceostatin is taken from Bonnal et al.[32] with permission from Nature Publishing Group.)

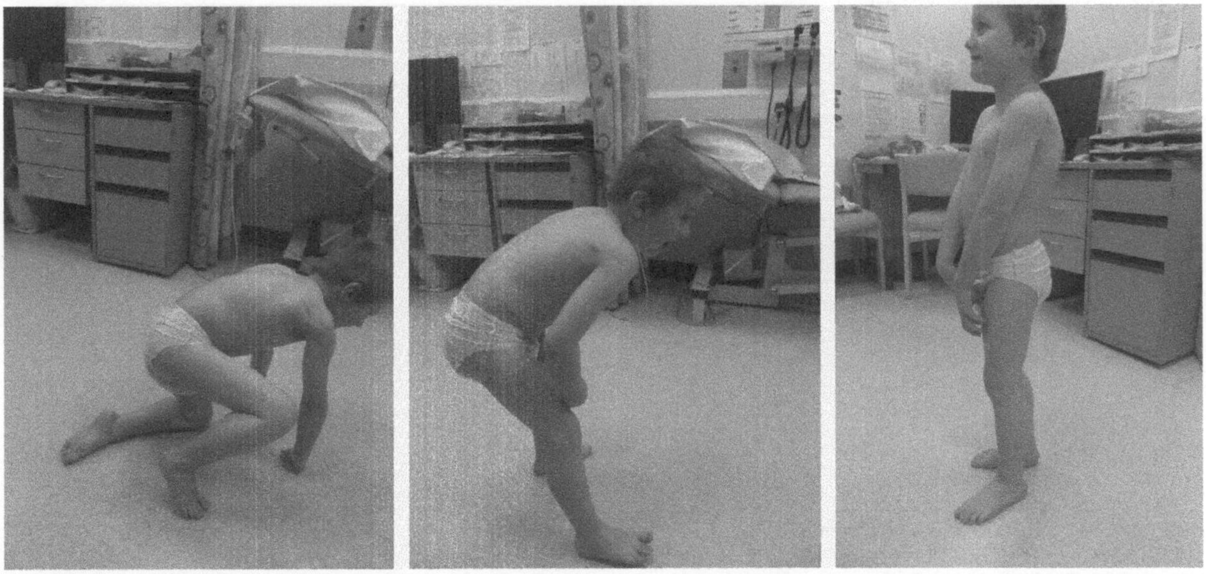

Figure 7.22 Duchenne muscular dystrophy. Photographs courtesy of Professor Volker Straub.

have a second gene copy on the other X chromosome to back this up. In general, DMD is more severe. One of the first characteristics of young boys with Duchenne muscular dystrophy is the use of hands and arms up the thighs to walk up their own bodies when getting upright from a squatting position or from the ground. This need to use the hands and arms is due to muscle weakness in the legs. The boy shown in Fig. 7.22 has DMD and is performing this typical movement, which is called Gowers' sign.

Dystrophin protein forms part of a complex that connects the cytoskeleton of a muscle cell to the extracellular matrix through the cell membrane, and this complex is very important for maintaining the structural integrity of muscle cells. DMD is the more severe of the two diseases, and is caused by mutations in which the translational reading frame encoding the 427kDa dystrophin protein is blocked by mutations that create stop codons. DMD develops between 2 and 6 years after birth, eventually involving all voluntary muscles and resulting in death by the time the patient is in his twenties. In contrast, BMD is caused by **missense mutations** within the *dystrophin* gene which lead to more subtle changes in the amino acid composition of the dystrophin protein, but a full-length protein is still made. BMD develops between 2 and 16 years after birth with similar symptoms to DMD but is much less severe, with survival well into middle age.

Hence mutations in the *dystrophin* gene cause a much less severe disease (BMD) if they maintain the open reading frame of the *dystrophin* mRNA to encode a full-length (albeit slightly different) protein, and a more severe disease if translation is stopped early. This is because although the dystrophin protein is very large and contains several domains, significant portions of it can be removed without destroying function (see Fig. 7.23). In particular,

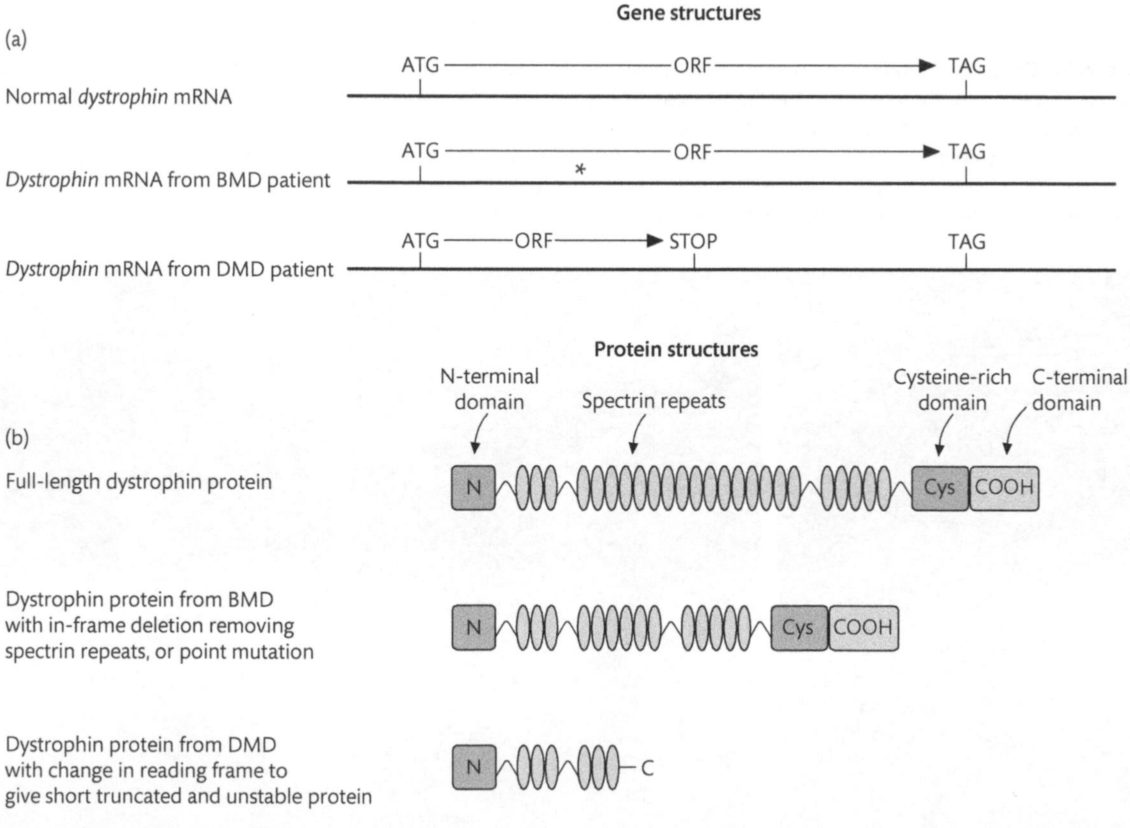

Figure 7.23 Duchenne and Becker muscular dystrophy are caused by different kinds of mutation in the *dystrophin* gene. (a) Becker muscular dystrophy is caused by missense mutation; a full-length dystrophin protein is still translated, but has a variant amino acid sequence as a result of the mutation (shown by an asterisk). Duchenne muscular dystrophy results if a premature stop codon (TAG) is introduced by mutation in the *dystrophin* gene to produce a short non-functional dystrophin protein from a truncated open reading frame. (b) The structure of the dystrophin protein. The form of dystrophin protein found in human muscles (called Dp427) has four domains encoded by 72 exons: an N-terminal actin-binding domain (amino acids 12–240), a large central domain (amino acids 253–3112, encoding 24 spectrin motifs), a cysteine-rich domain (amino acids 3113–3299), and a C-terminal complex (amino acids 3300–3685). The dystrophin protein maintains the structure and function of muscle fibres, while linking the extracellular matrix and cytoskeletal actin within muscle cells through a complex called the dystrophin–glycoprotein complex.

the central domain can be partially deleted (this central domain is made up of a number of spectrin repeats which might be somewhat redundant). This means that if it were possible to 'skip' a *dystrophin* exon in a DMD patient which contained a mutant stop codon (i.e. to prevent this exon from being spliced into the final transcript), a sufficiently functional version of the dystrophin protein should be restored if the downstream exon picks up the reading frame again.

In practice, exon skipping within the human *dystrophin* pre-mRNA has been achieved by transfecting cells from DMD patients with synthetic oligonucleotides which are complementary to the splice site sequences of mutated exons. On hybridization with *dystrophin* pre-mRNA within the nucleus, these oligonucleotides make the splice sites of the mutated exon invisible to the splicing apparatus (see Fig. 7.24). This leads to skipping of the mutated exon by the spliceosome, which instead joins together the upstream and downstream exons.

In order to be used as a therapy for treating DMD patients, splice-site-blocking oligonucleotides need to be delivered into patients' muscle cells. Clinical trials have shown that this approach might be possible in humans, but efficient delivery into muscles and stability of the oligonucleotides once they are in the body remain problematic.[42] In order to deliver splice-site-blocking nucleotides into humans, alternative chemistries have been used, particularly using 2′-O-methyl instead of 2′-OH on the ribose sugars, and using phosphothioate backbone linkages which contain sulphur atoms in the phosphodiester bonds. Both these chemistries improve the stability of the oligonucleotides in patients.

An important tool for testing therapeutic splicing inhibition has been the **Mdx mouse** (see Fig. 7.25).[41] Mice also have a *dystrophin* gene, and the *Mdx* mouse strain develops muscular dystrophy since it contains a stop codon in exon 23 of this gene which blocks translation of the encoded dystrophin protein (hence the *Mdx* mouse is a model of DMD). Splice-site-blocking oligonucleotides have been used to restore expression of dystrophin protein in the muscles of *Mdx* mice (see Fig. 7.25). To deliver the oligonucleotides within cells a hybrid gene was made that encodes U1 snRNA fused to sequences complementary to the 5′ and 3′ splice sites of mouse *dystrophin* exon 23 (the exon which contains the stop codon in the *Mdx* mouse). The U1 snRNA encoded by this hybrid gene acts both to stabilize the therapeutic RNA (since U1 snRNP is a very stable nuclear RNA) and to correctly localize the hybrid RNA within the nucleus. This hybrid gene was delivered into *Mdx* mouse muscle cells within an adenovirus injected into the bloodstream (this kind of adenoviral genome is called an adenoviral vector since it delivers a piece of foreign DNA). This kind of gene delivery proved to be very successful. Three months after adenovirus injection, skipping of the mutated *dystrophin* exon 23 could be detected using RT-PCR in multiple mouse tissues (see Fig. 7.25, which shows skipping of *dystrophin* exon 23 in muscle cells).

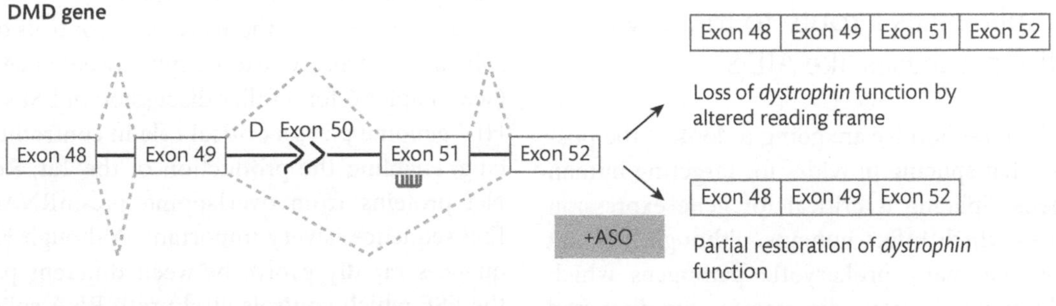

Figure 7.24 **Therapeutic oligonucleotides have been used to cause exon skipping in myotonic dystrophy.** One exon that has been targeted for skipping in in DMD patients is exon 51, which is frequently mutated in Duchenne muscular dystrophy. Antisense oligonucleotides complementary to exon 51 hybridize to the *dystrophin* pre-mRNA and block access of the spliceosome to important splicing signals. As a result of this, exon 50 is directly spliced onto exon 52, missing out exon 51. (From Singh and Cooper[17] with permission.)

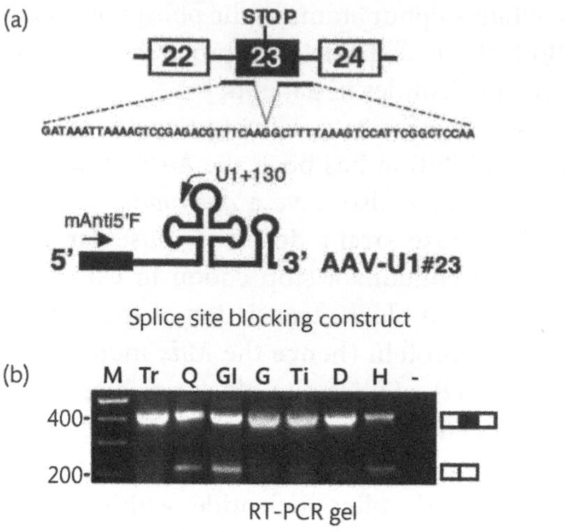

Figure 7.25 **Gene therapy can block splicing of exons containing stop codons in the *Mdx* mouse.** The *Mdx* mouse normally makes a short dystrophin protein since it has a stop codon in exon 23 of the *dystrophin* gene; this results in the mouse developing muscular dystrophy. Adenovirus vectors containing a gene encoding a hybrid U1 snRNA gene fused to sequences complementary to the 5′ and 3′ splice sites of *dystrophin* exon 23 have been constructed and injected into the tail veins of *Mdx* mice. (a) These adenovirus vectors infect cells and are transcribed within nuclei into chimaeric RNAs which can hybridize with the splice sites of *dystrophin* pre-mRNA exon 23 and block access of the splicing machinery. (b) The effect of this vector in restoring gene function in the *Mdx* mouse. After infection with adenovirus vector, RNA was prepared from different *Mdx* mouse tissues and analysed by RT-PCR to detect whether exon 23 was spliced into the *dystrophin* mRNA or not. These RT-PCR experiments identified exon 23 skipping in some muscle tissues (the smaller band towards the bottom of the agarose gel). Exon-23-skipped *dystrophin* mRNA was translated into dystrophin protein and restored some strength to the *Mdx* mouse muscle. As well as exon 23 skipping, some level of exon 23 was still spliced into *dystrophin* mRNA in each muscle tissue (upper band on the gel). (Reproduced with permission from Denti et al.[43])

7.13 Splicing as a route to therapy for infectious diseases like AIDS

In this final section we are going to look at the possibilities that splicing provides for targeting human pathogens. Splicing is critical for gene expression in viruses which infect humans, although it is not used by the many prokaryotic pathogens which cause human infectious diseases (as we discussed in Chapter 5, RNA splicing was first discovered in an adenovirus which infects human cells). A disease-causing virus for which splicing regulation has been shown to be critically important is **HIV** (the human immunodeficiency virus). Infection with HIV causes the disease AIDS (acquired immunodeficiency syndrome) to develop. HIV is a retrovirus; this means that it has an RNA genome which has to be converted into DNA by the enzyme reverse transcriptase during the life cycle of the virus in the cell (the gene structure of the HIV virus is shown in Fig. 7.26). Between infections, the DNA copy of the HIV virus becomes integrated into the human genome (this integrated genome is called a **provirus**).

Conventional therapies for HIV have targeted protein components of the virus. Particularly important therapeutic drugs are inhibitors of reverse transcriptase, such as the drug AZT which prevents copying of the HIV viral genome into cDNA, and protease inhibitors, which prevent the processing steps required for the assembly of viral protein into new viruses late in infection. These antiviral therapies are often used in combination with each other; the natural selection imposed by antiviral drugs can and does select for drug-resistant variants within the HIV virus population.

Another option that has been explored for HIV treatment is pre-mRNA splicing control. Splicing control is very important for this virus. The HIV genome has a maximum upper limit on size to enable the RNA genome to fit within the virus particle, and to cope with this small genome size the virus uses alternative splicing. From a genome of about 9kb the HIV virus produces a total of 40 different mRNAs. As well as alternative splicing of viral transcripts, it is also critical that late in the viral life cycle the virally encoded RNA is *not* spliced in order to provide full-length HIV genomic RNAs—these become packaged as virions to infect new cells.

For these reasons there are tight controls on HIV splicing patterns. An exonic splicing enhancer (ESE) (see Chapter 6 for a fuller discussion of ESEs) in the HIV genome plays a critical role in controlling splicing, enabling the production of the Tat, Rev, and Nef proteins from overlapping pre-mRNAs. This ESE sequence is very important. Although HIV sequences rapidly evolve between different patients, the ESE which controls viral pre-mRNA splicing of the *tat*, *env*, and *nef* genes does not change between HIV strains. This is because this ESE is needed to maintain a functional balance between splice isoforms needed during early infection for protein

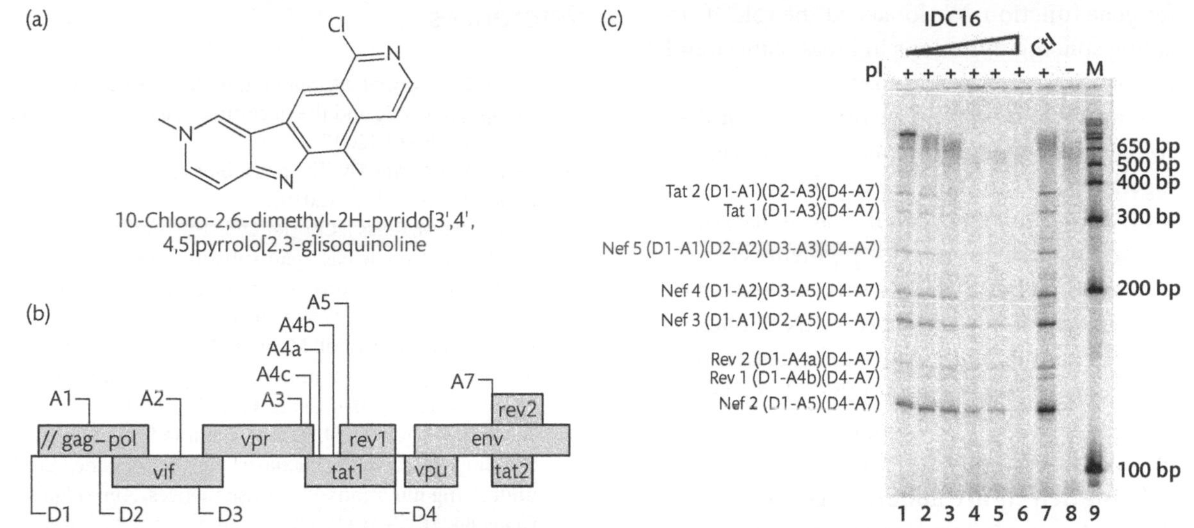

Figure 7.26 **RNA processing is needed to produce a full repertoire of proteins from the HIV genome and this has been targeted by small molecule inhibitors.** Full-length ORFs for the *tat*, *rev*, and *nef* genes are produced by alternative pre-mRNA splicing. An ESE binding the splicing activator protein SRSF1 plays a critical role in expression of these three genes. (a) The structure of the small molecule splicing inhibitor IDC16, which is effective at inhibiting the proliferation of the HIV genome. (b) A schematic map of the HIV genome showing the splice donor (D) and acceptor (A) sites with the different ORFs boxed. (c) The effect of increasing the concentration of IDC16 on the HIV splicing pattern. (From Bakkour et al.[44] ©2007 Bakkour et al.)

production and full-length genomic RNAs that are made in the final stages of infection.

This critical HIV ESE binds the SR protein splicing regulator **SRSF1** (the same protein discussed above that is over-expressed in some cancers). Small molecule inhibitors have been identified which bind to and inhibit SRSF1 function. The structure of one of these, called IDC16, is shown in Fig. 7.26. IDC16 strongly inhibits splicing of HIV pre-mRNAs and very potently inhibits HIV infection at low concentrations which do not affect the growth of the infected human cells.

SRSF1 is a very important splicing regulator in normal cells, so why do low concentrations of IDC16 have such a strong effect on splicing of HIV pre-mRNAs rather than affecting splicing in host cells? This is probably because the small size of the HIV genome means that this critical ESE can only bind SRSF1 and not other splicing regulators since it is so short. Most ESEs in host cell pre-mRNAs bind multiple splicing factors *including* SRSF1. Hence inhibition of SRSF1 would strongly inhibit the function of the critical HIV ESE, while host cell ESEs would additionally bind other splicing regulators which compensate for loss of SRSF1 function. In the future the use of splicing inhibitors might become important as a potential route for HIV therapy.

Key points

Point mutations which introduce stop codons into the *dystrophin* gene cause a much more severe form of muscular dystrophy than mutations which cause amino acid substitutions. This led to the idea of inducing exon skipping as a treatment option. The approach used is to block the splice sites of mutated exons containing translational stop codons, thereby preventing them from being spliced into the dystrophin mRNA. Although slightly shorter as a result of the missing exons, the dystrophin protein made after exon skipping should still be sufficient to restore function to muscle cells. Inhibition of splicing has also been carried out using small molecules. One application of this has been to block the production of the different splice isoforms needed by HIV.

Summary

In this chapter we have examined the important roles of pre-mRNA splicing in human disease and development.

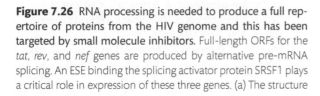

- **Diseases can be caused by mutations in the splicing code.** Mutations in splicing signals can be both frequent and particularly catastrophic for gene function. We looked at the role of *cis*-acting splicing mutations in breast cancer and premature ageing syndromes.

- **Diseases can be caused by mutations in the spliceosome** which recognizes and splices together exons in the nucleus. Mutations in these *trans*-acting splicing proteins cause diseases including a form of retinitis pigmentosa.
- **Diseases can be caused by mis-expression of splicing factors.** Normal pre-mRNA splicing patterns are maintained by a delicate balance of splicing regulators within the cell. Defects in expression levels of splicing factors contribute to the pathological properties of cancer cells.
- **Splicing defects are double-edged.** On the one hand splicing defects can cause devastating genetic diseases and on the other the manipulation of splicing within cells offers potential treatments. We saw how manipulation of splicing to induce exon skipping is being used as a therapy for treating muscular dystrophies.
- **Splicing offers a route to the treatment of infectious diseases like AIDS.** Small molecule inhibitors of splicing might also be important therapeutic tools for treating disease caused by viruses, such as HIV, which rely on compact splicing control elements to regulate alternative splicing of RNA made from their genomes.

Questions

7.1 Why are mutations that affect splicing often particularly serious for gene function?

7.2 What kinds of sequences within genes are mutated to cause splicing defects in their encoded pre-mRNAs?

7.3 What is the mechanism of disease in the premature ageing condition HGPS?

7.4 How are the mutations that affect splicing patterns detected?

7.5 Name two diseases in which there are mutations within splicing components.

7.6 Give three examples of splicing changes in cancer cells, and explain why they are important.

7.7 How is splicing used as a tool to treat disease? Give two examples.

References

1. **Wang GS, Cooper TA.** Splicing in disease: disruption of the splicing code and the decoding machinery. *Nat Rev Genet* **8**, 749–61 (2007).
2. **Ward AJ, Cooper TA.** The pathobiology of splicing. *J Pathol* **220**, 152–63 (2010).
3. **Krawczak M, Reiss J, Cooper DN.** The mutational spectrum of single base-pair substitutions in mRNA splice junctions of human genes: causes and consequences. *Hum Genet* **90**, 41–54 (1992).
4. **Ars E, Kruyer H, Morell M, et al.** Recurrent mutations in the NF1 gene are common among neurofibromatosis type 1 patients. *J Med Genet* **40**, e82 (2003).
5. **Teraoka SN, Telatar M, Becker-Catania S, et al.** Splicing defects in the ataxia-telangiectasia gene, ATM: underlying mutations and consequences. *Am J Hum Genet* **64**, 1617–31 (1999).
6. **De Sandre-Giovannoli A, Bernard R, Cau P, et al.** Lamin A truncation in Hutchinson-Gilford progeria. *Science* **300**, 2055 (2003).
7. **Eriksson M, Brown WT, Gordon LB, et al.** Recurrent de novo point mutations in lamin A cause Hutchinson-Gilford progeria syndrome. *Nature* **423**, 293–8 (2003).
8. **Scaffidi P, Gordon L, Misteli T.** The cell nucleus and aging: tantalizing clues and hopeful promises. *PLoS Biol* 3 e395 (2005).
9. **Scaffidi P, Misteli T.** Reversal of the cellular phenotype in the premature aging disease Hutchinson-Gilford progeria syndrome. *Nat Med* **11**, 440–5 (2005).
10. **Gordon LB, Cao K, Collins FS.** Progeria: translational insights from cell biology. *J Cell Biol* **199**, 9–13 (2012).
11. **Scaffidi P, Misteli T.** Lamin A-dependent nuclear defects in human aging. *Science* **312**, 1059–63 (2006).
12. **Cartegni L, Krainer AR.** Disruption of an SF2/ASF-dependent exonic splicing enhancer in SMN2 causes spinal muscular atrophy in the absence of SMN1. *Nat Genet* **30**, 377–84 (2002).
13. **Liu HX, Cartegni L, Zhang MQ, Krainer AR.** A mechanism for exon skipping caused by nonsense or missense mutations in *BRCA1* and other genes. *Nat Genet* **27**, 55–8 (2001).
14. **Cartegni L, Wang J, Zhu Z, et al.** ESEfinder: a web resource to identify exonic splicing enhancers. *Nucleic Acids Res* **31**, 3568–71 (2003).
15. **Baralle D, Baralle M.** Splicing in action: assessing disease causing sequence changes. *J Med Genet* **42**, 737–48 (2005).
16. **O'Driscoll M, Ruiz-Perez VL, Woods CG, et al.** A splicing mutation affecting expression of ataxia-telangiectasia and Rad3-related protein (ATR) results in Seckel syndrome. *Nat Genet* **33**, 497–501 (2003).
17. **Singh RK, Cooper TA.** Pre-mRNA splicing in disease and therapeutics. *Trends Mol Med* **18**, 472–82 (2012).
18. **Edery P, Marcaillou C, Sahbatou M, et al.** Association of TALS developmental disorder with defect in minor

splicing component U4atac snRNA. *Science* **332**, 240–3 (2011).

19. **He H, Liyanarachichi S, Akagi K, et al.** Mutations in U4atac snRNA, a component of the minor spliceosome, in the developmental disorder MOPD I. *Science* **332**, 238–40 (2011).
20. **McKie AB, McHale JC, Keen TJ, et al.** Mutations in the pre-mRNA splicing factor gene PRPC8 in autosomal dominant retinitis pigmentosa (RP13). *Hum Mol Genet* **10**, 1555–62 (2001).
21. **Liu MM, Zack DJ.** Alternative splicing and retinal degeneration. *Clin Genet* **84**, 142–9 (2013).
22. **Mordes D, Luo X, Kar A, et al.** Pre-mRNA splicing and retinitis pigmentosa. *Mol Vis* **12**, 1259–71 (2006).
23. **Cao H, Wu J, Lam S, et al.** Temporal and tissue specific regulation of RP-associated splicing factor genes PRPF3, PRPF31 and PRPC8—implications in the pathogenesis of RP. *PLoS One* **6**, e15860 (2011).
24. **Yoshida K, Sanada M, Shiraishi Y, et al.** Frequent pathway mutations of splicing machinery in myelodysplasia. *Nature* **478**, 64–9 (2011).
25. **Quesada V, Conde L, Villamor N, et al.** Exome sequencing identifies recurrent mutations of the splicing factor SF3B1 gene in chronic lymphocytic leukemia. *Nat Genet* **44**, 47–52 (2012).
26. **Bechara EG, Sebestyen E, Bernardis et al.** RBM5, 6, and 10 differentially regulate NUMB alternative splicing to control cancer cell proliferation. *Mol Cell* **52**, 720–33 (2013).
27. **Imielinski M, Berger AH, Hammerman PS, et al.** Mapping the hallmarks of lung adenocarcinoma with massively parallel sequencing. *Cell* **150**, 1107–20 (2012).
28. **Oltean S, Bates DO.** Hallmarks of alternative splicing in cancer. *Oncogene* **46**, 5311–18 (2014).
29. **Hanahan D, Weinberg RA.** Hallmarks of cancer: the next generation. *Cell* **144**, 646–74 (2011).
30. **Ladomery M.** Aberrant alternative splicing is another hallmark of cancer. *Int J Cell Biol* **2013**, 463786 (2013).
31. **Chen M, Zhang J, Manley JL.** Turning on a fuel switch of cancer: hnRNP proteins regulate alternative splicing of pyruvate kinase mRNA. *Cancer Res* **70**, 8977–80 (2010).
32. **Bonnal S, Vigevani L, Valcárcel J.** The spliceosome as a target of novel antitumour drugs. *Nat Rev Drug Discov* **11**, 847–59 (2012).
33. **Christofk HR, Vander Heide MG, Harris MH, et al.** The M2 splice isoform of pyruvate kinase is important for cancer metabolism and tumour growth. *Nature* **452**, 230–3 (2008).
34. **Harper SJ, Bates DO.** VEGF—A splicing: the key to anti-angiogenic therapeutics? *Nat Rev Cancer* **8**, 880–7 (2008).
35. **Ladomery MR, Harper SJ, Bates DO.** Alternative splicing in angiogenesis: the vascular endothelial growth factor paradigm. *Cancer Lett* **249**, 133–42 (2007).
36. **Cheng J, Zhou T, Liu C, et al.** Protection from Fas-mediated apoptosis by a soluble form of the Fas molecule. *Science* **263**, 1759–62 (1994).
37. **Manley JL, Krainer AR.** A rational nomenclature for serine/arginine-rich protein splicing factors (SR proteins). *Genes Dev* **24**, 1073–4 (2010).
38. **Larsson SH, Charlieu JP, Miyagawa K, et al.** Subnuclear localization of WT1 in splicing or transcription factor domains is regulated by alternative splicing. *Cell* **81**, 391–401 (1995).
39. **Karni R, de Stanchina E, Lowe SW, et al.** The gene encoding the splicing factor SF2/ASF is a proto-oncogene. *Nat Struct Mol Biol* **14**, 185–93 (2007).
40. **Das S, Anczukow O, Akerman M, Krainer AR.** Oncogenic splicing factor SRSF1 is a critical transcriptional target of MYC. *Cell Rep* **1**, 110–17 (2012).
41. **Ghigna C, Giordano S, Shen H, et al.** Cell motility is controlled by SF2/ASF through alternative splicing of the Ron protooncogene. *Mol Cell* **20**, 881–90 (2005).
42. **Jarmin S, Kymalainen H, Popplewell L, Dickson G.** New developments in the use of gene therapy to treat Duchenne muscular dystrophy. *Expert Opin Biol Ther* **14**, 209–30 (2014).
43. **Denti MA, Rosa A, D'Antona G, et al.** Body-wide gene therapy of Duchenne muscular dystrophy in the *mdx* mouse model. *Proc Natl Acad Sci USA* **103**, 3758–63 (2006).
44. **Bakkour N, Lin YL, Maire S, et al.** Small-molecule inhibition of HIV pre-mRNA splicing as a novel antiretroviral therapy to overcome drug resistance. *PLoS Pathog* **3**, 1530–9 (2007).

8 Co-transcriptional pre-mRNA processing

Introduction

In prokaryotes, the process of mRNA translation is co-transcriptional—in other words, ribosomes associate with a transcript while it is still in the process of being synthesized, and translation begins. The picture is rather different in eukaryotes; messenger RNA needs to be exported from the nucleus to the cytoplasm prior to translation, and there are several intervening steps between transcription and the eventual translation of an mRNA. This chapter will discuss the pre-mRNA processing events that occur co-transcriptionally (while transcription is still taking place), including the synthesis of the 7-methyl guanosine 5′ cap, pre-mRNA splicing, and the formation of the 3′ end of an mRNA by cleavage and polyadenylation. We will also consider briefly the connections that exist between chromatin structure and pre-mRNA processing. The chapter will end with a discussion of the special case of metazoan histone mRNA 3′ end formation.

8.1 Transcription and the RNA polymerases

8.1.1 RNA polymerases and the basics of transcription

Before we embark on this topic, it is worth revising the essential elements of the process of **transcription** and of the structure and function of the **RNA polymerases** (the enzymes that transcribe RNA from a DNA template). A detailed account of the processes that underpin transcription and its regulation are beyond the scope of this book. However, it is important to appreciate that transcription and post-transcriptional processes are intimately linked. In effect, most of the processes described in this chapter occur co-transcriptionally. Therefore a proper understanding of the process of transcription is required to appreciate the links between transcription and pre-mRNA processing events. Thus, the reader is strongly advised to become familiar with the process of transcription. The connection between transcription and post-transcriptional processes is illustrated very powerfully by the properties of the C-terminal domain (CTD) of RNA polymerase II, as we shall see later in the chapter.

RNA polymerase was independently discovered in 1960 by Jerard Hurwitz and Sam Weiss.[1] RNA polymerases are complex enzymes made up of several subunits.[2–4] In prokaryotes, there is a single RNA polymerase; in contrast, there are at least three in eukaryotes (RNA polymerases I, II, and III), each of which is responsible for transcribing particular characteristic RNAs. Transcription can be divided into three phases: initiation, elongation, and termination. RNA polymerases initiate transcription at sites in genes known as promoters.

RNA polymerase I transcribes the 47S precursor of the 28S, 18S, and 5.8S ribosomal RNA (rRNA). We shall cover the topic of rRNA processing in more detail in Chapter 14. As ribosomal RNA is very abundant in the cell (constituting up to 96% of all RNA), RNA polymerase I is very active. Transcription by

RNA polymerase I is initiated with the binding of UBF (upstream binding factor) to a sequence about 50 nucleotides upstream of the transcription start site. UBF then recruits another transcription factor, SL1, which in turn recruits RNA polymerase I through protein–protein interactions.

RNA polymerase III transcribes transfer RNA (tRNA), the 5S ribosomal RNA, the U6 small nuclear RNA (snRNA), the 7SL RNA which is part of the signal recognition particle, the RNA component of RNAse MRP (involved in RNA processing), and several repeated sequences such as the Alu elements in humans. The transcription of the 5S rRNA gene is well studied: it begins with the binding of the transcription factor TFIIIA to an 'internal control sequence' about 50 bases downstream of the transcription start site. (Recall from Chapter 4 that TFIIIA was the first zinc-finger protein to be characterized.) Two other transcription factors, TFIIIC and TFIIIB, then bind; these then recruit RNA polymerase III.

RNA polymerase II is responsible for the transcription of all protein-coding genes, microRNAs (miRNAs), and the snRNAs (U1, U2, U4, U5, and U7, but not U6). Most promoters responding to RNA polymerase II contain a conserved sequence element known as the TATA box, generally located close to the transcription start site. The essential steps in the initiation of RNA polymerase II transcription are summarized in Fig. 8.1. The TATA box is bound by the TFIID complex (TFIID contains the TATA-box-binding protein (TBP)) and TFIIA, followed by the recruitment of TFIIB. TFIIB then recruits RNA polymerase II and TFIIF. Next, TFIIE and TFIIH bind to the complex. TFIIH is particularly important as it phosphorylates serine 5 in the CTD of RNA polymerase II; this phosphorylation then allows RNA polymerase II to initiate transcription. However, this in itself is not enough for transcription to continue: another protein kinase, pTEF-b, is recruited, which phosphorylates serine 2 in the CTD. Only then can transcription continue through the gene (this is known as the *elongation of transcription*).

Some genes do not have a TATA box. In these cases the TATA box is substituted by another sequence element, the 'initiator' element. The initiator is bound by another transcription factor

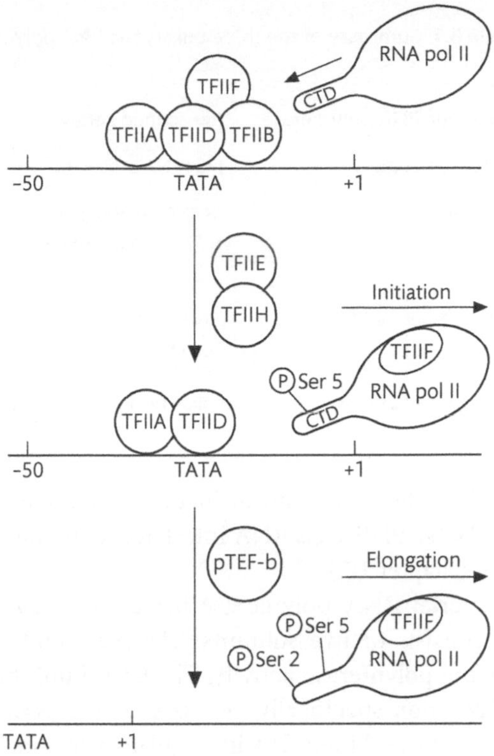

Figure 8.1 Essential steps in the initiation of transcription by RNA polymerase II. A complex of core transcription factors assembles at the TATA box (TFIIA, TFIIB, TFIIF, and TFIID; the latter contains the TATA-binding protein (TBP)). The TATA box is present in most promoters that are responsive to RNA polymerase II. RNA polymerase II is recruited, and initiation of transcription occurs, but only after phosphorylation of serine 5 in the C-terminal domain (CTD) by TFIIH. A further phosphorylation, this time of serine 2, by the protein kinase pTEF-b is required for transcriptional elongation to take place. Note that TFIIF remains bound to the polymerase.

which has the ability to recruit TFIID, and hence RNA polymerase II, through protein–protein interactions.

The three eukaryotic RNA polymerases can be differentiated by their sensitivity to the fungal toxin α-amanitin. RNA polymerase I is insensitive to it; RNA polymerase III is moderately sensitive; while RNA polymerase II is very sensitive, being inhibited at a toxin concentration of 1 $\mu g\ ml^{-1}$.

The main properties of the eukaryotic RNA polymerases are summarized in Table 8.1. It is also worth mentioning that additional RNA polymerases have been discovered in plants. Plant RNA polymerases IV and V are involved in the synthesis of short non-coding RNAs involved in RNA interference,

Table 8.1 Summary of the three eukaryotic RNA polymerases*

Eukaryotic RNA polymerase	Transcribed genes	Key transcription initiation factors	Sensitivity to α-amanitin
RNA polymerase I	47S rRNA precursor	UBF and SL1	Insensitive
RNA polymerase II	Protein-coding genes, miRNAs, most snRNA	TFIID and TFIIA; TFIIB, TFIIF, TFIIE; the kinases TFIIH and pTEF-b	High sensitivity
RNA polymerase III	tRNA, 5S rRNA genes, U6 snRNA	TFIIIA, TFIIIB, TFIIIC	Medium sensitivity

*Their target genes, the main core transcription factors involved, and their sensitivity to α-amanitin are shown

including the formation of inactive heterochromatin.[5] (We shall discuss RNA interference in more detail in Chapter 16.)

Bacterial RNA polymerase has a core structure that consists of five subunits. The β subunit contains the polymerase activity. The β1 subunit binds to DNA non-specifically, two α subunits assemble the enzyme and interact with regulatory factors, and the ω subunit promotes the assembly of the whole enzyme complex. A separate subunit, sigma (σ), increases binding specificity for correct promoter sequences.

When inactive, bacterial RNA polymerase binds non-specifically to DNA; but it can be recruited rapidly by the σ subunit to wherever transcription is required. Eukaryotic RNA polymerases share a similar core architecture with the prokaryotic RNA polymerase because they work through a common catalytic mechanism, but they have acquired added complexity with the presence of additional subunits. Thus, in the evolution of eukaryotes several proteins were incorporated into active RNA polymerase to become genuine subunits, giving greater functional and regulatory complexity.

The Archaea are a unique group of prokaryotic microorganisms that are thought to be the descendants of an ancient group of species that bridge the evolutionary gap between bacteria and eukaryotes. They are worth mentioning because their RNA polymerase system is distinct from that of bacteria. Like bacteria, the Archaea only have a single RNA polymerase.[6] However, the overall structure and subunit composition of archaeal RNA polymerase is reminiscent of eukaryotic RNA polymerase II.

In all RNA polymerases, the mechanism of addition of ribonucleotides to a nascent RNA transcript is essentially the same. A common feature is the role of Mg^{2+}. Aspartyl residues in the RNA polymerase catalytic subunit bind Mg^{2+} ions. The first Mg^{2+} holds the α-phosphate of the last nucleotide in the nascent transcript, while the second Mg^{2+} holds the pyrophosphate of the next NTP. The 3′ –OH group in the last nucleotide of the nascent transcript mediates a nucleophilic attack onto the next NTP. Figure 8.2 illustrates the process of transcriptional elongation

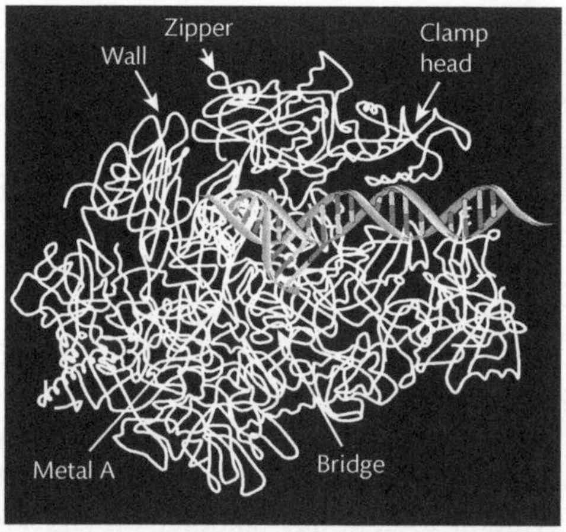

Figure 8.2 Crystal structure of RNA polymerase II in the process of transcribing pre-mRNA. Note that the DNA double helix enters from the right and makes an abrupt turn through the enzyme where it is unwound. Mg^{2+} (metal A) is involved in positioning the next NTP to be added to the nascent transcript. The identity of the next NTP is guided by hybridization to the DNA template strand. As the DNA exits the enzyme, it re-anneals and the nascent pre-mRNA transcript exits separately. (From Gnatt et al.[7])

by RNA polymerase II. It is taken from a seminal *Science* paper by Roger Kornberg.[7] Kornberg won the Nobel Prize in Chemistry in 2006 for his work on elucidating how DNA is converted to RNA, i.e. the process of transcription.

8.1.2 Transcription is regulated

Transcription is regulated in complex ways. For example, the initiation and elongation of transcription, described in this section, can be regulated through the direct phosphorylation of RNA polymerase II. Transcription factors play a major role in mediating transcriptional regulation. The transcription factors we describe in this section are known as the *core transcription factors*. However, there are several other transcription factors that bind to specific DNA elements in promoters. In Chapter 4 we saw that some transcription factors also bind RNA and work post-transcriptionally. Many of these transcription factors include transcription activation or repression domains, or alternatively interact with cofactors that either activate or repress transcription. Transcription is also regulated at a distance by transcriptional enhancers and silencers. Last but not least, epigenetic modifications of DNA and alterations in chromatin play a very important role in regulating transcription. (An epigenetic mechanism is one in which changes in gene expression or phenotype are due to changes *other than* direct changes in DNA sequence.) As we shall see in Chapter 15, RNA can also play a part in epigenetic modifications of DNA and chromatin.

So, why place this emphasis on transcription and its regulation in a chapter about co-transcriptional processing? Transcriptional and post-transcriptional processes are intimately connected, often through multifunctional proteins, so it is impossible to completely divorce the processes. It is also clear that specific promoters can influence post-transcriptional events, presumably through the recruitment of proteins that affect pre-mRNA processing, splicing, and so on. In Chapter 4 we encountered several proteins that can bind both DNA and RNA; it is likely that several of these coordinate transcriptional and post-transcriptional processes.

Key points

RNA polymerases are ancient enzymes that evolved to copy a DNA template strand into an RNA transcript. In prokaryotes, there is a relatively simple single RNA polymerase. In eukaryotes there are several RNA polymerases; of which RNA polymerase II is responsible for the transcription of protein-coding genes. In eukaryotes, several pre-mRNA processing steps take place co-transcriptionally. In prokaryotes, mRNA translation occurs co-transcriptionally.

8.2 Formation of the ends of an mRNA

8.2.1 Formation of the 7-methyl guanosine cap

While transcription is occurring, the very 5′ end of the nascent pre-mRNA is immediately modified. This happens once the first 25 nucleotides are synthesized and the nascent transcript has first emerged from the exit channel of the RNA polymerase. An extra guanosine (not encoded by the DNA template) is added at the very 5′ end of the transcript. (Remember that guanosine consists of the base guanine with a ribose sugar group attached.) However, the linkage is unusual in two ways. First, there are three phosphate groups between the two last ribose groups, and secondly they are joined 5′ to 5′ as shown in Fig. 8.3. The guanine group on the cap is methylated at position 7 by the enzyme methyl transferase. Therefore the 5′ cap is known as the **7-methyl guanosine cap** (usually abbreviated m^7G).[8]

The process of methylation takes place as follows. First, a terminal phosphate group is removed by a phosphatase enzyme. Then, GTP is added to the remaining two terminal phosphates by the enzyme guanylyl transferase. The addition of GTP creates the atypical 5′ to 5′ triphosphate linkage. A methyl transferase then methylates the nitrogen at position 7 to complete the cap structure. The enzymes required for the formation of the cap associate together in the **capping enzyme complex** (CEC). The CEC is actually bound to RNA polymerase II before transcription starts, so that RNA polymerase II transcripts are 5′ capped immediately.

What is the function of the 5′ cap? First, it prevents attack by 5′ exonuclease because it does not

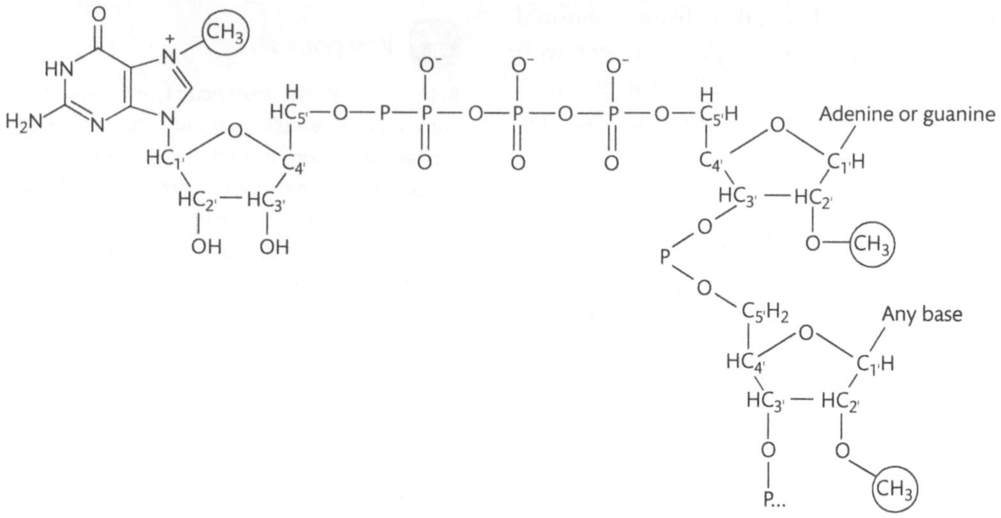

Figure 8.3 Structure of the 7-methyl guanosine cap. Note the unusual 5′–5′ linkage between the modified 7-methyl guanosine and the first nucleotide in the transcript.

resemble the proper substrate. As such, it protects the 5′ end of the nascent transcript from undergoing digestion. (As we shall see in Chapter 12, decapping of mRNA by decapping enzymes is an essential step in active mRNA degradation pathways.) The cap is also bound in the nucleus by the **cap-binding complex** (CBC). The CBC is in turn recognized by the nuclear pore complex; this combined interaction facilitates the export of the transcript from the nucleus. The CBC also protects the cap from decapping enzymes. After the first round of translation in the cytoplasm (the so-called *pioneer round*), CBC is replaced by the translation initiation factor and cap-binding protein eIF4E. The function of eIF4E in mRNA translation will be discussed in detail in Chapter 11.

8.2.2 Cleavage and polyadenylation at the 3′ end

While the 5′ end of a transcript is defined by where transcription starts, the 3′ end is defined by an active pre-mRNA cleavage process.[9,10] The vast majority of eukaryotic mRNAs are polyadenylated (in other words they receive a string of As) at the very 3′ end, giving rise to the **poly(A) tail**. The process of 3′ end formation by cleavage is functionally coupled to the process of polyadenylation. Two signals in the nascent RNA transcript define the polyadenylation site: AAUAAA and a downstream G/U signal. These signals can occur quite a long way after the stop codon—even at a distance of thousands of nucleotides—and define the end of the 3′ UTR (3′ untranslated region). Cleavage occurs about 10–30 nucleotides after the AAUAAA signal, and the poly(A) tail is then added to the free 3′ end of the pre-mRNA. Note that the poly(A) tail can only be added once cleavage has occurred. The architecture of a typical mature eukaryotic mRNA is summarized in Fig. 8.4.

The site of cleavage is determined by two multiprotein complexes, CPSF and CstF. In mammals CPSF contains five subunits: Fip1, CPSF-30, CPSF-73, CPSF-100, and CPSF-160. CPSF-160 recognizes the AAUAAA signal. Note that there are slight variations in the signal across species: in mammals, for example, almost 50% of mRNAs have a divergent AAUAAA signal (e.g. AUUAAA). Fip1 and CPSF-30 also bind the pre-mRNA, improving specificity, especially when there is not a perfect match to the consensus AAUAAA sequence. The G/U-rich signal is recognized by another complex, known as the cleavage stimulation factor (CstF). CstF contains three subunits: CstF-50, CstF-64, and CstF-77. The G/U-rich sequence is specifically recognized by CstF-64. Another protein, called symplekin, acts as a platform to help bring together the CPSF and CstF complexes.

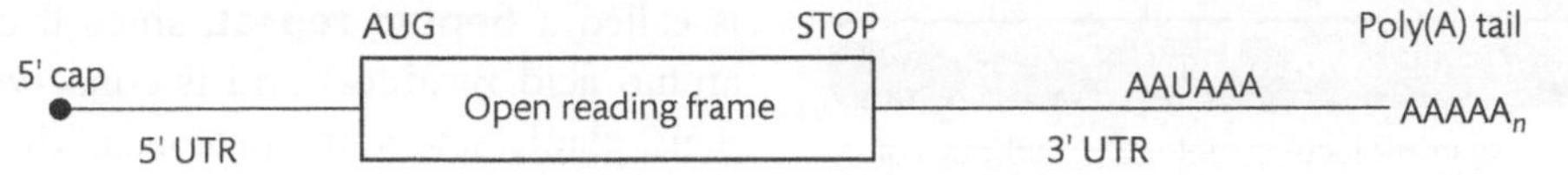

Figure 8.4 Architecture of a typical eukaryotic mRNA. After processing, a mature mRNA consists of a 7-methyl guanosine cap at the 5′ end, followed by the 5′ UTR (untranslated region). The start and stop codons define the open reading frame (ORF). The sequence after the stop codon is the 3′ UTR, which usually ends in a poly(A) tail. The first of the two cleavage signals, AAUAAA, is retained in the mature mRNA and precedes the poly(A) tail.

While CPSF and CstF act in concert to define the cleavage site, it is further defined by additional protein complexes—the cleavage factors CFI and CFII. In mammals, CFI binds to the sequence UGUAA proximal to the AAUAAA site, where it can help recruit CPSF if the AAUAAA signal is missing or divergent. Interestingly, efficient cleavage also requires the presence of the C-terminal domain of RNA polymerase II (the CTD). This is important because it helps to couple the process of transcription to 3′ end cleavage; and furthermore it is consistent with the role of the CTD in recruiting pre-mRNA processing factors, as we shall see later in this chapter.

The interaction of the 3′ end cleavage machinery with the CTD is mediated by CstF-50, a subunit of CstF. The protein that actually produces the cleavage is thought to be CPSF-73, whose defining feature is the presence of a metallo-β-lactamase domain. CPSF-73 is also involved in the cleavage of histone mRNA 3′ ends, as we shall see in Section 8.6.

Once the mRNA is cleaved, the 3′ end is finally defined and a poly(A) tail can be added. Like the 5′ cap, the poly(A) tail has several functions: to protect the mRNA from degradation (in this case by 3′ exonucleases), to promote nuclear export, and to promote translation. Polyadenylation starts as soon as cleavage has occurred and is catalysed by the enzyme polyadenylate polymerase, which is specialized in the addition of adenosine monophosphates. Once an initial short poly(A) tail has been formed, relatively slowly, it is bound by the RNA-binding protein PAB2. Binding by PAB2 increases the affinity of the polyadenylate polymerase for the RNA, which then extends the poly(A) tail more rapidly (this is known as fast polyadenylation). Once about 250 As have been added, the enzyme stops working as it can no longer be stimulated by CPSF upstream. The interlinked processes of 3′ end cleavage and polyadenylation are summarized in Fig. 8.5. Note that many mRNAs are *alternatively polyadenylated*,[11,12] as described in Box 8.1.

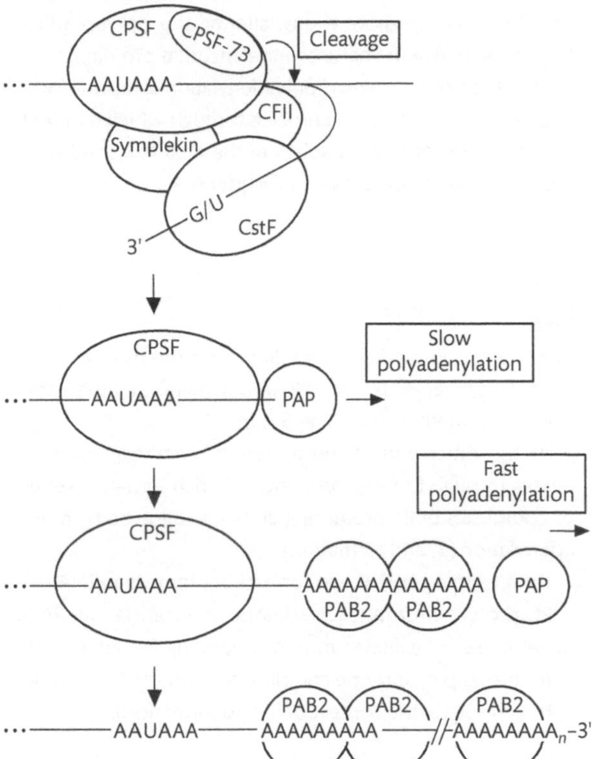

Figure 8.5 3′ end formation: cleavage and polyadenylation. The cleavage signal AAUAAA is recognized by CPSF and the downstream G/U-rich signal is recognized by CstF. The CTD of RNA polymerase II helps to recruit these complexes to the transcript. Symplekin acts as a platform to bring CPSF and CstF together. Additional complexes, CFI and CFII, help to define the site of cleavage. Cleavage is stimulated by the C-terminal domain of RNA polymerase II and is mediated by CPSF-73. Following cleavage, a short poly(A) tail is added by polyadenylate polymerase. This short tail is bound by PAB2, which then stimulates the polymerase to add further As, up to about 250 in total.

Box 8.1 Alternative polyadenylation

Hardly any process in molecular biology is without complications, not least polyadenylation. All aspects of gene expression have the potential to be regulated, providing evolution with additional tools to generate phenotypic diversity. In the human genome, more than half the genes are alternatively polyadenylated; this means that there may be more than one site of polyadenylation. The consequences of alternative polyadenylation for the mRNA depend on the site at which it occurs. If the alternative polyadenylation site lies in the 3′ UTR, the consequence will be a shortened (or lengthened) 3′ UTR, which may affect the stability, localization, and translation of mRNA, all of which can be influenced by sequences in the 3′ UTR. If the alternative site lies upstream of the usual stop codon (e.g. in an intron) the result could be a truncated transcript that lacks one or more exons, altering the open reading frame and hence the nature of the encoded protein.

This type of alternative polyadenylation is also a form of alternative splicing–the process through which different combinations of exons end up in the final exported processed mRNA, as described in Chapter 6.

Key points

The 5′ end is immediately modified once the nascent transcript first emerges from the RNA polymerase by the addition of a 7-methyl guanosine 5′ cap.

At the other end of the transcript, once transcription passes through the AAUAAA and G/U-rich signals, a series of complexes bind, promoting cleavage followed by polyadenylation (3′ end formation).

Both capping and 3′ end formation promote mRNA export, protection from exonucleases, and translation. RNA polymerase II facilitates mRNA processing by interacting with the capping enzyme complex (to form the 5′ cap) and with CstF-50 (in the context of 3′ end formation).

8.3 The C-terminal domain (CTD) of RNA polymerase II

In contrast with other RNA polymerases, RNA polymerase II has acquired during evolution a unique and repetitive structure at its C-terminus. The C-terminal end (the CTD) of the large subunit of RNA polymerase II has a very important function. It contains 52 repeats of the sequence YSPTSPS (this is called a **heptad repeat**, since it contains seven amino acid residues) and is conserved across animals, plants, and some protozoa. The CTD of RNA polymerase II plays a crucial role in linking transcription and RNA processing.[13–15] The CTD is also involved in the process of co-transcriptional surveillance[16] described in Box 8.2. In the crystal structure of RNA polymerase II the CTD lies at the exit point of the emerging transcript. This is a strategic position where it can influence the processing of newly synthesized pre-mRNA.

As we saw earlier, the transcription of a gene by RNA polymerase II into an mRNA copy involves the formation of a pre-initiation complex, followed by transcriptional initiation and elongation, and termination and release of RNA polymerase II. The CTD of RNA polymerase is phosphorylated differentially during each of these steps.

Each heptad Y**S**PT**S**PS repeat in the CTD contains two serine residues which are major sites of reversible phosphorylation (the phosphorylated serines are in bold type). As we saw in Section 8.1, phosphorylation of the fifth serine in the heptad is required for transcriptional initiation, and phosphorylation of the second serine is required for transcriptional elongation. Since there are 52 CTD repeats in the large subunit of human RNA polymerase II, there is a total of 104 potential phosphorylation sites in a single CTD!

Box 8.2 The role of the CTD in co-transcriptional surveillance

There are several 'checkpoints' in the maturation of mRNA. Having these checkpoints is important because if faulty mRNAs end up in the cytoplasm they can be translated into deleterious proteins. Thus mechanisms have evolved in the cell nucleus in which mRNA processing is interrupted if a problem occurs.

The CTD is thought to play a role in this mRNA surveillance. At transcription initiation, the CTD is phosphorylated at serine 5, but not at serine 2, so that it stalls (a process mediated by the inhibitory protein complex NELF) and allows capping to take place. The capping enzyme itself disables NELF-mediated stalling of transcription; thus correct capping is followed by transcriptional elongation. In other words, the RNA polymerase is allowed to proceed with the rest of the transcript only when an mRNA is correctly capped.

The phosphorylation status of the serines in the CTD controls the movement of the extending RNA polymerase as it moves along the gene. Changes in CTD phosphorylation to promote movement of RNA polymerase are caused by assembly of early splicing factors onto the nascent pre-mRNA. The splice machinery component U1 snRNP interacts with the core transcription factor TFIIH, which phosphorylates the CTD of RNA polymerase II on serine 5, promoting the formation of transcription complexes on genes. Hence the assembly of the splicing machinery has a direct and positive effect on transcription.

In the pre-initiation complex, serine residues within the CTD are not phosphorylated. This form of RNA polymerase II is called RNA polymerase IIA; it binds to the promoter but only makes short transcripts by what is called non-processive transcription (this means that transcription does extend the full length of the gene). This non-processivity is due to two inhibitory protein complexes called DSIF and NELF.

As discussed in Section 8.1, at the next stage, transcriptional initiation, serine 5 of the CTD is phosphorylated by TFIIH, a component of the transcription initiation complex. This leads to the recruitment of the capping enzyme complex and co-transcriptional 5′ capping of the transcript. The next stage is transcriptional elongation, whereby transcription becomes *processive*, and the RNA polymerase is able to extend the full length of the gene to make a complete pre-mRNA transcript. At this stage, serine 2 of the CTD is phosphorylated by pTEF-b to give hyperphosphorylated RNA polymerase IIo.

What drives the switch from non-processivity to processivity? DSIF and NELF, the inhibitory protein complexes, are unable to bind the fully phosphorylated CTD. Therefore, once hyperphosphorylation of the RNA polymerase has occurred, DSIF and NELF no longer bind, and the inhibition they cause is lifted.

Finally, when the RNA polymerase reaches the 3′ terminus of the gene, serine 5 becomes dephosphorylated; at this stage the CTD is only associated at serine 2, allowing association with polyadenylation factors.

Overall, the CTD makes an important contribution to recruitment to sites of transcription of proteins involved in 3′ end formation and 5′ capping. However, the impact of the CTD is felt beyond transcription; deletion of the CTD from RNA polymerase II also prevents efficient pre-mRNA splicing. The assembly of splicing components on the nascent RNA (still attached to RNA polymerase II) triggers reversible phosphorylation of the CTD, which has a positive effect on the transcriptional elongation of RNA polymerase II. The CTD is also required for co-transcriptional RNA editing of specific transcripts, including *auto-editing* of intron 4 of ADAR2, causing a change in a splice site.[17] We shall cover the process of RNA editing in Chapter 13. In the next section we discuss the links between transcription and splicing.

> **Key points**
>
> RNA polymerase II has acquired a unique repetitive structure at its C-terminus—the CTD. It has up to 52 repeats of a heptad sequence which contain two conserved serines. Phosphorylation of these serines influences the ability of RNA polymerase II to initiate and elongate transcription, and to recruit splice factors and other pre-mRNA processing components to the nascent transcript.

8.4 The links between splicing, transcription, and chromatin

It was originally thought that splicing takes place after the transcription of genes, possibly in a different compartment of the cell nucleus (model 1 in Fig. 8.6). However, the most recent information suggests that this is not usually the case, and that splicing takes place generally co-transcriptionally[18–21] (model 2 in Fig. 8.6; notice that the exons are spliced together as the RNA polymerase II is still in the process of transcribing the gene). It is also possible that the binding of components of the splicing machinery, such as the U1 snRNP (see Chapter 5), might take place while the gene is still being transcribed by RNA polymerase II, but the catalytic steps of splicing take place later (model 3 in Fig. 8.6).

Next we discuss the links between transcription and splicing in greater detail. We shall see that these links are temporal (in terms of timing), physical (the ways in which splicing components and

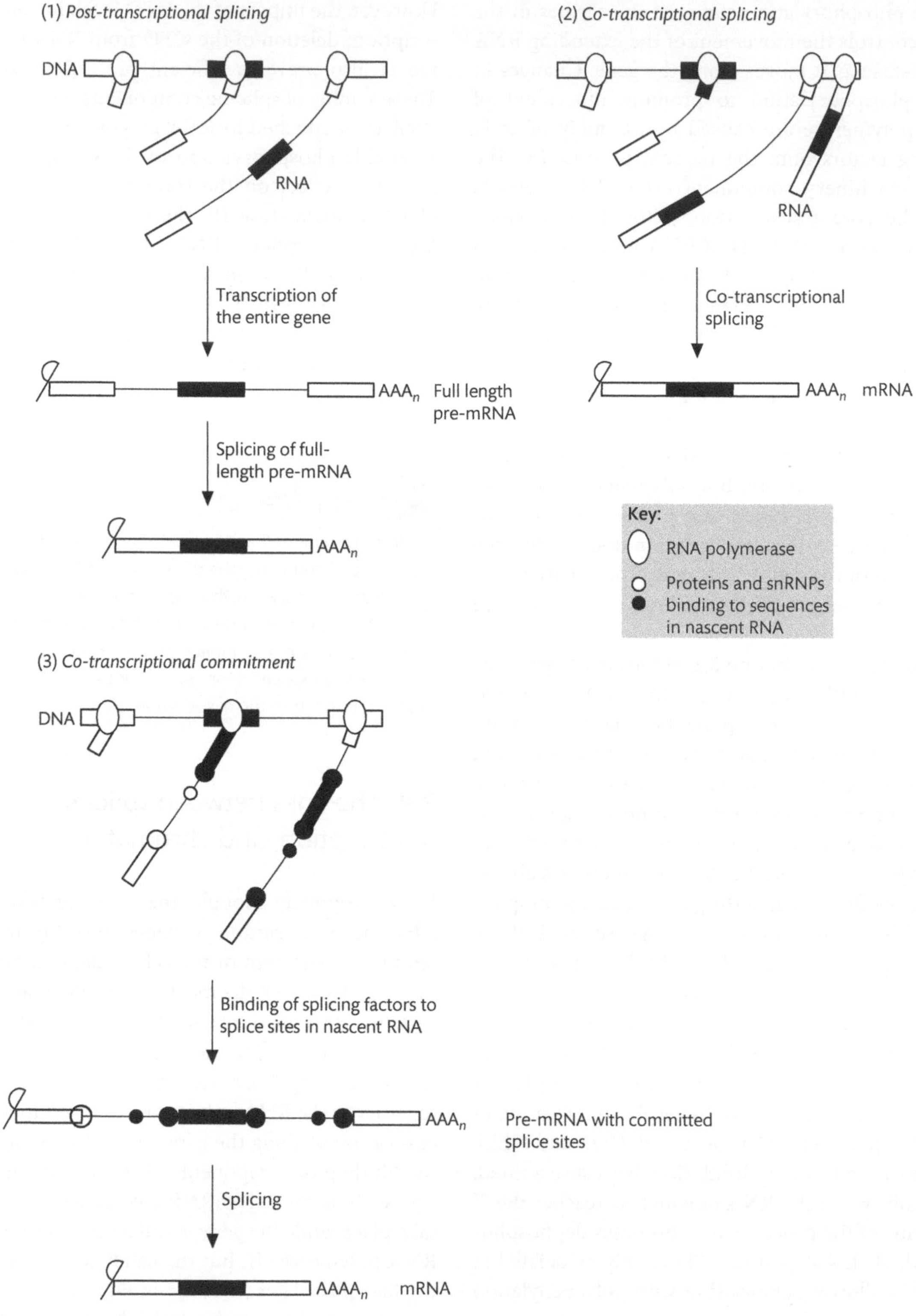

Figure 8.6 Possible models of when splicing takes place relative to transcription. Splicing can take place (1) post-transcriptionally after the entire pre-mRNA has been transcribed or (2) co-transcriptionally while transcription is still taking place, or (3) splice sites can be occupied by early components of the splicing machinery during transcription (co-transcriptional commitment) with the actual catalytic steps of splicing taking place later.

transcriptional components are physically linked), and spatial (splicing and transcription are compartmentalized in the cell nucleus).

Many pre-mRNAs are either spliced while they are being transcribed (model 2 in Fig. 8.6) or at least bind early splicing factors and start to assemble spliceosomes (spliceosomes are the molecular machines that carry out splicing—model 3 in Fig. 8.6). To compare the relative timings of when transcription and splicing take place it has proved useful to look at very long genes whose transcription can be induced. In this way it is possible to initiate transcription, and then look over the time-course to compare the progress of both splicing and transcription.

One of the first experiments to use this kind of approach is shown in Fig. 8.7. This experiment utilized what is apparently the longest gene in the human genome, *dystrophin*.[22] *Dystrophin* is induced during muscle differentiation and, with a length of 2.5Mb, takes 16 hours to transcribe! However, splicing is not delayed for the 16 hours which it takes for the full-length pre-mRNA to be made. Instead, it takes place while transcription is still happening—a process called **co-transcriptional splicing**. Other experiments suggest that co-transcriptional splicing is likely to be the rule rather than the exception. For example, data supporting co-transcriptional splicing have also been obtained in fruit flies; where splicing of upstream exons in the hormone-inducible *E74A* gene was detected before the 3′ end of the gene had been transcribed.[23]

The experiments which showed co-transcriptional splicing of *dystrophin* pre-mRNA (see Fig. 8.7) used a molecular biology approach; RNA was isolated from cells at various times after initiation of transcription by muscle differentiation, and then analysed by RT-PCR. The kinetics of transcription and pre-mRNA splicing have also been compared visually using cell biology approaches. An example of this is shown in Fig. 8.8, in which chromosomes isolated from embryos of the fruit fly *Drosophila melanogaster* were visualized microscopically.[24] In these microscope spreads, elongating pre-mRNAs can be seen as strands emanating from the chromatin as they are copied from their chromosomal template gene. The interpretation of spreads like the one shown in Fig. 8.8 suggests that spliceosomes start forming on splice sites within seconds of transcription, and that splicing might take place within three minutes of the intron being

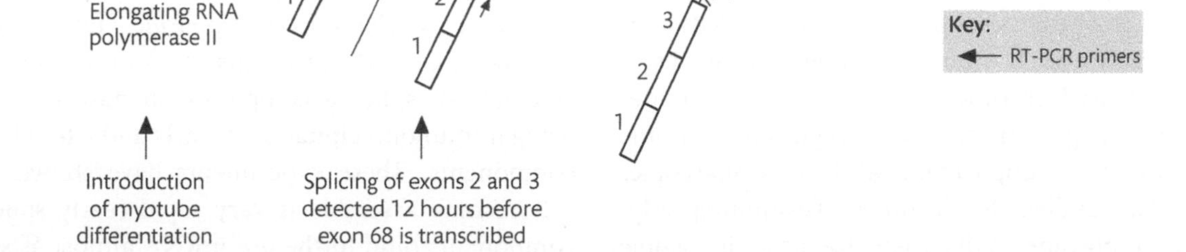

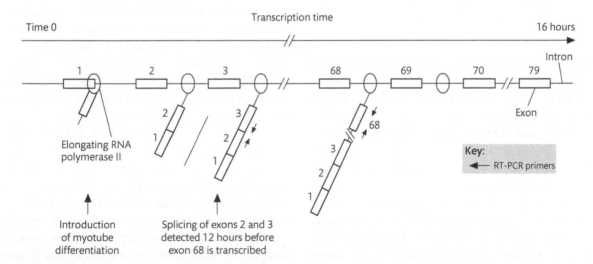

Figure 8.7 Splicing of the human *dystrophin* gene takes place co-transcriptionally. Co-transcriptional splicing was shown experimentally for the *dystrophin* gene by inducing cultured muscle precursor cells (myoblasts which do not express *dystrophin*) to differentiate into muscle cells (myotubes which do express *dystrophin*). This experiment is outlined as a cartoon, which shows the first wave of transcription of the *dystrophin* gene after differentiation is induced. The elongating RNA polymerase II tracks along the gene making the mRNA—this takes 16 hours in total. During this first wave of transcription, the splicing of exons in the newly synthesized (nascent) pre-mRNA was monitored using reverse transcription and primers in adjacent exons. Spliced exons from the 5′ end of the gene (exons 2 and 3) could be detected 12 hours before transcription of exon 68 was detected, suggesting that splicing occurs co-transcriptionally.

Figure 8.8 **Spliceosome assembly and splicing have been visualized microscopically on chromatin spreads of *Drosophila* embryos.** These chromatin spreads show chromatin prepared from the nuclei of fruit fly embryos, with attached nascent RNAs. Distinct loops corresponding to lariat intermediates and excising introns can be seen within the nascent RNAs (which are still undergoing transcriptional elongation). Nascent RNAs attached to the downstream part of the gene are shorter than predicted because of the removal of these intronic sequences. Granules (arrow) are present on nascent RNA and correspond to assembling spliceosomes. (Reproduced from Beyer and Osheim[24] with permission from Elsevier.)

transcribed while the pre-mRNA is still attached to the RNA polymerase.

Similar results have been obtained by looking at insect cells with particularly large chromosomes, called **polytene chromosomes**. Instead of having two copies of each chromosome per cell, cells containing polytene chromosomes appear to have a single copy of each chromosome when visualized under the microscope—in fact, each 'single' chromosome copy contains multiple identical chromosome copies aligned in parallel. The advantage of examining polytene chromosomes is that their size makes it possible for gene expression to be visualized microscopically; active genes on polytene chromosomes appear more diffuse because the DNA is less condensed. These less condensed regions are called **Balbiani rings**, on which extending RNA polymerases with their attached nascent RNAs can be seen. Experiments using three-dimensional reconstruction of electron microscope images to analyse in situ transcription and splicing of transcripts derived from Balbiani rings in the midge *Chironomus tentans* showed that splicing takes place within a kilobase of the intron being synthesized, and within 30 seconds of the intron being transcribed.

Dystrophin and the Balbiani ring genes are particularly long. This gives plenty of time for co-transcriptional splicing to take place; a longer gene will take longer to transcribe, and so the RNA polymerase will be associated with it for longer. However, not all introns are removed co-transcriptionally. Whether an intron is excised co-transcriptionally or post-transcriptionally also depends on its position in the gene. Upstream introns in genes are most likely to be co-transcriptionally removed since these are the first to be transcribed and so are associated with the elongating RNA polymerase for the longest time. Different genes are also spliced at different speeds. Introns which are more slowly spliced are more likely to be spliced post-transcriptionally.

There is also evidence to support model 3 (co-transcriptional splicing commitment) in Fig. 8.6. Although many pre-mRNAs are fully spliced co-transcriptionally, this is not always the case. Therefore another important process is co-transcriptional **splice site commitment**. Commitment in this context means that early splicing factors bind to splice sites on the pre-mRNA, which then commits the spliceosome to use these sites when splicing takes place once transcription has finished. Co-transcriptional splice site commitment—as opposed to complete co-transcriptional splicing—tends to happen for very short genes like those found in the yeast *Saccharomyces cerevisiae*. Experiments have been carried out in this yeast to monitor the assembly of splicing complexes on nascent RNAs using immunoprecipitation of early and late splicing components. These experiments have shown that spliceosome assembly is very rapid (early splicing components bind to the pre-mRNA almost as soon as the 3′ splice site, namely the 3′ end of the intron, is transcribed). However, only on longer yeast genes can splicing itself occur while transcription is still happening. Because of the short size of most yeast genes, however, most transcripts are spliced post-transcriptionally even though splicing commitment takes place much earlier.

The experiments described in this section suggest that either splicing or commitment to using particular splice sites by the spliceosome often happens while nascent transcripts are still being transcribed. Co-transcriptional splicing has important implications for alternative splicing—the process through which different exon combinations can be derived from mRNAs originating from the same gene. This is because the transcription machinery can communicate with, or in other words influence, the splicing machinery. The mechanisms and biology of alternative splicing is discussed in detail in Chapter 6.

There are also shared components in the transcription and splicing machinery; a number of RNA splicing components have been found in RNA polymerase II and in spliceosomes, including the protein CA150, a transcription elongation factor. CA150 contains six FF domains, present in several transcription and splicing factors. The FF domains recognize the phosphoserine motifs in the CTD of RNA polymerase II. The N-terminus of CA150 also contains three WW domains which interact with the splice factor SF1. Thus, CA150 may be directly involved in coupling transcription and splicing.[25] This property is not unique to CA150—as we saw in Section 8.3, the splice machinery component U1 snRNP actually interacts with the core transcription factor TFIIH, which phosphorylates the CTD of RNA polymerase II on serine 5, promoting the initiation of transcription.

Over recent years yet another layer of complexity in co-transcriptional splicing has emerged, namely that splicing is influenced not only by the machinery of transcription but also by the **chromatin** environment.[26–28] Chromatin is the material that makes up chromosomes: in essence chromosomal DNA packaged by proteins (notably the **histones**), as well as RNA (transcripts and non-coding RNAs involved in chromatin modifications). The histones together form regular packaging units called **nucleosomes**, around which a regular length of DNA is coiled.

There are three main mechanisms through which chromatin can affect co-transcriptional splicing, and they are listed in Table 8.2. One mechanism is nucleosome positioning (sometimes referred to as nucleosome occupancy). Genome-wide analysis of nucleosome positioning shows that it is not random—it varies considerably and has a significant role in regulating gene expression. Exon sequences tend to be more associated with nucleosomes compared with introns. This is due to the fact that exons tend to be GC-rich whereas introns tend to be AT-rich (GC-rich sequences are favoured by nucleosomes). It is not a coincidence that the average size of exons is 150 nucleotides—the length of DNA that coils around a nucleosome.

Nucleosome positioning also impacts on the progress of RNA polymerase II. Promoters are often depleted of nucleosomes, facilitating the initiation of transcription. Once transcription is under way, the presence of nucleosomes tends to slow down transcriptional elongation and as a result the density of RNA polymerase processing through exons is higher. Therefore this increases the concentration of CTDs on exons, enhancing the recruitment of pre-mRNA splicing factors where they are most needed. In fact the tendency of nucleosomes to slow down transcriptional elongation when traversing exons facilitates the process of exon recognition by the splicing machinery.

Histones are post-translationally modified in a myriad of ways. They can be acetylated, methylated, and phosphorylated, as well as other modifications. A detailed discussion of histone modifications is beyond the scope of this book. The consequence of these histone modifications is to affect nucleosome

Table 8.2 The effect of chromatin on alternative splicing

Process	Mechanism and effect on splicing
Nucleosome occupancy	A higher density of nucleosomes on GC-rich exons facilitates exon recognition
Histone modifications	Modified histones interact with adaptor proteins that affect transcriptional elongation rates or interact directly with splice factors
DNA methylation	DNA methylation aids exon recognition; exons that tend to be skipped are less methylated

structure and positioning; but they can also allow the histones to recruit several interacting proteins called *adaptor proteins* that in turn can affect gene expression. These adaptor proteins can influence the elongation rate of RNA polymerase II, and they can also facilitate the recruitment of splice factors. For example, histone H3 is trimethylated at lysine 36 by SETD2 during transcription, attracting the adaptor protein MRG15. MRG15 interacts with the splice factor PTB, a polypyrimidine tract binding protein whose recruitment generally induces exon skipping. Recruitment of MRG15 and PTB affects the alternative splicing of the FGFR-2 receptor which controls tumour growth and invasiveness.[29]

Another process that affects chromatin is the methylation of GC-rich sequences. DNA methylation plays a very important role in epigenetic modification; hypermethylated promoters are generally inactive. It turns out that exons also tend to be more methylated than introns, even when flanking intron sequences have a similar GC content. This suggests that DNA methylation might also contribute to exon recognition. In support of this idea, exons that tend to be skipped are generally less methylated. The methyl-CpG-binding protein MeCP2 is enriched in highly methylated exons that are alternatively spliced. Both the inhibition of methylation and the ablation of MeCP2 result in aberrant exon skipping.[30]

Key points

Splicing, namely the removal of introns from pre-mRNA, occurs mainly co-transcriptionally. The transcription and splicing machineries interact and share molecular components. The chromatin environment also influences alternative splicing, either by affecting transcriptional elongation rates or by recruiting specific splice factors.

8.5 The spatial organization of pre-mRNA processing

As we saw in Section 8.4, both the binding of early splicing factors and sometimes splicing itself can take place while the pre-mRNA is still being transcribed. In fact, other steps of pre-mRNA processing are also biochemically interlinked with transcription. This has led to the idea of a **gene expression factory**, in which all the molecular machines that carry out different steps of mRNA processing are physically linked together.[31] Because of this, pre-mRNA processing is physically likely to take place at sites of transcription within the nucleus.

A lot of research has been carried out to examine the physical location of splicing factors in eukaryotic nuclei and, hence, where splicing takes place. Figure 8.9 shows the localization of the splice factor SRSF2 (previously known as SC35) in **nuclear speckles**. Nuclear speckles are also known as **splicing factor compartments** (SFCs) and are sites of storage of splice factors.[32,33] SFCs occupy about 20% of the nuclear volume. SFCs are visualized by electron microscopy as **interchromatin granule clusters** (IGCs); they are also associated with **perichromatin fibrils** (PFs), which are sites of nascent transcripts. The association of large stores of splice factors with nascent transcripts underlines the co-transcriptional nature of splicing. Table 8.3 lists the subnuclear sites that are enriched in splicing factors.

Where does splicing take place in the cell? How do the SFCs correlate with the sites at which pre-mRNA transcripts are spliced in the nucleus? SFCs do not seem to be the actual physical sites of transcription. Instead, SFCs are closely adjacent to and partially overlap sites of active gene transcription, and act as a reservoir of splicing factors for processing newly synthesized pre-mRNAs. This makes a lot of sense; it is more efficient for the transcription machinery to draw on a nearby store of splice factors rather than rely on random collisions with free splice factors in the nucleoplasm.[34–37]

Although they look uniform under the light microscope, under the electron microscope SFCs can be resolved into structures called interchromatin granule clusters (ICGCs), which are thought to be primarily storage sites for splicing factors, and perichromatin fibrils, which are thought to be the nascent transcripts which are actively being spliced. A cartoon of this organization is shown in Fig. 8.10.

The localization of splicing factors has also been examined in living cells by visualizing splicing factor–GFP (green fluorescent protein) fusion proteins using a microscopic technique called **FRAP** (fluorescence recovery after photobleaching). FRAP experiments have shown that although SFCs appear to

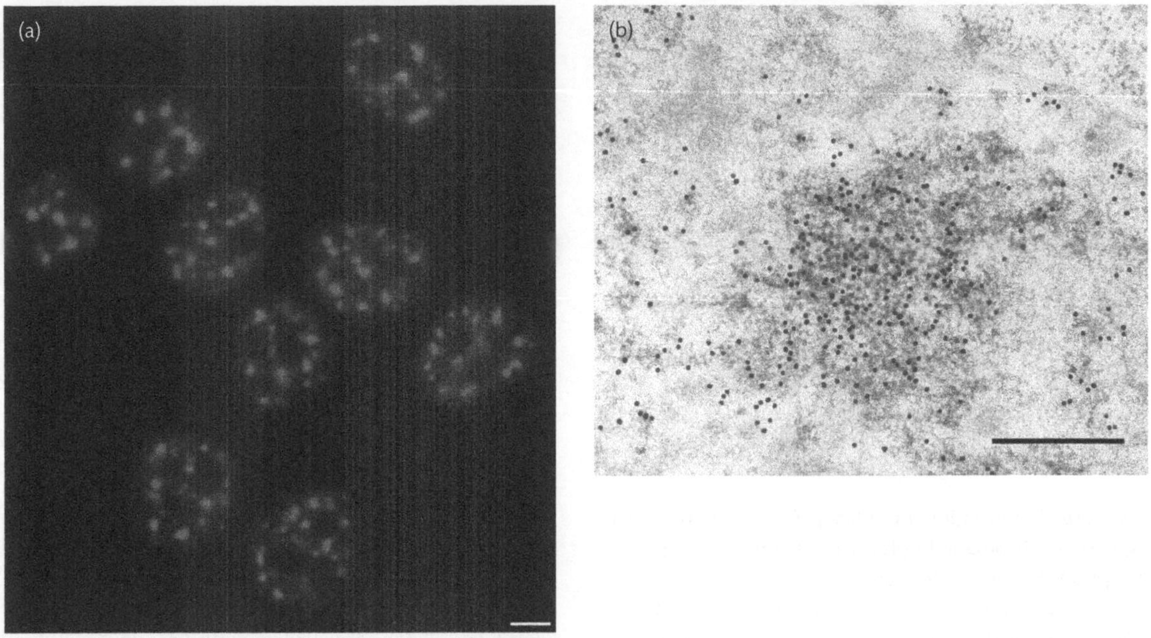

Figure 8.9 Presence of nuclear speckles in interchromatin space. (a) Splice factors such as SRSF1 are localized in both a speckled and diffuse pattern in the nucleus. Human HeLa cells, scale bar = 5μm. (b) Immunoelectron microscopy of a nuclear speckle using an antibody against SRSF2 and a secondary antibody conjugated with colloidal gold (15nm particles). Speckles are made up of particles 20–25nm in diameter connected by thin fibrils, giving a 'beaded chain' appearance. Scale bar = 500nm. (Adapted with permission from Spector and Lamond.[33])

be distinct morphological structures in fixed cells, within the living cell they are highly dynamic and are being continually reorganized. For example, a single molecule of the splicing factor SRSF1 (previously known as ASF/SF2) has been calculated to remain in a given SFC for only 45 seconds, and at least 10 000 molecules of SRSF1 leave SFCs per second per nucleus with approximately equal numbers

Table 8.3 Subnuclear sites that are enriched in splicing factors*

Subnuclear site	Function in RNA processing
Perichromatin fibril (component of the splicing factor compartment)	Splicing factors localized on a nascent transcript which is being spliced
Interchromatin granule cluster (component of the splicing factor compartment)	Storage compartment next to perichromatin fibrils; stores splice factors and other proteins involved in pre-mRNA processing
Cajal body	Contains snRNPs but not many other important splicing components. A site of snRNA and snoRNA modification, snRNP and snoRNP assembly
Paraspeckle	Nuclear retention of hyper-edited mRNAs (A→I editing)
Nucleolus	Transcription and processing of rRNA and assembly of ribosomal subunits

*The splicing factor compartments (also known as nuclear speckles) are thought to consist of two important components, the interchromatin granule clusters in which splice factors and other pre-mRNA processing components are stored, and the perichromatin fibrils in which splicing is taking place on nascent transcripts. snRNPs are complexes of small nuclear RNAs and several associated proteins; several snRNPs make up the spliceosome, the molecular machine that carries out splicing

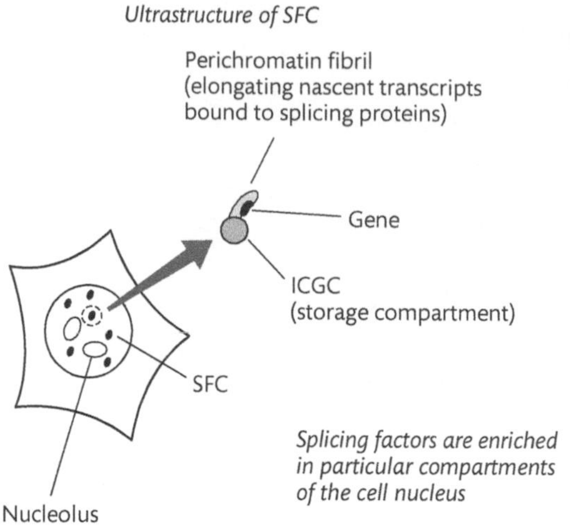

Figure 8.10 Organization of splicing factor compartments (SFCs) and perichromatin fibrils involved in active splicing and transcription. Splicing factors are enriched in particular structures of the cell nucleus known as SFCs. Ultrastructural analysis reveals these to contain two components. Perichromatin fibrils are sites of active genes laden with elongating transcripts packed with splicing factors. They are associated with interchromatin granule clusters (ICGC)—the sites of storage of splicing factors.

being recruited. The shapes of SFCs also change rapidly, suggesting a dynamic internal organization.

Hence, splicing proteins within the nucleus have a high degree of mobility, and so are able to 'roam' the nuclear space randomly and be available for interaction with protein partners or for target RNAs. It has been calculated from experiments using fluorescently labelled proteins that single splicing proteins can cross the nucleus in a few seconds, while larger structures such as spliceosomes only take a few minutes. Although fast, this movement of splicing complexes within the nucleus occurs by diffusion, and does not require energy. Movement within the nucleus is also strongly influenced by molecular interactions within the nucleus, both by specific interactions (i.e. with partner proteins to form slower-moving macromolecular complexes) and by the presence of physical obstacles (such as chromatin). The relatively random and dynamic motion of splice factors within the nucleus is referred to as the *stochastic model* and is described in Box 8.3.

How are these SFCs held together within the nucleus? In the rest of the cell, phospholipid membranes are very important for cellular compartmentalization. Although the nucleus is surrounded by the nuclear envelope, there are no membranes within it to generate subcompartments. Instead, the nuclear structures containing splicing factors are thought to be held together by protein interactions. Many proteins involved in RNA processing have protein interaction domains which cause them to interact through intermolecular forces both with themselves and with other splicing components, and these interactions may allow the assembly of SFCs within the nucleus. There is a lot of research interest in where specific proteins are located in the cell nucleus. Details of many of these sites of protein localization can be found online at the Nuclear Protein Database (NPD)[37] (http://npd.hgu.mrc.ac.uk/user/) (see Fig. 8.11).

The composition of SFCs has also been examined using a proteomics approach.[38] This showed that they contain at least 146 proteins, of which more than half are involved in pre-mRNA processing. As well as splicing factors, SFCs also contain some proteins involved in transcription, 3′ end formation,

Box 8.3 Stochastic model of nuclear organization

Splicing factors are not thought to occur in very stable long-lived complexes within the nucleus. Instead, current dynamic models of gene expression within the nucleus suggest that splicing factors and other nuclear components of gene expression are continually exchanging partners. The overall picture from these experiments is that the eukaryotic nucleus is in a constant state of flux, with random movement of proteins and the pairing and unpairing of protein interaction domains. Rather than being fixed in cells, splicing proteins in the cell (and other nuclear proteins involved in gene expression such as transcription factors) are constantly exchanging their partners. This model of nuclear organization is called a 'stochastic' model, and is important since it allows cells to respond to stimuli rapidly.

Because nuclear proteins are constantly changing partners, the cell is able to adjust to changing conditions much more rapidly than if proteins were fixed in long-term partnerships. If signalling pathways change either the properties (phosphorylation status of splicing proteins/transcription factors) or quantities (synthesis of new transcription factors) of nuclear components, this change will have an instant effect. In contrast, if proteins and nucleic acids were instead in long-term partnerships in cells it would take several hours or longer to respond to changes in the cellular environment.

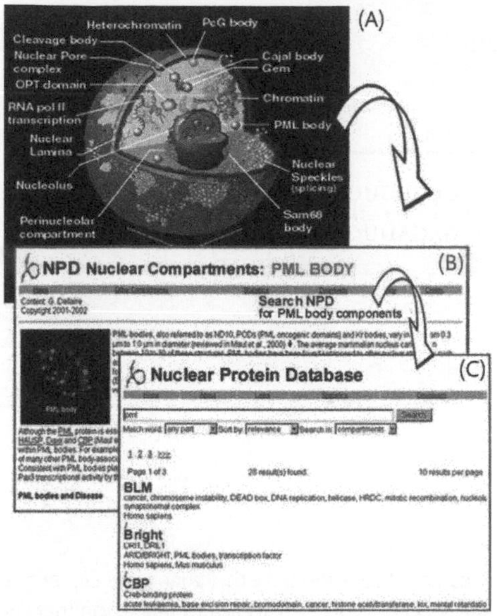

Figure 8.11 The spatial organization of nuclear proteins can be visualized online at the Nuclear Protein Database (http://npd.hgu.mrc.ac.uk/user/). The compartment browser provides an overview of each of the principal subnuclear compartments (A). An overview page (B) is generated with an image of that compartment and a short abstract describing its structure, function, and relevance to human disease. The database can be searched for proteins associated with a given domain (C) from the overview page. (Reproduced from Dellaire et al.[37] with permission from Oxford University Press.)

and even some ribosomal proteins. SFCs also contain the long non-coding RNA MALAT1, which is required for the retention of the splice factor SRSF1. Therefore they are quite complex in terms of their composition. There is the possibility of screening for compounds that might disrupt their formation.[39] One such compound is tubercidin, an analogue of adenosine. Tubercidin causes the dispersal of splice factors such as SRSF1 from SFCs. The consequence of disrupting SFC formation with tubercidin is to change alternative splicing patterns of several genes, including Clk1. Coincidentally, Clk1 encodes a splice factor kinase which modulates the release of SRSF1 from SFCs through hyperphosphorylation. Tubercidin causes skipping of Clk1's exon 4, and the resulting Clk1 isoform is catalytically inactive. This experiment further demonstrates how the regulation of splice factor availability through their sequestration into SFCs can contribute to the regulation of alternative splicing.

Key points

Splicing factors are concentrated in specific splicing factor compartments (SFCs) within the nucleus. SFCs (also known as nuclear speckles) contain both splicing factors actively involved in splicing pre-mRNAs and adjacent storage compartments. Splicing factors rapidly move around the cell nucleus and are dynamically recruited to SFCs by actively transcribing genes.

8.6 Histone mRNA 3′ end formation

In metazoans (multicellular animals), mRNAs that encode all five histones, the proteins responsible for packaging DNA into chromatin, are unusual in two ways: they do not contain introns and they do not end with a poly(A) tail. Experiments in the early 1980s showed that they end with a highly conserved stemloop structure that is recognized by a specialized 3′ processing machinery. The stemloop element is followed about 15–20 (and not more than 100) bases downstream by a purine-rich 'histone downstream element' (HDE). The cleavage of histone pre-mRNAs occurs between these two elements.[40]

The juxtaposition of these two conserved elements, which work together, is described as the **bipartite processing element**. The stemloop is recognized by the stemloop binding protein (SLBP) and the HDE interacts through base-pairing with U7 snRNP.[41] U7 is one of the snRNAs which is not involved in splicing. SLBP remains bound to histone mRNAs in the cytoplasm, where it stimulates their translation. This stimulation is particularly useful because the lack of a poly(A) tail may otherwise result in poor translation.

Histone mRNAs reach their highest levels during the S-phase of the cell cycle (when DNA synthesis takes place). This makes sense because newly copied DNA must be packaged by histones. Once the S-phase is completed, SLBP promotes the degradation of histone mRNAs. Thus SLBP coordinates three stages in the life of a histone mRNA: the formation of its 3′ end, its translation into histone proteins, and its eventual degradation.

The structure of the stemloop is remarkably conserved in terms of size, consisting of a six-base-pair stem and a four-nucleotide loop. The specific sequence of the stemloop is not always critical, whereas

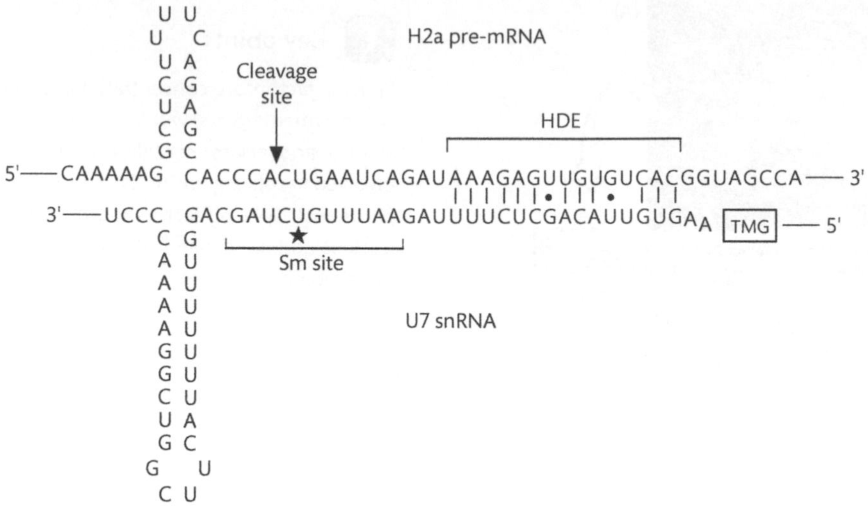

Figure 8.12 Base pairing interaction between mouse histone H2A mRNA and U7 snRNA. U7 snRNA recognizes the purine-rich element (HDE) downstream of the stemloop structure. The cleavage site, which is effectively the end of the histone mRNA, is also indicated. TMG is the trimethyl guanosine cap at the 5′ end of U7 snRNA and the asterisk indicates the region that UV cross-links with Lsm11, a U7 snRNP-specific protein. (Reprinted from Dominski and Marzluff[41] with permission from Elsevier.)

mutations in the HDE sequence are usually deleterious. The positioning of the two elements is also important, as shifting the HDE a few nucleotides away can compromise 3′ end formation. The cleavage site is typically an adenosine; in mammals this is generally the fifth nucleotide after the stemloop. Moving the HDE also shifts the position of the cleavage site, suggesting that U7 snRNA has a key role in defining the precise position of cleavage. The base pairing between U7 snRNA and the HDE is shown in Fig. 8.12 and the components of the histone 3′ end processing system are illustrated in Fig. 8.13.

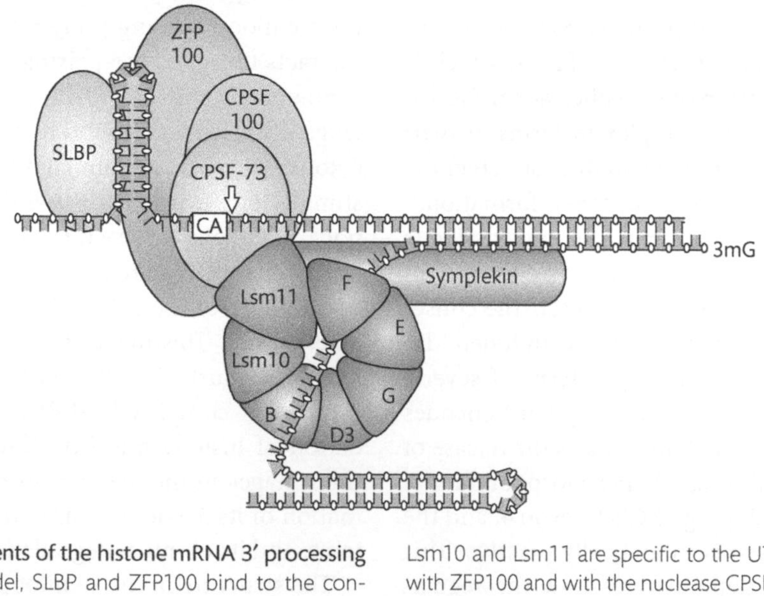

Figure 8.13 Components of the histone mRNA 3′ processing machinery. In this model, SLBP and ZFP100 bind to the conserved stemloop (six-base stem and four-base loop). The U7 snRNA base-pairs with the HDE. The U7 snRNP contains seven Sm proteins that form a ring bound to the Sm site. Five of these are found in other snRNAs (Sm proteins B, D3, G, E, and F), whereas Lsm10 and Lsm11 are specific to the U7 snRNP. Lsm11 interacts with ZFP100 and with the nuclease CPSF-73/CSPF-100 complex, ensuring that the cleavage occurs at an adenosine just downstream of the stemloop. (Reprinted from Dominski and Marzluff[41] with permission from Elsevier.)

In humans, SLBP is a 31kDa RNA-binding protein that migrates at about 45kDa in SDS–PAGE gels due to unusual electrophoretic properties. SLBP contains a 70 amino acid RNA-binding domain in the middle of the protein which does not have immediate similarities to other known RNA-binding domains. However, like other RNA-binding domains, it contains a number of conserved basic and aromatic amino acids that are critical for RNA binding.

The main role of the SLBP is thought to be the stabilization of the binding of U7 snRNA to the HDE. Thus, some histone mRNAs have a weaker HDE, in which case SLBP plays a critical role in recruiting U7 snRNA. At the end of the S-phase of mitosis SLBP is actively degraded; this follows its phosphorylation, an event triggered by a cyclin-dependent kinase (these are kinases that control the progression of the cell cycle). Thus levels of SLBP are in synchrony with the stage of mitosis during which it is required.

How might SLBP promote binding of U7 snRNA? The protein ZFP100 is a 100kDa zinc-finger protein which contains 18 zinc fingers of the Cys2His2 type (see Chapter 4 for a description of the zinc finger domain). ZFP100 is thought to work as a molecular adaptor because it interacts with the stemloop, with SLBP, and also with LSm11, a protein component of the U7 snRNP particle.

The U7 snRNA is transcribed by RNA polymerase II. It is a short RNA, ranging between 57 and 71 nucleotides in length. The 3′ end of the U7 snRNA forms a large stemloop that varies in size. The middle of the U7 snRNA contains an 'Sm site', a characteristic of other snRNAs excluding U6. As we saw in Chapter 5, the Sm site is bound by the Sm proteins, forming a ring structure known as the Sm core. However, there are also two U7-specific proteins in the U7 snRNP, namely Lsm10 and Lsm11 (the U7 snRNP protein that contacts ZFP100). The 5′ end of U7 snRNA is single-stranded and contains a pyrimidine-rich stretch that base pairs with the HDE in the histone mRNAs. The base-pairing between the HDE and U7 snRNA can be confirmed by making compensatory mutations in both the HDE and the 5′ end of the U7 snRNA. (A compensatory mutation is one that restores the original wild-type phenotype, in this context normal 3′ end formation of histone mRNAs.)

What, then, is responsible for the actual cleavage of the histone mRNA 3′ end? The cleavage is mediated by the nuclease CPSF-73 which, as we saw in Section 8.2, is a member of the metallo-β-lactamase family of nucleases. Within the U7 snRNP, Lsm11 is required for the recruitment of CPSF-73 to the site of cleavage. CPSF-73 provides the endonuclease activity responsible for cleaving the 3′ end, but it is also implicated in the exonucleolytic degradation of the released RNA. In order to work effectively, CPSF-73 is aided by the protein CPSF-100. These two proteins are also both found in CPSF, the complex involved in 3′ end formation of non-histone mRNAs. Thus there is a degree of overlap, in terms of molecular components, between histone 3′ end formation and the 3′ end formation of other pre-mRNAs.

Key points

Metazoan histone mRNAs process their 3′ ends in a different way. They do not possess an AAUAAA and G/U-rich signal, but instead contain a conserved stemloop and a purine-rich element–the HDE. The stemloop and HDE are bound by a series of proteins and by the U7 snRNA, resulting in the cleavage of the histone pre-mRNA. Histone mRNAs are not polyadenylated.

Summary

- The process of transcription was reviewed briefly, as was the structure and function of the main RNA polymerases. In bacteria and the Archaea there is a single RNA polymerase, whereas in eukaryotes there are three main RNA polymerases: I, II, and III. Each transcribes a different set of genes. RNA polymerase II is mainly responsible for the transcription of protein-coding genes. In eukaryotes, pre-mRNA processing is mostly co-transcriptional. In prokaryotes mRNA *translation* is co-transcriptional.
- mRNA is modified at both ends. Soon after initiation of transcription, a 7-methyl guanosine cap is added at the 5′ end. At the 3′ end, the signal AAUAAA followed by a

G/U-rich element dictates the site of cleavage and polyadenylation. Cleavage follows the binding of several multiprotein complexes: CPSF, CstF, CFI, and CFII. After cleavage, the enzyme polyadenylate polymerase adds the poly(A) tail.

- The C-terminal domain (CTD) of RNA polymerase II contains up to 52 repeats of a heptad sequence that contains two conserved serines. Phosphorylation of these serines influences the ability of the RNA polymerase to initiate and elongate transcription, and to recruit pre-mRNA processing factors.
- After intron-containing genes have been transcribed, an interplay between RNA polymerase II and the machinery of splicing facilitates both transcription and the different steps of pre-mRNA processing. Splicing is intimately linked to transcription, often occurring at the same time and place. The process of splicing enables both efficient transcription and exit from the nucleus into the cytoplasm.
- It is clear that the chromatin environment also contributes to the regulation of co-transcriptional pre-mRNA splicing. The positioning of nucleosomes and the methylation of DNA help to define the position of exons. The modification of histones allows the recruitment of molecular adaptors which in turn can help recruit specific splice factors to the nascent transcript.
- Splicing factors are enriched in distinct regions of the nucleus corresponding to sites of active transcription and storage sites from which they can be recruited. These storage sites are called splice factor compartments (SFCs); within them storage occurs in interchromatin granule clusters, and perichromatin fibrils are the sites where the SFCs interact with nascent transcripts.
- The 3′ ends of histone mRNA are different in that they are not polyadenylated. Cleavage at the 3′ end is mediated by a conserved stemloop and purine-rich sequence, the HDE. Key players are the RNA-binding proteins SLBP and the U7 snRNP which together with other factors define the precise cleavage point.

Questions

8.1 Describe the process of transcription and the properties and target genes of the different RNA polymerases.

8.2 What is the structure and function of the 7-methyl guanosine 5′ cap?

8.3 How are cleavage of pre-mRNA and polyadenylation achieved and coordinated?

8.4 What are the key features of the C-terminal domain (CTD) of RNA polymerase II?

8.5 How does the CTD help coordinate transcription and pre-mRNA processing?

8.6 How does the chromatin structure influence the process of pre-mRNA splicing?

8.7 What is the structure and function of a splice factor compartment?

8.8 How do the U7 snRNP and SLBP work together to process the 3′ ends of histone mRNAs?

References

1. **Hurwitz J.** The discovery of RNA polymerase. *J Biol Chem* 280, **42**, 477–85 (2005).
2. **Cramer P.** Multisubunit RNA polymerases. *Curr Op Struct Biol* **12**, 89–97 (2002).
3. **Werner F.** Structural evolution of multisubunit RNA polymerases. *Trends Microbiol* **16**, 247–50 (2008).
4. **Cramer P.** Recent structural studies of RNA polymerases II and III. *Biochem Soc Trans* **34**, 1058–61 (2006).
5. **Wierzbicki AT, Ream TS, Haag JR, Pikaard CS.** RNA Polymerase V transcription guides ARGONAUTE4 to chromatin. *Nat Genet* **41**, 630–34 (2009).
6. **Werner F.** Structure and function of archaeal RNA polymerases. *Mol Microbiol* **65**, 1395–404 (2007).
7. **Gnatt AL, Cramer P, Fu J, et al.** Structural basis of transcription: an RNA polymerase II elongation complex at 3.3 A resolution. *Science* **292**, 1876–82 (2001).
8. **Gonatopoulos-Pournatzis T, Cowling VH.** Cap-binding complex (CBC). *Biochem J* **457**, 231–42 (2014).
9. **Mandel CR, Bai Y, Tong L.** Protein factors in pre-mRNA 3′-end processing. *Cell Mol Life Sci* **65**, 1099–122 (2008).
10. **Proudfoot N, O'Sullivan J.** Polyadenylation: a tail of two complexes. *Curr Biol* **12**, R855–7 (2002).
11. **Tian B, Hu H, Zhang H, Lutz CS.** A large-scale analysis of mRNA polyadenylation of human and mouse genes. *Nucleic Acids Res* **33**, 201–12 (2005).
12. **Lutz CS.** Alternative polyadenylation: a twist on mRNA 3′ end formation. *ACS Chem Biol* **3**, 609–17 (2008).
13. **Bentley DL.** Coupling mRNA processing with transcription in time and space. *Nat Rev Genet* **15**, 163–75 (2014).

14. **Phatnani HP, Greenleaf A.** Phosphorylation and functions of the RNA polymerase II CTD. *Genes Dev* **20**, 2922–36 (2006).
15. **Chapman RD, Heidemann M, Hintermair C, Eick D.** Molecular evolution of the RNA polymerase II CTD. *Trends Genet* **24**, 289–96 (2008).
16. **De Almeida S, Carmo-Fonseca M.** The CTD role in cotranscriptional RNA processing and surveillance. *FEBS Lett* **582**, 1971–6 (2008).
17. **Laurencikiene J, Källman AM, Fong N, Bentley DL, Öhman M.** RNA editing and alternative splicing: the importance of co-transcriptional coordination. *EMBO Rep* **7**, 303–7 (2006).
18. **Kornblihtt AR, de la Mata M, Fededa JP, et al.** Multiple links between transcription and splicing. *RNA* **10**, 1489–98 (2004).
19. **Gornemann J, Kotovic KM, Hujer K, Neugebauer KM.** Cotranscriptional spliceosome assembly occurs in a stepwise fashion and requires the cap binding complex. *Mol Cell* **19**, 53–63 (2005).
20. **Listerman I, Sapra AK, Neugebauer KM.** Cotranscriptional coupling of splicing factor recruitment and precursor messenger RNA splicing in mammalian cells. *Nat Struct Mol Biol* **13**, 815–22 (2006).
21. **Neugebauer KM.** On the importance of being co-transcriptional. *J Cell Sci* **115**, 3865–71 (2002).
22. **Tennyson CN, Klamut HJ, Worton RG.** The human *dystrophin* gene requires 16 hours to be transcribed and is cotranscriptionally spliced. *Nat Genet* **9**, 184–90 (1995).
23. **LeMaire MF, Thummel CS.** Splicing precedes polyadenylation during *Drosophila* E74A transcription. *Mol Cell Biol* **10**, 6059–63 (1990).
24. **Beyer AL, Osheim YN.** Visualization of RNA transcription and processing. *Semin Cell Biol* **2**, 131–40 (1991).
25. **Sánchez-Alvarez M, Goldstrohm AC, Garcia-Blanco MA, Suñé C.** Human transcription elongation factor CA150 localizes to splicing factor-rich nuclear speckles and assembles transcription and splicing components into complexes through its amino and carboxyl regions. *Mol Cell Biol* **26**, 4998–5014 (2006).
26. **Luco RF, Allo M, Schor IE, et al.** Epigenetics in pre-mRNA splicing. *Cell* **144**, 16–26 (2011).
27. **Fernandes de Almeida S, Carmo-Fonseca M.** Reciprocal regulatory links between cotranscriptional splicing and chromatin. *Semin Cell Dev Biol* **32**, 2–10 (2014).
28. **Lee K-M, Tarn W-Y.** Coupling pre-mRNA processing to transcription on the RNA factory assembly line. *RNA Biol* **10**, 380–90 (2013).
29. **Sanidas I, Polytarchou C, Hatziapostolou M, et al.** Phosphoproteomics screen reveals akt isoform-specific signals linking RNA processing to lung cancer. *Mol Cell* **53**, 577–90 (2014).
30. **Maunakea AK, Chepelev I, Cui K, Zhao K.** Intragenic DNA methylation modulates alternative splicing by recruiting MeCP2 to promote exon recognition. *Cell Res* **23**, 1256–69 (2013).
31. **Misteli T.** The concept of self-organization in cellular architecture. *J Cell Biol* **155**, 181–5 (2001).
32. **Mao YT, Zhang B, Spector DL.** Biogenesis and function of nuclear bodies. *Trends Gen* **27**, 295–306 (2011).
33. **Spector DL, Lamond AI.** Nuclear speckles. *Cold Spring Harb Perspect Biol* **3**, 1–12 (2011).
34. **Misteli T.** Protein dynamics: implications for nuclear architecture and gene expression. *Science* **291**, 843–7 (2001).
35. **Matera AG, Shpargel KB.** Pumping RNA: nuclear bodybuilding along the RNP pipeline. *Curr Opin Cell Biol* **18**, 317–24 (2006).
36. **Sleeman JE, Trinkle-Mulchay L.** Nuclear bodies: new insights into assembly/dynamics and disease relevance. *Curr Opin Cell Biol* **28**, 76–83 (2014).
37. **Dellaire G, Farrall R, Bickmore WA.** The Nuclear Protein Database (NPD): sub-nuclear localisation and functional annotation of the nuclear proteome. *Nucleic Acids Res* **31**, 328–30 (2003).
38. **Saitoh N, Spahr CS, Patterson SD, et al.** Proteomic analysis of interchromatin granule clusters. *Mol Biol Cell* **15**, 3876–90 (2004).
39. **Kurogi Y, Matsuo Y, Mihara Y, et al.** Identification of a chemical inhibitor for nuclear speckle formation: implications for the function of nuclear speckles in regulation of alternative pre-mRNA splicing. *Biochem Biophys Res Comm* **446**, 119–24 (2014).
40. **Weiner AM.** E pluribus unum: 3′ end formation of polyadenylated mRNAs, histone mRNAs, and U snRNAs. *Mol Cell* **20**, 168–70 (2005).
41. **Dominski Z, Marzluff WF.** Formation of the 3′ end of histone mRNA: getting closer to the end. *Gene* **396**, 373–90 (2007).

9 Nucleocytoplasmic traffic of messenger RNA

Introduction

Eukaryotic cells are defined by the presence of a *nuclear membrane* (this is also referred to as the *nuclear envelope*) as well as having other membrane-bound organelles (subcellular compartments) (see Fig. 9.1). Eukaryotic genes are transcribed in the nucleus but mRNAs are translated in the cytoplasm. This means that most mRNAs need to be exported across the nuclear membrane into the cytoplasm in order to be translated. For mRNAs passage from the nucleus to the cytoplasm is one directional. Prokaryotes have no nuclear membrane, and transcription takes place alongside translation.

This chapter focuses on the nuclear export of mRNAs. mRNAs are encoded by ~21 000 genes in humans, each of which is transcribed by RNA polymerase II. Although protein-coding genes are numerous, it is important

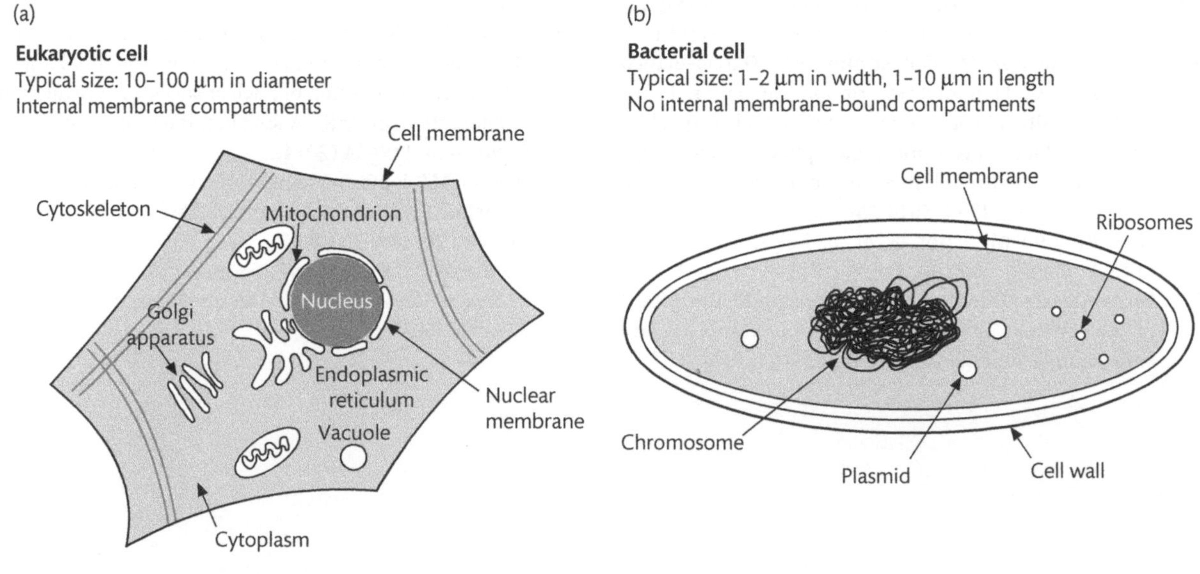

Figure 9.1 The organization of eukaryotic and prokaryotic cells imposes a requirement for mRNA nuclear export. (a) Eukaryotic cells are organized into a number of membrane-bound compartments. mRNA is made in the nucleus but needs to be exported into the cytoplasm for translation, which creates a need for nuclear mRNA export. (b) Prokaryotic cells are much smaller and do not have this same internal organization. The lack of a nuclear membrane in prokaryotes means that there is no equivalent need for mRNA nuclear export.

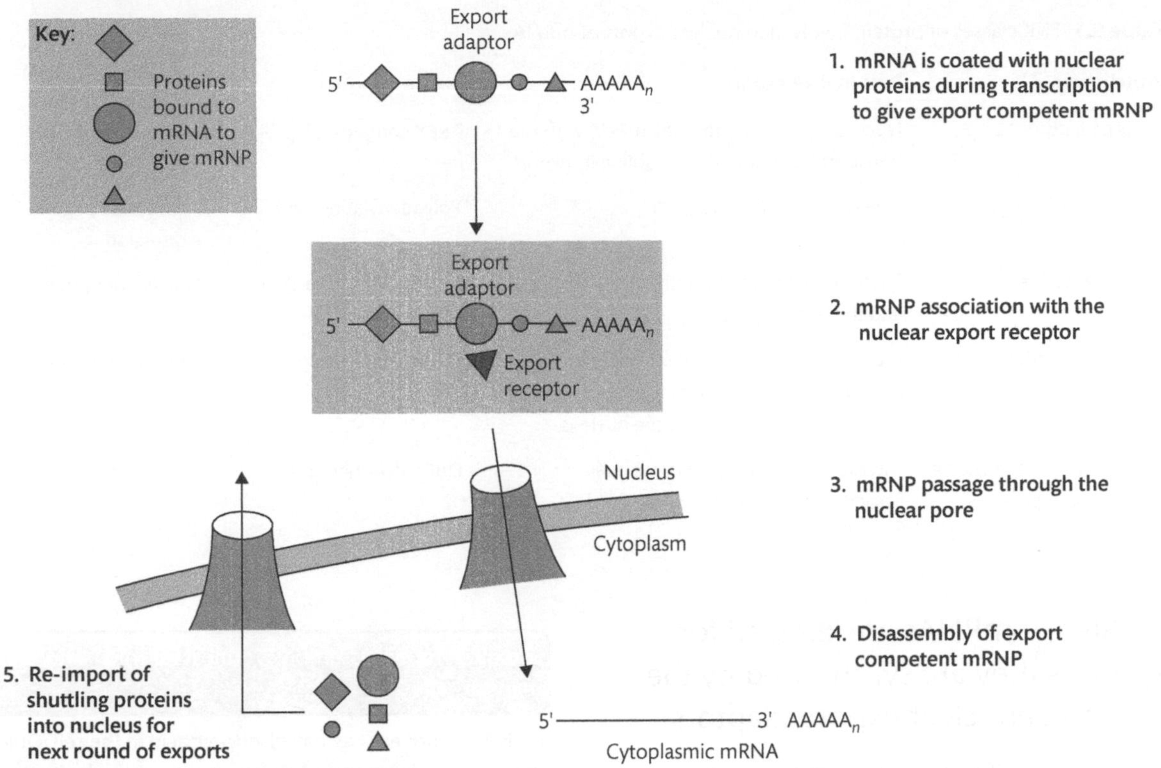

Figure 9.2 Major steps in the nuclear export of mRNA. (1) During transcription, mRNA becomes coated with nuclear proteins, including RNA export adaptors to create mRNPs. (2) These mRNPs cross the nucleoplasm by diffusion. (3) mRNPs associate with nuclear export receptors. (4) mRNP nuclear export receptors enable export through the nuclear pore. (5) mRNP export complexes are disassembled once they reach the cytoplasm, and their component proteins are re-imported into the nucleus for the next round of nuclear mRNA export.

to note that mRNA is exported from the nucleus by a different mechanism to that used for the non-coding RNAs, which actually make up the bulk of cellular RNA. Because of these differences, the mechanisms used to export non-coding RNAs from the nucleus are described separately in Chapter 14.

All current eukaryotes share the same common mechanisms to export mRNAs and each of the different classes of ncRNAs out of the nucleus, suggesting that these export challenges were solved very early in eukaryotic evolution.

For the purposes of discussion in this chapter, mRNA nuclear export is broken down into five major steps. A flow diagram of these steps is shown in Fig. 9.2. The first step starts very early in the lifetime of the mRNA as it is being copied from the gene, and the last is only completed after the fully mature mRNA is in the cytoplasm (see Fig. 9.2). We shall discuss each of these steps in turn in this chapter. The names of the proteins and protein complexes involved in nuclear export of mRNA are defined in the glossary, and highlighted in the text where they are first used.

Many proteins contribute to mRNA nuclear export. Five important types of protein involved in mRNA nuclear export are shown in Table 9.1 to make them easier to refer to during this chapter. It should be noted that these proteins have sometimes been given different names by different research groups, particularly between yeast and multicellular animals (metazoans). For simplicity, unless specifically mentioned, the metazoan names are used in this chapter. Some of the original pictures used as figures in this chapter also deviate from the main names used in the text, and where this happens we provide an explanation of which protein we are talking about.

Table 9.1 Five classes of protein involved in nuclear export of mRNA

Protein type	Role in RNA export	Name of protein	Cellular location
Export adaptors	Nuclear export adaptors link mRNA with the transport apparatus, thus enabling its export	TREX complex, SR proteins	Active sites of gene expression
3′ end formation	Release transcript from gene	Polyadenylation proteins	Active sites of gene expression
Export receptors	These proteins bind to both export adaptors and the nuclear pore	Mex67/Mtr2 (yeast), TAP/NXT (metazoans)	Nuclear periphery
Nuclear pore proteins	Proteins found on the inner channel of the nuclear pore which interface with the RNA export complex as it leaves the nucleus	FG repeat nucleoporins	Nuclear membrane
Disassembly and recycling of export components	Disassemble RNA export complexes once they reach the cytoplasm	DBP5 RNA helicase	Cytoplasm

9.1 Step 1: mRNAs are 'dressed for export' as they are synthesized by the addition of nuclear export adaptors

Nuclear export of mRNA starts at the gene and ends in the cytoplasm. The very first steps in mRNA export occur during transcription. Proteins are added to the pre-mRNA as soon as it starts to be made by some of the co-transcriptional processing events we discussed in Chapter 8 (you will remember these steps include capping, splicing, and polyadenylation). Although most mRNAs undergo each of these co-transcriptional steps, not all transcripts need necessarily follow the same pathway (e.g. histone mRNAs are neither spliced nor polyadenylated). By adding proteins to RNA, co-transcriptional processing steps convert naked mRNA into mRNP (see Box 9.1; mRNP is mRNA coated with protein).

Amongst the important nuclear export proteins added to mRNAs while they are being transcribed are *mRNA export adaptor proteins*.[1] The job of export adaptor proteins is to bind to mRNAs in the nucleus, and then connect them with the RNA export machinery. There are two subsets of important export adaptors called the **TREX complex** and the **SR proteins** (see Table 9.1).

An important feature of the TREX and SR proteins is that they can bind to many different mRNAs that have individually different sequences by (1) binding to degenerate sequences which are found in all transcripts, and (2) being loaded onto transcripts by processes which are shared between all mRNAs. This loading takes place during transcription of the mRNA (see Fig. 9.3).

We will look at these two different kinds of export adaptors in a little more detail next.

Box 9.1 mRNP assembly is critical for mRNA nuclear export

mRNAs do not exist as 'naked' nucleic acid in the cell. Quite the opposite is true—mRNAs are rapidly coated with proteins during their synthesis to give *mRNPs*, or messenger ribonucleoproteins. Once these steps have been completed properly the mRNA transcript will be in the form of an *export-competent mRNP*, which can be transported to the cytoplasm. Protein components of the mRNPs carry instructions to export cargo mRNA into the cytoplasm. Incorrectly processed RNAs do not enter the export pathway, and instead become degraded in the nucleus.

9.1.1 The TREX complex is an mRNA export adaptor

First, the name TREX complex is an acronym for transcription/export complex. As its name suggests, the TREX complex is important for both transcription of mRNA and nuclear export of mRNA. The TREX complex is loaded onto mRNAs just at their

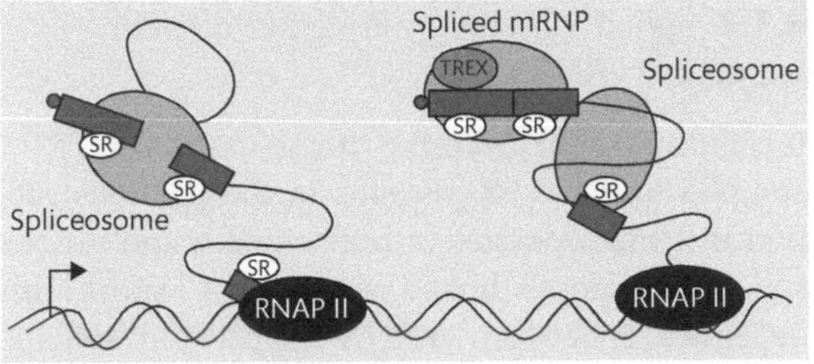

Figure 9.3 Important mRNA nuclear export adaptors. The TREX complex is made up of three components: the export adaptor REF, the RNA helicase UAP56 (also involved in splicing), and the THO complex (also involved in transcription). SR proteins are distributed across the length of the mRNA. (From Reed and Cheng[2] with kind permission from *Current Opinion in Cell Biology*.)

very 5′ ends through an interaction with the cap binding complex (CBC).[3] Notice that REF is the most 5′-located protein in the TREX complex.

As the name complex suggests, the TREX complex is made up of a number of individual protein components. Some important TREX components are shown in Fig. 9.4. The TREX complex is assembled as mRNA is being made by transcription and splicing, using energy conserved through ATP hydrolysis.[4, 5] This ATP hydrolysis is carried out by an ATP-dependent RNA helicase called **UAP56**. UAP56 protein is both a member of the TREX complex and part of the spliceosome (where it is involved in the recruitment of U2 snRNP to the spliceosome). This

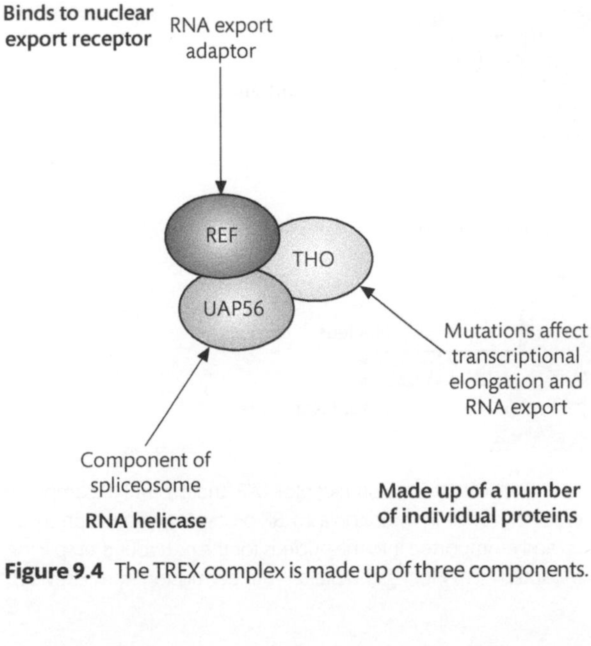

Figure 9.4 The TREX complex is made up of three components.

link between splicing and nuclear export of mRNA is important, and we will discuss it again later in this section. Full assembly of the TREX complex acts as a signal that nuclear mRNA processing is complete and the mRNA is now ready for export. The TREX complex is added to the 5′ end of mRNAs and is stabilized on the mRNA by protein interactions with the cap binding complex (CBC) at the 5′ end of the transcript.

Another critical component of TREX as far as export is concerned is a protein called **REF** (which stands for **R**NA **E**xport **F**actor) which is the actual export adaptor. The REF protein contains a central RNA recognition motif and so is able to bind RNA, but it normally does not bind to mRNA very strongly by itself. Instead, REF only binds to mRNA strongly when it interacts with UAP56 in the TREX complex.

The third component of the TREX complex is the **THO** complex of proteins. The THO complex itself contains several proteins, one of which is named Tho2, and is involved in transcriptional elongation as well as mRNA export.

9.1.2 A group of the SR proteins are mRNA adaptor proteins

The second set of RNA export adaptors are SR proteins which are a large group of proteins containing domains enriched in serine and arginine (RS domains). They are also important in in exon splicing (see Chapters 4 and 6–8). The SR protein family contains several proteins (see Chapter 4). A subset of three SR proteins (shown in Fig. 9.5) is involved in mRNA nuclear export.

There is an important logic to the way that SR proteins work as mRNA export adaptors. This logic is shown in Fig. 9.6. Phosphorylated SR proteins bind to the pre-mRNA before splicing as components of the spliceosome, but at this point cannot operate as export adaptors, and this prevents pre-mRNA from being inadvertently exported from the nucleus. SR proteins are only converted into RNA export adaptors by dephosphorylation during the process of splicing, so mRNA export from the nucleus is dependent on proper completion of RNA splicing. This is an efficient way of recruiting export adaptors since in metazoans most genes contain exons (see Chapter 6). This means that most transcripts go through the splicing process before they leave the nucleus.

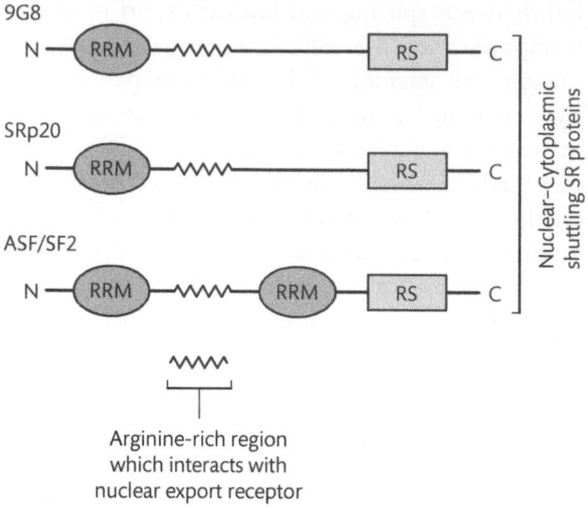

Figure 9.5 Three SR proteins function as mRNA export adaptors. These SR proteins interact with mRNA through their RRM sequences, and with the RNA export receptor through a domain which is rich in arginine amino acid residues.

9.1.3 Splicing is closely linked to nuclear export of mRNA

It should be clear from the discussion of the TREX complex and the SR proteins in the previous section that the processes of transcription and mRNA splicing are closely linked with mRNA export. This dependence between mRNA export and splicing was first demonstrated in experiments in the late 1990s carried out using *Xenopus* oocytes. In these experiments, nuclei were micro-injected with either unspliced pre-mRNA or fully spliced pre-mRNA made from a cDNA clone. Whether these RNAs were exported into the cytoplasm or stayed in the nucleus was then tested. The *Xenopus* oocytes were dissected to give nuclear and cytoplasmic fractions, and monitored to see where the transcripts were. Only the RNA which was spliced in the nucleus was efficiently exported. The RNA made from cDNA

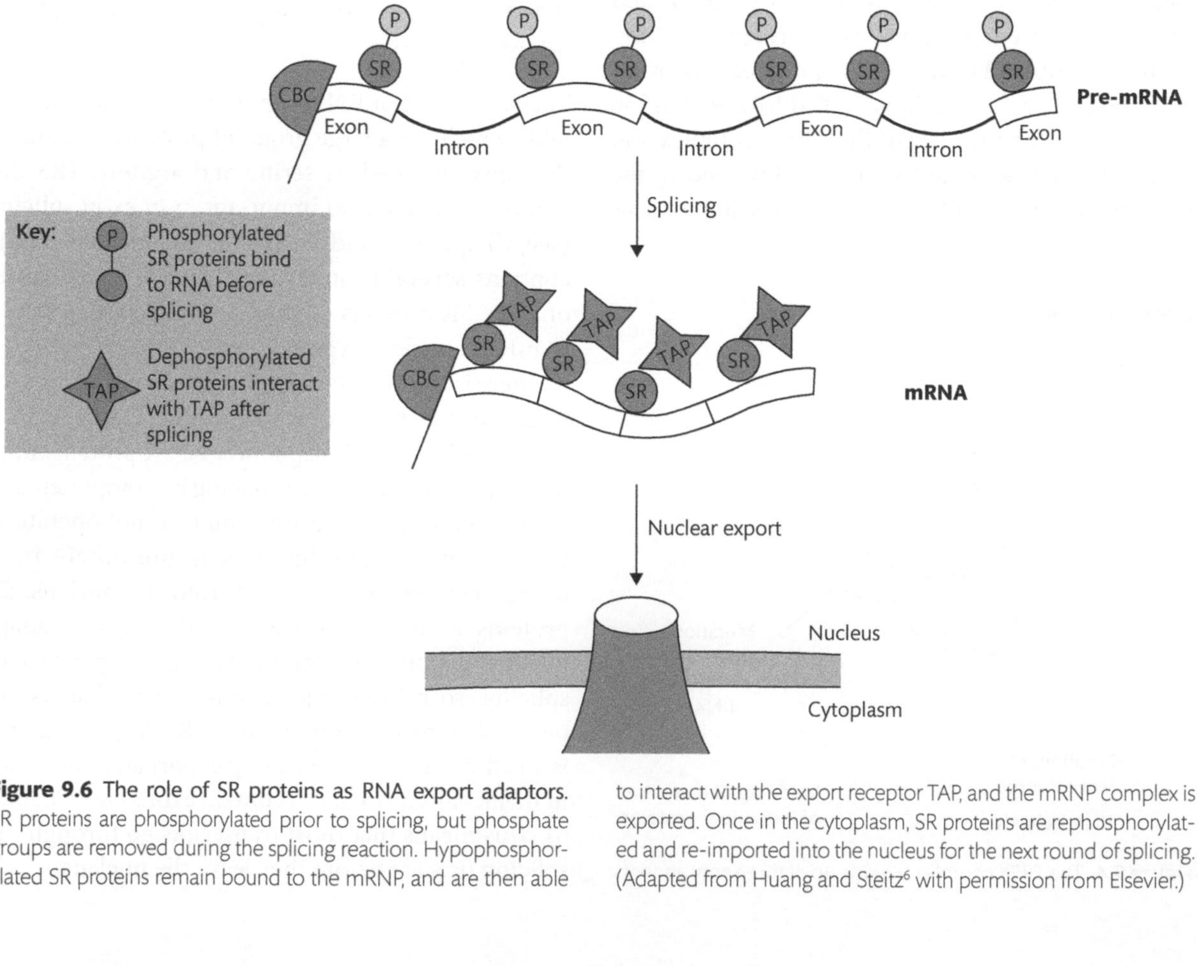

Figure 9.6 The role of SR proteins as RNA export adaptors. SR proteins are phosphorylated prior to splicing, but phosphate groups are removed during the splicing reaction. Hypophosphorylated SR proteins remain bound to the mRNP, and are then able to interact with the export receptor TAP, and the mRNP complex is exported. Once in the cytoplasm, SR proteins are rephosphorylated and re-imported into the nucleus for the next round of splicing. (Adapted from Huang and Steitz[6] with permission from Elsevier.)

which was not spliced stayed in the nucleus.[7] More recent experiments in cultured human cells have similarly shown that introns are very important for the nuclear export of human mRNAs. These experiments in human cells analysed the nuclear export of mRNA expressed from genes with and without introns. Data from one of these experiments carried out in human cells are shown in Fig. 9.7.

Splicing makes three important contributions to the nuclear export of mRNA.

1 Splicing physically adds both SR proteins and the TREX complex onto mRNAs.

2 Splicing modifies the SR protein mRNA export adaptors to remove phosphate groups.

3 Splicing is important in that it removes introns from transcripts. Introns can act as signals to retain RNAs within the nucleus. Experiments carried out in the 1980s in yeast cells showed that introns normally act as nuclear retention signals, preventing the escape of nuclear RNA into the cytoplasm.[9] In these experiments, a yeast gene which encoded a pre-mRNA that contained an ORF for the enzyme β-galactosidase was artificially constructed. This ORF would only be translated if the pre-mRNA was translated in the cytoplasm. In wild-type yeast cells, no β-galactosidase activity was detected because the unspliced pre-mRNA was restricted to the nucleus. However, in yeast cells which contained temperature-sensitive mutations in early splicing components or splice site mutations (see Chapter 6), pre-mRNA was released into the cytoplasm at the non-permissive temperature. Hence, sequestering mRNAs within spliceosomes acts to retain unspliced pre-mRNAs in the nucleus. This nuclear retention signal is released once the exons are spliced together.

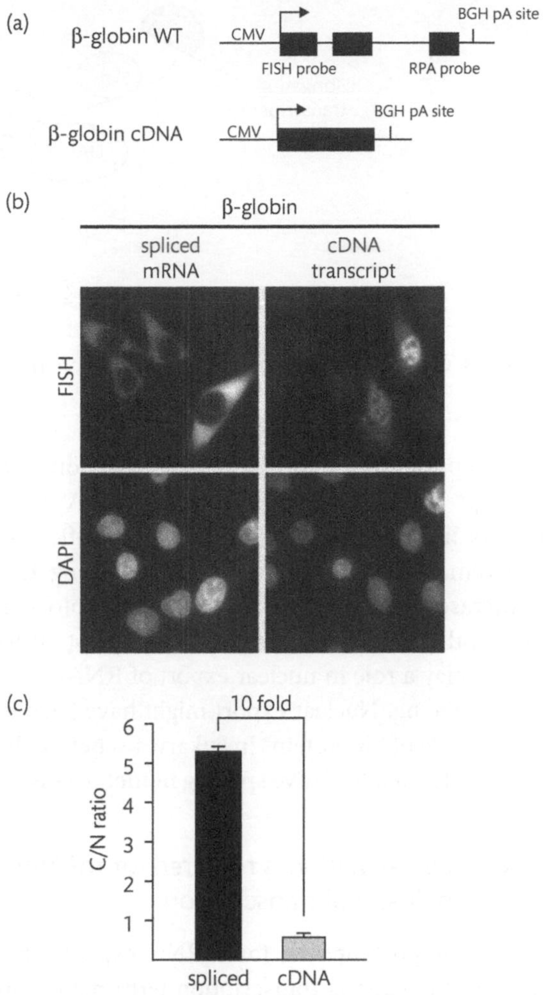

Figure 9.7 mRNA export in human cells is linked to splicing. In this experiment, the cellular location of a transcript was monitored using fluorescence in situ hybridization (FISH) after it had been expressed in cultured cells from either an intron-containing gene or a cDNA copy of this gene (lacking introns). Around six- to tenfold more mRNA was exported to the cytoplasm if introns were present in the gene, necessitating a splicing step before nuclear export of the mRNA, rather than when it was directly encoded by the cDNA. (From Valencia et al.[8] with permission from *PNAS*.)

How are mRNAs which are not spliced exported from the nucleus? For each of these three reasons, splicing usually has to be finished before an mRNA can be exported. Most metazoan mRNAs go through the splicing pathway before nuclear export. However, some important mRNAs are not encoded by intron-containing genes and so do not follow the splicing pathway. The histone mRNAs are not spliced, but SR proteins still bind to an element within an intronless histone transcript to promote its nuclear export. Viral transcripts without introns have also adapted direct ways to get out of the nucleus (see later) by binding nuclear export adaptors. Hence mRNAs that have not been spliced still use some of the same splicing-related proteins in their nuclear export.

What happens to nuclear mRNA export in species that have genes predominantly without introns? This is the case in the yeast *S. cerevisiae*, where most of the genes are intronless and so are not transcribed into pre-mRNAs that need splicing (see Chapter 6). In yeast cells the TREX complex is recruited transcriptionally as part of RNA polymerase II, and then

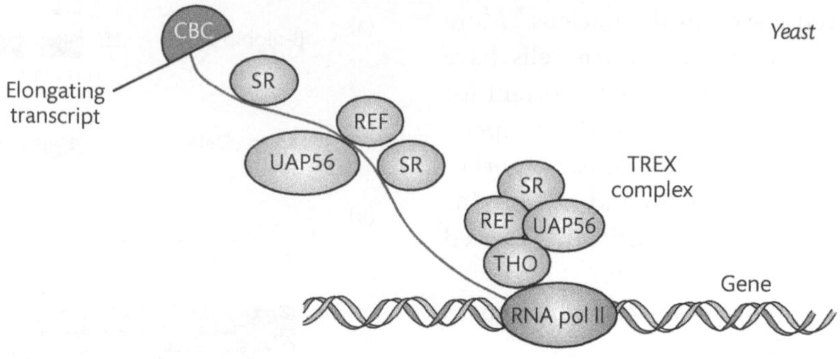

Figure 9.8 Coupling of mRNA export to transcription in yeast. (Adapted from Reed and Cheng[2] with permission from Elsevier.)

TREX deposits REF and UAP56 onto the elongating transcript. Transcriptional coupling of RNA export adaptors in yeast is shown in Fig. 9.8. Notice that THO remains associated with the elongating RNA polymerase, while the nuclear export adaptors are added to the nascent transcripts. Also in yeast SR-like proteins play a role in nuclear export of RNAs rather than in splicing. Nuclear export might have been the ancestral role of SR proteins in eukaryotes before they evolved roles in alternative splicing in metazoans.

9.1.4 Polyadenylation is required for mRNAs to leave their site of transcription

The next important step for mRNA export which occurs at the gene is transcription termination and release of the mRNA from the gene template; without this step the mRNA cannot be exported, or indeed move away from its template gene. Nascent transcripts have been observed to pause at or near their genes within the nucleus. Two things happen while mRNAs pause: (1) the nascent mRNA is released from the RNA polymerase, and around 200–250 adenosines are added to the 3′ end of new transcripts in a process called polyadenylation (see Chapter 8); (2) proteins bind to the poly(A) tail of the mRNA.

The mechanism of polyadenylation is covered in more detail in Chapter 8. Experiments in yeast have shown that the process of polyadenylation is important, but not essential, for RNA export.[10] In these experiments, versions of a gene encoding green fluorescent protein (GFP) were made which generated mRNAs with different 3′ ends (see Fig. 9.9). The cellular location of the mRNAs made from these genes were then analysed using FISH and by monitoring translation of GFP. These experiments demonstrated that normal mRNAs (3′ end made with a poly(A) tail) were normally exported into the cytoplasm and translated into protein. However, if the 3′ end of the mRNA was made by a ribozyme cleavage, the mRNA was unable to be efficiently exported. (Ribozymes are catalytic RNAs, and are described in Chapter 3. In this case the ribozyme cut the mRNA to release it from the RNA polymerase.) RNA export could be rescued by adding adenosine residues before the ribozyme cut site, showing that these poly(A) sequences also bind proteins that are important for nuclear export.

These experiments mean that it is the run of A residues themselves rather than some other aspect of the polyadenylation process which is important for nuclear export of mRNA. Similar experiments have shown that 3′ end formation is also important in mammalian cells. The probable role of poly(A) tails of newly made nuclear mRNAs is to bind to a protein called the nuclear poly(A)-binding protein which is important for nuclear export of mRNA.

Key points

mRNA is converted to mRNP while it is being transcribed. Amongst the proteins added, export adaptors provide a molecular tag for an mRNP, addressing it for export from the nucleus into the cytoplasm. Proper nuclear export is dependent on how mRNA is synthesized and processed in the nucleus. These processing steps include capping, splicing, and polyadenylation, and each adds new proteins to produce an *export competent mRNP*. Appropriate mRNP formation also ensures that mRNAs have undergone each of the nuclear processing pathways completely and properly, so that intron-containing mRNAs which could not be translated into full-length proteins are not exported into the cytoplasm.

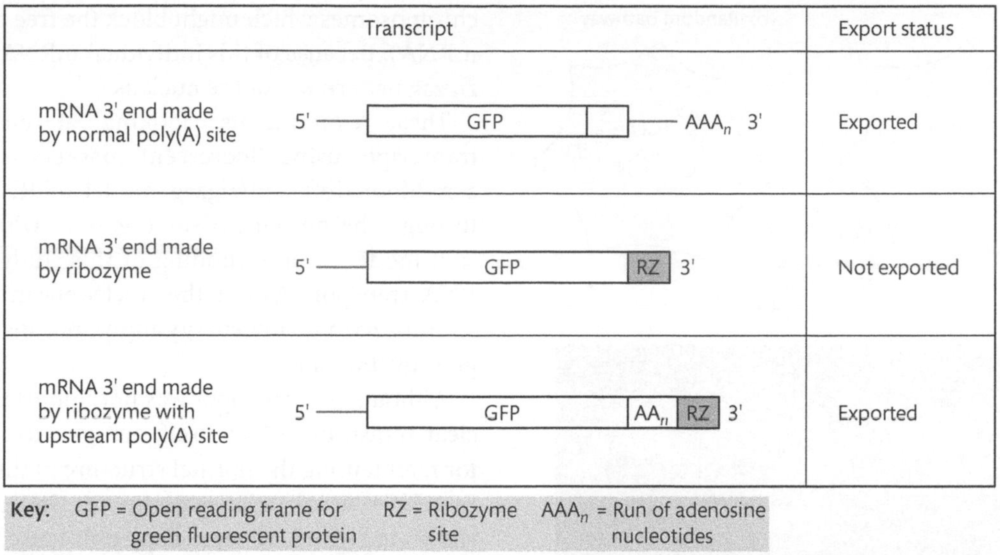
Transcript
Export status
mRNA 3' end made by normal poly(A) site
5'
GFP
AAA$_n$ 3'
Exported
mRNA 3' end made by ribozyme
5'
GFP
RZ
3'
Not exported
mRNA 3' end made by ribozyme with upstream poly(A) site
5'
GFP
AA$_n$
RZ
3'
Exported
Key: GFP = Open reading frame for green fluorescent protein
RZ = Ribozyme site
AAA$_n$ = Run of adenosine nucleotides

Figure 9.9 Polyadenylation is required for mRNAs to leave their site of transcription. Three different versions of a gene encoding green fluorescent protein (GFP) were expressed in yeast. These transcripts generated their 3′ ends in three different ways. **Transcript 1:** a gene terminating in a normal poly(A) site (pA) produced an mRNA through the normal 3′ end pathway which was polyadenylated. Formation of this transcript resulted in normal cytoplasmic export of the GFP transcript into the cytoplasm. **Transcript 2:** a gene which did not have a poly(A) site, but instead encoded a synthetic ribozyme (RZ), generated its 3′ end through cleavage at this RZ site. This transcript was not polyadenylated. This resulted in nuclear accumulation of RZ-terminated transcripts near the site of transcription. **Transcript 3:** a run of just 48 adenosine residues was placed upstream of the RZ site in the gene encoding the third transcript. Although the 3′ end of this transcript was generated by RZ cleavage, it contained a run of A residues at its 3′ end. This synthetic poly(A) tail was sufficient to enable cytoplasmic accumulation of the GFP transcript. (This figure summarizes data from Dower et al.[10])

9.2 Step 2: mRNA transcripts reach the nuclear pore by random nuclear diffusion

Intuitively, one might expect that active genes would be located at the nuclear membrane close to nuclear pores which are the points of mRNA exit. In fact almost the opposite seems to be true. Although a proportion of transcriptionally active genes are found at the nuclear periphery, most gene-rich chromosomes and transcriptionally active DNA are instead found within the nuclear interior. This raises the question of how transcripts are moved from their site of synthesis (genes) to their nuclear exit point at the nuclear pore. This is the topic of this section.

Over the years there has been a lot of controversy about whether mRNAs follow distinct pathways, or 'tracks', to leave the nucleus, or whether they randomly diffuse from the gene to the nuclear pore. These two models of mRNA transit are shown in Figs 9.10a and 9.10b. The *RNA tracking model* was originally suggested in the 1980s, based on observations of mRNA transcripts transcribed from integrated copies of the Epstein–Barr virus (EBV). In support of this kind of transit, in situ hybridization of cells with an EBV-specific probe detected a distinct 'track' containing EBV transcripts leading from the gene to the cytoplasm.[11] This evidence supporting the tracking model of nuclear export is shown in Fig. 9.10c; note that the EBV mRNA follows a straight track by the shortest available route to the NPC at the nuclear periphery.

These early observations of RNA export via tracks were based on virally infected cells. However, more recent experiments on endogenous genomic transcripts suggest that most cellular transcripts actually move through the nucleus via *random diffusion*.[12] The experimental evidence supporting diffusion of mRNA in the nucleus is shown in Fig. 9.11; note that

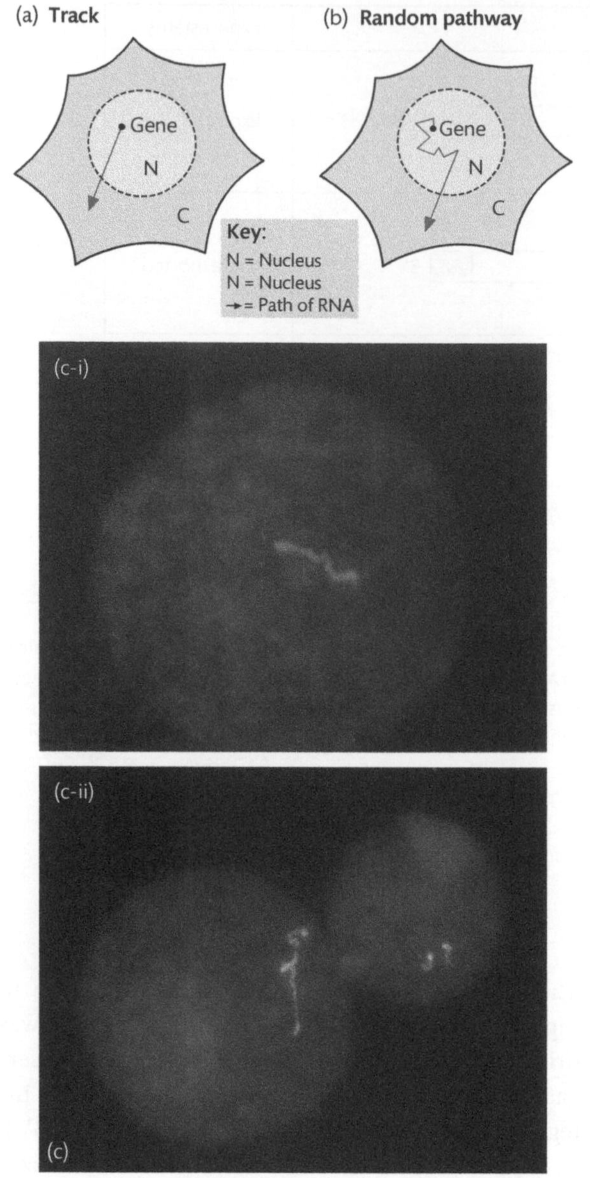

Figure 9.10 **Pathways of RNA export out of the nucleus.** RNAs have been proposed to follow either (a) specific tracks out of the nucleus or (b) random pathways within the nucleus, eventually leading to export of mRNA out of the nuclear pore. (c) Epstein–Barr virus mRNAs were shown to follow tracks out of the nucleus in cultured cells, but endogenous mRNAs may follow random pathways. (Part (c) reprinted from Lawrence et al.[11] with permission from Elsevier.)

after transcription at the gene, the mRNA follows an essentially random path within the nucleus until it leaves through a nuclear pore. Although mRNA movement is random in the nucleus, it is constrained by 'objects' floating in the nucleoplasm (such as chromosomes which might block the free passage of mRNA). Because of this individual mRNAs follow a zigzag pattern out of the nucleus.

These experiments, looking at endogenous transcripts using fluorescent markers, supported a random diffusion/zigzag model of RNA transit through the nucleus. Also consistent with a diffusion model, another finding of this study was that RNA transport within the nucleoplasm does not seem to require any energy input, but instead happens by diffusion.

Although energy input was not important for nuclear transit of mRNA itself, energy *was* important for maintaining the normal structure of the nucleus. In mRNA visualization experiments similar to those shown in Fig. 9.11, energy depletion caused condensation of nuclear structures, reducing the nuclear mobility of mRNAs; under these circumstances mRNAs hit nuclear obstructions more frequently.

Key points

Transcription of genes often takes place distant from the nuclear periphery. After transcription and release of mRNA from the gene, these mRNAs move through the nucleoplasm by diffusion rather than following straight tracks from the gene to the nuclear pore, although they are impeded by solid structures within the nucleus such as chromosomes.

9.3 Step 3: Transit through the nuclear pore requires addition of nuclear export receptors

The next step in mRNA export from the nucleus into the cytoplasm is mRNA passage across the nuclear membrane.[13,14] Transport across the nuclear pore requires a second group of proteins called **nuclear export receptors** which bind to the nuclear adaptors. Both the TREX complex and SR proteins bind to the same nuclear export receptor, a protein called TAP, and an associated protein called p15.

The job of the TAP mRNA nuclear export receptor is to move mRNAs through the nuclear pore. This is a critical job. Experiments have been carried out in fruit fly cells to show the important roles of both TAP and the p15 receptor subunits for mRNA

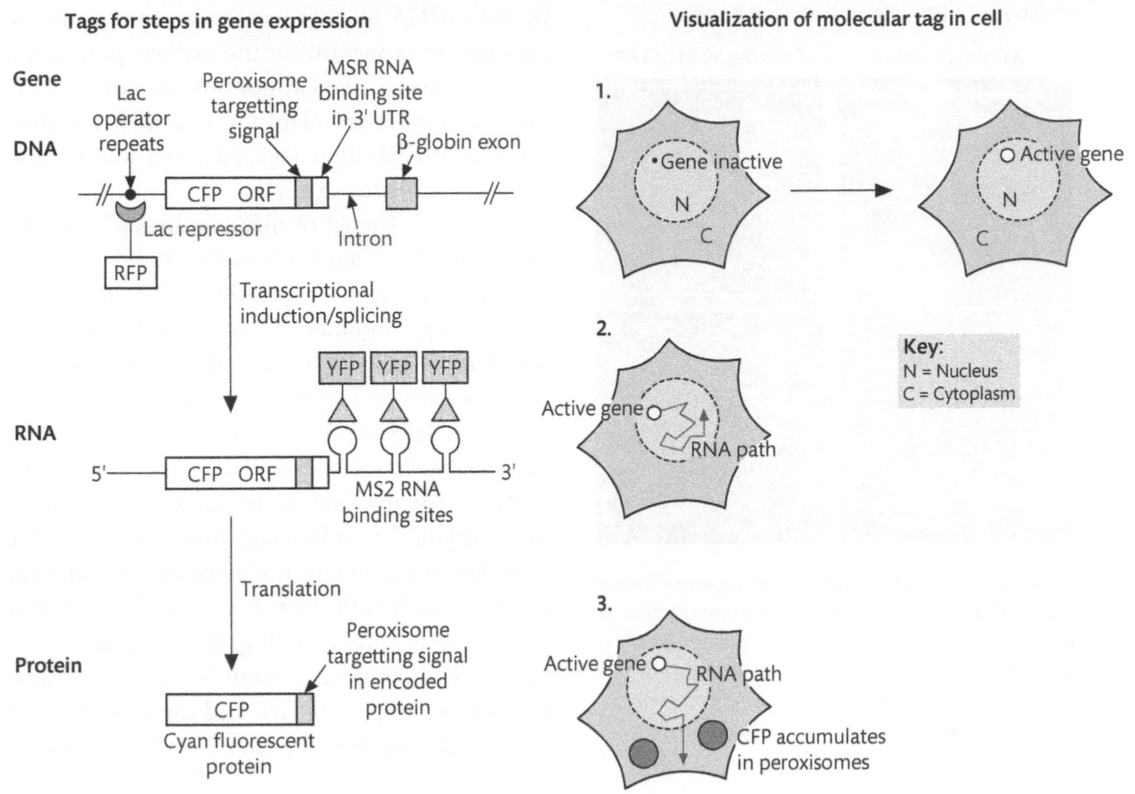

Figure 9.11 **Experiments in living cells suggest that RNA follows a random path through the nucleoplasm to reach the nuclear pore.** These experiments used a cell line for which different molecular stages in the expression of an integrated CFP (cyan fluorescent protein) gene could be followed with different fluorescent 'markers'. The gene itself encoded a CFP fused to a peroxisome-targeting sequence, repeats of the binding sequence for the bacteriophage MS2 RNA-binding protein, and a β-globin intron and exon. (Note that to make gene expression easier to detect, the cell line contained multiple tandem copies of the integrated CFP gene rather than a single copy.) **Detection of the *CFP* gene at the DNA level:** each integrated gene was tagged upstream with a binding site for the Lac repressor (*LacI*); co-expression in this same cell line of the Lac repressor fused to RFP (red fluorescent protein) revealed the position of the *CFP* gene in the cell nucleus as a bright dot. **Detection of *CFP* mRNAs at the RNA level:** *CFP* mRNAs were visualized within the cell by binding a yellow fluorescent protein (YFP) bound to MS2, which attached to the MS2-binding sites within the *CFP* mRNA. **Detection of the encoded CFP protein:** the eventual translation product was CFP protein targeted to peroxisomes. Each of these steps in gene expression was followed in real time after induction of the *CFP* gene to enable the route followed my mRNA out of the nucleus to be monitored. Firstly, **transcriptional activation** of the *CFP* gene led to the gene locus becoming larger as it unfolded. Secondly, mRNP traffic in the nucleus showed mRNPs following a random pathway out of the nucleus. Finally, the CFP protein was translated.in the cytoplasm after nuclear export of mRNAs. CFP expression of the final protein from these genes was present within peroxisomes in the cytoplasm. (Data summarized from Shav-Tal et al.[12])

export. These are shown in Fig. 9.12. mRNA accumulated very rapidly within the nucleus when levels of TAP or p15 were decreased by **RNA interference** (RNAi). Similarly, as shown in Fig. 9.13, yeast cells which contain temperature-sensitive mutations in the homologous mex67 protein (the yeast version of TAP; mex is an acronym for messenger RNA export) also rapidly accumulate nuclear mRNA when they are switched to the non-permissive temperature.

mRNA is only exported from the nucleus when it is fully processed. Why does the nuclear mRNA export receptor not bind RNA non-specifically in the nucleus and then export it out into the cytoplasm? Recent work in human cells has shown that an intramolecular interaction prevents the TAP protein from binding to RNA, even though it contains an RNA recognition motif. However, a physical interaction between TAP and nuclear export adaptors bound to RNA changes the structure of TAP to expose its RNA recognition motif (see Fig. 9.14). This increase in RNA binding strength by TAP means that at this point mRNA is firmly bound

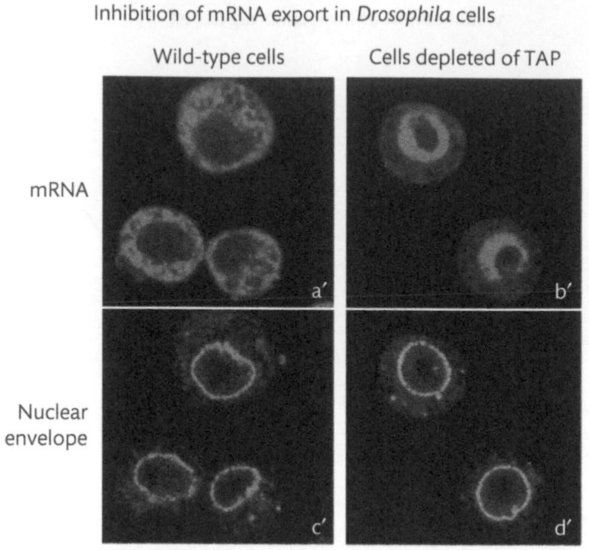

Figure 9.12 **Depletion of the nuclear export receptor in fruit fly cells blocks mRNA nuclear export.** These experiments were carried out in *Drosophila* cells depleted of TAP protein by siRNA (please note nomenclature here: *NXF1* is the *Drosophila* TAP gene). For a full colour version of this figure, see Colour Illustration 9. (From Stutz and Izaurralde[15] with permission from Elsevier.)

by the mRNA nuclear export receptor and is ready for nuclear export out of the nuclear pore complex. Note that since TAP cannot bind to mRNA by itself, but only via export adaptors, TAP does not start exporting mRNAs that lack adaptors and so are not properly ready for export.

The actual transit of mRNPs through the nuclear pore has been visualized in the midge *Chironomus tentans* by electron microscopy.[16] The salivary glands of this midge contain enormous chromosomes called **Balbiani rings**, which are visible under the electron microscope (see Fig. 9.15). The mRNPs transcribed from the Balbiani rings first make contact with the nuclear basket of the nuclear pore, and then pass through the nuclear pore in a 5′→3′ direction. This explains the important role of the cap-binding complex in stabilizing the association of the export adaptor TREX with the mRNP. This 5′→3′ transport of mRNA is shown in Fig. 9.16; notice the 5′-located TREX complex, containing the export adaptor REF, interacting with TAP and directing the mRNP through the nuclear pore in a 5′→3′ direction.

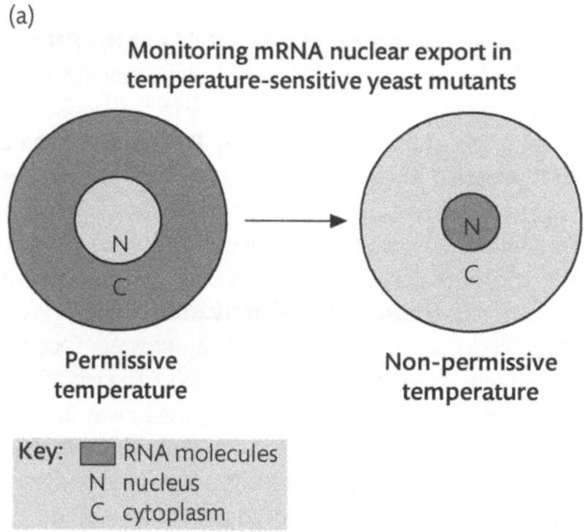

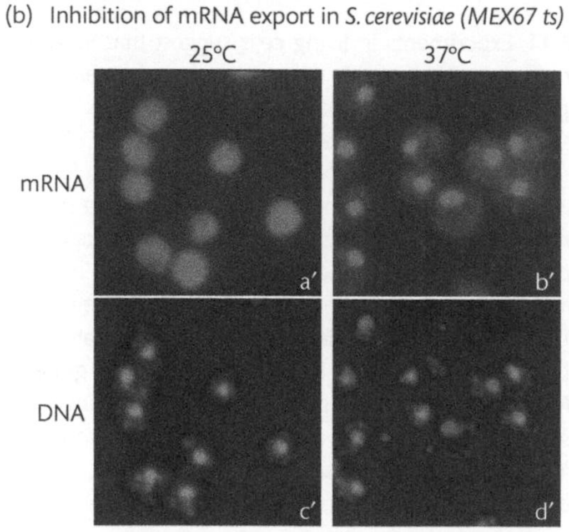

Figure 9.13 **Temperature-sensitive mutations in the yeast mRNA export receptor block mRNA nuclear export.** Many of the early screens for mRNA export proteins were carried out using genetics in yeast. Pools of temperature-sensitive mutants were screened for mRNA localization at the permissive temperature and the non-permissive temperature (the temperature at which the mutant phenotype kicks in). (a) mRNA export is normal when the mutant yeast cells are grown at the permissive temperature. However, note that the RNA accumulates in the nucleus when temperature-sensitive yeast mutant cells are switched to a non-permissive temperature. (b) Visualization of nuclear retention of polyadenylated RNA in yeast with a temperature-sensitive mutation in the *MEX67* gene. This gene encodes the yeast TAP protein. The image shows multiple yeast cells at either 25°C (permissive temperature, mRNA export normal) or at 37°C (non-permissive temperature, mRNA export blocked). The cellular location of most of the mRNA in a cell was visualized after in situ hybridization with a fluorescent oligo-d(T) probe. Oligo-d(T) base pairs to the poly(A) sequence in polyadenylated mRNA. In normal (wild-type) yeast cells, or in temperature-sensitive mRNA export mutants at the permissive temperature, most mRNA is detected in the cytoplasm and very little in the nucleus. In yeast cells containing temperature-sensitive mutations in RNA export components, mRNA fails to export and accumulates rapidly within the nucleus (within minutes of changing temperature). Yeast cells were also stained with DAPI (to show nuclear DNA). For a full colour version of Fig. 9.13b, see Colour Illustration 10. (Part (b) from Stutz and Izaurralde[15] with permission from Elsevier.)

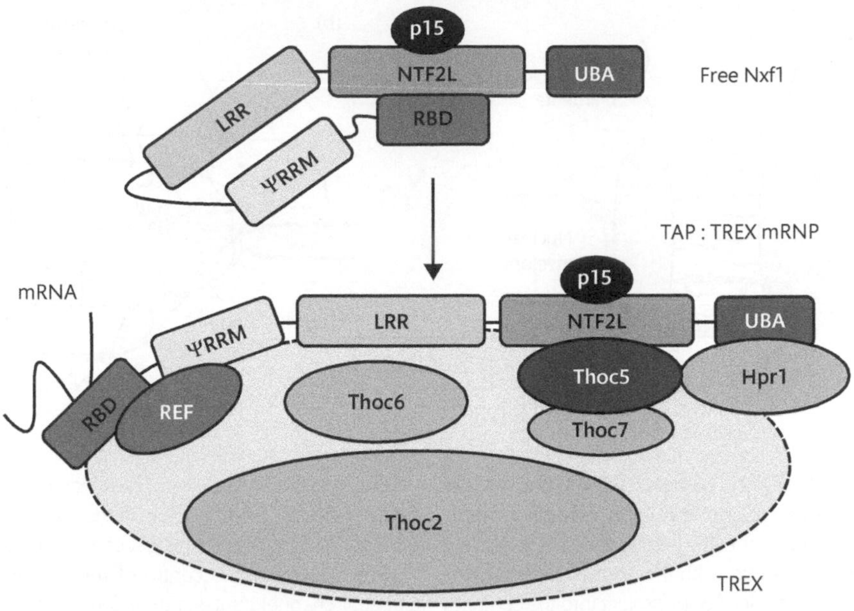

Figure 9.14 Interaction with the RNA export adaptor TREX increases the binding affinity of TAP for mRNA. The RNA binding domain (RBD) of TAP is normally sequestered by an intramolecular interaction. Interaction with the TREX complex (notice that the individual protein components are shown in the TREX complex) remodels the secondary structure of TAP so that the RBD is available to bind to RNA. Notice that the TAP mRNA export receptor is also associated with the p15 protein. (Adapted from Viphakone et al.[13] with permission from *Nature Communications*.)

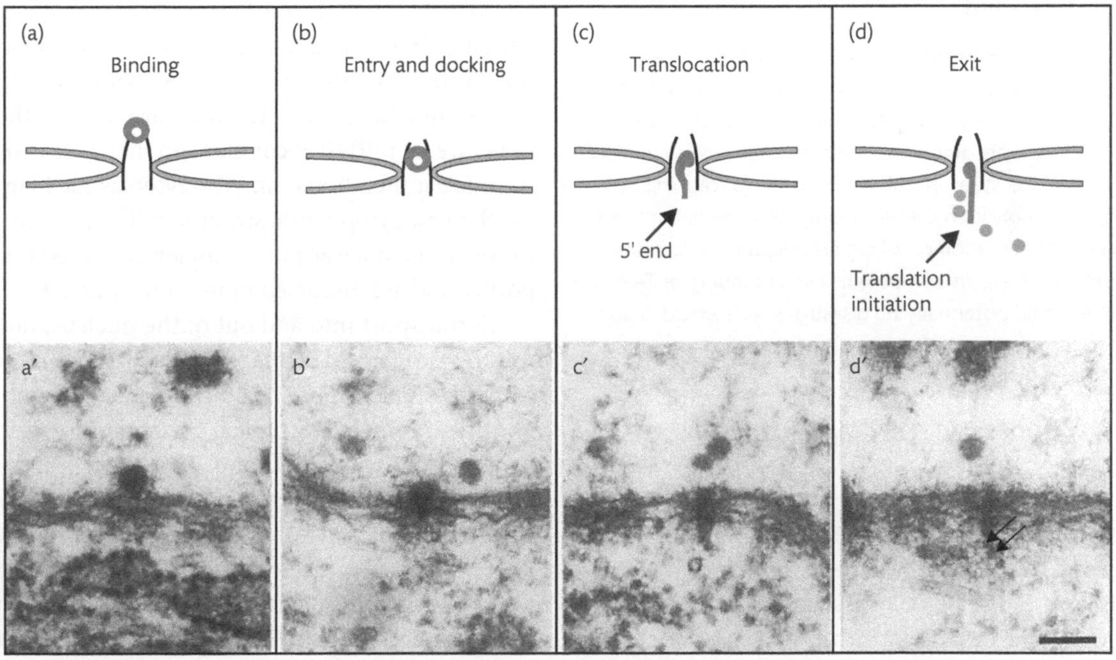

Figure 9.15 mRNPs have been visualized by electron microscopy going through the nuclear pore complex. These images were obtained using electron microscopy of cells from the salivary glands of the midge *Chironomus tentans*. Transcripts from genes present in the Balbiani rings were visualized at different stages of nuclear export ((a)–(d) shows schematic view; (a′)–(d′) shows electron micrographs) from initially binding to the nuclear pore on the nuclear side (a) to emerging into the cytoplasm (d). The mRNAs immediately start being translated by ribosomes once on the cytoplasmic side of the nuclear pore (arrowed in (d)). Scale bar = 100nm. (From Bjork and Wieslander[16] with permission from *Chromosoma*.)

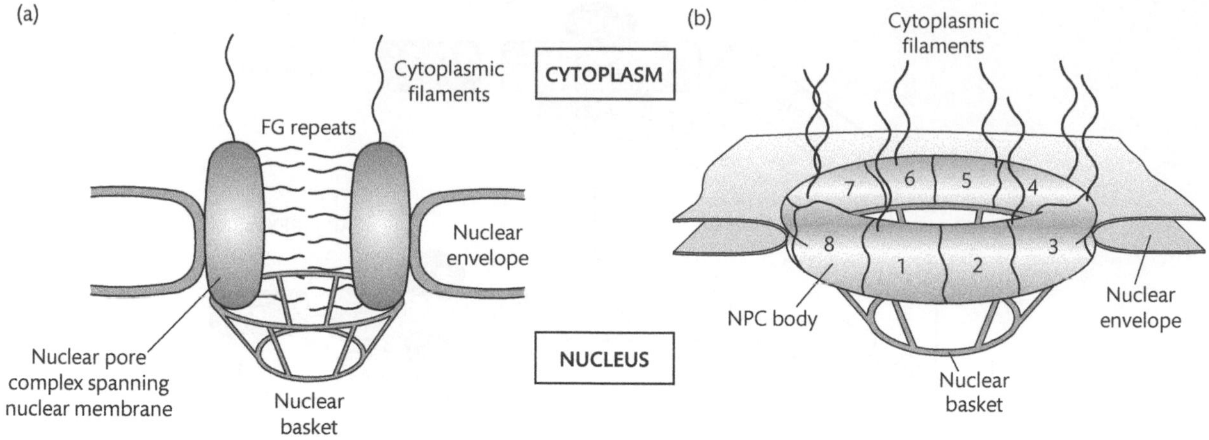

Figure 9.16 The nuclear pore complex (NPC). **The nuclear pore complex provides a channel through the nuclear membrane.** (a) Side view showing the NPC embedded in the nuclear membrane, with extensions into both the nucleus and the cytoplasm. Notice that the FG nucleoporins project into the centre of the NPC. (b) Viewed from above and at an angle the NPC looks like a doughnut embedded in the nuclear membrane. The NPC has eightfold rotational symmetry: eight subcomplexes are embedded in the nuclear membrane, each with a single cytoplasmic filament and a single nuclear filament contributing to the nuclear basket. Notice that the actual channel through which transport takes place is in the centre of these eight subcomplexes. The central channel contains the FG repeats. This central channel is normally 10nm wide, but can enlarge to 40nm to allow passage of large molecules. (Part (b) reprinted from Stewart[17] with permission from Macmillan Publishers Ltd.)

Key points

Why do nuclear export receptors depend on nuclear export adaptors? Why does TAP not instead directly bind to mRNA and thereby direct its nuclear export? The addition of mRNA export adaptors within the nucleus is coupled to each of the steps in mRNA biosynthesis (capping, splicing, and polyadenylation) being properly completed. Because of this, addition of export adaptors indicates that the mRNP is export competent. Direct binding of TAP to mRNA would potentially be disastrous, as it would enable the export of pre-mRNA before it was properly processed.

9.3.1 Movement of mRNPs through the nuclear pore

Once mRNPs have navigated their way across the nucleoplasm, the final stage in their export to the cytoplasm is the actual passage out of the nucleus and into the cytoplasm. This passage takes place through nuclear pores.[18,19]

Nuclear pores are not simple holes in the membrane. Instead, they are made up of multiple protein components called nucleoporins which form a **nuclear pore complex**. A single nuclear pore complex (abbreviated **NPC**) is shown in Fig. 9.16; notice that the NPC spans the nuclear membrane but also has extensions into both the nucleus (called the nuclear basket, since this resembles a basket-like structure under the electron microscope) and the cytoplasm (called cytoplasmic filaments). The nuclear basket is the last structure an mRNP encounters on the nuclear side of the nuclear membrane, and the cytoplasmic filaments are the first cytoplasmic structures. The proteins that make up the nuclear pore complex are called nucleoporins, and are discussed more fully in Box 9.2.

All transport into and out of the nucleus, not just mRNA export, takes place through nuclear pores. Multiple nuclear pores stud the nuclear membrane of a eukaryotic cell; a typical vertebrate somatic cell contains between 1000 and 10 000 nuclear pores. Nuclear pores are continually involved in moving molecules into and out of the nucleus. Each nuclear pore has been estimated to carry out 1000 translocation events every second. Exit or entry through the nuclear pore is selective. Small molecules can get through them without any extra help, but molecules larger than 30–40kDa need to bind to specific export receptors which export them through the pore.

This last step in mRNA nuclear export through the nuclear pore complex represents just the last 200nm of the voyage, but takes place in a different

Box 9.2 Nuclear pores are built up from protein subunits called nucleoporins

Nuclear pores are assembled from about 30 protein subunits called **nucleoporins**. Some of these nucleoporins are present once in each of the eight identical subcomplexes of the nuclear pore (which means that there are eight copies in total within an NPC), while others are present in up to seven copies per subcomplex (meaning that there are 56 copies in total in the nuclear pore). There are three classes of nucleoporin, which are divided up depending on their job within the nuclear pore.

- Some nucleoporins anchor the nuclear pore in the nuclear envelope. Looking at Fig. 9.16, the anchor nucleoporin proteins are those that are embedded in the nuclear envelope.
- Around a third of nucleoporins are **FG repeat nucleoporins**. The name FG reflects that these nucleoporins contain runs of amino acids rich in the hydrophobic amino acid phenylalanine (symbol F) and the neutral amino acid glycine (symbol G). FG repeat nucleoporins line the centre of the NPC channel. These FG repeats are critically important for nuclear pore function. The FG repeats within this class of nucleoporin are thought to be fairly unstructured but to stick out into the central channel of the nuclear pore (notice these projections in Fig. 9.16). Within this central channel, FG repeats are thought to form either a brush-like structure or a gel-like structure which provide two important functions: (i) it acts as a sieve, blocking the export of larger molecules from the nucleus while small molecules can still pass through; (ii) it provides binding sites for proteins involved in nuclear export. In this way FG repeats act as stepping stones out of the nucleus, thereby allowing passage of export proteins and their cargoes into the cytoplasm.
- The final class comprises structural nucleoporins, which play a role in building up the complex and positioning other proteins.

environment from the intranuclear transit we have discussed up to now. Rather than freely diffusing through the nucleoplasm, the mRNP has to pass through the constrained volume of the central channel of the NPC through a mesh of hydrophobic FG repeat nucleoporins. Movement of mRNP through the nuclear pore is thought to take place by Brownian (i.e. random) motion, with the hydrophobic nuclear export receptor TAP interacting with and stepping between proteins called FG repeat nucleoporins (see Box 9.2) inside the core channel of the nuclear pore. This is where the key role of the nuclear export receptor becomes important. Because of the hydrophobic nature of the nuclear pore, in order to move through its interior mRNPs need to be bound to the hydrophobic nuclear export receptor which is able to interact with the FG repeat nucleoporins.

9.4 Step 4: Disassembly of the export competent mRNP

Why is mRNP transit across the nuclear pore unidirectional? In other words, once mRNPs start to move across the nuclear pore, why does mRNP move out into the cytoplasm through nuclear pores and not slide back into the nucleus? The mechanism to ensure forward movement and prevent backsliding of mRNP involves an RNA helicase called **DBP5** (see Fig. 9.17). DBP5 enables directional mRNP transport by removing the export receptor TAP from mRNP once it reaches the cytoplasm. Because of removal of the nuclear export receptor, once the mRNP is in the cytoplasm and stripped it cannot re-enter the hydrophobic environment of the nuclear pore.

Intuitively, we would expect the protein which strips TAP from the mRNP to be on the cytoplasmic side of the NPC. In fact, the important DBP5 RNA helicase which carries out this function joins the mRNA as it is being transcribed at the gene and travels between the nucleus and the cytoplasm with the mRNP, but it is only activated to remove TAP on the cytoplasmic side. This activation is carried out by a protein called Gle1, which is anchored on the cytoplasmic side of the NPC, along with a Gle1 cofactor molecule called IP6.

Once activated by Gle1 attached to the cytoplasmic filaments of the nuclear pore, DBP5 uses ATP hydrolysis to strip off nuclear proteins, including TAP, from the mRNP. Because of the cytoplasmic location of DBP5 and IP6, mRNP export complexes are stripped off the mRNP as they pass out of the nuclear pore. Once they are disassembled, the RNA is unable to interact with FG nucleoporins and move back into the nucleus through the pore.

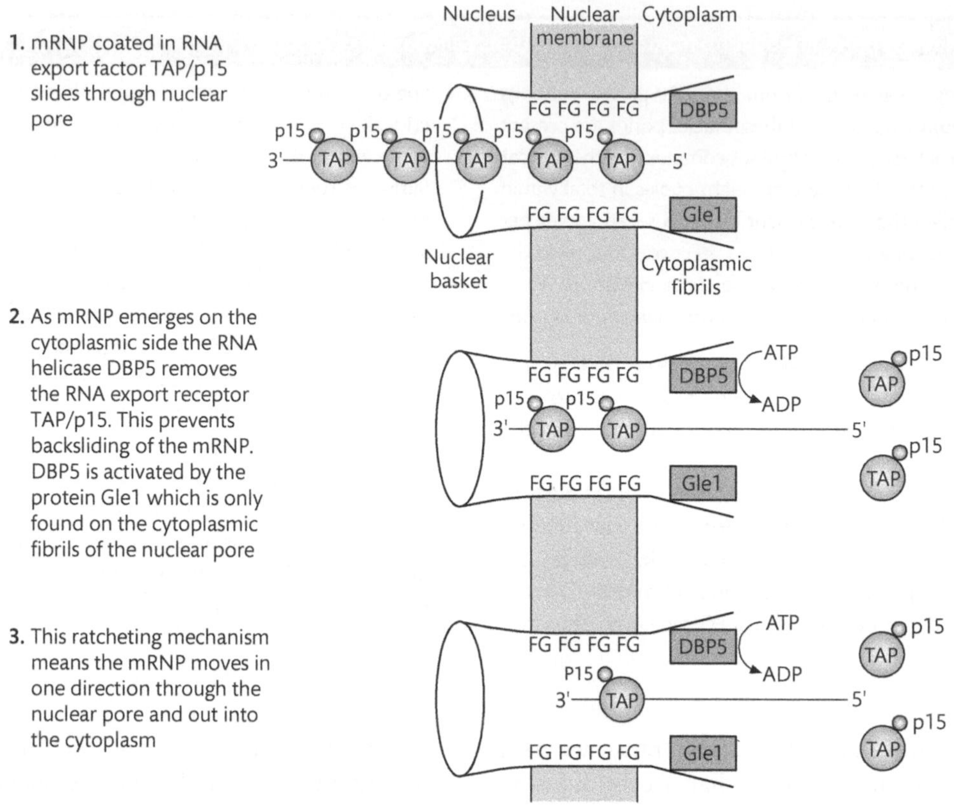
1. mRNP coated in RNA export factor TAP/p15 slides through nuclear pore
2. As mRNP emerges on the cytoplasmic side the RNA helicase DBP5 removes the RNA export receptor TAP/p15. This prevents backsliding of the mRNP. DBP5 is activated by the protein Gle1 which is only found on the cytoplasmic fibrils of the nuclear pore
3. This ratcheting mechanism means the mRNP moves in one direction through the nuclear pore and out into the cytoplasm
Nucleus
Nuclear membrane
Cytoplasm
FG FG FG FG
DBP5
Gle1
p15
TAP
3'
5'
Nuclear basket
Cytoplasmic fibrils
ATP
ADP
P15

Figure 9.17 **mRNP may exit through the nuclear pore by a ratchet-like mechanism.** (1) An mRNP is shown in the process of exiting through a nuclear pore complex (NPC). The mRNP is coated with export receptors (TAP/p15) bound to the RNA through export adaptors (REF and SR proteins—for simplicity the actual adaptors are not shown). TAP/p15 enables the RNA to pass through the FG nucleoporins which line the channel of the NPC. (2) On reaching the cytoplasmic side of the nuclear pore, the mRNP protein DBP5 is activated by the Gle1 protein to remove the TAP/p15 nuclear export receptor. Once stripped of the export receptor, mRNP can no longer pass through the channel of the NPC. In this way, once part of the mRNA has entered the cytoplasm, it does not slide back into the nucleus, and directional transport is achieved. (3) This process continues stepwise until the entire mRNP has moved through the nuclear pore. Since the action of DBP5 and Gle1 proteins holds mRNP that has moved into the cytoplasm, together they act as a ratchet to stop backsliding into the nucleus. (Adapted from Stewart[20] with permission from Elsevier.)

Key points

The mRNA export receptor TAP facilitates movement through the pore by interacting with the hydrophobic FG repeat nucleoporins. mRNPs move through the NPC as long linear molecules, a bit like snakes moving through a hole. Once it passes through the nuclear export receptor TAP is progressively removed from the mRNP as it enters the cytoplasm—remember that the mRNP on the cytoplasmic side is stripped of TAP while the nucleoplasmic mRNP still has TAP attached. Once TAP is removed in the cytoplasm the mRNP is no longer able to interact with the nucleoporins.

9.5 Step 5: Export receptors shuttle between the nucleus and the cytoplasm

Nuclear export adaptors are concentrated in sites of active gene expression in the nucleus (Fig. 9.18 shows an image of REF localized in the nucleus—notice that it has a speckled localization). However, even though they are most concentrated near active genes at any one time, each of the mRNA export adaptors actually continually shuttles between the nucleus and the cytoplasm. mRNA export adaptors are carried out of the nucleus with mRNP export complexes, and then transported back into the

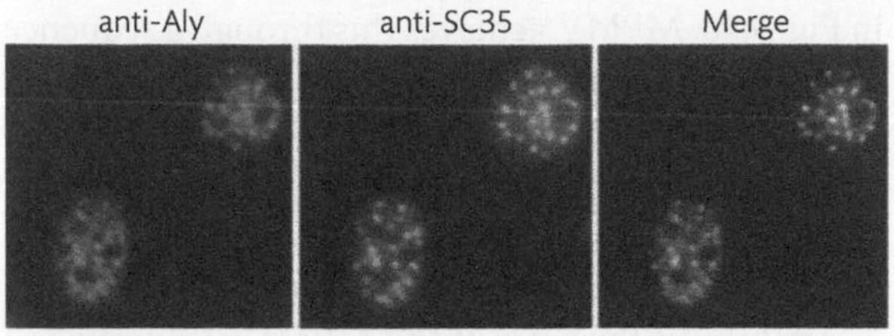

Figure 9.18 A REF protein (called ALY in this figure) is located in the nucleus within subnuclear regions called speckles which also contain the SR protein splicing factor SC35. This experiment used antibodies to localize proteins with indirect immunofluorescence. For this reason ALY is localized with anti-ALY antibodies, and SC35 with anti-SC35 antibodies. For a full colour version of this figure, see Colour Illustration 11. (Reprinted from Zhou et al.[21] with permission from Macmillan Publishers Ltd.)

nucleus. This shuttling of export components is very important, and is described here as step 5 of nuclear export. Without this return of the nuclear export components into the nucleus after a round of mRNA nuclear export, over time export components would accumulate in the cytoplasm such that no further nuclear mRNA export would be possible.

This shuttling of mRNA export adaptor proteins was demonstrated in *heterokaryon* experiments. In these, human cells were fused with mouse cells using a chemical agent called PEG. The resulting fused cells (called heterokaryons, i.e. cells that contain multiple different nuclei) contained two nuclei, one mouse and one human, which can be distinguished under the microscope. These heterokaryon experiments were carried out in the presence of cycloheximide, to block any new protein synthesis, and could directly visualize the movement of the shuttling export adaptors (see Fig. 9.19).

Figure 9.19b shows the results from one of these heterokaryon experiments. REF–GFP fusion protein leaves the human nucleus and then enters the mouse nucleus. On the other hand, proteins such as hnRNPC that do not shuttle remain in the human nucleus.

9.6 mRNA export can be hijacked by some viruses

Although export of mRNA from the nucleus does not usually take place through direct TAP binding, there are some important exceptions. These are provided by some viral RNAs which have evolved TAP-binding sites to ensure their efficient nuclear export. We shall discuss one of these examples in this next section.

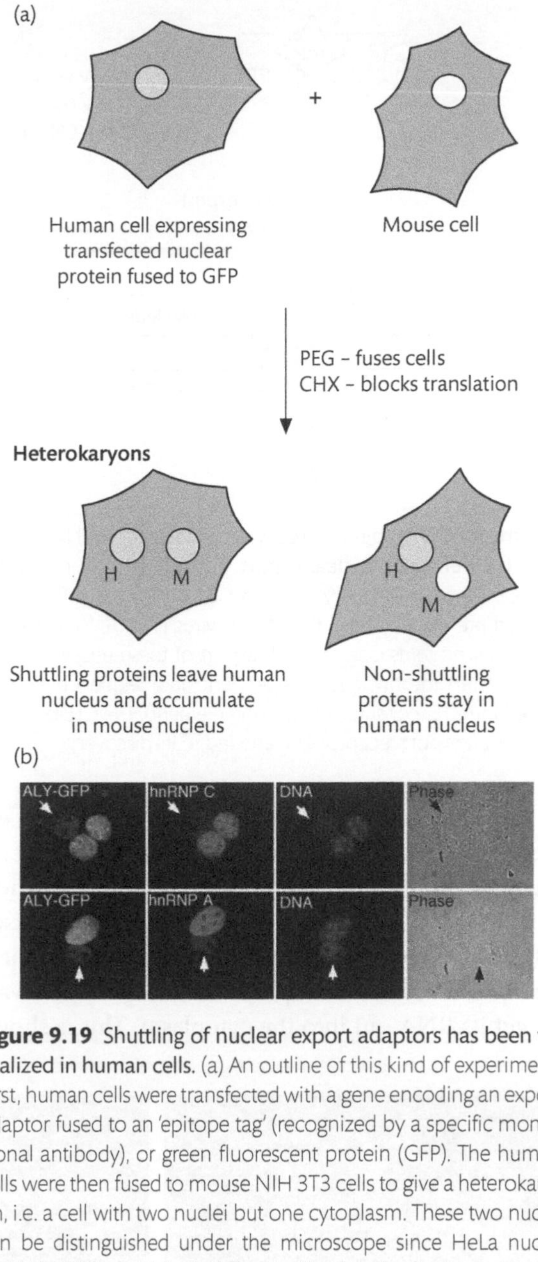

Figure 9.19 Shuttling of nuclear export adaptors has been visualized in human cells. (a) An outline of this kind of experiment. First, human cells were transfected with a gene encoding an export adaptor fused to an 'epitope tag' (recognized by a specific monoclonal antibody), or green fluorescent protein (GFP). The human cells were then fused to mouse NIH 3T3 cells to give a heterokaryon, i.e. a cell with two nuclei but one cytoplasm. These two nuclei can be distinguished under the microscope since HeLa nuclei and NIH 3T3 nuclei look different. The fused cells were cultured in medium containing cycloheximide to prevent any new protein synthesis. This experiment measures whether shuttling takes place, since any shuttling proteins will leave the human nucleus and then be re-imported into either the human but sometimes the mouse nucleus. Newly appearing fluorescent proteins in the mouse nucleus show that these must be shuttling proteins. Non-shuttling proteins remain in the human cell nucleus. (b) An actual experiment, where the REF adaptor protein fused to GFP (note that REF is named ALY-GFP in this figure) shuttled between the nucleus and the cytoplasm, as did another protein called hnRNP A. In contrast, another nuclear protein called hnRNPC did not shuttle, and so stayed put in the human nucleus. For a full colour version of this figure, see Colour Illustration 12. (Reprinted from Zhou et al.[21] with permission from Macmillan Publishers Ltd.)

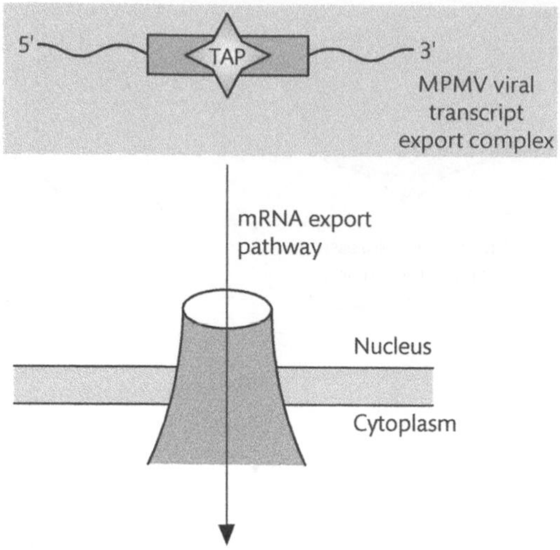

Figure 9.20 **TAP binds directly to some viral mRNAs to ensure their efficient nuclear export.** The MPMV virus hijacks the mRNA export machinery to export unspliced pre-mRNA. Unspliced pre-mRNAs from the MPMV virus encode for the structural viral proteins Gag and Pol. Export of these unspliced transcripts from the nucleus is dependent on a sequence element called the CTE (constitutive transport element). The CTE binds the metazoan export receptor TAP and led to its discovery.

A number of viruses hijack the cellular RNA export machinery to get their RNA molecules out of the nucleus and into the cytoplasm. HIV is one of these viruses, as we will discuss in Chapter 16. The Mason Pfizer monkey virus (MPMV) is a simpler retrovirus than HIV but it also exploits RNA export pathways to get its RNA out into the cytoplasm. This is shown in Fig. 9.20. MPMV achieves this through a sequence element called the *constitutive transport element* (CTE) within its own pre-mRNA. CTE is a high-affinity binding site for the mRNA export adaptor TAP. By binding TAP, the CTE directs unspliced MPMV pre-mRNAs down the mRNA export pathway.

Analysis of the CTE of the MPMV also provided scientists with new information on the mRNA nuclear export pathways followed by endogenous cellular RNAs. The micro-injection of RNAs containing CTEs into the nuclei of *Xenopus* oocytes blocked the general export of mRNA from the nucleus to the cytoplasm, suggesting for the first time that the CTE RNA sequence was accessing the mRNA export machinery. Co-injection of the human TAP protein overcame this dominant negative effect, showing that TAP is an mRNA export receptor.

9.7 mRNA export can become defective in human diseases

This process of mRNA nuclear export is absolutely critical, and is hijacked by viral RNAs which lack proper export signals.[23] Furthermore, mRNA export is also affected in several human genetic diseases. These include mytonic dystrophy (see Fig. 9.21), as we discussed in Chapter 6 with regard to splicing, and some other diseases (some important examples are shown in Table 9.2).

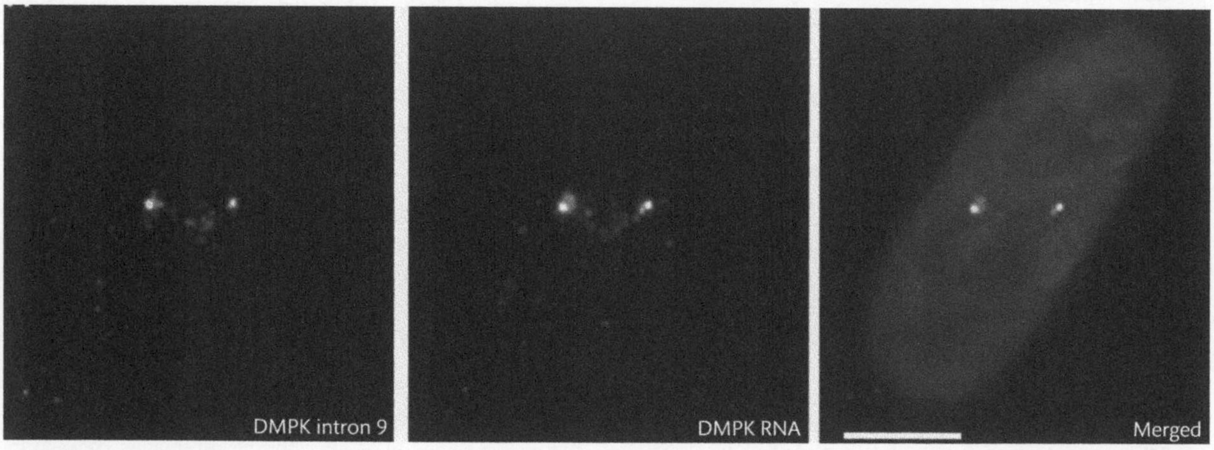

Figure 9.21 **mRNA export of the *DMPK* mRNA is inhibited in patients with myotonic dystrophy.** Notice that the FISH signal from the *DMPK* mRNA co-localizes with an intron signal from the *DMPK* gene which indicates where it is in the nucleus. Since human cells are diploid, there are two FISH signals in each nucleus. The expanded CAG repeat in the *DMPK* gene that occurs in muscular dystrophy probably prevents the mRNA from associating properly with components of the nuclear export machinery. Scale bar = 5μm. (From Smith et al.[22] with permission from *The Journal for Cell Biology*.)

Table 9.2 Mutations in the mRNA export pathway that cause human genetic diseases

Disease	mRNA export defect
Myotonic dystrophy (see Chapter 6 for more details of this disease)	*DMPK* mRNA with expanded CUG repeats accumulate at the *DMPK* gene; putting an mRNA export signal from a virus into these mutant mRNAs can help exacerbate the myotonic dystrophy phenotype at a cellular level
Osteogenesis imperfecta type I (a type of bone disease)	Nuclear retention of collagen mRNA caused by mutation in splicing sequences which prevents splicing and causes nuclear retention of mRNA
Lethal congenital contracture syndrome and lethal arthrogryposis with anterior horn cell disease (both are types of motor neuron disease)	Mutations in the mRNA export factor Gle1

Summary

In this chapter we have described how mRNAs are exported from the nucleus into the cytoplasm. The key points from this chapter are as follows.

- Export of mRNA involves at least five different classes of protein. These are shown in Fig. 9.2 and summarized in Table 9.1. These five classes of protein bind sequentially to the mRNA, leading to its export out of the nucleus and disassembly of the export complexes in the cytoplasm:
  1. **Nuclear assembly of mRNP:** nuclear export adaptors bind to RNA as it is being transcribed. The resulting mRNP–export adaptor complexes diffuse within the nucleoplasm.
  2. **Release of the fully formed mRNA from the gene by the 3′ end machinery:** release has to be completed before RNA can move out towards the nuclear periphery.
  3. **Association of mRNP with the nuclear export receptor:** on diffusing mRNPs, the nuclear export adaptors bind to mRNP export receptors located near the nuclear periphery.
  4. **Passage through the nuclear pore:** the mRNP nuclear export receptors in turn interact with the central channels of the NPC and mediate nuclear export of mRNP.
  5. **Disassembly of export competent mRNP:** once in the cytoplasm, the export complexes are disassembled and the carriage proteins are re-imported.
- Nuclear ↔ cytoplasmic traffic takes place through an aperture in the nuclear membrane called the nuclear pore complex (NPC).
- The NPC is a protein complex made up of a set of proteins called nucleoporins. Within the interior of the channel formed by the NPC are a group of nucleoporins rich in the hydrophobic amino acid phenylalanine and the neutral amino acid glycine. These are named the FG repeat nucleoporins after the single-letter symbols for phenylalanine and glycine.
- RNAs are exported from the nucleus as cargoes of RNA export complexes consisting of RNA and soluble carrier proteins.
- One class of soluble carrier protein are the nuclear export receptors, which contain hydrophobic regions which can interact with FG repeat nucleoporins. The TAP/p15 heterodimer is the mRNA export receptor.
- Nuclear export receptors are often linked via export adaptors with their carriage RNAs.
- Transport of RNA through the nucleus and across the nuclear pore occurs via diffusion. The proteins which assemble on mRNPs do not provide a motor for transport, but instead enable the mRNA to pass through the hydrophobic environment of the nuclear pore.
- mRNAs are coated with proteins as soon as they are made in the nucleus, to give mRNPs. The composition of these mRNPs is adjusted through the lifetime of the mRNA. Importantly, mRNPs are remodelled to remove the mRNA export

adaptor TAP/p15 as they are exported out of the nucleus. Hence nuclear and cytoplasmic mRNPs have different properties, with only the nuclear mRNP being able to cross the NPC.

Questions

9.1 What are the major steps in mRNA export from the nucleus?

9.2 Why has an mRNA export system evolved that requires adaptors and receptors to export mRNA? Why not just use a molecule that can act as both an adaptor and receptor?

9.3 How do the challenges of navigating through the nucleoplasm differ from transit through the nuclear pore?

9.4 Why does mRNA export move molecules out of the nucleus, rather than allowing them to slide back in?

References

1. **Walsh MJ, Hautbergue. M, Wilson SA.** Structure and function of mRNA export adaptors. *Biochem Soc Trans* **38**, 232–6 (2010).
2. **Reed R, Cheng H.** TREX, SR proteins and export of mRNA. *Curr Opin Cell Biol* **17**, 269–73 (2005).
3. **Cheng H, Dufu K, Lee CS, et al.** Human mRNA export machinery recruited to the 5′ end of mRNA. *Cell* **127**, 1389–1400 (2006).
4. **Dufu K, Livingstone MJ, Seebacher J, et al.** ATP is required for interactions between UAP56 and two conserved mRNA export proteins, Aly and CIP29, to assemble the TREX complex. *Genes Dev* **24**, 2043–53 (2010).
5. **Masuda S, Das R, Cheng H, et al.** Recruitment of the human TREX complex to mRNA during splicing. *Genes Dev* **19**, 1512–17 (2005).
6. **Huang Y, Steitz JA.** SRprises along a messenger's journey. *Mol Cell* **17**, 613–15 (2005).
7. **Luo MJ, Reed R.** Splicing is required for rapid and efficient mRNA export in metazoans. *Proc Natl Acad Sci USA* **96**, 14 937–42 (1999).
8. **Valencia P, Dias AP, Reed R.** Splicing promotes rapid and efficient mRNA export in mammalian cells. *Proc Natl Acad Sci USA* **105**, 3386–91 (2008).
9. **Legrain P, Rosbash M.** Some *cis*- and *trans*-acting mutants for splicing target pre-mRNA to the cytoplasm. *Cell* **57**, 573–83 (1989).
10. **Dower K, Kuperwasser N, Merrikh H, Rosbash M.** A synthetic A tail rescues yeast nuclear accumulation of a ribozyme-terminated transcript. *RNA* **10**, 1888–99 (2004).
11. **Lawrence JB, Singer RH, Marselle LM.** Highly localized tracks of specific transcripts within interphase nuclei visualized by in situ hybridization. *Cell* **57**, 493–502 (1989).
12. **Shav-Tal Y, Darzacq X, Shenoy SM, et al.** Dynamics of single mRNPs in nuclei of living cells. *Science* **304**, 1797–1800 (2004).
13. **Viphakone N, Hautbergue GM, Walsh M, et al.** TREX exposes the RNA-binding domain of Nxf1 to enable mRNA export. *Nat Commun* **3**, 1006 (2012).
14. **Hung ML, Hautbergue GM, Snijders AP, et al.** Arginine methylation of REF/ALY promotes efficient handover of mRNA to TAP/NXF1. *Nucleic Acids Res* **38**, 3351–61 (2010).
15. **Stutz F, Izaurralde E.** The interplay of nuclear mRNP assembly, mRNA surveillance and export. *Trends Cell Biol* **13**, 319–27 (2003).
16. **Bjork P, Wieslander L.** Nucleocytoplasmic mRNP export is an integral part of mRNP biogenesis. *Chromosoma* **120**, 23–38 (2011).
17. **Stewart M.** Molecular mechanism of the nuclear protein import cycle. *Nat Rev Mol Cell Biol* **8**, 195–208 (2007).
18. **Cole CN, Scarcelli JJ.** Transport of messenger RNA from the nucleus to the cytoplasm. *Curr Opin Cell Biol* **18**, 299–306 (2006).
19. **Stewart M.** Structural biology. Nuclear trafficking. *Science* **302**, 1513–14 (2003).
20. **Stewart M.** Ratcheting mRNA out of the nucleus. *Mol Cell* **25**, 327–30 (2007).
21. **Zhou Z, Luo MJ, Straesser K, et al.** The protein Aly links pre-messenger-RNA splicing to nuclear export in metazoans. *Nature* **407**, 401–5 (2000).
22. **Smith KP, Byron M, Johnson C, et al.** Defining early steps in mRNA transport: mutant mRNA in myotonic dystrophy type I is blocked at entry into SC-35 domains. *J Cell Biol* **178**, 951–64 (2007).
23. **Hurt JA, Silver PA.** mRNA nuclear export and human disease. *Dis Model Mech* **1**, 103–8 (2008).

10 Messenger RNA localization

Introduction

An essential part of the molecular architecture of cells is the intracellular distribution of its components—not least the localization of proteins. How then are proteins localized to specific subcellular compartments? Nature has found several strategies to achieve this. The simplest mechanism is general diffusion coupled to a docking mechanism. In this process proteins contain motifs within their amino acid sequences that allow them to be directed to specific subcellular compartments. A classic example is the signal recognition particle (SRP) which binds transiently to a signal in nascent polypeptides (five to ten hydrophobic amino acids in the N-terminus) and delivers them to the endoplasmic reticulum (ER) membrane. Other well-known examples include the nuclear localization signal (NLS), a short motif rich in basic amino acids required for the nuclear localization of several nuclear proteins, and the short hydrophobic nuclear export signal (NES) required for the export of proteins from the nucleus to the cytoplasm.

However, diffusion- and docking-based localization mechanisms are not always realistic or sufficiently efficient in some circumstances. In some cases it is better to localize the message—mRNA—because a single mRNA can be used in multiple rounds of protein synthesis. Moreover, the genetic information within an mRNA can also be stored in a particular cytoplasmic location until such a time as the protein it encodes is required.

10.1 The need for mRNA localization

This chapter is focused on the localization of mRNAs[1–6] (summarized in Fig. 10.1). Note, however, that other classes of RNA are also localized. In Section 10.4.2 we will also briefly describe the localization of microRNAs (small non-coding RNAs that regulate mRNA translation) in neurons.

Measuring mRNA localization is technically demanding. One of the main techniques used is **in situ hybridization**, in which a complementary antisense nucleic acid probe is labelled typically with a fluorescent probe. This technique is known as **FISH** (fluorescence in situ hybridization). 'In situ' refers to the fact that the localization of the mRNA is detected in intact cells. FISH techniques have been greatly improved, to the extent that it is now possible to examine the localization of thousands of mRNAs. The sensitivity of fluorescent probes used to detect localized mRNAs has also shown remarkable improvement, as described in Box 10.1.

The importance of mRNA localization is powerfully illustrated by a genome-wide analysis by Eric Lécuyer and colleagues of mRNA localization during *Drosophila* embryogenesis.[1,2] They found that during early development in *Drosophila* a surprising 71% of 2314 expressed genes express mRNAs that are localized, and these were grouped into 35 localization categories. These categories included the following: 15.8% of mRNAs showing subcellular

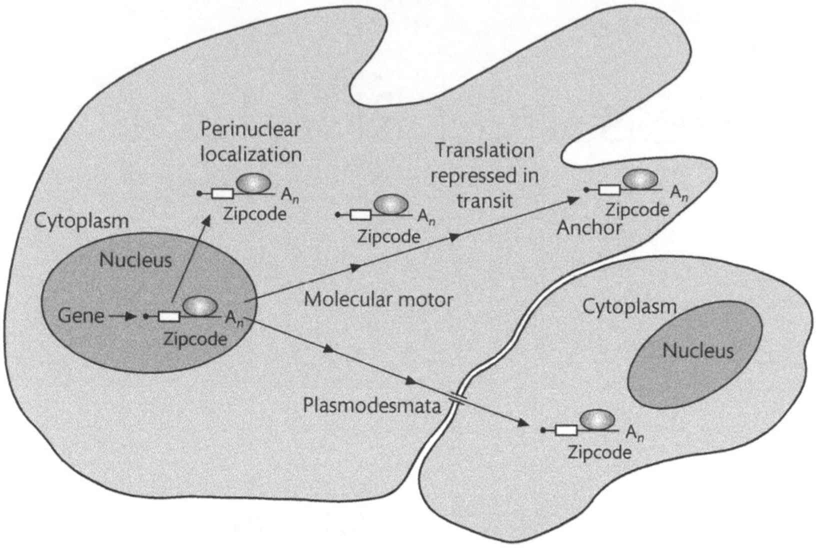

Figure 10.1 Interlinked processes in mRNA localization. The process of mRNA localization begins in the nucleus where the deposition of specific RNA-binding proteins onto pre-mRNAs can influence their ultimate localization. In some cases, mRNAs can even be trafficked from cell to cell. Once localized to the appropriate cellular compartment, mRNAs need to be anchored or they will simply diffuse away. mRNA translation can also be regulated at the site of localization, and mRNAs can be selectively degraded or protected from degradation depending on need. The net effect is the temporal and spatial regulation of protein synthesis.

localization; 8.4% germ cell (cells that will give rise to the next generation) localization; 8.6% posterior localization (the posterior part of the embryo); 3.5% perinuclear (close to the cell nucleus). This high proportion of localized mRNAs may be especially prominent in *Drosophila* because of their particular developmental features (not least the syncytium of *Drosophila* embryos in which several cells effectively fuse). However, the results indicate that mRNA localization may be a very common phenomenon in nature.

Incidentally, observing the pattern of an mRNA's localization has an added value. Localization of mRNA can help to predict or define the *function* of a particular gene. For example, if an mRNA is localized to germ cells, it is likely to encode a protein with a function in germ cells.

In the following sections we shall look at the machinery of mRNA localization including the role of the cytoskeleton, **RNA zipcodes**, and the influence of the nuclear history of a transcript on its localization. We will illustrate the phenomenon of mRNA localization in early development using three classic examples, and then move on to discuss the localization of mRNAs in differentiated animal and plant cells. We will briefly examine the special cases of perinuclear, endoplasmic reticulum, mitochondrial, and chloroplast localization. Examples of disrupted mRNA localization in human disease are also mentioned.

Box 10.1 Advances in the techniques used to examine mRNA localization

Initially the techniques used to examine mRNA localization were not particularly sensitive and were prone to several artefacts. These were due to non-specific hybridization of antisense nucleic acid probes and low signal-to-noise ratios. Nowadays techniques have advanced significantly, enabling more powerful and precise studies of mRNA localization. Probes have been developed to achieve sensitivity to even a single mRNA molecule (i.e. the ability to detect a single localized mRNA in a cell). Santangelo and colleagues describe a new detection method, MTRIP (multiply-labelled tetravalent RNA imaging probes), in an article published in the journal *Nature Methods*.[6] These probes consist of heavily modified nucleic acid ligands labelled with up to three fluorophores (molecules that generate a fluorescent signal). The ligands have been cleverly engineered to tetramerize, thus enhancing fourfold the strength of the signal. These advanced probes allow visualization in live cells in real time of the motion of mRNAs and of the granules that contain them.

Key points

In development and normal physiology there is a need to localize proteins to specific subcellular compartments. An efficient way of doing this is to localize mRNAs so that proteins can be synthesized at the correct cellular location.

10.2 The machinery of mRNA localization

It is now clear that evolution has developed a machinery that can localize mRNAs to specific areas in an oocyte or embryo, or even in subcellular compartments within a cell.[7–18] An emerging picture is that the process of mRNA localization is initiated in the nucleus where specific RNA-binding proteins bind mRNAs destined for localization. Once in the cytoplasm, **mRNP** complexes (mRNA packaged by proteins) are remodelled and acquire additional localization and intracellular transport factors.

10.2.1 The localization of *Xenopus Vg1* mRNA through cytoskeletal motors

Vg1 is a *Xenopus laevis* mRNA that is localized to the vegetal pole of the oocyte. The vegetal pole is the lower-hemisphere yolk-rich region, directly opposite the animal pole. Several proteins—XStau, hnRNP1, and Vera (also known as Vg1RBP)—bind an mRNA localization element (LE) in *Vg1* mRNA and are required for correct Vg1 localization: XStau binds to dsRNA and is localized, with *Vg1* mRNA, to the vegetal pole during mid-oogenesis. Xstau and *Vg1* mRNA are detected in a distinct RNP complex that sediments at 20S (S is the Svedberg unit; see Section 11.2) on sucrose gradients. XStau also co-sediments with the cytoskeletal motor protein kinesin, suggesting that it has a role in mRNP transport. Consistent with these observations, mutant versions of XStau block the correct localization of *Vg1* mRNA.

Whereas XStau only binds *Vg1* mRNA in the cytoplasm, hnRNP I and Vera bind *Vg1* and other mRNAs in the nucleus and remain bound to them in the cytoplasm. It is worth noting that hnRNP I was first identified as a generic pre-mRNA packaging protein; but it is also known as PTB (pyrimidine tract-binding protein), a protein that is often involved in the repression of 3′ splice sites. Once again, this underlines the multifunctional nature of many RNA-binding proteins which are therefore able to coordinate multiple processes. Prrp (proline-rich RNA-binding protein) is a fourth protein that binds the *Vg1* LE in the cytoplasm. The proline-rich domain of Prrp associates with profilin, a protein involved in promoting actin polymerization. Actin polymerization is a necessary step in the formation of actin filaments which are thought to be required in anchoring *Vg1* mRNA to the vegetal pole. Thus Prrp acts as a molecular adaptor that connects a localized mRNA to the cytoskeleton. In summary, several proteins bind to the *Vg1* LE, some of which are already in the nucleus, influencing the eventual fate of *Vg1* mRNA in the cytoplasm. This is summarized in Fig. 10.2.

Several other RNA-binding proteins bind to localization elements in the nucleus—for example, ZBP1 binds the mammalian β-actin localization element in the nucleus. The β-actin localization element is described as the 'zipcode'. In another example, Squid, the *Drosophila* orthologue of hnRNP A1 (a very abundant pre-mRNA packaging protein also involved in alternative splicing) is required for *oskar* mRNA localization and binds to *oskar* mRNA in the nucleus.

10.2.2 The exon junction complex and mRNA localization

In Chapter 12 we will discuss the **exon junction complex** (EJC) in the context of mRNA degradation; the EJC is also highly relevant to the process of mRNA localization. The EJC is deposited about 20 nucleotides upstream of exon junctions; and is removed after the first round of translation (the 'pioneer' round). Before its removal, the EJC is able to influence mRNA fates in several ways. First, by marking correctly spliced transcripts, the EJC facilitates their nuclear export. EJCs also direct newly transcribed mRNAs to regions of the cytoplasm that are enriched in polysomes, favouring their immediate translation.

In the *oskar* mRNA of *Drosophila*, all of its three introns are located in the open reading frame (ORF); because the *oskar* mRNA EJCs are required for its localization, they must not be removed before the mRNA has had a chance to become localized. Consequently *oskar* must be kept *translationally repressed* in transit

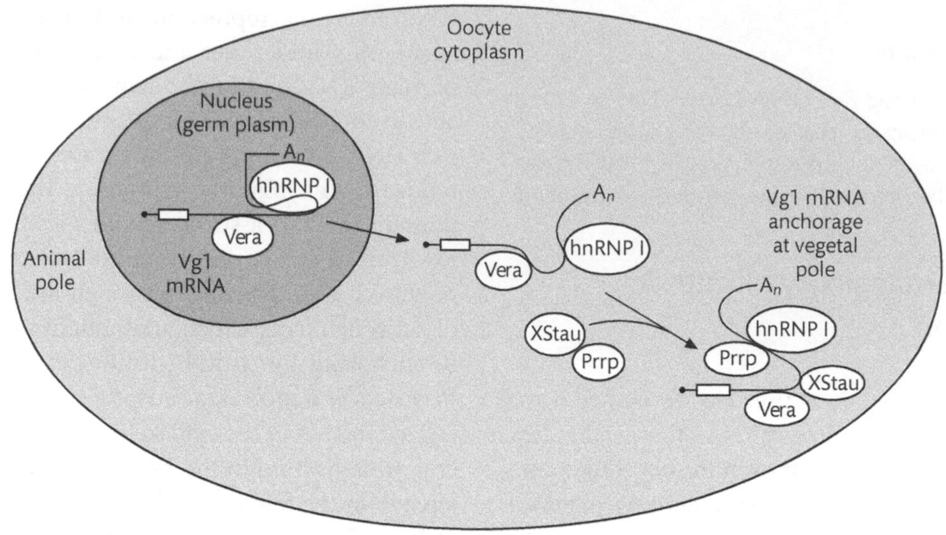

Figure 10.2 Model of the localization of *Vg1* mRNA. In the nucleus, the *Vg1* mRNA localization element (LE) is recognized by the proteins hnRNP I and Vera. After export to the cytoplasm, the mRNP complex is remodelled and hnRNP I and Vera no longer interact directly. XStau and Prrp join the *Vg1* mRNP complex, facilitating its transport to the vegetal pole where it is anchored.

to its eventual cytoplasmic destination in order to prevent removal of the EJC and premature translation. It is interesting to note that mammalian counterparts of Magoh and Y14 (components of the EJC required for localization of *oskar* mRNA in *Drosophila*) are also detected in the dendrites of mammalian neurons, suggesting that the EJC is involved in mRNA localization more generally.

How then are mRNAs trafficked in the crowded environment of the cytoplasm? The answer lies in the **molecular motors**; these are molecular machines that mediate the transport of vesicles, organelles, and macromolecules. There are three families of motors: myosins, which move along actin filaments, dyneins, and kinesins, which move along microtubules. mRNA cargos have been shown to move along both actin filaments (e.g. yeast *Ash1* mRNA) and microtubules (*β-actin* mRNA). Motion is achieved through conformational changes associated with ATP hydrolysis. The direction of travel of a molecular cargo is determined by the type of motor and the polarity of the cytoskeletal structure with which it interacts. Moreover, the direction and rate of travel can be regulated by environmental cues.

mRNAs are transported in large granules whose molecular composition is complex. In neurons, enormous granules (>1000S) associate with conventional kinesin and contain in excess of 40 proteins. These include proteins involved in RNA transport (e.g. Staufen and FMR1) or in protein synthesis (EF1α, eIF2α, eIF2β, eIF2γ, or ribosomal protein L3), RNA helicases (DDX1, DDX3), hnRNP proteins (hnRNP A1, hnRNP B2, hnRNP U), and other RNA-associated proteins and splice factors (Aly, NonO, nucleolin, and PSF). The importance and function of each of these proteins remains to be determined. Overall, their combined functions are to anchor mRNAs bearing appropriate 'zipcodes' to the molecular motors, to keep them translationally repressed while in transit, and to activate their translation at the appropriate time once the destination is reached.

The movement of the transport granules has been observed in real time in neurons by GFP (green fluorescent protein) tagging. The granules can move at different speeds with an average of approximately 34nm/s. The model that has emerged is described as the 'Venice model' of mRNA transport, and was proposed by Lopez de Heredia and Jansen.[11] It is explained in Fig. 10.3. According to this metaphor, messenger RNAs are exported from the nucleus via the nuclear pore complex (pre-marked by specific nuclear RNA-binding proteins and by the EJC); they then associate with different cytoskeletal tracks (canals) and are transported by specific motors

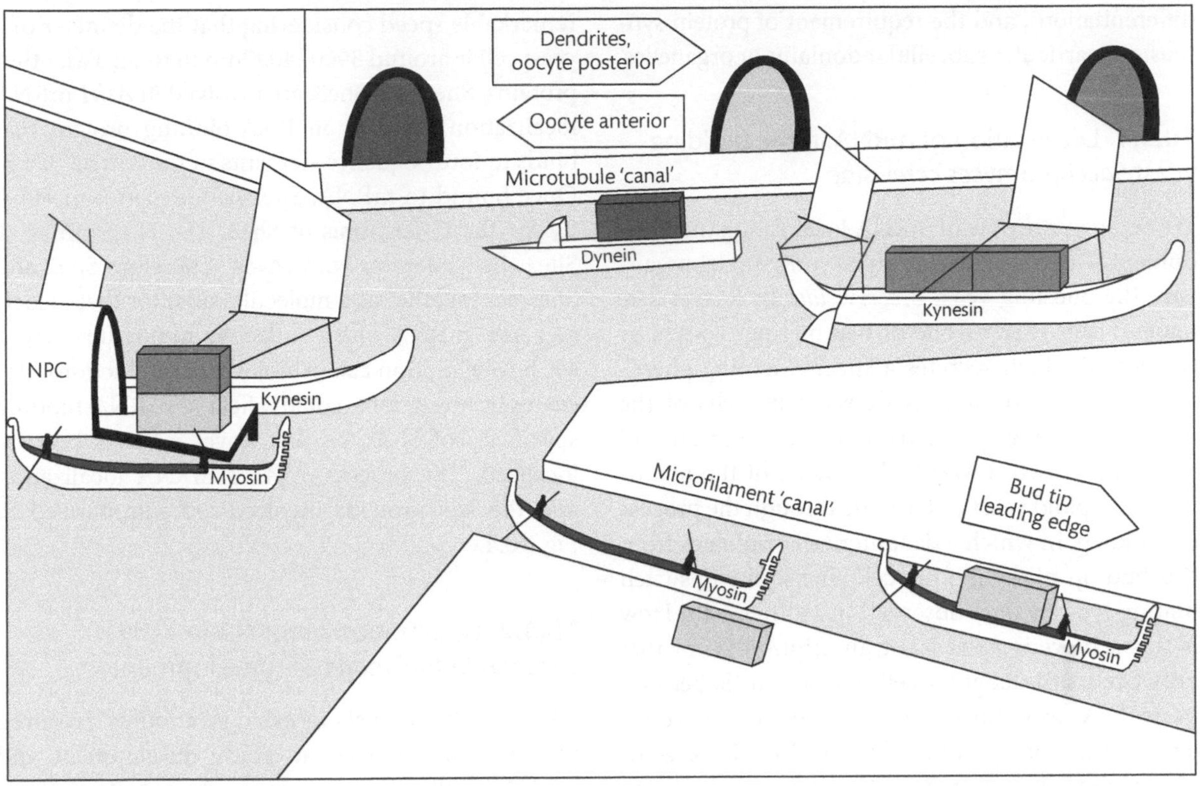

Figure 10.3 The 'Venice model' of mRNA traffic. According to this model, mRNPs are exported from the nucleus through the nuclear pore complex (NPC). In the cytoplasm they associate with different cytoskeletal filaments (the canals in the analogy) and are delivered to their destinations via specific 'motors' (gondolas and other boats in the analogy). (Reproduced from de Heredia and Jansen[11] with permission from Elsevier.)

(gondolas or boats) until they reach their cytoplasmic destination. At their destination, the mRNAs need to be anchored or they will float away!

Key points

mRNA localization invariably involves the specific interaction of RNA-binding proteins with mRNA sequence elements, generally in the 3′ UTR, which direct their localization. These elements have been referred to as 'zipcodes'.

It is also clear that the *nuclear history* of an mRNA influences its eventual localization, owing to the deposition of RNA-binding proteins on the nucleus that accompany the mRNAs into the cytoplasm. These proteins include components of the multifunctional exon junction complex (EJC).

In the cytoplasm the transport of mRNAs involves molecular motors that drive mRNAs along cytoskeletal filaments to their destination. mRNA is generally translationally repressed while in transit. Once localized, mRNAs need to be anchored to prevent them diffusing away.

10.3 Classical examples of mRNA localization in development

Developmental biology has to take a large proportion of the credit for the discovery of mRNA localization.[9,19–22] Well before the advent of modern molecular biology, developmental biologists working on ascidians and sea urchins were aware that specific cell lineages in early development arise from specific regions of the oocyte. We now know that, to a large extent, this compartmentalization of information in oocytes is due to localized mRNAs. As we shall see, mRNA localization is a widespread and ancient mechanism observed across eukaryotes in several settings in normal development and in differentiated somatic cells. The localization of mRNA is associated with the establishment of morphogen gradients in development (e.g. gradients involved in the definition of body axes), the segregation of cell fate determinants (proteins involved in initiating cellular differentiation), and the requirement of protein synthesis in particular subcellular domains or organelles.

10.3.1 Localization of mRNA in the budding yeast *Saccharomyces cerevisiae*

We begin our review of mRNA localization in development with a unicellular eukaryotic model organism, the budding yeast *S. cerevisiae*. In *S. cerevisiae* haploid cells exist as one of two mating types, a or α, each of which secretes a specific mating pheromone. The pheromone is detected by cells of the opposing mating type causing cell cycle arrest and polarized growth towards the source of the pheromone. Haploid yeast cells divide through the process of budding, in which a daughter cell originates from the 'bud tip' of the mother cell. The ability to switch mating types is only observed in mother cells. How is this achieved? Ash1 is a transcription factor that only occurs in daughter cell nuclei. This is because its mRNA accumulates in the bud tips. Ash1 represses the transcription of the *HO* nuclease gene. HO nuclease is required for the mating switch, and is expressed in mother cells which lack Ash1.

Several other mRNAs are localized in the yeast bud tip that gives rise to the daughter cells, but Ash1 is the best-studied example. The localization of *Ash1* mRNA requires the actin cytoskeleton and *cis*-acting RNA elements also known as mRNA zipcodes. The best-known zipcode is located in the *Ash1* 3′ UTR—the E3 element located at the end of the *Ash1* ORF. The localization of *Ash1* mRNA has been visualized directly as follows. An *Ash1* mRNA construct was designed to incorporate MS2 RNA stemloops (these stemloops are specifically bound by the prokaryotic RNA-binding protein MS2). The MS2 protein was in turn fused to GFP, enabling the movement and localization of *Ash1* mRNA to be observed and recorded in real time.

Using the power of yeast genetics (many mutants can be generated and screened in a relatively short time), several *trans*-acting factors have been found to be involved in *Ash1* mRNA localization. It is now clear that its localization is dependent on the protein Myo4 (also known as She1). Myo4 is a form of myosin, a cytoskeletal protein involved in molecular motors. Myo4 is required to transport *Ash1* mRNA along actin filaments. *Ash1* mRNA moves along the filaments at speeds of 200–400nm/s, quite a remarkable speed considering that the diameter of a yeast cell is around 3000–4000nm in total! Two other proteins, She2 and She3, are involved in *Ash1* mRNA localization. She2 is an RNA-binding protein that binds to four sequence elements in the *Ash1* 3′ UTR. Once bound to mRNA, She2 acquires greater affinity for the C-terminus of She3. The N-terminus of She3 then interacts with Myo4. Therefore She2 and She3 act together as a molecular adaptor that associates *Ash1* mRNA with a molecular motor. It is easy to see how evolution can exploit these interactions. For example, mutations could affect the RNA-binding specificity of She2, so that novel mRNAs become localized. The process of *Ash1* mRNA localization and the key proteins involved are summarized in Fig. 10.4.

10.3.2 Localization of mRNA in early *Drosophila melanogaster* development

The fruit fly *D. melanogaster* is another favoured model organism used to study development and mRNA localization. One of its greatest strengths is the ability to generate mutants. A specific category of genes important in early development are the *maternal effect genes*; these genes are expressed in oocytes or in the 'nurse cells' that surround and nurture them. Maternal effect genes are involved in early development, including the formation of the basic body plan including segmentation. Several mRNAs are transcribed and processed in nurse cells and then transported into oocytes, where several become localized. Some of the most widely studied include mRNAs that encode proteins involved in the generation of the antero-posterior axis, i.e. the front and back of the fly. Thus Bicoid and Hunchback are involved in the patterning of anterior structures (head and thorax) in the *Drosophila* embryo, whereas Nanos and Caudal are involved in the formation of posterior abdominal structures. Bicoid is a multifunctional transcription factor whose DNA-binding homeodomain binds to a specific RNA sequence in the 3′ UTR of *caudal* mRNA, blocking its translation.

Oskar mRNA is also synthesized in nurse cells and localizes in the posterior end of the oocyte, where it is required for the proper formation of the germplasm, a cytoplasmic structure required for germ

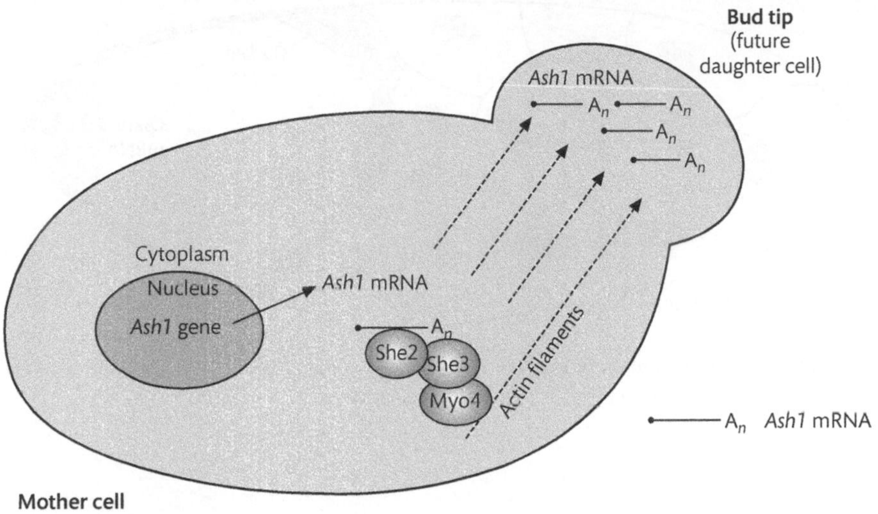

Figure 10.4 Mechanism of *Ash1* mRNA localization in budding yeast. The *Ash1* gene is transcribed in the nucleus of the mother cell *Ash1* mRNA and exported, fully processed, to the cytoplasm. Sequence elements in its 3′ UTR are recognized by the RNA-binding protein She2, which then binds She3. She3 in turn binds Myo4 (myosin) which enables directed transport of the mRNA along actin filaments towards the bud tip, the future daughter cell. Expression of Ash1 protein in the daughter cell results in repression of the HO nuclease, ensuring that a mating type switch does not occur in the daughter cell.

cell formation. Reminiscent of the *Vg1* example in *Xenopus* oocytes, *oskar* mRNP particles are transported in a translationally repressed state with the participation of the dsRNA-binding protein Staufen and the kinesin heavy chain motor protein Khc. Khc is very important as it drives the localization process. Other proteins required for the proper localization of *oskar* mRNA include Mago-nashi and Y14; these are components of the EJC that is deposited onto splice junctions, as discussed in Section 10.2. The deposition of EJCs facilitates nuclear export and marks mRNAs that have been correctly spliced; thus only correctly processed transcripts are localized in the oocyte.

The *oskar* mRNP is even more complex, and includes the following additional proteins. Yps is a Y-box protein that represses *oskar* mRNA translation. As we shall see in Chapter 11, Y-box proteins work in *Xenopus* oocytes as 'masking proteins' to aid the long-term storage of mRNAs and repress their translation. Dcp1, a conserved eukaryotic decapping enzyme, is also a component of the *oskar* mRNP and is required for its eventual degradation. Thus *oskar* mRNA carries with it all of the necessary requirements for its entire life cycle: localization, repression of translation, translation, and eventually degradation. Typical patterns of mRNA localization in *Drosophila* oogenesis are as summarized in Fig. 10.5.

10.3.3 Localization of mRNA in early *Xenopus laevis* development

A widely studied vertebrate model in which extensive mRNA localization takes place is the oocyte of the African clawed toad *X. laevis*. *Xenopus* oocytes afford unique advantages when used as model organisms; including their size, well-characterized developmental stages, clearly defined animal–vegetal axes, and the ease with which reagents can be micro-injected into either the cytoplasm or the germinal vesicle (nucleus). There are six stages in *Xenopus* oogenesis. Stage I is pre-vitellogenic (i.e. before the formation of yolk). Vitellogenesis starts in stage II and continues until stage V, when the animal–vegetal poles are clearly defined. By stage VI, oocytes are 1.2mm in diameter and are ready for fertilization after exposure to progesterone. The definition of the animal–vegetal axis is important because the highly pigmented 'animal pole' gives rise to the ectoderm and neuroectoderm; while the 'vegetal pole' gives rise to the endoderm.

Several mRNAs are known to localize in the animal pole, including *An1* which encodes a ubiquitin-like

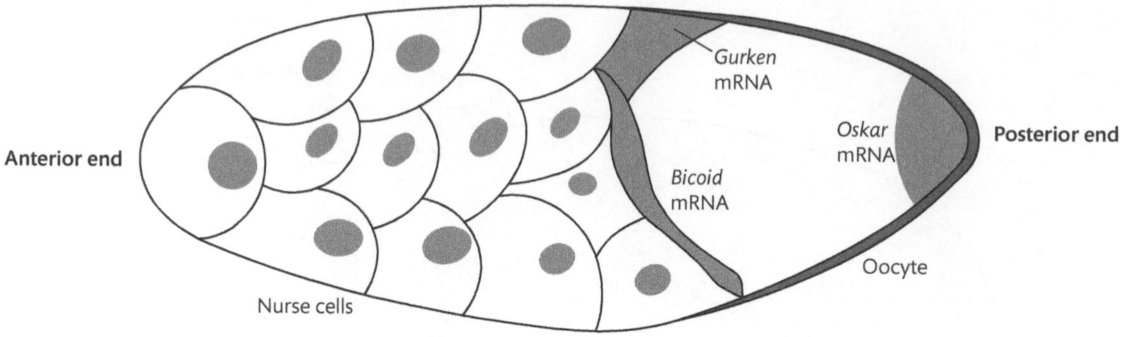

Figure 10.5 mRNA localization in *Drosophila* oogenesis. The oocyte is supported by the nurse cells at the anterior end. *Bicoid*, *gurken*, and *oskar* mRNAs are localized to different parts of the developing oocyte.

protein, *An2* (a mitochondrial ATPase subunit), *An3* (an RNA helicase), *fibronectin* (an extracellular matrix component), and *Tcf1* (a transcription factor). mRNAs that are localized in the vegetal pole include *Xcat2* (a zinc-finger protein related to *Drosophila* nanos), *VegT* (a T-box transcription factor), and *Xdazl* (an RNA-binding protein involved in spermatogenesis). *Vg1*, which we discussed in Section 10.2, is a well-studied example of an mRNA that is localized in the vegetal pole. It encodes a member of the transforming growth factor-beta (TGFβ) family. If Vg1 protein is mis-expressed in the animal pole, mesodermal cells are induced where there should normally be ectodermal cells. *Vg1* mRNA is transcribed early in oogenesis in stage I, and it becomes distributed throughout the cytoplasm. By mid-oogenesis *Vg1* mRNA is transported to the vegetal pole where it remains until the end of oogenesis.

Key points

Classical examples of mRNA localization in development include *Ash1* in the unicellular fission yeast, and in multicellular organisms *oskar* in early *Drosophila* and *Vg1* in early *Xenopus* development, among several others.

10.4 Localization of mRNA in differentiated somatic cells

So far we have concentrated on the localization of mRNAs in early development. However, in several somatic cell types mRNA localization also occurs in fully developed organisms.[23–41] Examples include the localization of mRNA in lamellipodia and neuronal axons. As we shall see, messenger RNA can also be localized around the nucleus and specific organelles including the endoplasmic reticulum, mitochondria, and chloroplasts.

10.4.1 Localization of *β-actin* mRNA

The *β-actin* mRNA is localized in fibroblast lamellipodia, a characteristic feature at the leading edge of motile cells. Its localization is dependent on targeted transport of the mRNA with the assistance of microfilaments. Microfilaments are found in the cytoplasm of all eukaryotic cells where they are involved in cellular architecture, intracellular transport, the amoeboid movement of cells, and muscle contractions. The principal ingredient of microfilaments is filamentous actin (F-actin) which forms through the polymerization of actin monomers. β-actin is one of the two non-muscle actins and is distinguished by its mRNA being localized in the leading edge of the lamellipodia of 'crawling cells', and in the filaments of the microvilli in epithelial cells. A series of calculations has shown that in crawling cells over three million molecules of actin per minute are used in the polymerization of F-actin! Thus a concentration of *β-actin* mRNA in the lamellipodia would cater for the local high demand for actin monomers. The localization of *β-actin* mRNA is even associated with metastatic potential in cancer (see Box 10.2).

The localization of *β-actin* mRNA is also dependent on the presence of a zipcode RNA

Box 10.2 Localization of mRNA and metastatic potential

There is an interesting connection between *β-actin* mRNA localization and metastatic potential,[28-30] i.e. the ability of cancer cells to migrate around the body generating secondary tumours. In one study, two cell lines were derived from a breast adenocarcinoma; one was highly metastatic and the other was weakly metastatic. The least metastatic cells exhibited an intrinsic cell polarity and polarized cell locomotion; that was not the case in the highly metastatic cells. One of the key differences is that the highly metastatic cells have lost the ability to localize *β-actin* mRNA. This is due to reduced expression of ZBP1, the RNA-binding protein that binds to the zipcode in the *β-actin* mRNA. Thus reduced ZBP1 expression is associated with a more metastatic phenotype, whereas over-expression of ZBP1 reduces the invasiveness of metastatic tumour cells.

Several other mRNAs have been described as being localized in the *pseudopodial protrusions* of cancer cells; for instance, mRNAs encoding Ras-related protein RAB13 and Plakophilin-4 in breast cancer cells. Furthermore, the tumour suppressor protein APC (adenomatous polyposis coli) has been shown to help target mRNAs to cancer cell protrusions. Loss of APC is associated with cancer progression, notably in the context of familial adenomatous polyposis (FAP). These examples clearly illustrate the importance of mRNA localization in cancer progression.

element in the 3′ UTR. Transplantation of the zipcode sequence to heterologous mRNAs can direct reporters into the leading edges of lamellipodia; furthermore, targeting the zipcode with antisense oligonucleotides prevents correct mRNA localization. The 3′ UTR zipcode contains the sequence ACACCC which is recognized by a specific protein ZBP1 (zipcode-binding protein 1, an RNA-binding protein with a KH-homology domain). A dominant negative mutant of ZBP1 blocks the correct localization of *β-actin* mRNA and specifically inhibits fibroblast cell motility. ZBP1 is also involved in the formation of dendritic filopodia and in synapses in the nervous system, indicating that the mechanisms that underlie *β-actin* mRNA localization are used in several cell types. The processes that underpin *β-actin* mRNA localization are summarized in Fig. 10.6.

The translation elongation factor EF1α provides an interesting link between the process of

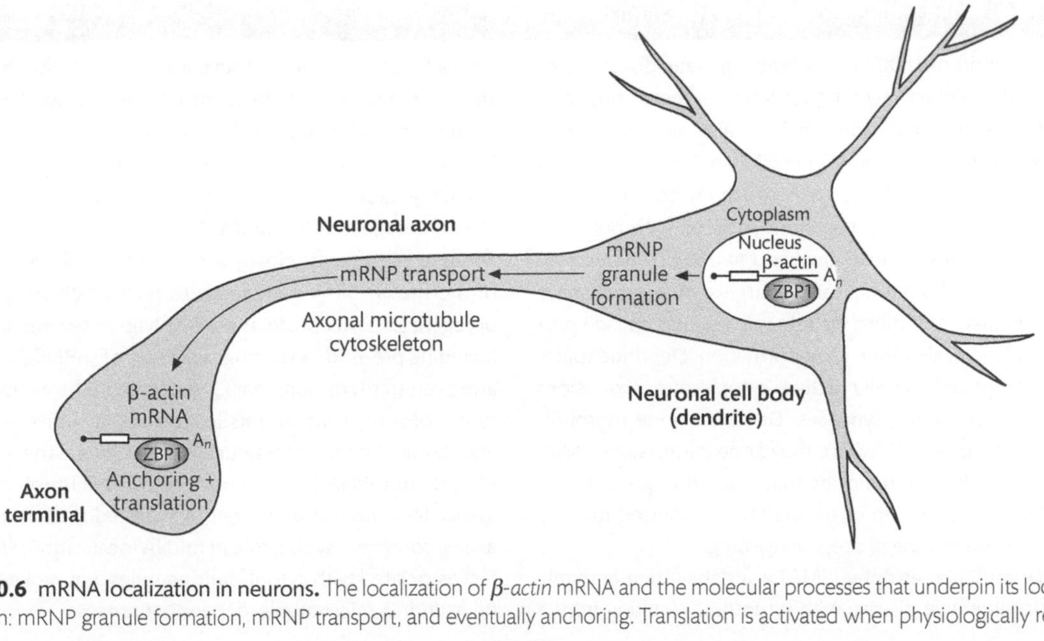

Figure 10.6 **mRNA localization in neurons.** The localization of *β-actin* mRNA and the molecular processes that underpin its localization are shown: mRNP granule formation, mRNP transport, and eventually anchoring. Translation is activated when physiologically required.

translation and the anchorage of *β-actin* mRNA to actin filaments. EF1α is a highly conserved protein that catalyses the binding of amino-acyl tRNA to ribosomes. In fact it is a translation elongation factor (the process of translation is covered in Chapter 11). In some circumstances up to 50% of EF1α is bound to the cytoskeleton. EF1α itself contains actin-binding sites and associates with those actin filaments with which mRNAs become localized. Thus EF1α is thought to contribute towards the localized translation of mRNAs such as *β-actin*.

10.4.2 Localization of mRNA in neurons

Neurons are fascinatingly complex cells. The neuron features a cell body which contains the nucleus and is surrounded by dendrites. Axons extend out of the cell body for remarkably long distances; they are protected by myelin sheaths and, at their ends, synaptic vesicles provide the basis for the neurotransmission of signals. The unique cellular architecture of neurons provides a perfect context for the localization of mRNAs because the physical distance between the nucleus and the axons and synapses is, relatively speaking, very great, and it is most efficient to transport mRNAs to sites where their translation is required. However, mRNA localization also occurs in cell types associated with neurons; for example, oligodendrocytes localize the mRNA for myelin basic protein (MBP) in the myelin compartment. Oligodendrocytes surround and insulate the axons through which neurons transmit electrical impulses. Faulty localization of mRNA in the human central nervous system has been associated with disease; two examples—fragile X syndrome and spinal muscular atrophy—are given in Box 10.3.

The localization of mRNA in neurons is thought to occur in stages. First, echoing the processes that occur in early development, mRNAs destined for localization are bound by specific proteins in the nucleus. Once in the cytoplasm, additional proteins associate to form a transport-competent mRNP particle. Then mRNP particles become associated with a molecular motor and are transported with the assistance of the cytoskeleton. Once they arrive at their destination, specific mRNAs are translationally repressed until the protein they encode is physiologically required. Their translation is then triggered by specific environmental cues, including stimulated synapses.

Box 10.3 Faulty mRNA localization in the human central nervous system

Fragile X syndrome (FXS) is a heritable genetic disease associated with X-linked mental retardation. Affected boys also exhibit developmental abnormalities and autism. Its incidence is 1 in 3600 males and 1 in 6000 females. FXS is caused by a mutation in the *FMR1* gene, which encodes the FMRP protein, an RNA-binding protein associated with polyribosomes. It contains an 'RGG box' and two KH (K homology) domains. Several potential mRNA targets, or 'cargoes', have been identified, and these include mRNAs that encode proteins involved in dendritic spine formation. Dendritic spines are electrophysiologically active cytoplasmic protrusions involved in excitatory synapses. Dendritic spine morphology is abnormal in FXS, where they appear immature. FMRP protein is localized in dendritic spines and thought to be involved in the regulation of translation of localized mRNAs, particularly during the process of learning.

Spinal muscular atrophy (SMA) is a debilitating heritable neurodegenerative disease with symptoms resulting from a loss of function of motor neurons associated with the brainstem and spinal cord; these include muscle weakness, difficulties in swallowing, and the accumulation of secretions in the lungs and throat. It occurs in 1 in 6000 births and is a leading cause of infant mortality. Recessive mutations in the homologous *SMN1* and *SMN2* (*survival of motor neuron*) genes in humans are clearly associated with SMA; knockout of the mouse *SMN* gene leads to fetal death early in embryogenesis. SMN proteins are involved in several processes including pre-mRNA splicing, assembly of snRNPs, snoRNPs, and even transcription. SMN is detected in the axons of mature motor neurons, and its loss results in shortened axons and dysfunctional motor neurons. What then is the relevance of SMN to mRNA localization? Loss of SMN protein is also associated with reduced levels of localized *β-actin* mRNA in axons, consistent with a role in mRNA localization. SMN is in fact associated with cytoplasmic granules that move through neurons in a microtubule-dependent process.

As well as mRNAs, several 'granules' have been shown to localize in neurons. These include **transport granules** (RNP complexes that contain localized mRNAs) and RNA-binding proteins involved in localization and translation. Neurons also contain **stress granules**, with mRNAs that are translationally arrested in response to a wide range of environmental stresses, and **P-bodies**, sites of mRNA degradation and translational repression. The presence of these granules suggests that the machinery of mRNA localization is highly compartmentalized and thus, by extension, more efficient. The reason for increased efficiency is the concentration, in the same granule, of all the proteins required to achieve translational repression and, when required, mRNA degradation, as we shall see in Chapter 12.

Specific zipcodes have evolved to determine the coordinated localization of classes of mRNA in specific cellular compartments. Several mRNAs and RNA-binding proteins are localized specifically in neuronal dendrites. As discussed earlier, β-actin localization occurs in fibroblasts. In dendrites, the localized translation of *β-actin* mRNA is associated with critical stages of neuronal development including synapse formation and axon guidance. CaMKIIα is a calcium-dependent protein kinase, highly expressed in neurons; it phosphorylates calmodulin, a calcium-binding protein involved in calcium-mediated signalling. As is the case for *β-actin* mRNA, its localization also requires a sequence element in the 3′ UTR which displays high homology to an RNA element in the 3′ UTR of *neurogranin* mRNA, another dendritically localized transcript. It is also worth noting that there are two cytoplasmic polyadenylation elements (CPEs) in the *CaMKIIα* mRNA which are, perhaps surprisingly, also required for its localization. As cytoplasmic polyadenylation influences the regulation of translation (Chapter 11), the implication is that both mRNA localization and the activation of translation can be coordinated.

A fascinating experiment in mice provides further confirmation of the importance of the correct localization of *CaMKIIα* mRNA. A transgenic mouse was produced in which the 3′ UTR was deleted, but the mRNA was still able to produce full-length protein. In these transgenic mice, CaMKIIα protein is produced normally but fails to localize to dendrites; the consequence is impaired memory, i.e. a very distinctive behavioural phenotype!

MicroRNAs are small non-coding RNAs that have the ability to regulate gene expression. They bind complementary target sequences in target mRNAs and will be described in detail in Chapter 16. Given the fact that mRNAs are localized in axons, might it be possible that microRNAs that target localized mRNAs are also localized in the same cellular compartment? This is indeed the case. FISH shows that miR181a-1* and miR-532 localize in distal axons and growth cones localize in distinct RNP granules. The implications are that, as well as in axons, microRNAs might be similarly localized in other cellular compartments together with their target mRNAs.

10.4.3 Perinuclear localization of mRNA

A special case of mRNA localization is the concentration of mRNAs around the nucleus (perinuclear localization), the rationale being to achieve a more efficient delivery of proteins such as transcription factors into the nucleus.

The proto-oncogene c-*myc* encodes a transcription factor associated with the regulation of genes involved in cellular proliferation. Its mRNA contains a localization signal (zipcode) within its 3′ UTR that directs a reporter sequence to the perinuclear cytoplasm and, in particular, to cytoskeletal-bound polysomes. A critical feature of the c-*myc* mRNA localization signal is an AUUUA sequence. As we shall see in Chapter 12, such AU-rich sequences are associated with mRNA instability; but this particular zipcode does not appear to influence mRNA stability.

In a similar vein, c-*fos* encodes a transcription factor involved in the regulation of genes associated with development, in particular that of the skeletal system. c-*fos* mRNA is enriched in polyribosomes (multiple ribosomes in the process of translating a given mRNA) that are associated with the cytoskeleton in the perinuclear compartment. This association is also dependent on sequence elements in the 3′ UTR. Consistent with its role in mRNA localization, sequences in the 3′ UTR of c-*fos* mRNA are highly conserved in evolution.

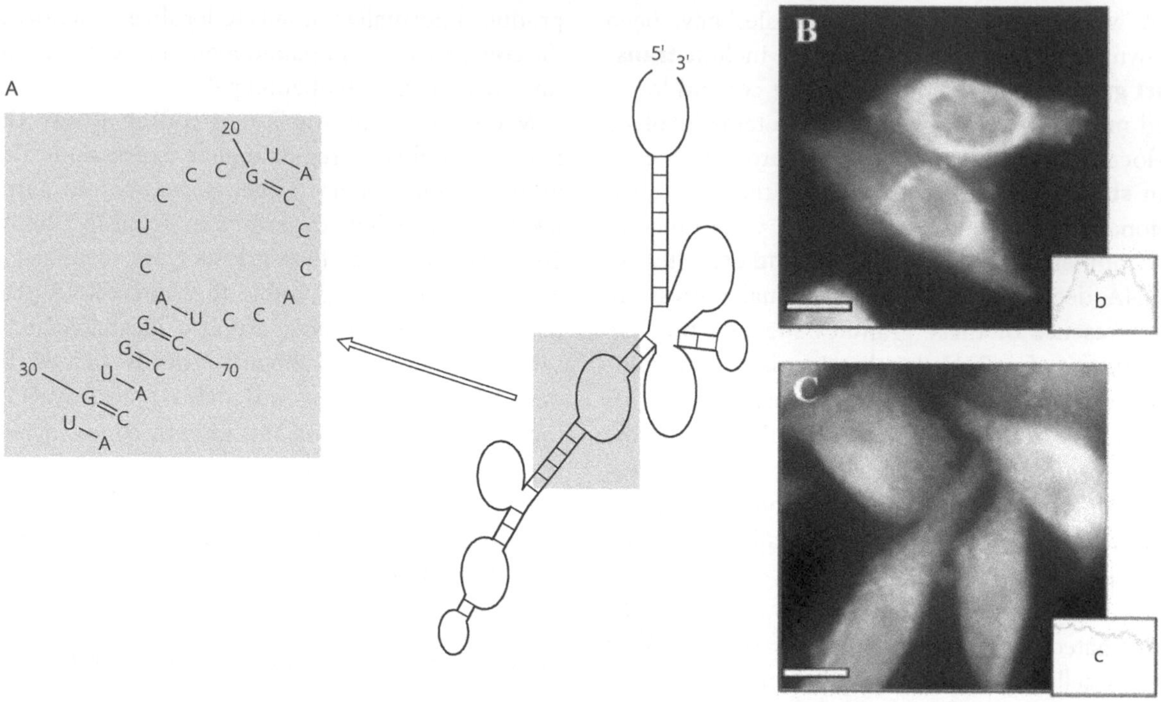

Figure 10.7 Perinuclear localization of *MT1* mRNA. Chinese hamster ovary cells were transfected with *MT1* messenger RNA constructs detected by in situ hybridization. The presence of a complete 3′ UTR (untranslated sequence after the stop codon) results in strong perinuclear localization (B), unless the perinuclear localization sequence shown in (A) is deleted (C). The extent of perinuclear localization is also illustrated in graph format (panels b and c). Scale bars = 10μm. (Reprinted from Nury et al.[34] ©The Biochemical Society, London.)

The retinoic-acid-binding protein CRABPI influences the metabolism and intracellular trafficking of retinoic acid. Retinoic acid is a metabolite of vitamin A that is involved in normal development and binds to specific transcription factors. Both CRABPI protein and its mRNA are localized to the perinuclear compartment and the localization is, once again, dependent on a zipcode sequence in the 3′ UTR.

Vimentin is a crucial component of intermediate filaments, whose synthesis is thought to initiate around the cell nucleus. For that reason, vimentin protein is required in the perinuclear compartment, and its mRNA indeed localizes there. Once again its targeting is dependent on a sequence in the 3′ UTR. In an elegant experiment, a coding sequence for GFP was fused to a chimeric 3′ UTR containing several sequences derived from the vimentin 3′ UTR. These sequences were sufficient to direct the GFP signal to the perinuclear compartment.

MT1 (*metallothionein 1*) mRNA encodes a small metal-binding protein involved in the sequestration of heavy metals. Although MT1 protein is mostly cytoplasmic, it can also be detected in the nucleus. A 10-base stretch of nucleotides in the 3′ UTR that contains a CACC repeat is required for targeting of *MT1* mRNA to the perinuclear compartment. The CACC repeat is involved in the formation of a secondary structure in the mRNA that is recognized by a hitherto unknown RNA-binding protein. The perinuclear localization of *MT1* mRNA is illustrated in Fig. 10.7.

10.4.4 Localization of mRNA in cellular organelles

Many proteins are destined for the ER (endoplasmic reticulum) because they are eventually secreted, or because they are integral membrane proteins. As we shall see in Section 11.4.3 a translation-dependent mechanism has evolved that directs nascent proteins to the ER; it makes use of a short polypeptide signal present in the N-terminus and the

signal recognition particle (SRP). However, there is strong evidence of an alternative translation-independent mechanism that is based on mRNA localization. The localization of mRNAs to the ER, once again, relies on the presence of sequence motifs in the mRNAs that are generally uracil-rich and adenine-poor.

ER localization of mRNAs has been described in yeast, plants, insects, and mammals, but it is best understood in *S. cerevisiae*. Examples of mRNAs that are localized to the ER in *S. cerevisiae* include *USE1*, encoding a tail-anchored membrane protein, and *SUC2*, encoding a soluble secreted enzyme. Their ER localization is independent of translation as they still localize to the ER when translation is blocked by removing the start codon or by using the translation inhibitor cycloheximide. It is also independent of the SRP as they also localize correctly when mutants in which the SRP is inactivated are used. Localization is achieved through the interaction of dedicated RNA-binding proteins with the ER localization sequences. Thus the deletion of yeast genes that encode two RNA binding proteins, SHE2 and PUF2, significantly reduces the ER localization of *USE1* and *SUC2* mRNAs.

Evidence also suggests that mRNAs can be localized to other cellular organelles including mitochondria and chloroplasts. The process of mitochondrial mRNA localization has also been studied in *S. cerevisiae*, and it is similarly translation-independent. Mitochondrial localization requires sequence elements in the 3′ UTR (untranslated region) of the localized mRNAs, the mitochondrial RNA-binding protein PUF3, and proteins involved in the *translocase of the outer membrane* (TOM) import machinery. The latter is a complex required for the import of proteins encoded by nuclear genes into the mitochondria.

In chloroplasts, mRNA localization has been observed in the eukaryotic unicellular green alga *Chlamydomonas reinhardtii*. An example is the chloroplast protein psbA, a thylakoid membrane protein. FISH was used to show that *psbA* mRNA is localized at the T zones, which are punctate regions adjacent to the pyrenoid (microcompartments of chloroplasts involved in CO_2 fixation, found in many algae). When cells are treated with the antibiotic lincomycin (a protein synthesis inhibitor) *psbA* localization was maintained, again suggesting that its localization is translation-independent.

Key points

The localization of mRNA is not confined to early development and is widely used in adult somatic cells. Well-known examples include the localization of *β-actin* mRNA in fibroblasts and the localization of several mRNAs in axons, synapses, and dendrites in neurons. A special case of mRNA localization occurs around the nucleus (perinuclear localization) in order to facilitate the accumulation of nuclear proteins in the nucleus, on the endoplasmic reticulum in order to facilitate the translation of mRNAs that encode secreted or membrane-associated proteins, and on mitochondria and chloroplasts.

10.5 Localization of mRNA in algae and plants

The need to localize mRNA applies to all eukaryotes, ranging from unicellular yeast to humans. Therefore mRNA localization is also observed in photosynthetic organisms. As we saw in the previous section, there is evidence of mRNA localization in chloroplasts. In this section we will discuss other examples of localized mRNA in algae and vascular plants,[42–49] including an example of intercellular traffic of mRNA through plasmodesmata (channels that connect adjoining cells).

10.5.1 Localization of mRNA in the algae *Acetabularia acetabulum* and *Fucus serratus*

The unusual properties of the green alga *Acetabularia acetabulum* were known early in the twentieth century. In the 1930s scientists chose it as an ideal model in which to study morphogenesis. By grafting together parts of the plant they showed that an agent, which they called 'morphogenetic substance', migrated from the nucleus to the cytoplasm and stimulated morphogenesis. The ability of these substances to drive morphogenesis could last for weeks, even when the nucleus was removed. The substances clearly existed in gradients because of the different morphogenetic properties of different cellular fragments. In the early 1960s, it became apparent that these substances were in fact mRNAs.

A. acetabulum is a unicellular organism whose large size and unique shape make it ideal for morphogenetic studies. This single-celled organism can be observed as a 60mm stalk; at its base is a rhizoid which contains the nucleus, and at the apical pole a whorl of gametophores is located in a cap-like structure. Clearly, this giant cell should provide another excellent context in which mRNA localization can be observed, and this is indeed the case.

The processes that underpin localization of mRNA in *A. acetabulum* are not dissimilar from what is observed in animal cells, so that specific mRNAs are localized in specific cellular compartments. Thus mRNA encoding Ran-G, a conserved GTPase involved in nucleocytoplasmic transport, is localized to the basal region in the rhizoid, i.e. close to the nucleus where it is required. Distributions of mRNA in *A. acetabulum* also change during development; thus *MAP kinase* mRNA is concentrated in the basal region in immature cells but accumulates in the cap region in mature cells. Several isoforms of calmodulin are expressed in *A. acetabulum*—calmodulin-2 and calmodulin-4 are the most evolutionarily divergent, with differences in amino acid sequences affecting their ability to bind calcium and interacting proteins. *Calmodulin-4* mRNA is concentrated in the apical region and *calmodulin-2* mRNA in the basal region; especially in more mature cells.

Similarly to animal cells, actin RNAs are localized in the brown alga *Fucus serratus*. In early zygotes of *F. serratus*, actin mRNA is evenly distributed, but it then becomes more concentrated in the thallus pole end until the end of the first cell division, when it redistributes to the cell plate. Within the space of 22 hours, actin mRNA is observed in three distinct distributions: evenly distributed (first 8 hours), localized to the thallus pole (at 12 hours), and finally localized to the cell plate (22 hours). Interestingly, the fertilized *F. serratus* egg is not polarized initially, but becomes irreversibly polarized 10 hours after fertilization in response to unilateral light.

10.5.2 Localization and traffic of mRNA in vascular plants

In vascular plants roots exhibit polarized tip growth that is mediated by specialized root epidermis cells (trichoblasts). Specific proteins and mRNAs accumulate there, including expansins which are associated with cell loosening. Root tips are also rich in filamentous actin (F-actin). Treatment of trichoblasts with latrunculin B, which depletes F-actin, stops the formation of root hairs. This suggests that actin-mediated mRNA localization occurs in trichoblasts, and more generally that microfilaments are also involved in mRNA localization in plants. The redistribution of microfilaments and *profilin* mRNA is shown in Fig. 10.8. (Profilin is a protein that promotes actin polymerization.)

Plants are unique in their extensive use of cell-to-cell transport of macromolecules via

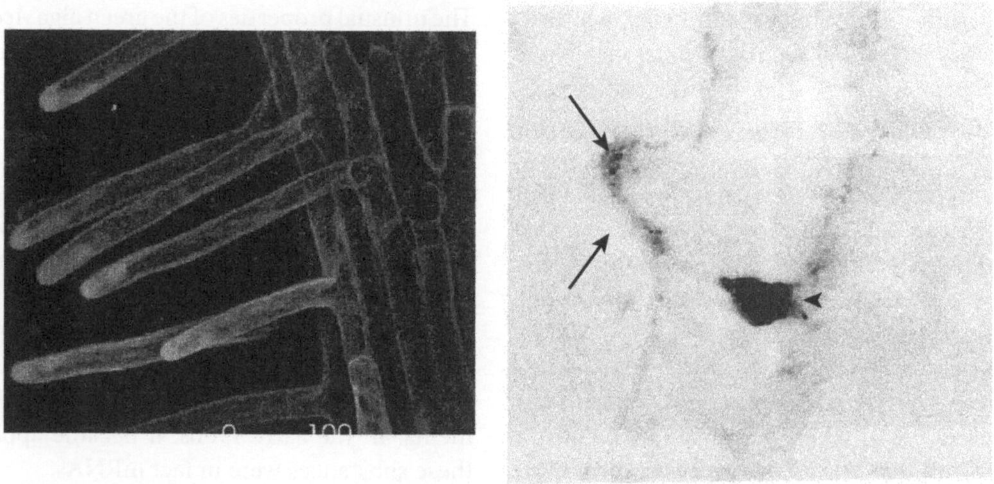

Figure 10.8 Redistribution of microfilaments and *profilin* mRNA in root hair formation. Actin filaments concentrate in the outgrowing bulge of the root hair (left panel). *Profilin* mRNA also becomes localized during root hair formation. (Reprinted from Baluska et al.[42] with permission from Elsevier.)

plasmodesmata (channels that connect the cytoplasm of adjoining cells) to the extent that plants have been defined as *supracellular organisms*. This suggests the potential for traffic of RNA from cell to cell—and this is indeed the case.

The *KNOX* genes are a class of homeobox gene involved in leaf development. KN1 is known to traffic, as a protein, from internal tissues of the leaf to the epidermis. It is involved in the initiation and maintenance of the meristem (undifferentiated plant tissue from which new cells are formed). Experiments show that KN1 tagged with GFP (green fluorescent protein) can traffic between cells across plasmodesmata. However, KN1 also mediates the trafficking of its own mRNA. When labelled sense (but not antisense) strand *KN1* mRNA is co-injected into mesophyll cells with KN1 protein, the mRNA moves rapidly from cell to cell. The effect is specific, because KN1 does not promote the traffic of the sense strand of cucumber mosaic virus (CMV) RNA. Instead, the intercellular movement of CMV RNA is facilitated by a virally encoded movement protein (plant viruses are known to traffic through plasmodesmata in order to infect adjoining cells). Thus the KN1 protein augments its intercellular traffic by facilitating the movement of its own message! Interestingly, the domain responsible for intercellular trafficking is the homeodomain itself—traditionally a DNA-binding domain, which in this context has acquired novel functions.

As well as movement across plasmodesmata, longer-distance mRNA traffic in plants can occur through the phloem. An example is the *CmNACP* mRNA from pumpkin (*Cucurbita maxima*). CmNACP is a member of a family of proteins involved in development of the apical meristem. The purpose of the phloem traffic is to deliver mRNA from the body of the plant to the shoot apex. This suggests that plants can direct genetic information from tissue to tissue even at long distances.

Key points

Localization of mRNA occurs in plants, ranging from algae to vascular plants. Localization can occur during early development or during mature plant growth. mRNA traffic can occur between cells (through plasmodesmata) and even tissues (through phloem).

Summary

- Intracellular localization of mRNA is a widespread phenomenon in nature. The rationale is simple and elegant: localize the genetic information to wherever it is required in the cell, so that proteins can be synthesized where they are spatially required.
- Localization of mRNA involves the recognition of particular sequences, usually in the 3′ UTR and generally called zipcodes, by RNA-binding proteins. Movement of mRNA to its destination generally requires the cytoskeleton and its molecular motors.
- There are strong indications that the fate of an mRNA, including its localization, is predetermined in the cell nucleus, where specific RNA-binding proteins, including the exon junction complex, are deposited on mRNAs.
- Localization of mRNA occurs extensively in early development, notably during oogenesis.
- Adult somatic cells also rely on mRNA localization, particularly neurons given their very specialized cellular architecture.
- MicroRNAs, small non-coding RNAs involved in the regulation of mRNA translation, can also be localized.
- Faulty mRNA localization can result in disease.
- Localization to specific subcellular compartments or organelles can also occur; this includes perinuclear, endoplasmic reticulum, mitochondrial, and chloroplast localization.
- mRNA localization is also extensive in plants, and there is even traffic of mRNAs between cells and tissues via plasmodesmata and the phloem.

Questions

10.1 How was mRNA localization discovered?

10.2 Discuss why mRNA localization might be needed.

10.3 How widespread is mRNA localization?

10.4 What techniques can be used to study mRNA localization?

10.5 Describe an example of mRNA localization in early development.

10.6 Describe an example of mRNA localization in a somatic cell.
10.7 How is mRNA localized in neurons?
10.8 What is the role of the exon junction complex in mRNA localization?
10.9 How are mRNA localization and mRNA translation coordinated?
10.10 Find an example of an RNA 'zipcode'. Which RNA-binding proteins are involved?
10.11 Discuss an example of aberrant mRNA localization in disease.

References

1. **Lécuyer E, Yoshida H, Parthasarathy N, et al.** Global analysis of mRNA localization reveals a prominent role in organising cellular architecture and function. *Cell* **131**, 174–187 (2007).
2. **Lécuyer E, Yoshida H, Krause HM.** Global implications of mRNA localization pathways in cellular organization. *Curr Opin Cell Biol* **21**, 409–15 (2009).
3. **Medioni C, Mowry K, Besser F.** Principles and roles of mRNA localization in animal development. *Development* **139**, 3263–76 (2012).
4. **Chabanon H, Mickleburgh I, Hesketh J.** Zipcodes and postage stamps: mRNA localisation signals and their trans-acting binding proteins. *Brief Funct Genomic Proteomic* **3**, 240–56 (2004).
5. **Martin KC, Ephrussi A.** mRNA localization: gene expression in the spatial dimension. *Cell* **136**, 719–30 (2009).
6. **Santangelo P, Lifland AW, Curt P, et al.** Single molecule-sensitive probes for imaging RNA in live cells. *Nat Methods* **6**, 347–51 (2009).
7. **Bullock SL.** Translocation of mRNAs by molecular motors: think complex? *Semin Cell Dev Biol* **18**, 194–201 (2007).
8. **Bullock SL.** Messengers, motors and mysteries: sorting of eukaryotic mRNAs by cytoskeletal transport. *Biochem Soc Trans* **39**, 1161–5 (2011).
9. **Gonsalvez GB, Urbinati CR, Long RM.** RNA localization in yeast: moving towards a mechanism. *Biol Cell* **97**, 75–86 (2005).
10. **Forrest KM, Gavis ER.** Live imaging of endogenous RNA reveals a diffusion and entrapment mechanism for nanos mRNA localization in *Drosophila*. *Curr Biol* **13**, 1159–68 (2003).
11. **Lopez de Heredia M, Jansen R-P.** mRNA localization and the cytoskeleton. *Curr Opin Cell Biol* **16**, 80–5 (2004).
12. **St Johnston D.** Moving messages: the intracellular localization of mRNAs. *Nat Rev Mol Cell Biol* **6**, 363–74 (2005).
13. **Kanai Y, Dohmae N, Hirakawa N.** Kinesin transports RNA: isolation and characterisation of an RNA-transporting granule. *Neuron* **43**, 513–25 (2004).
14. **Giorgi C, Moore MJ.** The nuclear nurture and cytoplasmic nature of localized mRNPs. *Semin Cell Dev Biol* **18**, 186–93 (2007).
15. **Kress TL, Yoon YJ, Mowry KL.** Nuclear RNP complex assembly initiates cytoplasmic mRNA localization. *J Cell Biol* **165**, 203–11 (2004).
16. **Lewis R, Mowry KL.** Ribonucleoprotein remodelling during RNA localization. *Differentiation* **75**, 507–18 (2007).
17. **Van de Bor V, Davis I.** mRNA localisation gets more complex. *Curr Opin Cell Biol* **16**, 300–7 (2004).
18. **Palacios I.** How does an mRNA find its way? Intracellular localisation of transcripts. *Semin Cell Dev Biol* **18**, 163–70 (2007).
19. **Yoon YJ, Mowry KL.** *Xenopus* staufen is a component of a ribonucleoprotein complex containing Vg1 RNA and kinesin. *Development* **131**, 3035–45 (2004).
20. **King ML, Messitt TJ, Mowry KL.** Putting RNAs in the right place at the right time: RNA localisation in the frog oocyte. *Biol Cell* **97**, 19–33 (2005).
21. **Lin MD, Fan SJ, Hsu WS, Chou TB.** *Drosophila* decapping protein, dDcp1, is a component of the *oskar* mRNP complex and directs its posterior localization in the oocyte. *Dev Cell* **10**, 601–13 (2006).
22. **St Johnston D.** The beginning of the end. *EMBO J* **20**, 6169–79 (2001).
23. **Bassell GJ, Kelic S.** Binding proteins for mRNA localization and local translation, and their dysfunction in genetic neurological disease. *Curr Opin Neurobiol* **14**, 574–81 (2004).
24. **Lapidus K, Wyckoff J, Mouneimne G, et al.** Systems analysis of RNA trafficking in neural cells. *J Cell Sci* **120**, 3173–8 (2007).
25. **Condeelis J, Singer RH.** How and why does β-actin mRNA target? *Biol Cell* **97**, 97–110 (2005).
26. **Dahm R, Kiebler M, Macchi P.** RNA localisation in the nervous system. *Semin Cell Dev Biol* **18**, 216–23 (2007).
27. **Dahm R, Macchi P.** Human pathologies associated with defective RNA transport and localization in the nervous system. *Biol Cell* **99**, 649–61 (2007).
28. **Kiebler MA, Bassell GJ.** Neuronal RNA granules: movers and makers. *Neuron* **51**, 1–6 (2006).
29. **Yasuda K, Zhang H, Loiselle D, et al.** The RNA-binding protein Fus directs translation of localized mRNAs in APC-RNP granules. *J Cell Biol* **203**, 737–46 (2013).
30. **Jakobsen KR, Sørensen E, Brøndum KK, et al.** Direct RNA sequencing mediated identification of mRNA localized in protrusions of human MDA-MB-231 metastatic breast cancer cells. *J Mol Signal* **8**, 9–23 (2013).
31. **Sasaki Y, Gross C, Zing L, et al.** Identification of axon-enriched microRNAs localized to growth cones of cortical neurons. *Dev Neurobiol*, **74**, 397–406 (2014).

32. **Bermano G, Shepherd RK, Zehner ZE, Hesketh JE.** Perinuclear mRNA localisation by vimentin 3′-untranslated region requires a 100 nucleotide sequence and intermediate filaments. *FEBS Lett* **497**, 77–81 (2001).
33. **Veyrune J-L, Campbell GP, Wiseman J, et al.** A localisation signal in the 3′ untranslated region of c-myc mRNA targets c-myc mRNA and β-globin reporter sequences to the perinuclear cytoplasm and cytoskeletal-bound polysomes. *J Cell Sci* **109**, 1185–94 (1996).
34. **Nury D, Chabanon H, Levadoux-Martin M, Hesketh J.** An eleven nucleotide section of the 3′-untranslated region is required for perinuclear localization of rat metallothionein-1 mRNA. *Biochem J* **387**, 419–28 (2005).
35. **Levadoux-Martin M, Li Y, Blackburn A, et al.** Perinuclear localisation of cellular retinoic acid binding protein I mRNA. *Biochem Biophys Res Commun* **340**, 326–31 (2006).
36. **Lerner RS, Seizer RM, Zheng T, et al.** Partitioning and translation of mRNAs encoding soluble proteins on membrane-bound ribosomes. *RNA* **9**, 1123–37 (2003).
37. **Jagannathan S, Nwosu C, Nicchitta, CV.** Analyzing subcellular mRNA localization via cell fractionation. *Methods Mol Biol* **714**, 301–21 (2011).
38. **Kraut-Cohen J, Gerst JE.** Addressing mRNAs to the ER: *cis* sequences act up! *Trends Biochem Sci* **35**, 459–69 (2009).
39. **Kraut-Cohen J, Afanasieva E, Haim-Vilmovsky L, et al.** Translation- and SRP-independent mRNA targeting to the endoplasmic reticulum in the yeast *Saccharomyces cerevisiae*. *Mol Biol Cell* **24**, 3069–84 (2013).
40. **Gadir N, Haim-Vilmovsky L, Kraut-Cohen J, Gerst JE.** Localization of mRNAs coding for mitochondrial proteins in the yeast *Saccharomyces cerevisiae*. *RNA* **17**, 1551–65 (2011).
41. **Uniacke J, Zerges W.** Chloroplast protein targeting involved localized translation in *Chlamydomonas*. *Proc Natl Acad Sci USA* **106**, 1439–44 (2009).
42. **Baluska F, Salai J, Mathur J, et al.** Root hair formation: F-actin-dependent tip growth is initiated by local assembly of profilin-supported F-actin meshworks accumulated within expansin-enriched bulges. *Dev Biol* **227**, 618–32 (2000).
43. **Bouget F-Y, Gerttula S, Shaw SL, Quatrano RS.** Localisation of actin mRNA during the establishment of cell polarity and early cell division in *Fucus* embryos. *Plant Cell* **8**, 189–201 (1996).
44. **Kim JY, Yuan Z, Cilia M, et al.** Intercellular trafficking of a KNOTTED1 green fluorescent protein fusion in the leaf and shoot meristem of *Arabidopsis*. *Proc Natl Acad Sci USA* **99**, 4103–8 (2002).
45. **Lucas WJ, Bouché-Pillon S, Jackson DP et al.** Selective trafficking of KNOTTED1 homeodomain protein and its mRNA through plasmodesmata. *Science* **270**, 1980–3 (1995).
46. **Kim J-Y, Rim Y, Wang J, Jackson DP.** A novel cell-to-cell trafficking assay indicates that the KNOX homeodomain is necessary and sufficient for intercellular protein and mRNA trafficking. *Genes Dev* **19**, 788–93 (2005).
47. **Okita TW, Choi SB.** mRNA localization in plants: targeting to the cell's cortical region and beyond. *Curr Opin Plant Biol* **5**, 553–9 (2002).
48. **Ruiz-Medrano R, Xoconostle-Cazáres B, Lucas WJ.** Phloem long-distance transport of CmNACP mRNA: implications for supracellular regulation in plants. *Development* **126**, 4405–19 (1999).
49. **Vogel H, Grieninger GE, Zetsche KH.** Differential messenger RNA gradients in the unicellular alga *Acetabularia acetabulum*. Role of the cytoskeleton. *Plant Physiol* **129**, 1407–16 (2002).

11 Translation of messenger RNA

Introduction

In this chapter we cover the remarkable process of mRNA translation. We begin by reviewing the structure and function of the essential machinery of translation, namely the ribosome and transfer RNA. We then outline the three phases of translation: initiation, elongation, and termination. Finally, we discuss several ways in which mRNA translation can be regulated.

11.1 What is translation?

Not to be confused with transcription, translation is the process whereby **ribosomes** read the genetic information encoded by **messenger RNAs** (mRNAs). Box 11.1 describes the discovery of mRNA in the middle of the twentieth century. Ribosomes catalyse the synthesis of polypeptide chains that form when amino acids are covalently linked through peptide bonds. As well as mRNA, translation requires non-coding RNAs, specifically ribosomal RNA (**rRNA**) and transfer RNA (**tRNA**). Transfer RNAis the molecular adaptor which associates a **codon** in the mRNA with a specific amino acid.

Translation consists of three phases: initiation, elongation, and termination. Several proteins are also involved in translation: ribosomal proteins, which together with rRNA constitute the ribosome, and several additional translation factors. The latter include the translation initiation factors (IFs in prokaryotes and eIFs in eukaryotes), the translation elongation factors (EFs and eEFs), and release (or termination) factors (RFs and eRFs).

11.2 The structure and function of the ribosome

11.2.1 Ribosome structure

The starting point in our overview of translation is the ribosome.[1–10] We now understand both prokaryotic and eukaryotic ribosome structure in exquisite detail. Ribosomes are large ribonucleoprotein complexes about 20nm in diameter. They are typically composed of ~65% ribosomal RNA (rRNA) and ~35% ribosomal proteins (the exact proportions vary from species to species). Ribosomes are made up of large and small subunits. Both subunits contain ribosomal RNA (rRNA) and a multitude of ribosomal proteins, denoted L1, L2, . . . (present in the large ribosomal subunit) and S1, S2, . . . (present in the small ribosomal subunit).

Ribosomal RNA is very complex, and extensive base-pairing networks and modified RNA bases add to its functional versatility. In Chapter 14 we will discuss the complex biosynthesis and processing of ribosomal RNA. The basic three-dimensional structure of the eukaryotic ribosome is shown in Fig. 11.1.

The different components of the ribosomal subunits are distinguished according to the rate at which they sediment, as measured in Svedberg units (S). In prokaryotes, the large ribosomal subunit contains 23S rRNA and 5S rRNA; and the small subunit

Box 11.1 The discovery of messenger RNA

Three scientists, Jacques Monod, André Lwoff, and François Jacob, shared the 1965 Nobel Prize in Physiology or Medicine for their elucidation of the nature of messenger RNA. Previously it had become clear that DNA is the repository of genetic information and that protein synthesis occurs in the cytoplasm, but how might nuclear DNA direct the synthesis of specific polypeptides?

Nobel laureates are rightly rewarded for their groundbreaking achievements, but it is important to note that several other scientists made vital contributions. Mahlon Hoagland and Paul Zimmerick of Harvard University had shown that the carrier of amino acids was a type of RNA that they called 'transfer' RNA. Elliot Volkin and Lazarus Astrachan, at the Oak Ridge National Laboratory, also discovered that, as a result of bacteriophage infection (a bacteriophage is a type of virus that attacks bacteria), a kind of RNA was formed that was similar in sequence to the bacteriophage's DNA. They found that the bacteriophage instructs the bacterial cells to synthesize its proteins by making a short-lived RNA that is a copy of its DNA. They called it 'DNA-like RNA' in 1956, three years after Watson and Crick had elucidated the double-stranded structure of DNA.

In the late 1950s, François Jacob, Jacques Monod, and Arthur Pardee performed a series of experiments that looked at the synthesis of specific proteins by a set of bacterial genes that are activated in the presence of lactose. They obtained evidence that led them to suggest that an RNA is produced as an exact copy of the DNA in a bacterial gene or 'operon'. In the famous experiment known as 'PaJaMo' (the first two letters of each of their surnames), they performed a series of conjugations (matings) between *Escherichia coli* bacteria. They generated a strain that had an inactive mutated *lac* operon (encoding the machinery of β-galactosidase synthesis). Wild-type bacteria could restore *lac* operon expression in the mutant strain almost immediately after mating, thanks to the action of the RNA copy of the relevant genes. They called this type of RNA *messenger RNA*.

Jacob and Monod were also famous for their pioneering work in the study of gene regulation. The addition of an 'inducer' of the *lac* operon caused an increase in its transcription. This was shown with pulse-labelling experiments (radiolabelling of newly synthesized mRNA). *lac* operon mRNA synthesis reached a maximum a mere two minutes after the addition of the inducer. The inducer they used was IPTG (isopropyl-β-d-thiogalactopyranoside), a stable derivative of galactose. IPTG is still routinely used in the laboratory to induce gene expression in genetically modified bacteria.

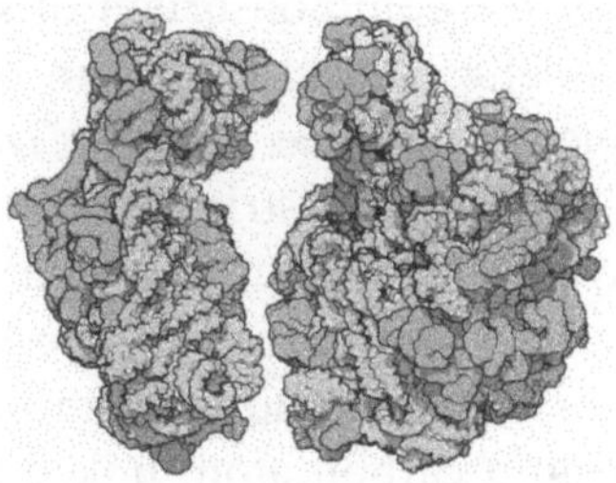

Figure 11.1 Ribosome structure. Structure of a eukaryotic ribosome. The small subunit is on the left, and the large subunit is on the right. The process of translation takes place in the intersubunit space. The small subunit is responsible for ensuring that each codon pairs with the anticodon in the correct tRNA (the decoding centre). The large subunit comprises the active site in which peptide bonds are created (the peptidyl transferase centre). Note the significant extent to which rRNA makes up the ribosomal subunit structure (lighter grey); numerous ribosomal proteins also contribute to the structure, helping the rRNA to adopt the correct structure (darker grey). Ribosomal RNA is responsible for the catalysis of chemical reactions, and is thus working as a ribozyme. (Reproduced courtesy of David S. Goodsell and the RCSB PDB under a CC-BY-3.0 licence.)

contains 16S rRNA. In eukaryotes, the large subunit contains 28S, 5.8S, and 5S rRNA, and the small subunit contains 18S rRNA. In prokaryotes, the small ribosomal subunit sediments at 30S, the large subunit at 50S, and the fully formed ribosome at 70S. In eukaryotes, the small subunit sediments at 40S, the large at 60S, and the fully formed ribosome at 80S. (Notice that S values are not additive; for example, the eukaryotic ribosome, 80S, is not the sum of 40S and 60S.)

The structure and catalytic activities of the ribosome involve complex interactions between ribosomal RNA and ribosomal proteins. The small ribosomal subunit contains the **decoding centre** required for reading the mRNA's genetic information, and the large subunit contains the **peptidyl transferase centre** where amino acids are added to the nascent polypeptide. In the next section we explore how catalysis is achieved through ribosomal RNA.

11.2.2 Catalysis in the ribosome is RNA based

The active site of catalysis in which peptide bond formation occurs has been mapped in ribosomes which have been stalled by antibiotics, and found to be composed entirely of RNA. No protein is found closer than 18Å to the reaction centre of the ribosome (see Fig. 11.2a). Elucidation of the mechanism of action of the ribosome led to the award of the 2009 Nobel Prize in Chemistry to Thomas Steitz (Yale University, USA), Venkatraman Ramakrishnan (MRC Laboratory of Molecular Biology, Cambridge, UK), and Ada Yonath (Weissman Institute of Science, Israel).

The chemistry of peptide bond formation is shown in Fig. 11.2b and takes place in the active site of the ribosome. This active site is called the peptidyl transferase centre. Two pockets in the ribosome active site called the A (aminoacyl) site and the P (peptidyl) site hold two tRNAs close together. The incoming tRNA in the A site is primed with a new amino acid to be added to the growing protein chain, while the P site contains a tRNA attached to the growing polypeptide chain.

To make a new peptide bond, the amine group of the incoming amino acid (notice the unpaired electrons in Fig. 11.2b) chemically attacks the carbon at the end of the growing polypeptide chain. (This attacked carbon is joined by an ester bond to the tRNA in the P site.) This chemical attack results in the incoming amino acid on the A-site tRNA becoming attached to the growing polypeptide, and at the same time releases the P-site tRNA.

RNA plays a critical role in peptide bond formation. The active site juxtaposes the tRNA molecules to enable peptide bond formation. The tRNAs in the P and A sites of the ribosome exactly position the reacting groups during peptide bond formation, but also have another important catalytic role. Notice in Fig. 11.2b that the 2′ –OH of the tRNA in the P site of the ribosome is very close to the amino acid chemical groups which react to form the new peptide bond. The P-site tRNA 2′ –OH group plays a central role in catalysis by donating a hydrogen atom to the released 3′ oxygen of the P-site tRNA. Addition of this 3′ hydrogen atom enables release of the negatively charged 3′ oxygen atom after peptide bond formation.

A number of hydrogen atoms move around the transition state during peptide bond formation; movement of these hydrogen atoms is shown in detail in Fig. 11.3. Notice that while the P-site tRNA 2′ –OH donates a hydrogen atom to stabilize the leaving 3′ oxygen of the P-site tRNA, a hydrogen of the amine (NH_2) group of the newly added amino acid moves onto the P-site 2′ oxygen at the same time. This movement results in the formation of an amide group (–NH) on the new amino acid residue added to the growing peptide chain, and restores the 2′ –OH group on the P-site tRNA.

This movement of hydrogen atoms to catalyse peptide bond formation is called a *proton shuttle*. It occurs in the active site of the ribosome because, at physiological pH, it is much more likely that –OH groups will accept and exchange rather than simply release hydrogen atoms.

11.3 Deciphering the genetic code

11.3.1 Essential features of the genetic code

Before we consider how the ribosome mediates polypeptide synthesis, it is worth reviewing the fundamental nature of the genetic code.[11–16] The conservation of the genetic code is one of the strongest arguments for the common origin of species. The code is essentially a continuous run of three-nucleotide triplets, known as codons. Translation is the process of reading a string of continuous codons and synthesizing a polypeptide whose primary structure is determined by the sequence of codons. There are 20 amino acids in nature, characterized by an α carbon atom linked to a hydrogen, a basic amino group (NH_3^+), an acidic carboxy group (COO^-), and a side chain R. The nature of R defines the properties of the amino acid. The 20 amino acids are illustrated in Fig. 11.4. Note that almost all amino acids are encoded by more than one codon, as illustrated in Table 11.1. Most amino acids are encoded by four codons, in which the first two bases are invariant. (For example, glycine is encoded by GGG, GGA, GGC, and GGT; these are known as **synonymous codons**.) On the other hand, arginine, serine, and leucine are encoded by six codons. Methionine (AUG) and tryptophan

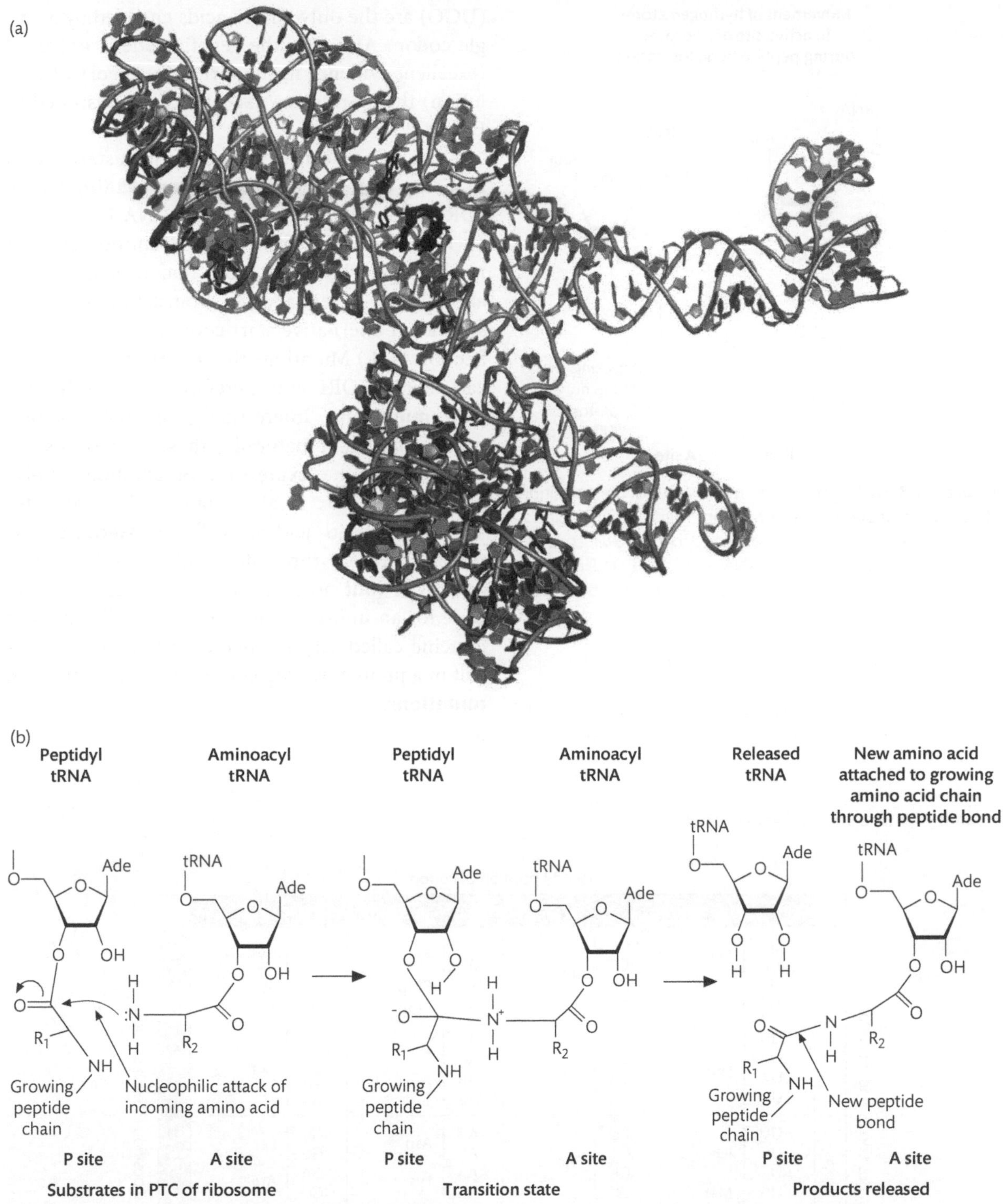

Figure 11.2 The ribosome catalyses peptide bond formation. (a) The crystal structure of the large subunit rRNA complexed with puromycin. This crystal structure shows that the active site of the ribosome has no protein side chains closer than about 18Å to the reaction centre, indicating that peptide bond formation must be catalysed by RNA. (Redrawn from the RCSB Protein Data Bank (ID RR0013) by Jonathan Crowe, Oxford University Press.) (b) The chemistry of peptide bond formation. PTC = premature termination codon. (Redrawn from Schmeing et al.[6] with permission from Elsevier.)

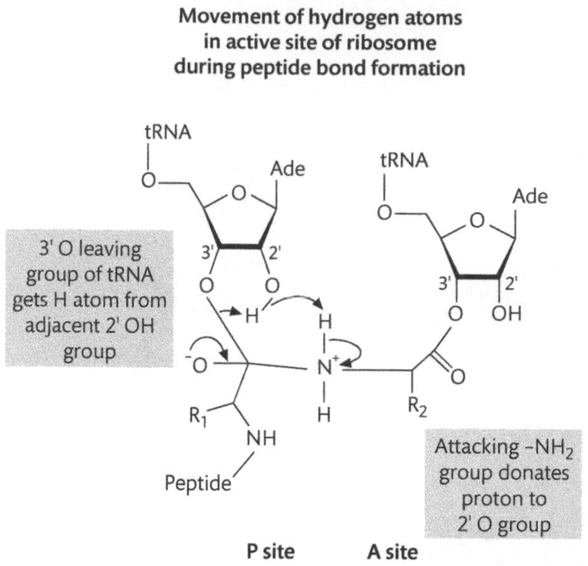

Figure 11.3 Hydrogen atoms move around the transition state during peptide bond formation. Note that the P site of the ribosome contains a tRNA connected to the growing peptide chain, and the A site contains a tRNA connected to the incoming amino acid. In the transition state of the reaction the ribose sugar 2′ -OH of the last nucleotide of the tRNA in the P site donates a hydrogen atom to its 3′ oxygen (this stabilizes the 3′ O^- leaving group after peptide bond formation). At the same time the amine group of the incoming amino acid donates a hydrogen atom to the 2′ O^- atom of the ribose sugar to regenerate the 2′ -OH group. (From Schmeing et al.[6] with permission from Elsevier.)

(UGG) are the only amino acids encoded by a single codon. Although the genetic code is universal (excellent evidence in support of the theory of evolution) there are some exceptions, as illustrated in Box 11.2.

A run of consecutive codons from a start codon to a stop codon is called an **open reading frame** (ORF). The first codon in any mRNA is normally an AUG, and therefore the first amino acid is a methionine. (However, note that in some cases, about 10% in eukaryotes, translation can also start from alternative start codons as described in Section 11.4.) Mutations that insert or delete bases can alter the ORF and therefore are called **frameshift mutations.** Interestingly, in some viruses, frameshifts occur naturally; these frameshifts are required for the expression of additional polypeptides (see Box 11.3). Mutations that only alter a specific amino acid are called **missense mutations.** There are three stop codons—UAA, UAG, and UGA (but note that in some cases UGA can code for an unusual amino acid, a modification of cysteine called selenocysteine). Mutations that result in a premature stop codon are called **nonsense mutations.**

Table 11.1 The universal genetic code

Second position of codon

First position (5′ end) of codon	U	C	A	G	Third position (3′ end) of codon
U	UUU, UUC: Phe UUA, UUG: Leu	UCU, UCC, UCA, UCG: Ser	UAU, UAC: Tyr UAA: **Stop** UAG: **Stop**	UGU, UGC: Cys UGA: **Stop** UGG: Trp	U C A G
C	CUU, CUC, CUA, CUG: Leu	CCU, CCC, CCA, CCG: Pro	CAU, CAC: His CAA, CAG: Gln	CGU, CGC, CGA, CGG: Arg	U C A G
A	AUU, AUC, AUA: Ile AUG: Met	ACU, ACC, ACA, ACG: Thr	AAU, AAC: Asn AAA, AAG: Lys	AGU, AGC: Ser AGA, AGG: Arg	U C A G
G	GUU, GUC, GUA, GUG: Val	GCU, GCC, GCA, GCG: Ala	GAU, GAC: Asp GAA, GAG: Glu	GGU, GGC, GGA, GGG: Gly	U C A G

Most amino acids are encoded by four codons. Methionine and tryptophan are generally encoded by one codon; whereas serine and arginine are encoded by six codons. There are three stop codons. Note that the stop codon UGA can also encode the rarer amino acid selenocysteine (Sec).

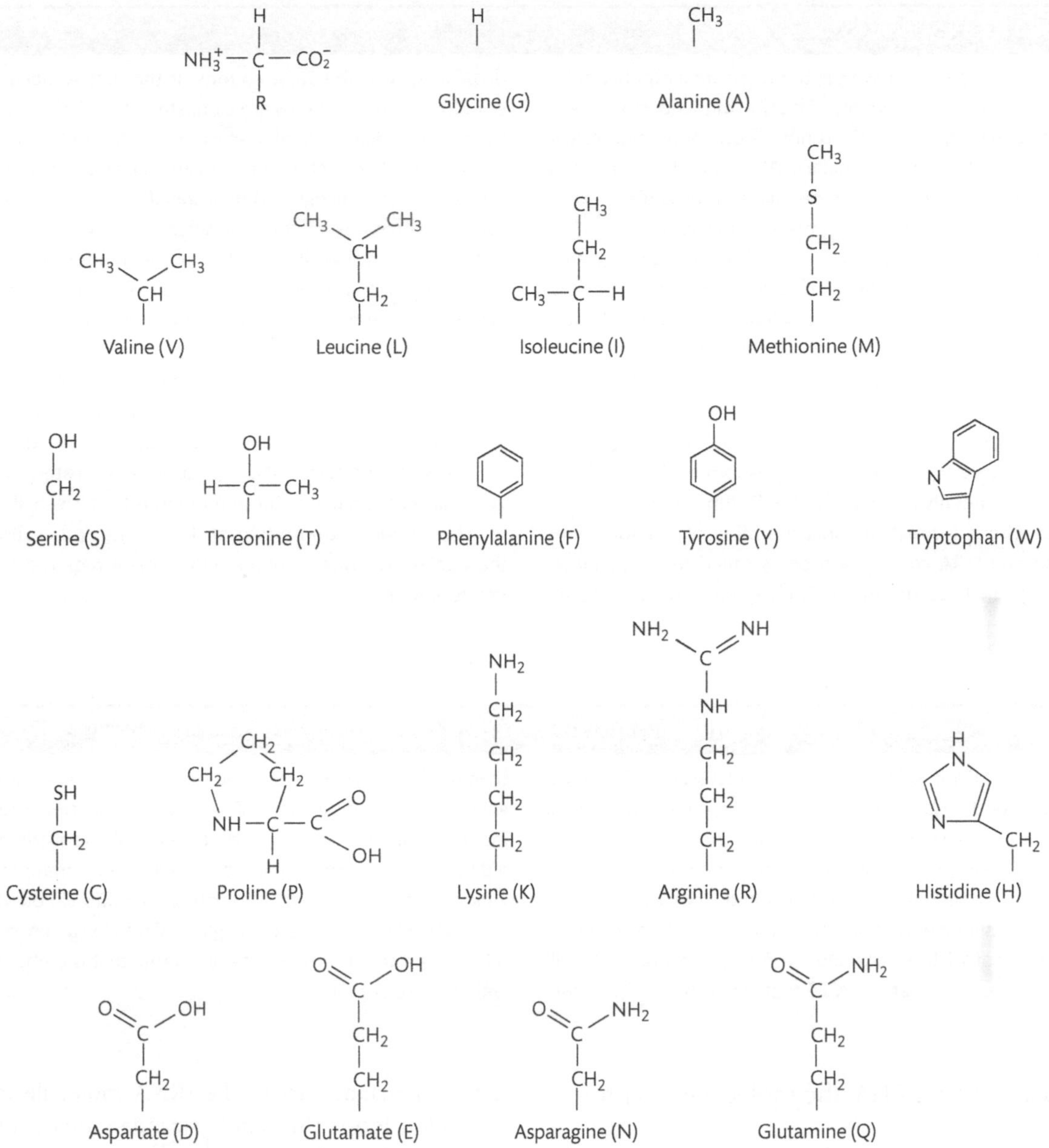

Figure 11.4 The 20 amino acids. Their chemical structure consists of the α carbon atom linked to a hydrogen, a basic amino group (NH_3^+), an acidic carboxy group (COO^-), and a side chain R. The chemical structures of R, the amino acid names, and the single-letter codes are shown. Glycine, alanine, and serine are *small* amino acids; alanine, valine, leucine, isoleucine, and methionine have *aliphatic* (hydrophobic) side chains; serine, threonine, and tyrosine have a free *hydroxyl group*; phenylalanine, tyrosine, and tryptophan have *aromatic* side chains; lysine, arginine, and histidine have a *basic* side chain; aspartate and glutamate have *acidic* side chains; asparagine and glutamine have *carboxamide* side chains. Cysteine contains *a sulfhydryl* (thiol) group that is characterized by the ability to form a covalent disulphide bond with another sulfhydryl group. Proline is an unusual amino acid; it also has an aliphatic side chain but it is bonded to both the α carbon and the nitrogen atom. Proline has the ability to create kinks in polypeptides.

Box 11.2 Exceptions to the genetic code

The genetic code is universal, but there are a number of exceptions in specific codons, notably in mitochondria. In vertebrate mitochondria UGA is not a stop codon, but instead it codes for tryptophan; similarly, AUA codes for methionine instead of isoleucine. In some ciliates (unicellular eukaryotes), UAG and UAA do not work as stop codons.

There is also the special case of the amino acid selenocysteine (abbreviated Sel or U), a modified version of cysteine, which is present in the selenoproteins. A selenium atom replaces the sulphur atom, changing the thiol group to a *selenol* group. Selenocysteine is encoded by UGA, normally a stop codon. However, UGA is not recognized as a stop codon in the presence of a SECIS element (selenocysteine insertion sequence). In eukaryotes the SECIS element is generally found in the 3′ UTR (untranslated region) from where it directs the incorporation of selenocysteine at several UGA codons (a process called *translational recoding*). In bacteria, the SECIS sequence is immediately downstream of the UGA codons. If the diet is lacking in selenium, selenoproteins are truncated at the UGAs, which then behave like normal stop codons. There is a specialized tRNA called tRNA(Sec) that can be charged with selenocysteine. It is recognized by a specific elongation factor (mSelB) which ensures that the selenocysteine is delivered to the ribosome's A site with the aid of an additional subunit called SBP-5 in eukaryotes. Thus SBP-5 binds to the SECIS RNA element and senses the need to incorporate selenocysteine.

Defects in the production of selenoproteins can be associated with disease. For example, mutations in the gene *SEPN1* are linked to early onset myopathy.[16] *SEPN1* encodes for the selenoprotein SelN, a glycoprotein located in the endoplasmic reticulum. These mutations have been shown to result in reduced insertion of selenocysteine because of the weakened structure of an RNA element required for its incorporation.

Box 11.3 Pseudoknots and viral translation

The pseudoknot RNA structures we discussed in Chapter 2 have been found to have important functions within viral mRNAs. Frameshifting sequences in viruses frequently contain a 'slippery' sequence (a sequence rich in U and A) in which the ribosome slips, and a pseudoknot which stimulates this slippage by preventing the mRNA from unwinding in front of the ribosome. Normally the ribosome will initiate translation at the AUG that starts the ORF, and will then read each triplet within the reading frame as an amino acid. However, some viral mRNAs contain additional reading frames if translation is started from a different triplet in a different frame. Ribosome slippages introduce frameshifts, which enable the ribosome to read extra ORFs encoded by the viral mRNA. As a result, a single viral mRNA can encode a different protein through a reading frame that is embedded within the major ORF.

11.3.2 Transfer RNA, the molecular adaptor

Transfer RNA (tRNA) is the molecule that allows the genetic code to be read. It acts as a molecular adaptor that associates an mRNA codon with a specific amino acid. Because of the existence of synonymous codons, amino acids can be recognized by more than one tRNA. Indeed, it is thought that there are up to 100 different tRNAs in eukaryotic cells.

Transfer RNAs are typically about 70 nucleotides long and fold into a series of four base-paired stems described as a 'cloverleaf' structure when drawn in two dimensions. In reality tRNA adopts an L-like shape in three-dimensional space. The structure of a typical tRNA molecule is shown in Fig. 11.5. The most important part of the tRNA molecule is the anticodon loop. This single-stranded area contains the **anticodon**, which hybridizes to a corresponding codon in the mRNA. The 5′ and 3′ ends of a tRNA molecule base pair together to form the acceptor stem—the site at which the amino acid is added. The 3′ end of the tRNA terminates in the residues CCA. The CCA terminus is 'charged' with the relevant amino acid. Each tRNA is charged with a specific amino acid by a family of enzymes called the **amino-acyl tRNA synthetases**. Their specificity ensures that tRNA is only charged with the appropriate amino acid.

Added complexity in tRNA structure arises from extensive base modifications, which change its

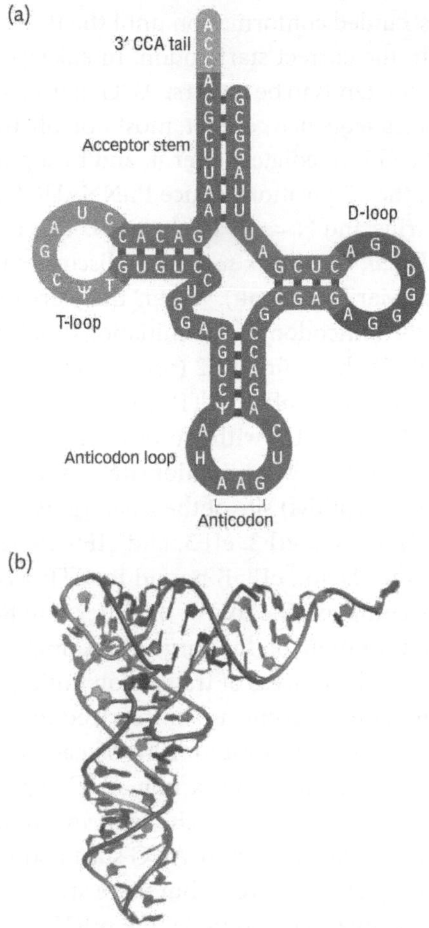

Figure 11.5 (a) A traditional 'cloverleaf' diagram of the folding pattern of a typical transfer RNA (tRNA). Abbreviations: A, adenosine; G, guanosine; C, cytidine; U, uridine; ψ, pseudouridine; D, dihydrouridine; H, hypermodified purine; T, ribothymidine. (b) Three-dimensional L-shaped structure of a typical tRNA. Note the single-stranded anticodon loop (which hybridizes to mRNA codons) and the 3′ acceptor end (charged with a specific amino acid). PDB code 1TRA (Reproduced from Craig et al.[17] with permission from Oxford University Press.)

nucleotides to generate pseudouridine, inosine, dihydrouridine, and ribothymidine. The biogenesis of tRNA will be described in more detail in Chapter 14. If the modified base is in the anticodon, base-pairing specificity can change. This results in the phenomenon called anticodon **wobble**. Generally, the first two bases in the anticodon–codon pair follow standard Watson–Crick base pairing. However, the third base of the codon affords a large degree of flexibility. For example, a U in the third base of an mRNA codon can be recognized by an A, G, or I (inosine) in the anticodon in a non-Watson–Crick base pairing. The consequence is that a given codon in mRNA can be recognized by more than one tRNA, ensuring that the process of translation is more efficient. Mutations that affect the third base in an mRNA codon are usually phenotypically neutral because they do not alter the resulting amino acid sequence.

Key points

Translation is the process whereby genetic information, transmitted from protein coding genes through mRNA, is read by the ribosome with the aid of the tRNA molecular adaptor leading to the synthesis of a polypeptide chain.

tRNA is charged with the appropriate amino acids by amino acid tRNA synthetases. The anticodon loop in tRNA is hybridized to mRNA codons. Flexibility in base pairing gives rise to the phenomenon known as wobble, in which given codons can be recognized by multiple tRNAs.

Although the genetic code is universal there are some notable exceptions in mitochondria; and the stop codon UGA can also code for the modified amino acid selenocysteine.

11.4 The three phases of translation

We now consider the three phases of translation: initiation, elongation, and termination.[2–4,6–8,10,18,19] Note that just as multiple RNA polymerases can work at the same time on a gene, an mRNA can be loaded with multiple ribosomes. mRNAs carrying multiple ribosomes are known as **polyribosomes** (also known as polysomes). Although the same principles apply in prokaryotes, this section mainly focuses on the process of translation in eukaryotes.

11.4.1 Translation initiation

In translation initiation, the aim is to recruit the small ribosomal subunit and an initiator tRNA charged with **N-formyl methionine**, the first amino acid in the nascent polypeptide, to the mRNA molecule, and to find the appropriate start codon in the mRNA (generally, but not always, an AUG). This process is guided by translation **initiation factors** (IFs in prokaryotes and eIFs in eukaryotes).

The first step in translation initiation is the formation of a ternary complex between eIF2, GTP, and

Met-tRNA$_i$ (the latter is the initiator tRNA charged with N-formyl methionine). Additional initiation factors (eIF1, eIF1A, and eIF3) then facilitate the association of the ternary complex with the 40S small ribosomal subunit to form the 43S pre-initiation complex. The eukaryotic small ribosomal subunit is recruited to the mRNA via the 7-methyl guanosine cap at the 5′ end of the mRNA. This relies on a complex of initiation factors, collectively called eIF4F, comprising three proteins, eIF4E + eIF4G + eIF4A. eIF4E is the cap-binding protein; it locks the eIF4F complex onto the cap. The mRNA associates with the 43S pre-initiation complex aided by the eIF3 complex, producing the 48S pre-initiation complex. Translation initiation does not only rely on eFI4F binding the 5′ end of an mRNA; poly(A) tails also stimulate translation by directing ribosomes that have just finished translating back to the 5′ end of the mRNA. How does this work? The eIF4G subunit of eIF4F interacts with the poly(A)-binding proteins which are bound at the 3′ end poly(A) tail of mRNA.

Once the 48S pre-initiation complex is formed, it is time to locate the appropriate start codon.A critical question is: How is the correct start codon found? In prokaryotes, there is a conserved seven-base sequence 13 bases upstream of the AUG start codon. This is the **Shine–Dalgarno sequence** (consensus TAAGGAG) which is complementary to the 3′ end of the prokaryotic small ribosomal subunit 16S rRNA. Its function is to bind the ribosome, thereby bringing it to the correct AUG start codon. In eukaryotes, once assembled at the 5′ end of mRNA, the small ribosomal subunit with its associated factors is thought to scan the 5′ UTR of the mRNA to find the first start codon. This process is known as cap-dependent translation as it relies on the presence of a cap at the 5′ end of mRNA in order to recruit the small ribosomal subunit. The process of initiation is aided by eIF4A (which is part of the eIF4F complex). eIF4A was the first RNA helicase to be discovered and characterized. The main role of eIF4A is to unwind stemloop structures in the 5′ UTR that otherwise impede translation initiation. Note that yet another initiation factor, eIF4B, is required to activate eIF4A, and that additional RNA helicases contribute to the unwinding of the 5′ UTR. The result of the unwinding process is that the mRNA can efficiently pass through the 40S mRNA binding channel in a single-stranded conformation until the P site associates with the correct start codon. In eukaryotes, the start codon tends to be the first AUG. It is further defined by its sequence context, most notably the presence of a G immediately after it, and by a purine (A or G) at the −3 position (hence PuNNAUGG, where Pu = purine and N = any nucleotide). This is known as the *Kozak consensus sequence* (discovered by the scientist Marilyn Kozak). When the start codon is found, the anticodon of the initiatior tRNA hybridizes to it. At this point eIF2 (remember, the ternary complex contains eIF2, GTP, and Met-tRNA$_i$) hydrolyses GTP to GDP with the aid of eIF5, a GTPase-activating protein. The initiatior tRNA is then lodged into the P (peptidyl) site of the small ribosomal subunit. eIF1, eIF1A, eIF2, eIF3, and eIF5 are released, and another factor, eIF5B bound to GTP, associates with the complex. The role of the latter is to facilitate the incorporation of the large ribosomal subunit, completing the process of translation initiation. This complex series of events is summarized in Fig. 11.6. Less frequently (in around 10% of cases) translation is also initiated from a non-AUG start codon: in eukaryotes CUG and UUG, and in prokaryotes GUG and AUU. These non-AUG start codons are a lot less frequent than AUG, but there are some interesting examples. For example, the mRNA encoding the growth factor FGF-2 can start translation from a typical AUG but also from three separate CUGs. The resultant isoforms of FGF-2 have differing physiological properties and can induce the expression of different sets of genes.

In eukaryotes there are cases in which the start codon is not the first AUG but is located further downstream in the mRNA. This is mediated by the **IRES** (internal ribosome entry site). IRES sequences were discovered in the poliovirus (in the Nahum Sonenberg laboratory) and in encephalomyocarditis mRNAs (in the Eckard Wimmer laboratory). In several RNA viruses the 5′ UTR contains an IRES, which allows cap-independent translation initiation. Viral IRESs allow ribosomes to initiate translation at a downstream start codon. In viral mRNAs ribosomes also find start codons through a process called *leaky scanning*, in which the first start codon can be skipped. Reasons that favour leaky scanning and skipping of the first start codon include a poor Kozak consensus, the presence of another proximal

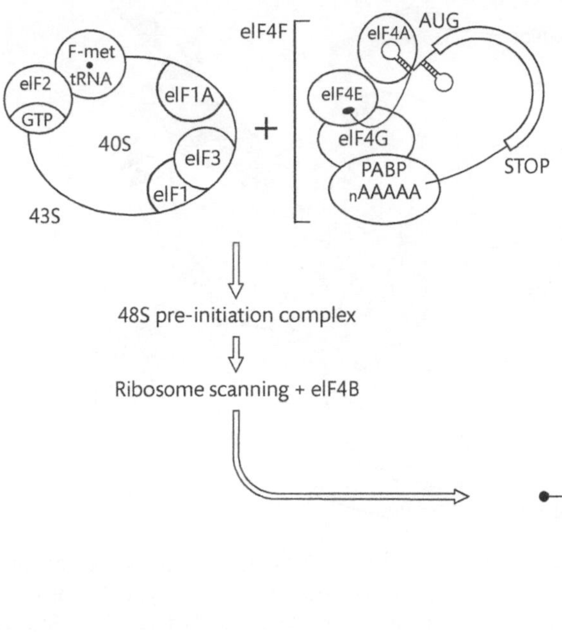

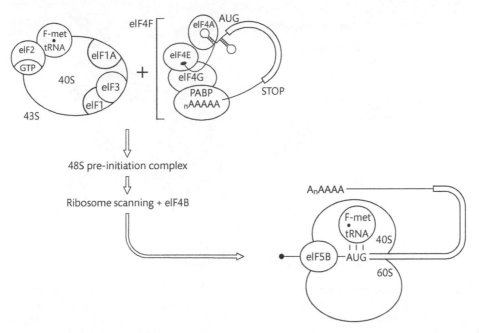

Figure 11.6 The process of eukaryotic translation initiation. The ternary complex (eIF2–GTP bound to f-Met $tRNA_i$) associates with the 40S ribosomal subunit together with eIF1A, eIF3, and eIF1 to form the 43S pre-initiation complex. The 43S complex associates with the mRNA, with the eIF4F complex bound to the 5′ m^7G cap on the mRNA. The cap-binding activity is provided by eIF4E. eIF4G interacts with the poly(A)-binding protein; thus the poly(A) tail, together with the 5′ m^7G cap, stimulate translation (*closed-loop model*). Hence the mRNA is loaded with a small ribosomal subunit (the 48S pre-initiation complex). The small ribosomal subunit scans until the correct start codon is found, with the RNA helicase eIF4A unwinding the 5′ UTR structure, the latter activated by eIF4B. The initiator tRNA hybridizes to the start codon, leading to the conversion of eIF2.GTP to eIF2.GDP. The initiator tRNA is then moved to the P (peptidyl) site. Next, eIF5B.GTP facilitates the joining of the large ribosomal subunit, completing the process of translation initiation.

AUG or stop codon, and the first start codon being too close to the 5′ cap.

Some viruses operate by shutting down host cell translation by destroying eIF4G, therefore preventing the interaction between the 5′ cap and the poly(A) tail (which normally facilitates translation initiation of most cellular mRNAs). The viral mRNAs then take over the translation machinery using their IRES sequences. IRESs are not confined to viruses; they are found in several cellular mRNAs associated with stress and mitosis. During mitosis, cells dephosphorylate eIF4E, reducing its affinity for the 5′ cap. In this scenario the translation machinery preferentially translates mRNAs that contain IRES sequences.

11.4.2 Translation elongation and termination

In translation elongation, the ribosome processes along the mRNA, adding amino acids to the nascent polypeptide chain as it does so. The process of translation elongation is summarized in Fig. 11.7. The arrival of the next amino acid at the ribosome and its addition to the nascent polypeptide through the formation of a peptide bond causes the ribosome to move on to the next codon. This is known as the process of **translocation**, and is aided by a number of **elongation factors** (eEFs).

There are three important sites in the large ribosomal subunit: the **A site** (*acceptor site*, where the amino-acyl tRNA lands, charged with the next amino acid to be added to the polypeptide chain), the **P site** (*peptidyl-tRNA site*, occupied by the most recent amino acid to be added on the growing polypeptide), and the **E site** (*exit site*, where the tRNA due to leave the ribosome following delivery of its cognate amino acid is located). Although the bulk of the A, P, and E sites lie in the large ribosomal subunit, the complete A, P, and E sites are formed only when the small and large ribosomal subunits associate.

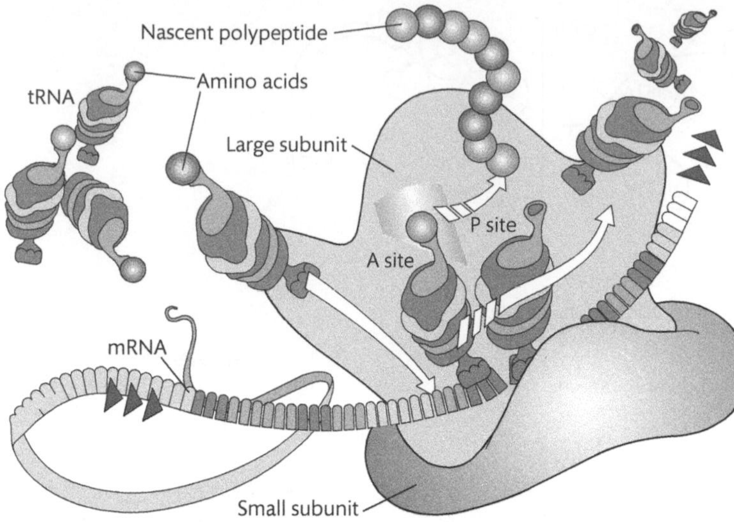

Figure 11.7 Process of translation elongation. As the mRNA is translated, tRNAs charged with specific amino acids use their anticodon loop to hybridize to their target codons on the mRNA in the A (acceptor) site. The new amino acid is then added to the nascent polypeptide chain into the P (peptidyl-tRNA) site. At each cycle a tRNA, now no longer charged with an amino acid, is released, ready for the next cycle. The process of elongation is stimulated by elongation factors (eEFs). (Reproduced with permission from Mariana Ruiz Villarreal.)

The essential aspects of the process in eukaryotes (and similarly in prokaryotes) are as follows. After translation initiation has taken place and the first methionine-charged initiator tRNA is placed in the P site, the next amino acid to be added to the polypeptide enters the A site (attached to its cognate tRNA) with the aid of eEF1α bound to GTP (EF-Tu in prokaryotes). The interaction of the codon with the anticodon triggers conformational changes in the 18S rRNA that promote GTP hydrolysis. After GTP hydrolysis, a peptide bond is formed between this second amino acid and the methionine that occupies the P site. Peptide bond formation is catalysed by the ribozyme activity of the rRNA itself (the peptidyl transferase reaction), aided by ribosomal proteins, and *translocation* occurs. Translocation is facilitated by hydrolysis of GTP bound to eEF2 (EF-G in prokaryotes). The initiator tRNA that has lost its amino acid is shifted to the E site (it will be the first tRNA to be released), and the tRNA carrying the second amino acid (which has just become covalently joined to the polypeptide) moves to the P site. This frees the A site for the next charged tRNA (carrying the third amino acid to be added), and the process is repeated until a stop codon is encountered.

Polypeptide synthesis ends with the process of translation termination once a stop codon enters the A site. The stop codons are UAG (named the *amber* codon in classical genetics), UGA (the *opal* codon), and UAA (the *ochre* codon). At first it was thought that there are tRNAs that specifically recognize the stop codons. Instead, it turns out that stop codons are recognized by **release factors** (eRFs in eukaryotes, RFs in prokaryotes) that help release the finished polypeptide from the mRNA. When the ribosome reaches a stop codon, the release factor eRF1 docks in the proximity of the A site in association with eRF3-GTP. eRF1 recognizes all three stop codons. GTP hydrolysis promotes the cleavage of the completed polypeptide from the last tRNA and the dissociation of the ribosomal subunits. Once dissociated, the ribosomal subunits are free to commence another round of translation.

The three-dimensional structure of release factors has been studied in detail. Interestingly, they actually resemble tRNAs, despite being proteins. This makes sense, as release factors need to dock with the same sites in the ribosomes that are recognized by tRNAs. The similarity between release factors and tRNA is an example of *molecular mimicry*. At the end of translation, ribosomes are recycled and dissociated into small and large subunits with the aid of the initiation factors eIF3 and eIF1A, ready to start a new round of translation.

11.4.3 The role of the signal recognition particle in translation

Several proteins are destined for either secretion from the cell or insertion into cellular membranes. The **signal recognition particle** (SRP) is an RNA–protein complex which binds to a short signal sequence present at the N-terminus of these proteins and mediates protein targeting. The signal sequence is fairly short, comprising 9–12 hydrophobic amino acids. The SRP binds to the signal sequence while the protein is still being translated. SRP binding leads to an arrest in translation, and targeting of the SRP/nascent peptide/ribosome to a network of membranes in the cell called the endoplasmic reticulum (ER) through which membrane-bound and secreted proteins pass. The SRP in turn interacts with a receptor on the surface of the ER called the signal receptor (SR), and transfers the nascent peptide/ribosome onto a *translocon protein* which is inserted in the ER membrane.

The translocon serves as a pore, allowing the newly synthesized protein to be inserted within the membrane. Future translation of the nascent polypeptide then takes place through the translocon, either to produce a membrane-bound protein or to insert the protein into the interior of the ER. The RNA component of the SRP provides a scaffold for proteins to bind which interacts with the ribosome, the SR, and the translocon, as shown in Fig. 11.8.

> **Key points**
>
> There are three steps in translation: initiation, elongation, and termination. The signal recognition particle is a ribonucleoprotein which delivers nascent polypeptides into cellular membranes or secretory pathways.

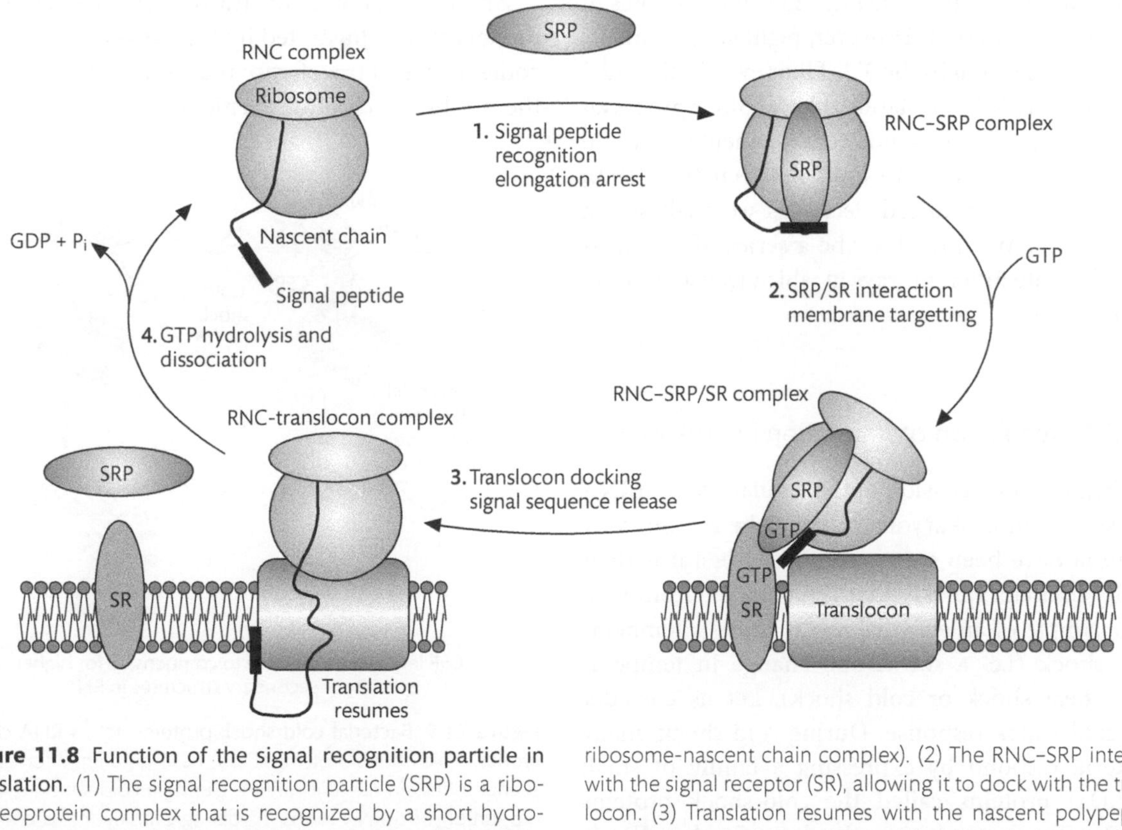

Figure 11.8 **Function of the signal recognition particle in translation.** (1) The signal recognition particle (SRP) is a ribonucleoprotein complex that is recognized by a short hydrophobic sequence at the N-terminus of proteins destined for secretion or insertion into cellular membranes. Binding of the SRP causes arrest of translation elongation (RNC is the ribosome-nascent chain complex). (2) The RNC–SRP interacts with the signal receptor (SR), allowing it to dock with the translocon. (3) Translation resumes with the nascent polypeptide passing through the translocon. (4) GTP hydrolysis facilitates the dissociation of the ribosome. (Reprinted from Egea et al.[19] with permission from Elsevier.)

11.5 Regulation of mRNA translation

We now turn our attention to examples of how mRNA translation can be regulated.[15,20–35] Ultimately, the amount of protein that can be produced from a given mRNA depends on the rate at which the mRNA can be translated. Thus, many regulatory processes are directed at influencing the rate at which ribosomes are loaded onto an mRNA and translation is subsequently initiated. An mRNA that is heavily translated will be loaded more densely with ribosomes, resulting in 'heavy polyribosomes'. On the other hand, an mRNA that is not translated much may be mostly present as 'free mRNP', or free messenger ribonucleoprotein particles (i.e. it is free of ribosomes). We will encounter a special case of free mRNP in Section 11.6 when we consider *masked messages*.

A general theme in the regulation of translation is the influence of secondary structures present in mRNA. Stemloop structures in the untranslated leader (known as the 5′ UTR) can influence rates of translation initiation. However, regulatory elements can also be present in the 3′ UTR or even in the ORF. Note that mRNA translation, like all other processes in gene expression, is not compartmentalized and isolated. Thus, there are links between translation and 'nonsense-mediated decay' as we shall see in Chapter 12. We now describe a series of examples that illustrate different ways in which translation can be regulated.

11.5.1 Regulation of translation in prokaryotes

We begin our discussion of the regulation of mRNA translation in prokaryotes. One of the reasons why bacteria have been so incredibly successful is their ability to adapt quickly to a changing environment. An example of an environmental stress is temperature shock (i.e. a substantial change in temperature—heat shock or cold shock). Let us consider the cold-shock response. During cold shock, many bacteria respond by expressing a family of small (~7kDa) proteins called the cold-shock proteins (CSPs). The first to be identified was CspA in *E. coli.* CspA consists of a highly conserved domain, the **cold-shock domain** (CSD), whose structure has been determined. The CSD is a β-barrel structure able to bind to single-stranded nucleic acids (including RNA) via conserved basic and aromatic amino acids that protrude from its surface. Interestingly, as we saw in Chapter 4, the CSD has been incorporated into the eukaryotic 'Y-box protein' family, a family of proteins involved in both transcription and mRNA packaging. What, then, is the function of the cold-shock proteins in bacteria?

One theory is that they work as **RNA chaperones**. According to this model they bind non-specifically to single-stranded RNA in messenger RNA. Colder temperatures favour the formation of a greater amount of double-stranded RNA, which hinders translation. This is because the hydrogen bonding in Watson–Crick base pairing is kinetically favoured at lower temperatures (conversely, dsDNA and dsRNA are denatured by heat). CSPs bind to mRNA during cold stress and therefore help to maintain mRNA in a single-stranded conformation. This model is illustrated in Fig. 11.9.

Another example of translation regulation in prokaryotes is illustrated by the *cat* gene, which encodes an acetyl transferase that confers resistance to the antibiotic chloramphenicol. The correct AUG

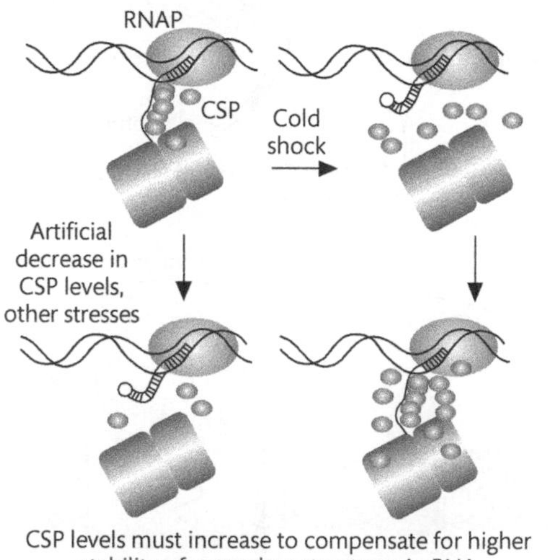
RNAP
CSP
Cold shock
Artificial decrease in CSP levels, other stresses
CSP levels must increase to compensate for higher stability of secondary structures in RNA

Figure 11.9 Bacterial cold shock proteins act as RNA chaperones. Cold-shock proteins (CSPs) have an affinity for single-stranded RNA in nascent mRNA. During a cold shock, a larger proportion of dsRNA forms, hindering translation. To counteract this problem, CSPs bind to nascent mRNA, helping to keep it single-stranded. RNAP = RNA polymerase. (Reprinted from Graumann and Marahiel[20] with permission from Elsevier.)

start codon within the mRNA is blocked as it is located in a stable stemloop structure. However, in *cat* mRNA, there is an additional upstream AUG which encodes a 'leader peptide'. In the presence of sublethal levels of chloramphenicol, ribosomes translate the leader peptides from the upstream AUG, but tend to stall (due to the action of chloramphenicol; see Table 11.2). Ribosomal stalling leads to a destabilization of the stemloop structure which then frees the correct AUG. This allows the expression of the acetyl transferase encoded by the *cat* gene. This phenomenon is called **translational attenuation** and is shown in Fig. 11.10.

A very ancient type of translation regulation is mediated by the riboswitch. We discussed the fundamental properties and functions of riboswitches in Chapter 2. Depending on their position within the mRNA in bacteria, riboswitches can control transcription or translation. The lysine-responsive riboswitch (L-box) has attracted significant interest because of its association with the development of resistance to lysine analogues (a family of antimicrobial agents). Figure 11.11 shows the structure of the L-box from the bacterium *Thermotoga maritima*. Originally isolated from geothermally heated marine sediment, this bacterium is an interesting model organism because it is part of a very ancient lineage. Riboswitches impede translation in two ways. They can occlude the ribosome-binding site, or alternatively they can cause a steric hindrance that blocks the elongation of translation.

Prokaryotic translation and the ribosomes that mediate this process are well-known targets of antibiotics. The aminoglycoside gentamicin binds to the decoding region of ribosomes which results in misreading of mRNA, generating aberrant or truncated proteins. The macrolide erythromycin binds to 23S rRNA, blocking the exit of the polypeptide chain.

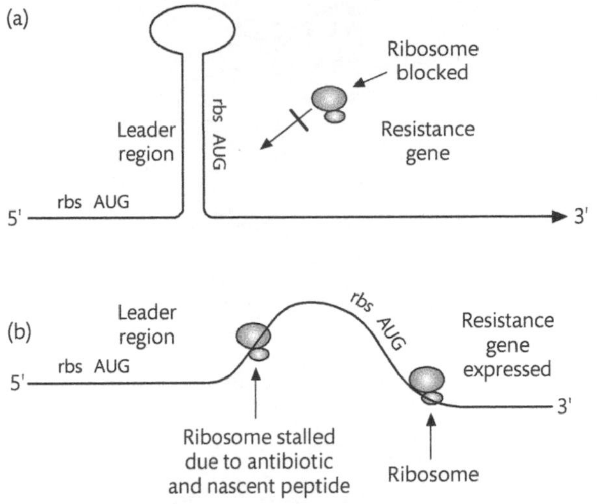

Figure 11.10 Translational attenuation of the prokaryotic *cat* mRNA. (a) The AUG at the start of the *cat* chloramphenicol antibiotic resistance gene is masked by a stemloop structure. (b) When initially exposed to sublethal doses of chloramphenicol, the upstream ribosome stalls—the consequence is that the downstream AUG becomes accessible and the resistance gene is expressed. As a result the bacteria acquire the ability to neutralize the action of chloramphenicol. rbs = ribosome-binding site. (Reproduced from Goldman.[27] ©2001 John Wiley & Sons Ltd.)

Tetracycline inhibits the codon–anticodon interaction, and the bacteriostatic agent chloramphenicol interferes with the formation of peptide bonds. The inhibition of translation by antibiotics is a favoured means of counteracting bacterial infections, as summarized in Table 11.2.

11.5.2 Regulation of translation in eukaryotes

We now move on to examples of translation regulation in eukaryotes. The first set of examples describes the role that translation factors can have in regulating translation. The second set of examples illustrates the action of *trans*-acting RNA-binding

Table 11.2 Common antibiotics and their effects on translation in bacteria

Antibiotic	Target	Consequence
Gentamicin	Decoding region of ribosomes	Misreading of mRNA code
Erythromycin	23S rRNA	Blocks exit of polypeptide chain
Tetracycline	Codon–anticodon interaction	Inhibition of protein synthesis
Chloramphenicol	Peptide bond formation	Bacteriostatic: peptide synthesis is stalled

Figure 11.11 Structure of the lysine riboswitch from *Thermotoga maritima*. Riboswitches in bacteria typically repress genes by a mechanism of transcriptional or translational attenuation. The overall structure of a lysine riboswitch is shown, and the lysine-binding pocket is highlighted in the inset. Note how the riboswitch produces a three-dimensional structure that envelops the lysine. Binding involves water molecules, hydrogen bonds, and K^+ cation coordination. For a full colour version of this figure, see Colour Illustration 13. (Figure kindly provided by Alexander Serganov, Memorial Sloan-Kettering Cancer Center.)

proteins that recognize specific RNA elements in mRNA.

The cap-binding protein eIF4E, which is a component of eIF4F, is bound by a protein called 4E-BP (eIF4E-binding protein). When bound to eIF4E, 4E-BP prevents the formation and action of eIF4F. (As we saw in Section 11.4, eIF4F is the complex between eIF4E, eIF4G, and eIF4A.) Therefore 4E-BP represses cap-dependent translation. However, when phosphorylated, 4E-BP is released, thus freeing eIF4E and allowing eIF4F to form.

Some mRNAs have a very large amount of secondary structure in their 5′ UTR, and in their case eIF4F activity is particularly important. mTOR (mammalian target of rapamycin) is a serine/threonine protein kinase that regulates cell proliferation and protein synthesis. mTOR is thought to integrate the signals received from several growth factors including insulin, IGF-1, and IGF-2. The activation of mTOR leads to the phosphorylation of 4E-BP which in turn releases eIF4E and results in enhanced translation of mRNAs. This model is illustrated in Fig. 11.12.

Interestingly, the aberrant expression and activity of eukaryotic translation initiation factors is increasingly linked to cancer. We now consider another example of how the activity of translation initiation factors can regulate translation. There are two main isoforms of CCAAT/enhancer-binding protein (C/EBP). The large isoform has a DNA-binding domain and a transactivation domain (a domain that helps to recruit RNA polymerase and therefore initiate transcription). It activates the transcription of a set of target genes, leading to proliferation arrest and terminal differentiation. It results from translation starting at a start codon denoted AUG1. The smaller isoform lacks the N-terminal *trans*-activation domain and arises from the use of a downstream start codon denoted AUG2. The smaller isoform binds to the same target genes as the larger isoform, but instead *represses* their expression because it lacks a transcriptional *trans*-activation domain. This repression results in the induction of cell proliferation. Not surprisingly, the short form of C/EBP is upregulated in some cancers.

The driving force behind the regulation of expression of the C/EBP isoforms is the relative activity of the translation initiation factors eIF2 and eIF4E. So how does this work? As well as AUG1 and AUG2, there is a third, additional, upstream AUG start codon. When eIF2 and eIF4E activities are high, the upstream AUG is used preferentially, resulting in the translation of a short upstream open reading frame (uORF). When the uORF is used the ribosome tends to scan on to AUG2 but not AUG1. Conversely, when eIF2 and eIF4E activities are relatively lower, only AUG1 tends to be used. This example illustrates the concept that the relative levels and activities of translation factors in cells have profound effects on gene expression. This phenomenon reminds us of the regulation of alternative splicing, in which the relative levels and activities of splice factors also have a determining influence on outcome.

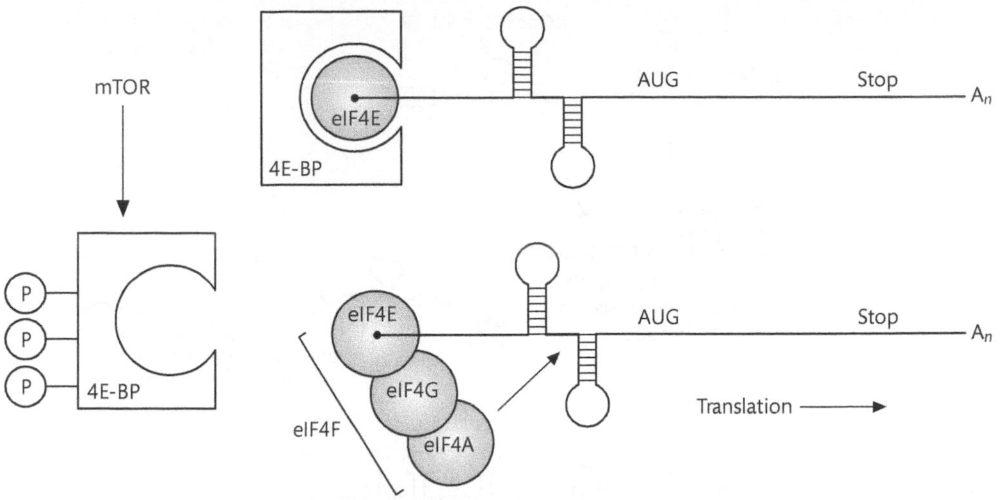

Figure 11.12 Regulation of eIF4E activity by 4E-BP. The presence of active eIF4E-binding protein (4E-BP) prevents eIF4E from interacting with eIF4G, which normally recruits the RNA helicase eIF4A forming the eIF4F complex and initiating translation. As a result of the activation of mTOR, after the cell signalling cascade initiated by growth factors, 4E-BP is hyperphosphorylated and then separates from eIF4E. As a result, eIF4F forms, allowing translation initiation to occur.

Next we consider the regulation of translation by RNA-binding proteins as opposed to translation initiation factors. An elegant and widely cited example of the regulation of mRNA translation relates to the control of iron levels in cells, the mechanism of which is illustrated in Fig. 11.13. Iron concentrations are normally tightly regulated, as iron plays a critical role in the active site of many enzymes, but excess iron can oxidize cellular components. When iron levels are high, it is necessary to prevent toxic levels from accumulating. When iron levels are low, however, available free iron needs to be released in order for iron-dependent enzymes to operate. Ferritin is an intracellular iron-binding protein which sequesters iron. The 5′ UTR of *ferritin* mRNA contains **IREs** (iron response elements, not to be confused with IRES, the internal ribosome entry site described in Section 11.4). IREs are hairpin stemloop structures. As discussed in Chapter 2, hairpin stemloops are RNA secondary structure motifs comprising an RNA helix with a terminal loop, and often containing unpaired nucleotides within the stem, which form bulges. RNA hairpin loops provide important binding sites for RNA-binding proteins.

The IRE acts as a binding site for two very similar proteins called the iron response binding proteins 1 and 2 (IRP1 and IRP2). IRP1 and IRP2 are iron-binding proteins, and binding of iron to these proteins affects their overall structure. In the absence of iron, IRP1 and IRP2 can bind to the IRE with very high affinity. By binding to the IRE, IRPs prevent the small ribosomal subunit from binding to *ferritin* mRNA. At high iron levels, IRPs are rendered inactive and do not bind the IRE. As a result the initiation of translation can proceed normally. This makes sense; if iron levels are high, ferritin protein is required to sequester the iron and regulate its concentration in the cell.

Another component of the iron metabolism system is the transferrin receptor (TfR). This receptor binds to transferrin, the protein that carries iron in the circulation. TfR is able to import transferrin that is complexed to iron into the cell. IREs are also present, but in this case in the 3′ UTR of the *TfR* mRNA. These IREs are quite rich in AU, and AU-rich sequences are often associated with mRNA instability as we shall see in Chapter 12. If IRPs are not bound (as is the case in the presence of high iron levels), *TfR* mRNA is unstable and therefore less TfR protein is made. When iron levels are low, however, *TfR* mRNA is stabilized via bound IRPs and more TfR protein is made, favouring the import of iron into the cell.

These elegant examples illustrate the fact that RNA-binding proteins (e.g. the IRPs) and their RNA targets (e.g. the IRE) have been evolutionarily

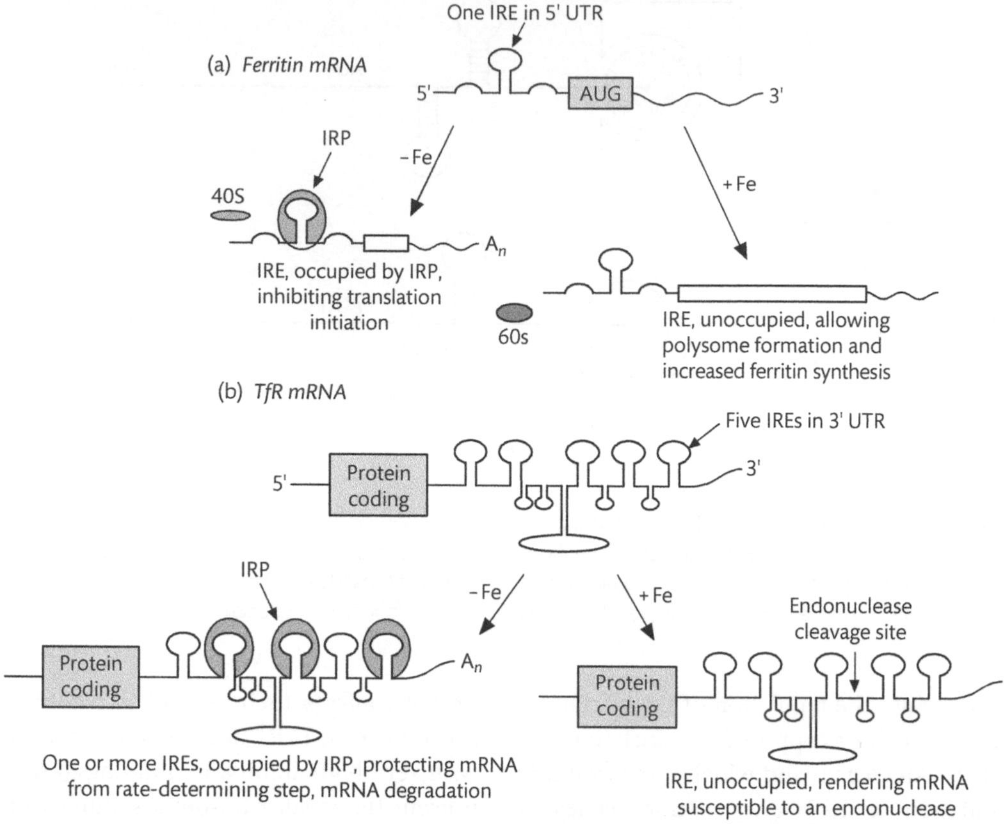

Figure 11.13 Translational regulation of iron metabolism. (a) A single IRE (iron response element) stemloop is present in the 5′ UTR of *ferritin* mRNA. In the absence of iron, IRP binds to the IRE, hindering the initiation of translation. In the presence of iron, however, IRP dissociates and translation of *ferritin* can occur. (b) There are five IREs in the 3′ UTR of the transferrin receptor mRNA. In the presence of iron, IRP does not bind them and the mRNA is susceptible to degradation. In the absence of iron, IRP binds to the five IREs stabilizing the mRNA, resulting in the production of transferrin receptor. (Reprinted from Rouault[24] with permission from Macmillan Publishers Ltd.)

conserved to regulate, in a coordinated fashion, two mRNAs that are involved in the same physiology—in this case the regulation of intracellular iron levels. It is also worth pointing out that this mechanism allows a rapid response to changes in iron concentration. In contrast, the regulation of ferritin and TfR expression at the transcriptional level will be a lot slower (because genes need to be transcribed, and the transcripts then need to be processed and exported).

The importance of the IRPs has been demonstrated in mouse models. Mice with homozygous deletions of both genes encoding these proteins die embryonically, and mice missing just IRP2 undergo neurodegeneration and anaemia, and develop iron-containing inclusions in their brains.

11.5.3 Regulation of translation by microRNAs

Genomes encode a multitude of non-protein-coding RNAs (ncRNAs) which have a regulatory function; as we shall see in Chapter 16, many of these are called microRNAs. MicroRNAs deserve a mention in this chapter because it is now known that they can regulate target genes at the level of mRNA translational repression. Among the first microRNAs to be discovered was the lin-4 microRNA in the nematode model organism *Caenorhabditis elegans*. Lin-4 binds via sequence complementarity to the 3′ UTR of *lin-14* and *lin-28* mRNAs, whose genes are involved in developmental timing. Binding of lin-4 to its target mRNAs results in their translational repression, and this regulatory event is required in normal development.

There are several ways in which microRNAs can repress translation. Evidence shows that in order to be translationally repressed by microRNAs, many target mRNAs require an m^7G cap. MicroRNAs are packaged into ribonucleoprotein (miRNPs); the proteins AGO2 and GW182 are components of the miRNP thought to interact with translation initiation factors, thereby preventing the recruitment of ribosomal subunits. MicroRNAs can also repress translation post-initiation by decreasing the rate of translation elongation, by causing premature ribosomal *drop-off*, or even by promoting the degradation of nascent polypeptides.

Thus, small non-coding RNA molecules (in this case, microRNAs) can regulate the translation of mRNA targets. This exciting development in the field of translation has opened the door to an even greater range of mechanisms that confer the ability to regulate mRNA translation.

11.5.4 Coupling translation with other post-transcriptional processes

In prokaryotes, translation is co-transcriptional; polypeptides are synthesized while RNA polymerase is still actively transcribing the gene. (Note that this is not the case in eukaryotes, where pre-mRNA splicing, capping, and 3′ end formation are co-transcriptional.) Transcription, be it in prokaryotes or eukaryotes, is coupled to later steps in gene expression, making the process of gene expression more efficient and coordinated. The existence of multifunctional RNA-binding proteins, described in Chapter 4, underpins the coordination of multiple steps in gene expression, including translation. Thus it is not unusual to see that splice factors, or even transcription factors, remain associated with mRNAs in the cytoplasm where they can contribute to the regulation of translation.

Work on the well-studied splice factor SRSF1, a member of the SR family of proteins, underlines these connections. SRSF1 is a shuttling protein, i.e. it can move continuously between the nucleus and the cytoplasm. Its import into the nucleus is facilitated through phosphorylation by the splice factor kinase SRPK1. In the nucleus SRSF1 is a key regulator of alternative splicing, as discussed in Chapter 9. It remains bound to mRNA as it is exported into the cytoplasm where it associates with translating ribosomes and promotes translation, especially when bound to exonic splice enhancer sequences.

SRSF1 can promote translation through the m^7G cap at the 5′ end of mRNA. It does so as follows. As we saw in Section 11.5.2, eIF4E-bp binds to the cap-binding protein eIF4E, preventing the latter from activating translation; this does not occur when eIF4E-bp is phosphorylated. The protein kinase mTOR is the 'mammalian target of rapamycin', a drug generally used as an immunosuppressant. mTOR regulates cell growth, proliferation, transcription, and protein synthesis; eIF4E-bp is one of the targets of mTOR. In contrast the protein phosphatase PP2A, whose substrates include proteins involved in cell signalling cascades which contribute to the regulation of cell proliferation, dephosphorylates eIF4E. SRSF1 binds to both mTOR and PP2A. Therefore SRSF1, initially deposited on target mRNAs during pre-mRNA processing, could enhance translation by helping to recruit mTOR to mRNAs or inhibiting the function of PP2A, helping to maintain eIF4E-bp in a phosphorylated state.

Given that SRSF1 regulates both alternative splicing and translation, an interesting question is whether or not the pre-mRNAs whose alternative splicing is altered by SRSF1 are the same target mRNAs whose translation it enhances. The identification of genuine in vivo RNA targets is not a trivial task; the gold standard approach is to use CLIP (crosslinking of RNA:protein complexes followed by immunoprecipitation). CLIP was used to define SRSF1 RNA targets in both the nucleus and cytoplasm. A significant degree of overlap was found, suggesting that SRSF1 binds to pre-mRNA affecting alternative splicing and then potentially remains bound as the mRNA is exported into the cytoplasm. To illustrate this idea, SRSF1 co-regulates the alternative splicing and the translation of transcripts derived from the *PABC1*, *NETO2*, *ENSA*, and *SRSF1* genes. The latter demonstrates that SRSF1 autoregulates both the alternative splicing *and* the translation of its *own* mRNA. These findings provide evidence that the regulation of alternative splicing and of translation may be connected. The same principle can be extended to other processes; for example, the Y-box proteins described in Chapter 4 can facilitate both the transcription and translational repression of maternally expressed mRNAs in oocytes. It is evident that many RNA binding proteins can do this, suggesting that the coordination of different steps of gene expression is a widespread phenomenon.

Key points

Translation is highly regulated in both prokaryotes and eukaryotes. The activity and levels of translation factors and RNA-binding proteins can determine the amount, or even the isoform, of a protein translated from an mRNA. MicroRNAs are naturally occurring short non-coding RNAs that hybridize to mRNAs, resulting in reduced translation.

11.6 Masked messages

A very special example of the regulation of mRNA translation is the **masked message.**[36–42] Although the term is not widely used nowadays, we define masked messages here as a special class of fully processed mRNA whose translation is significantly delayed, potentially for a very long time. Masked messages are stored in association with mRNA-binding proteins, sometimes termed 'masking proteins', in a cytoplasmic pool of translationally repressed mRNAs. Masked messages were first described in the freshwater fish *Misgurnus fossilis* (see Box 11.4).

11.6.1 Masked messages in germ cells

Oocytes are well known for possessing masked messages; but which mRNAs are masked? The answer to this question was provided by a classic experiment shown in Fig. 11.14. Nancy Standart and colleagues extracted mRNP from clam oocytes, deproteinized them, and added the naked mRNA to rabbit reticulocyte lysate to obtain in vitro translation. To their surprise, ribonucleotide reductase and cyclins A and B were produced, yet these proteins are not normally present in oocytes. Clearly then, their corresponding mRNAs must have been present, but in an

Box 11.4 Discovery of masked messages

The masked message hypothesis started in the early 1960s with the work of the Russian scientist Alexander Spirin and colleagues on embryos of the loach *Misgurnus fossilis*, a freshwater fish. The irradiation of fish embryos by X-rays was, strangely, intermittently lethal, and lethality depended on the developmental stage at which irradiation took place. If the embryos were irradiated during stages in which embryonic nuclei were transcriptionally active, such as between 6 and 8.5 hours after fertilization, the embryos did not develop to gastrula. However, if they were irradiated between 8 and 14.5 hours after fertilization, when transcription is inactive, they progressed to gastrula. This strange observation was explained as follows.

During the transcriptionally active phase nuclei produce the mRNAs that convey the information required for the embryos to progress to gastrulation. X-rays are lethal during this phase because they destroy the DNA to the extent that no further mRNA synthesis is possible. In contrast, during the inactive phase of transcription, genetic information exists in the form of multiple copies of stored cytoplasmic mRNAs. As there are multiple copies of the mRNAs, genetic information encoded by them is more resistant to irradiation.

The analysis of cytoplasmic extracts fractionated on sucrose gradients revealed the presence of a broad spectrum of mRNP particles sedimenting between 40 and 60S. When combined with ribosomes in vitro, these mRNPs formed polyribosomes. These mRNPs were named 'informosomes' in the sense that they carry the information required to direct the synthesis of proteins necessary for development to progress. However, the term informosome did not catch on. Instead, they became known as 'masked messages'—mRNAs whose translation is delayed. Masked messages were then discovered in the sea urchin *Lytechinus pictus*, and in various other animal tissues such as sheep thyroid, rat brain and liver, and HeLa cells (human cervical cancer), and in plants including wheat grain.

Whereas the size of the masked messages (mRNP particles) is heterogeneous because of the variable length of mRNA molecules contained within them, their density on caesium density gradients is remarkably uniform, at about 1.4g/cm^3, indicating a protein to mRNA ratio of 3:1. This suggested that the types of protein present in masked messages are quite universal. In fact, they could be described as 'mRNA histones' by analogy with the uniform packaging of DNA by histones. Alexander Spirin further developed the concept of masked messages by suggesting that they carry translation initiation and elongation factors, as well as components which repress and mask mRNA. He described the concept with the Latin proverb *omnia mea mecum porto*, meaning 'I carry all my things along with me'.

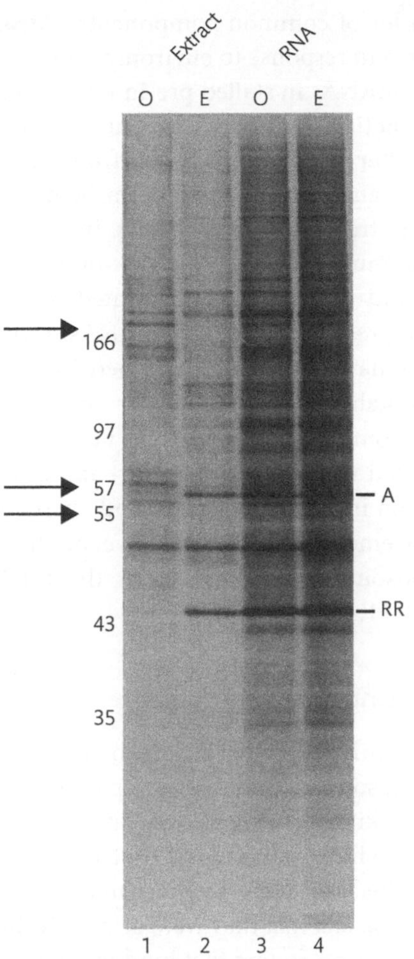

Figure 11.14 Masked messages in the oocytes of the clam *Spisula solidissima*. Before and after fertilization different sets of proteins are translated in oocytes and embryos. After fertilization, early development requires rapid cell division, and therefore proteins such as cyclin A (marked as A) and ribonucleotide reductase (RR) are produced only when an embryo (track 2) but not an oocyte (track 1) extract is added to an in vitro translation reaction. When mRNP proteins are removed by phenol extraction, and *naked* mRNA is added to an in vitro translation reaction, the same proteins are synthesized (tracks 3 and 4). This experiment shows that unfertilized oocytes contain a set of translationally repressed, masked mRNAs whose translation is repressed by mRNA-bound proteins. (Reproduced with permission from Standart et al.[39])

untranslated state. This makes sense; the developing oocyte does not need these proteins, because they are required for cell proliferation and division *after* fertilization has occurred.

The need for masked messages is dramatically illustrated in the early development of the African clawed toad *Xenopus laevis*. In *Xenopus*, oogenesis is a lengthy process, lasting up to eight months. During this period the diameter of oocytes increases from 50 to 1200μm, and a large store of macromolecules is accumulated—a staggering 10^{12} ribosomes and 2×10^{11} masked mRNAs per single oocyte, with the latter containing approximately 20 000 distinct mRNA species. At oocyte maturation, induced by the hormone progesterone, fundamental changes occur in preparation for fertilization. Individual masked mRNAs are recruited for translation; this occurs at different time points depending on the mRNAs. In other words, their 'unmasking' is temporally regulated.

After fertilization, a remarkable process follows: embryonic cells divide rapidly to the extent that, within three days, a swimming tadpole has developed and the original fertilized egg has given rise to 10^6 cells. During this rapid development, embryonic cells have no time to transcribe and process transcripts. Instead they rely on the pool of **maternal mRNAs** which encode the proteins required for rapid proliferation, including DNA synthesis enzymes, proto-oncogenes, and histone mRNAs.

A well-studied case of a *Xenopus* oocyte masked message is the proto-oncogene c-*myc*. The transcription factor Myc is required for cell proliferation and is often upregulated in tumours. c-*myc* mRNA accumulates from early oogenesis, and up to 5×10^6 c-*myc* mRNAs have accumulated in, and are stored in, a masked state by late oogenesis. Myc protein is then unmasked as it is required during early development; once used, c-*myc* mRNA is actively degraded so that, by gastrulation, there are only 10 transcripts per cell!

Masked messages have also been detected in mammalian oocytes, but the number is not as large as in *Xenopus* oocytes. A mammalian context in which masked messages are important is spermiogenesis. In mice, spermiogenesis takes two weeks, during which time spermatocytes give rise to haploid spermatids, which in turn develop into mature sperm cells. To do this they need to develop specialized structures, including the flagellum and the sperm head, in which the genome is highly compacted by the protamines (which replace histones). In spermatids, transcription eventually ceases once the protamines have condensed and packaged the DNA. Thus, mRNAs required for spermatid development are synthesized in early spermatids and stored for several days until required in late spermatids.

11.6.2 Masked messages in other contexts

Masked messages are not restricted to germ cells; they are well documented in somatic tissues. In mammals, the development of reticulocytes into red blood cells, which transport oxygen attached to haemoglobin around the body, involves the loss of the nucleus and mitochondria. Messenger RNAs encoding globins are very abundant. Another abundant mRNA encodes the enzyme *LOX* (15-lipo-oxygenase) required for the degradation of mitochondria. *LOX* mRNA is transcribed in bone marrow, but is only translated when the reticulocytes reach the peripheral blood.

Masked messages are also important in plants. Protein synthesis in wheat grains declines during seed ripening, and a significant proportion of mRNA is found in a non-translating masked state. In the plant alfalfa, mRNAs encoding storage proteins are expressed in the early stages of the 'somatic embryo' and are only recruited for translation at the beginning of cotyledon development.

11.6.3 Structure and function of the P-body

There is evidence that mRNAs that are not translating accumulate in cytoplasmic foci. These foci are known as processing bodies (**P-bodies**). P-bodies were first described in the late 1990s when scientists were looking at the intracellular localization of proteins associated with the regulation of mRNA translation or decay. They noticed that several of these antigens could be seen in numerous discrete foci. Whereas some mRNAs enter a P-body in order to be degraded, others can be translationally repressed long term (i.e. masked messages).

P-bodies contain mRNAs that are not translating; and as a result they do not contain (other than eIF4E) translation initiation factors and ribosomes. P-bodies also contain a set of core proteins associated with mRNA decay (mRNA decay will be described in Chapter 12), including decapping enzymes and nucleases. They also contain proteins that are associated with microRNA function (microRNAs will be described in Chapter 16), translational repression, and nonsense-mediated decay.

As well as P-bodies, several cytoplasmic granules have been described that appear distinct (but share a lot of common components). **Stress granules** form in response to environmental stress. They contain mRNAs in stalled pre-initiation complexes. Their function is still unclear; but they are thought to sequester and protect mRNAs from environmental stress and prevent them from being translated until the stress trigger disappears. In order to form, stress granules require the RNA-binding proteins TIA-1 and TIA-R, both associated with translational repression of specific mRNAs. Cytoplasmic granules have also been described in specific cell types, notably germ granules (or maternal mRNA storage granules) in oocytes and neuronal granules (associated with mRNA transport and translational repression in neurons). Parker and Sheth[41] describe the movement of mRNAs betweem the nucleus, polyribosomes, and granules as the *mRNA cycle* (see Fig. 11.15).

11.6.4 Unmasking messages

How are mRNAs masked and then unmasked? In germ and somatic cells, they are packaged by abundant 'masking proteins'. These include the 'Y-box proteins', which are required (but not sufficient) to repress their translation. The name Y-box protein refers to the fact that they were originally discovered as transcription factors that bind to a DNA element called the Y box. The Y-box proteins bind both DNA and RNA, and serve as the main mRNA packaging proteins. As you will recall from Chapter 4, they are not the only proteins that can bind both DNA and RNA. A classic example is TFIIIA, a multi-zinc-finger protein that binds both the *5S* rRNA gene and its gene product 5S rRNA.

The Y-box proteins bind and package mRNA via an evolutionarily conserved domain called the cold-shock domain. As we saw earlier, the cold-shock domain binds single-stranded RNA and is thought to act as an 'RNA chaperone'. Thus, Y-box proteins are an integral part of mRNP particles and have in fact been referred to as the 'RNA histones' originally predicted by Alexander Spirin (see Box 11.4). The mRNA packaging proteins are also associated with an RNA helicase in humans called p54/RCK, which is closely related to eIF4A. p54/RCK is functionally complex because it is thought to be involved not only in the formation

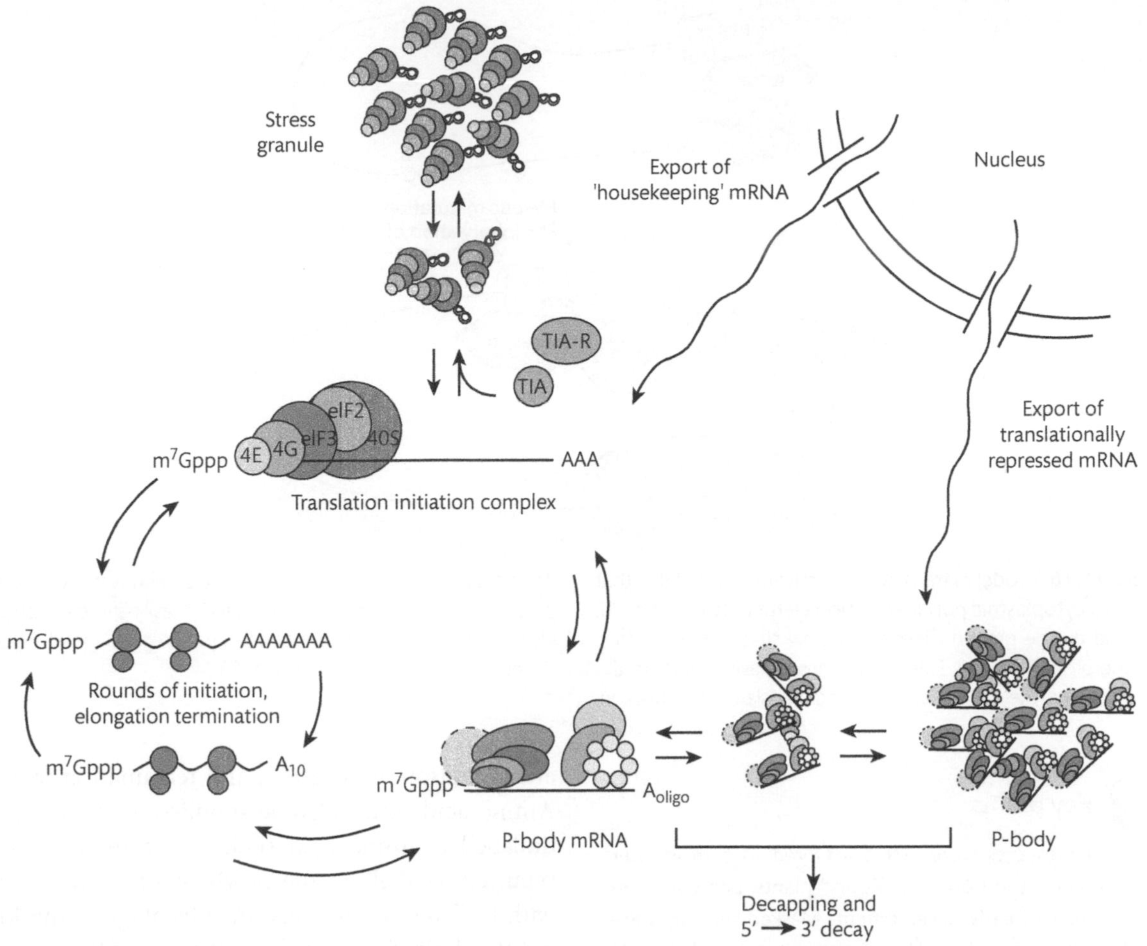

Figure 11.15 The mRNA cycle. According to the model proposed by Parker and Sheth, as a result of defects in translation initation or terminaton, mRNAs assemble into P-bodies. Once in a P-body an mRNA can be degraded or translationally repressed, or it can re-enter into translation. mRNAs which, upon exiting the nucleus, need to be translated immediately do not enter P-bodies. In contrast mRNAs (including masked messages) that need to be translationally repressed immediately assemble into P-bodies. TIA-1 and TIA-R promote the formation of stress granules in response to environmental stress, sequestering mRNAs away from the translation machinery. (Reproduced from Parker and Sheth.[41])

of translationally repressed mRNP, but also in their translational recruitment, and even degradation. p54/RCK is, in fact, also an essential component of P-bodies.

The mechanism through which messages are masked (probably co-transcriptionally) is not yet fully understood. However, there is a good understanding of how messages might be recruited for translation, or *unmasked*, in the cytoplasm. Masked messages in *Xenopus* oocytes have a U-rich *cytoplasmic* polyadenylation element (CPE) bound by CPEB (CPE-binding protein). CPEB interacts with the protein maskin, which also interacts with the cap-binding protein eIF4E. When bound by maskin, eIF4E is unable to bind to translation initiation factors. At the appropriate time, CPEB is phosphorylated and, in this form, it then releases maskin. Maskin then does two things: it frees eIF4E, and causes the recruitment of CPSF (a cytoplasmic form of cleavage and polyadenylation specificity factor) and PAP (poly(A) polymerase). This recruitment stimulates **cytoplasmic polyadenylation**, i.e. further elongation of the poly(A) tail. The presence of a poly(A) tail then promotes translation after it is bound by PABP 1 (PABP 1 is the poly(A) binding protein I in the cytoplasm; whereas PABPN1 is involved in nuclear polyadenylation). This model is illustrated in Fig. 11.16.

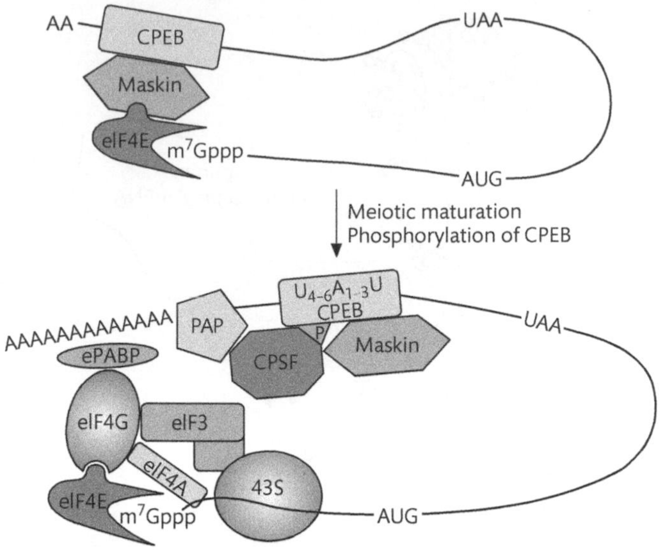

Figure 11.16 Model of translational activation of mRNAs that contain a cytoplasmic polyadenylation element (CPE). The CPE is bound by the protein CPEB at multiple sites. CPEB interacts with the protein maskin. During oogenesis maskin binds to the cap-binding protein eIF4E and prevents translational activation. At fertilization CPEB is phosphorylated, and the cytoplasmic polyadenylation machinery is recruited. Maskin's interaction with eIF4E is abrogated and translation initiation goes ahead. (Reproduced from Colegrove-Otero et al.[40])

Key points

Masked messages were first discovered in a freshwater fish. They exist in both animals and plants, particularly in the context of early development. Masked messages are mRNAs that are packaged by proteins into a translationally inactive state, where they can remain for months until they are required. When required they are unmasked and can then be rapidly translated.

11.7 Manipulating translation

11.7.1 In vitro translation

In the final section of this chapter we consider the manipulation of translation,[42–44] beginning with **in vitro translation**. An in vitro translation system is a means of synthesizing a protein of choice in a cell-free extract. The reasons for doing this might be to produce high levels of a particular protein so that its properties can be investigated (as opposed to isolating a rare protein out of a complex mix). In order to do this, the extract must contain all the machinery of translation, including ribosomes, tRNAs, and the corresponding amino-acyl tRNA synthetases and initiation, elongation, and termination factors. Amino acids must also be supplied, including a radiolabelled amino acid such as [^{35}S]methionine if required. (Labelling the newly synthesized protein with [^{35}S]methionine has the advantage of marking out the desired protein from the rest of the proteins present in the cell-free extract, but note that the protein of interest should contain a number of methionines, otherwise the labelling is weak.) The cell-free extract must also be devoid of any endogenous mRNAs originally present in the source from which it was derived, or they will also be translated.

How is in vitro translation achieved? There are two approaches. One is to transcribe a full-length mRNA in vitro, generally from a plasmid vector. The in vitro transcribed mRNA must comprise the 5′ UTR with an m^{7}G cap, the native start codon and ORF, and the 3′ UTR, ideally with a poly(A) tail included. The mRNA can simply be added to the cell-free extract, together with an amino acid master mix and the translation process can begin. However, an easier approach is to use a coupled in vitro transcription–translation system. In this approach, the cell extract includes a prokaryotic RNA polymerase, for example T7, T3, or SP6 RNA polymerase. You simply add a plasmid construct, and the mRNA of choice is first transcribed from the plasmid driven by a T7, T3, or SP6 promoter. The resulting mRNA is then translated in the same reaction tube.

Three types of commercially available cell-free lysates are commonly in use—rabbit reticulocyte, wheat germ, and *E. coli*—all of which were developed from cells in which protein synthesis occurs at a high rate. Rabbit reticulocyte lysate is one of the most popular. As the name suggests, it is derived from the reticulocyte, a precursor of the enucleated red blood cell (erythrocyte). Reticulocytes produce high levels of haemoglobin and their mRNA translation machinery is very active. Indeed, over 90% of the protein synthesized in a reticulocyte is haemoglobin. When preparing the extract, endogenous mRNA (most of which is globin mRNA) is generally eliminated using a Ca^{2+}-dependent micrococcal nuclease. Failure to eliminate endogenous mRNAs from the reticulocyte lysate results in high background translation (i.e. the translation of mRNAs other than the target mRNA). Wheat germ lysate can be used instead of reticulocyte lysate. However, this type of lysate is more dependent on the presence of 5′ mRNA cap structures than reticulocyte lysate, and is also prone to higher drop-off rates.

An alternative approach is to use an *E. coli* lysate. These are best suited for coupled transcription–translation systems and are very efficient because they are less dependent on the more complex translation initiation process in eukaryotes. However, their disadvantage is that they tend to contain more endogenous ribonucleases compared with reticulocyte or wheat germ lysates.

There are a number of applications in which in vitro translation can be particularly useful. For instance, in some cases the protein that you wish to translate is toxic to the cell, particularly insoluble, or very susceptible to degradation by intracellular proteases. In all these cases in vitro protein synthesis may be helpful. You may also wish to label your protein of interest, including truncated or mutated versions, and study their properties. Researchers often use in vitro translated proteins in a co-immunoprecipitation assay to confirm an interaction between two proteins. A [^{35}S]methionine-labelled protein can also be added to an affinity chromatography column to see if it binds to a given molecule. It can also be added to a specific extract or subcellular fraction, after which it is possible to test whether it co-immunoprecipitates with its candidate interacting partner using an antibody to the latter.

11.7.2 Blocking translation with antisense oligonucleotides

Another way of manipulating translation is to block it. When investigating the function of genes it is often useful to 'silence' or 'knock down' genes. (This is not the same as a gene knockout, where the expression of a gene is entirely removed, for example by deleting it.) The reason for knocking down genes is to study gene function in the laboratory; or even to counteract the expression of deleterious genes in the context of gene therapy.

One of the ways in which genes can be knocked down is to block mRNA translation. An ingenious way of doing this is to exploit the fact that mRNA needs to become single-stranded where translation is occurring, which makes it amenable to targeting by **antisense oligonucleotides**.

One antisense approach is to use long stretches of antisense RNA, the theory being that the translation of the target mRNA can be blocked by base pairing with its complementary RNA. In practice, this approach is problematic; long antisense RNA is prone to degradation and is more difficult to transfect into target cells. It can also lead to non-specific effects, because a longer stretch of antisense RNA is more likely to have areas of sequence complementarity with unintended target mRNAs. In addition, antisense RNA is prone to attack by endogenous ribonucleases, which reduce its effectiveness. A better approach is the use of short antisense DNA oligodeoxyribonucleotides (often referred to simply as oligos).

A shorter antisense oligo (typically 20–25 bases long) is easier to transfect into cells, while being long enough to confer sequence target specificity. However, when designing antisense oligos, care must be taken to ensure that the target sequence is only found in the intended target mRNA. DNA oligonucleotides are also prone to degradation by endogenous nucleases. The oligonucleotides are degraded by nucleases that target single-stranded

DNA and they also attract the attention of RNAse H, a ribonuclease that degrades RNA in DNA–RNA hybrids.

How might it be possible to stabilize oligonucleotides and increase their half-life and effectiveness? The solution is to introduce chemical modifications into the oligonucleotides either in the sugar–phosphate backbone or even in the bases themselves (without impairing their ability to hybridize to complementary bases, i.e. their base pair specificity). The most successful and popular DNA-based antisense oligonucleotides are the **phosphorothioates**, in which an oxygen in the backbone phosphate is replaced by a sulphur atom.

Phosphorothioates have been particularly useful in the clinical setting. A notable example is fomivirsen, a drug that is in fact a 21-nucleotide antisense phosphorothioate oligo complementary to a sequence of the mRNA encoded by immediate early region 2 (IE2) of human cytomegalovirus. Hybridization of fomivirsen to its target viral mRNA results in the inhibition of IE2 protein synthesis which, in turn, results in the inhibition of viral replication. Fomivirsen is injected into the eyes of AIDS patients to counteract CMV-induced retinitis.

There are additional problems underlying the ability of antisense oligonucleotides to be effective. Effectiveness is influenced by accessibility of the mRNA target. mRNAs are not static molecules; they fold into complex structures through the intrinsic ability of single-stranded RNA to form intramolecular base pairs. Thus, the intended target sequence may be hidden from the antisense oligo by virtue of being part of an extensive double-stranded RNA structure such as a stemloop. On the other hand, the stemloop might exist in a dynamic state. For example, a stemloop in the 5′ UTR might be frequently unwound by the RNA helicase eIF4A during the process of translation initiation, and at this point any antisense oligo in the proximity may be able to find its target.

Morpholinos are modified oligonucleotides, in which the normal phosphodiester bond in the backbone is replaced by a non-ionic phosphorodiamidate linkage. A six-membered morpholine ring also replaces the deoxyribose group, giving the morpholinos their name. The modification of the backbone renders the morpholinos resistant to nucleases. Consequently, they have a longer half-life and therefore are more effective at knocking down gene expression. In addition, their backbone is non-ionic, which reduces the chances that they will interact with proteins, resulting in additional non-specific effects.

The thinking behind morpholinos is essentially the same as for other antisense oligo approaches; they bind a target sequence in an mRNA and thereby block its translation. An important difference between morpholinos and conventional antisense oligonucleotides is that, owing to their modified chemistry, they do not result in RNAse H cleavage of the RNA–DNA duplex when hybridized to target mRNA. If they are bound to targets within the ORF, they are prone to being displaced by the ribosome and associated machinery as it travels along the mRNA. For this reason, morpholinos are usually targeted to the 5′ UTR, or frequently to sequences around the native AUG start codon, with the aim of preventing translation from being initiated in the first place.

Morpholinos have become favourite tools in gene knockdown experiments in the study of development in model organisms, including most notably zebrafish, *Xenopus*, and sea urchins (particularly during early development).[45] Morpholinos can be injected directly into eggs or embryos. They are also effective in cell culture and have also been used in mammalian models. Morpholinos can be so effective that the protein encoded by the target mRNA actually becomes undetectable!

As with any antisense oligonucleotide experiment, appropriate controls must be incorporated into the experimental design. Ideally, the same effect should be obtained with more than one discrete morpholino. The same point applies to the use of small interfering RNAs (siRNAs) in gene knockdown experiments, as we shall see in Chapter 16. Control morpholinos are also used, including one with one or more base mismatches and a scrambled sequence (the morpholino's base sequence randomly rearranged).

There is considerable hope that morpholinos will be useful in the clinical setting. Examples include the treatment of cancer, cardiovascular disease, muscular dystrophy, and bacterial and viral diseases including hepatitis C, influenza A virus, West Nile virus, SARS, and Ebola.

Key points

Translation can be achieved in vitro in the laboratory using translation-competent extracts from plant, animal, yeast, or bacterial cells. Thus, specific proteins can be synthesized efficiently and their properties studied. Translation can also be blocked by using the principle of antisense technology. Antisense RNA, or antisense DNA oligodeoxyribonucleotides (oligos), can hybridize to an mRNA target and block translation. Oligonucleotides can be modified chemically, rendering them more stable and effective; examples include the phosphorothioates and morpholinos which can be used as experimental tools or even therapeutic agents.

Summary

- Messenger RNA was discovered in the 1950s in a series of classical experiments. It was shown to be the intermediate between DNA (the repository of genetic information) and polypeptide synthesis. Translation of mRNA occurs in ribosomes, with the aid of transfer RNAs, the molecular adaptors that associate mRNA codons with amino acids. Transfer RNAs are charged with specific amino acids by the amino-acyl tRNA synthetases.
- Ribosomes are exquisitely complex ribonucleoprotein molecular machines. They consist of large and small subunits, both of which contain ribosomal RNA (65%) and ribosomal proteins (35%).
- There are some minor exceptions to the genetic code, notably in mitochondria and the incorporation of the modified amino acid selenocysteine at the UGA stop codon.
- The signal recognition particle (SRP) is a ribonucleoprotein complex that binds to hydrophobic leader sequences on proteins destined for cell membranes or secretory pathways, temporarily stalling translation. The SRP then interacts with the signal receptor which brings the ribosome and nascent polypeptide in contact with the translocon on the endoplasmic reticulum (ER) membrane. The rest of the polypeptide is then synthesized through the translocon and into the ER space.
- Translation of mRNA is divided into three phases, all of which can be regulated: initiation, elongation, and termination. Protein factors facilitate each of these steps in both prokaryotes and eukaryotes, but rRNA also contributes key catalytic activities.
- There are several ways in which mRNA translation can be regulated. The presence of stemloops in the 5′ UTR can prevent translation, as can the modulation of levels and activities of initiation factors. The poly(A) tail is thought to stimulate initiation of translation through an interaction between the poly(A) binding protein and the initiation factor eIF4G. Some viral and cellular mRNAs use an IRES sequence to direct cap-independent translation initiation. MicroRNAs generally target the 3′ UTR to repress translation.
- Masked messages are an extreme example of translation regulation, in which mRNAs can be stored in a translationally inert state in animals and plants for periods as long as months. The process of unmasking is characterized by cytoplasmic polyadenylation.
- mRNA can be translated in vitro, with numerous applications. mRNA translation can be blocked with antisense oligonucleotides to study gene function. Antisense oligonucleotides can be modified to increase their half-life and efficiency (e.g. phosphorothioates and morpholinos). The inhibition of mRNA translation with antisense oligonucleotides can be also used in a therapeutic setting.

Questions

11.1 Describe the three phases of mRNA translation.

11.2 What is the function of the large and small ribosomal subunits?

11.3 How does rRNA work as a ribozyme in mRNA translation?

11.4 How does the structure of tRNA facilitate its job as a molecular adaptor?

11.5 What is the genetic code and how is it deciphered?

11.6 How is translation initiated, and how do translation elongation and termination occur?

11.7 Describe the role of the signal recognition particle in translation.

11.8 Give three examples of how translation can be regulated, in both prokaryotes and eukaryotes.

11.9 What is the structure and function of P-bodies and how does it relate to the regulation of mRNA translation?

11.10 What are masked messages, how were they discovered, and where might they be found?

11.11 How can translation be manipulated in vivo?

References

1. **Frank J, Agrawal RK, Verschoor A.** Ribosome structure and shape. In *Encyclopedia of Life Sciences*. Chichester: John Wiley (2001).
2. **Joseph S.** After the ribosome structure. How does translocation work? *RNA* **9**, 160–4 (2003).
3. **Stahl G, McCarty GP, Farabaugh PJ.** Ribosome structure: revisiting the connection between translational accuracy and unconventional decoding. *Trends Biochem Sci* **27**, 178–83 (2002).
4. **Ramakrishnan V.** Ribosome structure and the mechanism of translation. *Cell* **108**, 557–72 (2002).
5. **Fedor MJ, Williamson JR.** The catalytic diversity of RNAs. *Nat Rev Mol Cell Biol* **6**, 399–412 (2005).
6. **Schmeing TM, Huang KS, Kitchen DE, et al.** Structural insights into the roles of water and the 2′ hydroxyl of the P site tRNA in the peptidyl transferase reaction. *Mol Cell* **20**, 437–48 (2005).
7. **Brunelle JL, Shaw JJ, Youngman EM, Green R.** Peptide release on the ribosome depends critically on the 2′ OH of the peptidyl-tRNA substrate. *RNA* **14**, 1526–31 (2008).
8. **Erlacher MD, Polacek N.** Ribosomal catalysis: the evolution of mechanistic concepts for peptide bond formation and peptidyl-tRNA hydrolysis. *RNA Biol* **5**, 5–12 (2008).
9. **Green R, Lorsch JR.** The path to perdition is paved with protons. *Cell* **110**, 665–8 (2002).
10. **Simonovic M, Steitz TA.** A structural view on the mechanism of the ribosome-catalyzed peptide bond formation. *Biochim Biophys Acta* **1789**, 612–23 (2009).
11. **Goldman E.** Transfer RNA. *Encyclopedia of Life Sciences*. Chichester: John Wiley (2008).
12. **Arnez J, Moras D.** Aminoacyl-tRNA synthetases. *Encyclopedia of Life Sciences*. Chichester: John Wiley (2009).
13. **Rich A, Kim SH.** The three-dimensional structure of transfer RNA. *Sci Am* **238**, 52–62 (1978).
14. **Tuite M.** Transfer RNA in decoding and the wobble hypothesis. *Encyclopedia of Life Sciences*. Chichester: John Wiley (2001).
15. **Hinnebusch AG.** Molecular mechanism of scanning and start codon selection in eukaryotes. *Microbiol Mol Biol Rev* **75**, 434–67 (2011).
16. **Maiti B, Arbogast S, Allamand V, et al.** A mutation in the SEPN1 selenocysteine redefinition element (SRE) reduces selenocysteine incorporation and leads to SEPN1-related myopathy. *Hum Mutat* **30**, 411–16 (2009).
17. **Craig N, Cohen-Fix O, Green R, et al.** *Molecular Biology: Principles of Genome Function* (2nd edn). Oxford: Oxford University Press (2014).
18. **Sonenberg N, Hinnebush A.** Regulation of translation initiation in eukaryotes: mechanisms and biological targets. *Cell* **136**, 731–45 (2009).
19. **Egea PF, Stroud RM, Walter P.** Targeting proteins to membranes: structure of the signal recognition particle. *Curr Opin Struct Biol* **15**, 213–20 (2005).
20. **Graumann P, Marahiel M.** A superfamily of proteins that contain the cold-shock domain. *Trends in Biochem Sci* **23**, 286–90 (1998).
21. **Phadtare S, Severinov K.** Nucleic acid melting by *Escherichia coli* CspE. *Nucleic Acids Res* **33**, 5583–90 (2005).
22. **Serganov A, Huang L, Patel DJ.** Structural insights into amino-acid binding and gene control by a lysine riboswitch. *Nature* **455**, 1263–8 (2008).
23. **Wouters BG, Van den Beucken T, Magagnin MG, et al.** Control of the hypoxic response through regulation of mRNA translation. *Semin Cell Dev Biol* **16**, 487–501 (2005).
24. **Rouault TA.** The role of iron regulatory proteins in mammalian iron homeostasis and disease. *Nat Chem Biol* **8**, 406–14 (2006).
25. **Kozak M.** Regulation of translation via mRNA structure in prokaryotes and eukaryotes. *Gene* **161**, 13–37 (2005).
26. **Reynolds N, Cooke H.** Role of the DAZ genes in male fertility. *Reprod Biomed Online* **10**, 72–80 (2005).
27. **Goldman E.** Translation control by RNA. *Encyclopedia of Life Sciences*. Chichester: John Wiley (2008).
28. **Spilka R, Ernst C, Kuldeep Mehta A, Haybaeck J.** Eukaryotic translation initiation factors in cancer development and progression. *Cancer Lett* **340**, 9–21 (2013).
29. **Wilkie GS, Dickson KS, Gray NK.** Regulation of mRNA translation by 5′- and 3′-UTR-binding factors. *Trends Biochem Sci* **28**, 182–8 (2003).
30. **Pasquinelli AE.** MicroRNAs: deviants no longer. *Trends Genet* **18**, 171–3 (2002).
31. **Cannell IG, Kong YW, Bushell M.** How do microRNAs regulate gene expression? *Biochem Soc Trans* **36**, 1224–31 (2008).
32. **Liu J.** Control of protein synthesis and mRNA degradation by microRNAs. *Curr Opin Cell Biol* **20**, 214–21 (2008).
33. **Sanford JR, Gray NK, Beckmann K, Cáceres JF.** A novel role for shuttling SR proteins in mRNA translation. *Genes Dev* **18**, 755–68 (2004).
34. **Michlewski G, Sanford JR, Cáceres JF.** The splicing factor SF2/ASF regulates translation initiation by enhancing phosphorylation of 4E-BP1. *Mol Cell* **30**, 179–89 (2008).
35. **Sanford JR, Coutinho P, Hackett JA, et al.** Identification of nuclear and cytoplasmic mRNA targets for the shuttling protein SF2/ASF. *PLoS ONE* **3**, e3369 (2008).

36. **Aitkozhina MA, Belitsina NV, Spirin AS.** Nucleic acids in the early stages of development of fish embryos (based on the loach *Misgurnus fossilis*). *Biokhimia* **29**, 169–75 (1964).
37. **Masters AK, Shirras AD, Hetherington AD.** Maternal mRNA and early development in *Fucus serratus. Plant J* **2**, 619–22 (1992).
38. **Meric F, Searfoss AM, Wormington M, Wolffe API.** Masking and unmasking maternal mRNA. *J Biol Chem* **48**, 30 804–10 (1996).
39. **Standart N, Dale M, Stewart E, Hunt T.** Maternal mRNA from clam oocytes can be specifically unmasked in vitro by antisense RNA complementary to the 3′-untranslated region. *Genes Dev* **4**, 2157–68 (1990).
40. **Colegrove-Otero L, Minshall N, Standart N.** RNA-binding proteins in early development. *Crit Rev Biochem Mol Biol* **40**, 21–73 (2005).
41. **Parker R, Sheth U.** P bodies and the control of mRNA translation and degradation. *Mol Cell* **25**, 635–46 (2007).
42. **Minshall N, Kress M, Weil D, Standart N.** Role of the p54 RNA helicase activity and its C-terminal domain in translational repression, P body localization and assembly. *Mol Biol Cell* **20**, 2464–72 (2009).
43. **Corey DR, Abrams JM.** Morpholino antisense oligonucleotides: tools for investigating vertebrate development. *Genome Biol* **2**, reviews1015, 1–3 (2001).
44. **Audic Y, Boyle B, Slevin M, Hartley RS.** Cyclin E morpholino delays embryogenesis in *Xenopus. Genesis* **30**, 107–9 (2001).
45. **Thummel R, Bai S, Sarras MPJr, et al.** Inhibition of zebrafish fin regeneration using *in vivo* electroporation of morpholinos against fgfr1 and msxb. *Dev Dynamics* **235**, 336–46 (2006).

12 Stability and degradation of mRNA

Introduction

The half-life of RNA molecules varies greatly (the half-life is the time it takes for half of a population of molecules to degrade). In this chapter we shall consider the stability of mRNA, which is highly regulated.[1,2] When mRNA was first discovered in the mid-twentieth century (as described in Chapter 11), one of its defining properties was its transient nature. In yeast, the half-life of mRNA varies from 1 to 100 minutes; and in mammalian cells it varies from less than 20 minutes to as much as 50 hours. Regulating the stability of mRNA provides another means of controlling gene expression; if an mRNA is more stable, more protein molecules can be synthesized from it.

12.1 Messenger RNAs have a half-life

A classical way to demonstrate the stability of mRNA is to block transcription with the poison α-amanitin, a cyclic eight amino acid peptide found in the *Amanita* genus of mushrooms, which includes the famous 'death-cap', *Amanita phalloides*. α-Amanitin specifically blocks the activity of RNA polymerase II. As a result, no new mRNA molecules are transcribed. When pre-mRNA transcription is stopped, the existing pool of mRNAs will progressively decay over several hours. Those mRNAs that are less stable disappear faster. The amount of protein produced from an mRNA, which gives a measure of its productivity and biological activity, is a function of the rate of translation initiation and also of the half-life of an mRNA. An mRNA that only survives for a short time may nonetheless produce a substantial amount of protein if rates of translation are high. In contrast, a more stable mRNA that is poorly translated may produce less protein overall. Thus, the productivity of each mRNA is different and is influenced by mRNA stability. Figure 12.1 illustrates a classical experiment that demonstrates the variable stability of an mRNA. The mRNA shown in Fig. 12.1 is *IL-3* (interleukin 3, a cytokine involved in stimulating immune response[3]). *IL-3* expression can be stimulated by the drug ionomycin, an ionophore that raises intracellular calcium. Ionomycin treatment also causes the stabilization of *IL-3* mRNA.

Other examples of mRNAs with differing half-lives include *casein* and c-*myc*. Casein is expressed in the mammary gland; the half-life of *casein* mRNA increases from 1 to 40 hours in response to the hormone prolactin. This makes sense, because the function of prolactin is to promote the production of milk, of which casein is a major component. At the other extreme, the stability of the proto-oncogene c-*myc* mRNA (which encodes a transcription factor that promotes cell proliferation) decreases from 35 minutes to less than 10 minutes in response to stimuli that promote cellular differentiation. In both cases, an environmental or physiological cue specifically affects the stability of the mRNA by either promoting or preventing **mRNA decay** or degradation.

How, then, is the stability of mRNA regulated? Is it purely a function of its secondary structure

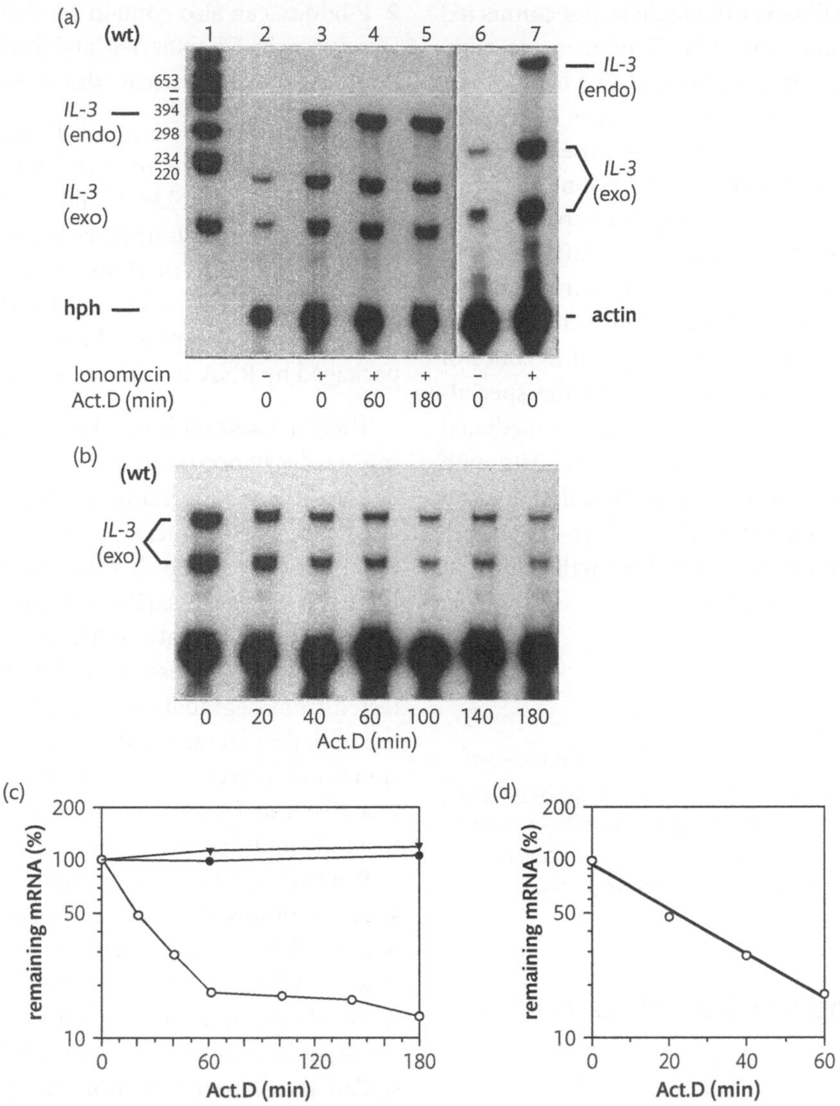

Figure 12.1 Expression and stability of *IL-3* (interleukin 3) **mRNA.** Panels (a) and (b) show the results of an RNAse protection assay in which a specific fragment of an mRNA is protected from degradation by RNAse by hybridizing to a radiolabelled DNA probe of specific length. Fragments labelled endo represent the *endogenous* expressed *IL-3* mRNA, whereas the two fragments labelled exo are derived from *exogenous IL-3* expressed from an engineered plasmid construct transfected into a cell line. (a) Lane 1, size marker; lanes 2 and 6, RNA from unstimulated cells; lanes 3 and 7, cells treated with the drug ionomycin to induce *IL-3*; lanes 4 and 5, actinomycin D was also added to block transcription. Hph and actin serve as a loading control. (b) Decay of exogenous *IL-3* mRNA can be clearly observed in unstimulated cells following actinomycin D treatment. (c), (d) Quantitation of results from (a) and (b), with *IL-3* signals normalized to hph. Exogenous *IL-3* (open circles) decays quite rapidly in unstimulated cells, indicating that it is actively degraded. In stimulated cells, endogenous *IL-3* (closed triangles) and exogenous *IL-3* (closed circles) are stabilized as mRNA levels do not drop even after transcription is blocked by actinomycin D. (With permission from Stoecklin et al.[3])

or of the length of the poly(A) tail, for example? Research focused on the stability of several mRNAs has revealed the existence of specific RNA elements that can govern mRNA stability. Not surprisingly, RNA-binding proteins recognize specific stability elements, and in turn interact with the mRNA degradation machinery as either single **ribonucleases** (RNAses) or complexes of proteins that contain ribonuclease activity. Ribonuclease activities can be subdivided into exonucleases (which work 5′→3′ or 3′→5′ from the free ends of an RNA molecule) or endonucleases (which cleave an RNA internally).

This chapter will cover the main issues connected with mRNA decay—an area of intense research. Broadly speaking, there are two reasons for the evolution of regulated mRNA decay. One is to regulate the amount of protein produced, and the other is to eliminate faulty mRNAs that could potentially produce toxic proteins—a type of quality control (also known as **mRNA surveillance**). Special aspects of mRNA decay will be presented, including the connection between extracellular stimuli and mRNA decay, the role of P-bodies as cytoplasmic zones of concentrated mRNA degradation, and the specialized decay processes known as 'nonsense-mediated decay', 'no-go decay', and 'non-stop decay'. Although the chapter focuses on the stability and degradation of mRNA, the reader should be aware that other classes of RNA molecules also have half-lives that are, in many cases, regulated.

Key points

mRNAs, like all macromolecules, have specific half-lives. Their stability is regulated and their degradation is spatially organized within the cell, providing yet another means of regulating gene expression. Faulty mRNAs are actively degraded, minimizing the production of toxic proteins.

12.2 Sites and mechanisms of mRNA degradation

It has become increasingly clear that the degradation of mRNA is spatially organized in the cell—in other words it does not happen to an equal extent throughout the cell. The best-studied sites of mRNA degradation are the **P-bodies**, also known as the Dcp (decapping) or GW bodies. These are cytoplasmic foci, often described as granules, where there is a very high concentration of several proteins involved in mRNA degradation.[4–7]

P-bodies can be subdivided into three parts.

1 The core of P-bodies is a series of proteins involved in mRNA decay, and includes the decapping complex Dcp1/Dcp2, the RNA helicase Dhh1 (RCK/p54 in vertebrates), the 5′→3′ exonuclease Xrn1, and the deadenylase Ccr4, a 3′→5′ exonuclease.

2 P-bodies can also contain microRNA components (discussed in Chapter 16) and proteins involved in nonsense-mediated decay (discussed in Section 12.5).

3 The third component of P-bodies is mRNA—in fact, mRNAs are required for the assembly of P-bodies. The mRNAs found in P-bodies are not engaged in translation; there are no ribosomes and translation factors in P-bodies so translation cannot occur. Thus P-bodies are filled with aggregates of translationally repressed mRNP particles (mRNA packaged by RNA-binding proteins).

There are several lines of evidence to suggest that mRNA decay occurs in P-bodies. When the mRNA decay pathways are stalled, P-bodies change in size and mRNA decay intermediates are observed within them. However, not all mRNAs that have entered P-bodies are necessarily degraded. Instead, some can return to translation. Messenger RNAs that are translationally repressed in P-bodies may be protected from degradation by binding of the 5′ cap to the initiation factor eIF4E. This translational repression is not irreversible and so the mRNAs can return to active translation. The proposed traffic of mRNAs to and from P-bodies is shown in Fig. 12.2.

Whereas P-bodies are present in unstressed cells, **stress granules** (SGs) are cytoplasmic structures in which mRNAs accumulate in response to cellular stress (including oxidative stress, heat shock, osmotic shock, and viral infection). Specifically, SGs are thought to be sites of accumulation of mRNAs in stalled translation initiation complexes.[8] However, SGs also contain proteins involved in the regulation of mRNA stability (HuR and TTP, which will be discussed in Section 12.3). Indeed, P-bodies and SGs are thought to interact physically, and they share several protein components. For example, RAP55/SCD6 is an evolutionarily conserved protein first detected in humans as an autoantigen in primary biliary cirrhosis, and is associated with translationally repressed mRNAs. RAP55 is detected in both P-bodies and SGs. Depletion of RAP55 also inhibits the formation of P-bodies.[9]

SGs may represent the site of systematic translational silencing of mRNAs which need to be repressed in the context of cellular stress so that the stressed cell can prioritize the translation of only those mRNAs that encode proteins required to

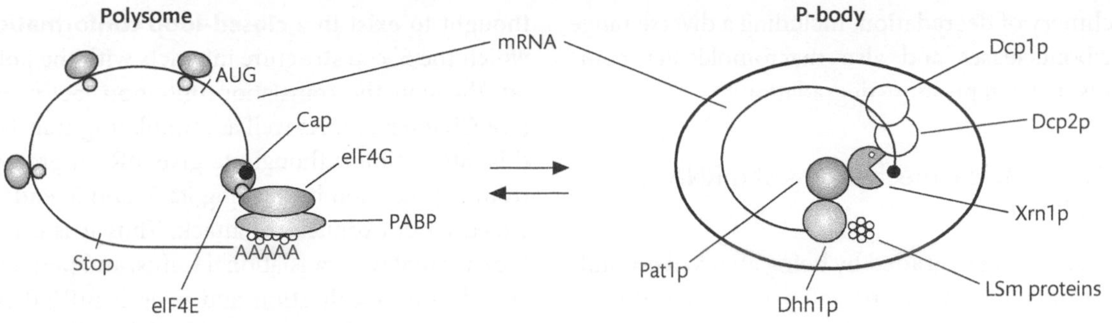

Figure 12.2 Transport of mRNAs from polyribosomes to P-bodies. Actively translating mRNAs form polyribosomes, aided by the interaction between the poly(A) tail bound by poly(A)-binding protein (PABP) and the translation initiation factor eIF4G. However, when translation is repressed, the mRNAs move to the P-bodies to be degraded or stored. P-bodies contain the decapping enzymes (Dcp1p and Dcp2p), the 5′→3′ exonuclease Xrn1, the RNA helicase Dhh1, and other components. (From Newbury[10] with permission from Portland Press.)

adapt to the environmental stress. The interaction of SGs with P-bodies represents a coordinated effort to repress translation and also promote mRNA decay of specific classes of stress-associated transcripts.

A main trigger for the formation of SGs is the phosphorylation of the eukaryotic translation initiation factor eIF2α. This initiation factor is phosphorylated during cellular stress, which leads to a decrease in the levels of the ternary complex eIF2–tRNA–GTP and hence reduced translation initiation. Two RNA-binding proteins, the translation regulators TIA-1 and TIAR, then bind the pre-initiation complex, resulting in polysome disassembly and formation of the SGs.

The process called 'RNA interference' and small non-coding RNAs known as 'microRNAs' are associated with a special form of targeted mRNA degradation that is covered in detail in Chapter 16. MicroRNAs and the RNA interference pathway repress the expression of genes by selective degradation or translational silencing of target mRNAs. These processes are briefly mentioned here because there is evidence to suggest that they interact with P-bodies; microRNAs and mRNAs that are targeted for translational repression or even degradation by RNA interference are both directed to P-bodies, where they are effectively sequestered from the translation machinery and come into contact with the general machinery of mRNA decay.[11] What is the evidence for this interaction? Some of the proteins associated with the machinery of RNA interference (namely, the Argonaute proteins discussed in Chapter 16) and also microRNAs can be detected in P-bodies. Blocking the RNA interference pathway or microRNA biogenesis pathways prevents the formation of P-bodies.

> **Key points**
>
> mRNA decay is spatially organized in the cytoplasm, particularly in P-bodies and stress granules.

12.3 The process of mRNA degradation

How does mRNA degradation happen? Degradation occurs in the cytoplasm, most likely in dedicated structures such as the P-bodies, and requires ribonucleases, i.e. enzymes that specialize in the degradation of RNA. There are several mechanisms through which degradation of mRNA might occur. Perhaps some mRNAs have a half-life that is influenced by their structure—the three-dimensional shape they adopt as a result of secondary structure in the mRNA and the way that it is packaged by RNA-binding proteins into an mRNP particle. Some mRNP particles might, by implication, be inherently more stably packaged. The degradation machinery could simply strip from an mRNA the proteins that otherwise protect it from degradation, and then rely on random collisions with ribonucleases for it to be degraded. However, this mechanism of degradation is not likely to be efficient. Instead, evolution has generated a complex and highly mRNA-specific machinery of degradation, including a diverse range of ribonucleases and also macromolecular complexes that can promote degradation.[1–3]

12.3.1 Principal mechanisms of mRNA degradation

mRNAs can be degraded by being attacked on multiple fronts. In other words, all parts of an mRNA—the 5′ cap, the 5′ UTR, the coding region, the 3′ UTR, and the poly(A) tail—can contribute to the regulation of mRNA stability. The principal mechanisms of mRNA degradation are summarized in Fig. 12.3.

mRNA degradation is also functionally linked with the repression of translation. This is because there is evidence that actively translating mRNAs are protected from degradation, whereas mRNAs that are not yet being translated can be targeted for degradation to prevent further translation taking place. mRNAs that are actively translated are thought to exist in a **closed-loop conformation** in which the 5′ cap structure interacts with the poly(A) tail through the translation initiation factor eIF4G (see Chapter 11). As well as stimulating translation, this interaction is thought to give mRNA protection from degradation by keeping its 5′ and 3′ ends protected from exonuclease attack. Thus it is certainly the case that the translational status, and perhaps the cytoplasmic localization and overall mRNP structure, may contribute to an mRNA's half-life.

The best-studied examples of regulated mRNA decay involve a highly sophisticated degradation machinery that works on specific mRNA sequence elements. There are several essential processes in mRNA degradation: **decapping** (removal of the cap at the 5′ end of the mRNA), **endonucleolytic cleavage** (breaking the bond between adjoining ribonucleotides within an RNA), 5′→3′ and 3′→5′ exonucleolytic degradation, for example **deadenylation** (removing the poly(A) tail at the 3′ end of the mRNA). As for the order in which they occur,

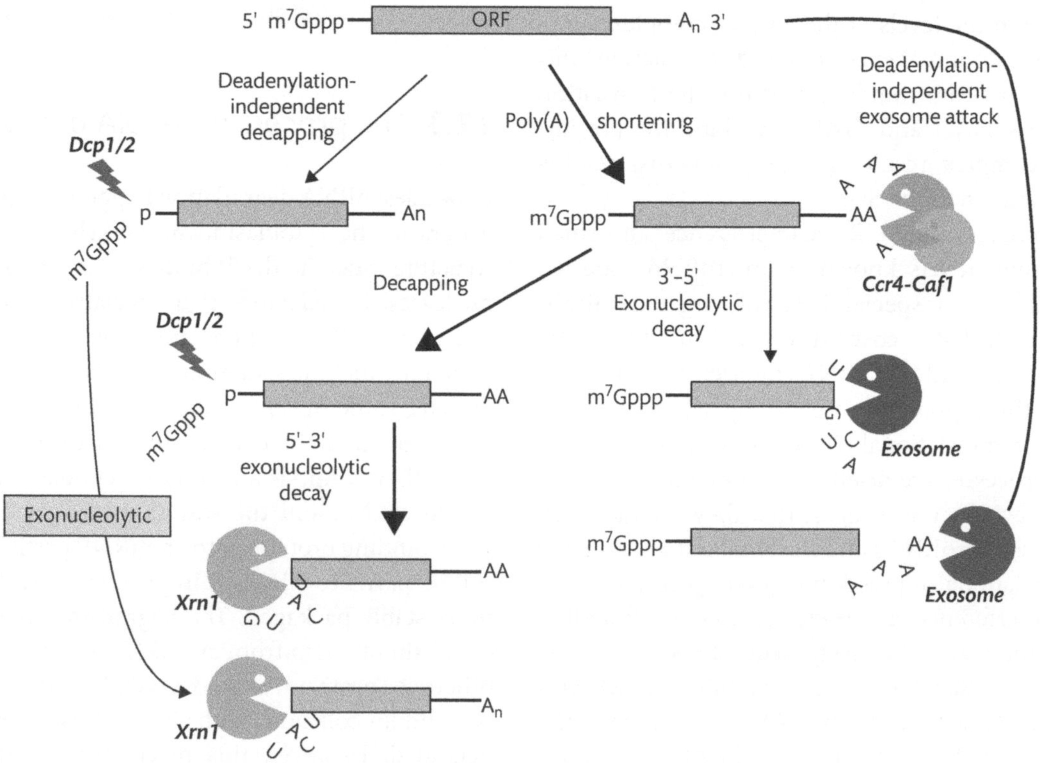

Figure 12.3 Eukaryotic mRNA decay pathways in the cytoplasm. The large arrows indicate the major pathway for mRNA degradation, triggered by deadenylation. (With permission from Chen and Shyu.[2])

evidence shows that most mRNA decay pathways that have been characterized in eukaryotes begin with deadenylation.[3] These processes have been studied in great detail in yeast. However, the main principles also apply to other eukaryotes. Let us consider each of these aspects in turn.

Decapping of the 5′ cap structure is achieved by the decapping proteins Dcp1 and Dcp2, which exist as a complex. This decapping complex has a preference for substrates that are more than 25 nucleotides long, which therefore protects mRNAs on which the translation initiation complex has assembled because translation initiation factors mask the very 5′ end of an mRNA.

Once the cap is removed, the 5′ end is attacked by the exonuclease Xrn1—a highly conserved processive exoribonuclease with homologues in all eukaryotes.[12] A processive nuclease moves in a given direction, digesting one nucleotide after another; the prefix 'exo' indicates that its activity begins from a free end of the mRNA. Xrn1 works from the 5′ end, releasing mononucleotides as it processes in a 5′→3′ direction. The *Drosophila* Xrn1 orthologue is known as Pacman; mutants exhibit low viability and display several developmental defects including faulty spermatogenesis. Xrn1 proteins are not only involved in standard mRNA decay pathways, but are also involved in RNA interference and nonsense-mediated decay and are detected in P-bodies, the sites of concentrated mRNA decay.

In contrast with 5′ degradation, deadenylation is required to begin the attack at the 3′ end of an mRNA. There are several deadenylases which specialize in the destruction of the poly(A) tail.[13] The best studied deadenylase activity in mammalian cells is provided by PARN (polyadenylate ribonuclease). Once the poly(A) tail is removed, the 3′ end of an mRNA is available for exoribonucleases that work in a 3′→5′ direction. Note that there are several deadenylases with distinct functional properties and expression patterns, which are thus able to confer tissue-specific regulation by controlling poly(A) tail lengths. The human deadenylases CCR4, CAF1, and PARN shuttle between nucleus and cytoplasm; CCR4 and CAF1 are mostly cytoplasmic and can be recruited to P-bodies.[13]

3′→5′ degradation is not mediated by a single exonuclease, but rather by a protein complex called the **exosome**[14,15] whose composition is illustrated in Fig. 12.4. The exosome is highly conserved in eukaryotes, and consists of a core of nine subunits that form a ring-like structure which envelops the target RNA. It contains several ribonucleases in association with the SKI complex of RNA helicases, which aid mRNA degradation by unwinding higher-order structures. The exosome has multiple RNA-processing functions in both the nucleus (where it is involved with the processing and degradation of snRNAs, snoRNAs, and rRNAs) and the cytoplasm, where it is principally involved in mRNA turnover. In some circumstances, most notably in RNA interference (Chapter 16) and nonsense-mediated decay (the destruction of mRNAs with premature stop codons, covered in Section 12.5), mRNA degradation is initiated by endonucleolytic cleavage. Internal cleavage of mRNA produces free 5′ and 3′ ends, which can in turn be attacked by Xrn1 or the exosome, respectively. Figure 12.3 summarizes the principal mechanisms of mRNA degradation in eukaryotes. Exosomes are also enriched in cytoplasmic granules known as *exosome granules*, whereas exosome components are not detected in P-bodies. Exosome granules are enriched with mRNAs that contain **AU-rich elements** (AREs);[16] these are described in the next section.

12.3.2 RNA elements that influence mRNA stability: the ARE

Most of the known RNA elements that govern mRNA stability are found in the 3′ UTR. The most widely studied is the AU-rich element (ARE), which is present in up to 8% of human mRNAs.[17] AREs are often, but not exclusively, present in the 3′ UTR. The ARE was discovered in the 1980s when its removal from the 3′ UTR of c-*fos* mRNA enhanced the ability of c-*fos* to transform cells (i.e. to make cells divide uncontrollably).[18] Fos is an oncogenic transcription factor whose expression is highly regulated. Why does the removal of the ARE from the 3′ UTR enhance the transformation potential of c-*fos*? The answer lies in the way it acts as a destabilizing element of the c-*fos* mRNA. When AREs are transplanted into a heterologous mRNA, the recipient mRNA acquires the instability (or short half-life) associated with the parent mRNA.

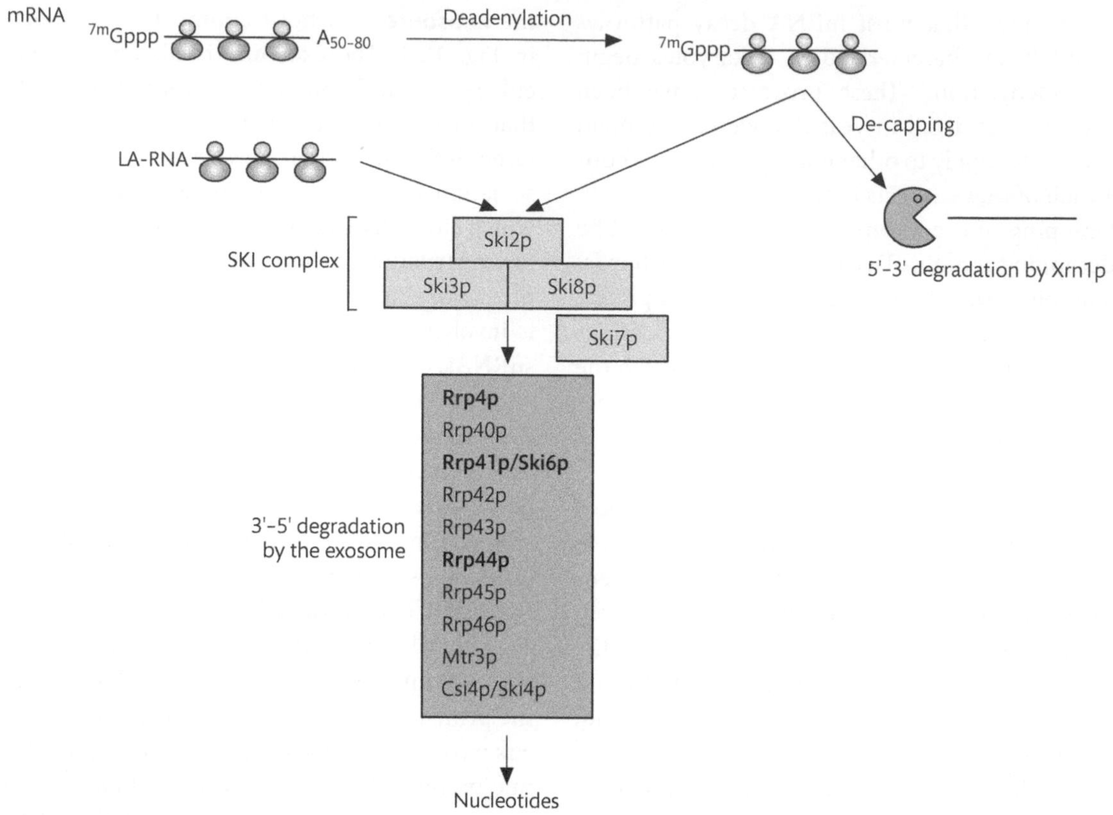

Figure 12.4 **Composition, substrates, and products of the cytoplasmic exosome.** The composition of the exosome has been worked out in great detail in yeast. In yeast, the predominant pathway for mRNA decay starts with deadenylation and is followed by decapping. When decapped, mRNAs can be attacked at the 5′ end by Xrn1p. Note that LA virus mRNAs are naturally unadenylated; like other mRNAs they can interact with the SKI complex (Ski2p, Ski3p, and Ski8p). The latter contains RNA helicase activity (Ski2p) which helps to accelerate degradation and delivers the substrate to the exosome via Ski7p. (Note that the suffix p, as in Xrn1p in yeast denotes the protein.) The core exosome contains several proteins; essentially an assortment of ribonucleases (boxed). Note that the nuclear exosome is slightly different in composition. (Reprinted from Butler[14] with permission from Elsevier.)

Several mRNAs that contain AREs encode oncogenes and cytokines, whose products are necessarily short-lived or are required in bursts of expression (see Table 12.1). Therefore their stability needs to be tightly controlled. The specific sequences in the AREs are varied, but typically comprise the pentamer AUUUA and the nonamer UUAUUUA(U/A)(U/A). AREs are divided into three classes: class I comprises one to three copies of the pentamer AUUUA next to a U-rich region; class II comprises two or more overlapping copies of the nonamer also located next to a U-rich region; class III contains a U-rich region without the pentamer or nonamer. These three ARE classes are listed in Table 12.2. It is worth noting that AREs are very important in mRNAs associated with cancer, a notable example being the proto-oncogene c-*myc*.[19] Cell signalling pathways that are often activated in cancer also have a direct effect on mRNA turnover.[20,21]

How, then, do AREs influence mRNA stability? AREs are recognized by RNA-binding proteins, examples of which are tristetraprolin (TTP), AUF1, and Hu antigen R (HuR). TTP is a zinc-finger protein whose expression is rapidly induced by insulin, serum, phorbol esters, and other signals. A mouse knockout of TTP resulted in the accumulation of tumour necrosis factor alpha (TNFα) mRNA. Biochemical experiments then demonstrated that

Table 12.1 Messenger RNAs that contain AREs (AU-rich elements)

Gene	Function	ARE and notable observations
c-fos	Transcription factor activated in several cancers	AREs in the 3′ UTR make it unstable
		Viral fos mRNA lacks the ARE
c-myc	Transcription factor activated in several cancers	Loss of AREs reported in cancer
Cyclins	Control checkpoints in cell cycle progression	Increased stability of cyclin mRNA mediated by HuR binding to AREs
VEGF	Growth factor required for angiogenesis	Hypoxia stabilizes its mRNA by HuR binding to AREs
Cox-2	Catalyses the production of prostaglandins	mRNA stabilized by binding of CUGBP2 to ARE
	Over-expressed in cancer	
Bcl-2	Inhibitor of apoptosis	mRNA stabilized by nucleolin binding to ARE
	Over-expressed in several leukaemias	

Table 12.2 Different types of *cis*-element associated with mRNA stability

cis-element	Motif	Example of mRNA
ARE class I	Dispersed AUUUA motifs within or near U-rich regions	*c-fos*
ARE class II	Overlapping AUUUA motif within or near U-rich regions	GM-CSF
ARE class III	U-rich region but not AUUUA	*c-jun*
Iron-responsive element (IRE)	Stemloop structure	Transferrin receptor (*TfR*)
Poly(C)-binding element	Poly(C) region	*α-globin*

GM-CSF, granulocyte-macrophage colony-stimulating factor.

There are three types of ARE (AU-rich regions), but also different types of elements, including the iron-responsive element (IRE) and the C-rich sequence in α-globin mRNA.

Adapted from Miller et al.[18]

TTP binds to an ARE in the 3′ UTR of TNFα mRNA in some way, causing its destabilization. AUF1, also known as hnRNP D, binds AREs via RNA recognition motifs (RRMs) (RNA-binding domains described in Chapter 4) and promotes mRNA decay. AUF1 interacts with the translation initiation factor eIF4G and with poly(A)-binding protein; this interaction is interesting because it suggests that AUF1 can sense the translational status of an mRNA and then promote mRNA decay accordingly.

In contrast, HuR proteins generally have the opposite effect; their binding to AREs *stabilizes* mRNA. HuR proteins belong to a family of RNA-binding proteins first characterized in *Drosophila*—the ELAV proteins (embryonic-lethal abnormal vision)—which are associated with neural development (Box 12.1).[22] HuR contains three RRMs, of which two are involved specifically in binding AREs. How do HuR proteins stabilize mRNA? A possible explanation is that they compete for binding to AREs with proteins that destabilize mRNAs. Thus, when HuR levels are low the AREs are normally bound by proteins that promote degradation. But when HuR levels increase, they mask the AREs, preventing degradation and stabilizing the mRNA. Examples of HuR at work are given in Box 12.1.

The principal catalytic activity associated with ARE-mediated mRNA decay is thought to be

Box 12.1 Multiple targets in mRNA decay for a multifunctional HuR

Human HuR is a member of the ELAV (embryonic-lethal abnormal vision) family of RNA-binding proteins which includes HuB, HuC, and HuD. HuR is involved in differentiation, the regulation of cell division, the response to nutrient availability, and promoting malignant phenotypes and the expression of proteins that regulate apoptosis. HuR is well known for promoting the stability of several mRNAs, including VEGF (vascular endothelial growth factor), c-*fos*, and various cyclin mRNAs.

HuR is a shuttling protein which, like several other RNA-binding proteins, works in both the nucleus and cytoplasm. Its roles are not confined to the regulation of mRNA stability; HuR is also involved in the regulation of mRNA translation and localization. When its cytoplasmic levels are high, it promotes the expression of anti-apoptotic genes by regulating both mRNA stability and translation. HuR is also involved in the genotoxic response. It accumulates in the cytoplasm in response to ultraviolet radiation and other genotoxic stimuli, where it stabilizes mRNAs that encode proteins involved in DNA repair. HuR's ability to block apoptosis and promote DNA repair, taken together, explain how it works in a complex way to promote cell survival.

provided by the exosome, as immunodepletion (using an antibody specific to an exosome component to precipitate out and thereby reduce levels of the exosome) stabilizes ARE-containing mRNAs. TTP also co-precipitates with the exosome, suggesting that it helps to recruit the exosome to mRNAs that contain AREs, as shown in Fig. 12.5.

12.3.3 Other RNA elements that influence mRNA stability

Every part of an mRNA can influence its stability. For example, the 5′ UTR of interleukin 2 (*IL-2*) mRNA contains an mRNA stability element known as the JRE (JNK-responsive element). JNK is a mitogen-activated protein kinase that phosphorylates Jun, a partner of the oncogenic transcription factor Fos. The JRE is required for the stabilization of *IL-2* mRNA during T-cell activation.

Some mRNA even contains elements in the coding region that promote mRNA decay. An example is c-*fos*; while this mRNA includes AREs in the 3′ UTR, as previously discussed, it also includes a purine-rich element in the ORF that contributes to its instability. c-*myc* also contains mRNA instability elements within its ORF. Furthermore, the ability of the c-*myc* RNA elements to induce instability is dependent on translation, further illustrating the connections that exist between mRNA turnover and translation.

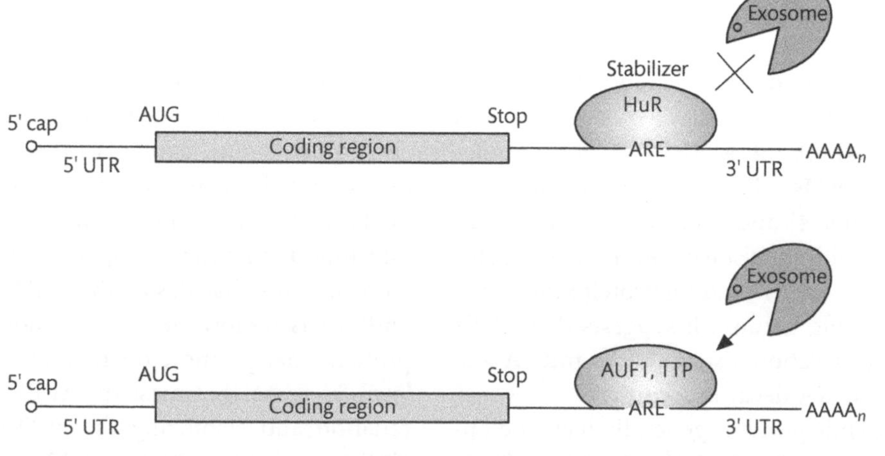

Figure 12.5 AU-rich elements (AREs) and mRNA decay. AREs are often, but not exclusively, present in the 3′ UTR. When bound by proteins that promote stability (HuR) the mRNA is protected from the exosome. In contrast, proteins that promote instability (AUF1 and TTP) recruit the exosome.

The presence of stability elements within the ORF is perhaps surprising, in the sense that the ORF encodes amino acids. In other words an AU-rich element in the ORF may lead to the incorporation of specific amino acids in the protein which are not needed by the protein. However, the stability element may be embedded in a part of the ORF where the amino acid sequences may be less crucial.

Key points

There are three essential steps in the initiation of mRNA decay: decapping, deadenylation, and cleavage. Cleavage is mediated by endonucleases. Exonucleases digest RNA free 5′ ends (Xrn1) or 3′ ends (deadenylases and the exosome). AU-rich elements (AREs) are present in several mRNAs whose stability is regulated, including oncogenes and cytokines. Other classes of RNA element also exist and can be located in all the parts of an mRNA.

12.4 Extracellular stimuli influence the stability of mRNA

By now it should be clear that at the heart of post-transcriptional gene regulation is the ability of RNA-binding proteins (as ***trans*-acting factors**) to recognize specific RNA sequences and structures (as ***cis*-acting elements**). Once a protein–RNA interaction is defined, it is possible to understand the binding and recognition in exquisite detail, and to explore the functional consequences of the interaction. We saw in the preceding section how the binding of different RNA-binding proteins to characteristic RNA sequences mediates the control of mRNA stability. But what regulates the interaction between protein and RNA? In other words, what makes an RNA-binding protein able to bind to its intended target, and how does it 'know' when to do so?

The answer is that there are well-characterized cell signalling pathways which are able to induce changes in mRNA stability.[23] The protein kinase PKC is involved in regulation of the stability of several mRNAs involved in apoptosis (*Bcl-2*), nitric oxide (NO) mediated signalling (i-*NOS*), and inflammation (*IL-1α*). Hypoxia results in the stabilization of VEGF, which promotes cellular growth and angiogenesis. Calcium affects the stability of mRNAs involved in haematopoiesis (*GM-CSF*) and in the immune response (*IL-2* and *IL-3*). The MAP kinase p38 also contributes to the stabilization of mRNAs involved in the immune response (*IL-3*, *IL-6*, and *IL-8*) and inflammation (*Cox*-2 and *TNFα*).

Having determined the signalling pathways involved, the question then arises: what are the mechanisms that result in altered mRNA stability? Clearly, there must be mechanisms through which cell signalling alters the properties of RNA-binding proteins. To illustrate this point, consider tristetraprolin (TTP), one of the targets of the MAPK pathway.[24] As described in Section 14.3, TTP binds AREs and promotes instability. TTP binds the ARE more avidly when hypophosphorylated. (Phosphorylation is the addition of a phosphate group commonly to a serine, threonine, or tyrosine residue within a polypeptide; *hypo*phosphorylation indicates that phosphorylation occurs to a very limited degree.) Hence, the phosphorylation of TTP via the MAPK pathway inhibits its ability to bind target AREs, resulting in the stabilization of mRNAs that contain them. Therefore the phosphorylation can directly impede the ability of TTP to bind to RNA.

The example of insulin further illustrates how extracellular stimuli can affect the stability of mRNA in response to physiological needs. At the heart of diabetes is a failure of appropriate insulin production or secretion. Rates of transcription, translation, and splicing all influence insulin levels, but so does mRNA stability, which is regulated via the 3′ UTR of *preproinsulin* mRNA.[25] (Preproinsulin is the precursor polypeptide from which insulin is derived in pancreatic β-cells.) The pyrimidine-tract binding protein (PTB), which is also a repressor of splicing, binds to the 3′ UTR of *preproinsulin* mRNA, promoting its translation and stability. PTB also binds to the 3′ UTRs of other mRNAs that encode proteins associated with the machinery of insulin secretion. Thus, PTB achieves the coordinated post-transcriptional regulation of a set of genes involved in the same pathway. Insulin expression is stimulated by the presence of glucose. In fact, PTB expression is also stimulated by glucose in a coordinated fashion. How does PTB respond to glucose-mediated cell signalling? mTOR is a protein kinase that regulates cell growth and protein synthesis in response to nutrient status. It promotes the ability of PTB to bind to the *preproinsulin* mRNA. Rapamycin is a drug that blocks mTOR, which stops the glucose-induced binding of PTB to *preproinsulin* mRNA, thus reducing its stability and translation into insulin.

It should be noted that, in general, several RNA-binding proteins may be involved in the regulation of mRNA stability, especially proteins that compete for similar binding sites. For example, like PTB, the protein TIAR also binds to the *preproinsulin* mRNA 3′ UTR. TIAR is involved in the formation of stress granules (cytoplasmic structures that form to concentrate mRNAs that are translationally repressed, as described in Section 12.2). Nutrient deprivation results in increased binding by TIAR, as a result of which *preproinsulin* mRNA is not translated but is instead degraded. This makes sense, as insulin is not required during conditions of nutrient starvation.

The combined influence of PTB and TIAR on *preproinsulin* mRNA is summarized in Fig. 12.6. Thus,

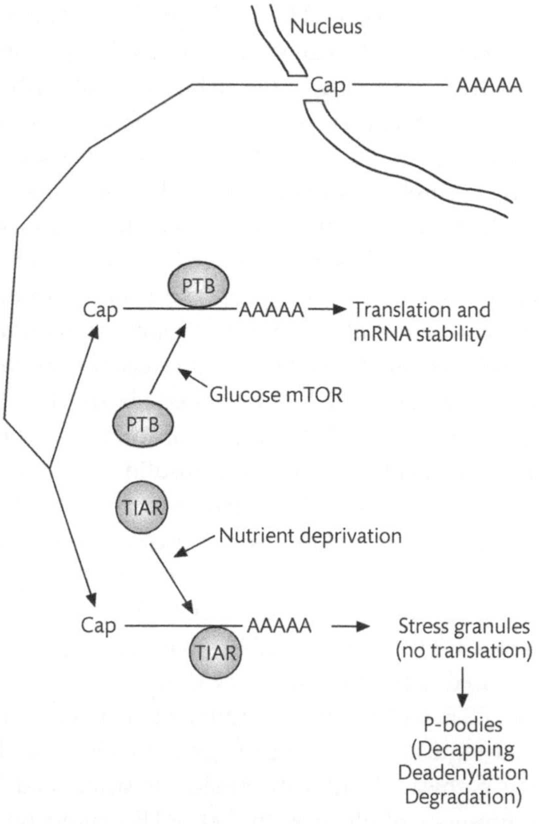

Figure 12.6 Model for the control of *preproinsulin* mRNA stability in insulin-producing cells. Binding of PTB to the 3′ UTR promotes translation and stability. PTB binding is stimulated by glucose and nutrients via the protein kinase mTor. In contrast, nutrient deprivation promotes binding of TIAR to the same 3′ UTR region, promoting translational repression and degradation in P-bodies. (Reprinted from Freda and Welsh[25] with permission from Elsevier.)

the interplay of the levels, localization, and activity of PTB and TIAR in pancreatic β-cells in response to extracellular signalling exerts an effect on *preproinsulin* mRNA translation and stability.

You may ask: why go to this trouble? Why not simply regulate insulin transcription (i.e. why not simply control the synthesis of the mRNA itself, rather than waiting until the mRNA has been produced and then regulating its stability)? The point of insulin is that it needs to be produced rapidly in response to dietary conditions and blood sugar. Post-transcriptional regulation provides the ability to achieve rapid bursts of translation of a given mRNA. In contrast, transcription and nuclear pre-mRNA processing steps take time.

Key points

Extracellular signals can influence mRNA stability. This often happens through the activation of protein kinases that modulate the ability of RNA-binding proteins to bind to their mRNA targets. A classical example is *preproinsulin* mRNA.

12.5 Nonsense-mediated, non-stop, and no-go mRNA decay

In this section we describe very particular forms of mRNA decay: nonsense-mediated, non-stop, and no-go mRNA decay. We will mainly focus on nonsense-mediated decay as it is more prevalent and better understood. **Nonsense mutations**, in which an inappropriate stop codon is generated, result in a **premature termination codon** (PTC). Premature termination codons result in the synthesis of frequently non-functional and deleterious truncated proteins.

Nonsense-mediated decay (NMD) is a remarkable eukaryotic process that eliminates mRNAs carrying PTCs.[26–30] The mechanism was first observed when the half-life of mRNAs was reduced (but only when a specific mutation resulted in a premature stop and not some other missense mutation). NMD research has clinical relevance because up to one-third of all known human genetic diseases involve mutations that generate PTCs.[31,32] Box 12.2 describes how PTCs can even be the target of therapy by using compounds that make the translation machinery bypass stop codons.[33]

Box 12.2 Premature stop codons as therapeutic targets

Although some nonsense mutations can have a primary effect on the transcript by interfering with exonic splicing signals (either ESEs or ESSs), in many of the point mutations causing genetic diseases the primary effect will be a premature termination of translation. Gentamycin belongs to a group of antibiotics called aminoglycosides, which affect the fidelity of ribosomes, enabling stop codons to be read as amino acids. In 1997 a group at the University of Birmingham in Alabama, USA, reported that gentamycin could enable full-length translation of *CFTR* transcripts in bronchial cells from cystic fibrosis (CF) patients which contained stop codons. As well as exhibiting normal *CFTR* transcript stability, these patients produced a normal functional CFTR protein.

Premature stop codons are usually only found in around 3% of patients with CF, but this number rises to 60% amongst Ashkenazi Jews. In a study of this group, gentamycin nasal sprays increased nasal expression of the *CFTR* transcript, although this was not enough to provide a significant clinical improvement. Most recently, small molecule drugs have been developed which interfere with nonsense-mediated decay. One of these, PTC124 (ataluren), is now under clinical trial and has been found to have fewer side effects than gentamycin (whose side effects include kidney failure and deafness).

12.5.1 Key properties of nonsense-mediated decay

At face value, this process was quite mysterious. How could the cell 'sense' mutations that give rise to a PTC? The process was first investigated in depth in yeast. NMD requires two ingredients: a ribosome stalled at a stop codon, and a downstream *cis*-acting sequence. Ribosomes were the key to unlocking the mystery—it is not the stop codon sequence per se that is the problem (after all, UGA, UAG, and UAA may occur in other reading frames) but rather the act of ribosomes stalling that is somehow sensed by the NMD machinery.

The precise pathway of NMD and the *cis*-acting sequences varies across species. In yeast, the *cis*-acting sequence is either a downstream sequence element (DSE) or the 3′ UTR itself. Plant NMD is also dependent on sequence elements downstream of PTCs, and on abnormally long 3′ UTRs. In yeast, the decay of PTC-containing mRNAs is initiated by decapping followed by 5′→3′ decay mediated by Xrn1. In *Drosophila*, decay starts with endonucleolytic decay proximal to the PTC, followed by the action of Xrn1 and the exosome. In contrast, NMD in mammals is dependent on a complex known as the **exon junction complex** (EJC), deposited approximately 20 nucleotides upstream of exon–exon boundaries. We encountered the exon junction complex in the context of mRNA localization in Chapter 10. It is further described in Box 12.3 and Fig. 12.7.

As well as targeting mRNAs with PTCs, NMD is thought to work in a broader context of mRNA surveillance—not only in the cytoplasm, but also in the nucleus. A striking fact is that up to a third of all inherited and acquired diseases are associated with a PTC; therefore NMD has a lot to guard against. For example, NMD targets the products of aberrant splicing—when an intron that lies in between two coding exons is not correctly spliced out, it is likely to introduce a PTC which needs to be eliminated by NMD. NMD also targets non-functional RNAs that are inappropriately expressed by transposon and retroviral sequences in the genome, or by upstream open reading frames (uORFs) in 5′ UTRs. As such, NMD has been described as a molecular 'vacuum cleaner' or 'Swiss army knife' which contributes significantly to the control and surveillance of correct gene expression.[29] Therefore it is not surprising that mouse knockouts of components of the NMD pathway are generally lethal.

Box 12.3 The exon junction complex

Two observations suggested that the splicing machinery leaves a marker on the mRNA at the site of splicing. One was the fact that cytoplasmic nonsense-mediated decay (NMD) in mammals is dependent on the presence of a 'spliceable' intron downstream of a PTC. The second was the fact that spliced mRNAs can be more productive in terms of translation than intronless transcripts. Biochemical analysis of exon–exon junctions revealed the presence of a specific and relatively uniform protein complex about 20 nucleotides upstream of exon–exon junctions, which was termed the exon

... *continued*

junction complex (EJC). The EJC is now known to play a role in several processes, not only in NMD but also in mRNA export, translation, and mRNA localization.

The structure of the EJC and the nature of its binding to mRNA are now understood.[34,35] Four proteins form the core of the EJC: MLN51, also known as hBarentsz, Magoh, Y14, and the RNA helicase eIF4AIII. These proteins help to anchor the EJC to the appropriate site on the mRNA. Y14 and Magoh form a heterodimer and are found originally in spliceosomes after the first *trans*-esterification reaction has occurred. More transiently associated factors include splicing-associated proteins such as SRm160 and SAP18, mRNA export factors UAP56, REF/Aly, and TAP/p15, and NMD-associated proteins UPF2 and UPF3. Most EJCs are removed after the first 'pioneering' round of translation (except any EJCs deposited on exon–exon junctions in untranslated regions).

Interestingly, EJCs are present in flies, but are not required for NMD. In flies the proteins Y14 and Magoh are required for the proper localization of *oskar* mRNA. (We saw in Chapter 10 how *oskar* mRNA localizes in the posterior end of the oocyte, where it is required for the proper formation of the germplasm, a cytoplasmic structure required for germ cell formation.) Thus the EJC probably evolved to assist the coordination of mRNA export, localization, translation, and mRNA decay. Therefore pre-mRNAs that are correctly spliced would be marked by the EJC to promote all the downstream steps in the life cycle of an mRNA in the cytoplasm.

It has been suggested that the new function for the EJC in NMD arose in mammals because of the explosion of alternative splicing–and therefore the increase in numbers of potentially mis-spliced transcripts harbouring PTCs.

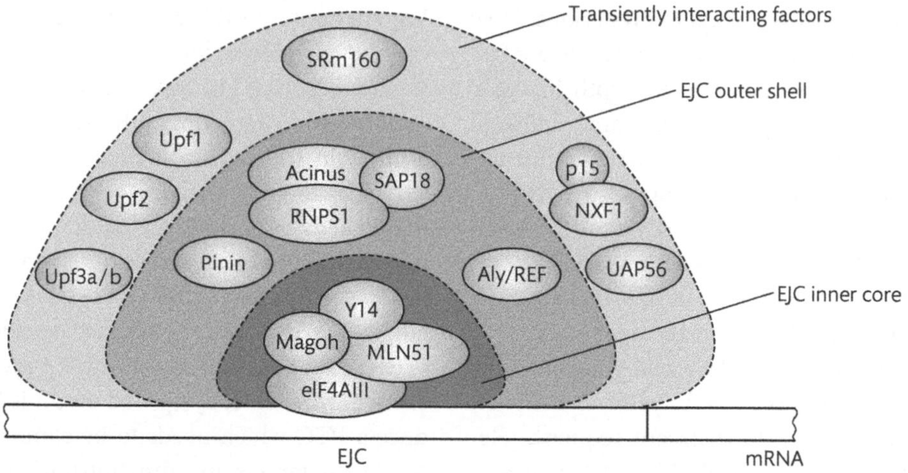

Figure 12.7 Components of the exon junction complex (EJC). There are three layers to the EJC. The inner layer is a tetrameric complex that contains eIF4AIII, MLN51, Magoh, and Y14 in which eIF4AIII provides direct contact to the mRNA. Several proteins are found in the outer shell: RNPS1, Acinus, and SAP18 can stably associate as a trimeric complex. SAP18, RNPS1, and Aly/REF are shuttling proteins, whereas Acinus and Pinin are nuclear restricted. Transiently interacting factors interact dynamically with either the EJC core or outer-sphere protein; these include the UPF proteins involved in NMD. (From Tange et al.,[35] with permission from Cold Spring Harbor Laboratory Press.)

12.5.2 Mechanism of nonsense-mediated decay

Initial research into NMD in yeast, primarily through genetic screens, led to the definition of a specific 'surveillance complex' which monitors for the presence of mRNAs with inappropriate stop codons. The surveillance complex includes the UPF proteins, which are conserved throughout eukaryotes. In mammals, the UPF proteins are actually present in the EJC itself. UPF1 is an RNA helicase that interacts with translation release factors; this interaction links NMD to mRNA translation.

NMD works in mammals as follows: UPF1 is recruited by translation release factors after the ribosome has encountered a stop codon. It then interacts with UPF2 and UPF3. UPF2 and UPF3 are part of a downstream EJC, forming an active surveillance complex. The UPF proteins interact and recruit factors that are involved in both 5′→3′ and 3′→5′ decay.

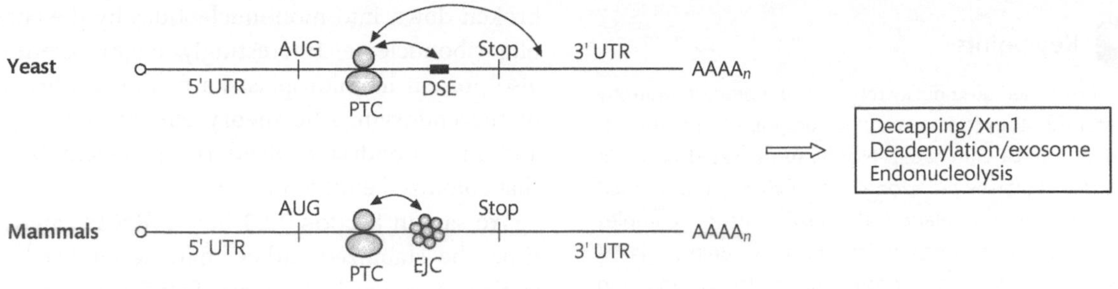

Figure 12.8 Mechanisms of nonsense-mediated decay (NMD) in yeast and mammals. In yeast, when a ribosome stalls at a premature termination codon (PTC), a downstream sequence element (DSE) or abnormally long 3′ UTR promotes NMD. In mammals, the presence of an exon junction complex (EJC), normally removed after the pioneer round of translation, triggers NMD. Once NMD is triggered by the UPF proteins, standard mRNA decay pathways are activated.

These interactions can explain the mechanism of NMD in mammals; after the first or **pioneer round of translation**, EJCs are removed, thus also rapidly removing UPF2 and UPF3. As a result of the pioneer round of translation, NMD does not occur, because the UPF proteins are unable to recruit factors that promote mRNA degradation. However, when translation stalls, due to a PTC, UPF1 is recruited and there is time to form the surveillance complex with UPF2 and UPF3, resulting in mRNA degradation. Yeast and mammalian models of NMD are compared in Fig. 12.8.

However, there is an unavoidable question: why don't real stop codons result in NMD? A probable answer is that exon–exon boundaries, at least in mammals, are not generally juxtaposed to proper stop codons. Exon–exon boundaries are simply parts of the mRNA where exons had been spliced together; these are normally marked by the EJC. This simple model would answer the above question. However, by the same token, some genuine PTCs may escape detection in mammals because by chance there may not be a downstream exon–exon junction with an EJC deposited on it. Therefore it is likely that other forms of PTC recognition also exist in mammals; and that NMD may not necessarily be wholly dependent on the EJC system.

12.5.3 Non-stop and no-go mRNA decay

We now describe two other specialized forms of mRNA decay. The first of these is *non-stop decay* (NSD), in which rather than a PTC causing premature translation termination, the defect is the absence of a stop codon.[36] This results in the ribosome translating through the 3′ UTR and the poly(A) tail. The result of translating the poly(A) tail is the addition of polylysine at the C-terminal end of the resulting polypeptide (the codon AAA encodes lysine).

The mechanism of non-stop decay has been studied in yeast.[36] After the ribosome has translated through the poly(A) tail it eventually arrives at the end of the mRNA, when it stalls because the ribosomal A site is left empty (whereas stop codons result in the recruitment of the release factor eRF3). A protein called Ski7p recognizes the empty ribosomal A site, causing the recruitment of the exosome and dissociation of the ribosome. The exosome then initiates degradation from the 3′ end, eliminating the offending mRNA. Any resulting proteins are also actively degraded, as the polylysine tract promotes proteolysis.

No-go decay (NGD) is the degradation of mRNAs that stall during the process of translation elongation.[37] This can occur when rare codons are used, or when highly stable secondary RNA structures are encountered. The mechanism of NGD has also been researched in yeast. It involves endonucleolytic cleavage of the faulty mRNA at the stall site by the protein Dom34.[38] The resulting mRNA fragments are then mopped up by Xrn1 and the exosome. Evidence shows that NGD also occurs in insect cells, suggesting that it is a conserved process in eukaryotes. Specifically, pelota, the *Drosophila* homologue of Dom34 involved in mitosis, is able to complement (rescue) Dom34-deficient yeast cells.[38]

Key points

Nonsense-mediated decay (NMD) is the process whereby premature stop codons can be recognized and the offending mRNA eliminated. Nonsense-mediated decay in mammals involves the exon junction complex, deposited at exon–exon boundaries. The exon junction complex recruits the mRNA degradation machinery unless it is removed during the pioneer round of translation. Non-stop decay (NSD) occurs in the absence of a stop codon, and no-go decay (NGD) targets mRNAs that stall during translation elongation.

12.6 Degradation of mRNA in bacteria and plants

Historically, a large proportion of research into mRNA degradation pathways has focused on yeast, invertebrate, and mammalian systems. However, it is no less important in other organisms. In this section, we briefly consider the state of play of mRNA decay research in bacteria and plants.

12.6.1 Degradation of mRNA in bacteria

Turnover of mRNA is an important pathway for gene regulation in bacteria, because bacteria need to adapt very quickly to changes in the environment.[39,40] mRNAs are frequently very short-lived in bacteria, with half-lives ranging from 30 seconds to 20 minutes. In *Escherichia coli*, the average half-life of mRNAs is a mere 2.4 minutes! mRNA degradation in bacteria is generally initiated by the enzyme RNAse E, particularly in AU-rich areas with little secondary structure. RNAse E is part of a complex known as the **degradosome** which also contains the exoribonuclease polynucleotide phosphorylase (PNPase) and an RNA helicase, RhlB, which facilitates unwinding of mRNA destined for degradation. The degradosome mediates the initial cleavages of target mRNAs, which are therefore immediately unavailable for translation. Once the degradosome has chopped up mRNAs into fragments, further degradation proceeds 3′→5′ via the exoribonucleases PNPase, RNAse II, and RNAse R. (Note that degradation is not thought to occur 5′→3′ in bacteria.) Once fragments become very short, they are finally broken down into mononucleotides by the enzyme oligoribonuclease. Interestingly, degradosomes are also present in chloroplasts, a further confirmation of the endosymbiotic theory wherein chloroplasts and mitochondria evolved from ancient bacteria that colonized eukaryotic cells.

We saw in Section 12.3 how mRNAs can sometimes be stabilized rather than destabilized. This is also observed in bacteria. Stabilizing motifs are found in the 5′ end of some mRNAs that are degraded slowly in bacteria. The 5′ element in the *ompA* mRNA encodes a hairpin structure. The existence of this hairpin structure agrees with the finding that RNAse E, a main constituent of the degradosome, interacts primarily with the 5′ end of mRNA; the hairpin structure may impede access to a free single-stranded 5′ end. Interestingly, RNAse E cleaves its own mRNA at the 5′ end, thus generating an autoregulatory loop.

Other features that counteract mRNA degradation in bacteria are the presence of a 5′ triphosphate at the 5′ end of an mRNA, or of a ribosome. Thus mRNAs that are actively translating are more stable, presumably because the association with ribosomes prevents access to the degradosome. Once the degradosome has acted, chopping up an mRNA towards its 5′ end, a new free 5′ end is generated ready for additional cleavage by the degradosome. Progressive fragmentation then presents additional substrates for the 3′→5′ exonucleases so that they keep on digesting away. Hairpin structures have also been documented at the 3′ end of certain bacterial mRNAs, where they can act as barriers to degradation by the 3′→5′ exonucleases.

12.6.2 Degradation of mRNA in plants

Messenger RNA decay processes are in fact universal, and apply to plants as well as bacteria and animals.[41] The sequencing of the *Arabidopsis thaliana* genome has facilitated the analysis of changes in gene expression globally in response to specific mutations or environmental challenges. In a pioneering experiment, researchers undertook a systematic inhibition of transcription in seedlings with the drug cordycepin.[42] Once transcription is inhibited, mRNA half-life can be measured over the course of time. Of 7800 mRNAs analysed, about 1% were observed to be relatively unstable, with half-lives of less than an hour. These included, disproportionately, mRNAs that encode transcription factors (reminiscent of the instability of mammalian c-*myc* or c-*fos*) and proteins associated with circadian (day/night) rhythm. This makes sense, because these mRNAs are precisely those that one would expect to have to be degraded rapidly, because the transition between day and night is fast.

The machinery of mRNA degradation is conserved in plants. Most of the key players described in yeast or animals have been identified, including decapping enzymes, 5′→3′ exonucleases, the exosome, NMD components, and PARN. The analysis of an *Arabidopsis* PARN mutant showed an elevated expression of genes that are normally induced by the plant hormone abscisic acid (ABA).[43] ABA is involved in the regulation of stress-associated genes. This suggests that changes in mRNA stability arise as a result of ABA signalling. Thus mRNA stability can be regulated through cell signalling in both plants and animals.

Key points

Degradation of mRNA also occurs in bacteria. The enzyme RNAse E, which is part of the degradosome, is a key player. The degradation of several plant mRNAs, including those involved in circadian rhythms or the stress response, is also regulated by an evolutionarily conserved machinery.

Summary

- mRNAs have a characteristic half-life which can be regulated by *cis*-acting RNA elements that are themselves bound by *trans*-acting RNA-binding proteins. A well-studied example is the AU-rich element (ARE) which promotes mRNA decay.
- P-bodies and stress granules are cytoplasmic granules where translationally repressed mRNAs can be degraded or stored. They contain high levels of proteins involved in mRNA degradation.
- *Cis*-acting RNA sequences and *trans*-acting RNA-binding proteins regulate mRNA decay. RNA-binding proteins can either promote or prevent mRNA decay. Messenger RNA decay is universally present in bacteria, single-celled eukaryotes, plants, and animals.
- Decay pathways involve the attack of the 5′ and 3′ ends of mRNAs by exonucleases, once the 5′ cap and poly(A) tail have been removed by specialized enzyme complexes. In contrast, endonucleases also cleave mRNAs internally. The machinery of mRNA degradation includes individual ribonucleases (Xrn1, PARN) or complexes with ribonuclease activity (exosome, degradosome).
- The levels and activities of RNA-binding proteins involved in mRNA decay can be regulated by extracellular stimuli, generally through protein kinases. These kinases modulate the activities of RNA-binding proteins involved in decay.
- Nonsense-mediated decay is a specialized form of decay that protects against premature termination codons, and in mammals is dependent on the exon junction complex. Exon junction complexes have evolved to coordinate several post-transcriptional processes, including translation, mRNA export, localization, and decay.
- Non-stop decay is the process whereby mRNAs that lack stop codons are degraded. No-go decay is the removal of mRNAs that stall during the translation elongation.

Questions

12.1 What is the evidence that mRNAs have a half-life?

12.2 Where in the cell does mRNA degradation take place?

12.3 What are the principal mechanisms of mRNA degradation?

12.4 How do AU-rich elements influence mRNA stability?

12.5 How do extracellular stimuli affect mRNA stability?

12.6 What is nonsense-mediated decay and how does it work?

12.7 What is the function of the exon junction complex in the context of mRNA degradation?

12.8 How do non-stop and no-go mRNA decay work?

12.9 How is mRNA stability regulated in plants and bacteria?

References

1. **Schoenberg DR, Maquat L.** Regulation of cytoplasmic mRNA decay. *Nat Rev Genet* **13**, 246–59 (2012).
2. **Chen C-Y, Shyu A-B.** Mechanisms of deadenylation-dependent decay. *Wiley Interdiscip Rev RNA* **2**, 167–83 (2011).
3. **Stoecklin G, Hahn S, Moroni C.** Functional hierarchy of AUUUA motifs in mediating rapid interleukin-3 mRNA decay. *J Biol Chem* **269**, 28 591–7 (1994).
4. **Parker R, Sheth U.** P-bodies and the control of mRNA translation and degradation. *Mol Cell* **25**, 635–46 (2007).
5. **Eulalio A, Behm-Ansmant I, Schweizer D, Izaurralde E.** P-body formation is a consequence, not the cause, of RNA-mediated gene silencing. *Mol Cell Biol* **27**, 3970–81 (2007).
6. **Anderson P, Kedersha N.** RNA granules: post-transcriptional and epigenetic modulators of gene expression. *Nat Rev Mol Cell Biol* **10**, 430–6 (2009).
7. **Adjibade P, Mazroui R.** Control of mRNA turnover: implication of cytoplasmic granules. *Semin Cell Dev Biol* **34**, 15–23 (2014).
8. **Kedersha N, Anderson P.** Stress granules: sites of mRNA triage that regulate mRNA stability and translatability. *Biochem Soc Trans* **30**, 963–9 (2002).
9. **Marnef A, Ladomery MR.** RAP55: insights into an evolutionarily conserved protein family. *Int J Biochem Cell Biol* **41**, 977–81 (2009).
10. **Newbury SF.** Control of mRNA stability in eukaryotes. *Biochem Soc Trans* **34**, 30–4 (2006).
11. **Liu J, Valencia-Sanchez MA, Hannon GJ, Parker R.** MicroRNA-dependent localization of targeted mRNAs to mammalian P-bodies. *Nat Cell Biol* **7**, 719–23 (2005).
12. **Jones CI, Zabolotskaya MV, Newbury SF.** The 5′→3′ exoribonuclease XRN1/Pacman and its functions in cellular processes and development. *Wiley Interdiscip Rev RNA* **3**, 455–68 (2012).
13. **Yan Y-B.** Deadenylation: enzymes, regulation, and functional implications. *Wiley Interdiscip Rev RNA* **5**, 421–43 (2014).
14. **Butler JS.** The yin and yang of the exosome. *Trends Cell Biol* **12**, 90–6 (2002).
15. **Mukherjee D, Gao M, O'Connor JP, et al.** The mammalian exosome mediates the efficient degradation of mRNAs that contain AU-rich elements. *EMBO J* **21**, 165–74 (2002).
16. **Lin W-J, Duffy A, Chen C-Y.** Localization of AU-rich element-containing mRNA in cytoplasmic granules containing exosome subunits. *J Biol Chem* **282**, 19 958–68 (2007).
17. **Bakheet T, Frevel M, Williams BRG, et al.** ARED: human AU-rich element-containing mRNA database reveals an unexpectedly diverse functional repertoire of encoded proteins. *Nucleic Acids Res* **29**, 246–54 (2003).
18. **Miller AD, Curran T, Verma M.** *c-fos* protein can induce cellular transformation: a novel mechanism of activation of a cellular oncogene. *Cell* **36**, 51–60 (1984).
19. **Wisdom R, Lee W.** The protein-coding region of *c-myc* mRNA contains a sequence that specifies rapid mRNA turnover and induction by protein synthesis inhibitors. *Genes Dev* **5**, 232–43 (1991).
20. **Benjamin D, Moroni C.** mRNA stability and cancer: an emerging link? *Expert Opin Biol Ther* **7**, 1–15 (2007).
21. **Hollams EM, Giles KM, Thomson AM, Leedman PJ.** mRNA stability and the control of gene expression: implications for human disease. *Neurochem Res* **27**, 957–80 (2002).
22. **Gorospe M.** HuR in the mammalian genotoxic response. *Cell Cycle* **2**, 412–15 (2003).
23. **Shim J, Karin M.** The control of mRNA stability in response to extracellular stimuli. *Mol Cell* **14**, 323–31 (2002).
24. **Carballo E, Cao H, Lai WS, et al.** Decreased sensitivity of tristetraprolin-deficient cells to p38 inhibitors suggests the involvement of tristetraprolin in the p38 signaling pathway. *J Biol Chem* **276**, 42 580–7 (2001).
25. **Freda R, Welsh N.** The importance of RNA binding proteins in preproinsulin mRNA stability. *Mol Cell Endocrinol* **15**, 28–33 (2009).
26. **Conti E, Izaurralde E.** Nonsense-mediated mRNA decay: molecular insights and mechanistic variation across species. *Curr Opin Cell Biol* **17**, 316–25 (2005).
27. **Maquat L.** Nonsense-mediated mRNA decay in mammals. *J Cell Sci* **118**, 1773–6 (2005).
28. **Alonso C.** Nonsense-mediated RNA decay: a molecular system micromanaging individual gene activities and suppressing genomic noise. *BioEssays* **27**, 463–6 (2005).
29. **Neu-Yilik G, Gehring NH, Hentzem MW, Kulozik AE.** Nonsense-mediated mRNA decay: from vacuum cleaner to Swiss army knife. *Genome Biol* **5**, 218–22 (2004).
30. **Popp MWL, Maquat LE.** The dharma of nonsense-mediated mRNA decay in mammalian cells. *Mol Cell* **37**, 1–8 (2014).
31. **Kuzmiak H, Maquat LE.** Applying nonsense-mediated mRNA decay research to the clinic: progress and challenges. *Trends Mol Med* **12**, 306–16 (2006).
32. **Nguyen LS, Wilkinson MF, Gecz J.** Nonsense-mediated mRNA decay: inter-individual variability and human disease. *Neurosci Biobehav Rev* **46**, 175–86 (2014).
33. **Kerem E, Konstan MW, De Boeck A, et al.** Ataluren for the treatment of nonsense-mutation cystic fibrosis: a randomised, double-blind, placebo-controlled phase 3 trial. *Lancet Respir Med* **2**, 539–47 (2014).
34. **Le Hir H, Rom Andersen G.** Structural insights into the exon junction complex. *Curr Opin Struct Biol* **18**, 112–19 (2008).
35. **Tange T, Nott A, Moore MJ.** Biochemical analysis of the EJC reveals two new factors and a stable tetrameric protein core. *RNA* **11**, 1869–83 (2005).

36. **Vasudevan S, Peltz SW, Wilusz CJ.** Non-stop decay—a new mRNA surveillance pathway. *BioEssays* **24**, 785–8 (2002).
37. **Doma MK, Parker R.** Endonucleolytic cleavage of eukaryotic mRNAs with stalls in translation elongation. *Nature* **440**, 561–4 (2006).
38. **Passos DO, Doma MK, Shoemaker CJ, et al.** Analysis of Dom34 and its function in no-go decay. *Mol Biol Cell* **20**, 3025–32 (2009).
39. **Regnier P, Arraiano CA.** Degradation of mRNA in bacteria: emergence of ubiquitous features. *BioEssays* **22**, 235–44 (2000).
40. **Deutscher MP.** Degradation of RNA in bacteria: comparison of mRNA and stable RNA. *Nucleic Acids Res* **34**, 659–66 (2006).
41. **Belostotsky DA.** State of decay: an update on plant mRNA turnover. *Curr Topics Microbiol Immunol* **326**, 179–99 (2008).
42. **Gutierrez RA, Ewing RM, Cheery JM, Green PJ.** Identification of unstable transcripts in *Arabidopsis* by cDNA microarray analysis: rapid decay is associated with a group of touch- and specific clock-controlled genes. *Proc Natl Acad Sci USA* **99**, 11 513–18 (2002).
43. **Nishimura N, Kitahata N, Seki M, et al.** Analysis of ABA hypersensitive germination revealed the pivotal functions of PARN in stress response in *Arabidopsis*. *Plant J* **44**, 972–84 (2005).

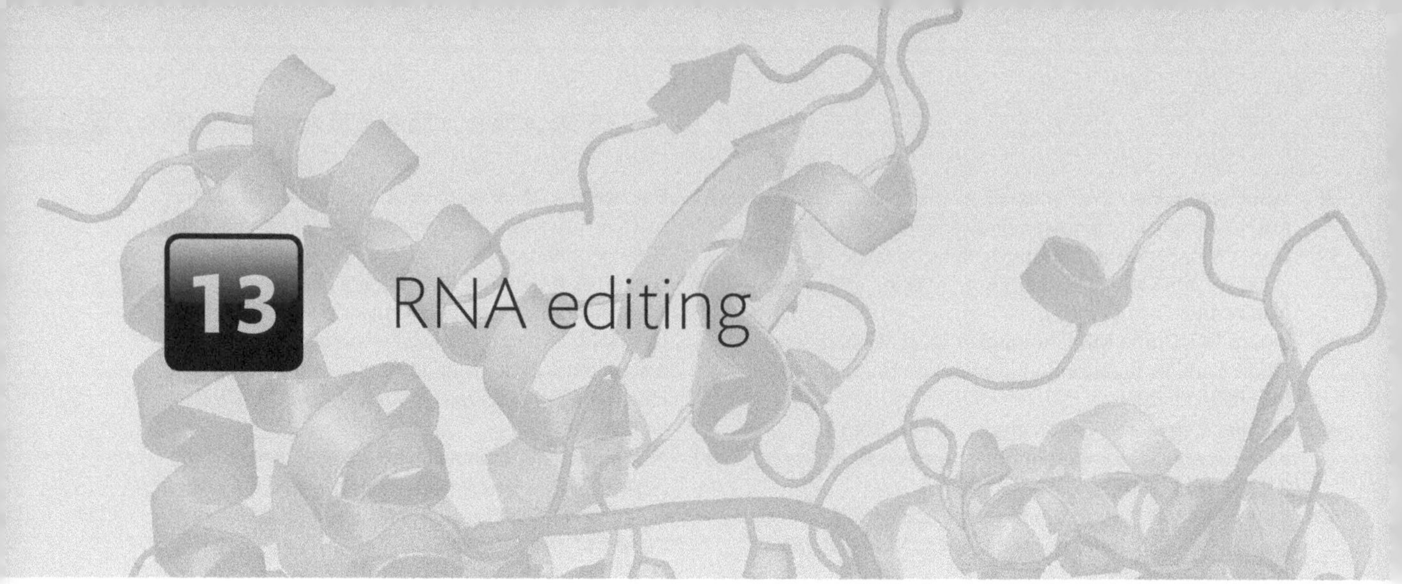

13 RNA editing

Introduction

A major surprise from molecular biology has been that the **open reading frames** (ORFs) of some RNAs can be altered after transcription by **RNA editing**, a process that does not affect the sequence of their parent genes. Many of the targets of RNA editing are also non-coding RNAs, and RNA editing is also thought to play an important role in keeping selfish DNA elements in the genome in check. RNA editing also plays a significant role in enabling tRNAs to translate mRNAs efficiently—a process that is conserved between bacteria and eukaryotes.

RNA editing in its different forms is the topic of this chapter. Terms introduced in this chapter to describe RNA editing are highlighted in bold and defined in the Glossary.

13.1 What is RNA editing and why might it exist?

RNA editing changes the sequence of RNAs once they have already been transcribed. Two general types of RNA editing occur through quite different mechanisms (see Fig. 13.1).

1 RNA editing through base modification changes the chemical identity of nucleotides already present within the transcript, and the length of the edited RNA stays the same. RNA editing through base modification is akin to changing the letters in words on a printed page (e.g. the conversion of the word 'edit' to the word 'exit' by changing the second letter from d to x). This type of RNA editing through base modification happens in higher eukaryotes.

2 Insertion/deletion RNA editing adds or subtracts uridine nucleotides to RNAs and so changes the length of the edited RNA compared with the pre-edited RNA. RNA editing through nucleotide insertion and deletion is important in a protozoan parasite, where it restores reading frames to mitochondrial transcripts made from scrambled genes called cryptogenes. Before RNA editing these cryptogenes do not have clear ORFs, and so cannot encode full-length proteins.

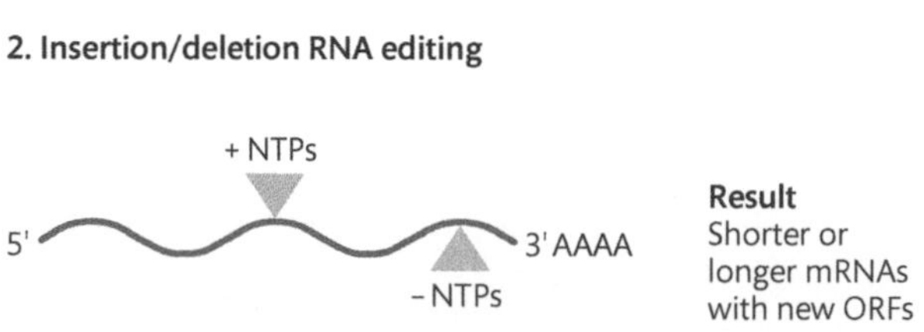

Figure 13.1 RNAs can be edited via two types of mechanism. (1) Base modification of nucleotides within mRNAs changes the sequence of transcripts without affecting their overall length. Notice that the two forms of editing shown here are A→I and C→U; we will be discussing both these kinds of editing later in the chapter. (2) Nucleotide addition or deletion (usually of U residues) changes the length of the edited mRNA. This second type of RNA editing is found in trypanosome mitochondria, and is covered at the end of the chapter.

Although these types of editing occur through different mechanisms, a common theme which crops up in both kinds of RNA editing is the role of RNA structure in guiding editing events (see Table 13.1). Historically, RNA editing through nucleotide insertion and deletion in trypanosome mitochondria was discovered first,[1] followed by RNA editing through base modification.[2,3] In this chapter we start by describing **A→I editing** through base modification since this form of editing is the most widely distributed in nature.

Why edit a transcript instead of changing the nucleotide sequence of the corresponding gene?[4] One reason might be that editing increases the options for making different mRNAs from the same gene—both edited and non-edited versions can be made. Editing also introduces a step through which gene expression can be controlled, since the number of edited sites per transcript and the fraction of editing which takes place at each site can vary between cells and tissues.

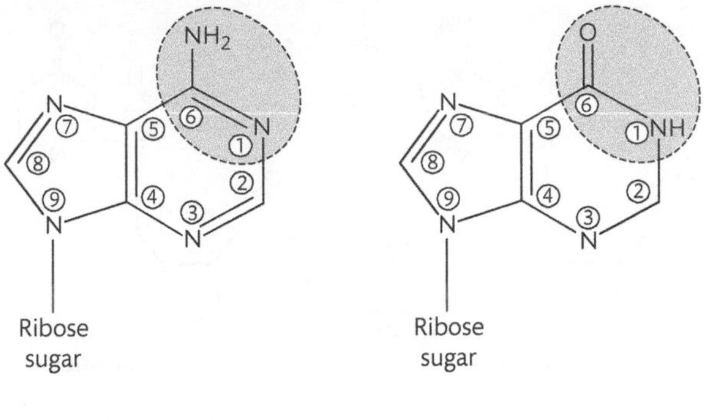

Figure 13.2 Adenosine (A) and inosine (I) contain slightly different purine bases (circled). Inosine contains the non-standard base hypoxanthine, while adenosine contains the base adenine. Both adenine and hypoxanthine are purine bases (containing two rings; see Chapter 2), but adenine has an amino group attached to N6 of the purine ring. Hypoxanthine has an oxygen at the equivalent position; and nitrogen N1 of hypoxanthine has a hydrogen side group.

13.2 A→I editing takes place by modification of adenosine through removal of an amino group

In Chapter 2 we discussed the structure of the four main nucleotides found in RNA. Although these four standard nucleotides are most frequently found in RNA, several other nucleotides also exist, including inosine (I). Inosine is similar to adenosine—these nucleotides are shown side by side in Fig. 13.2.

The bases of some nucleotides can be modified to form new bases, effectively changing the sequence of RNA (see Chapter 2). Remember that it is the bases that are important for nucleotide sequence;

Table 13.1 Types of RNA editing covered in this chapter

Type of RNA editing	Examples	Is RNA editing guided by double-stranded RNAs?
RNA editing through base modification	A→I editing in nervous system development	Yes. RNA editing is targeted by intramolecular RNA helices recognized by the ADAR enzymes
	A→I editing of tRNAs	No. RNA editing is carried out by ADAT enzymes
	C→U editing in mammalian intestine	RNA editing is targeted by RNA-binding ApoBec enzymes, but target RNA forms stemloop
	cDNA editing (this kind of editing takes place on DNA but is evolutionarily related to RNA editing)	No
RNA editing through deletion or insertion of uridine nucleotides	RNA editing found in trypanosome mitochondria	Yes. RNA editing is targeted by small guide RNAs separate from the RNA molecule which is the editing substrate

ADAR, adenosine deaminase acting on RNA; ADAT, adenosine deaminase acting on tRNA; ApoBec, carries out C→U RNA editing of the *APOB* transcript.

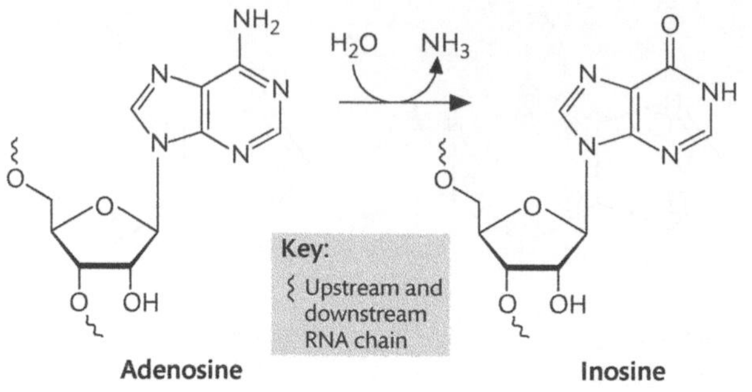

Figure 13.3 A→I editing converts adenosine to inosine through a deamination reaction (removal of an amino group). In the A→I editing reaction carbon C6 of adenosine is deaminated and replaced by an oxygen, and nitrogen N6 is protonated. The ribose sugar is connected upstream and downstream to the rest of the transcript (shown by a squiggly line) by phosphodiester bonds. (Reproduced from Nishikura[9] with permission from Macmillan Publishers Ltd.)

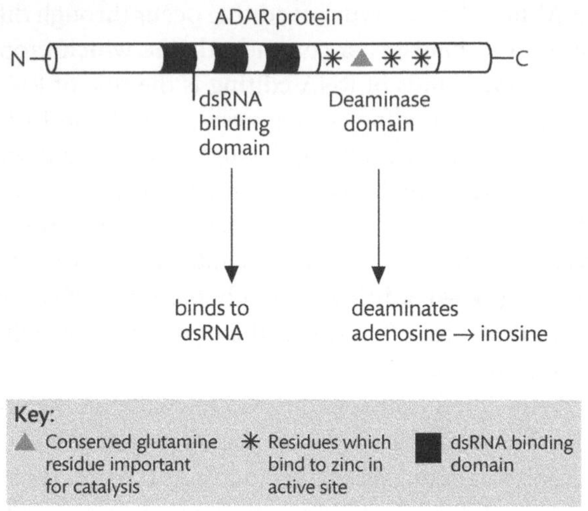

Figure 13.4 ADAR enzymes contain a double-stranded RNA-binding domain and a catalytic deaminase domain. How these important amino acid positions in the deaminase domain function catalytically is shown in Fig. 13.5.

the phosphodiester backbone of RNA is invariant. In the case of inosine, this is made from adenosine by removal of the amino group on the adenine base and its replacement with an oxygen atom (see Fig. 13.3). This kind of reaction is called a **deamination** because it involves loss of an amino group. RNA editing actually takes place while the edited RNA is still being transcribed in the nucleus (i.e. it is co-transcriptional).[5]

A→I editing of mRNAs is catalysed by a group of enzymes called **ADARs** (adenine deaminases which act on RNA).[6–8] ADAR enzymes edit double-stranded (ds) RNAs which contain adenosine. Short dsRNA helices more than about 20 nucleotides long are selectively A→I edited at specific positions. Longer dsRNAs are extensively and promiscuously A→I edited at multiple positions (this is called **pan editing**).

ADAR enzymes all contain two important domains (see Fig. 13.4).

1 A double-stranded RNA-binding domain (abbreviated dsRBD) (see Chapter 4). The dsRBD binds to the double-stranded RNAs that are editing substrates, thus explaining why RNA secondary structure is so important in RNA editing. Atomic structures derived by X-ray crystallography show that dsRBDs bind to the sugar phosphate backbone of the RNA (particularly the 2′ -OH groups on the ribose sugar) rather than to the bases.[10] This means that dsRBDs do not have a high sequence specificity in terms of RNA target, and instead selectively bind to dsRNAs based on how extensively they are base paired.

2 A deaminase catalytic domain that carries out the catalytic conversion of adenine to inosine.[11] ADAR catalytic domains have important features. First, the active sites of ADAR enzymes are positively charged because of their amino acid contents. This positive charge facilitates interactions between the enzyme and the negatively charged dsRNA. Secondly, three particular amino acid residues in the catalytic deaminase domain play specific roles in catalysis. These are the cysteine and histidine residues, which bind to the zinc atom in the active site of ADAR enzymes, and a conserved glutamic acid residue. The catalytic deaminase domain catalyses a chemical reaction between adenine and water to give inosine and ammonia (see Fig. 13.5).

Key points

ADAR enzymes catalyse the chemical modification of adenosine to inosine. ADAR enzymes contain dsRNA-binding domains, which bind to secondary structure in RNA. ADAR enzymes contain deaminase domains which carry out catalysis, and special RNA-binding domains which bind dsRNAs.

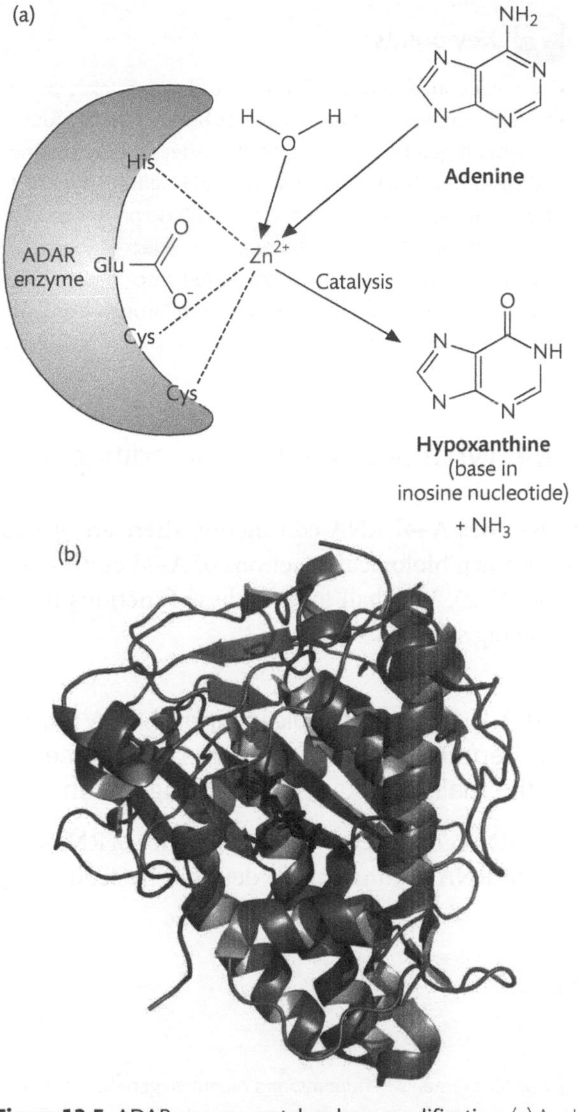

Figure 13.5 **ADAR enzymes catalyse base modification.** (a) A→I RNA editing takes place as the result of a chemical reaction between adenine and water to give inosine and ammonia. A glutamic acid residue within the active site of the deaminase domain of ADAR enzymes is key for catalysis. Glutamic acid has a carboxyl group side chain. This carboxyl group reacts with an incoming water molecule in the presence of zinc to generate a hydroxyl ion and a proton ($H_2O \rightarrow HO^- + H^+$). The proton ($H^+$) attaches to the adenine base at the N1 position. This destabilizes the double bond between N1 and C6 (which contains the amino group), which is then attacked by the hydroxyl ion (HO^-), resulting in the production of inosine and ammonia (NH_3). Notice that all the atoms in the water (H_2O) and adenine molecules going into the reaction are accounted for in the ammonia and inosine products. (b) The protein structure of the catalytic domain of the human ADAR2 protein, which folds around a molecule of inositol hexakisphosphate needed for enzyme activity.[11] (Part (b) redrawn from the RCSB Protein Data Bank accession 1ZY7/pdb by Jonathan Crowe, Oxford University Press.)

13.3 The biological consequences of A→I RNA editing: adenosine and inosine form different base pairs in RNA secondary structure

A→I RNA editing is important because it changes the base-pairing interactions of RNAs. Adenosine base pairs with uridine (to give A–U base pairs). Inosine base pair best with cytidine (to give an I–C base pair) (see Fig. 13.6), but can also base pair more weakly with adenosine (I–A) and uridine (I–U). This flexible base-pairing property of inosine is important for tRNAs, as we shall see later in this chapter.

Base pairing is important in RNAs for forming secondary structures where strands of RNA are held together (see Chapter 2). A→I RNA editing was first discovered through effects on such base pairing.[2] In these experiments, RNA helices formed by A–U base pairing were incubated in cell extracts. Surprisingly this led to unwinding of the

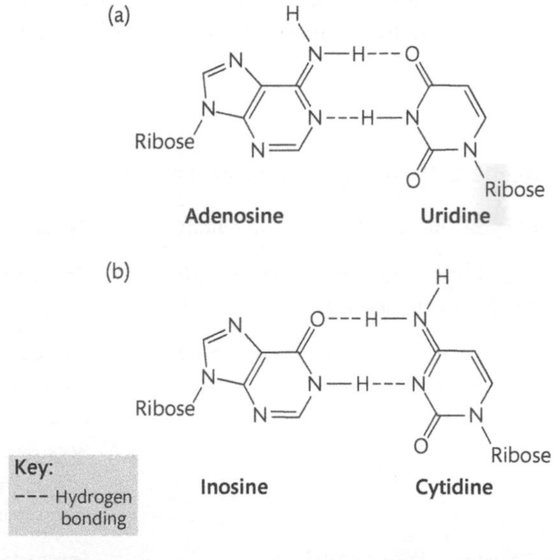

Figure 13.6 **A→I editing modifies base pairing between nucleotides.** Base pairs in double-stranded RNA are held together by hydrogen bonds between complementary bases (see Chapter 2). (a) Adenosine forms Watson–Crick base pairs with uridine. (b) Inosine forms strong Watson–Crick base pairs with cytidine. This change in base pairing occurs because of the positions of the charged groups which contribute to hydrogen bonding. The amino group ($-NH_2$), which is removed from adenosine during A→I RNA editing, normally acts as a hydrogen bond donor. (Reproduced from Nishikura[9] with permission from Macmillan Publishers Ltd.)

Figure 13.7 **A→I RNA editing was first discovered because it caused double-stranded RNAs to partially unwind.** This unwinding effect occurred because the A→I edited nucleotides were no longer able to base pair with their normal partners. (This figure summarizes the conclusions of Bass and Weintraub.[2])

dsRNA duplexes (this unwinding phenomenon is shown in Fig. 13.7). The unwinding of the RNA helices was caused by A→I RNA editing destabilizing the initial base pairing that created the helices. As a result of editing, the A→I edited nucleotides could no longer strongly base pair with their original partner uridine, leading to unwinding of the helix.

Key points

Chemical conversion of adenosine to inosine takes place through removal of an amino group. Before editing, adenosine hydrogen bonds with thymine. After editing, inosine hydrogen bonds with cytosine. A→I RNA editing was first discovered because it destabilized A-U base pairing within molecules of RNA. Base pairing was affected because adenosine nucleotides were converted into inosine nucleotides, with concomitant effects on hydrogen bonding between bases in the dsRNA.

13.4 What does A→I mRNA editing do?

What does A→I RNA editing do? There are at least five known biological functions of A→I editing (see Table 13.2). We shall look at these functions in the following sections.

13.4.1 Editing of double-stranded RNA made from repetitive genetic elements: Alu elements are the main targets of A→I editing in humans

Since RNA editing changes the sequence of RNAs, the sites of RNA editing can be detected by sequencing

Table 13.2 The biological functions of A→I RNA editing within mRNAs

Biological function of A→I mRNA editing	Explanation
1. Editing of double-stranded RNA made from repetitive genetic elements	Controls the activity of Alu elements in humans, and repetitive genetic elements in fruit flies
2. Modifying the reading frames of already transcribed mRNAs	Each codon in an mRNA that encodes an amino acid has a complementary tRNA with an anticodon which recognizes it by base-pairing rules. Adenosines are complementary to uridines, and inosines to guanosines. Because of this change A→I edited codons will be recognized by tRNAs with different anticodon sequences and can result in different amino acids being incorporated into the polypeptide chain
3. Changing splicing signals	Splice site selection depends on sequence-specific RNA–RNA and RNA–protein interactions. By changing critical sequences A→I mRNA editing can change these interactions and so regulate alternative pre-mRNA splicing
4. Nuclear retention of specific mRNAs	Some nuclear RNA-binding proteins, including p54nrb and PSF, show strong affinity for inosine-containing RNA, and this leads to selective retention of certain edited mRNAs in the nucleus
5. Competition with the RNAi processing machinery	A→I editing can affect the dsRNA substrates recognized by the RNAi pathways. Thus editing and RNAi production can compete with each other

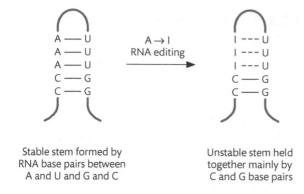

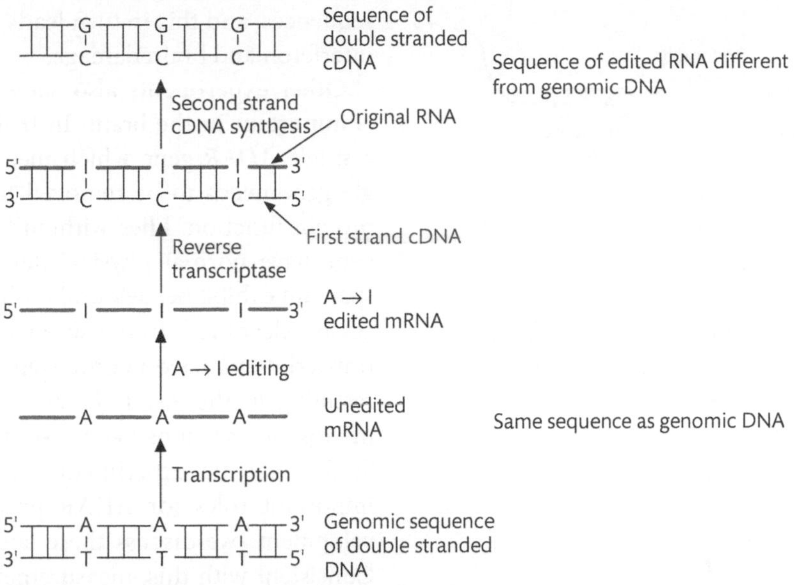

Figure 13.8 Detecting sites of A→I editing through sequencing. Sites of A→I editing can be detected by comparing genomic and cDNA sequences. If A→I RNA editing post-transcriptionally changes the sequence of RNA, this will mean that edited transcripts and genomic DNA will have slightly different sequences. Inosine nucleotides in an edited transcript will be copied as a C by reverse transcriptase (since I is complementary through base pairing to C). Non-edited adenosines are copied into thymidines in cDNA (since A is complementary to T). Hence the sequences of genomic DNA and unedited RNAs will be the same, but the sequences of edited RNAs will be different.

RNAs and comparing them with genomic DNA (see Fig. 13.8). Such experiments indicate which RNA sequences undergo most A→I RNA editing. RNA editing is particularly targeted at repetitive elements in the genome called Alu sequences.[12] More than 90% of A→I editing in the transcriptome takes place on transcribed Alu sequences. There are around a million Alu elements in the human genome, and over 75% of human genes contain Alu elements (mainly within introns). All of these Alu elements have similar sequences; the organization of a single Alu sequence is shown in Fig. 13.9.

Why should Alu elements be such frequent targets for A→I RNA editing? The reason is that Alu elements provide the secondary structures that RNA editing requires. Alu elements are frequently inserted in multiple positions and in both orientations within genes. Once transcribed into pre-mRNAs, Alu elements present within the same RNAs but in different orientations are complementary, and these complementary sequences can base pair and form the double-stranded RNA structures required for RNA editing (see Fig. 13.10).

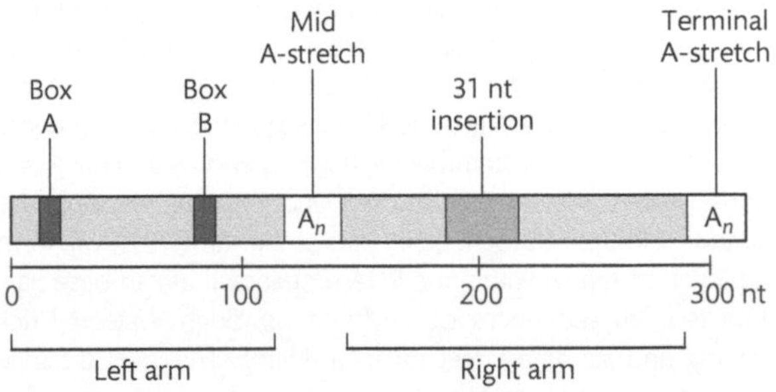

Figure 13.9 Alu elements. There are more than a million Alu copies in the human genome (corresponding to 10% of the human genome). Alu elements are mobile elements, and specifically retrotransposons since they need to be transcribed into RNA in order to transpose in the genome. The genetic structure of each Alu element has two similar arms and is about 300 nucleotides (nt) long. The left arm contains A and B boxes which are required for internal initiation by RNA polymerase III. The right arm terminates in a stretch of A residues. (From Hasler and Strub.[13])

Is A→I RNA editing of Alu sequences important? The answer seems to be yes. Studies have examined the effects of human mutations in the *ADAR1* gene. These mutations cause an inflammatory disease called Aicardi–Goutières syndrome (AGS) which

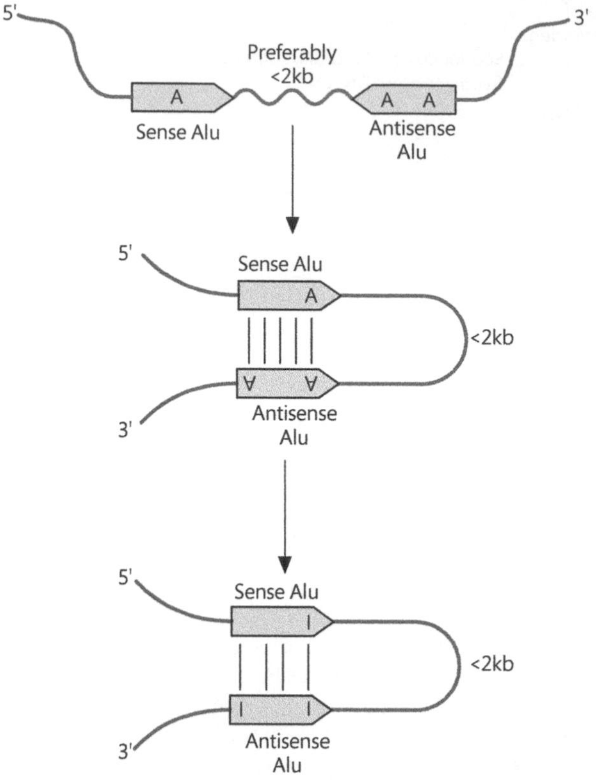

Figure 13.10 The secondary structures formed by oppositely orientated Alu elements act as targets for A→I RNA editing. Because of their high levels of insertion into the human genome and their propensity for inserting into coding regions, on average human pre-mRNA will include 15 Alu sequences. Alu sequences can insert into either strand of the human genome and have very conserved sequences. Two Alu sequences inserted on opposite strands and then transcribed as part of the same RNA will form an inverted repeat within the RNA sequence that can base pair. Inverted Alu sequences can hybridize through Watson–Crick pairing and act as a target for A→I editing. The ideal distance between two inverted Alu sequences to enable maximal editing is around 2kb. (Redrawn from Hasler and Strub.[13])

affects the brain and skin.[14] The presence of double-stranded RNAs in a cell causes expression of molecules called interferons. The brains of AGS sufferers resemble those of embryos that have been infected with RNA viruses like rubella, which cause a massive and damaging immune response by inducing interferon expression. These disease symptoms suggest that by editing repetitive Alu sequences after they are transcribed, ADAR1 protein might limit levels of double-stranded RNA in the cell and so reduce the expression of interferons. Mutations in the *ADAR1* gene lead to increased levels of double-stranded Alu sequences, and this in turn leads to the induction of interferon and brain damage.

Other experiments also show that RNA editing is important in the brain. In fruit flies there is just a single *ADAR* gene which means that only a single gene needs to be removed to eliminate ADAR protein function. Flies without a functional *ADAR* gene have normal physical development and lifespan but exhibit behavioural problems, including excessive cleaning.[15] In the absence of editing of these transcripts massive neurodegeneration takes place, resulting in the visible holes in the brain seen on microscope sections like those shown in Fig. 13.11. Similar genetic experiments in mice also suggest important roles for ADAR proteins in brain development (we discuss these later in this chapter). Consistent with this, measurements in different tissues show that inosine nucleotides are particularly enriched in the brain.[16]

13.4.2 Modifying the reading frames of already transcribed mRNAs: RNA editing can create new proteins

While almost all cellular mRNA editing is targeted to Alu elements located in pre-mRNAs (and often within introns before splicing takes place), some biologically very important examples of RNA editing still take place within the coding portion of mRNAs. Known target mRNAs for A→I editing in fruit flies encode ion channels which are important for normal development of the nervous system. RNA secondary structure is still important for editing of mRNAs. Frequently this RNA secondary structure comes from base pairing between the exon (containing the A nucleotide that becomes edited) and a region of nearby intron sequence that is called the exon complementary sequence (ECS) (see Fig. 13.12). RNA editing events within mRNAs can change the reading frames of mRNAs, resulting in new protein sequences being translated from them.

mRNA editing events are important in the brains of other animals as well as the fruit fly. One of the best characterized A→I mRNA editing events takes place in the mouse brain, within the mRNA encoding the glutamine receptor GluR2 subunit.[4,18] The GluR2 protein is an important component of a neurotransmitter receptor that functions in

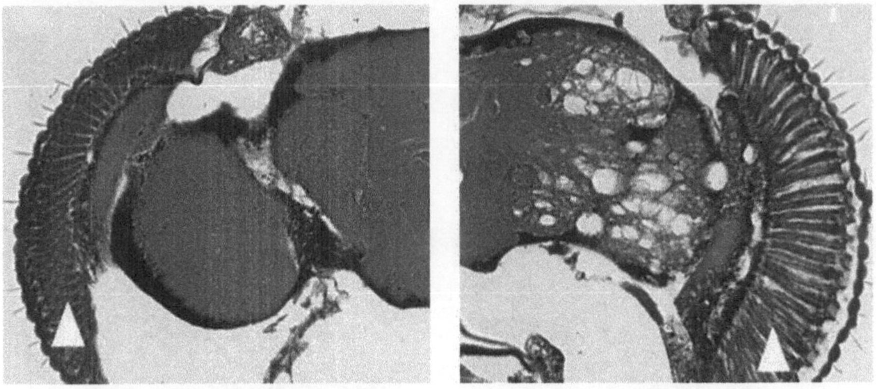

Figure 13.11 Analysis of fruit flies without functional *ADAR* genes indicates an important function for A→I editing in the brain. An important genetic approach is to inactivate or mutate genes and then to analyse what happens. Any defects observed in mutant cells or organisms can then be assigned directly or indirectly to the function of the missing protein. Fruit flies are particularly useful for this kind of analysis. Fruit fly genes can be either mutated or inactivated by a set of transposable elements called P elements. P elements located close to genes can be induced to 'hop' out of the genome, removing neighbouring fragments of genomic DNA including adjacent genes. Here these techniques were used to analyse the effects of mutating the single fruit fly *ADAR* gene. Notice the large holes in the brain of the fly without any ADAR protein (right) compared with the wild-type brain (left). (Reprinted from Palladino et al.[15] with permission from Elsevier.)

learning and memory (see Fig. 13.13). A→I RNA editing catalysed by the RNA editing enzyme ADAR2 converts a CAG codon (encoding glutamine) into a CIG codon (read as CCG, encoding arginine). Because of this amino acid change, editing of the *GluR2* mRNA transcript is sometimes called *Q/R editing* (the symbol for glutamine is Q, and that for arginine R). As a result of the Q→R amino acid change, receptors containing a GluR2 subunit translated from an A→I edited mRNA transcript do not allow the passage of calcium ions.

This *GluR2* RNA editing event is very important.[4] In mice, Q/R editing of the GluR2 receptor is critical for normal development of the nervous system (see Fig. 13.14). Experiments have shown that mice with a point mutation in *GluR2*, which prevents A→I editing, develop seizures and die within three weeks of birth. Similarly the *ADAR2* gene that carries out the editing of the *GluR2* mRNA is essential for mouse development. Mice that do not have ADAR2 protein (these mice lack a functional *ADAR2* gene, and are called *ADAR2*$^{-/-}$ mice in Fig. 13.14) could not edit the *GluR2* mRNA, and similarly suffered neurological defects and early death. However, surprisingly, the deaths of

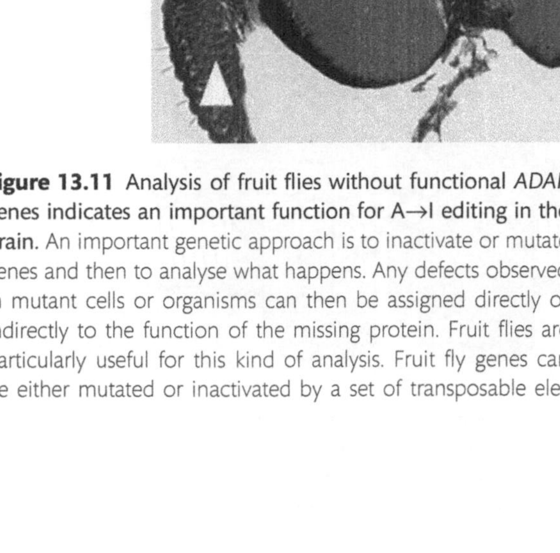

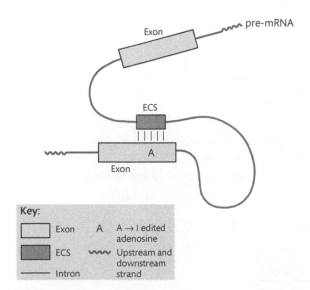

Figure 13.12 ADAR enzymes catalyse A→I mRNA editing. RNA editing by ADAR enzymes is targeted to RNA double helices which form between the portion of RNA sequence containing the adenosine to be edited and a complementary sequence which can base pair with it. Because of their complementary nature, base-pairing partners which target editing are called editing complementary sites (ECSs). An ECS can be located either close to or very distant from the nucleotide sequence to be edited.

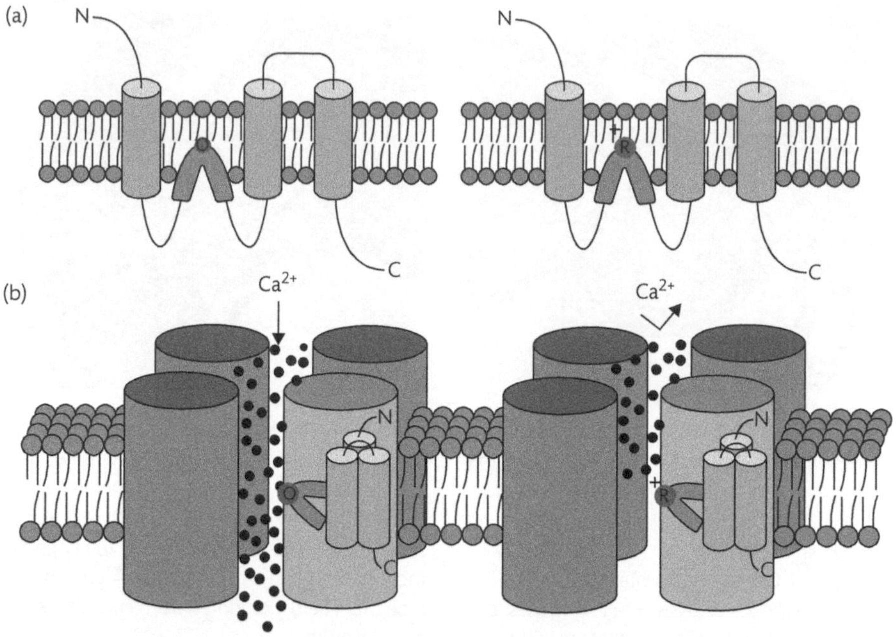

Figure 13.13 A→I mRNA editing changes the open reading frame of the GluR2 receptor subunit so that two different proteins are made. (a) Conversion of a CAG codon into a CIG codon by ADAR2 replaces arginine (R) with glutamine (Q) in the GluR2 subunit protein. Because of the amino acid changes produced, this kind of editing is sometimes called Q/R editing, and creates receptor subunits with different permeability to calcium ions. (Reproduced from Slotkin and Nishikura[17] with kind permission.)

(a)

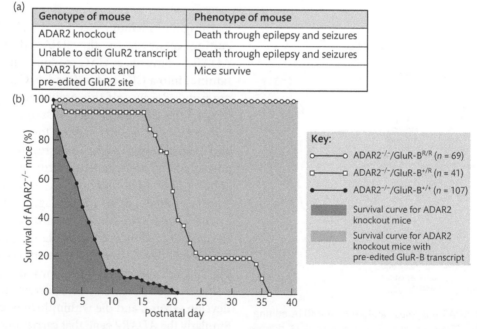

Genotype of mouse	Phenotype of mouse
ADAR2 knockout	Death through epilepsy and seizures
Unable to edit GluR2 transcript	Death through epilepsy and seizures
ADAR2 knockout and pre-edited GluR2 site	Mice survive

(b)

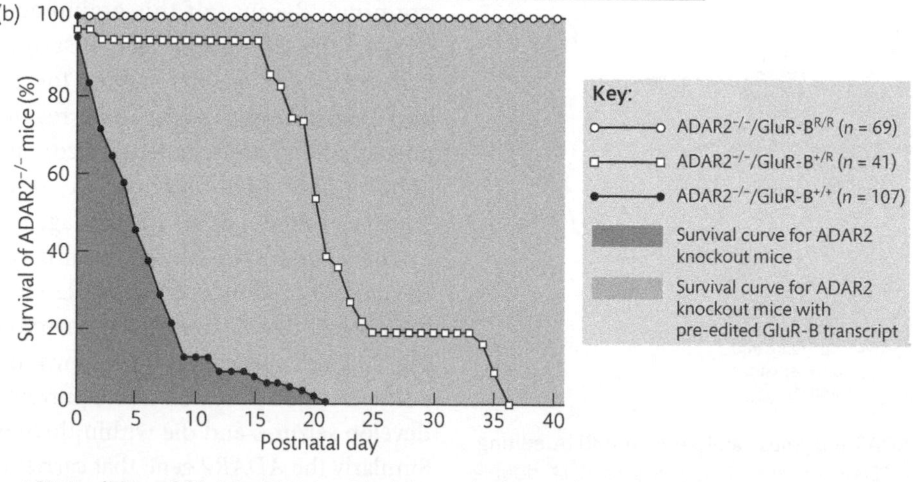

Figure 13.14 A→I RNA editing of GluR2 mRNA is critical for normal development of the nervous system in the mouse. (a) The effects of various mouse genotypes tabulated to summarize the results of the genetic experiments. (b) Mice with homozygous deletions of the *ADAR2* gene (*ADAR2*$^{-/-}$ mice) cannot carry out Q/R editing of the *GluR2* mRNA and die as a result of nervous system defects. This lethality is shown by the decline in survival in the graphs—notice the steep drop in the survival curve after birth. The lethality of the *ADAR2*$^{-/-}$ mutation is rescued by a nucleotide change in the *GluR2* gene. This 'pre-edited' nucleotide within *GluR2* introduces an arginine in the Q/R edited position. *ADAR2*$^{-/-}$ mice with two copies of the pre-edited nucleotide fare better. (Part (b) reprinted from Higuchi et al.[18] with permission from Macmillan Publishers Ltd.)

the $ADAR2^{-/-}$ mice could be prevented by introducing a point mutation into the *GluR2* gene which pre-edits the encoded mRNA. This point mutation changes the normal CAG codon into CCG. In terms of tRNA recognition this change is equivalent to a CIG, so it 'pre-edits' the Q/R site, removing the need for ADAR2 for A→I editing of *GluR2* mRNA (these pre-edited mice are *GluR*-$B^{+/R}$ and *GluR*-$B^{R/R}$ in Fig. 13.14).

13.4.3 Changing splicing signals: A→I editing of pre-mRNAs can affect splicing

By changing the sequence of important splicing signals in pre-mRNAs, A→I RNA can affect how these are recognized by the spliceosome (see Chapters 5–7). The connection between A→I mRNA editing and the regulation of splicing was first shown in 1999 for the pre-mRNA encoding the ADAR2 protein which catalyses A→I editing.[19]

Two different mRNA isoforms are made from the mouse *ADAR2* gene. One *ADAR2* splice isoform encodes a full-length ADAR2 protein and the other encodes just a short non-functional ADAR2 protein (see Fig. 13.15). These two mRNA isoforms are made by alternative splicing of pre-mRNAs encoded by the *ADAR2* gene via selection of an alternative 3′ splice site by the spliceosome.

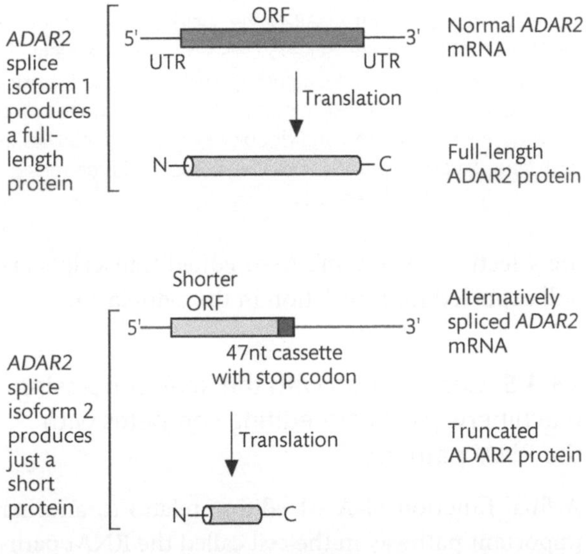

Figure 13.15 There are two splice isoforms of the *ADAR2* mRNA. Two alternative splice isoforms of the *ADAR2* mRNA transcript are made. One *ADAR2* mRNA splice isoform contains a full-length ORF encoding a functional ADAR2 protein. A second alternative splice isoform contains an insertion of 47 nucleotides through alternative splicing. Insertion of these 47 nucleotides changes the reading frame of the *ADAR2* transcript so that a short non-functional ADAR2 protein is made by translation.

The ADAR2 protein itself regulates this alternative event, within its own pre-mRNA, through A→I RNA editing controlling splice site recognition. Most introns in the human and mouse genomes start with the nucleotides GT and end with AG, and this is required for recognition by the spliceosome (see Chapters 5 and 6). In contrast, the genomic sequence of the upstream 3′ splice site for ADAR2 exon 5 is an AA rather than an AG. AA is not a functional 3′ splice site and would not be recognized by the spliceosome, but the genomic AA is edited to AI in the *ADAR2* transcript by the ADAR2 protein itself. AI is read as AG by the spliceosome which enables selection of the upstream 3′ splice site. An experiment illustrating the effect of ADAR2 protein on splicing of its own pre-mRNA is shown in Fig. 13.16; notice that the splicing pattern of ADAR2 changes when cells express increased levels of ADAR2 protein.

Why is this important? The reason is that editing of the *ADAR2* mRNA by the ADAR2 protein itself sets up a classical feedback loop that is critical for maintaining ADAR2 protein levels in the cell (see Fig. 13.17). If the level of ADAR2 protein increases in the cell, it edits the upstream 3′ splice site in the *ADAR2* mRNA more efficiently so that it is recognized by the spliceosome, resulting in the non-functional ADAR2 protein being made. When functional levels of ADAR2 protein fall, this reduces A→I editing and increases production of ADAR2 protein.

13.4.4 Nuclear retention of specific mRNAs: A→I editing can affect nuclear retention of transcripts

Another function of A→I editing is concerned with the export of RNA from the nucleus; A→I editing can block export of nuclear RNAs into the cytoplasm.[20] Nuclear-retained A→I edited transcripts include some viral RNAs (notably polyoma virus transcripts late in infection) and also some endogenous cellular mRNAs. Some inosine-containing transcripts become anchored by attachment to nuclear proteins (see Fig. 13.18). These nuclear retention mechanisms of adenosine-containing mRNAs

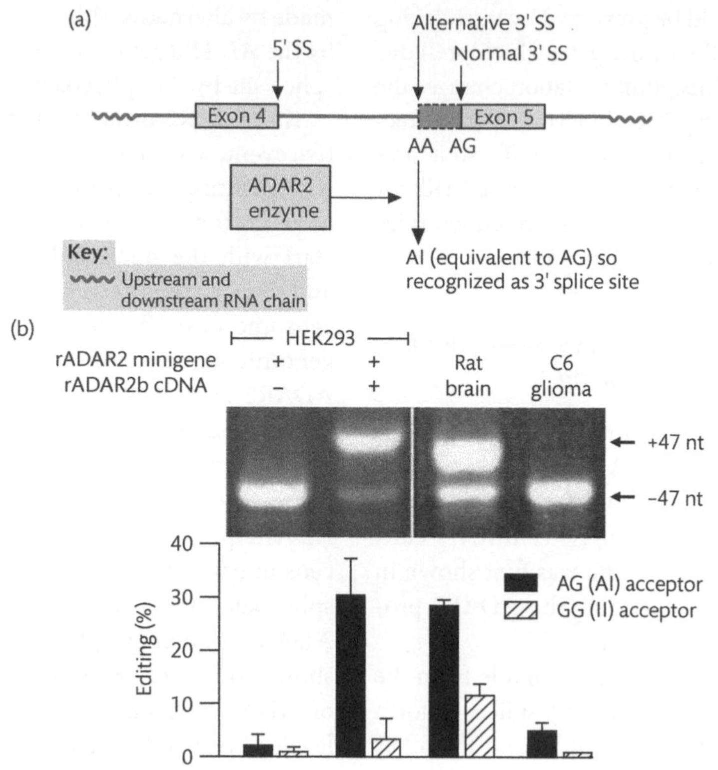

Figure 13.16 **ADAR2 protein autoregulates its own splicing through A→I mRNA editing.** (a) Two alternative splice versions of exon 5 using different 3′ splice sites (SS) are spliced into *ADAR2* mRNAs. The downstream exon 5 splice site sequence is used in the absence of editing, and is the usual AG sequence used at the many other 3′ splice sites. The upstream exon 5 alternative splice site sequence is AA, which is edited to AI (equivalent to AG which is a 3′ splice site) by the ADAR2 protein itself. (b) Experiment to show that ADAR2 protein catalyses A→I editing in its own pre-mRNA to create the alternative upstream 3′ splice site. Patterns of splicing for this part of the *ADAR2* mRNA are shown here on an agarose gel. RNA was analysed by RT-PCR in an experiment in which HEK293 cells had been transfected with a minigene containing exon 5 of the rat *ADAR2* gene. Increasing expression of rat ADAR2 protein in the HEK293 cells led to the alternative 3′ splice site being selected preferentially so as to include the extra 47 nucleotides in the *ADAR2* transcript. The histogram shows the levels of single-edited (AI) and double-edited (II) adenosines in each lane. (Part (b) reprinted from Rueter et al.[19] with permission from Macmillan Publishers Ltd.)

are selective, since some A→I edited transcripts are still exported for translation in the cytoplasm.

13.4.5 Competition with the RNAi processing machinery: A→I RNA editing competes with the RNAi pathway

A final function of A→I editing relates to another important pathway in the cell called the RNAi pathway (see Chapter 16).[21-23] *ADAR* genes have been inactivated in the nematode worm *Caenorhabditis elegans*. *ADAR* mutant worms look normal but have defects in chemotaxis; they fail to smell and to crawl towards food sources. This means that in worms the A→I RNA editing pathway must be involved in modifying transcripts involved in chemotaxis.

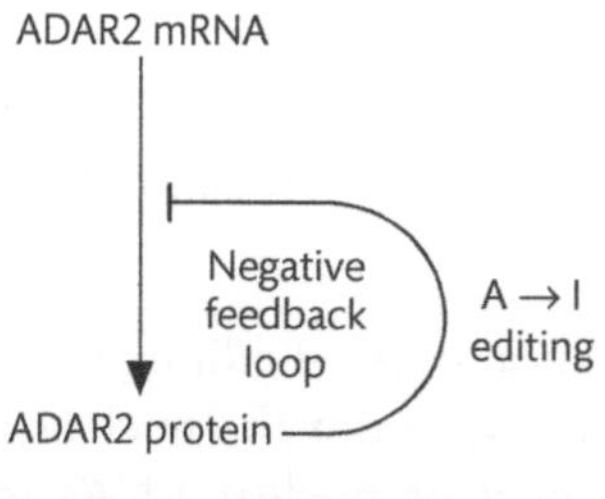

Figure 13.17 **ADAR2 regulates its own production through a negative feedback loop involving alternative RNA splicing.** The alternative 3′ splice site, which leads to an extra 47 nucleotides being inserted into the *ADAR2* mRNA, prevents production of a functional ADAR2 protein. The result of this is that levels of ADAR2 protein fall in the cell. As the ADAR2 protein levels fall, there is correspondingly less editing to make the *ADAR2* exon 5 upstream 3′ splice site. As a result of this more functional (full-length ORF) *ADAR2* mRNA is made, resulting in more ADAR2 protein.

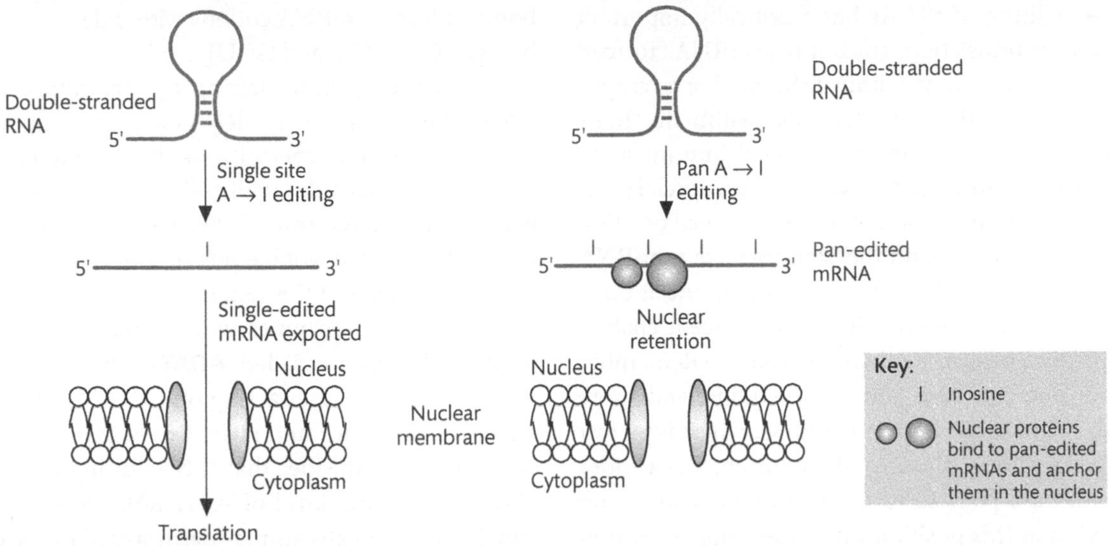

Figure 13.18 Inosine-containing transcripts may be held in the nucleus. (Adapted from DeCerbo and Carmichael.[20])

Surprisingly, the chemotaxis defects in these worms can be rescued by secondary mutations which deactivate the RNA interference (RNAi) pathway (Chapter 16). This means that the RNA editing and RNAi pathways must be connected, probably because both the RNAi and microRNA pathways operate on dsRNAs.

Key points

The substrates for A→I RNA editing are dsRNA helices which form within RNA molecules. A frequent source of dsRNA in the transcriptome is Alu elements present in inverted orientations within single RNA transcripts. RNA editing of these Alu elements in pre-mRNAs reduces the level of double-stranded RNAs formed by these elements which could induce harmful immune responses.

A→I RNA editing also modifies exon sequences (exons are sometimes spliced into mRNAs to encode proteins). A→I edited sites in mRNAs are recognized by editing ADAR enzymes since they are located in double-stranded RNA structures which form between exonic sequences and exon complementary sequences (ECSs) within introns. Thus, A→I RNA editing takes place before the intronic ECSs are removed by splicing. A→I editing in mRNAs can rewrite ORFs. Because inosine has different base-pairing properties from adenosine, otherwise equivalent codons containing adenine and inosine will bind to anticodons in different tRNAs. This can lead to the insertion of new amino acids at specific locations within proteins corresponding to the position of A→I edited sites in the parent mRNA. A→I RNA editing can also change splicing patterns of edited pre-mRNAs by modifying splicing signals, and addition of inosines can lead to nuclear retention of edited RNAs. Finally, the dsRNAs which are targeted by the A→I editing machinery can also be substrates for the microRNA processing machinery. By disturbing base pairing, A→I editing in dsRNA can compete with other microRNA processing pathways.

13.5 A→I editing plays an important role in the function of tRNAs

We have discussed the ability of A→I editing to rewrite the reading frames within mRNA molecules by changing codons. Before we leave A→I editing there is one further important topic to consider. This is the topic of A→I tRNA editing.[24–28]

tRNAs are a frequent target for A→I editing. Inosine-containing tRNAs are also found in all species: eukaryotes, bacteria, and even the Archaea. This means that inosines are much more widespread in tRNAs than they are in mRNAs. The implication of this is that A→I editing probably first evolved to modify tRNAs, i.e. before A→I editing became adapted to edit mRNAs as well.

A→I editing of tRNAs has a critically important function in translation, since it helps tRNAs to read the full set of codons within mRNAs. For example, Fig. 13.19 shows that two adenosines within the tRNA for alanine (tRNA^{ala}) are A→I edited into inosines. One of these edited adenosines is at a very special position within the anticodon of the tRNA called position 3, which base pairs with the **wobble base** in mRNA (see Box 13.1). A→I editing of the adenosine at position 3 of the anticodon is important since it enables the tRNA to read more than one codon. (Remember that inosine can base pair with cytidine, and more weakly with adenosine and uridine, whereas A just base pairs with U. Thus, A→I editing enables a wider range of base pairs to form between the codon and anticodon at this position.) In the example shown in Fig. 13.19, the A→I edited tRNA^{ala} is able to hydrogen bond with three mRNA codons with different wobble bases (GCC, GCA, and GCU).

By expanding base pairing opportunities, A→I RNA editing changes in tRNAs are very important to enable the full repertoire of the genetic code to be read by a relatively small collection of tRNA molecules. The tRNAs from all species on the planet are A→I edited at the wobble position of the anticodon to enable full use of the genetic code.

The group of enzymes which catalyse A→I editing of tRNAs are called **ADATs** (abbreviation of adenosine deaminases acting on tRNA). ADATs are related in sequence and structure to ADARs, again consistent with the idea that tRNA editing was probably the ancestral kind of A→I RNA editing from which enzymatically similar pathways of mRNA editing later evolved.

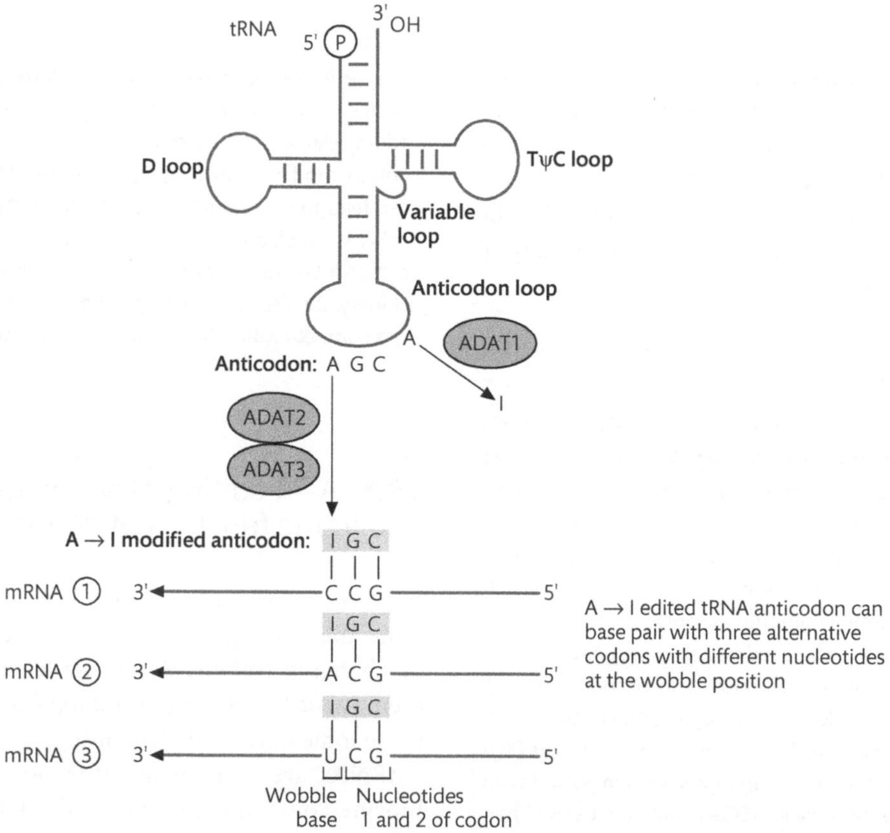

Figure 13.19 **Two critical adenosines within tRNAs are edited to inosines.** Transfer RNAs are important targets for A→I RNA editing. Two adenosines within tRNA^{ala} are converted into inosine by A→I editing. One of these A→I RNA changes is in anticodon position 1 which is complementary to the wobble position of the mRNA codon. This anticodon A→I editing is catalysed by a heterodimer of proteins called ADAT2 and ADAT3. Inosine at the wobble position within the anticodon enables tRNA^{ala} to bind to three different codons. A second adenosine in the tRNA anticodon loop, but not within the anticodon itself, is A→I edited by the enzyme ADAT1. Although use of the wobble base has evolved a particularly important role in eukaryotes, ADATs are found in *all* living organisms on the planet. Therefore ADAT enzymes probably evolved before ADAR enzymes.

Box 13.1 The wobble base has important consequences

Synonymous codons are similar codons in the mRNA which encode the same amino acid. The third position in the codon of the mRNA is called the wobble base, as synonymous codons often vary at this position rather than at positions 1 and 2 of the codon.

1. **The wobble base allows a limited set of tRNAs to read a group of synonymous codons.** Although there are 64 triplets in the genetic code (see Table 13.1), there are only around 30 tRNAs. This is possible because each tRNA can recognize more than one codon that usually differs in the wobble position. Hence 64 tRNAs are not required to read 64 codons.
2. **The wobble position of mRNAs also has implications for molecular evolution by providing a molecular clock.** Different synonymous codons can be used in different species to encode the same amino acid. These differences are often predicted to be neutral in terms of natural selection and, because of the wobble base, tend to accumulate at position 3 of the codon. The extent to which synonymous codon changes have accumulated in mRNA coding sequences has been used as a measure of divergence between species.

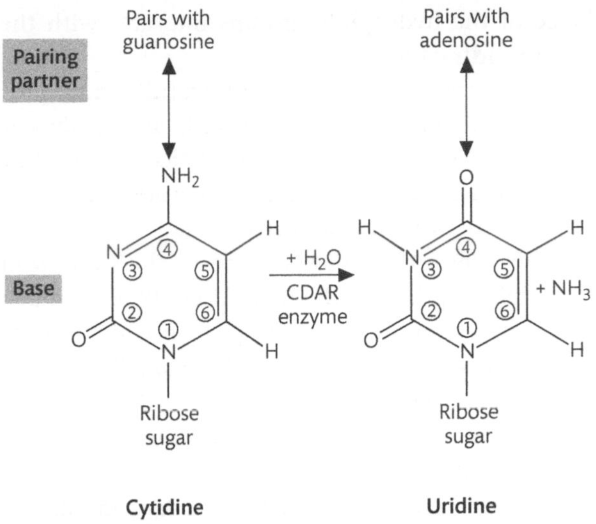

Figure 13.20 Cytidine can be converted to uridine by a deamination reaction. Both cytidine and uridine have single pyrimidine rings, but while cytidine has an amino group attached to position 4 of the ring, uridine has an oxygen. Cytidine is converted into uridine by removal of this amino group, and its replacement by an oxygen atom. The chemical reaction of cytidine with water removes the amino group to make uridine; this reaction is catalysed by an enzyme called **CDAR** (abbreviation of cytosine deaminase acting on RNA). This reaction releases an ammonia (NH_3).

Key points

A→I RNA editing of tRNA has an important role in translation. Adenosines at position 3 within the tRNA anticodon can be A→I edited, and it is position 3 of the anticodon that base pairs with the wobble base of the codon. This enables the nucleotide at the wobble base position within the mRNA codon to be more flexible. Non-standard (and slightly weaker) base-pairing combinations, including G-U and most importantly for this discussion A-I, C-I, and U-I, occur at the wobble position.

13.6 C→U RNA editing takes place through base deamination (removal of an amino group) of cytidine

A second important kind of RNA editing also relies on base modification. This is C→U RNA editing, and it is the topic of this section. Underlying this form of RNA editing, the bases cytosine (C) and uracil (U) are very similar chemically (see Fig. 13.20 and Chapter 2), and can be interconverted by removal of an amine group. C→U editing is important in both RNA and in DNA, as we shall see in the next three sections.

13.7 C→U RNA editing creates two different forms of the *APOB* mRNA in different tissues, and was the first RNA editing reaction to be discovered in animals

In 1987 scientists at the MRC Clinical Research Centre in London identified a tissue-specific C→U nucleotide change between the *APOLIPOPROTEIN B* (*APOB*) mRNA and its encoding gene in the intestine. In all other tissues the *APOB* mRNA had exactly the same sequence as its gene.[3] The reason for the interest in this gene is that APOB protein transports cholesterol and lipids around the body (see Box 13.2). ApoB is needed for these specific transport pathways because both lipids and cholesterol are hydrophobic yet need to be transported in the aqueous environment of blood. APOB enables blood transport of cholesterol and lipids by having both hydrophobic and hydrophilic surfaces. The hydrophobic groups interact with and envelop the fats and lipids, and an external face of the hydrophilic groups interacts with the surrounding blood.

There are two forms of APOB: a full-length protein of 100kDa called APOB100, and a shorter 48kDa protein called APOB48. Notice in Fig. 13.21 that these differently sized APOB proteins are both translated from transcripts encoded by the same *APOB* gene. APOB100 is made in the liver from the full-length *APOB* ORF. APOB48 is a much shorter version of the APOB protein which is only made in cells lining the wall of the small intestine. APOB48 is identical in sequence to APOB100 at its N-terminus, but lacks the C-terminus of the full-length protein.

The experiments reported from London in 1987 explained how the full-length APOB100 and the much shorter APOB48 protein are made from a single gene. The liver *APOB* mRNAs are not edited, and encode a full-length ORF translated to a 100kDa APOB protein (upper pathway in Fig. 13.21). In contrast, intestinal *APOB* mRNAs are C→U edited at a single site within the transcript which converts a CAA glutamine codon to a UAA stop codon. This shortened ORF is translated within the intestine to make a shorter APOB48 isoform (lower pathway in Fig. 13.21).

Box 13.2 Different isoforms of apolipoprotein B are involved in cholesterol and lipid transport

Two different isoforms of apolipoprotein B (APOB) are made: APOB100 in the liver and APOB48 in the intestine. The APOB100 protein is made in the liver from the full-length *APOB* ORF. The liver is one of the main sites of synthesis of cholesterol in the body. Cholesterol is transported through the bloodstream in low-density lipoproteins (LDLs) containing APOB100. APOB100 has two domains: a lipoprotein assembly domain and a LDL receptor-binding domain. APOBl00 encases lipids for transport in the LDLs via its C-terminal lipoprotein-binding domain. The N-terminal LDL receptor-binding site on APOB100 is recognized by LDL receptors on cells which need to take up cholesterol. As a result of this, cholesterol is efficiently absorbed from the bloodstream. In the intestine the shorter APOB48 protein coats fat globules absorbed from food in the intestine to make vesicles called chylomicrons, which transport the fat globules around the bloodstream. An important difference in the shorter APOB48 protein is that it lacks the C-terminal LDL receptor. Therefore chylomicrons coated in APOB48 are not absorbed by cells in the same way as LDLs. Instead, they are unloaded by lipoprotein lipases in muscles and other cells which require fats as a source of energy.

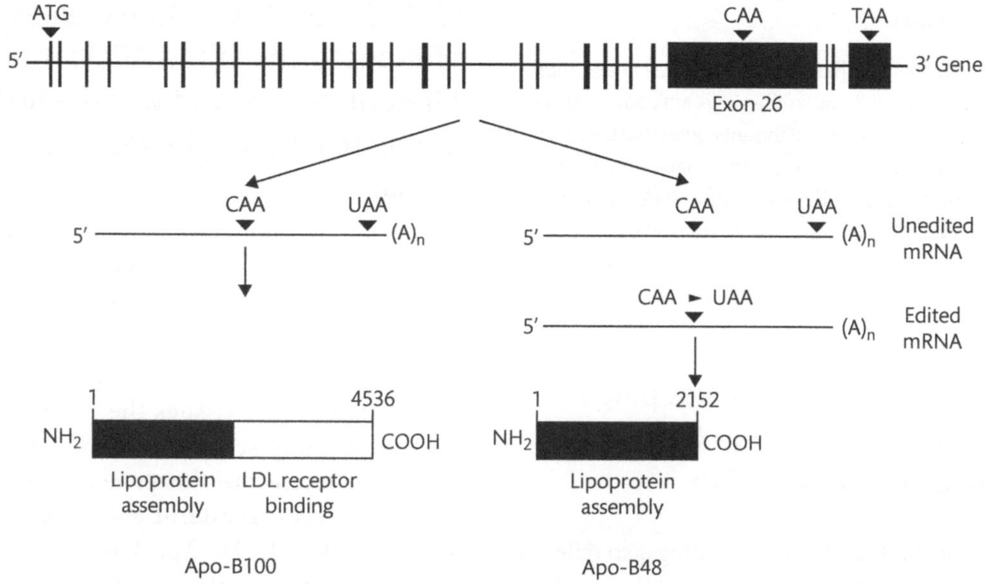

Figure 13.21 C→U RNA editing produces two forms of the apolipoprotein B (APOB). The same *APOB* gene is transcribed in liver and intestine. However, in the liver, the *APOB* transcript is translated to give a 4536 amino acid APOB100 protein which transports cholesterol made in the liver to other tissues. In the intestine, the same *APOB* transcript is transcribed, but is C→U edited at nucleotide 6666 to replace a glutamine-encoding CAA codon with a stop UAA codon. The intestinal *APOB* transcript is translated to give a shorter protein APOB48 containing only the lipoprotein assembly domain. APOB48 transports lipids from the intestine to where they are needed elsewhere in the body. (Reprinted from Chester et al.[29] with permission.)

13.8 *APOB* mRNAs are edited by an RNA editing complex containing the cytidine deaminase ApoBec1

The mechanism through which the *APOB* transcript is edited has been worked out, and involves an editing complex comprising a set of two interacting proteins which attach to RNA[30–33] (see Fig. 13.22). The enzyme which catalyses C→U RNA editing is a cytidine deaminase called ApoBec1 (abbreviation of apolipoprotein B editing enzyme, catalytic polypeptide-like). The other component of the editing complex is an RNA-binding protein called **ACF** (ApoBec complementing factor) and is needed for **ApoBec** activity.

The ApoBec1 and ACF proteins assemble on the *APOB* mRNA as a complex that carries out the RNA editing reaction which converts the genetically encoded CAA into the edited UAA triplet in the intestine (see Fig. 13.22). Notice from this figure that the core editing complex is made up of two interacting copies of ApoBec1 protein and a single copy of the interacting accessory protein ACF. The ApoBec editing complex binds to specific conserved nucleotides within the ApoB RNA sequence including AU-rich elements (these AU-rich elements are shown as boxes in the hairpin loop in Fig. 13.22). These boxed sequence elements are known to be important since their mutation prevents C→U RNA editing from taking place. The region edited in the *APOB* transcript also base pairs to form a specific stemloop secondary structure. This structure contains the edited cytidine residue within its loop and positions it for editing.

Although they both carry out base deamination reactions, the ApoBec1 cytidine deaminase is somewhat simpler in structure than the ADAR enzymes we discussed earlier—compare Figs 13.23 and 13.4. ApoBec1 does contain a deaminase domain with histidine and cysteine residues coordinated with zinc atoms, and a conserved glutamic acid residue that is important for catalysis. However, ApoBec1 does not contain an RNA-binding domain and binds to the AU-rich target RNA sequences fairly non-specifically (see Fig. 13.22).

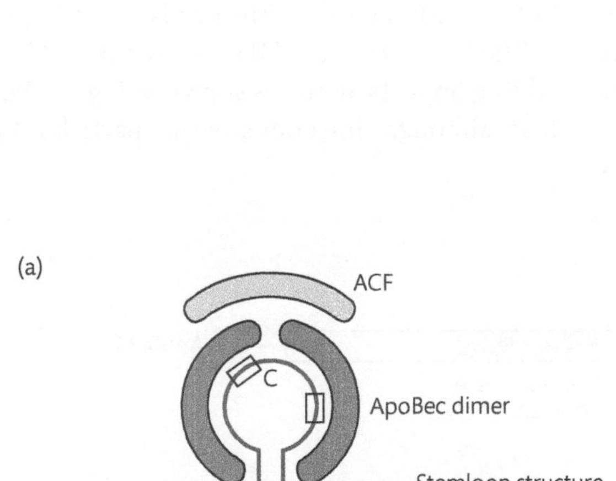

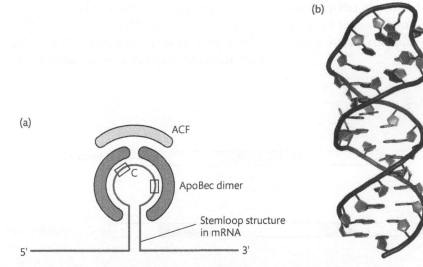

Figure 13.22 A multicomponent editing complex assembles on the *APOB* mRNA. (a) The editing complex includes two copies of the cytidine deaminase enzyme ApoBec1 and a single copy of the accessory factor ACF. ApoBec1 and ACF interact with each other and with RNA. The *APOB* mRNA forms a hairpin stemloop secondary structure, with the C to be edited in the loop of the hairpin. ApoBec1 protein is critically required for RNA editing since mice from which the *ApoBec1* gene has been deleted cannot carry out the editing reaction to produce the ApoB48-encoding mRNA. In humans, ApoBec1 is only expressed in cells in the intestine—the tissue where C→U *APOB* RNA editing takes place—so *APOB* mRNA is only edited in the intestine. (b) The structure of the stemloop RNA sequence recognized by ApoBec1 is shown, based on NMR data. The editing base (cytidine nucleotide 6666 in the *APOB* mRNA) is buried in the loop, and accessibility to ApoBec1 is achieved through binding of the ACF protein. (Part (b) redrawn from the RCSB Protein Data Base ID 1YNC by Jonathan Crowe, Oxford University Press.)

> **Key points**
>
> C→U RNA editing plays an important role in the production pathway making different forms of APOB protein in human liver and intestine. Full-length APOB protein is a 100kDa protein made in the liver and is involved in the transport of cholesterol through the bloodstream. A shorter form of APOB called APOB48 is made in the intestine and transports lipids through the bloodstream. The shorter form of APOB is made as the result of a CAA triplet (encoding glutamine) being edited into a UAA stop codon in the *APOB* mRNA. C→U RNA editing of the *APOB* mRNA is carried out by a cytidine deaminase called ApoBec1 and an accessory protein called ACF.

13.9 ApoBec proteins play an important role in innate immunity to retroviruses like HIV and in generating an antibody response

Although this chapter is primarily concerned with RNA editing, it is important to note that a related system of C→U editing takes place in DNA, also through base modification and involving ApoBec enzymes.[34,35] Since the C→U editing in DNA is more widespread than in RNA, the implication is that such DNA editing provided the evolutionary source from which RNA editing developed.

ApoBec1—the catalytic subunit which edits the apolipoprotein transcript—was the first animal CDAR to be identified and is the founding member of a larger ApoBec family of cytidine deaminases. Surprisingly, while ApoBec1 is primarily implicated in C→U RNA editing, other ApoBec proteins (including ApoBec3) carry out the same reaction on DNA. Each ApoBec family member contains either one or two cytidine deaminase domains (ApoBec1 and ApoBec3 are aligned in Fig. 13.23—notice the important amino acid residues in the cytidine deaminase domains).

C→U editing on DNA plays an important role in the function of the immune system in animals, functioning in both *innate immunity* (the part of the innate immune system which does not adapt after infection) and *adaptive immunity* (which uses antibodies made by the immune response and becomes more potent after repeated antigen stimulation).

13.9.1 C→U DNA editing in innate immunity

Two different ApoBec3 proteins help T cells avoid infection by retroviruses like HIV (which causes AIDS) and also inhibit the activity of retrotransposable elements, which are similar to retroviruses but are mobile only within cells (see Chapter 16).

The life cycle of the HIV retrovirus and how RNA editing impacts on this is shown in Fig. 13.24. Notice that although infectious virus particles (virions)

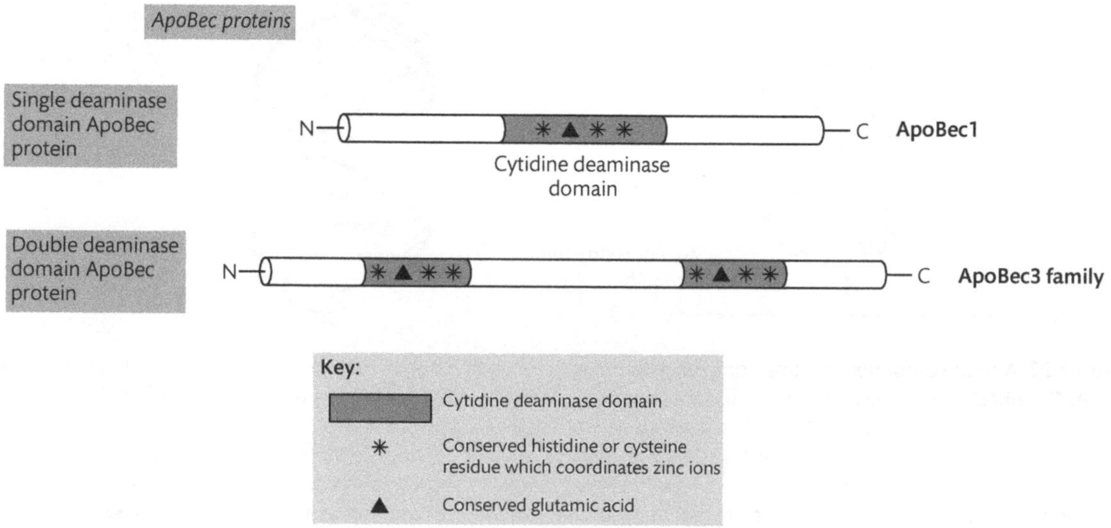

Figure 13.23 ApoBec1 is part of a family of proteins that contain cytidine deaminase domains. This figure shows two of these proteins: ApoBec1, which carries out C→U RNA editing on RNA, and ApoBec3, which carries out similar reactions on DNA.

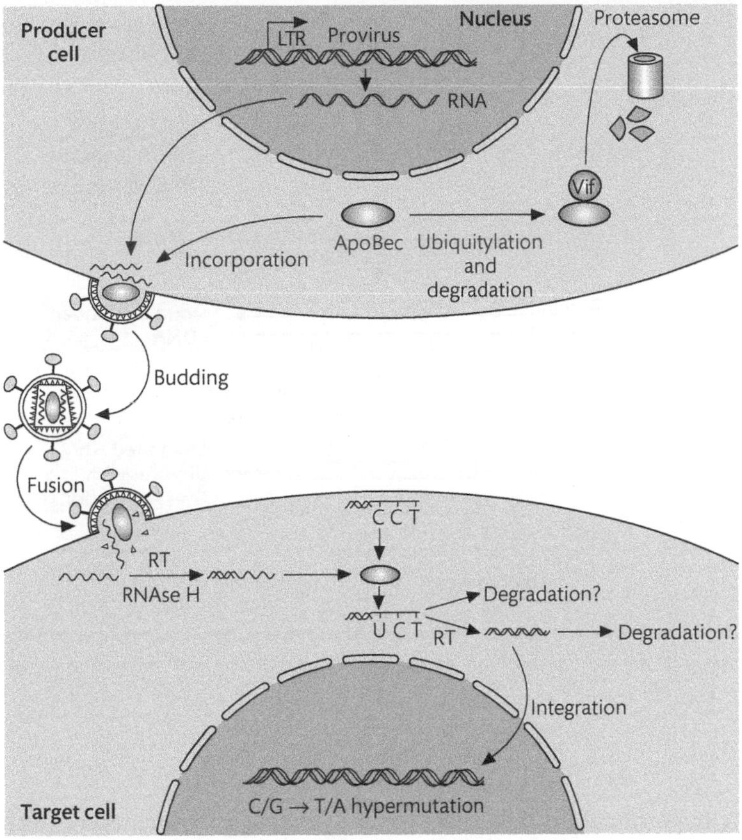

Figure 13.24 **Life cycle of a retrovirus like HIV showing the role of ApoBec proteins in limiting infectivity.** After infection, a complementary DNA (cDNA) copy of the retroviral RNA genome is made by reverse transcriptase (RT). The result of this is a first-strand cDNA hybrid with the original retroviral RNA. This first-strand cDNA is called the negative strand since this first strand is complementary in terms of sequence to the original retroviral RNA genome. The RNA in this hybrid is then digested with RNAse H, and replaced with a DNA strand. This double-stranded DNA virus then integrates into the genome to give a provirus (a viral genome inserted into the chromosome), which can be latent for many years before being transcribed into a new infectious viral RNA. (Reprinted from Harris and Liddament[35] with permission from Macmillan Publishers Ltd.)

contain an RNA genome, after infection the RNA genome converts into dsDNA which integrates as a provirus into a host chromosome. Conversion of the RNA genome of retroviruses into DNA is carried out initially by a viral enzyme called reverse transcriptase which makes **cDNA** (complementary DNA). ApoBec3 proteins make C→U DNA editing changes in this cDNA to create mutated proviruses (see Fig. 13.25). Such C→U DNA editing of viral sequences will introduce crippling mutations into important viral genes, and these can functionally inactivate the provirus. Notice that a C which is edited into a U in the first-strand cDNA will be copied as an A in the second-strand cDNA. The end result will be an inserted provirus with an A–T instead of a G–C base pair.

C→U DNA editing of first-strand cDNAs is very potent in cells expressing ApoBec proteins, changing up to 25% of the C residues into U residues. Notice in Fig. 13.25 that the ApoBec protein also inserts into the exiting HIV particle. In this way ApoBec proteins actually hitch a ride within the HIV virion to carry out further cDNA editing in the next cell that the virus infects. In fact, this form of innate immunity is so important that HIV has evolved a counter-attack mechanism against the ApoBec proteins. An HIV gene called *vif* (*virion infectivity factor*) recognizes and tags the ApoBec proteins with ubiquitin, thereby targeting them for degradation and limiting further mutation of proviruses (see Fig. 13.24).

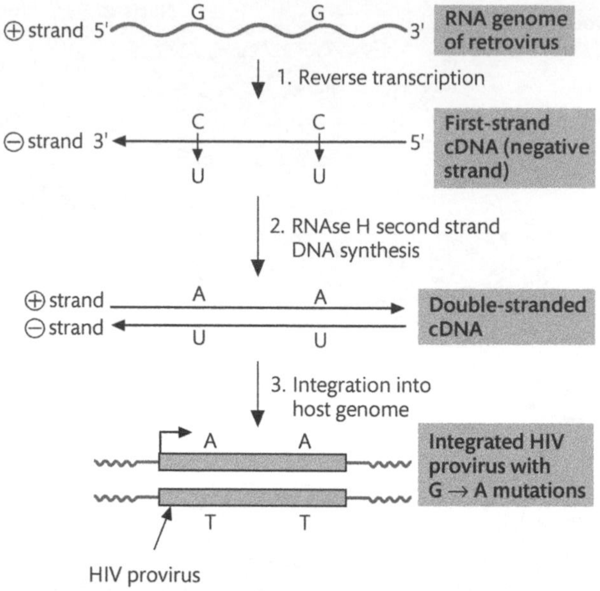

Figure 13.25 **RNA editing can hypermutate the RNA made from retroviruses.** This figure shows the production of viral genomic RNA in a producer cell, and its release and insertion into a target cell. (Redrawn from Harris and Liddament[35] with permission fropm Macmillan Publishers Ltd.)

The C→U DNA hypermutation that takes place in infecting HIV first-strand cDNAs could be very dangerous for the cell if it happened more generally to genomic DNA. However, compared with the massive C→U editing which takes place on HIV first-strand cDNA comparatively little C→U editing takes place on genomic DNA. The reason for this is that the ApoBec proteins are normally sequestered away from genomic DNA. The DNA-editing ApoBec enzymes are found within the cytoplasm, and in particular within cytoplasmic bodies called **P-bodies.** This means that while ApoBec proteins do not have access to nuclear DNA, they can mutate viral cDNAs made in the cytoplasm.

13.9.2 C→U DNA editing in adaptive immunity

ApoBec proteins also play an important role in the adaptive immune response, which uses antibodies to neutralize invading pathogens. Better antibodies are made as the immune system responds to an infection (see Box 13.3) through a process called *somatic mutation*. This rearranges the genes encoding the antibody proteins so that they have different heavy chains, and mutates the part of the genes that encode the part of the antibody that binds to the antigen. Somatic mutation of antibody genes is carried out by an ApoBec protein called **AID** (activation-induced cytidine deaminase).

Box 13.3 The ApoBec protein AID helps make higher-affinity antibodies during the immune response

The first antibodies produced in response to an infection tend to be of low affinity. Higher-affinity antibodies with different properties are produced later on in infection. The low-affinity antibodies, which are called immunoglobulin Ms (IgMs) and have a μ heavy chain, are produced early in infection. Later in infection, or after a repeat infection, soluble IgG antibodies which have high affinity for their target antigens are produced. C→U DNA editing of antibody-encoding genes plays an important role in these changes. IgGs have a γ heavy chain, and the cytidine deaminase AID functions in the switch from the μ to the γ heavy chain, possibly by introducing Us into DNA which stimulate the DNA recombination machinery to rearrange the antibody encoding genes. AID also plays a role in the somatic hypermutation of antibody genes to generate higher-affinity antibodies.

Interestingly, sequence comparisons show that *AID* is one of the most ancient *ApoBec* genes, suggesting that all C→U editing cytidine deaminases (working on either DNA or RNA) might have evolved from an early function in the animal immune system.

Key points

The ApoBec proteins behave as part of the innate immune system of the cell in carrying out hypermutation of invading retroviruses. Hypermutated retroviruses become incapacitated through changes to nucleotides in important viral genes, and as a result they are less able to mount a subsequent new infection. The AID ApoBec protein is involved in the adaptive immune response, where it mutates the genes encoding antibodies to increase antibody diversity and change antibody properties.

13.10 Trypanosome mitochondrial RNA is edited by base insertions and deletions to create ORFs from frameshifted transcripts

Each type of editing we have discussed so far in this chapter takes place through base deamination. The next, and final, editing topic in this chapter involves a different mechanism. This is the uridine addition/deletion type of RNA editing which takes place in trypanosome mitochondria.[36,37] Uridine addition/deletion editing changes the length of the edited RNA; however, it is still classed as RNA editing since it changes the sequence of the edited RNA compared with the genetically encoded version.

Before we go into more detail about the mechanism of RNA editing through uridine insertion and deletion, let us first take a moment to look at the context of where it takes place—within the mitochondria of trypanosomes which are protozoan parasites (see Box 13.4).

All mitochondria contain their own genetic material. The mitochondrial genome is much smaller than the nuclear genome, and is limited to a small set of genes encoding ribosomal RNAs and tRNAs, and mRNAs for a number of proteins required for oxidative phosphorylation.

Box 13.4 Trypanosomes

Trypanosomes are protozoan parasites which infect humans and cause sleeping sickness. They derive from an early branch in the eukaryotic evolutionary tree, so contain several unique features. Like other eukaryotes their cellular energy is largely provided by oxidative phosphorylation within subcellular organelles called mitochondria. Trypanosomes contain a single disc-shaped mitochondrion near the basal body of their flagellum which provides the locomotive power of the cell.

The trypanosome mitochondrial DNA is called the **kinetoplast** (see Fig. 13.26). The kinetoplasts of trypanosomes are unusual in two respects.

1 Kinetoplasts contain two different kinds of DNA circle: about 50 identical *maxicircles* (22kb in size) and about 10 000 smaller and much more heterogeneous *minicircles* (1kb in size). (In contrast, human mitochondrial DNAs are all equivalent.)

2 Maxicircles encode genes for subunits of the respiratory complex and are the functional equivalent of mitochondrial genomes in other species. However, many of the genes encoded on maxicircles are unrecognizable compared with the mitochondrial genes in other species. This is because the genes are either missing uridine nucleotides or in other cases have extra uridines inserted into and disrupting their reading frames. We say that the coding information of these unrecognizable genes is *encrypted*. The apparently non-functional genes are called *cryptic genes* or *cryptogenes*.

After transcription the mRNAs encoded by cryptogenes are converted by the addition or deletion of multiple uridine residues into their transcripts. Addition or deletion of uridine (U) from transcripts can create AUG start codons needed for the initiation of translation, correct frameshifts in reading frames, or in some cases entire ORFs via **pan editing** (where multiple uridines are added or deleted). The result of this form of editing is that the final edited mRNAs have different lengths and sequences from the starting mRNA, and are able to encode functional protein sequences.

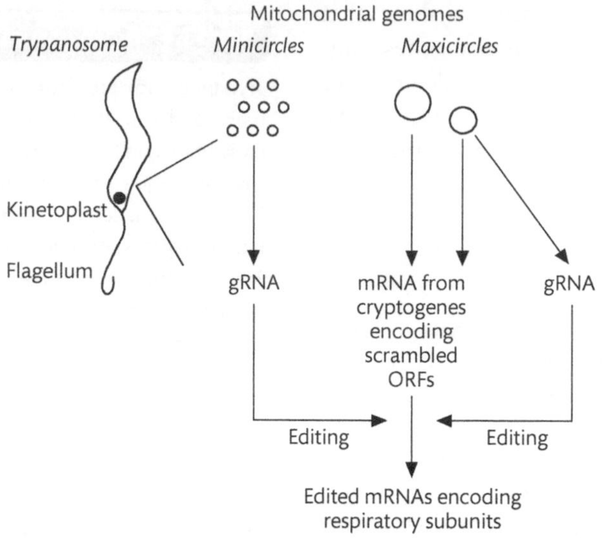

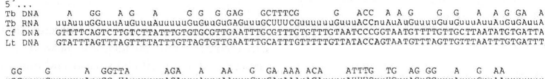

```
5´...
Tb DNA     A   GG   A G   A     G G  G GAG   GCTTTCG      G   ACC  A A G    G  G   A  A G GA  A
Tb RNA   uuAuuuGGuuuAuGuuuAuuuuuGuGuuGuGAGuuuGCUUUCGuuuuuuGuuuACCnuAuAuGuuuuGuuGuuuAuuAuGuGAuuA
Cf DNA   GTTTTCAGTCTTGTCTTATTTGTGTGCGTTGAATTTGCGTTTGTGTTTGTAATCCCGGTAATGTTTTGTTGCTTAATATGTGATTA
Lt DNA   GTATTTAGTTTAGTTTTATTTGTTAGTGTTGAATTTGCATTTGTTTTTGTTATACCAGTAATGTTTAGTTGTTTAATTTGTGATTT

 GG    G      A  GGTTA     AGA   A   AA   G  GA AAA ACA    ATTTG TG  AG GG   A   G  AA
uGGuuuuGuuuuuuAuuGG UAuuuuuuAGAuuuAuuuAAuuuGuuGAuAAAuACAuuuuAUUGuuUGuuAGuGGuuuAuuuGuuAAuuuuuuu
TGGTTTTGTATTTTACTGA TATTTTATTGTCGTATTTAATTTATTAATTAATACATTTTTATTATTTGTGATCGGATTATTCGTAAATTTTGTT
TGGTTTTGTATTTTATTGA TATTTTATTGATATTTTTAATTTACTAATTAATACATTTTTGTTATTTGTGAGTGGTTTATTCGTAAATTTTGTT
                                                                              COIII-6

G   G GTTTTTGG  AGG       G  GTTTTG  G   G A  A GA  GAG  G  G   GTTTTG      G      G GAAACCA
GuuuuGuGUUUUUGGuuuAGGuuuuuuuGuuG  UUGuuGuuuuGuAuuAuGAuuGAGuuuGuuGuuuG    GuuuuuuGuuuuuGuGAAACCA
TTATTTTTATTTTGATTTCGTTTTTTTTTATG  TGTATTATTTGTGCTTTGATCCGCTATATTTATTG    GTTTTTTATTTTTATGAAATCA
TTGTTTTTATTTTGATTTCGTTTTTTTTTATG  TGTTTTATTTATGTTATGAGTAGGAATATTATTTG    GTTTTTTATTTTTGTGAAATCA

G  ATGAGAGTTTTGCA  G  A    A  ACA  AAG  G GGTTTTTG     GG  C A     A     A  GGA    A TACA     ATTT
GuuAUGAGAG  UUUGCAuuGuuAuuuAuuACAuuAAGuuGuGG    UGuuuuuGGuuCuAuuuuAuuuuuAuuGGAuuuAuUACAuuuuA  U
GGTATGAGAA  TTTGCATTATTGTTTGTAACGTGCAGCTGTGG    TGTGTTTGGATCAATTTTATTTTTAATAGATTTACTACATTTTA  G
AGTATGAGAA  TTTGCATTATTGTTTGTAACATGTAGCTGTGG    AGTGTTTGGGTCAATATTATTTTTAATAGATTTATTACATTTTA  G
COIII-5

GCA G        AGG G   G  G  G   A   G    A GCG   G   AA      G G A GGA ACACG    G        G A  G G
GCAuGuuuuuuuAGGuGuuuuGuuGuuGuuuAuuuGuuuuAuGCGuuuGuuuAAuuuuuuGuGuAuGGAuACACGuuuuGuuuuuuuGuAuuGuG
TCATGTATTATTGGGTATATTTTTATTATTTATATGTTTTGGTAGATGCTTTAATTTTTTAAGTATGGACACACGTTTTGTTTTTTTATATGTTG
TCATGTGTTTTTAGGTATATTTTTATTATTTTTATGTTTTAGTCGTTGCTTTAATTTTTTGTGTATGGACACACGTTTTGTTTTTTTATATGTAG
                             COIII-4

   G   A A  GACA    G  GATTTAG   GA         A  GCGA   G   A    GA G    A G G  A G A   G G G G AA
uuuGuuuAuAuuGACAuuuuGuuGAUUUAGuuuGAuuuuuuuuuAuuGCGAuuuGuuuAuuuuGAuGuuuuAuGuGuuAuGuAuuuGuGuGuGuAA
TATGTTTATATTGACATTTTGTTGATTGTGTTTGATTTTTTTTATTACGTTTCGTATATTTTGATGTGTTATGCGTAGTGTATTTATGTGCATAA
TCTGTTTATATTGACATTTTGTAGATTGTGTTTGATTTTTTTTATTACGATTTGTATATTTTGATGTGTTAAGTGTAGTATACTTATATGCATAA
              COIII-3

    A  GG G    TTTAGTTG  GA   ATG  AA   G A  GGTAGTTTGTAGGAAG
uuuuAuuGGuGuuuuUUUAGUUGuuGAuuA GuuAAuuuGuAuuGGUAGUUUGUAGGAAGuuuuuuuuAuAAAAAAAAAA
ATATTTAATCCACACAAAT
GAAATAA
```

Figure 13.26 The mitochondrial genome has a distinct subcellular organization in trypanosomes. (a) Each trypanosome contains a single mitochondrion near the base of its flagellum. Within the mitochondrion there are multiple copies of circular DNA molecules called minicircles and maxicircles. In this diagram the minicircles and maxicircles are shown separately for clarity, but in vivo they are concatenated (enmeshed). Maxicircles encode components of the respiratory chain and also contain scrambled versions of genes called cryptogenes with no obvious protein-coding capacity. Both minicircles and maxicircles encode small RNAs called guide RNAs (gRNAs) which are used as partially complementary templates to edit cryptogene-encoded mRNAs. (b) Comparison of the sequences of the *COXIII* gene from trypanosome RNA and genomic DNA (Tb RNA and Tb DNA, respectively). Notice that the lower-case Us are uridines which are added post-transcriptionally. *COXIII* sequences from two other species are also added for comparison—in these species (Cf and Lt) nucleotides within the genome directly encode U residues in the RNA. ((b) is reprinted from Feagin et al.[38] with permission from Elsevier.)

13.11 RNA editing was discovered in trypanosomes by sequencing cDNAs encoded by mitochondrial genes

Trypanosome mitochondrial RNA editing was first discovered in 1986 by Rob Benne of the University of Amsterdam in the mRNA encoding mitochondrial cytochrome *c* oxidase subunit II (COXII).[1] This *COXII* gene was somewhat unusual compared with that in other sequenced species in that the reading frame was disrupted by a −1 frameshift (i.e. a nucleotide was missed from the ORF). As shown in Fig. 13.27, the −1 frameshift observed in the trypanosome *COXII* gene was towards the downstream end of the gene and as a result changed the reading frame for the C-terminal 40 amino acids.

At the time there were two possible explanations for this result. One was that there was another unframeshifted version of the *COXII* gene elsewhere in the genome that still had to be found. The other possibility, which proved to be correct, was that the frameshifted mRNA was corrected at the RNA level. Rob Benne found that four uridine residues (UUUU) are inserted into the encoded *COXII* mRNA to correct the −1 frameshift encoded by the mitochondrial genome.

Two years later, in 1988, a similar but even more elaborate form of RNA editing was reported in the trypanosome *COXIII* transcript by Jean Feagin and Ken Stuart in Seattle.[38] While *COXIII* is normally a highly conserved gene in eukaryotes the trypanosome *COXIII* gene is so extensively altered by RNA editing that it does not have a recognizable ORF before editing. Instead it has a cryptic sequence which can be converted into a *COXIII* ORF by the insertion of the multiple uridines that are added during RNA editing. Over 50% of the trypanosome *COXIII* reading frame is inserted as uridines by RNA editing. This kind of extensive editing is called pan editing. It is now known that 9 out of the total of 18 maxicircle protein-encoding genes are cryptic genes, converted after transcription by RNA editing either by base insertion or deletion.

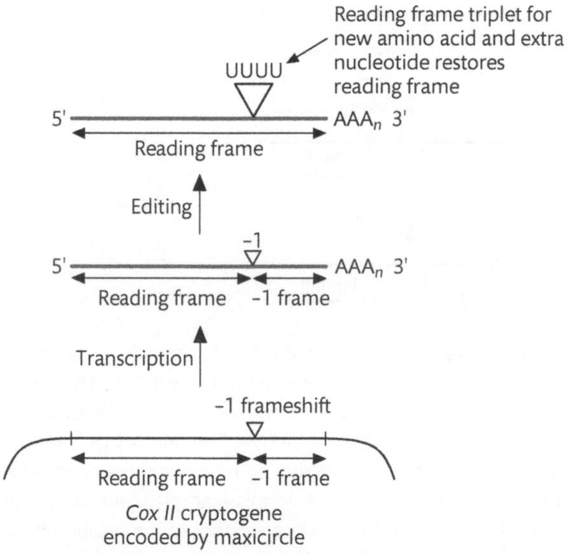

Figure 13.27 RNA editing was discovered in the *COXII* transcript of trypanosome mitochondrial RNA. The trypanosome *COXII* gene contains a frameshift in its reading frame at codon 170 caused by the loss of one nucleotide which changes the downstream reading frame to −1. As a result the *COXII* gene encodes the N-terminal 170 amino acids of COXII protein, but coding information for the remaining ~40 amino acids is out of frame. To correct the frameshift the mRNA 4 uridine residues (UUUU) are inserted at the site of the frameshift to restore the *COXII* reading frame by a U insertion RNA editing mechanism.

Key points

RNA editing was discovered in *COXII* mRNAs made in trypanosome mitochondria where they correct frameshifts encoded by the mitochondrial genome.

13.12 Short RNAs called guide RNAs target trypanosome mitochondrial RNA editing

The mechanism of RNA editing in trypanosome mitochondria involves a class of small non-coding mitochondrial RNAs called **guide RNAs** (gRNAs).[39–42] Guide RNAs were discovered in 1990 at the laboratory of Larry Simpson at the University of California, Los Angeles, and are encoded by both the minicircle and maxicircle mitochondrial genomes. Minicircles do not contain any protein-coding genes; instead each minicircle encodes three or four distinct gRNAs. Together, the heterogeneous group of minicircles within each trypanosome mitochondrion encode ~12 000 distinct gRNAs.

The sequences of gRNAs are partially complementary to target unedited mRNAs from maxicircle-encoded cryptogenes. As a result gRNAs can form base pairs with complementary cryptogene-encoded mRNAs (see Fig. 13.28). There are four important features to notice about the RNA helices which form between gRNAs and their cryptogene targets.

1 The gRNAs are much shorter than their complementary mRNAs.
2 Base pairing between the gRNA and its complementary mRNA is not continuous along the length of the gRNA.
3 The 5′ part of the gRNA is more extensively base paired with the mRNA. This part of the gRNA

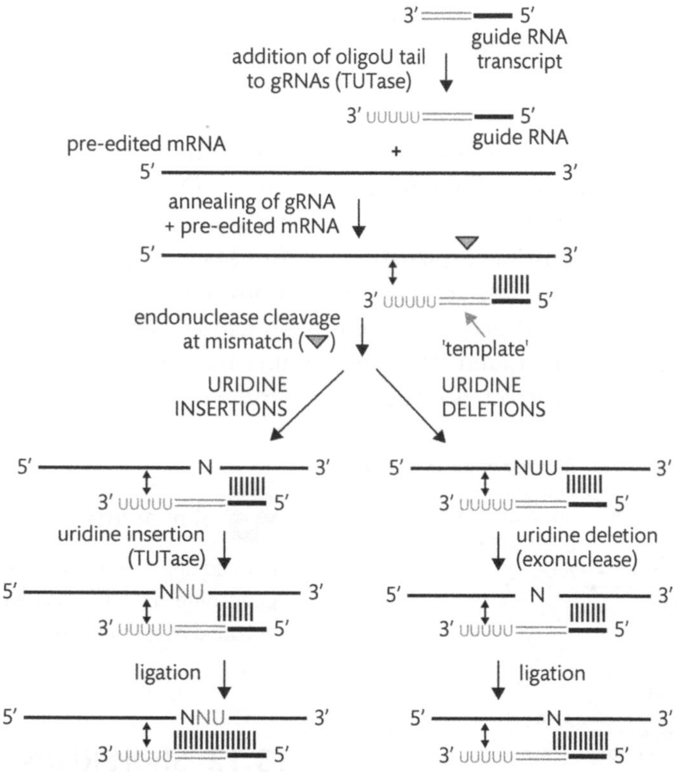

Figure 13.28 mRNA editing of trypanosome mitochondrial RNAs is directed by guide RNAs. (1) Guide RNA. The guide RNA transcript is partly complementary to the target RNA, and has a series of U residues added at the 3′ end by an enzyme called TuTase. These extra U residues help base pairing of the guide RNA to the pre-edited mRNA. (2) Guide RNA hybridization and RNA cleavage. After the guide RNA base pairs with the pre-edited mRNA, the mRNA in this duplex is cut by an endonuclease to give two fragments (an *endonucleolytic* cut is within an RNA molecule, as opposed to *exonucleolytic* digestion which is digestion from an RNA end). As a result of cutting there are now three RNA molecules in the duplex: the upstream and downstream fragments of mRNA and the gRNA holding these together. Only the mRNA is cut, the gRNA remains intact. (3) The mRNA is then edited by either U addition or U deletion. Left-hand pathway—U addition: an enzyme called terminal uridylyl transferase (TUTase) adds U residues to the 5′ phosphate group of the cut mRNA. The number of U residues added corresponds to the number of unpaired A residues present in the gRNA; after their insertion into the mRNA, these new U residues base pair with their template As within the gRNA. This increases the length of complementarity of the anchor sequence and the mRNA. The number of U residues added to an mRNA is controlled by how many unpaired A residues are present in the gRNA. In this example there are two unpaired A residues, and so this corresponds to the addition of two U residues into the mRNA. Once the As specified for insertion by the gRNA are in place, editing at this position in the mRNA is complete. Right-hand pathway—U deletion: the rule for deletion of uridine from mitochondrial cryptogenes is opposite to that for addition. Uridine residues that are present in the mRNA just upstream of the anchor site are *not* complementary to A residues in the gRNA. In this case, a U-specific 3′→5′ exonuclease removes the unpaired U residues from the mRNA. (4) Ligation. Once U addition is complete at this site the enzyme RNA ligase finally joins the upstream and downstream fragments of mRNA. (From Gott[43] with permission.)

is called the 'anchor'. Through this more extensive base pairing the anchor efficiently attaches the gRNA to the unedited cryptogene-encoded mRNA. (Notice in Fig. 13.28 that the anchor anneals downstream of the editing site in the mRNA.)

4 The 3′ part of the gRNA is modified after it is transcribed by the addition of a small number of uridine residues. This run of uridine residues is called an oligo-U tail and functions to help base pairing of the gRNA with mRNA upstream of the editing site.

13.13 Guide RNAs are used as a template for RNA editing through uridine insertions and deletions

RNA editing in trypanosome mitochondria usually takes place by addition of uridine to cryptogene-encoded transcripts, and less frequently by uridine deletion.[41,42] Exactly how editing takes place has been worked out in in vitro systems which carry out these reactions in the test tube. Uridine addition and deletion take place by similar four-step pathways that share some steps and enzymes, but each has some unique steps catalysed by different enzymes. These four-step pathways involve sequential cutting of the mRNA, addition of uridine nucleotides to or cleavage from the mRNA, and finally putting the RNA back together (ligation).

Conceptually, the pathway of RNA editing is similar to that which operates in the DNA repair pathway, and these two pathways might have some connection in terms of evolution. While there is no need for hybridization of a guide RNA in DNA repair, DNA repair involves sequential DNA cleavage, and nucleotide excision or addition followed by ligation of the repaired DNA strand. The pathways of RNA editing through U addition and deletion are described next.

RNA editing by uridine addition is achieved through the four-step pathway shown in Fig. 13.28. Notice that this pathway includes gRNAs, the mRNAs to be edited, and also protein enzymes which carry out the catalytic reactions on the mRNA. Although gRNAs are partially homologous to their target RNAs, the rule for RNA editing through uridine addition is that there are A residues present in the gRNA which are not in the mRNA. These unpaired A residues then act as templates for addition of uridine to the mRNA.

A given cryptogene mRNA may require multiple editing changes to fully restore its reading frame. Editing of a mRNA takes place in a 3′→5′ direction. This polarity is achieved because each step of RNA editing produces new mRNA sequences which can then be bound by the anchor sequence of the next gRNA used to modify the mRNA.

13.14 Trypanosome mitochondrial RNA editing requires nuclear-encoded proteins which might be useful therapeutic targets

The RNA editing pathway may be an Achilles' heel in the trypanosome; since the mitochrondria of their animal hosts do not carry out such RNA editing reactions, drugs targeted against RNA editing components may be useful therapeutic agents against trypanosome parasites.[44] Identification of the protein components involved in trypanosome mitochondrial RNA editing has enabled their expression to be targeted by siRNA (see Chapter 16) and through genetic knockout. The elimination of mitochondrial RNA-editing proteins blocks RNA editing and leads to trypanosome cell death, showing that RNA editing is essential for viability. In the absence of editing, trypanosomes lack key components of their mitochondrial respiratory chain.

Experiments have shown that trypanosome mitochondrial RNA editing takes place within a large complex called the *editosome*. Editosomes have been fractionated biochemically and contain three types of component: mitochondrial cryptogene-encoded mRNAs, gRNAs, and proteins required for the editing reaction (see Box 13.5). Because of the limited coding information capacity of the mitochondrial genome, these editing proteins are encoded by the nucleus and the mRNAs translated in the cytoplasm before the proteins are imported into mitochondria.

Box 13.5 Dissection of the molecular components of editosomes

For the RNA editing reactions shown in Fig 13.28 to take place four enzymatic activities are required.

- An *RNA endonuclease* is needed to cut the mRNA at the editing site during step 2. A candidate endonuclease for cutting the pre-edited mRNA has been identified which contains an RNAse III motif. RNAse III-type enzymes like DICER are also involved in RNA cleavage of RNA duplexes for microRNAs (see Chapter 16). This endonuclease probably binds to RNA through a zinc finger and dsRNA-binding domain.
- A *terminal uridylyl transferase* (TUTase) is needed to add U residues for step 3 of insertional editing.
- An *exo-UTPase* is needed to remove unpaired Us from the mRNA for step 3 of deletional editing.
- An *RNA ligase* is needed to join the 5′ and 3′ fragments of mRNA together at step 4 of the editing reaction.

Proteomic experiments have also shown that a much more extensive number of proteins than those providing the four basic enzymatic activities are involved in RNA editing in trypanosome mitochondria. These proteins include RNA helicases which play a role in RNA annealing and disassociation.

Key points

RNA editing reactions are targeted by a group of short RNAs encoded by the mitochondrial genome called guide RNAs (gRNAs). Guide RNAs are partially complementary to their target mRNAs, and control addition or deletion of uridine residues into mRNA. RNA editing takes place in a complex called the editosome. Protein components of the editosome are essential for creating a functional respiratory system in trypanosome mitochondria, and might also provide a potential target for therapeutic drugs.

Summary

- Two RNA editing mechanisms exist in nature.
  (1) Base modification reactions, which are found in A→I and C→U RNA editing in animals. Chemically, these base modifications occur by removal of amino groups from adenine and cytosine bases, respectively. These reactions are called base deaminations.
  (2) Base insertion/deletion reactions are found in the mitochondria of trypanosomes.
- Base modification changes the genetic code of mRNAs. The amino groups removed from bases by RNA editing through base modification are important hydrogen bond donors in nucleotide base pairing (see also Chapter 2); adenosine base pairs with uridine, but inosine base pairs best with cytidine. This means that inosine behaves exactly like guanosine in terms of base pairing. Because of this, inosine is also read as a G by the translation machinery and the enzyme reverse transcriptase.
- Transfer RNAs are also A→I edited in bacteria, the Archaea, and eukaryotes. Inosine can also base pair more weakly with A and U, which is important for decoding the wobble base in mRNAs. This enables a smaller group of tRNAs to read the full group of codons in the genetic code. Its evolutionary conservation suggests that tRNA editing may have been the ancestral form of A→I RNA editing.
- C→U editing of mRNA plays important roles in making different forms of the APOB protein in liver and intestine. C→U RNA editing may have evolved from similar reactions which occur on DNA and are important in the immune system.
- Double-stranded RNAs are important in RNA editing. Although there are mechanistic differences between different kinds of editing, double-stranded RNAs are a frequent feature of RNA editing reactions. Guide RNAs are important for guiding the specificity of trypanosome mitochondrial RNA editing. Intramolecular base pairing is important for A→I editing. RNA secondary structures also form at the site of C→U editing by ApoBec1.

Questions

13.1 Does RNA editing change the sequence of the parent gene?

13.2 Why are double-stranded RNA structures important for guiding RNA editing?

13.3 Name five biological outcomes of A→I RNA editing.

13.4 What is the reaction called that is used to edit adenosine to inosine, and cytidine to uridine?
13.5 Did RNA editing evolve first in mRNA or tRNA, and why is this thought to be the case?
13.6 In what biological system did C→U editing first evolve?
13.7 What is the purpose of editing in trypanosome mitochondria?

References

1. **Benne R, Van den Burg J, Brakenhoff JP, et al.** Major transcript of the frameshifted *COXII* gene from trypanosome mitochondria contains four nucleotides that are not encoded in the DNA. *Cell* **46**, 819–26 (1986).
2. **Bass BL, Weintraub H.** An unwinding activity that covalently modifies its double-stranded RNA substrate. *Cell* **55**, 1089–98 (1988).
3. **Powell LM, Wallis SC, Pease RJ, et al.** A novel form of tissue-specific RNA processing produces apolipoprotein-B48 in intestine. *Cell* **50**, 831–40 (1987).
4. **Reenan RA.** The RNA world meets behavior: A—>I pre-mRNA editing in animals. *Trends Genet* **17**, 53–6 (2001).
5. **Rodriguez J, Menet JS, Rosbash M.** Nascent-seq indicates widespread cotranscriptional RNA editing in *Drosophila*. *Mol Cell* **47**, 27–37 (2012).
6. **Gerber AP, Keller W.** RNA editing by base deamination: more enzymes, more targets, new mysteries. *Trends Biochem Sci* **26**, 376–84 (2001).
7. **Keegan LP, Leroy A, Sproul D, O'Connell MA.** Adenosine deaminases acting on RNA (ADARs): RNA-editing enzymes. *Genome Biol* **5**, 209 (2004).
8. **Savva YA, Rieder LE, Reenan RA.** The ADAR protein family. *Genome Biol* **13**, 252 (2012).
9. **Nishikura K.** Editor meets silencer: crosstalk between RNA editing and RNA interference. *Nat Rev Mol Cell Biol* **7**, 919–31 (2006).
10. **Ryter JM, Schultz SC.** Molecular basis of double-stranded RNA-protein interactions: structure of a dsRNA-binding domain complexed with dsRNA. *EMBO J* **17**, 7505–13 (1998).
11. **Macbeth MR, Schubert HL, Vandemark AP, et al.** Inositol hexakisphosphate is bound in the ADAR2 core and required for RNA editing. *Science* **309**, 1534–9 (2005).
12. **Ulbricht RJ, Emeson RB.** One hundred million adenosine-to-inosine RNA editing sites: Hearing through the noise. *Bioessays* **36**, 730–5 (2014).
13. **Hasler J, Strub K.** Alu elements as regulators of gene expression. *Nucleic Acids Res* **34**, 5491–7 (2006).
14. **Rice GI, Kasher PR, Forte GM, et al.** Mutations in ADAR1 cause Aicardi-Goutières syndrome associated with a type I interferon signature. *Nat Genet* **44**, 1243–8 (2012).
15. **Palladino MJ, Keegan LP, O'Connell MA, Reenan RA.** A-to-I pre-mRNA editing in *Drosophila* is primarily involved in adult nervous system function and integrity. *Cell* **102**, 437–49 (2000).
16. **Paul MS, Bass BL.** Inosine exists in mRNA at tissue-specific levels and is most abundant in brain mRNA. *EMBO J* **17**, 1120–7 (1998).
17. **Slotkin W, Nishikura K.** Adenosine-to-inosine RNA editing and human disease. *Genome Med* **5**, 105 (2013).
18. **Higuchi M, Maas S, Single SN, et al.** Point mutation in an AMPA receptor gene rescues lethality in mice deficient in the RNA-editing enzyme ADAR2. *Nature* **406**, 78–81 (2000).
19. **Rueter SM, Dawson TR, Emeson RB.** Regulation of alternative splicing by RNA editing. *Nature* **399**, 75–80 (1999).
20. **DeCerbo J, Carmichael GG.** Retention and repression: fates of hyperedited RNAs in the nucleus. *Curr Opin Cell Biol* **17**, 302–8 (2005).
21. **Heale BS, Keegan LP, McGurk L, et al.** Editing independent effects of ADARs on the miRNA/siRNA pathways. *EMBO J* **28**, 3145–56 (2009).
22. **Heale BS, Keegan LP, O'Connell MA.** ADARs have effects beyond RNA editing. *Cell Cycle* **8**, 4011–12 (2009).
23. **Tonkin LA, Bass BL.** Mutations in RNAi rescue aberrant chemotaxis of ADAR mutants. *Science* **302**, 1725 (2003).
24. **Gerber AP, Keller W.** An adenosine deaminase that generates inosine at the wobble position of tRNAs. *Science* **286**, 1146–9 (1999).
25. **Keegan LP, Gallo A, O'Connell MA.** Development. Survival is impossible without an editor. *Science* **290**, 1707–9 (2000).
26. **Keller W, Wolf J, Gerber A.** Editing of messenger RNA precursors and of tRNAs by adenosine to inosine conversion. *FEBS Lett* **452**, 71–6 (1999).
27. **Maas S, Gerber AP, Rich A.** Identification and characterization of a human tRNA-specific adenosine deaminase related to the ADAR family of pre-mRNA editing enzymes. *Proc Natl Acad Sci USA* **96**, 8895–900 (1999).
28. **Wolf J, Gerber AP, Keller W.** tadA, an essential tRNA-specific adenosine deaminase from *Escherichia coli*. *EMBO J* **21**, 3841–51 (2002).
29. **Chester A, Scott J, Anant S, Navaratnam N.** RNA editing: cytidine to uridine conversion in apolipoprotein B mRNA. *Biochim Biophys Acta* **1494**, 1–13 (2000).
30. **Hersberger M, Patarroyo-White S, Arnold KS, Innerarity TL.** Phylogenetic analysis of the apolipoprotein B mRNA-editing region. Evidence for a secondary structure between the mooring sequence and the 3′ efficiency element. *J Biol Chem* **274**, 34 590–7 (1999).

31. **Lellek H, Kirsten R, Diehl I, et al.** Purification and molecular cloning of a novel essential component of the apolipoprotein B mRNA editing enzyme-complex. *J Biol Chem* **275**, 19 848–56 (2000).
32. **Mehta A, Kinter MT, Sherman NE, Driscoll DM.** Molecular cloning of apobec-1 complementation factor, a novel RNA-binding protein involved in the editing of apolipoprotein B mRNA. *Mol Cell Biol* **20**, 1846–54 (2000).
33. **Navaratnam N, Bhattacharya S, Fujino T, et al.** Evolutionary origins of apoB mRNA editing: catalysis by a cytidine deaminase that has acquired a novel RNA-binding motif at its active site. *Cell* **81**, 187–95 (1995).
34. **Gallois-Montbrun, S Kramer B, Swanson CM, et al.** AT Antiviral protein APOBEC3G localizes to ribonucleoprotein complexes found in P bodies and stress granules. *J Virol* **81**, 2165-78 (2007).
35. **Harris RS, Liddament MT.** Retroviral restriction by APOBEC proteins. *Nat Rev Immunol* **4**, 868–77 (2004).
36. **Estevez AM, Simpson L.** Uridine insertion/deletion RNA editing in trypanosome mitochondria—a review. *Gene* **240**, 247–60 (1999).
37. **Stuart KD, Schnaufer A, Ernst NL, Panigrahi AK.** Complex management: RNA editing in trypanosomes. *Trends Biochem Sci* **30**, 97–105 (2005).
38. **Feagin JE, Abraham JM, Stuart K.** Extensive editing of the cytochrome *c* oxidase III transcript in *Trypanosoma brucei*. *Cell* **53**, 413–22 (1988).
39. **Blum B, Bakalara N, Simpson L.** A model for RNA editing in kinetoplastid mitochondria: 'guide' RNA molecules transcribed from maxicircle DNA provide the edited information. *Cell* **60**, 189–98 (1990).
40. **Blum B, Simpson L.** Guide RNAs in kinetoplastid mitochondria have a nonencoded 3′ oligo(U) tail involved in recognition of the preedited region. *Cell* **62**, 391–7 (1990).
41. **Madison-Antenucci S, Grams J, Hajduk SL.** Editing machines: the complexities of trypanosome RNA editing. *Cell* **108**, 435–8 (2002).
42. **Simpson L, Sbicego S, Aphasizhev R.** Uridine insertion/deletion RNA editing in trypanosome mitochondria: a complex business. *RNA* **9**, 265–76 (2003).
43. **Gott JM.** Two distinct roles for terminal uridylyl transferases in RNA editing. *Proc Natl Acad Sci USA* **100**, 10 583–4 (2003).
44. **Simpson L.** Uridine insertion/deletion RNA editing as a paradigm for site-specific modifications of RNA molecules. In: Gesteland R, Cech T, Atkins JF (eds), *The RNA World* (3rd edn), pp. 401–18. Cold Spring Harbor, NY: Cold Spring Harbor Laboratory Press (2006).

14 The biogenesis and nucleocytoplasmic traffic of non-coding RNAs

Introduction

Up to now, we have focused primarily on the processing of pre-mRNA precursors. However, non-coding RNAs (ncRNAs) are also processed during their biogenesis (formation). These include the processing and maturation of ribosomal RNA (rRNA), small nuclear RNA (snRNA), transfer RNA (tRNA), mitochondrial transcripts, and even telomerase. In order to thrive and multiply, cells must be able to generate large numbers of all of these molecules in the nucleus, process them, if necessary export them to the cytoplasm, and if required re-import them back into the nucleus. Elaborate machineries have evolved to achieve all of this processing. The purpose of this chapter is to give an overview of these very important aspects of RNA processing. More often than not, the RNA processing machinery involves the action of other RNA molecules. Examples are shown in Table 14.1.

In previous chapters we have described the structure and function of ribosomes. In this chapter we review the main features of the biogenesis of ribosomal RNA—the generation of those RNAs that form part of the structure of the ribosome—a process that is facilitated by a family of small RNA molecules known as the **snoRNAs** (small nucleolar RNAs). We then describe the organization of rRNA processing in an important multifunctional nuclear organelle, the **nucleolus**. We overview the processing of snRNAs (the small non-coding RNAs that are part of the pre-mRNA splicing machinery) by **scaRNAs**, another class of small non-coding RNAs that are localized in **Cajal bodies** (named after the Spanish cell biologist Santiago Ramón y Cajal who first described them in 1903). We discuss the involvement of non-coding RNAs in the biogenesis and function of **telomerase** (required for the formation of **telomeres**). We describe the processing of **transfer RNA (tRNA)**, another important class of non-coding RNA, and we also briefly discuss the processing of mitochondrial transcripts. Note that microRNA biogenesis will be discussed in Chapter 16. An integral part of the biogenesis of non-coding RNAs is their transport in and out of the nucleus. In this chapter we also consider in some detail the complex machinery that has evolved to facilitate the process called *nucleocytoplasmic traffic*. We end the chapter by describing how retroviruses cleverly hijack the machinery of RNA export.

Table 14.1 Examples of macromolecules that contain RNA, with the corresponding non-coding RNAs involved in their biogenesis

Macromolecule/complex involved in processing	Cellular function	Small non-coding RNA involved in its biogenesis
Ribosomal RNA (rRNA)	mRNA translation	Small nucleolar RNA (snoRNA)
Small nuclear RNA (snRNA)	Pre-mRNA splicing	Small Cajal body-associated RNA (scaRNA)
RNA component of telomerase (hTR)	Formation of telomeres	scaRNA-like component of hTR

14.1 The snoRNAs and scaRNAs: multiple roles in RNA biogenesis

As we saw in Chapter 11, ribosomes are exceedingly complex macromolecules, made of both RNA and proteins. There are approximately 70 known eukaryotic ribosomal proteins, encoded by mRNAs transcribed by RNA polymerase II. Ribosomal RNA makes up about 65% of the mass of ribosomes and is central to their structure and function. It provides, to a large extent, the catalytic functions required for mRNA translation. The fact that the essential machinery of protein synthesis contains catalytic RNA provides further evidence in favour of the RNA world hypothesis (see Chapter 1). The biogenesis of ribosomes is a highly intricate process. This topic is important because the production of ribosomes is a requirement for cell proliferation. In this section we shall focus on the biogenesis of eukaryotic rRNA.

14.1.1 Pre-rRNA cleavage and the U3 snoRNA

In Chapter 8 we saw that eukaryotic rRNA is first transcribed by RNA polymerase I as a ~7000 nucleotide 47S precursor (pre-rRNA), which gives rise to the 28S, 18S, and 5.8S rRNAs. These rRNAs are extensively processed post-transcriptionally.[1,2] A fourth rRNA, the 5S rRNA, is transcribed from a separate gene by RNA polymerase III. The 18S rRNA ends up in the small ribosomal subunit (40S in eukaryotes), whereas the other three rRNAs are incorporated into the large subunit (60S in eukaryotes).

There are three stages in rRNA synthesis: (1) the transcription of pre-rRNA; (2) its processing (including cleavage and removal of spacer sequences within the pre-rRNA); and (3) the modification of rRNA (up to 100 sites are modified in eukaryotic rRNA). The modification of rRNA is required for its proper folding and structural and catalytic function in mature ribosomes.

The transcription and processing of rRNA occurs in the nucleolus.[3] First, a large complex known as the 90S pre-ribosome is formed. This consists of the 47S precursor rRNA (transcribed by RNA polymerase III), the 5S rRNA (transcribed by RNA polymerase I, but not in the nucleolus), ribosomal proteins, additional non-ribosomal proteins, and a class of non-ribosomal RNAs known as the snoRNAs.[4,5] A critical initial step is the splitting of the 90S complex into two smaller particles: the 66S precursor, which is eventually processed into the 60S large ribosomal subunit; and the 43S precursor, which is processed into the 40S small ribosomal subunit. These essential steps in ribosome formation are summarized in Fig. 14.1.

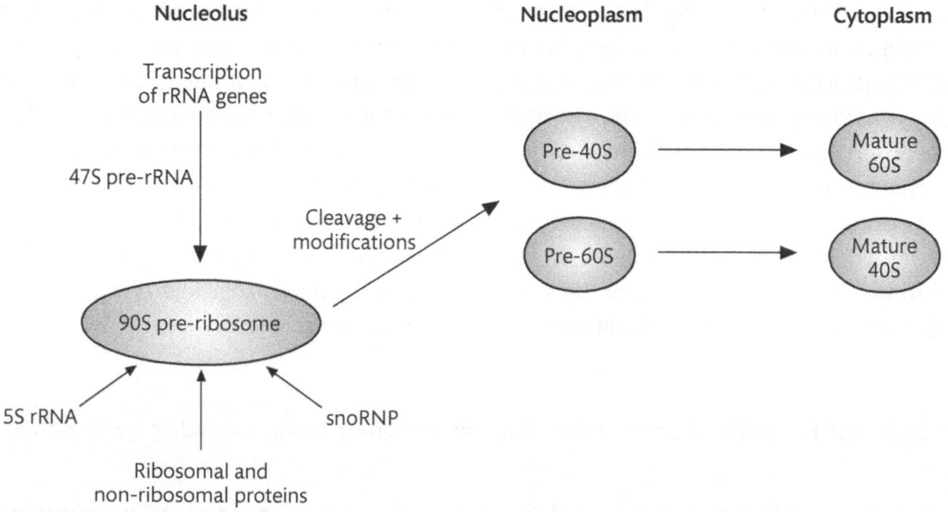

Figure 14.1 Essential steps in eukaryotic ribosome formation. Transcription of the 45S rRNA gene occurs in the nucleolus, producing a 47S pre-rRNA transcript. This combines with the 5S rRNA expressed by a separate gene. Ribosomal and non-ribosomal proteins contribute to the formation of the 90S pre-ribosome. Modifications of rRNA are facilitated by the snoRNPs in the nucleolus. Cleavage of pre-rRNA eventually gives rise to the pre-40S and pre-60S subunits which are released into the nucleoplasm. After further processing steps the mature small and large ribosomal subunits are released into the cytoplasm.

In the late 1960s, an electron microscopy technique known as the **Miller spread** (named after the scientist who invented it) allowed the direct visualization of chromatin and nascent transcripts. Miller found that electron-dense terminal 'blobs' were apparent at the 5′ end of all nascent pre-rRNAs. The composition of these blobs remained a mystery for several years. We now know that the main component of these blobs is the U3 snoRNP (U3 snoRNA, i.e. the U3 small nucleolar RNA plus associated proteins). Studies have shown that U3 snoRNA is required for early ribosomal RNA processing.[6] Specifically, it helps to cleave and remove the spacer regions in the primary transcript. Depletion of U3 snoRNA results in faulty pre-rRNA cleavage and the loss of the blobs present at the 5′ end of the pre-rRNA, as shown in the Miller spreads in Fig. 14.2.

The U3 snoRNP particle, which is also known as the **SSU processome**, is required for the formation of the 18S rRNA in the small subunit of the ribosome. It contains several U3-specific proteins in addition to the U3 snoRNA, forming a large complex (>2Mda). In fact, the U3 snoRNP is surprisingly large, sedimenting at 80S—the same sedimentation rate as a fully formed ribosome! The SSU processome is assembled co-transcriptionally onto the nascent pre-rRNA and is required for the early cleavages of the pre-rRNA. The overall structure and processing of a pre-rRNA transcript are shown in Fig. 14.3.

There are up to 28 proteins in the U3 snoRNP, including proteins involved in nuclear **RNA surveillance**. One of these is the 5′ exonuclease XRN2; its role is to help degrade faulty rRNA transcripts.[7] Surprisingly, five U3 snoRNP-associated proteins are actually small ribosomal subunit proteins. Why should these be present? A probable explanation is that they are part of the U3 snoRNP because they bind to the pre-rRNA at this early stage of processing and facilitate the folding of the pre-rRNA into a conformation that aids its correct cleavage. Interestingly, defects in this biogenetic process are linked to a human congenital syndrome called Diamond–Blackfan anaemia, as described in Box 14.1. One of the lesions associated with this type of anaemia is in fact a mutation that affects the structure of a small ribosomal subunit protein.[8]

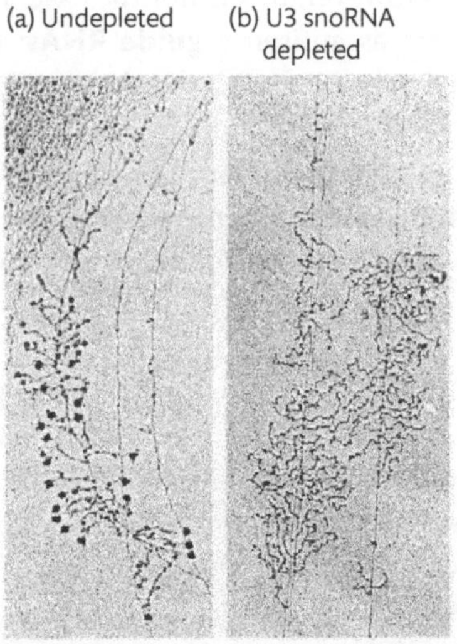

Figure 14.2 Presence of U3 snoRNA on nascent pre-rRNA. Nascent ribosomal RNA transcripts are visualized via a Miller spread. (a) Electron-dense blobs can be seen at the 5′ end of the nascent transcripts. (b) The spread is derived from a yeast strain in which the U3 snoRNA has been depleted, resulting in the loss of the terminal blobs. (Reprinted from Dragon et al.[6] with permission from Macmillan Publishers Ltd.)

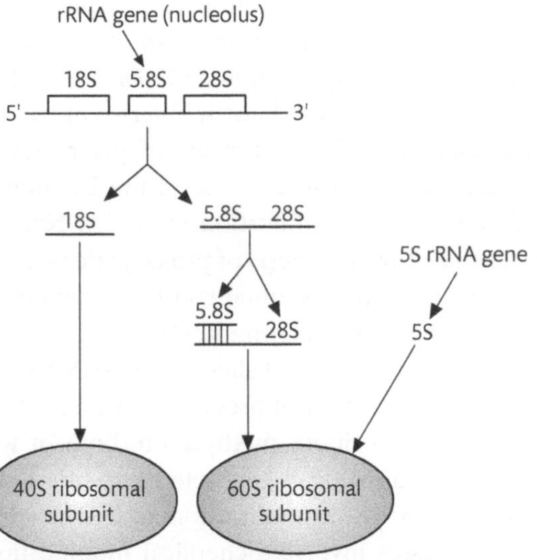

Figure 14.3 Pre-rRNA transcript structure and cleavage steps. The primary transcript (pre-rRNA) contains, starting at the 5′ end, the 18S precursor followed by the 5.8S and 28S precursors separated by spacer sequences. The first step is the cleavage of the precursor generating the 18S rRNA. The rest of the transcript is then cleaved, producing the 5.8S and 28S rRNA. The 5S rRNA is expressed from a different transcription unit. The 18S rRNA ends up in the small ribosomal subunit and the 28S and 5.8S rRNAs end up in the large ribosomal subunit.

Box 14.1 Abnormal ribosomal biogenesis in Diamond–Blackfan anaemia

Defects in ribosomal biogenesis have been linked to a human congenital syndrome called Diamond–Blackfan anaemia (DBA). DBA is characterized by underproduction of red blood cells (resulting in anaemia). It is a very rare condition, arising in just 1 in 200 000 live births. Twenty-five per cent of cases are associated with a mutation in the *RPS19* gene that encodes the small ribosomal subunit protein S19. A total lack of a given ribosomal protein is lethal—not surprisingly, the mouse knockout of *RPS19* is not viable. DBA is a result of haploinsufficiency, in which having only one wild-type copy of a gene is not lethal (but at the same time not sufficient to maintain a normal phenotype).

Knockdown of *RPS19* reduces the synthesis of 18S rRNA, with a resulting accumulation of rRNA precursors in the nucleus. These immature precursors can be observed as perinuclear dots (visible around the nucleus).

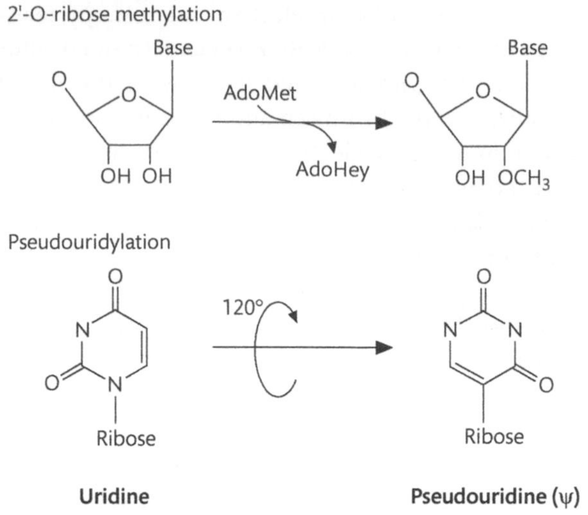

Figure 14.4 The two principal modifications of rRNA. Pseudouridine is an isomer of the nucleotide uridine, and is formed in a reaction called pseudouridylation by the enzyme ψ synthase. A second type of common modification is the addition of a methyl group to the 2′ hydroxyl group in the ribose sugar by the enzyme 2′-*O*-methylase.

14.1.2 snoRNAs and the modification of rRNA

Small nucleolar RNAs are abundant non-coding RNAs. Between 100 and 200 snoRNAs are known, and the number varies from species to species.[9] They are bound by proteins to form snoRNPs, referred to as 'snorps' (as opposed to snRNPs, referred to as 'snurps'). They are best known in terms of two complementary functions: cleavage of pre-rRNA precursors, as we saw in the case of the U3 snoRNA, and site-specific modification of rRNA. Note that, in contrast, the biogenesis of prokaryotic ribosomes is thought to occur spontaneously, i.e. without the contribution of non-coding RNAs.

There are two types of site-specific rRNA modification: the formation of pseudouridine and the formation of 2′-*O*-ribose methylation, both of which are illustrated in Fig. 14.4. Note the subtle distinction between RNA modification and RNA editing. Both processes involve a chemical modification of RNA. Editing generally results in changes in the coding potential, i.e. a change in the biological information encoded by the RNA (in the case of mRNA, for example). However, ribosomes are non-coding and therefore their RNA is modified but not edited, because changes in the primary structure of an rRNA do not result in the downstream modification of a polypeptide.

It is essential that the correct RNA nucleotides are modified. But how can this be achieved? One possibility is that the modification enzymes contain RNA-binding domains that recognize particular RNA sequences. However, this is not the case. In fact, the snoRNAs act as antisense **guide RNAs**; they align through hybridization with the sites in rRNA where the modifications are required.

There are two main families of snoRNAs, corresponding to the two main types of modification. The methylation guide snoRNAs are known as the **Box C/D snoRNA** family (SNORDs) and all contain two short sequence motifs, known as Box C (5′-PuUGAUGA-3′, where Pu stands for purine, an A or a G) and Box D (5′-CUGA-3′). These two motifs are brought together in close proximity through a single stemloop structure. Two similar but less well-conserved motifs, labelled C′ and D′, lie between the C and D motifs. Upstream of both Box D and Box D′ there are antisense elements of 10–21 nucleotides. These antisense elements hybridize to the intended target in the rRNA precursors, defining the site of methylation.

Like all RNAs, Box C/D snoRNAs are associated with proteins. The most important of these is fibrillarin, a nucleolar protein. Fibrillarin is the methylase responsible for the methylation of the 2′ hydroxyl group in the ribose sugar. The methyl group is provided by *S*-adenosyl methionine (AdoMet), a molecule derived from adenosine triphosphate (ATP) and the amino acid methionine. The methyl group attached to the sulphur atom is chemically reactive.

The second family of snoRNAs, the pseudouridylation guide snoRNAs, also possess a shared structure of two hairpins linked by a hinge. The hinge region contains the H box (ANANNA, where N represents any nucleotide), whereas the short tail contains the trinucleotide ACA, which always occurs three nucleotides from the very 3′ end. These snoRNAs are known as the **Box H/ACA snoRNAs** (SNORAs). Guide sequences that hybridize around the intended modification target are present in the internal loops of the two hairpins. Proteins that associate with the Box H/ACA snoRNAs include dyskerin, which is essential for the pseudouridylation reaction.

The basic structure of both classes of snoRNA is shown in Fig. 14.5. Like the snRNAs, the snoRNAs associate with a set of core snoRNP proteins, but more transiently with other proteins. Both fibrillarin and dyskerin are core snoRNP proteins. They can provide the catalytic functions of the snoRNPs again and again, facilitating the fast and efficient production of mature ribosomal RNA.

Interestingly, in vertebrates, snoRNAs are encoded within introns. Thus, they are not independently transcribed but are themselves processed out of intron lariats.[10] Because introns lariats are recycled, it is important to process the snoRNAs from them swiftly, otherwise they get degraded. By contrast, most snoRNAs in yeast are encoded by independent monocistronic or polycistronic mRNAs and are processed by endonucleases and exonucleases.

Note that snoRNAs are often encoded within introns of genes associated with ribosome biogenesis.[11] This makes sense because it means that snoRNA expression is coordinated with the expression of the components of ribosomes. There are also several snoRNAs (defined in terms of the presence of the conserved sequence motifs that characterize them) which do not possess the antisense sequences that would target them to rRNA. These are known as **orphan snoRNAs**, in the sense that their physiological targets are not yet known. But this does not mean that they have no function. These orphan snoRNAs may be involved in processes other than RNA modification. That this should be the case is illustrated by the SNORD115 family of snoRNAs.[4] They are expressed in the mammalian brain and direct the modification of a serotonin receptor subtype 2c pre-mRNA, influencing its alternative splicing through two possible mechanisms. This can be directly by

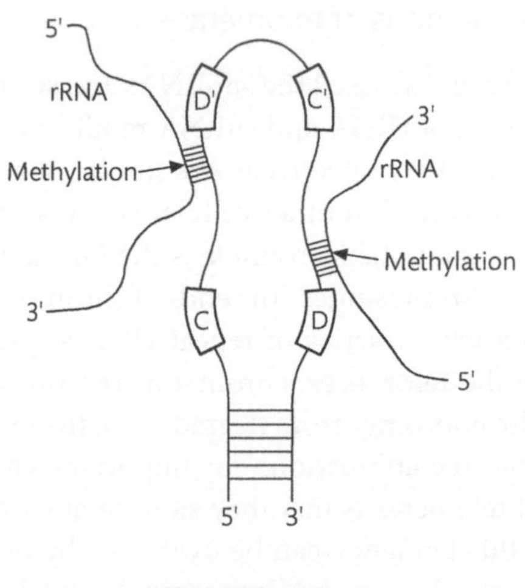

Box C/D snoRNA

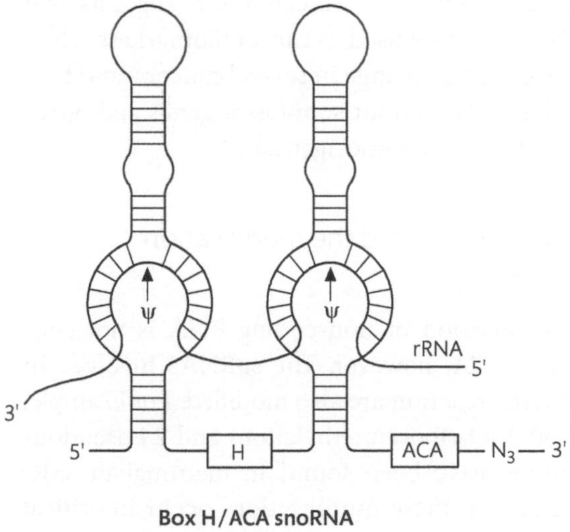

Box H/ACA snoRNA

Figure 14.5 Structures of typical Box C/D and Box H/ACA snoRNAs complexed to rRNA. Conserved sequences (Boxes C, C′, D, and D′) define the Box C/D family of snoRNAs. In the Box C/D snoRNA, the guide sequences that direct methylation are situated upstream of Box D and Box D′. Box H/ACA snoRNAs are also defined by conserved sequences and form two hairpin structures. The guide sequences that direct the site of pseudouridylation (′) are provided by the proximal loops at the beginning of the two hairpins.

Box 14.2 snoRNA-like molecules in the Archaea

The Archaea are a phylogenetic group of prokaryotes that are genetically distinct from bacteria and eukaryotes (one of the three domains of life). Prominent examples include the halophiles and the sulphur-dependent thermophilic acidophiles such as *Sulfobulus solfataricus*. They are thought to bridge bacteria and eukaryotes in the sense that they share features of both. Thus, their basic metabolism, genome organization, and cellular structure are reminiscent of bacteria. However, features of their transcription and translation appear closer to the equivalent processes in eukaryotes.

For example, the extent of 2′-*O*-ribose methylation of their ribosomal RNA is similar to that found in eukaryotes. Several distinct Box C/D-like RNAs have also been identified, as well as what looks like an archaeal version of fibrillarin, the methylase associated with Box C/D snoRNAs. Thus, this machinery evolved a long time ago, before the divergence of eukaryotes from prokaryotes.

physically binding to and masking a splicing silencer element, or indirectly by interfering with an A→I editing event mediated by the RNA editing enzyme ADAR2 in a critical regulatory sequence in an alternatively spliced exon.[12,13]

From an evolutionary perspective, snoRNAs would appear to pre-date eukaryotes, as similar molecules are present in the Archaea.[14] This is illustrated in Box 14.2. Research also suggests that snoRNAs could be used as cancer biomarkers. Their expression levels change in several cancers, and they can behave like tumour suppressor genes and oncogenes, influencing tumorigenesis.[15]

14.1.3 scaRNAs and the modification of snRNA

The modification of non-coding RNA is not confined to rRNA, however. The snRNAs involved in the splicing reaction are also modified. For example, up to 30 2′-*O*-ribose methylations and 24 pseudouridylations have been found in mammalian snRNAs. Many of these modifications occur in critical sequences in the snRNAs that are involved in the assembly and function of the spliceosome; therefore these modifications are very important. For example, the modification of the 5′ end of U2 snRNA is essential for the splicing reaction. Thus, a very precise machinery must exist to modify snRNAs. Might non-coding RNAs also be involved in the processing of snRNAs? The answer is yes: there is a special class of non-coding RNAs that concentrate in Cajal bodies and their function is to modify snRNA precursors.

Cajal bodies (CBs) are multifunctional structures in the nucleus, best known for their ability to store pre-mRNA processing factors. However, they have additional functions, including the processing of snRNAs and of snoRNAs themselves. SnRNAs are transcribed in CBs and must transit through CBs before being finally matured. A small non-coding RNA, U85, is required for both 2′-*O*-ribose methylation and pseudouridylation of U5 snRNA.[16] The fact that U85 can direct both modifications is explained by the fact that it contains both Box C/D and Box H/ACA sequences. In situ hybridization studies show that U85 co-localizes with the U5 snRNA in CBs. U85 and other small RNAs that localize to the CB are known as **scaRNAs** (small CB-associated RNAs).

How does U85 localize to CBs? Mutational analysis of U85 (i.e. a systematic introduction of mutations followed by in situ hybridization to determine its ability to localize to CBs) defined a CB-specific localization signal, the CAB box (sequence UGAG). These signals are located as two copies in the Box H/ACA domains of scaRNAs.[17]

14.1.4 Biogenesis of telomerase

So far we have discussed the snoRNAs and scaRNAs in the context of rRNA and snRNA modification in nucleoli and CBs, respectively. Are similar non-coding RNAs involved in other cellular processes? The answer is yes: a notable example is the formation of the ends of chromosomes. The ends of chromosomes are capped with a series of repeat DNA sequences that form the **telomeres**, chromatin structures that protect chromosomes from degradation, fusion, and undesirable recombination. An important characteristic of telomeres is that they shorten at each cell division; this tendency can be overcome by the enzyme complex known as **telomerase**. High telomerase activity is a characteristic of immortalized cells, which are able to renew their telomeres.

Telomerase is a ribonucleoprotein consisting of two main components: an RNA template embedded within the hTR RNA, and a reverse transcriptase (TERT) which produces the telomeric DNA sequences from the hTR template.[18] The hTR RNA, encoded by the *TERC* gene, consists of two main structural features: a pseudoknot at the 5′ end and a snoRNA H/ACA-like sequence at the 3′ end. The structure of the hTR RNA is illustrated in Fig. 14.6.

The hTR RNA has a snoRNA-like sequence and is complexed by proteins that are normally associated with Box H/ACA snoRNAs. The precise function of the snoRNA-like sequence in hTR is not yet fully understood. However, rather than being required for pseudouridylation (as is the case for typical Box H/ACA snoRNAs) this part of the hTR RNA is thought to be essential for the biogenesis of telomerase. To illustrate its importance, mutations of the H/ACA domain of hTR are associated with dyskeratosis congenita (DC). DC is an inherited disorder distinguished by abnormal nails and pigmentation, white patches in the mouth, bone marrow failure, premature ageing, and cancer. The mutations in the H/ACA domain cause defects in the assembly of the pre-RNP (i.e. the hTR precursor).[19]

There is a sequence in the 3′ terminal stemloop in hTR (known as the CR7 sequence) that is essential for the accumulation of hTR RNA. Thus hTR RNA promotes its own accumulation. The hTR snoRNA-like sequence also contains a CB localization motif (the CAB box described earlier, with the consensus sequence UGAG in the CR7 sequence).[20] Consistent with this finding, hTR RNA is detected in CBs. Accumulation of hTR in CBs is more pronounced during the S phase (DNA synthesis phase) of the cell cycle. (The S phase is the stage in the cell cycle when telomere synthesis also takes place.) Therefore the snoRNA-like sequence in hTR is required for the assembly of telomerase in the CBs of dividing cells.

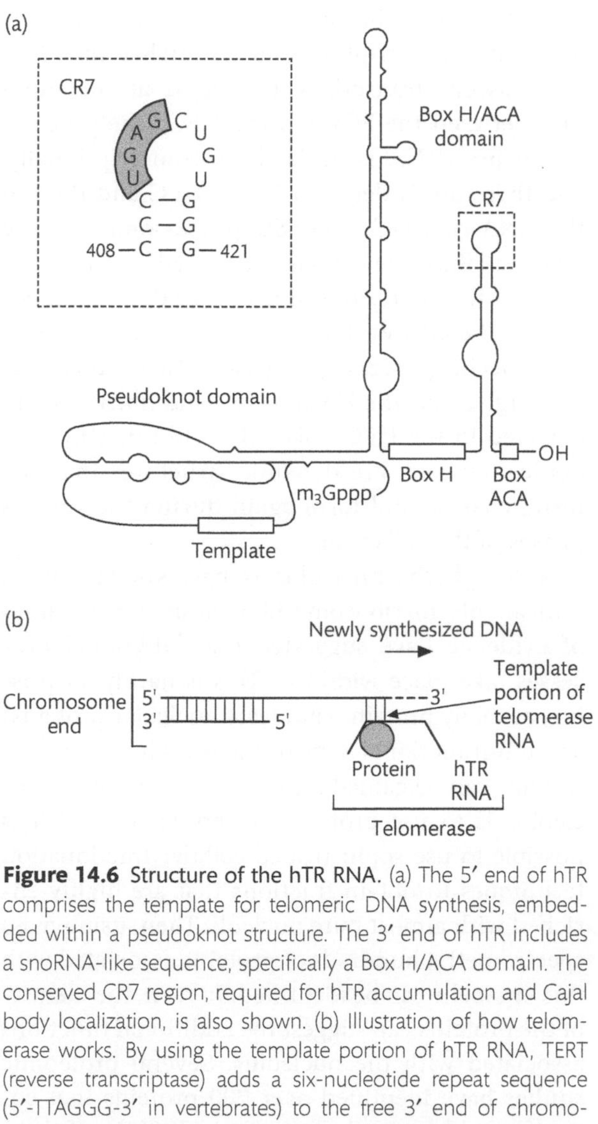

Figure 14.6 Structure of the hTR RNA. (a) The 5′ end of hTR comprises the template for telomeric DNA synthesis, embedded within a pseudoknot structure. The 3′ end of hTR includes a snoRNA-like sequence, specifically a Box H/ACA domain. The conserved CR7 region, required for hTR accumulation and Cajal body localization, is also shown. (b) Illustration of how telomerase works. By using the template portion of hTR RNA, TERT (reverse transcriptase) adds a six-nucleotide repeat sequence (5′-TTAGGG-3′ in vertebrates) to the free 3′ end of chromosomes. The repeat sequences are complexed by proteins, forming the telomere. (Reprinted from Jády et al.[20] with permission from the Rockefeller University Press.)

Key points

Ribosome biogenesis is a complex process that involves the transcription, cleavage, and modification of the main rRNA precursor. Pre-rRNA cleavage requires the U3 snoRNA and takes place in nucleoli. Ribosomal subunit assembly commences in the nucleoli.

Small nucleolar RNAs (snoRNAs) guide the modification of rRNA bases: 2′-O-ribose methylation is facilitated by the Box C/D snoRNAs, and pseudouridylation is facilitated by the Box H/ACA snoRNAs.

Small nuclear RNAs (snRNAs) are modified in Cajal bodies (CBs) by the scaRNAs. ScaRNAs are related to the snoRNAs and contain both Box C/D and Box H/ACA sequences.

Telomerase is a ribonucleoprotein required for the formation of telomeres. Its RNA component, hTR, contains a snoRNA-like sequence. hTR also localizes to CBs and its snoRNA-like sequence is required for the biogenesis of telomerase.

14.2 Structure and function of the nucleolus

When observing cells by phase contrast microscopy, nucleoli (literally 'little nuclei') are the most visible structures in the nucleus. Nucleoli were observed as far back as the nineteenth century, but their function did not become known until the twentieth century. The scientist Barbara McClintock proposed that nucleoli form through the activity of a **nucleolar organizing region** (NOR). In the 1950s, RNA was detected in nucleoli, and, in the 1960s, the NOR was shown to contain DNA for ribosomal genes. In fact, NORs are specific chromosomal sites which we now know correspond to clusters of ribosomal RNA genes. In humans, each NOR contains 30–50 rRNA gene copies in tandem, separated by non-transcribed spacer DNA. These observations led to the realization that one of the main functions of nucleoli is the synthesis of ribosomal RNA (and also ribosomes).[21–25]

It is clear that nucleoli have evolved, as organelles, to make ribosome synthesis efficient. In fact, it has been estimated that in a yeast cell growing at its optimal rate, in which rRNA synthesis is proceeding efficiently, up to 40 ribosomes can leave the nucleolus *every second*! However, as we shall see, nucleoli have functions other than ribosome biogenesis.[26]

The number of nucleoli varies both from cell to cell and according to the stage of the cell cycle that the cell is in. In humans, most cells will contain one or only a few nucleoli. In contrast, higher numbers are detectable in rapidly proliferating cells, including tumour cells. This makes sense; proliferating cells need to grow and this requires high rates of protein synthesis, hence ribosomes. Nucleoli disassemble and then reassemble during the cell cycle. The shape and diameter of nucleoli also varies, ranging from 0.5μm in differentiated lymphocytes to 9μm in proliferating cells—presumably a reflection of more intense activity in the latter. The positioning of nucleoli is also organized within the nucleus; they are often associated with the nuclear envelope, most probably to facilitate the export of maturing ribosomes and other RNP complexes into the cytoplasm.

Historically, nucleoli have been analysed in great detail using electron microscopy (EM), in which electron-dense areas appear darker. EM has been useful in defining the three main compartments within the nucleolus. These are the **fibrillar centres** (FCs) and the **dense fibrillar component** (DFC), both of which are embedded within the **granular component** (GC). The GC consists of several granules with a diameter of 15–20nm. How, then, is ribosomal RNA synthesis organized within these compartments?

The compartments of the nucleolus and the processes in ribosome biogenesis that occur within them are shown in Fig. 14.7. The FCs are associated with transcription and are richer in DNA, and nascent transcripts are found at the junction between the FCs and the DFC. The processing of pre-rRNA is initiated co-transcriptionally, and therefore is detected in the DFC and then in the RNA-rich GC. The SSU processome and the 90S pre-ribosome are also localized in the DFC, whereas the later processing of the 60S subunit occurs in the GC. Proteins and snoRNAs associated with specific processing steps are also localized accordingly; thus the U3 snoRNA and fibrillarin are detected in the DFC. Like CBs and other nuclear bodies, nucleoli are dynamic structures that can form, change, and form again during the various phases of the cell cycle.

Although the nucleolus is best known for its critical role in ribosome biogenesis, several lines of evidence have suggested that additional processes take place within it. This is hardly surprising, as many proteins (not least nucleolar proteins) are multifunctional. One of the ways in which new insights can be gained into the functions of the nucleolus is to use proteomics approaches.[27–30] It is possible to use sophisticated cellular fractionation techniques to obtain fractions that are highly enriched with almost pure nucleoli. Then, using mass spectrometry, hundreds of proteins associated with this organelle can be identified. In turn, the identity of the proteins can suggest the additional functions associated with the nucleolus. Several proteomic studies have identified over 700 proteins that copurify, stably, with human nucleoli; 90% of these can also be found in yeast nucleoli, indicating that

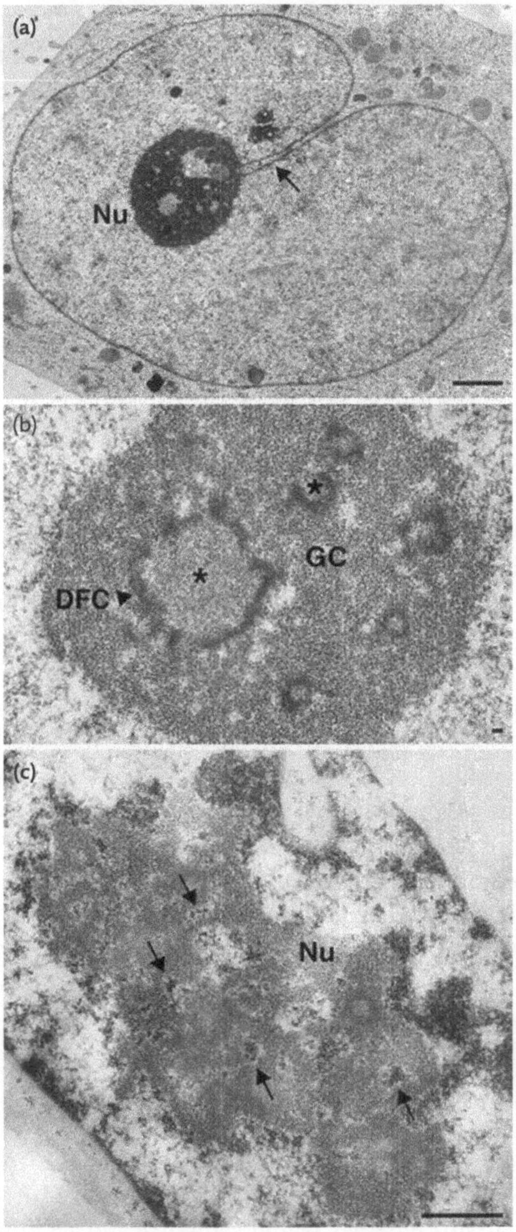

Figure 14.7 Compartmentalization of the nucleolus. Nucleoli are the result of the activation of ribosomal DNA genes, concentrated in the fibrillar centres (FC). The dense fibrillar centres (DFCs) surround the fibrillar centres and are the site of pre-rRNA synthesis and early processing events. The RNA-rich granular component (GC), generally the largest of the nucleolar compartments, is associated with late RNA processing and ribosome formation. (a) Electron micrograph demonstrating the invagination of the nuclear envelope towards a nucleolus (Nu) in a HeLa tumour cell line. (b) The three main compartments: asterisks denote the variable-size fibrillar centres, which are surrounded by DFCs and then the GC. (c) The highly contrasted chromatin around and inside the nucleolus (arrows). (Reprinted from Henandez-Verdun[21] with permission from Springer Science and Business Media.)

the various nucleolar functions occur throughout eukaryotes.

Table 14.2 lists the results of a typical proteomic analysis of a mammalian nucleolus. Perhaps surprisingly, only 25–30% of these proteins are known to be associated with the biogenesis of ribosomes. The reason for the presence of proteins that are not involved in ribosome biogenesis is because the nucleolus is involved in other processes. These include the assembly and maturation of non-ribosomal RNP, the cell cycle, tRNA maturation, mRNA transport, RNA interference, and the sensing of cellular stress.

To illustrate this point, the presence of the glycolytic enzyme GAPDH in the nucleolus of both plants and mammals is perhaps a surprising find. However, GAPDH is multifunctional; in addition to its role in glycolysis, it is involved in transcriptional regulation and in the regulation of mRNA splicing and stability. Thus GAPDH might be linking the metabolic status of a cell to the regulation of gene expression. However, why should GAPDH (or other unexpected proteins) be present in the nucleolus? The nucleolus may be a storage site for some of these proteins, and it might play an active part in several aspects of non-rRNA gene expression processes. For example, small nuclear RNAs also transit

Table 14.2 Proteomic analysis of mammalian nucleoli*

Functional group of proteins	Number of proteins	Overall percentage of nucleolar proteins
Ribosome biogenesis	168	25.2
Ribosomal proteins	71	10.7
mRNA metabolism	128	19.2
Translation	30	4.5
Chromatin structure	35	5.2
Fibrous proteins	8	1.2
Chaperones	13	1.9
Other	214	31.8

*Only about a third of proteins are clearly involved in ribosome biogenesis (including ribosomal proteins). The presence of proteins involved in a wide range of nuclear processes emphasizes the multifunctional nature of the nucleolus.[28]

Box 14.3 Targeting RNA polymerase I in cancer

Researchers in Australia have explored the potential of targeting RNA polymerase I in cancer. As a basis for their work, they used a mouse model of spontaneous B-cell lymphoma driven by the potent oncogene *c-myc*. Among its several targets, the Myc transcription factor upregulates the transcription of 47S rRNA and genes involved in rRNA processing and ribosome assembly, helping proliferating cells to increase protein synthesis. The small molecule CX-5461 is a highly selective inhibitor of RNA polymerase I. It can induce B-lymphoma cell death through the activation of the *nucleolar surveillance pathway*. This pathway is activated in conditions of severe perturbation of ribosome biogenesis.[33] It involves the protein p53, known as the 'guardian of the genome'—p53 has the ability to induce apoptosis (programmed cell death). Crucially, normal B-cells were spared by the treatment with CX-5461.

through the nucleolus. Thus the U6 snRNA, transcribed by RNA polymerase III, transits through the nucleolus wherein it acquires modifications. (Therefore snRNAs are processed in both CBs and in nucleoli.)

Nucleoli are particularly prominent and numerous in cancer cells, in which they are often seen to be abnormal in shape and structure. Abnormal numbers and shapes of nucleoli can even inform diagnosis, prognosis, and response to chemotherapeutic drugs. Since an increased rate of rRNA production is required by cancer cells, it has been suggested that it may be possible to target nucleoli in cancer therapy.[31] Box 14.3 describes one such approach, in which RNA polymerase I (active in nucleoli) is targeted with a small molecule, selectively killing B-lymphoma cells in vivo.[32]

Key points

Ribosome biogenesis is initiated and takes place mostly in the nucleolus. The nucleolus is divided into three compartments: the DNA-rich fibrillar centre (FC) is associated with rRNA genes; the dense fibrillar centre (DFC) surrounds the FC and is associated with rRNA gene transcription and early pre-rRNA processing steps; and the RNA-rich granular component (GC) is associated with the later steps of rRNA biogenesis.

Proteomic analysis of the nucleolus has shown that there are several hundred proteins involved in functions other than ribosome biogenesis. This is because the nucleolus has additional functions associated with the processing and regulation of non-ribosomal RNAs. Abnormalities of nucleolar numbers and structure are associated with cancer.

14.3 Processing of tRNA and of mitochondrial transcripts

We explored the process of mRNA translation in Chapter 11 and saw that tRNA is the critical adaptor molecule that reads the genetic code by bringing a particular amino acid to an mRNA codon. In tRNA, the anticodon loop hybridizes to a corresponding codon in the mRNA. If you recall, the 5′ and 3′ ends of a tRNA molecule base pair together to form the acceptor stem, and the 3′ end terminates in the residues CCA. It is the terminal A of the CCA that is 'charged' with the relevant amino acid. The three-dimensional L-shape of a tRNA includes several stemloop structures including the D-loop, the TψC-loop, and the anticodon loop. A mature tRNA must possess all of these structural features in order to work.

14.3.1 Critical steps in pre-tRNA processing

In eukaryotes, transfer RNA precursors (**pre-tRNA**) are transcribed by RNA polymerase III. The precursors have a 5′ leader sequence and a 3′ trailer sequence which need to be cleaved and processed in order to trim the tRNA down to the appropriate length.[34–36] This is achieved, as we shall see, by a set of specific ribonucleases. One of the trimming events includes the removal of an internal sequence. This is a form of splicing, albeit totally unrelated to pre-mRNA splicing. Several modifications are also made to the RNA nucleotides, and the CCA 3′ end is also formed. In general, the CCA sequence is not encoded in the tRNA gene itself and instead needs to be added separately. The

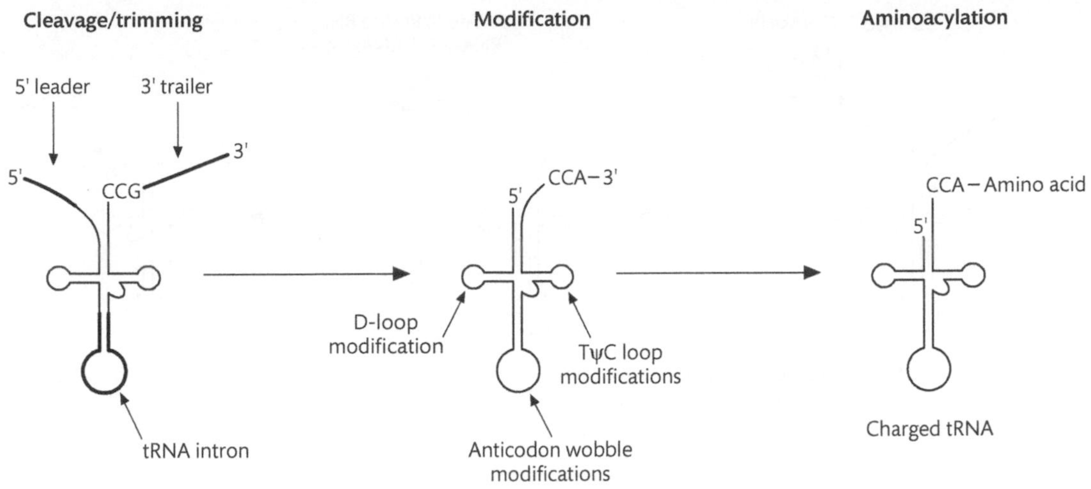

Figure 14.8 Summary of steps in pre-tRNA processing. The precursor tRNA is first cleaved and trimmed, finishing with the addition of CCA. Where applicable, intervening sequences are also spliced out. Several modifications are then applied, including in the anticodon loop. Once these steps have taken place, the tRNA can be charged with the appropriate amino acid generating the amino-acyl tRNA.

sequence of events in pre-tRNA processing is summarized in Fig. 14.8.

The biogenesis of a tRNA proceeds as follows: first it is cleaved and trimmed by ribonucleases; it then undergoes modification of the nucleotides in much the same way as rRNA nucleotides are modified. Modifications are essential for the tRNA to adopt its correct three-dimensional structure. However, beyond the modifications that influence structure, a special group of modifications occur in the anticodon itself; these facilitate the phenomenon known as wobble, in which a particular anticodon can recognize more than one codon. Thus, for example, the base A, in the third position of the anticodon, can be deaminated to give rise to an I (inosine) which can base pair with all four nucleotides. Once these modifications have taken place, the tRNA is ready to be charged with the appropriate amino acid.

Processing of tRNA is a sequential process that starts with trimming of the 5′ leader by the enzyme RNAse P.[37] The involvement of RNAse P in 5′ end pre-tRNA trimming is an ancient process, as it occurs in prokaryotes[37] as well as eukaryotes.[38] The next step is the cleavage of the middle of the 3′ end trailer by RNAse E and RNAse III, followed by exonucleolytic trimming by RNAse II, RNAse BN, and RNAse PH/tRNAse Z. If the CCA sequence is trimmed away, or is absent, a CCA-adding enzyme restores the CCA terminus at the 3′ end. The RNA cleavage and trimming steps and the enzymes involved in the processing of tRNA are summarized in Fig. 14.9.

Note that pre-tRNA processing in bacteria shares the same principal features, but is slightly different. In bacteria, pre-tRNA transcripts are usually part of a polycistronic transcript (in which multiple copies are present) and generally the CCA sequence is included. The first step in bacterial pre-tRNA processing is the endonucleolytic cleavage downstream of the CCA sequence in the 3′ trailer, followed by a first round of exonucleolytic trimming of the 3′ end. Then the bacterial RNAse P cleaves the 5′ leader, after which a second round of trimming of the 3′ end completes the process.

In eukaryotes, a quality control mechanism (RNA surveillance) exists to verify the correct processing of tRNA; it involves a complex of proteins called TRAMP.[39,40] The role of the TRAMP complex in nuclear RNA surveillance is discussed in Box 14.4.

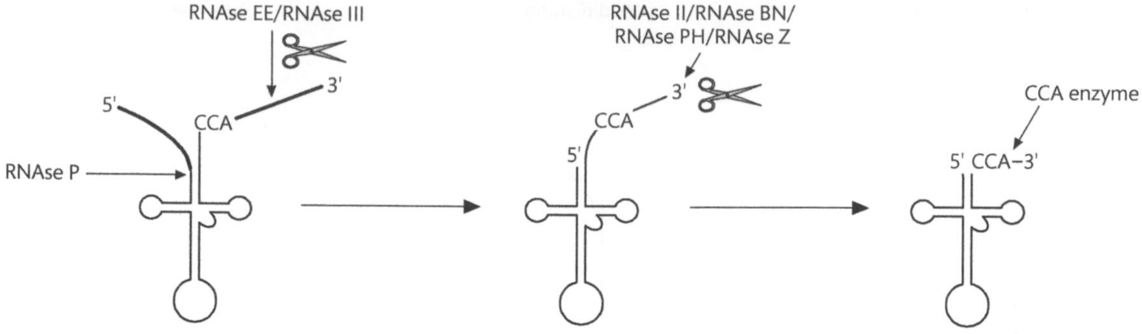

Figure 14.9 Sequence of pre-tRNA cleavage and trimming steps in eukaryotes. The 5′ leader is trimmed by the evolutionarily conserved ribozyme RNAse P. The 3′ end trailer is first cleaved endonucleolytically by RNAse E and RNAse III, and then trimmed exonucleolytically. Finally, the CCA sequence is restored at the 3′ end, which is the eventual site of amino-acylation.

Box 14.4 TRAMP and RNA surveillance in the nucleus

As they have evolved, cells have acquired the ability to eliminate non-functional transcripts, be they coding or non-coding. This is part of the process called *RNA surveillance*. In the nucleus, incorrectly processed or faulty mRNAs are targeted for degradation by the nuclear exosome. Incorrectly processed non-coding RNAs could also cause serious problems and need to be removed. Thus, faulty tRNA and rRNA molecules are also degraded by the exosome, first being marked for degradation by a complex called TRAMP (Trf4/Air2/Mtr4 polyadenylation complex).

TRAMP is a complex of proteins that becomes associated with faulty tRNA and rRNA precursors, to which it adds a poly(A) tail. (Note that tRNA and rRNA are not normally polyadenylated.) TRAMP contains a poly(A) polymerase (Trf4), an RNA-binding protein (Air2), and an RNA helicase (Mtr4). In this case polyadenylation is part of the process by which faulty RNAs are marked for degradation. TRAMP interacts directly with the exosome, which mediates RNA degradation.

Remarkably, TRAMP is able to sense even subtle mutations, including the absence of a methylation modification in tRNA. It does not necessarily sense the lack of specific modifications or other abnormalities (of which there are many), but rather the tendency of the faulty RNA to be improperly folded and less structurally stable.

There is also a process (called retrograde tRNA nuclear import) that involves the transport of faulty tRNAs from the cytoplasm back into the nucleus. This added layer of defence ensures that any faulty tRNAs that have ended up in the cytoplasm are destroyed.[41]

14.3.2 Key enzymes involved in pre-tRNA processing

In bacteria, RNAse P is a ribonucleoprotein that consists of a ~400 nucleotide RNA subunit with catalytic properties (i.e. a ribozyme) and a 15kDa small protein subunit, RnpB. Box 14.5 describes the experiments which identified RNAse P as a ribozyme, resulting in the joint award of a Nobel Prize. In eukaryotes, RNAse P is more complex as there are up to ten additional proteins associated with the H1 RNA. Some of these eukaryotic proteins are also present in Archaea which, as we saw earlier, can be considered chimaeras with features of both prokaryotes and eukaryotes. These RNAse-P-associated proteins are required for substrate recognition and contribute to the enzyme activity provided by the ribozyme moiety in the H1 RNA. However, the H1 RNA can bind to the tRNA substrate on its own, at least in vitro. RNAse P components can be detected in the nucleoplasm and the nucleolus.

Rather than having a single substrate, RNAse P is able to process each of the different tRNAs present in the cell, as well as some other cellular RNAs. The pathway through which RNAse P catalyses cleavage of the phosphodiester bond is shown in Fig. 14.10.

Three magnesium ions are found in the active site of RNAse P. *Magnesium ion A* deprotonates the attacking –OH group (this –OH group is provided by the solution), activating it as a nucleophile which attacks the phosphorus atom of the phosphodiester bond. *Magnesium ion B* stabilizes the negatively charged transition state of the reaction and balances

Box 14.5 The discovery that bacterial RNAse P operates as a ribozyme

During the same time period in which Tom Cech's laboratory was identifying catalytic RNAs in the rRNA of *Tetrahymena*, Sidney Altman was investigating how tRNA is made in bacterial cells. In the 1970s he had made an intriguing observation while working at the Laboratory of Molecular Biology in Cambridge, UK. By rapidly extracting tRNA from bacterial cells using phenol, he found that tRNA was generated from a precursor molecule, which was longer at both ends than the final tRNA molecule itself.

By the 1980s the enzyme that carries out the 5′ processing reaction (called RNAse P for RNAse precursor) had been purified and shown to be composed of RNA and protein components. The advent of recombinant DNA technology, and the possibility of making RNAs in the laboratory through in vitro transcription, meant that it was possible to clone the gene for the RNA component of RNAse P and express it in the laboratory. Surprisingly, this pure RNAse P RNA was catalytically active by itself in cleaving the 5′ terminal extension on pre-tRNA. This catalytic RNAse P RNA was a true enzyme: it was unchanged by the reaction, exhibited multiple turnover of substrates, was only needed in small amounts to catalyse reactions, and was stable.[42] While the RNA component of RNAse P was active catalytically, the protein component of RNAse P increased the catalytic rate of the enzyme—that is, it acted as a cofactor. It is now known that many ribozymes use additional protein cofactors to facilitate their catalytic reactions.

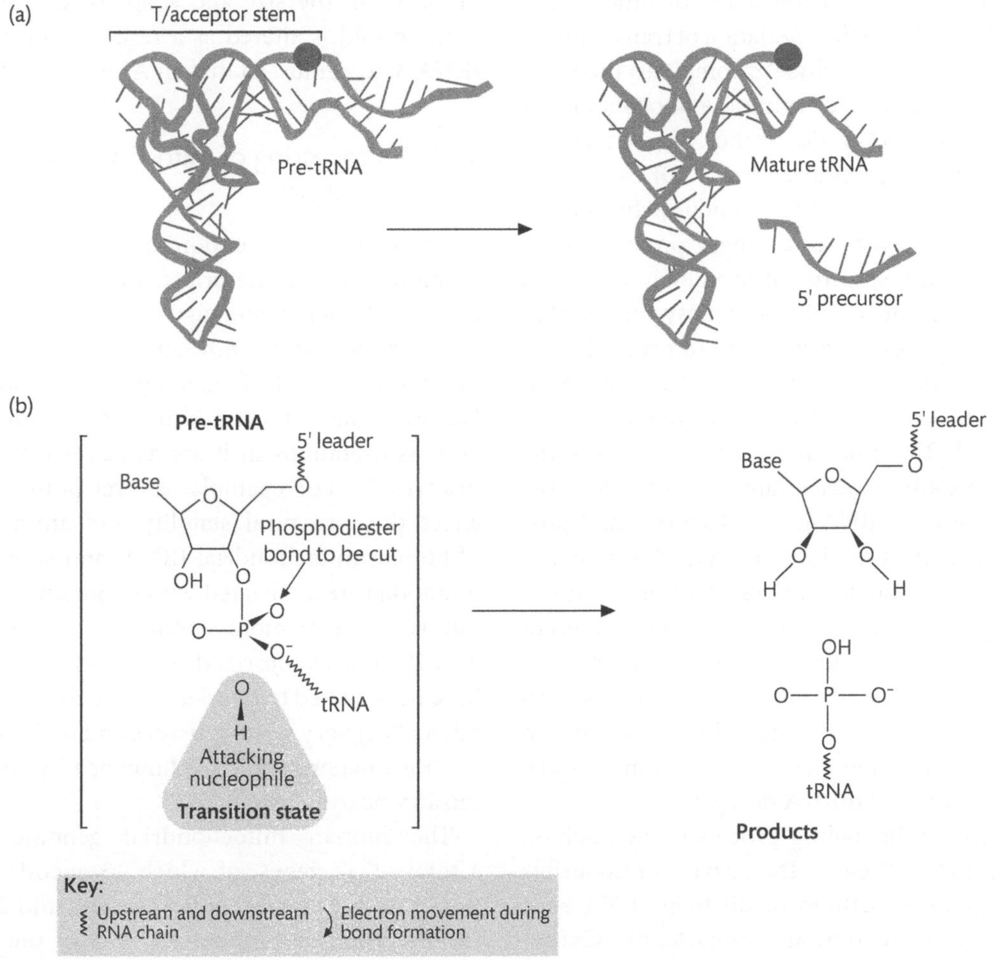

Figure 14.10 **RNAse P is a ribozyme which processes the 5′ terminus of bacterial tRNA.** (a) The conversion of pre-tRNA to tRNA by cutting a phosphodiester bond (the site of cleavage is shown as a sphere). (b) The transition state of this reaction. The attacking nucleophile group is the oxygen of an -OH group (arrowed), which breaks the phosphodiester bond and displaces the 3′ oxygen of the upstream ribose sugar. (From Kazantsev and Pace[37] with permission from Macmillan Publishers Ltd.)

the negative charge of the leaving 3′ oxygen ion. *Magnesium ion C* interacts with and protonates the 2′ –OH group of the released 5′ leader. These three magnesium ions also work as electrophilic catalysts to stabilize the high negative charges of the five oxygen atoms in the transition centre. Each of the magnesium ions interacts with non-bridging oxygens in the transition state, both directly and indirectly (through water coordination; see Chapter 2).

There is an added twist to the function of RNAse P. RNAse P subunits bind to the chromatin of active tRNA genes in mammalian cells and yeast. Such binding is thought to contribute to the activation of expression of the genes whose products are processed by RNAse P. This type of concerted action once again suggests the ability to coordinate processes and echoes the involvement of other RNA-associated proteins in the regulation of transcription.

The endonuclease tRNAse Z completes the trimming of the 3′ end by cleaving the phosphodiester bond that precedes the site of the CCA sequence.[43] In other words, it generates a free 3′ OH end ready for the addition of the CCA residues. In eukaryotes, tRNAse Z exists as a longer protein whose N-terminus can help direct it to mitochondria and chloroplasts. If you recall, the endosymbiotic theory states that these organelles were originally derived from prokaryotes, and they still encode, to some extent, their own tRNAs, which also need to be processed. The more ancient C-terminus of the eukaryotic tRNAse Z essentially corresponds to the shorter version of tRNAse Z in bacteria, and possesses a catalytic function. tRNAse Z contains a metallo-β-lactamase domain, responsible for its endonucleolytic activity. A similar domain is present in CPSF-73, which, as described in Chapter 8, is involved in pre-mRNA 3′ end formation. It has been suggested that tRNAse Z might also be involved in the processing of other classes of RNAs in bacteria, and in the process of mRNA decay.[44]

A particularly intriguing process is the addition of CCA to tRNA 3′ ends. The enzymes responsible must recognize a plethora of different tRNA ends and then add, unaided by any template, two Cs (pyrimidines) and switch to an A (purine). There are two classes of CCA-adding enzymes, class I and class II.[45] The class I family is defined by the archaeal *Archaeoglobus fulgidus* enzyme AfCCA, which has structural homology with RNA recognition motifs and eukaryotic poly(A) polymerase. The class II family is defined by the prokaryotic *Bacillus stearothermophilus* enzyme BstCCA. Perhaps unexpectedly, eukaryotes use class II enzymes to add CCA to their tRNAs. Both classes of enzymes recognize the substrate by detecting tRNA shape and charge complementarity; thus they are able to bind to all tRNAs based on shape and not sequence per se.

An interesting structural issue surrounds how these enzymes achieve the incorporation of specific nucleotides (CCA) when they do not utilize a template. The solution is very elegant; they use parts of the tRNA phosphate backbone and an arginine to mimic the Watson–Crick base pairing interactions that then attract the required nucleotides (two CTPs first). Then, the size and shape of the nucleotide-binding fold is altered as a result of the elongating tRNA 3′ terminus, so that an ATP is added last.

14.3.3 Processing of mitochondrial tRNA transcripts

We now turn our attention briefly to the processing of mitochondrial transcripts. Their compact circular genome (16.5kb in humans) underlines their prokaryotic origin. Mitochondria do rely on nuclear genes to express several of their important components. However, the mitochondrial genome is also important, as mutations in it are associated with human diseases.[46,47] For example, several point mutations affect the structural stability and aminoacylation of human mitochondrial tRNA, and some of these mutations are associated with ophthalmoplegia (paralysis of one or more extraocular muscles). In fact over 200 mitochondrial tRNA (mt-tRNA) mutations have been linked to a variety of disease states.[48] These often affect very specific tissues and include diabetes mellitus, hypertrophic cardiomyopathy, and neurosensory hearing loss.

The human mitochondrial genome contains a total of 37 genes, of which 13 encode polypeptides, two are ribosomal RNA genes, and 22 encode tRNAs. The polypeptides are part of the oxidative phosphorylation machinery through which ATP is synthesized.

In humans, three polycistronic transcription units are expressed from the mitochondrial genome.

Transfer RNA precursors are processed no differently than elsewhere; the process involves RNAse P, tRNAse Z, and CCA-adding enzymes, which are imported into the mitochondria. The protein-coding mRNAs are polyadenylated, an event that appears to confer greater stability to the mRNAs, as is the case with nuclear-encoded eukaryotic mRNAs. A mitochondrial-specific poly(A) polymerase performs this function. The mitochondrial mRNAs tend to have a very short 5′ UTR and they do not contain the canonical prokaryotic translation initiation site (the Shine–Dalgarno sequence). Thus, at least in some ways, mitochondrial mRNAs are quite unlike prokaryotic mRNAs, despite their prokaryotic origin.

Key points

Transfer RNA processing in all organisms involves a sequential step of endonucleolytic cleavage and exonucleolytic trimming of the 5′ leader and 3′ trailer of pre-tRNAs. The sequence CCA (the site of aminoacylation) is added at the 3′ end.

RNA nucleotides are extensively modified, including the anticodon loop. Once all the modifications have been made, tRNA can be charged with the appropriate amino acid. An RNA surveillance mechanism ensures that faulty tRNAs are degraded.

Mitochondria possess their own compact genome, which comprises rRNA, tRNA, and protein-coding genes. Polycistronic transcripts are processed in mitochondria, and defects in their processing are associated with mitochondrial dysfunction in several diseases.

14.4 SMN proteins and snRNP assembly

We now turn our attention to the SMN proteins and their role in snRNP assembly. As we saw in Section 14.1.3, snRNAs are modified through the action of scaRNAs in the nucleus. In fact snRNPs are assembled in *both* the nucleus and cytoplasm in a very intricate process. U1, U2, U4, U4atac, U5, U11, and U12 snRNA precursors are transcribed by RNA polymerase II, whereas U6 is transcribed by RNA polymerase III. The snRNAs are capped (with an m^7G), marking them for export from the nucleus. In the cytoplasm the snRNAs interact with SMN protein and Gemins 2–8 to form the *SMN complex.*[49] In the SMN complex a set of Sm proteins arranged in a ring-like structure associates with the snRNAs at the Sm site (see Chapter 5). Further processing occurs, including trimming of the 3′ end and hypermethylation of the 5′ terminal nucleoside. This latter modification is recognized by a protein called snurportin, facilitating the reimport of the snRNA into the nucleus. Additional maturation steps occur in the Cajal body, including snRNA base modifications and the annealing of U4 to U6 snRNA.

14.4.1 A protein important for snRNP assembly is affected by mutations causing spinal muscular atrophy (SMA)

The **SMN** proteins are of particular interest due to their involvement in spinal muscular atrophy (SMA). SMA (see Box 14.6) is a severe muscle disease caused by low levels of the ubiquitously expressed SMN (survival of motor neuron) protein. The SMN protein has two important connections to RNA processing,[50,51] shown in Fig. 14.11. Why do these connections to RNA processing lead to a defect in a particular motor neuron?

There are two possible answers to this: the SMN1 protein is essential for (1) *assembly of snRNPs in the cytoplasm* or (2) *recycling in the nucleus.* snRNAs are coated in proteins within the cell to form snRNPs. SMN protein is part of a complex in the cell which assembles snRNPs by adding a ring of Sm proteins around them within the cytoplasm (see Chapter 5). SMN protein is also found in Cajal bodies where it helps reassemble nuclear snRNP complexes *after* splicing. Because of the reduction in SMN protein in SMA patients, there is likely to be a resulting shortage of assembled snRNPs, leading to neuronal cell death.

Box 14.6 Spinal muscular atrophy (SMA)

SMA is a reasonably frequent genetic disease with an occurrence of 1 in 6000 live births, and with between 1 in 25 and 1 in 60 people carrying the mutation as heterozygotes. SMA is caused by autosomal recessive mutations within a gene called *SMN1*. In SMA, neurons at the anterior horn of the spinal cord degenerate, leading to weakness in the muscles that they innervate.

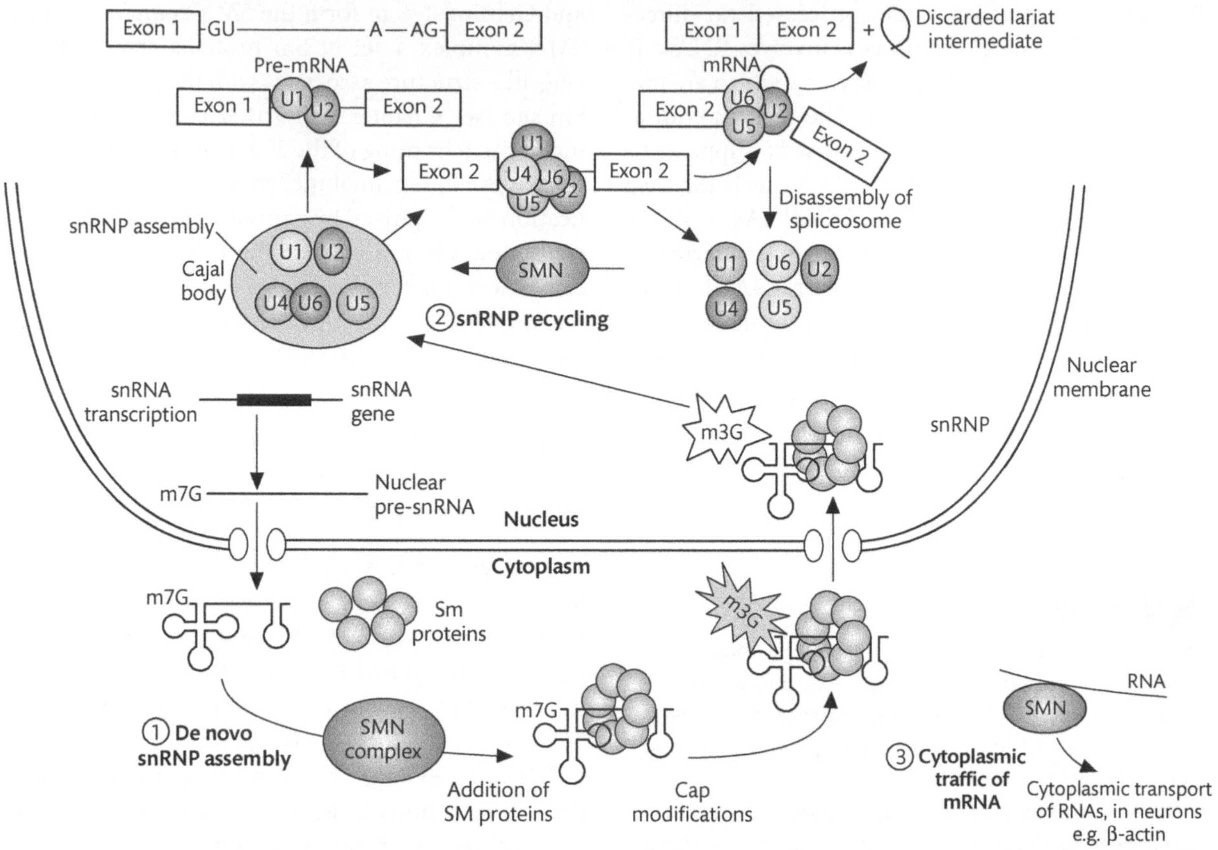

Figure 14.11 The SMN protein is involved in nuclear and cytoplasmic assembly of snRNPs and cytoplasmic transport of mRNAs. Although not shown here, as well as roles in snRNP assembly and RNA transport, SMN1 protein might also have a role within cells in transcription. (Redrawn from Eggert et al.[50] with permission form Elsevier.)

snRNPs are needed in spliceosomes for splicing in every cell in the body (Chapter 5). The tissue-specific symptoms of SMA patients raise a similar question to those of patients with retinitis pigmentosa. Why should snRNP assembly defects result in neuronal disease rather than problems elsewhere in the body? Particular neurons might need very high levels of assembled snRNPs, either to process specific pre-mRNAs or because of high overall levels of gene expression, making the snRNP deficiencies in SMA patients only acute in neurons in the anterior horn of the spinal cord.

SMN protein is involved in the *cytoplasmic transport of mRNAs*. Cytoplasmic mRNA transport might be particularly important in neurons, which are extremely long cells and so have to transport RNA long distances through their cytoplasm.

As well as the relation between SMN protein and snRNP assembly, and so with splicing, SMA has another connection with splicing: the severity of SMA is modulated by the expression of a second gene called *SMN2*, which is normally poorly expressed since it is inefficiently spliced.[52] *SMN1* and *SMN2* are almost identical multi-exon genes found on human chromosome 5q13, located within a region of the chromosome which has undergone an inverted duplication—the *SMN1* and *SMN2* genes can be seen side by side in Fig. 14.12. This gene duplication occurred during the fairly recent evolutionary past after the mouse and human lineages diverged—there are two *SMN* genes in humans, but only one in the mouse. However, in humans, while the *SMN1* gene is required for cellular function, the *SMN2* gene does not seem to be under any particular selection pressure in most people as it is starting to decay by accumulating mutations.

Although it might not normally have a function, the *SMN2* gene *does* become very important in SMA patients who lack the *SMN1* gene. This is because the *SMN2* gene could theoretically provide a genetic

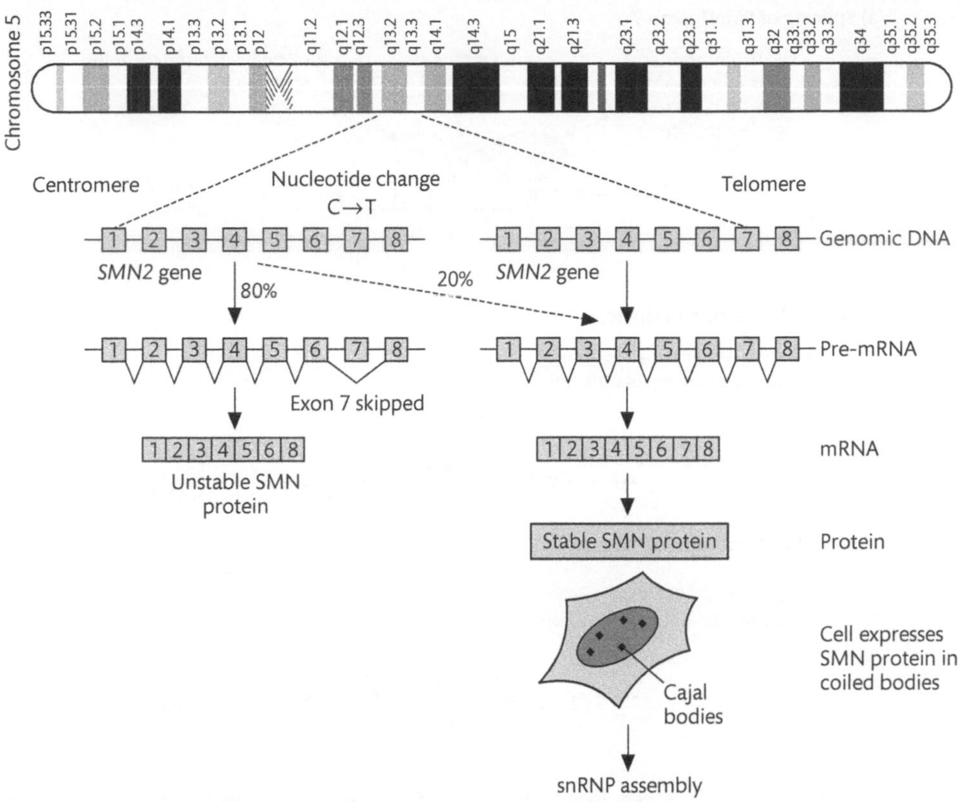

Figure 14.12 Humans have two *SMN* genes called *SMN1* and *SMN2* on the long arm of chromosome 5 but only *SMN1* is fully functional. *SMN1* encodes an eight-exon pre-mRNA (exons 1–7 are protein coding, while exon 8 encodes the 3′ UTR). These exons are spliced together into an mRNA, which is translated into the 294 amino acid SMN protein. The SMN protein is important for snRNP biogenesis and localizes to subnuclear structures, called Cajal bodies, within the cell nucleus. SMA is caused by deletions or mutations of the *SMN1* gene. As a result of the lack of functional SMN protein, the regions of the nucleus which play an important role in snRNP assembly (the Cajal bodies) are absent in cells from SMA patients. The second *SMN* gene is also on the long arm of chromosome 5 and is called *SMN2*. Exon 7 of the *SMN2* transcript is very inefficiently spliced. The 80% of *SMN2* mRNAs which are without exon 7 encode a shorter SMN protein which is unstable in the cell. Because of its inefficient splicing pattern *SMN2* does not fully compensate for the loss of the *SMN1* gene in SMA patients. (Adapted from Khoo et al.[52] with permission from Elsevier.)

backup for the missing *SMN1* gene, but with one important caveat. Although it is still almost identical to *SMN1* in its protein-coding sequence, *SMN2* has picked up one particularly crippling mutation. A single point mutation within exon 7 of *SMN2* means that this coding exon is no longer spliced efficiently into the *SMN2* mRNA (only 20% of the *SMN2* mRNAs include exon 7). Hence the splicing defect in *SMN2* pre-mRNA and its correction are of great importance (see the next two sections).

14.4.2 There is scientific controversy about why exon 7 of the *SMN2* gene is inefficiently spliced

Physiologically normal exon splicing patterns are controlled by a balance between positive and negative influences which affect recognition by the spliceosome—these are splicing enhancers and silencers, respectively (see Chapters 5 and 6). The point mutation within the *SMN2* gene affects an RNA sequence which controls splicing of exon 7 and leads to its being missed out by the spliceosome.[53] Two competing hypotheses have suggested that the C→U mutation in *SMN2* prevents splicing of exon 7 by either creating an exonic splicing silencer or disrupting an exonic splicing enhancer (these competing hypotheses are shown in Fig. 14.13).

Because of its importance in the pathology of SMA, a lot of work has gone into resolving the controversy of why splicing of *SMN2* exon 7 is defective.[54–57] While the precise molecular mechanisms

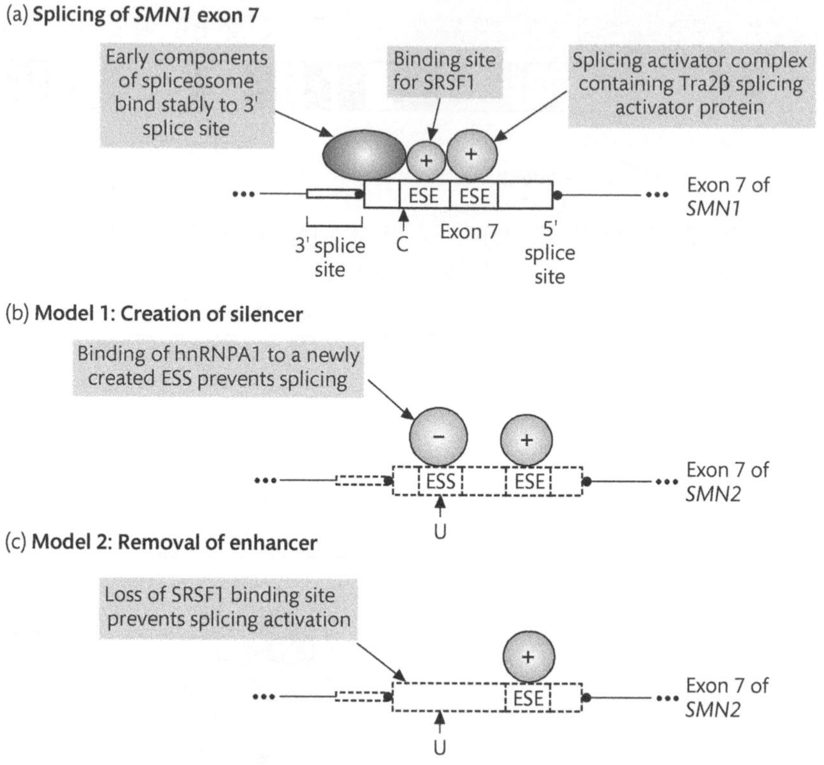

Figure 14.13 Exon 7 of SMN2 pre-mRNA is very inefficiently spliced because of mutations affecting the splicing code. (a) Exon 7 of the *SMN1* gene is efficiently spliced into *SMN1* mRNA since it contains the binding sites for a number of nuclear RNA-binding proteins which together create a splicing activator complex. These splicing activator proteins include Tra2β and SR (in particular the SR protein SRSF1) which bind to an exonic splicing enhancer (ESE) sequence within exon 7, and activate splicing of this exon by stabilizing the binding of early spliceosome components to the 3′ splice site of *SMN1* exon 7. Binding of U2AF and U2 are among the earliest steps in spliceosome assembly and are stabilized by this splicing activator complex (splicing activators are shown as a + sign; see Chapter 7 for a fuller explanation of how splicing activators work). A single point mutation (changing a C to a U in exon 7 of the *SMN2* gene) changes the splicing code and thereby prevents *SMN2* exon 7 from being spliced into the mRNA. Since it is not spliced into the mRNA very efficiently, *SMN2* exon 7 is shown with a broken line rather than the solid line of *SMN1*. Two competing theories have been proposed to explain exactly what this C→U change does. (b) Model 1 proposes that the C→U transition in *SMN2* creates an exonic splicing silencer (ESS). This splicing silencer site binds the splicing repressor hnRNP A1 and prevents the splicing activator complex bound to the ESE from working. (c) Model 2 proposes that the C→U transition in *SMN2* destroys a binding site for the splicing activator SRSF1. For the splicing activator complex to work it needs to bind a group of splicing activators (see (a)). Loss of the SRSF1 binding site results in the splicing activator complex being unable to activate splicing of *SMN2* exon 7.

are still somewhat controversial, the current picture is a hybrid of two models.

1 The silencer model. *SMN2* exon 7 is thought to contain a binding site for hnRNP A1 which acts to stop the exon being recognized by the splicing machinery. However, this exon 7 silencer sequence is not directly affected by the point mutation.

2 The activator model. On the other hand, a high affinity binding site for SRSF1 (previously known as ASF/SF2) is disrupted by the point mutation in *SMN2* exon 7, and this prevents proper activation of this exon.

Although the precise molecular mechanisms preventing splicing of *SMN2* exon 7 are still controversial, splicing manipulation of this exon provides the basis for strategies to treat SMA, as we shall see in the next section.

14.4.3 Molecular therapy for SMA is targeted at correcting the splicing of *SMN2* exon 7

As we have seen, patients with SMA are missing the *SMN1* gene through either mutation or deletion, although the adjacent *SMN2* gene may still be present.

While *SMN2* is inefficiently spliced it can still provide a limited genetic backup for *SMN1* function in some patients. Many therapies under development for treating SMA are based on improving the cellular expression level of *SMN2*. One approach has been to use a class of molecule called histone deacetylase (HDAC) inhibitors, such as valproic acid and 4-phenylbutyrate, to increase gene expression. HDACs increase general transcription of the *SMN2* gene and cause an increase in the splicing of *SMN2* exon 7 (possibly through increasing the expression of positive splicing factors in the cell, like Tra2β, which activate exon 7 splicing).

A disadvantage of using HDACs is that they are not very specific (an estimated 2% of cellular genes are upregulated by HDACs, although HDACs are under clinical trial for other diseases including some cancers). Furthermore, because of the inefficiency of exon 7 splicing, increasing normal *SMN2* gene expression only goes some way towards increasing the level of expression of the SMN2 protein. Because of this, other strategies are aimed at increasing the splicing efficiency of *SMN2* exon 7.[58–61]

Normally only around 20% of SMN2 exon 7 is spliced into SMN2 mRNA (see Fig. 14.14a). Two different strategies have been developed which increase the splicing efficiency of *SMN2* exon 7 in slightly different ways.

1 Targeted oligonucleotide enhancers of splicing (TOES) use oligonucleotides to anneal to *SMN2* exon 7 and bring in splicing activator proteins (Fig. 14.14b). This strategy was developed by a group of UK scientists led by Professors Ian Eperon and Francesco Muntoni, and uses hybrid oligonucleotides to restore splicing to *SMN2* exon 7. These hybrid oligonucleotides are made up of two parts: a sequence complementary to *SMN2* exon 7 which hydrogen bonds through Watson–Crick base pairing to the inefficiently spliced exon, and an RNA sequence (repeats of GGA which is an ESE consensus sequence) bound by splicing activator proteins like

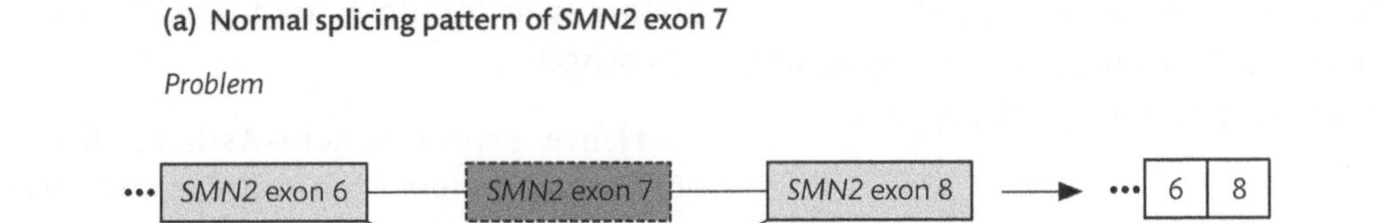

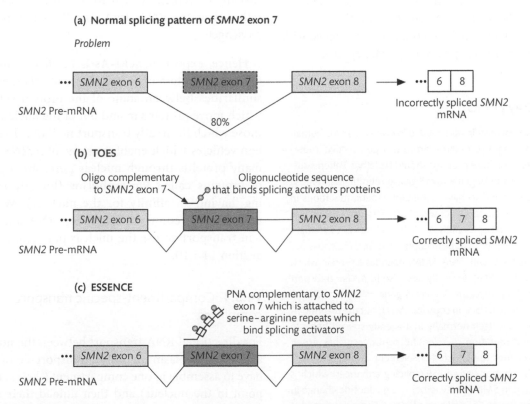

Figure 14.14 Therapeutic strategies have been developed to increase the splicing of *SMN2* exon 7. In principle, the approaches which have been developed to treat SMA could also be used to correct other defects where *cis*-acting splicing control sequences are mutated, and where the disease caused by these mutations prevents normal pre-mRNA processing.

SRSF1 which can rescue splicing of *SMN2* exon 7. This technique was very successful in increasing the incorporation of *SMN2* exon 7 in fibroblasts from SMA patients, and could also partially restore Cajal bodies, showing that the SMN protein was functional in these cells.

2 Exon-specific splicing enhancers by chimaeric effectors (ESSENCE) uses peptide nucleic acid complexes to bind to *SMN2* exon 7 and activate splicing (Fig. 14.14c). This strategy was developed by a group of US scientists led by Professor Adrian Krainer, and uses a protein nucleic acid (PNA) oligonucleotide fused to 10 alternating serine and arginine residues (RS_{10}). PNAs are made by the chemistry which is normally used to make short peptides. They have a peptide-like backbone, but also have bases as side groups which can anneal with RNA molecules. These side groups can act as landing sites for positive regulators of splicing such as SRSF1. PNAs have several advantages over conventional oligonucleotides: they are stable in the cell, they do not target RNAs for RNAse H digestion (normally target RNAs hybridized to DNA oligonucleotides would be digested by RNAse H within the cell), and they can cross the cell membrane when coupled with positively charged peptides or special peptides.

Key points

Spliceosomal proteins are needed in every cell in the human body to splice pre-mRNAs. From this perspective, mutations affecting these proteins should be lethal. Patients suffering from the degenerative disease retinitis pigmentosa have been found to have dominant genetic mutations in the genes encoding spliceosomal proteins—this means that one of the two copies of the gene in each cell of the patient is defective rather than both copies. The *SMN1* gene is deleted in patients with SMA. *SMN1* encodes a protein which functions in snRNP assembly and also in mRNP transport through the cytoplasm. A second gene called *SMN2* provides a potential backup copy for *SMN1*, but the pre-mRNA from *SMN2* is itself normally inefficiently spliced because of a point mutation. A scientific controversy has arisen about whether the affected *cis*-element in *SMN2* exon 7 normally functions as an ESE (splicing enhancer) which is inactivated in *SMN2*, or whether the nucleotide change in *SMN2* creates an ESS (splicing silencer). Therapies aimed at curing SMA are being developed which rescue splicing of the defective exon in the *SMN2* pre-mRNA.

14.5 Nucleocytoplasmic traffic of non-coding RNA

In Chapter 9 we looked at the route and mechanisms by which mRNA is transported out of the nucleus. In this section we are going to look at how non-coding (nc) RNAs are exported from the nucleus. Although ncRNAs and mRNAs follow the same route out of the nucleus (via the nuclear pore), traffic is achieved by somewhat different mechanisms.[62–64] However, nuclear export of ncRNAs follows the same general themes that we discussed in Chapter 9.

1 Nuclear pores provide a transit point between the nucleus and the cytoplasm.

2 Cargo proteins carry RNA transcripts from the nucleus to the cytoplasm. Non-coding RNAs bind proteins which shuttle between the nucleus and the cytoplasm. RNAs destined for nuclear cytoplasmic export are transported as the cargo or freight of these shuttling proteins. It is the shuttling proteins that are the vehicles which are actually transported, rather than the RNA itself. The RNA is a passive passenger.

Hence, export of ncRNAs is in effect a branch of nuclear protein export. Because of this there are similarities between some of the protein vehicles which move proteins in and out of the nucleus and those which indirectly transport ncRNAs. The protein vehicles which enable passage of ncRNAs and many proteins through nuclear pores are a group of proteins called **karyopherins** (literally meaning 'having an affinity for the nucleus'). We will return to the connection between RNA and protein transport across the nuclear pore complex in Section 14.5.10.

14.5.1 Compartment-specific transport complexes

For directional RNA transport between the nucleus and the cytoplasm to work, transport complexes have to assemble in one compartment (their starting point in the nucleus) and then unload their cargo in another (their destination in the cytoplasm). Transport complexes monitor exactly where they are in the cell, and respond to this information. For ncRNAs this is achieved by changing the stability of the transport complex depending on cellular location:

- Nuclear export complexes are only stable in the nucleus and then disassemble once they reach the cytoplasm.
- Nuclear import complexes are stable in the cytoplasm, and break down once they get back into the nucleus.

Non-coding RNAs monitor their position in the cell using a small molecule called **RAN**. In the next sections we will first describe in general how ncRNAs are exported using RAN as a location guide, and then look in more detail at the specific mechanisms which are used to export each class of ncRNA into the cytoplasm.

14.5.2 Nuclear transport of rRNA, tRNA, snRNAs, and microRNAs is dependent on the RAN-GTPase protein

RAN plays a central role in enabling nuclear RNA export complexes to monitor their cellular location (nuclear or cytoplasmic) and so to either assemble or disassemble as appropriate. RAN is central to the nuclear export of rRNA, tRNA, snRNA, and microRNAs (mRNAs use a different system—see Chapter 9). Since these ncRNAs are much more abundantly synthesized in the nucleus than mRNA, RAN is in effect responsible for regulating the nuclear export of most cellular RNA.

Experiments have shown that at any one time most RAN protein is in the nucleus of the cell. However, although RAN accumulates in the nucleus, in living cells it is actually dynamically and rapidly moving between the nucleus and cytoplasm and back again. Hence, although at any one time most RAN is in the nucleus, around 10^5 molecules of RAN are transported per second from the nucleus in transport complexes.

What kind of protein is RAN? RAN is one of a group of small **GTPases** which provide critical switches in the cell, and was first identified through its homology to the related RAS protein (Fig. 14.15). The name RAN itself is an acronym for RAS-related nuclear protein. RAS is one of the best known examples of a GTPase-thrown switch, and provides an on–off switch in important signalling pathways downstream of growth factor receptors. The gene encoding RAS can be mutated in several cancers, resulting in signalling pathways being permanently on and inappropriate cell growth.

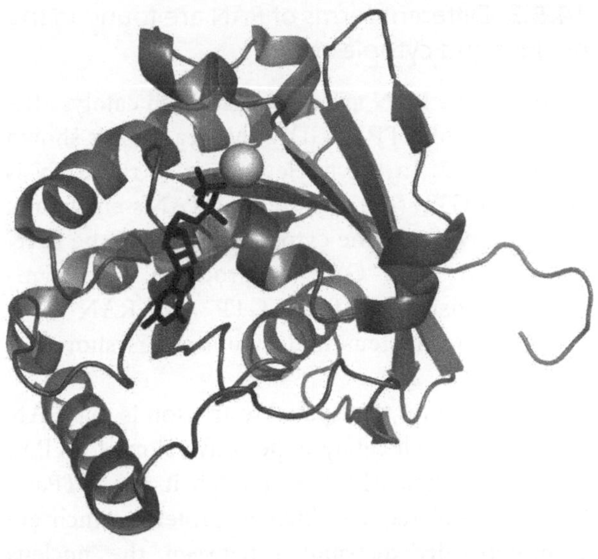

Figure 14.15 RAN is a GTPase which binds to its substrate GTP and then converts GTP to GDP. RAN is present in two forms in the cell. In the nucleus it is primarily bound to GTP, and in the cytoplasm it is primarily bound to GDP. (Protein structure redrawn from the RCSB Protein Data Bank ID 3GJ0 by Jonathan Crowe, Oxford University Press.)

How the RAN system works to provide spatial information in the cell depends on two things: (1) having different forms of RAN protein in the nucleus and the cytoplasm, and (2) having nuclear cytoplasmic transport complexes which are able to respond to these differences and either assemble or disassemble in the right place (Fig. 14.16).

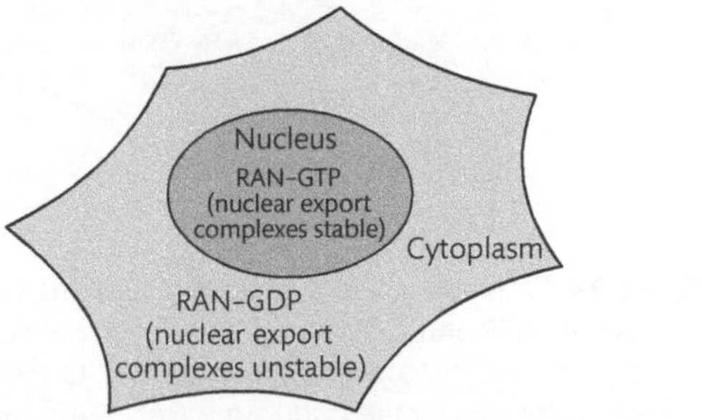

Figure 14.16 Cells contain a 100-fold concentration gradient of RAN protein.

14.5.3 Different forms of RAN are found in the nucleus and cytoplasm

GTPases like RAN are enzymes which catalyse the conversion of GTP to GDP. This reaction is shown in Fig. 14.17. In the nucleus most of the RAN is bound to GTP (the substrate of RAN's enzymatic activity), while in the cytoplasm most of the RAN is bound to RAN–GDP (its product). This asymmetric division of RAN–GTP and RAN–GDP between the nucleus and cytoplasm is shown in Fig. 14.16.

The reason for this spatial restriction is that RAN does not by itself catalyse the conversion of GTP to GDP very efficiently, even though it is a GTPase. Instead, RAN requires 'helper' proteins which are asymmetrically distributed between the nucleus and the cytoplasm. These proteins are shown in Fig. 14.17 and discussed next.

In the nucleus RAN–GTP acts as a component of RNA export complexes. The nuclear assembly of RAN–GTP is due to the presence in the nucleus of a protein called **RCC1**. RCC1 is anchored in the nucleus through binding to histones, the small basic proteins attached to chromosomal DNA.

Nuclear RCC1 creates RAN–GTP since it replaces any GDP bound to RAN with GTP (see Fig. 14.17).

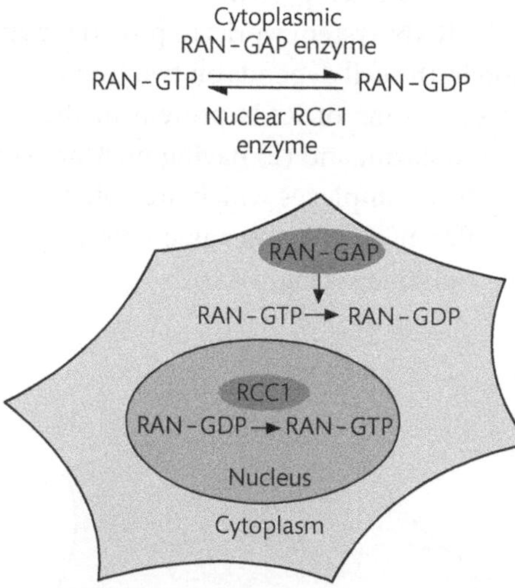

Figure 14.17 **Nuclear and cytoplasmic proteins switch RAN between its GTP and GDP forms.** RAN works inefficiently as a GTPase. The catalytic conversion of RAN–GTP to RAN–GDP by GTPase is activated by the cytoplasmic RAN–GAP protein. After catalysis, RAN does not release GDP. The nuclear RCC1 protein replaces GDP bound to RAN with GTP.

For this reason, RCC1 is termed a guanine nucleotide exchange factor (GEF). Once GTP has been added to RAN it is stable in the nucleus, because of the normal inefficiency of RAN as a GTPase.

Although RAN is normally very inefficient as a GTPase, it is highly active in the cytoplasm where it converts RAN–GTP to RAN–GDP. This increase in GTPase activity is because of the presence of a cytoplasmic protein called RAN–GTPase-activating protein (**RAN–GAP**), which stimulates the GTPase activity of RAN by a factor of about 10^5. Hence RAN–GDP is found in the cytoplasm because of the cytoplasmic location of RAN–GAP, as shown in Fig. 14.17.

RAN–GAP is concentrated on the cytoplasmic fibrils of the NPC (see Chapter 9) where RNA emerges from the nucleus. RAN–GAP is too big to diffuse into the nucleus, and does not contain any nuclear import signals. Production of RAN–GDP in the cytoplasm is also assisted by two other proteins, called **RANBP1** and **RANBP2**, which also help this catalysis to occur. Like RAN–GAP, RANBP1 and RANBP2 are exclusively cytoplasmic.

14.5.4 Non-coding RNA nuclear export complexes contain adaptors and receptors

A schematic overview of a nuclear RNA export complex for ncRNA is shown in Fig. 14.18. Notice that this is made up of three protein components.

1 An RNA export receptor (a karyopherin), which is the vehicle which interacts with and allows passage through the nuclear pore.
2 RAN in the GTP (nuclear) form, which provides the switch which tells the complex that it is in the nucleus.
3 Cargo RNA, which is usually connected to the complex via an adaptor protein.

The enzymatic activity of RAN is very important in determining its shape. Hydrolysis of RAN–GTP to RAN–GDP switches the conformation of the RAN protein in two key regions. These switch regions are important since they are in the parts of RAN which interact with karyopherins, the transport receptors for ncRNAs. We are going to look at these karyopherin proteins in more detail in the next section.

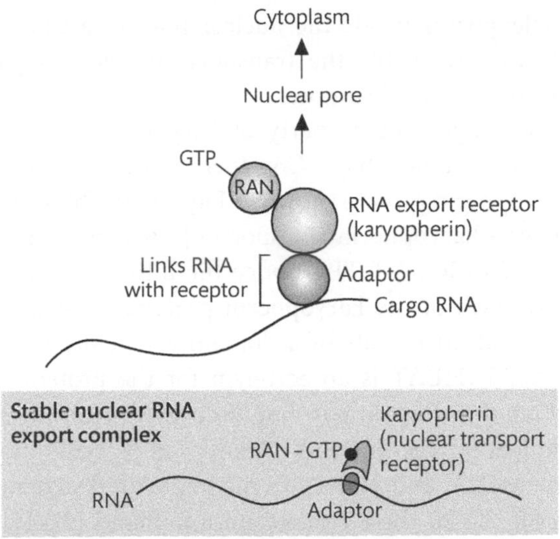

Figure 14.18 Nuclear RNA export complexes are composed of cargo RNA and protein vehicles. The proteins are adaptors which bind to RNA, and receptors which mediate passage through the nuclear pore complex.

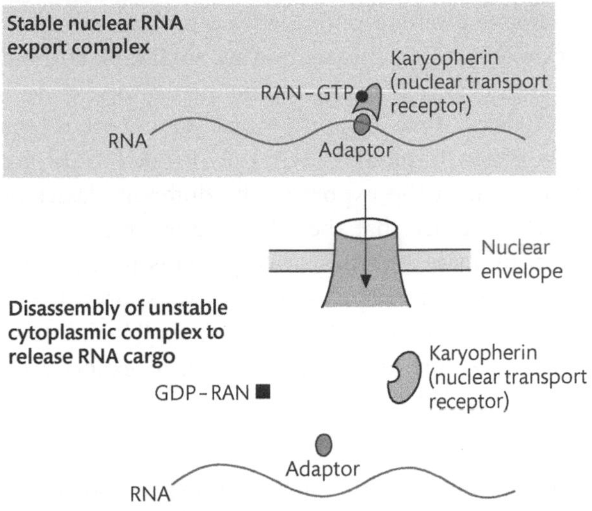

Figure 14.19 Stable nuclear export complexes contain RAN-GTP, but disassemble in the cytoplasm when RAN is converted to the GDP form. Once export complexes move into the cytoplasm, RAN-GTP is activated to convert to RAN-GDP, which causes the complex to disassemble. All non-coding RNAs are exported from the nucleus by export complexes which contain RAN-GTP but use different associated proteins. The actual proteins which fill these roles for each type of RNA are shown in Table 14.3.

Figure 14.19 shows how RAN controls directional nuclear export through the NPC. RNA nuclear export receptors assemble into RNA export complexes in the nucleus with RAN-GTP. RAN-GTP is the critical component holding this complex together. Once the RNA export complex moves into the cytoplasm, the GTPase activity of RAN is activated by RAN-GAP. This catalytic activation results in the catalytic conversion of GTP to GDP and a change in conformation of RAN which is now unable to bind stably to the RNA export receptor. This in turn changes the conformation of the nuclear export receptor. As a result of these changes, the nuclear RNA export complex breaks up, releasing the RNA cargo in the cytoplasm.

The functions of RAN and its associated proteins are not limited to RNA export. They also have important roles in spindle assembly during mitosis and in cell cycle control.[65] In fact, RCC1 was first discovered because it is also important for chromosome condensation and organization of the microtubules before cell division by mitosis. This explains its full name, regulator of chromosome condensation (RCC1). Likewise, the karyopherin **CRM1** stands for chromosome maintenance.

Key points

ncRNAs follow a distinct nuclear export pathway from that of mRNAs. Newly transcribed ncRNAs are exported from the nucleus into the cytoplasm as components of RNA export cargo complexes. The small GTPase RAN enables transport complexes which move ncRNAs out of the nucleus to monitor their position in the cell. RAN normally remains bound to its substrate (GTP) or product (GDP) and so can exist in two states in the cell. RAN-GTP is found only in the nucleus, and RAN-GDP is found only in the cytoplasm. Depending on whether it is in the RAN-GTP or RAN-GDP form, the RAN protein switches its conformation. These changes in shape either stabilize RNA export complexes (by RAN-GTP) or destabilize them (by RAN-GDP), releasing RNA cargo.

14.5.5 Nuclear export of ncRNA is dependent on nuclear export adaptor and receptor proteins

While RAN, export adaptors, and export receptors play an essential role in the nuclear export of all classes of ncRNA, different adaptors and receptor components are responsible for exporting each class of ncRNA.[66,67] In this section we shall look at these nuclear export adaptors and their receptors.

At least three important nuclear export receptors which respond to RAN–GTP/RAN–GDP are important in the export of the different classes of ncRNA (notice that the RNA export receptor column in Table 14.3 has three proteins in it). These nuclear export receptors are CRM1, exportin-t, and exportin-5. In the next section, we shall look at how these receptors and adaptors work to export each class of ncRNA.[68]

14.5.6 Karyopherins are an important group of nuclear export receptors which respond to positional information provided by RAN

Each of the nuclear receptors involved in the export of ncRNAs from the nucleus (see Table 14.3) belong to a larger family of proteins, the karyopherins. There are around 20 different karyopherins in humans and around 14 in yeast. As a group, karyopherins are involved in both export of nuclear proteins (and in some cases attached RNAs) and import of cytoplasmic proteins into the nucleus. Members of the karyopherin family of nuclear export receptors important for RNA transport include CRM1 (the receptor for the export of rRNA, snRNA, and some HIV transcripts), exportin-t (the receptor for the nuclear export of tRNA), and exportin-5 (the receptor for the nuclear export of microRNAs).

Karyopherins act as the protein scaffolds on which transport complexes assemble. Karyopherins also have domains which can interact with the FG nucleoporins inside the nuclear pore (see Chapter 9), and so enable the transport complex to pass through the nuclear pore.

An important property of karyopherins is that they can change shape. Consider the structure of the karyopherin protein shown in Fig. 14.20. This example of a karyopherin is importin β, which is shown bound to RAN–GTP—notice the intimate connection. Each of the karyopherin proteins is made up of about 20 repeats of a domain called the HEAT domain (HEAT is an acronym for the proteins in which this domain was first discovered: Huntingtin, Elongation factor 3, 'A' subunit of protein phosphatase 2A, and TOR1). In the importin β structure in Fig. 14.20 these repeats are numbered H1–H19. Individual HEAT domains are composed of two α-helices, and in the karyopherin molecule these repeats together form a superhelical spiral structure like the coils of a spring. This spring-like structure is flexible. All karyopherins bind to RAN–GTP (located in the nucleus) but not to RAN–GDP (located in the cytoplasm). Distortions are introduced into the spring-like karyopherin structure by RAN–GTP binding.

By affecting the superhelical structure of the karyopherin protein, binding of RAN–GTP affects the ability of the karyopherin to interact with other partner proteins. Binding RAN–GTP in the nucleus enables assembly of nuclear RNA export complexes (see Figs 14.18 and 14.19). Similarly, once these export complexes reach the cytoplasm, hydrolysis of RAN–GTP to RAN–GDP leads to a change in structure and disruption of RNA export complexes, releasing RNA.

Table 14.3 Components of the karyopherin-based nuclear export complexes responsible for nuclear export of ncRNAs*

Non-coding RNA class	RNA export adaptor	RNA export receptor (karyopherin)	RAN bound as part of complex
snRNA	PHAX	CRM1	RAN-GTP
rRNA	NMD3	CRM1	RAN-GTP
HIV-RRE	Rev	CRM1	RAN-GTP
tRNA	None	Exportin-t (direct binding to RNA)	RAN-GTP
Pri-microRNA	None	Exportin-5 (direct binding to RNA)	RAN-GTP

*Nuclear export complexes are made up of RNA and usually three proteins: export adaptors which bind to RNA, export receptors which interact with the nuclear pore, and RAN-GTP. Each of these export complexes has a similar modular design, but contains different protein components depending on the class of ncRNA.

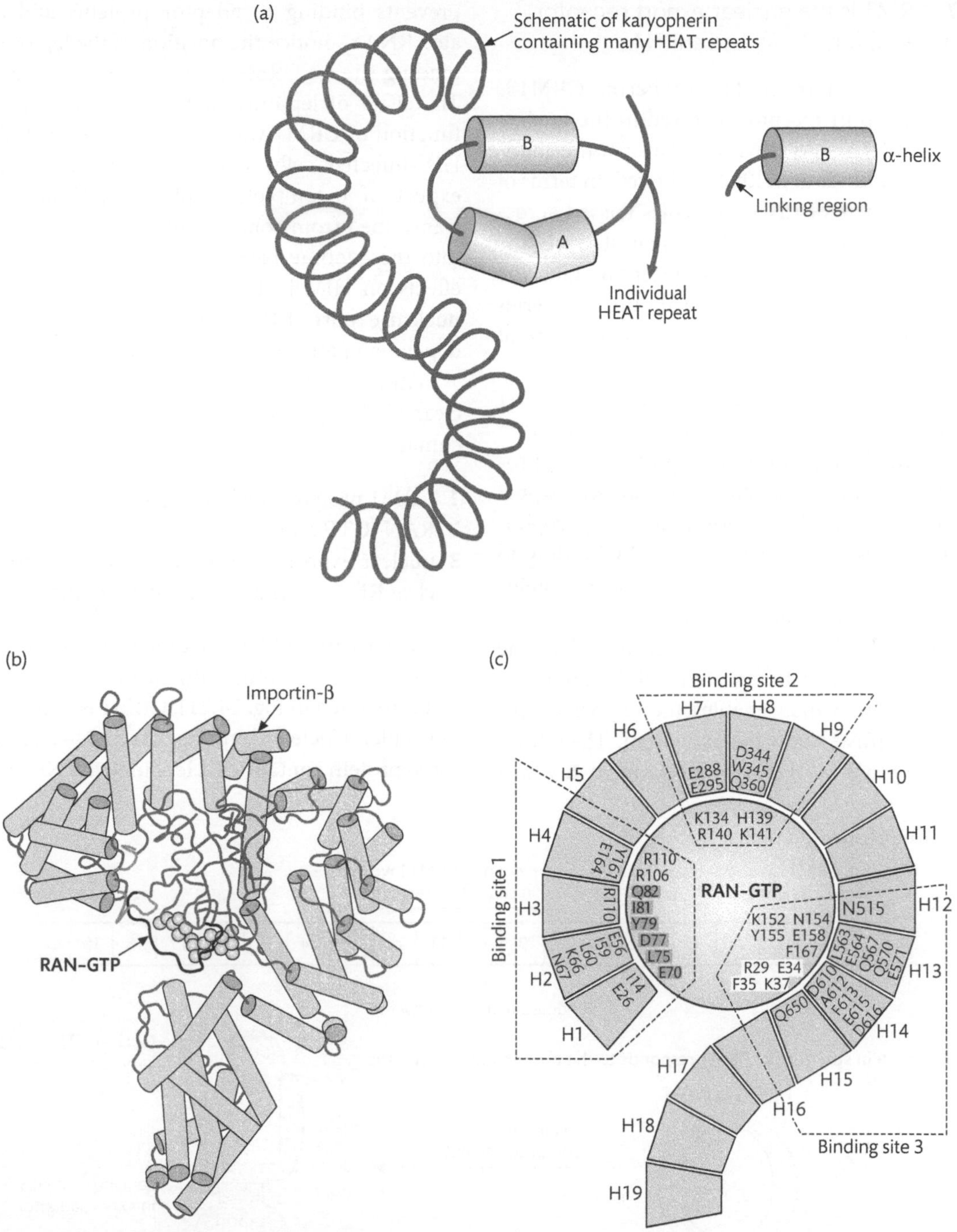

Figure 14.20 **Karyopherins are shape-changing molecules which interact with RAN–GTP.** The atomic structures of several karyopherins have been solved. The structure of the karyopherin importin β is shown here. (a) Karyopherins are made up of multiple repeats of a protein sequence called HEAT repeats which look like the coils in a spring. Individual HEAT repeats are made up of α-helices joined by linking regions. When not bound to RAN–GTP, the importin β molecule has an S shape. (b) The HEAT repeats of importin β (shown as helices and linking regions in this protein structural diagram) coil around RAN–GTP and interact through binding sites 1–3. (c) A schematic view—notice that there are 19 repeats in importin β numbered H1–H19. This spring-like structure is both flexible and provides a large potential interaction surface for protein partners. Binding of RAN–GTP distorts the structure of importin β, and affects interactions with cargo proteins and FG-repeat nucleoporins. (Reprinted from Stewart[67] with permission from Macmillan Publishers Ltd.)

14.5.7 CRM1 is the nuclear export receptor (karyopherin) for rRNAs and snRNAs

While there are a number of karyopherins, CRM1 is the nuclear export receptor involved in the nuclear export of snRNAs and rRNAs. Because of the abundance of these kinds of ncRNA in the cell, in terms of absolute numbers of RNA molecules passed across the nuclear membrane CRM1 is quantitatively the most important nuclear export receptor. Like the other karyopherins, CRM1 is made up of a series of HEAT repeats. The 19 HEAT repeat structure of CRM1 is illustrated in Fig. 14.21.

Like most other nuclear export receptors, CRM1 does not bind directly to the RNAs it transports.[69,70] Instead, CRM1 binds to RNA export adaptor proteins which contain a leucine-rich **nuclear export sequence** (NES).[71] NESs are rich in hydrophobic amino acids such as leucine, and by binding to CRM1 they act as address tags for nuclear proteins to join the nuclear export pathway.

Leucine-rich NESs were first discovered in the HIV Rev protein (the Rev protein will be discussed in Section 14.6) which enables the export of unspliced pre-mRNAs into the cytoplasm. The CRM1 protein is bound by the drug **leptomycin B** which prevents binding to adaptor proteins and associated RNAs—notice the position of the leptomycin B binding site on CRM1 protein shown in Fig. 14.21. The effect of leptomycin B on the nuclear export function of CRM1 was first identified in studies on HIV-infected cells. Leptomycin B blocks nuclear export of incompletely spliced HIV pre-mRNAs transcribed from copies of the HIV virus integrated into the genome (see Section 14.6). As well as its effects on HIV, leptomycin B generally blocks the nuclear export of RNA moving through all CRM1-dependent pathways, but not mRNA nuclear export.

Within the nucleus CRM1 forms tripartite nuclear RNA export complexes (see Fig. 14.22) which contain

1 CRM1 protein itself, bound to
2 RAN–GTP and
3 nuclear export adaptors which are bound to nuclear RNAs destined for export into the cytoplasm.

The structure of CRM1 protein has been worked out at atomic resolution (this structural information is summarized in Fig. 14.21). CRM1 is a ring-shaped molecule. Nuclear binding of RAN–GTP and a cargo protein containing a leucine-rich NES induces

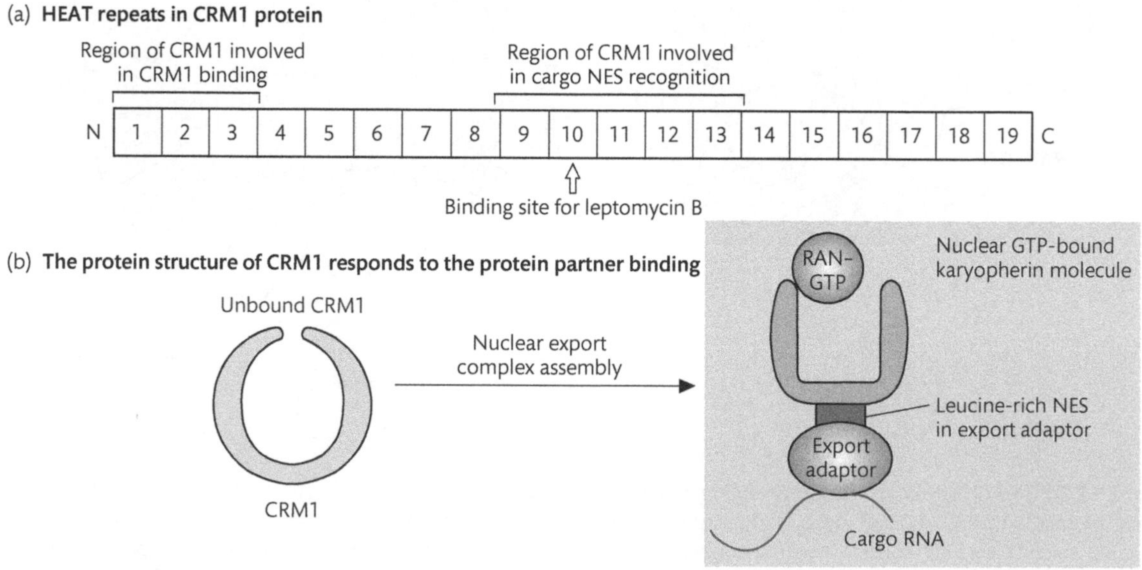

Figure 14.21 The karyopherin CRM1 changes shape in nuclear export complexes. (a) CRM1 is made up of 19 HEAT repeats which bind RAN–GTP, cargo proteins containing leucine-rich NESs, and FG-repeat nucleoporins. These interactions can be blocked by drugs; the antibiotic leptomycin B binds to HEAT repeat 10 and blocks interaction of CRM1 with cargo proteins which contain NESs. (b) Binding of partner proteins shifts the shape of CRM1; notice that CRM1 has a doughnut shape. (Part (b) is based on information in Debler et al.[69])

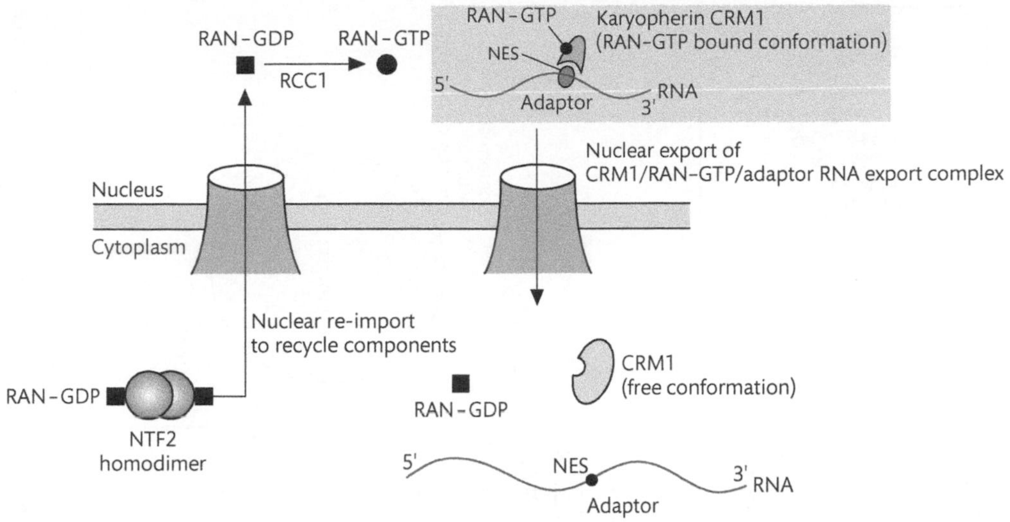

Figure 14.22 **RNA export complexes and their components shuttle between the nucleus and cytoplasm.** Tripartite export complexes containing CRM1 move RNAs out of the nucleus. These complexes disassemble in the cytoplasm. The transported RNAs stay in the cytoplasm, while the nuclear export components are transported back into the nucleus for the next round of nuclear RNA export.

a shift in the structure of CRM1 which facilitates the formation of a nuclear export complex. Notice the opening of the CRM1 ring structure in Fig. 14.21 on binding RAN–GTP and the leucine-rich NES in the adaptor protein. After nuclear export RAN–GTP hydrolyses in the cytoplasm, and RAN–GDP is released from CRM1. This release of RAN–GDP in the cytoplasm similarly causes a structural shift in CRM1, leading to disassembly of the export complex (notice that the CRM1 ring-like protein structure closes).

14.5.8 Different karyopherins act as export receptors for tRNA and microRNAs

While CRM1 acts as the nuclear export receptor for rRNA and snRNA, two other important nuclear export receptors are responsible for the nuclear export of tRNA[72] and microRNAs[73,74] (see Fig. 14.23) (microRNAs will be discussed in more detail in Chapter 16).

- **Exportin-t** is the nuclear export receptor for tRNA. Exportin-t will only bind to tRNA once it has been properly processed in the nucleus. Atomic structures have been worked out for nuclear exportin-t (bound to RAN–GTP), where it wraps around the 5′ and 3′ ends of the tRNA so that it can distinguish fully processed tRNAs. Once in the cytoplasm, exportin-t releases RAN–GDP and undergoes a conformational change so that it also releases its cargo tRNA.
- **Exportin-5** is the nuclear export receptor for exporting microRNA precursors (primary microRNAs, or pri-microRNAs) into the cytoplasm (see Chapter 16). The nuclear complex of exportin-5–RAN–GTP–microRNA is also understood at atomic resolution.

Notice from Fig. 14.23 that both exportin-t and exportin-5 directly bind to their target RNAs rather than needing an adaptor protein.

14.5.9 After releasing their loads in the cytoplasm nuclear RNA export components are moved back into the nucleus

The final destination of many ncRNAs, including tRNAs and rRNAs, is the cytoplasm. However, a different situation applies to the protein components of nuclear ncRNA export complexes—these need to be moved back into the nucleus. Obviously, if nuclear RNA export proteins were only flowing out into the cytoplasm from the nucleus, further RNA export would be impossible once their nuclear levels were depleted.

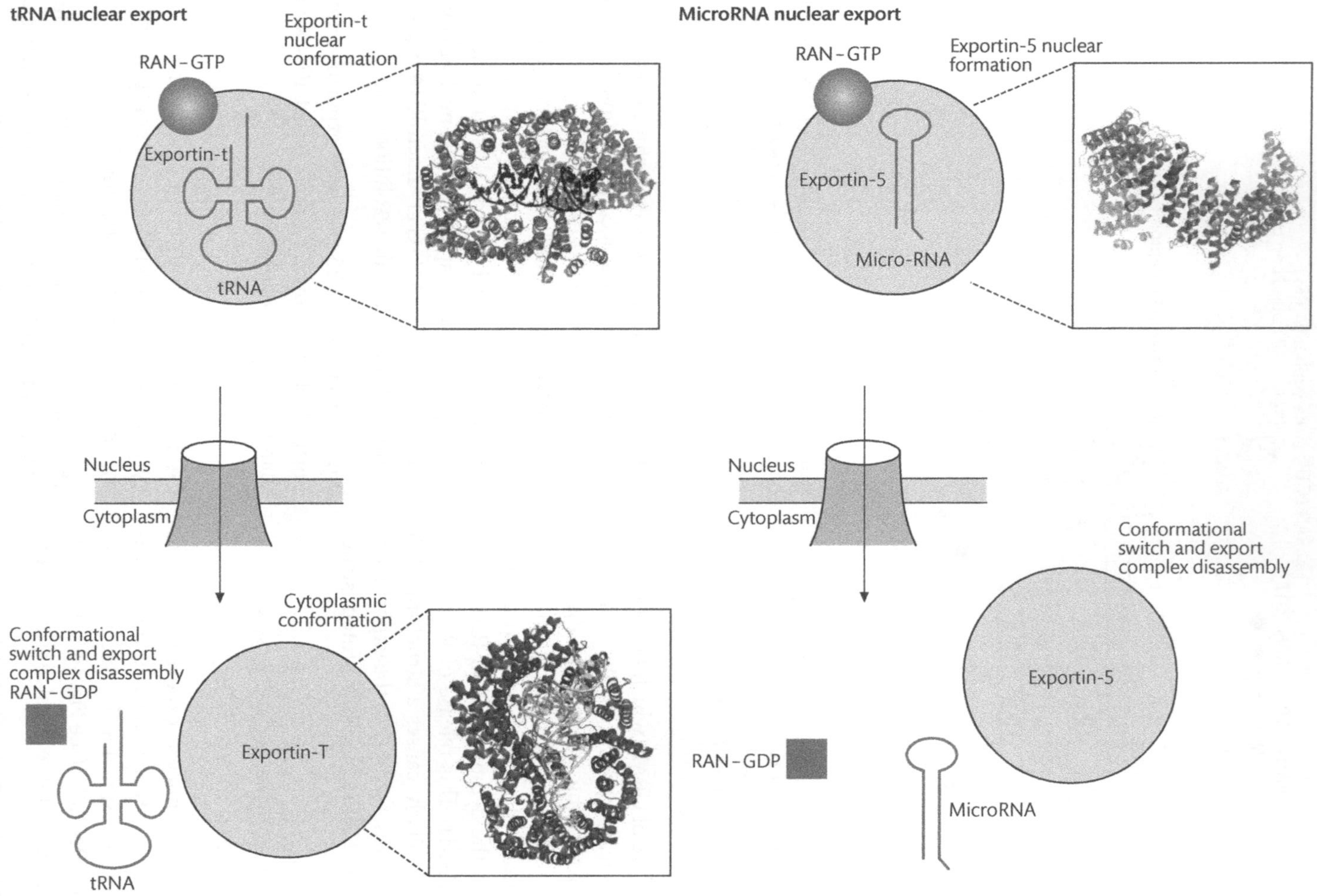

Figure 14.23 The karyopherins exportin-t and exportin-5 export tRNA and microRNAs from the nucleus. Both exportin-t and exportin-5 directly bind RNA in nuclear export complexes (note that this is different from CRM1). Notice the change in shape in exportin-t in the nucleus and in the cytoplasm once the RAN-GTP is converted to RAN-GDP; this releases the cargo RNA. (Protein and RNA structures redrawn from accessions at the RCSB Protein Data Bank (IDs NA0142, 3ICQ, and 3IBV) by Jonathan Crowe, Oxford University Press.)

Instead of a gradual accumulation of RNA export components in the cytoplasm, components of nuclear RNA export complexes are moved back into the nucleus for the next round of RNA export. As an example of this process, Fig. 14.22 shows the import of RAN–GDP back into the nucleus from the cytoplasm ready for another round of export. RAN–GDP is transported by a nuclear import receptor protein called **NTF2**, as part of a nuclear import complex which passes through the nuclear pore back into the nucleus.[75] Once in the nucleus RAN–GDP is converted into RAN–GTP by RCC1 and is released by NTF2. Released NTF2 then returns to the cytoplasm to pick up another RAN–GDP molecule for nuclear import.

Key points

The nuclear export receptors which enable ncRNAs to get through the nuclear pore belong to a group called the karyopherins. Karyopherin molecules are made up of motifs called HEAT repeats. The changes in shape of karyopherins when they bind and release RAN–GTP control the nuclear assembly and cytoplasmic disassembly of RNA export complexes. Karyopherins bind to RAN–GTP in the nucleus, and this causes them to either assemble export complexes within the nucleus or disassemble them in the cytoplasm. Karyopherins do not bind to RAN–GDP, and so can monitor where they are in the cell. Some karyopherins like CRM1 interact with RNA cargo through adaptor molecules with leucine-rich nuclear export sequences. Other karyopherins which transport tRNAs and microRNAs bind directly to cargo RNAs.

14.5.10 Nuclear transport of snRNAs

Up to this point we have considered nuclear *export* of RNA. This involves the transport into the cytoplasm of newly synthesized RNA, i.e. movement in an outward direction. However, sometimes RNA can also traffic in the opposite direction, from the cytoplasm back into the nucleus. A key example of nuclear imported RNAs are the snRNAs—these first move out of the nucleus and then later in the opposite direction from the cytoplasm back into the nucleus. We will now look at the intracellular journeys of snRNAs.

SnRNAs are first transcribed from their parent genes in the nucleus. After transcription, some snRNAs (U1, U2, U4, and U5 snRNAs) are exported into the cytoplasm for processing. Fully processed snRNAs are finally re-imported into the nucleus. Most snRNAs undergo this two-way traffic across the nuclear membrane. U6 snRNA is different in that it always stays in the nucleus. The four major spatial steps as snRNPs move through the cell are shown in Fig. 14.24 and described in detail next.

All but one of the snRNAs are transcribed in the nucleus by RNA polymerase II; the exception is U6 snRNA (see Section 14.5.12). Transcription of the snRNAs by RNA polymerase II takes place in the Cajal body (CB).[76,77] These primary snRNA transcripts have a 5′ cap structure typical of RNA polymerase II derived transcripts containing 7-methyl guanosine (see Chapters 5 and 8).

After transcription in Cajal bodies, snRNAs are exported from the nucleus into the cytoplasm using the CRM1 nuclear export receptor. CRM1 does not bind directly to snRNAs. Instead, an adaptor called PHAX attaches to the cap-binding complex (CBC) at the 5′ end of the snRNA.[78] PHAX contains a leucine-rich NES which is recognized by CRM1. PHAX then interacts with CRM1 to mediate nuclear export of the snRNP cargo. This snRNA nuclear export complex is shaded in Fig. 14.24.

After export snRNAs in the cytoplasm are trimmed and modified as follows.

1 The addition of seven Sm proteins: addition of Sm proteins is carried out by a protein called SMN (discussed in Section 14.4) and enhanced by a second complex called PMRT5 which methylates the Sm proteins. These Sm proteins assemble to form a ring-like structure around the snRNAs—notice this ring-like structure in Fig. 14.24.

2 Trimming the 3′ end of the snRNAs.

3 Addition of an snRNA 5′ cap structure: the final stage of cytoplasmic maturation of snRNAs is hypermethylation, to convert the 7-methyl guanosine cap to a trimethyl guanosine cap structure at their 5′ end (m_1Gppp to m_3Gppp in Fig. 14.24).

Addition of the trimethyl guanosine cap provides a signal for nuclear re-import of snRNPs (since modification of the cap structure is the last step, it indicates that the cytoplasmic processing steps are completed).

4 snRNPs are moved back into the nucleus by the nuclear protein import pathway which also actively targets a number of different proteins to the nucleus in eukaryotic cells.

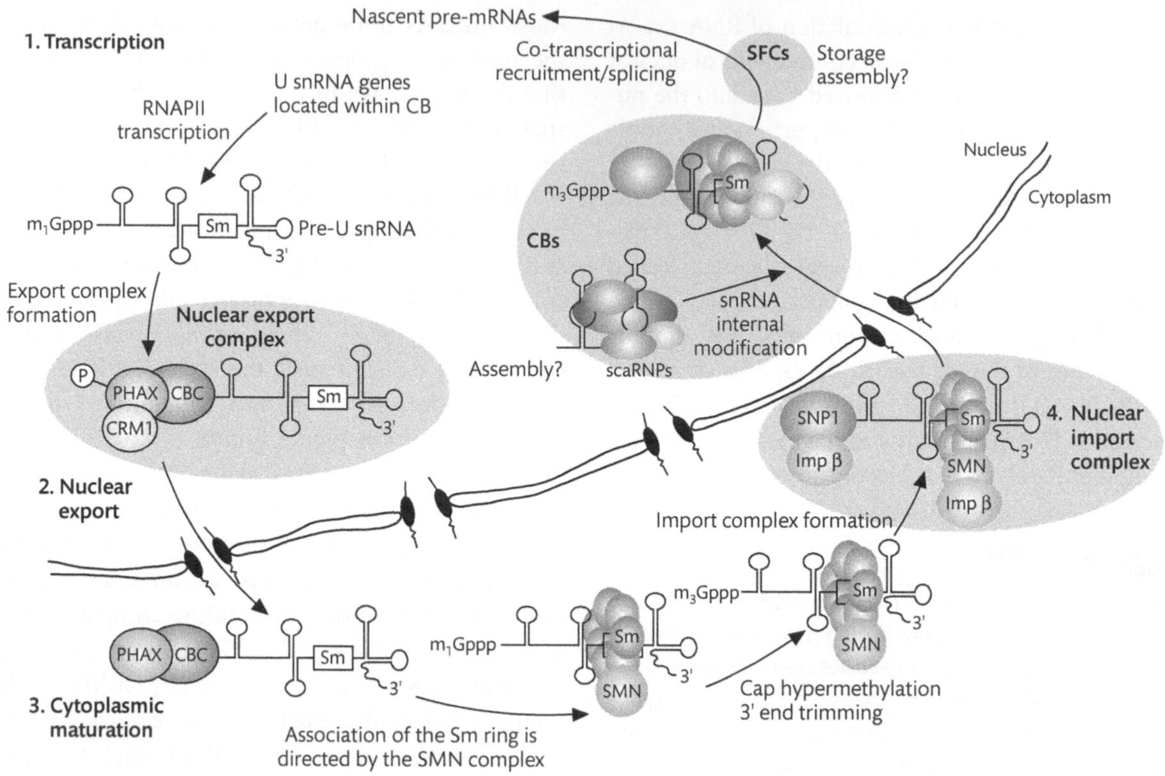

Figure 14.24 Most snRNPs transit between the nucleus and the cytoplasm as they mature. Maturation of snRNPs involves four major steps. (1) Transcription: U1, U2, U4, and U5 snRNAs are first transcribed in the nucleus by RNA polymerase II. This takes place within nuclear structures called Cajal bodies (CBs). (2) Nuclear export: snRNAs are then exported from the nucleus. (3) Cytoplasmic maturation: in the cytoplasm snRNAs associate with a set of seven Sm proteins which form a circle around a sequence called the Sm motif, and the 5′ cap of the snRNA is hypermethylated. (4) Nuclear import: proper cytoplasmic maturation provides a signal for snRNP nuclear import, after which final snRNP assembly and modification takes place at the CB before snRNPs move to sites of storage and splicing. (From Patel and Bellini[79] with permission from Oxford University Press.)

Once back in the nucleus snRNPs move into the Cajal bodies, where they were first transcribed. As an analogy, this return of snRNAs to their site of synthesis has been compared with the return of adult salmon to their original spawning grounds, and probably involves feats of (molecular) navigation of similar complexity. The final steps of snRNP synthesis take place within Cajal bodies; snRNP-specific proteins are added, and the snRNA component of the snRNPs is modified by methylation and pseudouridinylation. These snRNP modifications are mediated by the scaRNPs (see Section 14.1.3).

Since Cajal bodies are the site of both transcription and processing of snRNPs, they are a bit like 'nucleoli for snRNPs', although the transcription and processing of snRNAs in the Cajal body is separated by a cytoplasmic phase. Cajal bodies are conserved in evolution, and are found in plants, flies, and fungi as well as mammals, although not all the component proteins are conserved, particularly coilin; this species restriction of the coilin protein initially made it appear that Cajal bodies were specific to mammals (see Fig. 14.25). Cajal bodies are also dynamic structures, with their size being dependent on the amount of RNA that the cell is making. Cells which have a high metabolic rate, like cancer cells, tend to have large numbers of Cajal bodies.

Cajal bodies are also the sites where snRNP complexes are reassembled. As we discussed in Chapter 5, the splicing reaction involves the disassembly of the U4/U6 di-snRNPs, and these are put back together in the Cajal body by a protein called SART/PRP24. The U4/U6•U5 tri-snRNP, which is added

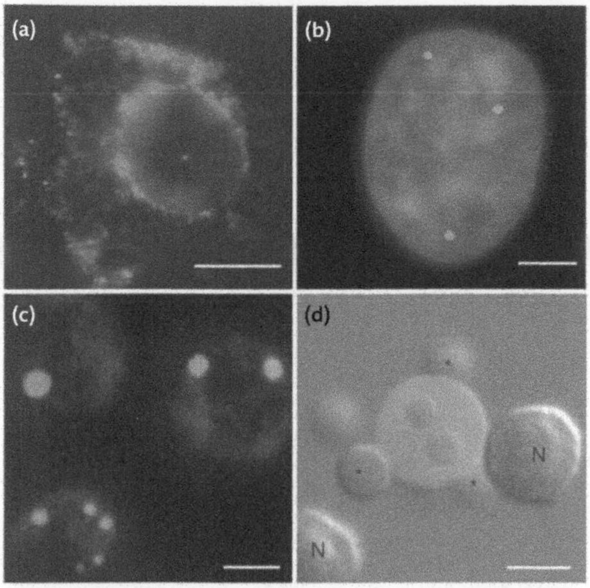

Figure 14.25 Cajal bodies in cells from (a) the fruit fly, (b) humans, (c) root tissue from the pea plant, and (d) *Xenopus*. Note that the Cajal bodies are dot-like structures within the nucleus, but can be very large in *Xenopus* oocytes. For a full colour version of this figure, see Colour Illustration 14. (Reprinted from Matera and Shpargel[76] with permission from Elsevier.)

to a unit of the developing spliceosome, is also assembled in the Cajal body. As well as their role in snRNP assembly, Cajal bodies play a role in the assembly of other nuclear RNPs, including U7 snRNP (required for histone RNA processing) and telomerase—an enzyme which adds telomere repeats onto the end of chromosomes composed of the telomerase RNA, the reverse transcriptase hTERT, and some Sm proteins. During the S phase, telomerase and telomeres are spatially associated at Cajal bodies.

Key points

Most ncRNAs undergo one-way traffic across the nuclear pore from the nucleus to the cytoplasm. However, some of the snRNAs which have critical roles in splice site selection and the spliceosome (see Chapter 5) have a more complex life cycle which includes transit in both directions across the nuclear pore. Fully splicing functional snRNPs are matured and assembled during a fascinating life cycle which involves a journey through different compartments of the cell. This journey is driven by sequential modifications which enable passage to the next destination in the processing pathway.

14.5.11 Mature snRNPs are re-imported into the nucleus using the nuclear protein import machinery

In this section we are going to discuss the nuclear import of the shuttling snRNPs in a bit more detail[69].

Similar to the export of ncRNA which relies on adaptor proteins which contain nuclear export sequences (NESs), the nuclear import of snRNPs relies on a short protein sequence called a **nuclear localization sequence (NLS)** (see Table 14.4 for the amino acid sequences of some NLSs) within an adaptor protein called snurportin (SNUPN). The NLS is recognized by a nuclear import receptor called importin β (see Fig. 14.20). Importin β is a karyopherin which mediates nuclear import rather than export.

The way that snRNPs move back into the nucleus is shown in Fig. 14.26—notice that importin β is shown as an S-shaped molecule. The four stages of nuclear import of snRNPs are as follows.

1 The first step involves the formation of a nuclear import cargo complex which contains (1) the snRNP destined for nuclear import which has acquired a trimethyl guanosine (TMG) cap, (2) the adaptor protein snurportin which binds to the TMG cap of the cargo snRNP, and (3) the nuclear import receptor importin β which binds to the NLS of snurportin.

2 Nuclear import: this nuclear snRNP import cargo complex is transported across the nuclear membrane via importin β, which can interact with the FG repeats in the channel of the nuclear pore. These FG-repeat interactions act like stepping stones, allowing the passage of the cargo complex through the nuclear pore.

3 Disassembly of the import complex. Once inside the nucleus, the nuclear import complex breaks up to release its cargo snRNP. Break-up is triggered by

Table 14.4 Sequences of two classical nuclear localization sequences (NLS) found in the nuclear SV40 T antigen and nucleoplasmin proteins

NLS type	Sequence	Protein
Monopartite	PKKKRRV	SV40 T antigen
Bipartite	KRPAATKKAGQAKKKK	Nucleoplasmin

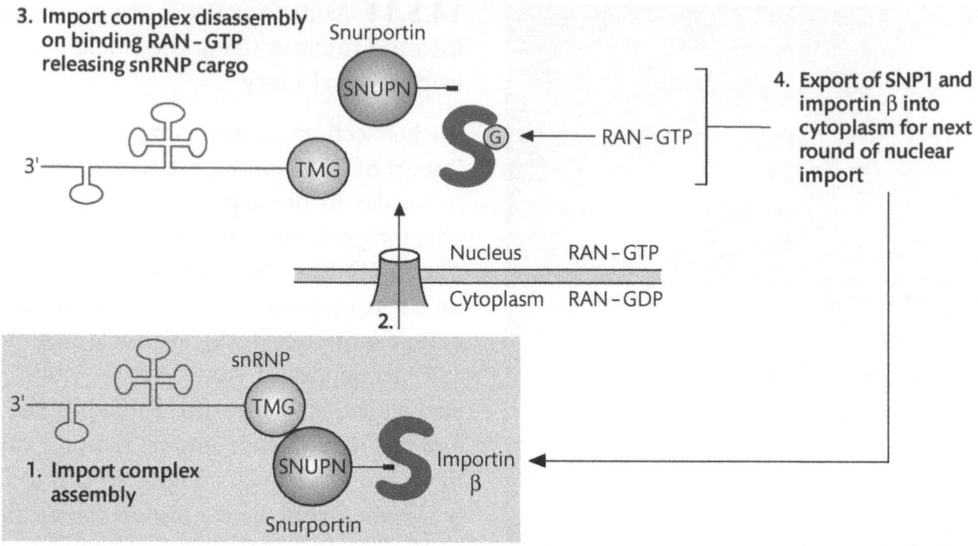

Figure 14.26 The nuclear import of snRNPs follows the route used for the nuclear localization of many nuclear proteins. This nuclear import route involves the use of the importin β karyopherin which recognizes nuclear localization sequences (NLSs). (1) Import complex assembly takes place in the cytoplasm between the snRNP to be imported, the adaptor snurportin (SNUPN) which interacts with the TMG cap of the snRNP, and the nuclear import receptor importin β which recognizes the NLS on snurportin. (2) Nuclear import, complex disassembly, and release of importin β: the cytoplasmic nuclear import complex moves into the nucleus through the nuclear pore. Once in the nucleus, since it is a karyopherin, importin β binds to RAN–GTP and changes shape, disassembling the import complex and releasing the snRNP. (3) Nuclear export of snurportin and importin β: the importin β and snurportin are exported back out into the cytoplasm ready for the next round of nuclear import.

importin β binding to RAN–GTP in the nucleus (since karyopherins bind to RAN–GTP), changing the conformation of importin β and so destabilizing the import complex. This change in conformation causes importin β to let go of the adaptor snurportin and the snRNP.

4 Recycling of import components into the cytoplasm. In the final stage, the import receptor importin β, RAN–GTP, and snurportin are both exported into the cytoplasm ready for a new round of nuclear import.

The pathways which are used to export nuclear RNA are similar to those used to import and export nuclear proteins. Importin β is also used to move other nuclear proteins into the nucleus and is known as the *classical protein import pathway*. As shown in Table 14.4, there are two types of NLS which are involved in moving proteins into the nucleus: a monopartite NLS, which is composed of a single short sequence, and a bipartite NLS, which is composed of two short sequences separated by a linker.

Key points

The karyopherin group also contains important nuclear import receptors responsible for moving proteins *into* the nucleus. Proteins destined for nuclear import contain a specific cellular spatial sorting code called a nuclear import sequence. The best-known nuclear import sequence is a basic sequence called a nuclear localization sequence or NLS. snRNPs which have undergone cytoplasmic processing use this protein import machinery to get into the nucleus.

14.5.12 Mature U6 snRNP is made exclusively in the nucleus

In the previous sections we have considered the cellular pathways followed by the shuttling snRNAs.[79] In this section we will complete our examination of the snRNAs by looking at U6 snRNA. Unlike the other snRNAs, U6 snRNA remains within the nucleus at all stages of its life cycle.

Like the other spliceosomal U snRNAs, U6 is transcribed in the nucleus and associates with proteins to generate U6 snRNP which functions in the spliceosome (see Chapter 5). However, despite this similarity of function, U6 snRNP has four distinct properties in terms of how it is made:

1 Unlike the other splicing snRNPs, U6 snRNP is made and matures exclusively in the nucleus—there is no cytoplasmic shuttling step.

2 U6 snRNA is transcribed by RNA polymerase III—the same polymerase that makes tRNAs (the other splicing snRNAs are transcribed by RNA polymerase II).

3 U6 snRNA does not have the same 5′ cap structure as the other snRNAs. The gamma phosphate of the first nucleotide of U6 snRNA is methylated. The other snRNAs have a 5′ TMG cap.

4 Although it does not go into the cytoplasm, U6 snRNA undergoes processing within the nucleolus before it is ready to work in the spliceosome within the nucleoplasm. During its time in the nucleolus, specific bases of U6 snRNA are modified by a group of nucleolar snoRNPs.

The exclusively nuclear processing pathways followed by U6 snRNA are shown in Fig. 14.27.

1 Transcription. Unlike the other snRNAs, U6 snRNA is transcribed by RNA polymerase III and then stays inside the nucleus. U6 snRNA molecules are bound at the 3′ end by a protein called the La protein—this increases the stability of U6 snRNA.

2 Nucleolar modification. Next the La protein is replaced by LSm proteins which form a ring structure around U6 snRNA (notice this LSm ring structure in Fig. 14.27).[80] Both the Sm and LSm proteins are very ancient in heritage, and are related to proteins which are involved in RNA metabolism in the Archaea and eubacteria. There are two types of modern eukaryotic LSm protein complex: one in the nucleus composed of LSm2–8 (which form a ring structure around U6 snRNA), and one composed of LSm1–7, which play an important role in RNA stability in the cytoplasm. Nucleotides within U6 snRNA are modified by snoRNPs.

3 Assembly of snRNPs within Cajal bodies. U6 becomes assembled into U4/U6, and U4/U6•U5 snRNP complexes ready for splicing.

4 Movement to splicing factor compartments (SFCs) to play a role in splicing. These snRNP complexes move to SFCs ready for use in splicing.

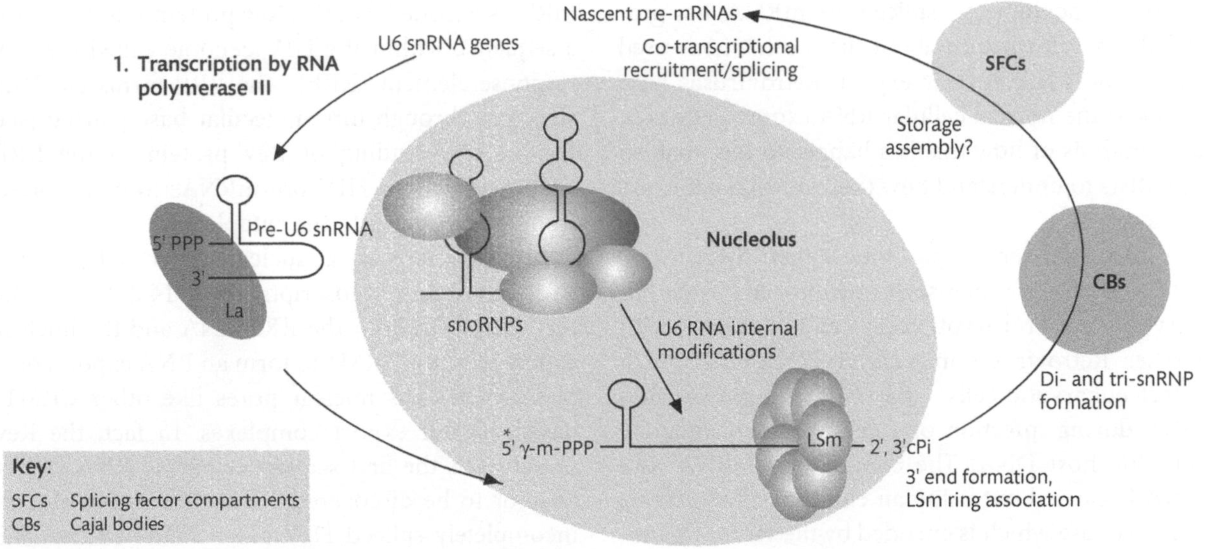

Figure 14.27 An exception to snRNP cytoplasmic maturation: U6 snRNP has an entirely nuclear life cycle. U6 snRNP is transcribed by RNA polymerase III, has a 5′ γ-monomethyl cap and does not leave the nucleus. U6 snRNPs become associated with a ring structure of LSm proteins that are added to the snRNA in the nucleolus. (From Patel and Bellini[21] with permission from Oxford University Press.)

Key points

U6 snRNA is different from the other spliceosomal snRNAs in that it remains in the nucleus. U6 snRNA associates with proteins to produce a U6 snRNP. An important group of U6 snRNP proteins are the LSm proteins, which form a ring around the U6 snRNA.

14.6 Retroviruses have hijacked the RNA export machinery to assist in the export of partially processed mRNAs

In this chapter we have discussed the nuclear export of ncRNAs. The nuclear export of ncRNAs generally follows a different molecular pathway to the nuclear export of mRNAs that we discussed in Chapter 9. However, there is an important group of cellular mRNAs which in fact follow an ncRNA nuclear export pathway. These important mRNAs are virally encoded, and are discussed in this final section of the chapter.

Analysis of the mechanisms which viruses use to export their mRNAs, and how they avoid the normal constraints imposed by the cell to limit the nuclear export of incompletely spliced pre-mRNA, has provided key information about the mechanisms used in normal RNA nuclear export. Retroviruses have hijacked the normal cellular RNA export pathways, and analysis of how this has happened has enabled scientists to understand how these normal pathways work.

Viruses are genetic parasites which infect, replicate, and amplify themselves within cells. Both eukaryotic and prokaryotic cells can be infected with viruses. Retroviruses are a class of RNA virus which infect eukaryotic cells—they copy themselves into DNA during infection of a cell and then integrate into the host DNA. The conversion of RNA into DNA is carried out using an enzyme called reverse transcriptase which is encoded by the virus. This integrated form of the virus is called a **provirus**, and it can remain latent for many years until the infectious cycle is initiated again.

The coding capacity of retroviruses is restricted. The retroviral genome needs to be small because smaller viruses will replicate faster within an infected cell and because the final RNA genome has to fit within a viral particle of finite size. For this reason, retroviruses try to fit maximal coding information into their genomes without increasing genome size. Some retroviruses such as the HIV virus, which causes AIDS, encode multiple transcripts for different proteins, including spliced and unspliced versions of transcripts made from the same gene.

Early in infection fully spliced HIV mRNAs are exported from the nucleus into the cytoplasm where they are translated.[81] At these early stages unspliced HIV pre-mRNA transcripts are restricted to the nucleus by anchoring to splicing components. However, later on in HIV infection something quite different happens; partially and even totally unspliced HIV RNAs are exported from the nucleus into the cytoplasm. These unspliced or partially spliced transcripts encode viral structural proteins and RNA copies of the full-length HIV genome. Nuclear export of unspliced viral transcripts is possible since HIV hijacks the normal RNA export machinery.

A key protein in the hijacking process is called Rev; this is translated from one of the early spliced mRNAs encoded by HIV. Rev protein interacts with a sequence within the HIV genome called the Rev response element (RRE). The RRE forms an RNA structure through intramolecular base pairing (see Fig. 14.28). Binding of Rev protein to the RRE enables unspliced HIV pre-mRNAs to be exported from the nucleus into the cytoplasm.

Rev functions as a nuclear export adaptor for RRE-containing transcripts (Fig. 14.28). Rev directly binds to both the RRE RNA and the nuclear export receptor CRM1 to form an RNA export complex which exits nuclear pores like other CRM1-based nuclear export complexes. In fact, the Rev protein was the first sequence-specific RNA export adaptor to be discovered. Hence the study of how incompletely spliced HIV pre-mRNAs entered the cytoplasm of infected cells provided important information about how RNAs are normally exported from the nucleus.

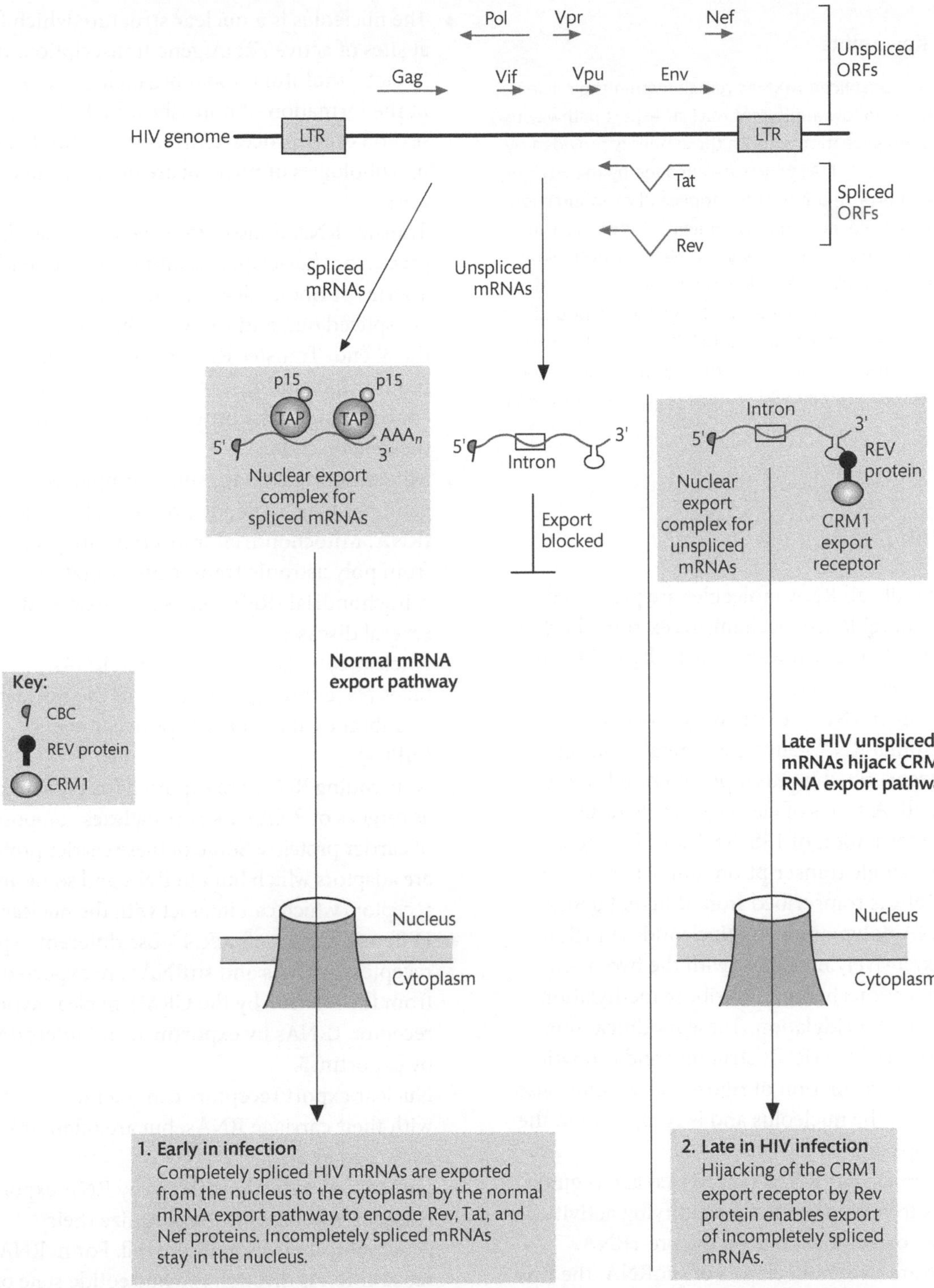

Figure 14.28 The HIV life cycle is critically dependent on the export of unspliced pre-mRNA mediated by Rev protein. The integrated HIV provirus encodes two classes of protein, spliced early transcripts (*Tat*, *Rev*, and *Nef*) and incompletely spliced late transcripts (encoding the viral genome and viral structural proteins). For export these unspliced transcripts require Rev protein binding to a structured RNA sequence called the RRE. The Rev protein binds to both the RRE and the CRM1 through its leucine-rich nuclear export sequence. In doing this, HIV hijacks the nuclear RNA export pathway used by snRNAs and rRNAs to enable nuclear export of unspliced pre-mRNAs.

Key points

In general, unspliced mRNAs remain within the nucleus, and mRNAs follow a different nuclear export pathway to ncRNAs. An exception to both these rules is provided by unspliced pre-mRNAs which are encoded by the HIV genome and exported into the cytoplasm late in infection. These unspliced pre-mRNAs contain a structured RNA sequence called the Rev response element (RRE) which binds a protein called Rev. Rev is a nuclear export adaptor, and contains a leucine-rich NES which is bound by the nuclear export receptor CRM1. This export complex leaves the nucleus and enters the cytoplasm to encode viral structural proteins which are only needed late in viral infection.

Summary

- Essentially all RNA molecules are processed. RNA surveillance mechanisms exist to check the quality of processing and any faulty RNAs are actively eliminated.
- Ribosomal RNA is extensively processed. Ribosomal RNA genes are generally clustered and transcribed by RNA polymerase I in the nucleoli. A series of cleavage events results in the formation of 18S, 5.8S, and 28S rRNA from a single transcription unit, whereas the 5S rRNA is transcribed from different genes by RNA polymerase III. Nucleotides in rRNA are extensively modified, with the two main modifications being 2′-*O*-ribose methylation and pseudouridylation. These modifications are essential for rRNA structure and catalytic function. Formation of ribosomal subunits also occurs in the nucleolus and is completed in the nucleoplasm.
- Small nucleolar RNAs (snoRNAs) act as guide RNAs to recruit the RNA-modifying activities to the correct locations on the pre-rRNA. There are two main classes of snoRNA: the Box C/D snoRNAs are responsible for 2′-*O*-ribose methylation, and the Box H/ACA snoRNAs are responsible for pseudouridylation.
- Similar small RNAs, the scaRNAs, are concentrated in the Cajal bodies where they are involved in the modification of snRNAs.
- The nucleolus is a nuclear structure which forms at sites of active rRNA gene transcription. It is in fact a multifunctional organelle, also involved in the formation of non-ribosomal RNP and in several other processes. Abnormal numbers and morphologies of nucleoli are observed in cancer cells.
- Transfer RNA is also extensively processed. The precursor transcript is trimmed at both ends by a series of ribonucleases, intervening sequences are spliced out, and CCA residues are added to the 3′ end. Transfer RNA nucleotides are also modified, including those in the anticodon loop contributing to the process of wobble in reading the genetic code.
- Mitochondria contain small compacted genomes which encode proteins, rRNA, and tRNA. Mitochondrial transcripts are processed from polycistronic transcripts. Mutations in mitochondrial tRNA genes are associated with several diseases.
- Nuclear cytoplasmic traffic of ncRNAs takes place through apertures in the nuclear membrane called nuclear pore complexes (NPCs).
- Non-coding RNAs are exported from the nucleus as cargoes of RNA export complexes composed of carrier proteins. Some of these carrier proteins are adaptors which bind to RNA and some are receptors which can interact with the nuclear pore.
- Different classes of ncRNA use different export receptors. rRNAs and snRNAs are exported from the nucleus by the CRM1 nuclear export receptor, tRNAs by exportin-t, and microRNAs by exportin-5.
- Nuclear export receptors can interact directly with their carriage RNAs, but are often linked via export adaptors.
- Directional traffic is achieved by RNA export complexes being able to recognize their geographical position in the cell. For ncRNAs this recognition is through the nucleotide state of the protein RAN which is an inefficient GTPase. In the nucleus RAN is in the GTP-bound form, and in the cytoplasm RAN is in the GDP-bound form.
- The continual nuclear re-import of export complex components is critical for continued nuclear export of RNAs.

- ncRNAs follow nuclear export and import pathways which are a branch of those used for the nuclear export and import of proteins.

Questions

14.1 What are the three stages of rRNA synthesis?
14.2 What does the U3 snoRNP do?
14.3 What are the principal chemical modifications of rRNA?
14.4 How do snoRNAs find their targets?
14.5 What is the function of scaRNAs?
14.6 What is the structure of a nucleolus?
14.7 What else, other than rRNA transcription and processing, do nucleoli do?
14.8 What are the critical steps in pre-tRNA processing?
14.9 List the key enzymes involved in pre-tRNA processing and their function.
14.10 How are mitochondrial pre-tRNAs processed?
14.11 How does the GTPase RAN work to facilitate nuclear transport?
14.12 How do nuclear export adaptors and receptors work?
14.13 How do karyopherins interact with RAN–GTP?
14.14 Describe the complex processing pathway of the snRNAs.
14.15 What is the function of a Cajal body?
14.16 How does the biogenesis of U6 snRNA differ from that of other snRNAs?
14.17 How and why do retroviruses hijack the RNA export machinery?

References

1. **Henras AK, Soudet J, Gérus M, et al.** The post-transcriptional steps of eukaryotic ribosome biogenesis. *Cell Mol Life Sci* **65**, 2334–59 (2008).
2. **Tshochner T, Hurt E.** Pre-ribosomes on the road from the nucleolus to the cytoplasm. *Trends Cell Biol* **13**, 255–63 (2003).
3. **Yip WS, Vincent NG, Baserga SJ.** Ribonucleoproteins in archaeal pre-rRNA processing and modification. *Archaea* 614735 (2013).
4. **Bachellerie J-P, Cavaille J, Huttenhofer A.** The expanding snoRNA world. *Biochimie* **84**, 775–90 (2002).
5. **Bratkovič T, Rogelj B.** The many faces of small nucleolar RNAs. *Biochim Biophys Acta* **1839**, 438–43 (2014).
6. **Dragon F, Gallagher JE, Compagnone-Post PA, et al.** A large nucleolar U3 ribonucleoprotein required for 18S ribosomal RNA biogenesis. *Nature* **417**, 967–70 (2002).
7. **Sloan KE, Bohnsack MT, Schneider C, Watkins NJ.** The roles of SSU processome components and surveillance factors in the initial processing of human ribosomal RNA. *RNA* **20**, 540–50 (2014).
8. **Choesmel V, Bacqeville D, Rouquette J, et al.** Impaired ribosome biogenesis in Diamond-Blackfan anemia. *Blood* **109**, 1275–83 (2007).
9. **Reichow SL, Hamma T, Ferré-D'Amaré AR, Varani G.** The structure and function of small nucleolar ribonucleoproteins. *Nucleic Acids Res* **35**, 1452–64 (2007).
10. **Decatur WA, Fournier MJ.** RNA-guided nucleotide modification of ribosomal and other RNAs. *J Biol Chem* **278**, 695–8 (2003).
11. **Richard P, Kiss T.** Integrating snoRNP assembly with ribosome biogenesis. *EMBO Rep* **7**, 590–2 (2006).
12. **Vitali P, Basyuk E, Le Meur E, et al.** ADAR-2 mediated editing of RNA substrates in the nucleolus is inhibited by C/D small nucleolar RNAs. *J Cell Biol* **169**, 745–53 (2005).
13. **Kishore S, Stamm S.** The snoRNA HBII-52 regulates alternative splicing of the serotonin receptor 2C. *Science* **311**, 230–2 (2006).
14. **Omer AD, Ziesche S, Decatur WA, et al.** RNA-modifying machines in Archaea. *Mol Microbiol* **48**, 617–29 (2003).
15. **Mannoor K, Liao J, Jiang F.** Small nucleolar RNAs in cancer. *Biochim Biophys Acta* **1826**, 121–8 (2012).
16. **Jády BE, Kiss T.** A small nucleolar guide RNA functions both in 2′-*O*-ribose methylation and pseudouridylation of the U5 spliceosomal RNA. *EMBO J* **3**, 541–51 (2001).
17. **Richard P, Darzacq X, Bertrand E, et al.** A common sequence motif determines the Cajal body-specific localisation of box H/ACA scaRNAs. *EMBO J* **22**, 4283–93 (2003).
18. **Zhang Q, Kim NK, Feigon J.** Architecture of human telomerase RNA. *Proc Natl Acad Sci USA* **108**, 20 325–32 (2011).
19. **Trahan C, Dragon F.** Dyskeratosis congenita mutations in the H/ACA domain of human telomerase RNA affect its assembly into a pre-RNP. *RNA* **15**, 235–43 (2009).
20. **Jády BE, Bertrand E, Kiss T.** Human telomerase RNA and box H/ACA scaRNAs share a common Cajal body-specific localisation signal. *J Cell Biol* **164**, 647–52 (2004).
21. **Henandez-Verdun, D.** The nucleolus: a model for the organisation of nuclear functions. *Histochem Cell Biol* **126**, 135–48 (2006).
22. **Raska I, Shaw PI, Cmarko D.** Structure and function of the nucleolus in the spotlight. *Curr Opin Cell Biol* **18**, 325–34 (2006).

23. **Lo SJ, Lee C-C., Lai H-J.** The nucleolus: reviewing oldies to have new understandings. *Cell Res* **16**, 530–8 (2006).
24. **Brown JW, Shaw PJ.** The role of the plant nucleolus in pre-mRNA processing. *Curr Top Microbiol Immunol* **326**, 291–311 (2008).
25. **Gerbim SA, Borovjagin AV, Lange TS.** The nucleolus: a site of ribonucleoprotein maturation. *Curr Opin Cell Biol* **15**, 318–25 (2003).
26. **Boisvert F-M, van Koningsbruggen S, Navascués J, Lamond AI.** The multifunctional nucleolus. *Nat Rev Mol Cell Biol* **8**, 574–85 (2008).
27. **Pederson T.** Proteomics of the nucleolus: more proteins, more functions? *Trends Biochem Sci* **27**, 111–12 (2002).
28. **Couté Y, Burgess JA, Diaz JJ, et al.** Deciphering the human nucleolar proteome. *Mass Spectrom Rev* **25**, 215–34 (2006).
29. **Andersen JS, Lam YW, Leung AK, et al.** Nucleolar proteome dynamics. *Nature* **433**, 77–83 (2005).
30. **Pendle AF, Clark GP, Boon AR, et al.** Proteomic analysis of the *Arabidopsis* nucleolus suggests novel nuclear functions. *Mol Biol Cell* **16**, 260–9 (2005).
31. **Quin JE, Devlin JR, Cameron D, et al.** Targeting the nucleolus for cancer intervention. *Biochim Biophys Acta* **1842**, 802–16 (2014).
32. **Bywater MJ, Poortinga G, Sanij E, et al.** Inhibition of RNA polymerase I as a therapeutic strategy to promote cancer-specific activation of p53. *Cancer Cell* **22**, 51–65 (2012).
33. **Deisenroth C, Zhang Y.** Ribosome biogenesis surveillance: probing the ribosomal protein-Mdm2-p53 pathway. *Oncogene* **29**, 4253–60 (2010).
34. **Agris PF, Vendeix FA, Graham WD.** tRNA's wobble decoding of the genome: 40 years of modification. *J Mol Biol* **366**, 1–13 (2006).
35. **Nakanishi K, Nureki O.** Recent progress of structural biology of tRNA processing and modification. *Mol Cells* **19**, 157–66 (2005).
36. **Morl M, Marchfelder A.** The final cut. The importance of tRNA processing. *EMBO Rep* **2**, 17–20 (2001).
37. **Kazantsev AV, Pace NR.** Bacterial RNase P: a new view of an ancient enzyme. *Nat Rev Microbiol* **4**, 729–40 (2006).
38. **Jarrous N, Reiner R.** Human RNAse P: a tRNA-processing enzyme and transcription factor. *Nucleic Acids Res* **35**, 3519–24 (2007).
39. **Andersen KR, Heick Jensen T, Brodersen DE.** Take the 'A' tail—quality control of ribosomal and transfer RNA. *Biochim Biophys Acta* **1779**, 532–7 (2008).
40. **Vaňáčová S, Wolf J, Martin G, et al.** A new yeast poly(A) polymerase complex involved in RNA quality control. *PLoS Biol* **3**, e189 (2005).
41. **Kramer EB, Hopper AK.** Retrograde transfer RNA nuclear import provides a new level of tRNA quality control in *Saccharomyces cerevisiae*. *Proc Natl Acad Sci USA* **110**, 21 042–47 (2013).
42. **Guerrier-Takada C, Gardiner K, Marsh T, et al.** The RNA moiety of ribonuclease P is the catalytic subunit of the enzyme. *Cell* **35**, 849–57 (1983).
43. **Spath B, Canino G, Marchfelder A.** tRNase Z: the end is not in sight. *Cell Mol Life Sci* **64**, 2404–12 (2007).
44. **Parwez T, Kushner SR.** RNase Z in *Escherichia coli* plays a significant role in mRNA decay. *Mol Microbiol* **60**, 723–37 (2007).
45. **Xiong Y, Steitz TA.** A story with a good ending: tRNA 3′-end maturation by CCA-adding enzymes. *Curr Opin Struct Biol* **16**, 12–17 (2006).
46. **Montoya J, Lopez-Perez MJ, Ruiz-Persini E.** Mitochondrial DNA transcription and diseases: past, present and future. *Biochim Biophys Acta* **1757**, 1179–89 (2007).
47. **Kelley SO, Steinberg SV, Schimmel P.** Functional defects of pathogenic human mitochondrial tRNAs related to structural fragility. *Nat Struct Biol* **7**, 862–5 (2000).
48. **Abbott JA, Francklyn CS, Robey-Bond SM.** Transfer RNA and human disease. *Front Genet* **5**, 158 (2014).
49. **Paushkin S, Gubitz AK, Massenet S, Dreyfuss G.** The SMN complex, an assemblyosome of ribonucleoproteins. *Curr Opin Cell Biol* **14**, 305–12 (2002).
50. **Eggert C, Chari A, Laggerbauer B, Fischer U.** Spinal muscular atrophy: the RNP connection. *Trends Mol Med* **12**, 113–21 (2006).
51. **Wirth B, Brichta L, Hahnen E.** Spinal muscular atrophy: from gene to therapy. *Semin Pediatr Neurol* **13**, 121–31 (2006).
52. **Khoo B, Akker SA, Chew SL.** Putting some spine into alternative splicing. *Trends Biotechnol* **21**, 328–30 (2003).
53. **Lorson CL, Hahnen E, Androphy EJ, Wirth B.** A single nucleotide in the SMN gene regulates splicing and is responsible for spinal muscular atrophy. *Proc Natl Acad Sci USA* **96**, 6307–11 (1999).
54. **Hofmann Y, Lorson CL, Stamm S, et al.** Htra2-beta 1 stimulates an exonic splicing enhancer and can restore full-length SMN expression to survival motor neuron 2 (SMN2). *Proc Natl Acad Sci USA* **97**, 9618–23 (2000).
55. **Kashima T, Manley JL.** A negative element in SMN2 exon 7 inhibits splicing in spinal muscular atrophy. *Nat Genet* **34**, 460–3 (2003).
56. **Cartegni L, Hastings ML, Calarco JA, et al.** Determinants of exon 7 splicing in the spinal muscular atrophy genes, SMN1 and SMN2. *Am J Hum Genet* **78**, 63–77 (2006).
57. **Kashima T, Rao N, David CJ, Manley JL.** hnRNP A1 functions with specificity in repression of SMN2 exon 7 splicing. *Hum Mol Genet* **16**, 3149–59 (2007).
58. **Cartegni L, Krainer AR.** Correction of disease-associated exon skipping by synthetic exon-specific activators. *Nat Struct Biol* **10**, 120–5 (2003).
59. **Skordis LA, Dunckley MG, Yue B, et al.** Bifunctional antisense oligonucleotides provide a *trans*-acting

splicing enhancer that stimulates SMN2 gene expression in patient fibroblasts. *Proc Natl Acad Sci USA* **100**, 4114–19 (2003).
60. **Hua Y, Vickers TA, Baker BF, et al.** Enhancement of SMN2 exon 7 inclusion by antisense oligonucleotides targeting the exon. *PLoS Biol* **5**, e73 (2007).
61. **Hua Y, Vickers TA, Okunola HL, et al.** Antisense masking of an hnRNP A1/A2 intronic splicing silencer corrects SMN2 splicing in transgenic mice. *Am J Hum Genet* **82**, 834–48 (2008).
62. **Cullen BR.** Nuclear RNA export. *J Cell Sci* **116**, 587–97 (2003).
63. **Kohler A, Hurt E.** Exporting RNA from the nucleus to the cytoplasm. *Nat Rev Mol Cell Biol* **8**, 761–73 (2007).
64. **Rodriguez MS, Dargemont C, Stutz F.** Nuclear export of RNA. *Biol Cell* **96**, 639–55 (2004).
65. **Dasso M.** Ran at kinetochores. *Biochem Soc Trans* **34**, 711–15 (2006).
66. **Stewart M.** Structural biology. Nuclear trafficking. *Science* **302**, 1513–14 (2003).
67. **Stewart M.** Molecular mechanism of the nuclear protein import cycle. *Nat Rev Mol Cell Biol* **8**, 195–208 (2007).
68. **Petosa C, Schoehn G, Askiaer P, et al.** Architecture of CRM1/Exportin1 suggests how cooperativity is achieved during formation of a nuclear export complex. *Mol Cell* **16**, 761–75 (2004)
69. **Debler EW, Blobel G, Hoelz A.** Nuclear transport comes full circle. *Nat Struct Mol Biol* **16**, 457–9 (2009).
70. **Dong X, Biswas A, Chook YM.** Structural basis for assembly and disassembly of the CRM1 nuclear export complex. *Nat Struct Mol Biol* **16**, 558–60 (2009).
71. **Dong X, Biswas A, Süel KE, et al.** Structural basis for leucine-rich nuclear export signal recognition by CRM1. *Nature* **458**, 1136–41 (2009).
72. **Cook AG, Fukuhara N, Jinek M, Conti E.** Structures of the tRNA export factor in the nuclear and cytosolic states. *Nature* **461**, 60–5 (2009).
73. **Bohnsack MT, Czaplinski K, Gorlich D.** Exportin 5 is a RanGTP-dependent dsRNA-binding protein that mediates nuclear export of pre-miRNAs. *RNA* **10**, 185–91 (2004).
74. **Okada C, Yamashita E, Lee SJ, et al.** A high-resolution structure of the pre-microRNA nuclear export machinery. *Science* **326**, 1275–9 (2009).
75. **Lange A, Mills RE, Lange CJ, et al.** Classical nuclear localization signals: definition, function, and interaction with importin alpha. *J Biol Chem* **282**, 5101–5 (2007).
76. **Matera AG, Shpargel KB.** Pumping RNA: nuclear bodybuilding along the RNP pipeline. *Curr Opin Cell Biol* **18**, 317–24 (2006).
77. **Stanek D, Neugebauer KM.** The Cajal body: a meeting place for spliceosomal snRNPs in the nuclear maze. *Chromosoma* **115**, 343–54 (2006).
78. **Ohno M, Segref A, Bachi A, et al.** PHAX, a mediator of U snRNA nuclear export whose activity is regulated by phosphorylation. *Cell* **101**, 187–98 (2000).
79. **Patel SB, Bellini M.** The assembly of a spliceosomal small nuclear ribonucleoprotein particle. *Nucleic Acids Res* **36**, 6482–93 (2008).
80. **Beggs JD.** LSm proteins and RNA processing. *Biochem Soc Trans* **33**, 433–8 (2005).
81. **Leblanc J, Weil J, Beemon K.** Posttranscriptional regulation of retroviral gene expression: primary RNA transcripts play three roles as pre-mRNA, mRNA, and genomic RNA. *Wiley Interdiscip Rev RNA* **4**, 567–80 (2013).

15 The 'macro' RNAs: long non-coding RNAs and epigenetics

Introduction

The topic of this chapter is long non-coding RNAs (ncRNAs). There are between 5000 and 15 000 individual long ncRNAs transcribed from separate genes.[1] As a group long ncRNAs are transcribed by RNA polymerase II and are often spliced to give the final ncRNA. However, unlike mRNAs, long ncRNAs do not have any open reading frames to encode proteins.[1] An important defining feature of this group of ncRNAs is their length, which is >200 nucleotides. This distinguishes the long ncRNAs from the short ncRNAs that are the topic of Chapter 16. This chapter focuses almost entirely on the long class of ncRNAs—this group is referred to as macroRNAs in the chapter title to distinguish them from the shorter microRNAs.

Long ncRNAs are functionally diverse, but many have roles in controlling gene expression. The long ncRNAs we discuss in this chapter are involved in epigenetic regulation of gene expression. In Chapter 17 we go on to discuss transcribed pseudogenes and circular RNAs.

A broad group of ncRNAs are involved in the epigenetic control of gene expression. These include the ncRNA *AIR*, which is 108kb long, the *XIST* RNA that is 17kb long, and the *H19* ncRNA which is just over 2kb long. Important groups of shorter ncRNAs also have partially overlapping functional roles in directing epigenetics—these include the siRNAs and rasiRNA (RITS complex-associated siRNA) that we discuss in the next chapter. Because of this overlap in epigenetic regulation, some of the important concepts relevant for the next chapter are introduced in the next section.

15.1 Epigenetic regulation and the epigenetic code

The first group of long ncRNAs we are going to discuss in this chapter are involved in epigenetic regulation of gene expression. However, before we discuss the specific roles of RNA molecules in controlling gene expression through epigenetic mechanisms, we shall say a few words about what epigenetics means.

The DNA within a cell nucleus does not exist as a 'naked' double helix, but instead is packaged with proteins to form a DNA–protein complex called **chromatin**. Packaging of the DNA double helix in chromatin has important implications for transcriptional control. In chromatin, DNA is wound around disc-like protein complexes called nucleosomes. The structure of a nucleosome is shown in Fig. 15.1. Notice that each nucleosome is assembled from four basic histone proteins called H2A, H2B, H3, and H4, with two copies of each histone within a nucleosome. The histone proteins and the DNA double helix are the basic components of chromatin which are targeted by epigenetic processes to control gene expression.

Epigenetic gene regulation works by controlling DNA storage within the nucleus and how easily DNA is accessed by RNA polymerase to be transcribed into RNA (see Box 15.1). There are two types of chromatin. More loosely packed DNA which is available for transcription is called **euchromatin**, and tightly packed and transcriptionally silent DNA is called **heterochromatin**.

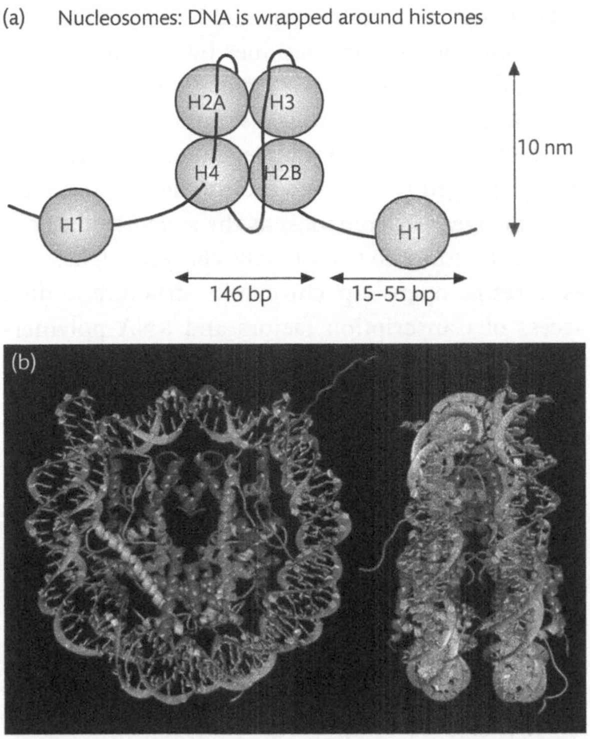

Figure 15.1 **Nuclear DNA is wrapped around nucleosomes in chromatin.** (a) A cartoon of a nucleosome. Each nucleosome is made up of two copies of each individual histone H2A, H2B, H3, and H4 which group together to form a histone octamer. Just under two turns of DNA are wrapped around a nucleosome. Notice that the DNA is on the outside of the nucleosome, not buried within it. A fifth histone, H1, binds to the linker sequence between histones. (b) A molecular model of a nucleosome with DNA wound around it. Within a nucleosome, DNA and nucleosomes are held together by ionic interactions between the positively charged histone proteins and the negatively charged phosphodiester backbone of the DNA. Therefore nucleosomal packaging of DNA is independent of the DNA sequence. Individual nucleosomes are joined together by linker regions of DNA giving them a 'beads on a string' appearance. Histone H1 binds to the linker regions between nucleosomes and helps to pack adjacent nucleosomes together to form a coiled structure called the *30nm fibre*. These supercoiled 30nm fibres result in the very tight packing of DNA within the nucleus. (Part (b) reprinted from Luger et al.[2] with permission from Macmillan Publishers Ltd.)

One conceptual way of thinking about storage of DNA in the cell as either euchromatin or heterochromatin is to compare this to a filing cabinet in an office (see Fig. 15.2). To keep the office tidy and working efficiently most unused files will be closely packed together within a filing cabinet until they are needed. However, some files which are actively in use will lie open on a desk, taking up proportionally

Box 15.1 Chromatin has two important functions in the cell

1. **DNA storage.** Genomes are split up into individual molecules of DNA called chromosomes. The physical length of each chromosome can exceed the length of the actual cells in which they reside. For example, the DNA in a typical human chromosome is 5cm long, but needs to be packed into a considerably smaller cell. Storage of DNA in chromatin enables this packaging to happen.
2. **Gene expression.** The second function of chromatin is to provide a mechanism to regulate transcription. Histones are designed so that the DNA wrapped around nucleosomes can be tightly packed together (in transcriptionally inactive heterochromatin) or much more loosely packed (in transcriptionally active euchromatin). The tightness of packing affects how easily the DNA can be expressed, and is dynamically regulated through chemical modification of histones to move chromatin from being inactive to active and vice versa.

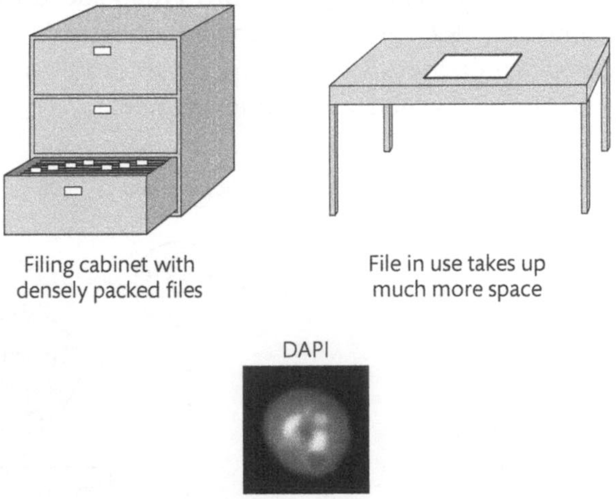

Figure 15.2 **Epigenetics provides a flexible storage system for DNA between actively transcribed and stored sequences.** The nucleus is used to store 99.5% of the cell's DNA within an organized 'filing system', conceptually analogous to the type used in offices. According to this analogy, transcriptionally active DNA will correspond to files which have been selected and are in use. Transcriptionally inactive DNA is archived in filing cabinets, but can be retrieved into an active folder when necessary. Notice that the files within the filing cabinet are stored a lot more densely than the open file on the desk. In the cell nucleus very tightly packed DNA is called heterochromatin, and is transcriptionally inactive. Less condensed DNA is more accessible to the transcriptional machinery, and is called euchromatin. The micrograph shows a mouse cell stained for DNA using the dye DAPI. Note that some regions of the nucleus stain very strongly since they contain high concentrations of very condensed DNA in heterochromatin. (Micrograph reprinted from Peters et al.[3] with permission from Elsevier.)

more space. In this analogy, the carefully stored files are equivalent to the heterochromatin in the cell nucleus, and the actively used files are the euchromatin. More tightly packaged chromosomal DNA (heterochromatin) is not so readily accessed by RNA polymerases, and so is less transcribed. And, just as files in an office can be moved between being active and being in storage, so DNA can move between being euchromatin or heterochromatin. More tightly packed chromatin stains stronger with dyes like DAPI that stain DNA (see Fig. 15.2).

Histone proteins are an important target for epigenetic modification. Each histone protein consists of a globular region and a tail. It is the histone tail which is modified epigenetically. In Fig. 15.3 histones H3 and H4 are shown, indicating some of the amino acid modifications which either activate (above the histones) or repress (below the histones) transcription. The symbol for lysine is K; hence mH3K9 means histone 3 methylated on lysine residue 9. Histone H3 has a trimethyl group added to amino acid K27 and is subsequently bound by a group of proteins called the *polycomb repressive complex 2* (PRC2) which repress transcription.

Histones can also be activated by the addition of acetyl groups to lysine residues by a group of proteins called histone acetyl transferases (HATs), which leads to the activation of transcription through two mechanisms. First, histone acetylation converts positively charged lysine residues into neutral residues. This change in charge loosens the general affinity of the nucleosome for negatively charged DNA, and as a result opens up chromatin structure, aiding access of transcription factors and RNA polymerase. Secondly, histone acetylation provides a binding site for proteins containing a bromodomain. Bromodomains are found in many transcriptional activator proteins, including transcription factors needed for transcriptional initiation. While euchromatin is acetylated by HATs, the histones in heterochromatin are deacetylated by **histone deacetylases** (HDACs). Histone deacetylation leads to transcriptional repression.

DNA can also be modified epigenetically by addition of methyl groups to cytosine residues. DNA methylation takes place on cytidine residues immediately adjacent to guanosine residues, and so is called CpG methylation.[4] The correlation between

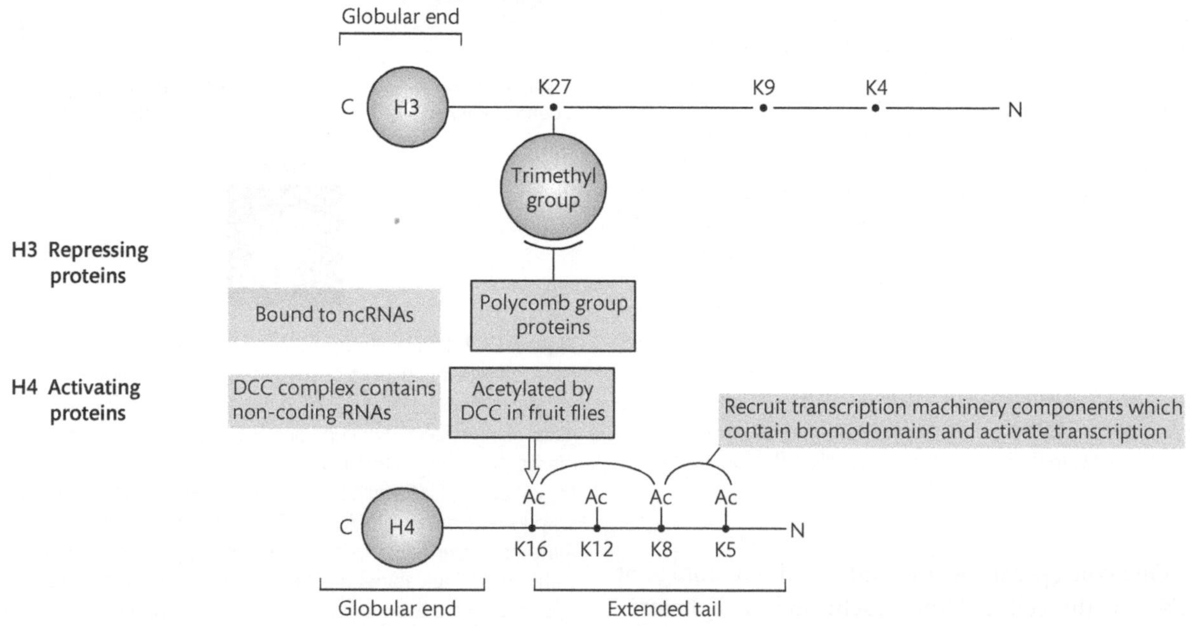

Figure 15.3 Some important histone modification patterns in histones H3 and H4 that are controlled by ncRNAs. Notice that the region of the histone proteins that is modified is the extended N-terminal tail. This is the region that projects from the chromosome in Fig. 15.1. Histone H3 can be modified by the addition of a trimethyl group to amino acid K27 by an important complex called the polycomb repressive complex 2 (PRC2). PRC2 contains both histone methyl transferases (HMTs) and proteins that bind to methylated histones and repress transcription. Histone H4 is acetylated during dosage compensation in male flies, described in Section 15.6.

histone modification and the effect on gene expression is captured by what is known as the **histone code**. The histone code deals specifically with modifications to histones, while the epigenetic code also covers cytosine modifications of DNA.

The key features of euchromatin and heterochromatin described by the epigenetic code are shown in Fig. 15.4. Notice that heterochromatin is very highly compacted, and contains both methylated DNA and deacetylated histones. This highly compacted DNA is illustrated by the more compact zigzag pattern. On the other hand, euchromatin is very loosely packed, with non-methylated DNA wrapped around nucleosomes with acetylated histones (notice the linear extended representation of euchromatin in Fig. 15.4).

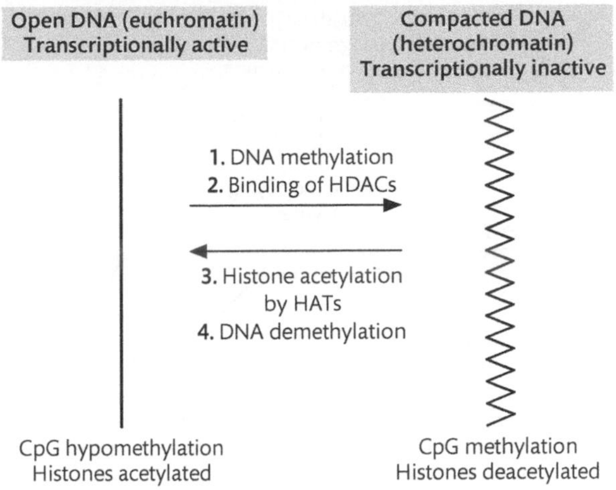

Figure 15.4 A set of chemical modifications to chromatin, called the epigenetic code, provides a switch between transcriptionally active and inactive chromatin. DNA in chromatin can be loosely packaged in euchromatin, or much more tightly packaged in heterochromatin. Addition and removal of two types of simple chemical groups control whether chromatin is packaged as transcriptionally active euchromatin or transcriptionally inactive heterochromatin. These chemical modifications are **methylation** (addition of methyl $-CH_3$ groups) and **acetylation** (addition of acetyl $-COCH_3$ groups). DNA itself can be methylated (on the base cytosine), and histones can be both methylated and acetylated (histone methylation and acetylation). Histone methylation and acetylation can both be inherited at cell division, and form the basis of stable patterns of gene expression controlled by epigenetics.

Key points

Within the eukaryotic nucleus DNA is wrapped around protein complexes called nucleosomes, a bit like the way thread is wound around a spool (note that there are only two turns of DNA per nucleosome, so a full chromosome will be packaged by wrapping around many nucleosomes). Nucleosomes are made up of octamers of positively charged proteins called histones, which have globular C-termini and extended N-terminal tails. Histone tails affect the way that the nucleosomes bind to DNA, and also act as landing sites for proteins. The properties of histones are modified by the addition of chemical groups to the tails. These chemical groups are acetyl groups, which activate transcription, and methyl groups, which can either activate or repress transcription. Histone modification is carried out by enzymes called HATs and HMTs, some of which are components of larger protein complexes which then bind to the modified histones and help to repress or activate chromatin.

15.2 Long ncRNAs are involved in epigenetic gene regulation of gene expression

The important link between RNA and epigenetics is encapsulated in the words of a review by Bernstein and Allis:[5] 'Two worlds have recently collided: those of RNA and chromatin'.

The epigenetic marks in chromatin that we discussed in the previous section are written, read, and erased by proteins. However, several of the proteins involved in epigenetic control are closely associated with RNA molecules (see Fig. 15.5). For example, the polycomb repressive complex PRC2 that silences gene expression by modifying histones is associated with multiple long ncRNAs.[6] The ncRNAs associated with PRC2 include important *XIST* long ncRNA that turns off one copy of the X chromosome in female cells (*XIST* is discussed in more detail in the next section).

Why should there be this association between proteins involved in epigenetics and RNA molecules? One reason might be that local transcription of ncRNAs could help to target epigenetic regulator proteins to particular genomic regions.[7] In other words, RNA molecules act as guidance systems for directing epigenetic changes onto chromatin.

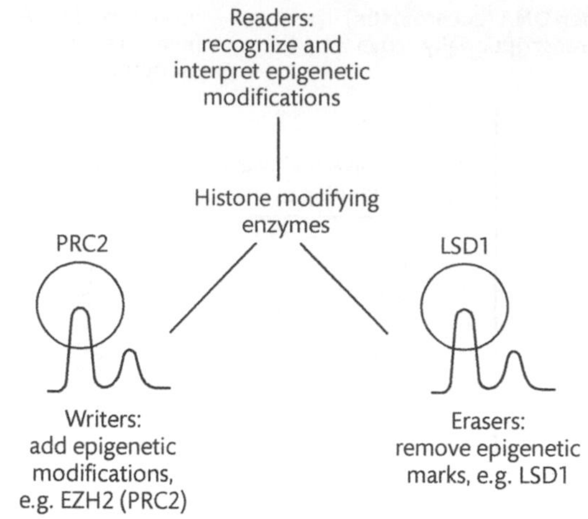

Figure 15.5 **Epigenetic regulator proteins can be associated with long ncRNAs.** Proteins involved in the epigenetic control of gene expression can be divided into the three functional groups: readers, writers, and erasers of the epigenetic code. Several readers, writers, or erasers of the genetic code are bound to long ncRNAs, and particularly to RNA secondary structures within these long ncRNAs. These include the epigenetic writer protein EZH2 (part of the polycomb repressive PRC2 protein complex) and the LSD1 epigenetic eraser protein. As we discuss later in this chapter, PRC2 is recruited to the long ncRNAs *XIST* and *HOTAIR*, and LSD1 also binds to *HOTAIR*.

Transcription of long ncRNAs might lead to binding of epigenetic regulator proteins to the RNA. Base pairing between ncRNAs and complementary sequences in the genome might also provide an efficient addressing system to localize chromatin-modifying complexes to their target sequences.

15.3 A long ncRNA called *XIST* epigenetically regulates the inactive X chromosome in female mammals

Most cells in female mammals contain two X chromosomes, but only one of these X chromosomes is active. This is to achieve parity of gene expression levels with male cells that contain a single X chromosome (Box 15.2). One of the first long ncRNAs to be identified was *XIST*.[8,9] The job of *XIST* is to equalize gene expression levels between male and female animals by turning off expression of the inactive X chromosome.

Box 15.2 Human males and females have different numbers of chromosomes

Human males have 22 pairs of autosomes, plus an X chromosome and a Y chromosome (Fig. 15.6). Females have 22 pairs of autosomes plus two X chromosomes. The Y chromosome is significantly smaller than the X chromosome. Since the X chromosome contains many more genes than the Y chromosome, each cell in a female will have correspondingly more genes than those present in male cells, leading to a genetic imbalance. Genetic imbalances have serious implications for development. In normal development, dosage compensation mechanisms exist to equalize sex chromosome gene expression between the sexes. These dosage compensation mechanisms make use of long ncRNAs to epigenetically equalize differences in gene expression between male and females.

The reason why one copy of the X chromosome is turned off in female cells is explained by looking at the relative sizes of the X and Y chromosomes (see Fig. 15.6). The X chromosome is much larger than the Y chromosome, and contains many more genes. Since a whole chromosome becomes dramatically compacted in each female cell, X inactivation provides one of the most dramatic examples of epigenetic effects targeted by long ncRNAs (Fig. 15.7).

A region of the X chromosome called the *X inactivation centre* (*Xic*) is responsible for X inactivation. When moved from the X chromosome to another chromosome, *Xic* will induce ectopic silencing of its new chromosome of residence. When searching for genes within *Xic* that caused X chromosome inactivation, the initial expectation was that a protein encoded by this region would be responsible. Therefore the identification of the *XIST* long ncRNA gene within *Xic* as controlling X inactivation was a complete surprise. *XIST* is an acronym for 'X-inactive specific transcript'. The reason for this name is that *XIST* ncRNAs are transcribed only from otherwise transcriptionally inactive X transcripts. Expression of *XIST* from the inactive X chromosome leads to X chromosome inactivation.

The *XIST* ncRNA is extremely long. Human *XIST* ncRNA is 19kb long, and is expressed from an intron-containing gene by RNA polymerase II

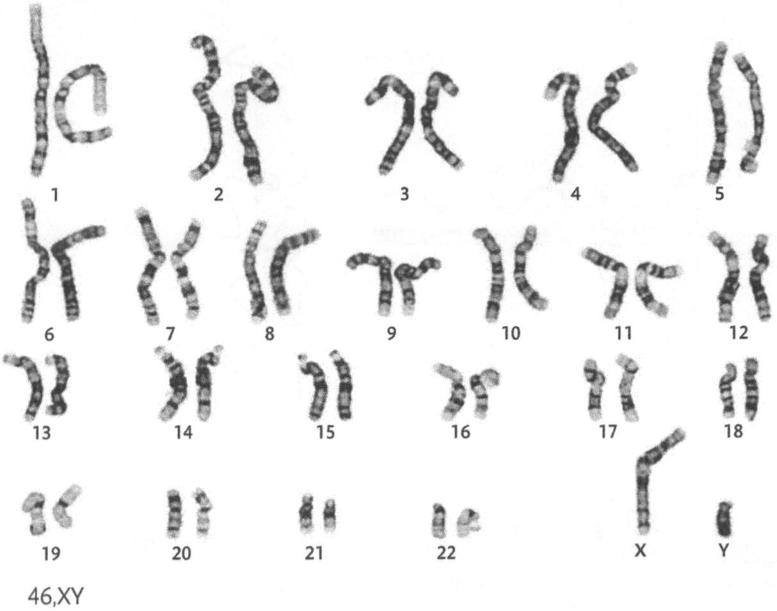

Figure 15.6 The human X chromosome is much larger than the Y chromosome. Even though almost every human cell contains 46 chromosomes, male and female cells have slightly different chromosomal contents. Males have one X and one Y chromosome, and females have two X chromosomes. Since this group of chromosomes contains a Y chromosome, it is from a male. (This image of a male karyotype was kindly provided by Alison Hammersley of the NHS Cytogenetics Unit, Newcastle.)

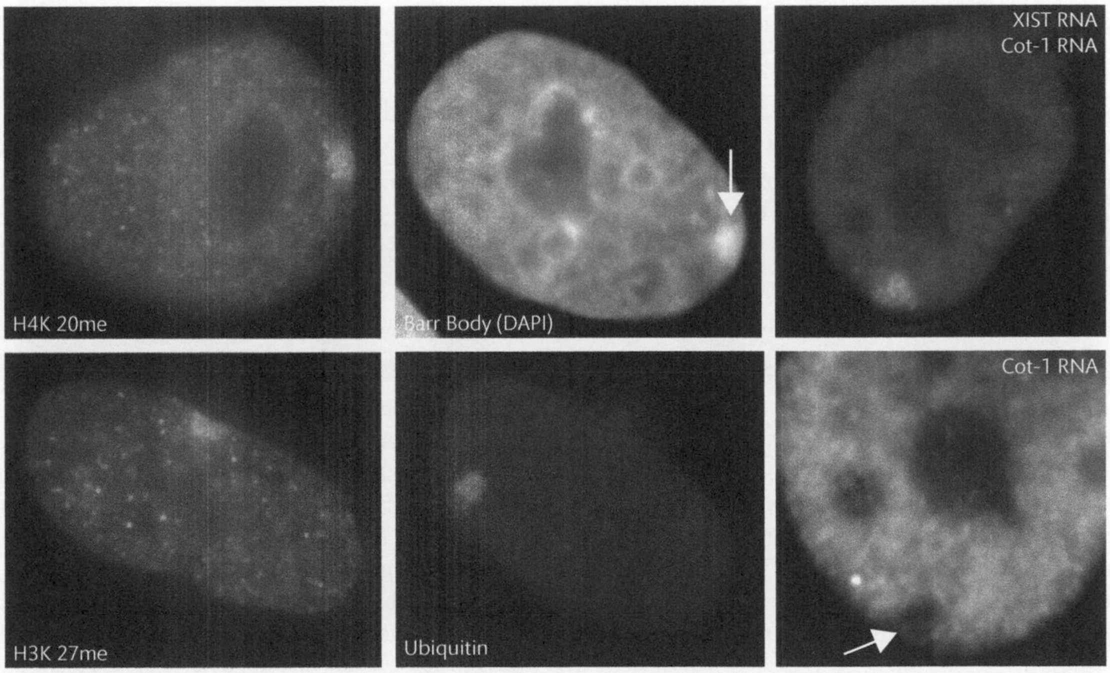

Figure 15.7 The *XIST* ncRNA coats the inactive X chromosome. This figure shows female nuclei stained for different markers. The epigenetic marks H4K20me and H3K27me are associated with the inactive X chromosome. The inactive X chromosome can be visualized using the DNA dye DAPI; notice the stronger staining with DAPI over the inactive X chromosome, within a structure called the Barr body. The Barr body is the inactive X chromosome. The inactive X chromosome is enriched for ubiquitin, but has low levels of Cot-1 RNA. *XIST* RNA can be visualized as a highly localized RNA within female cells by fluorescent hybridization. For a full colour version of this figure, see Colour Illustration 15. (From Hall and Lawrence.[10])

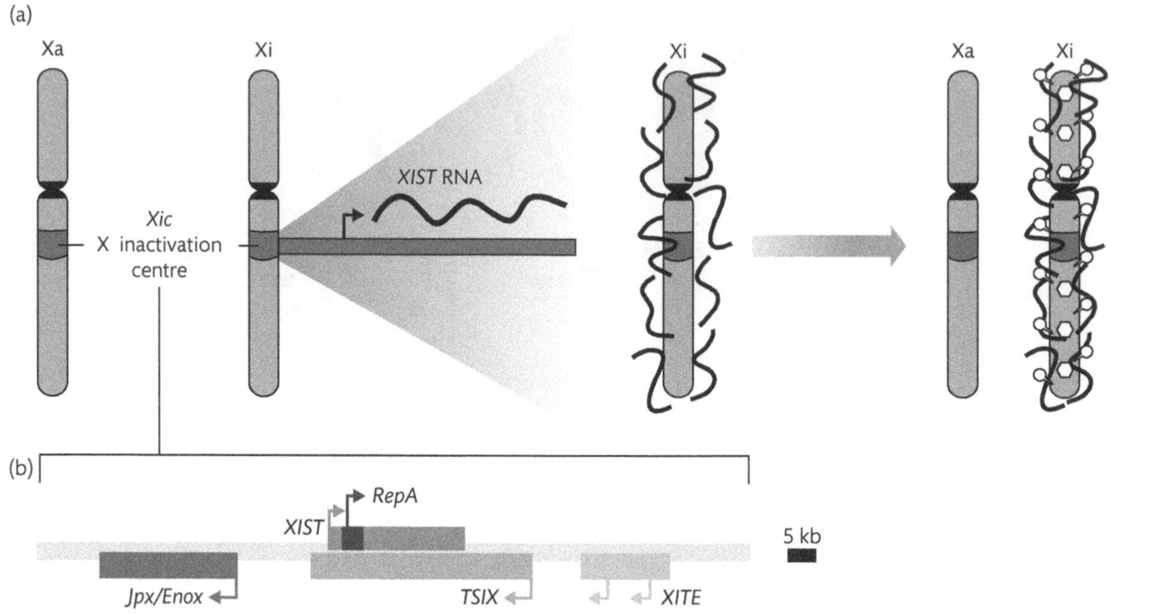

Figure 15.8 **The X inactivation centre (*Xic*) expresses a cluster of ncRNAs that control X chromosome inactivation.** (a) The location of the *Xic* on the long arm of the X chromosome. *XIST* is a long ncRNA expressed from *Xic*, that represses transcription of the rest of the X chromosome. *XIST* coats the surface of the inactive X chromosome, leading to its epigenetic modification including histone and DNA methylation. (b) the *Xic* region includes four ncRNA genes: (1) the long ncRNA *XIST*, which coats the inactive X chromosome and initiates epigenetic X chromosome inactivation; (2) the *TSIX* ncRNA which is transcribed from the opposite DNA strand from *XIST* and so is antisense; (3) a ncRNA encoded by the *XITE* gene; (4) the *Jpx/Enox* gene. *XITE* is thought to be an enhancer for the *TSIX* gene, and *XITE* transcription is associated with its function as an enhancer, while *Jpx/Enox* is a diffusible RNA that acts in *trans*. (From Lee.[13])

transcription. *XIST* is 5′ capped and polyadenylated. However, once transcribed from the inactive X chromosome, *XIST* RNA is not exported from the nucleus, or even away from the inactive X chromosome. Instead, the *XIST* RNA very specifically localizes to the X chromosome from which it was transcribed (see Fig. 15.8). As a result, the inactive X chromosome becomes 'painted' with *XIST* RNA, whereas the active X chromosome has no association with *XIST* at all (see Box 15.3).

Box 15.3 Is there an advantage of using a long ncRNA to inactivate a whole chromosome rather than a protein?

Because the *XIST* RNA is transcribed from the inactive X chromosome only and then coats this same chromosome, it does not have to search out a target chromosome—it is already in position. In contrast, if the inactive X chromosome encoded a protein responsible for X inactivation, such an encoded protein would have to be translated in the cytoplasm and then moved back into the nucleus and find its target—thus muddling which chromosome encoded it in the first place.[13]

How does *XIST* RNA cause transcriptional downregulation of the inactive X chromosome once it has painted it? *XIST* RNA acts as a 'magnet' to draw in proteins that add negative epigenetic marks to the inactive X chromosome. RNA secondary structures are important for binding these proteins (see Chapter 2 for more details about the formation of RNA secondary structures). A region of stemloop secondary structures within *XIST* long ncRNA called Repeat A (Rep A in Fig. 15.8) binds to an epigenetic writer protein complex called the polycomb repressive complex (abbreviated PRC2 in Fig. 15.8).

Once bound to the inactive X chromosome, this polycomb repressive complex then methylates histone H3K27 (this is an abbreviation of histone H3, lysine 27) to inactivate the inactive X chromosome.[14,15] The result of these modifications and other negative epigenetic marks that accumulate (including DNA methylation) is that the inactive X chromosome becomes highly compacted and transcriptionally inactive.

Coverage of the X chromosome by *XIST* starts from the *XIST* gene itself, and then goes on to include the entire X chromosome. The end result is that the entire inactive X chromosome is covered by *XIST* RNA. Association of *XIST* with the X chromosome takes place through interactions with RNA-binding proteins called hnRNP U and YY1.[11,12]

15.4 The X inactivation centre contains a number of non-coding RNAs as well as *XIST*

Surprisingly, *XIST* is not the only long ncRNA encoded by the X chromosome inactivation centre. The X inactivation centre also contains genes for other long ncRNAs including *TSIX* and *XITE*.[13] *TSIX* and *XITE* are also involved in inactivation of the X chromosome.

The *TSIX* gene actually partially overlaps with the *XIST* gene, but is transcribed from the opposite DNA strand. In other words, *TSIX* ncRNA is an antisense transcript. (This is cleverly indicated by the gene names: *TSIX* is the reverse of *XIST*. We discuss antisense transcripts later in the chapter.) *TSIX* directly regulates expression of *XIST*. *TSIX* is only expressed from the active X chromosome and not from the inactive X chromosome. On the active X chromosome, *TSIX* ncRNA expression negatively regulates *XIST* gene expression by directing negative epigenetic marks onto the *XIST* promoter. The inactive X chromosome does not express *TSIX*, which means that it expresses *XIST* instead.

These opposite expression patterns of *TSIX* and *XIST* are established early in development when both X chromosomes are active in females, but then one X becomes inactive. This expression pattern is shown in Fig. 15.9 as arrows pointing in opposite directions (indicating their opposite directions of

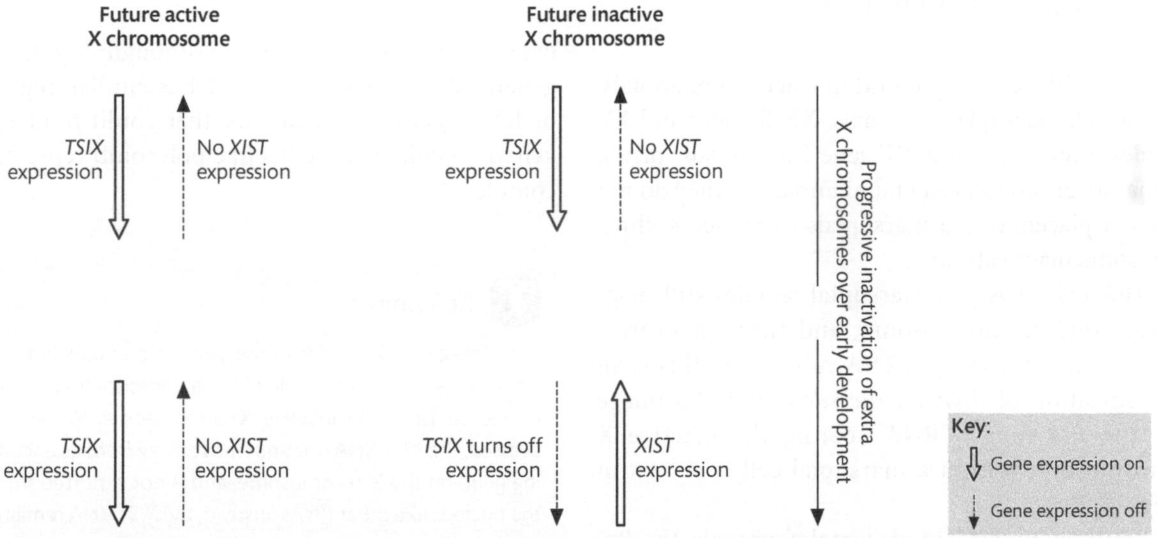

Figure 15.9 ***XIST*** **gene expression is regulated during establishment of the inactive X chromosome by the antisense transcribed** ***TSIX*** **gene.** The expression of the *XIST* and *TSIX* ncRNA genes is shown with their relative expression levels indicated by the width of the arrows (strong expression = wide arrow, weak expression = thin arrow). Notice that *TSIX* becomes inactive on the future inactive X chromosome and this enables expression of *XIST*. On the active X chromosome *TSIX* remains active, and represses *XIST*. These early developmental patterns of gene expression were identified in experiments in differentiating female embryonic stem (ES) cells. ES cells represent early developmental cells which can be maintained in culture but have the potential to differentiate into any other cell type in the body. Undifferentiated female ES cells contain two active X chromosomes, but one of these is inactivated in the first 48 hours after the ES cells undergo differentiation. This in vitro system (using cultured cells) is thought to be similar to X chromosome inactivation as it would normally happen in very early embryos.

transcription), with the width of the arrow indicating the level of transcription. Before X inactivation, *TSIX* is highly expressed from both active X chromosomes (indicated by wide arrows) and *XIST* is not transcribed (shown as a broken arrow). However, during development, *TSIX* is downregulated on the future inactive X chromosome (the *TSIX* arrow disappears and is replaced by a broken arrow), enabling *XIST* to be transcribed and coat the inactive X to induce heterochromatin formation. *TSIX* expression is maintained on the active X chromosome (the wide *TSIX* arrow remains), thus ensuring that the second X chromosome remains active by continuing to repress *XIST*.

The *XITE* (*X inactivation intergenic transcription element*) RNA is transcribed from the same DNA strand as *TSIX*, and within an enhancer from the *TSIX* gene. We discuss how enhancer RNAs operate in more detail later in the chapter.

15.5 Non-placental mammals also use a long non-coding RNA to inactivate an X chromosome in females

The *XIST* gene is only found in placental mammals. Although marsupials also have XX females and XY males, they have no *XIST* gene (marsupials form a separate group to placental mammals as they do not have a placenta). Do marsupials carry out X chromosome inactivation?

The answer is yes. Marsupial females still inactivate one X chromosome, and through expression of another long ncRNA called *Rsx* (this is an abbreviation of *RNA on the silent X*).[16] An image of the *Rsx* long ncRNA coating the inactive X chromosome within a marsupial cell is shown in Fig. 15.10.

Similarly to *XIST* in placental mammals, the *Rsx* ncRNA coats the inactive X chromosome of female marsupials only. When ectopically inserted into other chromosomes, the *Rsx* gene will also lead to their inactivation.

Hence marsupials and placental mammals each use a similar ncRNA strategy to inactivate one X chromosome/cell within female cells. This shows the utility of long ncRNAs for inactivating gene expression across evolution. Although *Rsx* is not genetically related to *XIST*, it has similar regions of RNA secondary structure that could bind epigenetic regulator proteins like polycomb repressive complex 2.

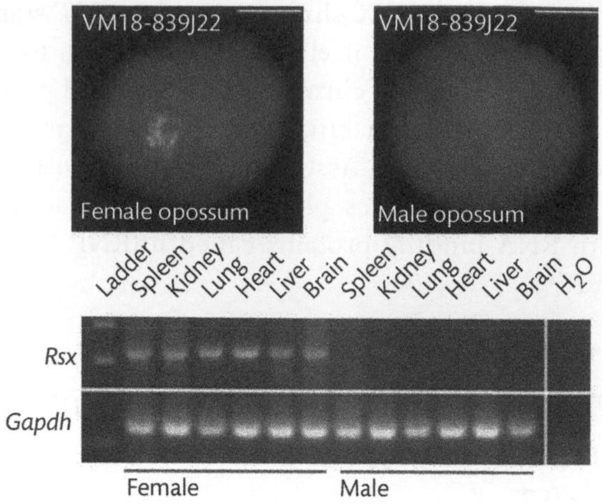

Figure 15.10 Expression of the *Rsx* long ncRNA from the marsupial inactive X chromosome. The upper panels show marsupial cell nuclei stained with the DNA dye DAPI and hybridized with a probe to detect *Rsx* expression by fluorescence in situ hybridization (bright spots). A single hybridization signal is observed in the female (XX) marsupials which will have one inactive X chromosome, and there is no signal in a male cell (XY) in which the single X chromosome will be active. The lower panel shows expression of the *Rsx* RNA monitored in different tissues using RT-PCR. Expression is only detected in female cells, which is expected since only female cells have an inactive X chromosome. (From Grant et al.[16])

Key points

The dosage of X chromosome gene expression is controlled by a long ncRNA called *XIST*, a transcript that is only expressed from the inactive X chromosome. When expressed, the *XIST* RNA does not move away from its encoding gene on the X chromosome, and is not exported from the nucleus like other RNAs. Instead, the *XIST* RNA remains local and coats the inactive X, targeting it for epigenetic repression. *XIST* gene expression is regulated by the ncRNA *TSIX*, which is an antisense RNA. When *TSIX* is expressed from the active X it represses *XIST*. *TSIX* itself is under the control of an enhancer called *XITE* which expresses ncRNAs. As well as their role in epigenetics, the *TSIX* and *XITE* ncRNA genes may also have roles in counting X chromosomes in female cells. Marsupials also use a long ncRNA to inactivate a single X chromosome in female cells.

Box 15.4 Dosage compensation is needed in both mammals and insects

Sex chromosomes are called X and Y in humans, mice, and fruit flies, with the X chromosome being significantly larger. Female cells contain two copies of the large X chromosome. Most male cells have one large X chromosome and one smaller Y chromosome. Despite their equivalent names, the X and Y chromosomes of fruit flies and mammals are not directly evolutionarily related to each other. Instead, both pairs of sex chromosomes arose through a similar process in which the male Y chromosome 'lost' most of its extra genetic material and so contracted in size. The reason for this jettisoning of genes by the Y chromosome is thought to be due to intragenomic competition between the X and Y chromosomes. According to this hypothesis, over evolutionary time the X chromosome has targeted genes on the Y chromosome to try to promote the inheritance of the X rather than the Y chromosome and so to selfishly propagate itself at the cost of the Y. In response, the Y chromosome of fruit flies and mammals has reduced its size in order to limit the number of potential genetic targets for interference by the X.

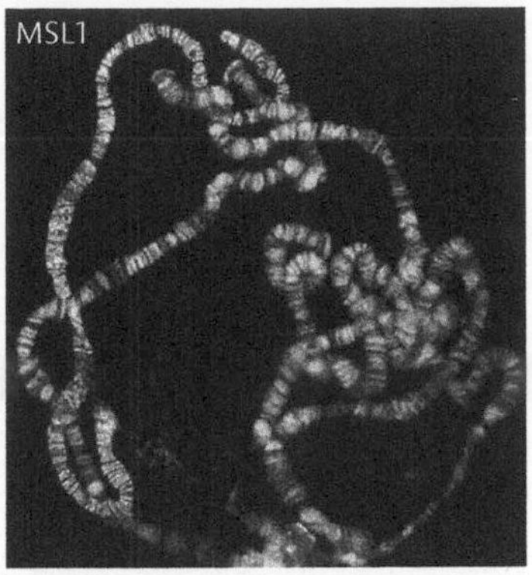

Figure 15.11 A complex called the dosage compensation complex (DCC) localizes to the single X chromosome of male fruit flies. The DCC contains both proteins and long ncRNAs. Here fruit fly chromosomes from a male cell are visualized using the DNA stain DAPI (shown in blue), and one of the protein components of the DCC called MSL1 (shown in red). Notice that the red signal only appears over one chromosome. This is the X chromosome. For a full colour version of this figure, see Colour Illustration 16. (Image from Ilik and Akhtar.[18])

15.6 Fruit flies use a long ncRNA to upregulate expression from a single male X chromosome

Insects also have sex chromosomes of different sizes named X and Y chromosomes (despite the shared name, the fly and mammalian sex chromosomes are not evolutionarily related—see Box 15.4). As in mammals, male flies are XY and female flies are XX. Since the fly X chromosome is larger, female flies have more genes in each cell. This raises the same problem that mammals have, of females having more genes than males.

Fruit flies solve the problem of equalizing gene expression between males and females in the opposite way to mammals—by doubling transcription from the single X chromosome present in the cells of male flies (XY) compared with female flies (XX).

Despite using opposite strategies, long ncRNAs also play a central role in dosage compensation in flies just as in mammals. Fruit fly males coat their single X chromosome with two long ncRNAs.[17] These long ncRNAs are part of an epigenetic reprogramming complex called the dosage compensation complex (DCC) which controls dosage compensation in male fruit flies.

As well as ncRNAs, the DCC complex also contains several proteins. An image of male fruit fly chromosomes stained with an antibody that recognizes one of the protein components of the DCC is shown in Fig. 15.11. Notice this protein component specifically localizes to the X chromosome, and not the other chromosomes.

The two long ncRNAs in the fruitfly DCC are called *roX1* and *roX2*, and only transcribed from the single male X chromosome (*roX* stands for *RNA on the X*).[19] Male flies need at least one of the two *roX* genes for survival. As in the case of *XIST*, RNA secondary structures are important in the *roX* RNAs. The *roX* RNAs have stemloop secondary structures that are reshaped by an RNA helicase called MLE, and this remodelling leads to the assembly of the full DCC complex containing several protein components that bind to the single male X chromosome.[20,21] The targeted DCC then spreads between these original chromosomal 'entry' sites to coat the entire male X chromosome.

The *roX* ncRNAs act as a scaffold for the assembly of the protein components of the DCC. These protein components epigenetically modify the male X chromosome.[22] The protein components of the DCC include a histone acetyl transferase (HAT) which helps

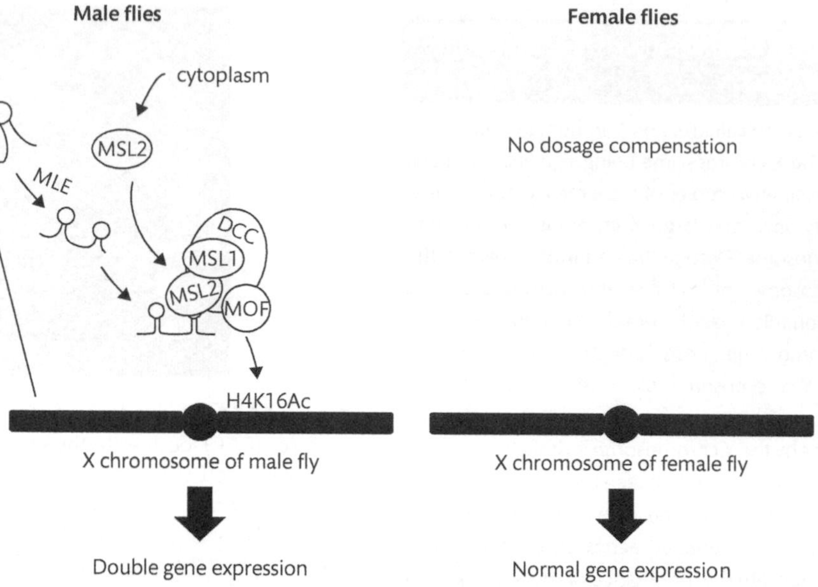

Figure 15.12 Male fruit flies assemble a dosage compensation complex to double gene expression levels from their single X chromosome. Male flies transcribe long non-coding *roX* RNAs from their single X chromosome. The secondary structures of the *roX* RNAs are remodelled by an RNA helicase called MLE. This enables the *roX* RNAs to act as a scaffold to associate with a group of proteins to form a dosage compensation complex (DCC). The DCC is an epigenetic reprogramming machine and contains several proteins essential for male development. Once targeted to multiple sites on the male X chromosome, the DCC spreads to coat the entire male X chromosome and thereby upregulate male X transcription twofold. The DCC proteins which carry out this transcriptional upregulation include a histone acetyl transferase called MOF (males absent on fourth), which acetylates histone H4 on the single male X chromosome (see also Fig. 15.3), and several other MSL (male-specific lethal) proteins. Fruit fly female X chromosomes both provide single-fold levels of gene expression, which together add up to the twofold expression levels from the single male X chromosome. (Adapted from Wutz.[23])

to upregulate gene expression on the male X chromosome by acetylating histone H4. This epigenetic modification of histone H4 makes the single X chromosome in males more transcriptionally active, so it is comparable to the expression from two X chromosomes in female fruit flies (see Fig. 15.12).

Thus upregulation of gene expression from the single male X chromosome provides dosage compensation in male flies so that it is equal to that of female fruit flies.

15.7 The logic of dosage compensation strategies used in flies and mammals

Although mammals and flies have solved the problems of dosage compensation between males and females in different ways, both use long ncRNAs to direct epigenetic changes.

The key differences between the strategies used for dosage compensation by mammals and flies are shown in Table 15.1. Notice that mammalian *XIST* only works in *cis* (it is transcribed from the inactive X chromosome, and paints that same inactive X chromosome). The fact that the inactive X chromosome of female mammals mediates its own inactivation by coating itself with one of its own RNA products simplifies inactivation, because the *XIST* RNA does not need to move away for protein translation and then find its way back to the parent X chromosome (and thus avoids painting the active X by mistake).

In male flies there is a single X chromosome which has to be upregulated so that there is no chance of mistakenly activating an extra X within the same cell. Hence, the insect *roX* RNAs can also work in *trans*. This means that the insect *roX* RNAs will upregulate expression from the single male X chromosome even if their genes are moved away from the X chromosome onto an autosome.

Table 15.1 Similarities and differences between X chromosome dosage compensation in flies and mammals

Similarities between dosage compensation in fruit flies and mammals	Differences between dosage compensation in fruit flies and mammals
1. Both use ncRNAs	1. Sex. Dosage compensation in flies takes place in males, while dosage compensation in mammals takes place in females
	2. Inducing euchromatin or heterochromatin. The dosage compensation machinery upregulates the single X chromosome in male flies, but downregulates one of the two X chromosomes in female mammals
	3. Mammalian *XIST* only works in *cis*, but the *roX* RNAs can work in *trans*

Key points

Mechanisms of dosage compensation operate to equalize gene expression levels between XX females and XY males. In mammals one of the X chromosomes in each cell of a female is inactivated, so only one X chromosome remains transcriptionally active, placing the XX female genetically on a par with XY males. In fruit flies the male upregulates gene expression from its single X chromosome. Although the strategies are slightly different, the dosage compensation mechanisms in both mammals and flies use long ncRNAs. The otherwise inactive female X chromosome in mammalian cells uses a long ncRNA called *XIST* to downregulate its gene expression. The single X chromosome in male flies uses two long ncRNAs called *roX* to upregulate its expression so that it is equivalent to the two X chromosomes in females.

15.8 Genetic imprinting uses long non-coding RNAs

Another kind of epigenetic regulation that involves long ncRNAs is called *imprinting*. The term imprinting refers to a kind of gene regulation in which maternal and paternal genes are differentially expressed. Imprinting just affects a subset of genes, yet is very important for development. Around 100 genes are imprinted in mammals.

Before we discuss imprinting we need to distinguish between the two copies of each chromosome in a cell. Most human cells are diploid. Diploid means that human chromosomes come in pairs, i.e. there are two copies of chromosome 1 in each cell etc. One copy of each chromosome originates in each original parent of the individual (e.g. there is one chromosome 1 from mum and one from dad etc.). For this reason, individual chromosomes in a pair can be distinguished as maternal and paternal chromosomes (see Box 15.5 for further explanation). This distinction does not usually matter; most of the many genes on a chromosome are equally expressed between the maternal and the paternal copies of the chromosome. The exception is for a subset of genes called *imprinted genes*.

Imprinted genes are usually involved in growth and development, and also occur in large clusters that contain some paternally imprinted and some maternally imprinted genes. In an imprinted locus, a paternally imprinted gene is silent on the paternal allele, and a maternally imprinted gene is silent on the maternal allele. Imprinted gene clusters can be huge, of the order of a megabase (10^6 base pairs) in size.

Imprinting is thought to be caused by genome conflict. This was first observed in experiments carried out in the 1980s (see Fig. 15.13) that manipulated

Box 15.5 Maternal and paternal chromosomes

Human cells are diploid. The two pairs of each chromosome in a diploid organism like the human can be traced back to when its first cell was formed by fertilization of an egg by a sperm (this first cell is called the zygote). Both the egg and sperm contain a single rather than a double set of chromosomes, i.e. they are haploid. The single set of chromosomes carried by the egg are called the maternal chromosomes (from the mother) and those from the sperm the paternal chromosomes (from the father). The normal diploid number of chromosomes is restored by the fusion of the maternal and paternal genomes immediately after fertilization (in the zygote). Although now present in pairs, each of the chromosomes in a cell from a diploid organism is directly descended from either the original maternal or the original paternal chromosome.

Box 15.6 Why does genetic imprinting exist?

Imprinting is thought to result from competition between 'male and female genomes'. Mammals are diploid, which means that they have two copies of each chromosome, one originally from their mother and one originally from their father, each descended from a single fertilized egg. These two sets of chromosomes are thought to retain somewhat different purposes during life. The genomic competition hypothesis suggests that the mother and father have slightly different 'hopes' for the fertilized egg and so programme their genes to influence development accordingly. The female wants to invest resources in the fertilized egg and maintain embryonic development, but not to a degree which might compromise either her own health or the health of any future offspring she might have. In contrast, the male would prefer the female to make a very large investment in supplying nutrients to the embryo which he has fertilized (and never mind about any future offspring from different fathers). Hence there is a conflict: males want large healthy embryos giving rise to robust offspring, while females want to conserve resources. The upshot of this genomic competition is that some genes will be alternatively expressed depending on their parent of origin.

Is imprinting important? The answer is yes, for reasons related to 'conflict' occurring between chromosome pairs, getting two chromosomes in either the male or female parent (see Box 15.5 for further explanation) can lead to disease.

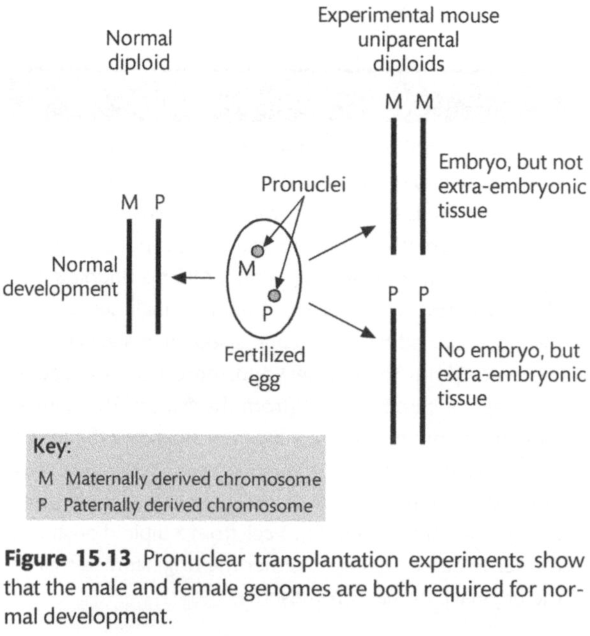

Figure 15.13 Pronuclear transplantation experiments show that the male and female genomes are both required for normal development.

events which occur very early in development (Box 15.6). For a brief period after fertilization of an egg by a sperm, mouse eggs contain two haploid nuclei (called pronuclei), usually one from the sperm and one from the egg. These two haploid nuclei fuse to give a diploid cell, which is the first cell of the new organism. Because of the equal contribution of the sperm and the egg nuclei, a normal fertilized egg will contain one set of paternal chromosomes (P) and one set of maternal chromosomes (M). The genes on these M and P chromosomes are responsible for making the embryo and also the extra-embryonic tissue (placenta) through which the embryo is nourished. Making both the embryo and the placenta is important for normal embryonic development.

Experiments in which fertilized eggs contain either two maternal pronuclei (only containing maternal chromosomes) or two paternal pronuclei (only containing paternal chromosomes) gave startling results. In the case of an egg containing two maternal genomes an embryo was produced normally but there was no extra-embryonic material. However, in the case of an egg containing two paternal genomes no embryo was produced, but just extra-embryonic tissue. The implication is that the production of the extra-embryonic tissue is programmed by the paternal genome, while the production of the embryo itself is programmed by the maternal genome.

The implication of these experiments is that the paternal genome is pre-programmed before fertilization to try to ensure that the embryo is well nourished. In normal development this pre-programming by the male is balanced by programming of the female's chromosomes, which try to ensure normal embryonic development. Although the maternal and paternal chromosomes are competing, the end result of this is a balance which ensures normal growth and development. Only when an organism is experimentally manipulated to contain just maternal or just paternal chromosomes (see Fig. 15.13), or in some genetic diseases where imprinted maternal or paternal alleles are deleted and this gives rise to different symptoms, does this conflict become apparent.

Long ncRNAs are particularly enriched in imprinted gene clusters, as are protein-coding genes that control embryonic growth. In some cases, these enriched long ncRNAs have been shown to have an important role in regulating imprinting. How this

Table 15.2 Two separate roles for ncRNAs in directing imprinting

Imprinted gene cluster	Role of ncRNA
The *IGF2/H19* imprinted gene cluster on human chromosome 11	The *H19* ncRNA is transcribed from a decoy promoter to IGF2, and provides a source of important microRNAs that can control growth
The *IGF2/AIR* imprinted gene cluster on human chromosome 17	The *AIRN* ncRNA directs epigenetic silencing of genes on the paternal chromosome 17

happens is the topic of the next two sections. We are going to look at two different long ncRNAs which each epigenetically regulate imprinting in different ways (Table 15.2).

15.9 Transcription of the *H19* long non-coding RNA acts as a decoy for transcription of the *IGF2* gene

One of the best-understood examples of genetic imprinting takes place on human chromosome 11, and is the topic of this section.

The *H19* long ncRNA is expressed at high levels from the maternal chromosome 11, and from one of the very first imprinted genes to be discovered.[24] The name *H19* is associated with the history of how this ncRNA gene was identified in 1991: *H19* was detected by hybridization, and this name reflects its position on the filter when it was first cloned. *H19* promotes embryonic growth.[25]

H19 expression has two separate effects on growth. First, *H19* is processed into shorter microRNAs that control growth.[26,27] Secondly, the promoter of *H19* competes for the transcriptional enhancer sequence that jointly controls the protein-coding *IGF2* (*insulin-like growth factor 2*) gene. *IGF2* is expressed from the paternal chromosome—the chromosome that does not express *H19*. This is called *enhancer competition*.

Figure 15.14 shows how this enhancer competition mode of *IGF2*/*H19* imprinting works. Remember when looking at Fig. 15.14 that both maternal (♀) and paternal (♂) chromosomes 11 are present in the same cell. The *H19* and *IGF2* genes are controlled by the same positive DNA control element, called an enhancer. This joint enhancer is adjacent to the *H19* gene, and activates either *H19* or *IGF2* expression (activation is indicated by the arrow between the enhancer and the relative genes). Whether the *H19* or *IGF2* gene is selected by the enhancer is controlled by a DNA sequence between the *IGF2* and *H19* genes called the **imprinting control region (ICR)**. The ICR is methylated on the paternal chromosome 11, and as a result the enhancer element activates transcription

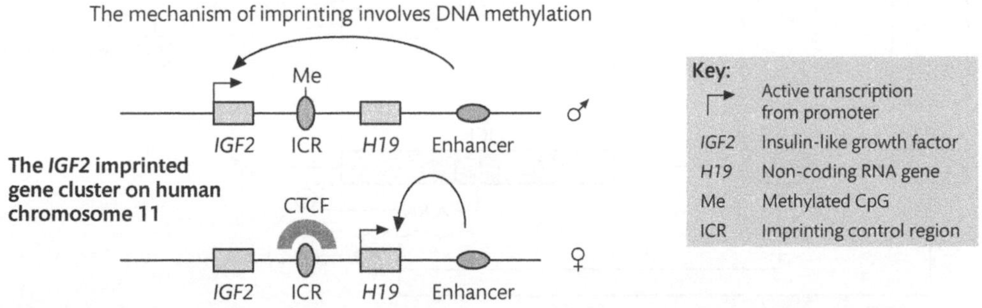

Figure 15.14 An imprinted cluster of genes on human chromosome 11 express either the ncRNA *H19* gene from the maternal chromosome (♀), or the protein-coding *IGF2* gene from the paternal chromosome (♂). The imprinting control region (ICR) is within the promoter of the *H19* long ncRNA. A downstream enhancer sequence controls activation of both *H19* and the protein-coding *IGF2* gene, but not both at the same time. This is called enhancer competition. In the ♀ copy of chromosome 11 the ICR is not methylated, and binds a protein called CTCF which blocks the enhancer sequence from activating *IGF2*. As a result, the shared enhancer sequence activates transcription of the *H19* ncRNA gene. On the ♂ chromosome, methylation of the ICR blocks both the *H19* promoter and binding of the CTCF protein. As a result the shared enhancer element activates expression of *IGF2*. (Adapted from Reik and Murrell[28] with permission from Macmillan Publishers Ltd.)

of *IGF2*. The ICR is not methylated on the maternal chromosome 11, and as a result binds to a protein called CTCF. When bound to the ICR, CTCF protein creates a barrier so that the enhancer element is no longer able to activate the *IGF2* gene. The result of this is that on the maternal chromosome 11, the joint enhancer element activates transcription of the *H19* long ncRNA instead.

15.10 The *AIRN* ncRNA epigenetically represses *IGF2R* gene expression by directing epigenetic chromatin modification

The second imprinted locus we are going to discuss is on mouse chromosome 17. This locus contains imprinted genes for a long ncRNA called *AIRN*, the protein-coding gene *IGFR2* (involved in embryonic growth), and two other protein-coding genes (encoding proteins involved in cation transport).

This chromosome 17 gene cluster is shown in Fig. 15.15. Active gene expression is indicated as a bent arrow at the start of the genes. Notice that the protein-coding genes in this gene cluster are only expressed from the maternal ♀ copy of chromosome 17, while the gene encoding the *AIRN* ncRNA is expressed from the paternal chromosome ♂.

Expression of the *AIRN* long ncRNA on the paternal copy of chromosome 17 silences activity of the imprinted protein-coding genes in this cluster. The *AIRN* long ncRNA is transcribed from a sequence that is actually within the *IGFR2* gene, but in the opposite orientation, making *AIRN* an antisense transcript (*AIRN* is an acronym for *antisense IGF2R RNA*). Since transcription of *AIRN* is antisense to the *IGF2R* gene, this antisense transcription might directly repress *IGF2R* expression.[29,30] (Antisense transcripts are discussed further in Section 15.13.)

After it is transcribed, the *AIRN* ncRNA is not exported from the nucleus, but instead remains localized near its site of transcription to repress gene expression of the imprinted cluster. The *AIRN* ncRNA represses expression of *IGFR2* and the other genes in this cluster by directing epigenetic changes on the paternal chromosome 17, particularly addition of negative epigenetic marks to the promoters of *IGF2R* and the other protein-coding genes encoding solute carrier proteins.

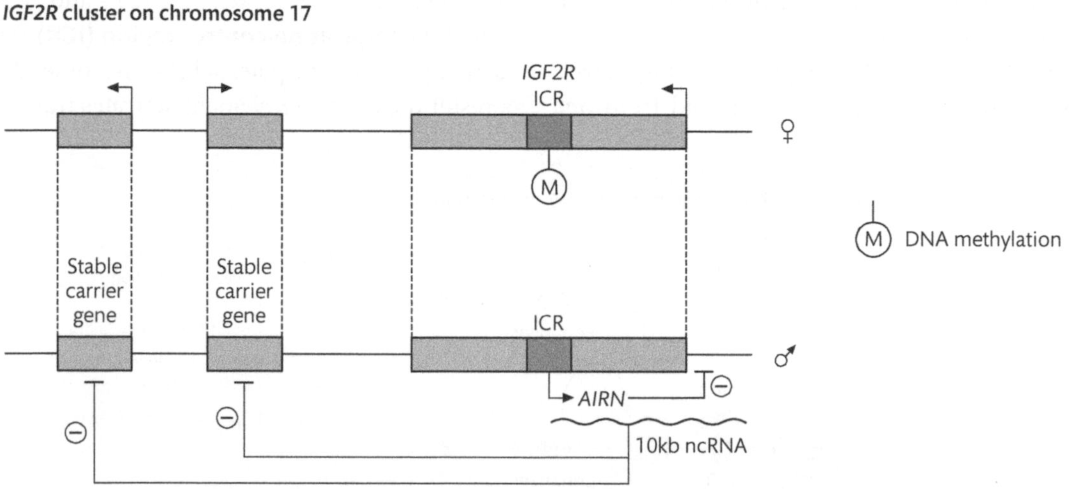

Figure 15.15 An imprinted gene cluster on human chromosome 17 includes the *AIRN* long ncRNA. The *AIRN* long ncRNA is antisense to the protein-coding *IGF2R* gene (notice the bent arrows indicating transcription, with the direction of the arrow being the direction of transcription). The male (♂) and female (♀) chromosomes of the cell are lined up to show differences in gene expression from these two chromosomes in the same cell. Expression of *AIRN* ncRNA on the paternal chromosome silences *IGF2R* and two other adjacent protein-coding genes; this repression is indicated by (–). On the maternal chromosome the promoter for the *AIRN* gene is methylated and so is inactive. As a result the *IGF2R* gene is transcribed from this chromosome, as are the adjacent genes which encode solute-carrier genes. The promoter for *AIRN* is in a part of the *IGFR2* gene called the imprinting control region (ICR).

How is this differential expression between the maternal and paternal chromosomes in this imprinted gene cluster controlled? DNA methylation again plays a key role. A region called the imprinting control region (ICR) is methylated on the maternal copy (labelled M in Fig. 15.15). This blocks access of RNA polymerase to the promoter of *AIRN* on the maternal chromosome 17, which as a result is not expressed. On the paternal chromosome the ICR is not methylated, and so *AIRN* is transcribed from the paternal chromosome. This then turns off the protein-coding genes in the cluster.

Many ncRNAs have been recently identified with important roles in gene expression (see Chapter 17 for more discussion). In the next section we are going to discuss the function of one of the more recent of such long ncRNAs to be identified and characterized, a long ncRNA called *HOTAIR*. *HOTAIR* has a key role in development, and regulates switches between euchromatin and heterochromatin of the *HOX* genes that control developmental organization in animals.

Figure 15.16 ***HOX* genes encode transcription factors which play an important role in establishing the body plan during embryonic development.** *HOX* genes were first identified by the analysis of so-called homeotic mutations in fruit flies which fail to establish a correct body plan. The phenotypes of homeotic mutant flies include appendages growing from inappropriate body segments. A classic homeotic mutation is *Ultrabithorax*, in which a fly has a second pair of wings on its thorax, rather than a single pair of wings and a pair of vestigial wings—a picture of an *Ultrabithorax* mutant is shown here. Fruit flies normally have a single full pair of wings and a pair of vestigial wings. Flies with mutations in the *Ubx* gene have two full pairs of wings instead. *HOX* genes were later identified in other animals, including mice and humans. Each *HOX* transcription factor contains a conserved DNA-binding domain called the homeobox domain (*HOX* being short for homeobox). Different *HOX* genes are responsible for controlling the correct position for the development of the eyes, limbs, and heart in the body. (Reprinted from Wagner[31] with permission from Macmillan Publishers Ltd.)

15.11 Long ncRNAs play an essential role in establishing animal body plans

Long ncRNAs epigenetically regulate a group of important developmental control genes called the homeotic or *HOX* genes. These homeotic genes get their name since parts of the body develop in the wrong places when they are mutated in flies (see Fig. 15.16).

HOX genes are clustered in animal genomes, with two clusters of *HOX* genes in flies and four clusters in vertebrates. Each cluster contains several *HOX* genes. Within clusters, *HOX* genes have very tight expression patterns that are important for establishing normal body plans. The *HOX* genes encode transcription factors that regulate the expression of the downstream transcription factors needed for proper development.

Transcription of particular *HOX* genes is first activated early in embryonic development in response to morphogen gradients. These morphogen gradients exist only transiently in the early embryo. Tight

Key points

Genetic imprinting describes where gene expression is controlled according to which parent the gene has been inherited from. Imprinted genes often occur in clusters. Many imprinted gene clusters include long ncRNAs. The *IGF2* region of chromosome 11 includes both genes for an ncRNA (called *H19*) and protein coding genes (*IGF2*). In this case the *H19* long ncRNA is transcribed from a decoy promoter to the *IGF2* protein-coding gene which controls placental growth, and also provides the source of important microRNAs that control placental growth. A long ncRNA called *AIRN* is expressed from the paternal copy of mouse chromosome 17, where it represses expression of a protein-coding gene called *IGFR2* that controls fetal growth.

patterns of *HOX* gene expression are maintained later in development through epigenetic mechanisms that activate and repress *HOX* gene expression as necessary.

The *HOX* protein-coding genes have been intensively studied. However, it came as a surprise that, in experiments to analyse patterns of *HOX* gene expression, there were high levels of intergenic RNA expression between human *HOX* genes.[32] This intergenic transcription gives rise to long ncRNAs.

One of the best understood of the long ncRNAs that is transcribed from between the human *HOX* genes is called *HOTAIR*.[33] *HOTAIR* is a long ncRNA (around 2kb) and is spliced and polyadenylated, but it does not encode a protein. The *HOTAIR* ncRNA (an abbreviation of *HOX antisense intergenic RNA*) is transcribed from the *HOXC* locus but in the opposite direction to *HOXC* genes.

Once transcribed, the *HOTAIR* long ncRNA acts as a scaffold on which epigenetic regulator proteins including PRC2 assemble.[34] This *HOTAIR* RNA–protein complex then targets and represses transcription of genes in another cluster of *HOX* genes called the *HOXD* cluster. The proteins scaffolded around *HOTAIR* change epigenetic chromatin marks that alter transcription patterns (see Fig. 15.17).

The *HOX* genes are critical for normal development—their genetic removal from mice causes severe abnormalities. What then is the role of *HOTAIR*? Is *HOTAIR* also important for development?

The answer from genetic experiments is that expression of *HOTAIR* is very important for normal patterns of development. Deletion of the *HOTAIR* gene in mice leads to loss and modification of specific skeletal structures.[35] An example of one of these anatomical changes in a mouse without the *HOTAIR*

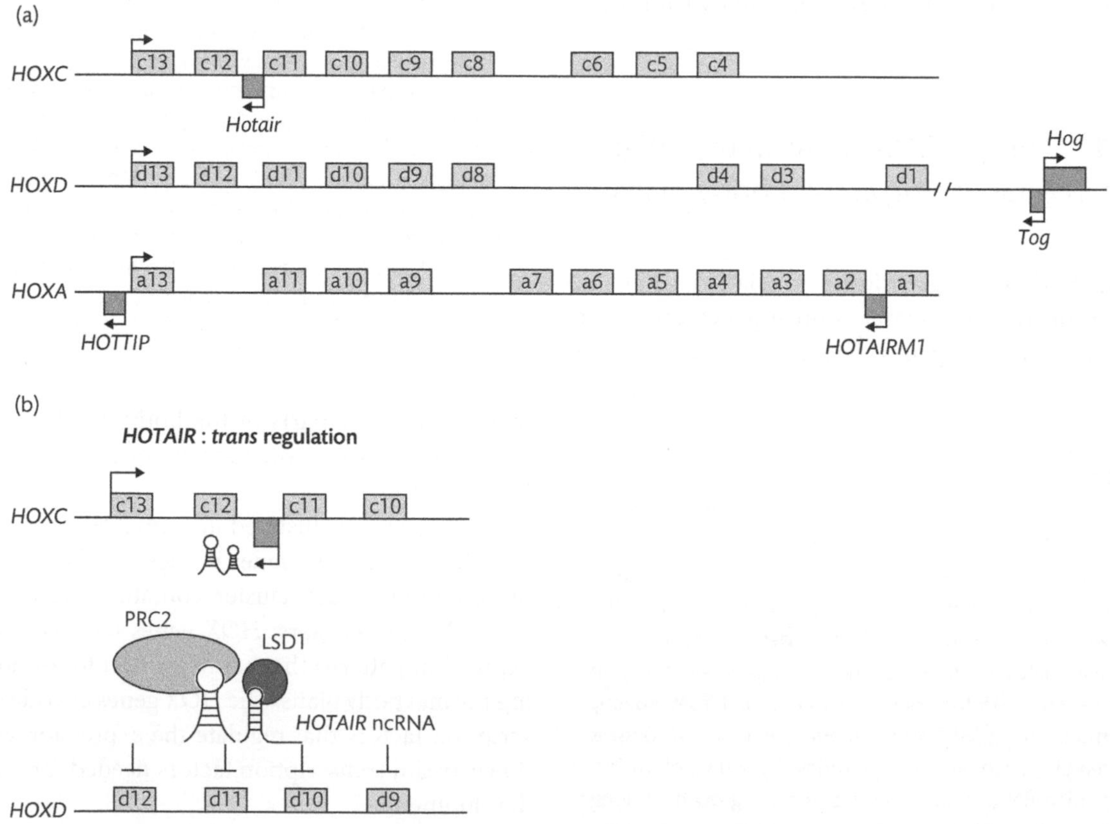

Figure 15.17 Transcription from the human HOXD locus is repressed by a long ncRNA called *HOTAIR*. *HOTAIR* is transcribed from the *HOXC* locus, but in the opposite direction to the other *HOXC* genes. *HOTAIR* represses transcription by binding to polycomb group (PRC2) and LSD1 epigenetic writer proteins. Notice that there are also some other long ncRNAs within these regions of the genome; these include *Hog/Tog*, *HOTTIP*, and *HOTAIRM1*. (From Dasen.[33])

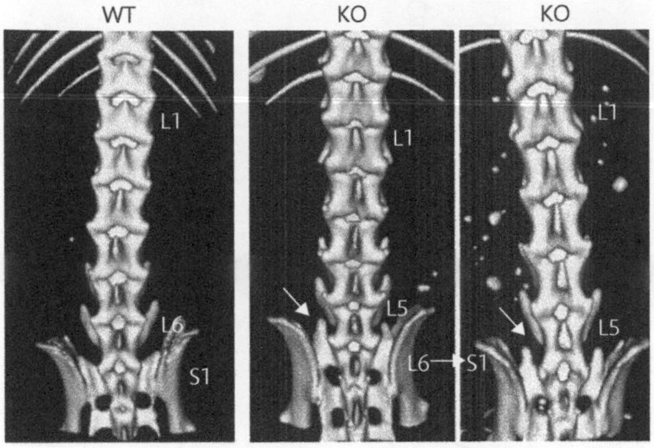

Figure 15.18 Removal of the gene encoding the *HOTAIR* ncRNA from mice causes morphological defects. A micro-CT scan shows skeletal abnormalities in the knockout mice (KO) compared with wild-type mice (WT), with loss of the sixth lumbar vertebra (L6) and structural malformation of the first sacral vertebra (S1). (From Li et al.[35])

gene is shown in Fig. 15.18. This CT scan shows complete loss of vertebra L6 in the *HOTAIR* knockout mouse. (Notice that the first sacral vertebra S1 directly follows lumbar vertebra L5 in the knockout, while in wild-type mice there is an L6 vertebra between L5 and the sacrum.)

Hence *HOTAIR*, like the *HOX* genes, has a key role in development. *HOTAIR* has important roles in disease as well as in normal development, particularly in cancer. In a third of breast cancers the levels of *HOTAIR* expression can be increased by up to 2000-fold. Increased *HOTAIR* expression correlates with a poor prognosis for breast cancer patients (see Box 15.7), with increased metastasis and mortality.

Why is increased *HOTAIR* expression so harmful in cancer cells? The answer is that inappropriate *HOTAIR* expression in breast cancer cells relocalizes epigenetic regulator proteins in cancer cells. This in turn changes the chromatin signatures of important

Box 15.7 Long ncRNAs in cancer prognosis

Another aspect of long ncRNAs that could be important for cancer is that they might be useful in prognosis. Long ncRNAs might be useful as prognostic markers in diseases like cancer since they are stable in body fluids because of their extensive secondary structures.

genes to affect the properties of the cancer cells. These changed cell properties include increased metastasis—the movement of tumours from their site of origin to other sites in the body.[36] Metastasis is one of the major killers in cancer—even more than the original primary tumours. The repressive effect of *HOTAIR* expression on three tumour suppressor genes is shown in Fig. 15.19; notice that *HOTAIR* binding leads to decreased expression from these genes, and so to increased metastasis.

Key points

HOX genes encode important transcriptional regulators of body plans in animals. *HOX* genes are activated or repressed epigenetically during development, and ncRNAs play an important role in establishing these epigenetic marks. The human *HOTAIR* ncRNA which is transcribed in the opposite direction to *HOXC* has an important role in repressing transcriptional activity at a separate *HOX* locus called *HOXD*. *HOTAIR* expression is important in both normal development and cancer.

15.12 Long ncRNAs are involved in transcriptional enhancer function

Enhancers are transcriptional control elements that are typically located at a distance from the genes that they regulate, and act over long distances to activate levels of gene expression. Enhancers have long been known to bind high concentrations of transcription factors that act to enhance gene expression on distant promoters through protein interactions that create DNA loops. These loops place the enhancer close to their regulated promoters within the three-dimensional organization of the nucleus, even though they might be separated by considerable distances in the actual DNA sequence.

Recent data show that some enhancers are transcribed to produce long ncRNAs.[37,38] This local transcription at enhancer sequences is important to how enhancers function. One function of enhancer RNAs might be to bind directly to epigenetic regulator proteins, and for these proteins then to modify chromatin in and around the enhancer (see Fig. 15.20a). Another function of enhancer RNAs is to provide a bridging molecule to establish contact

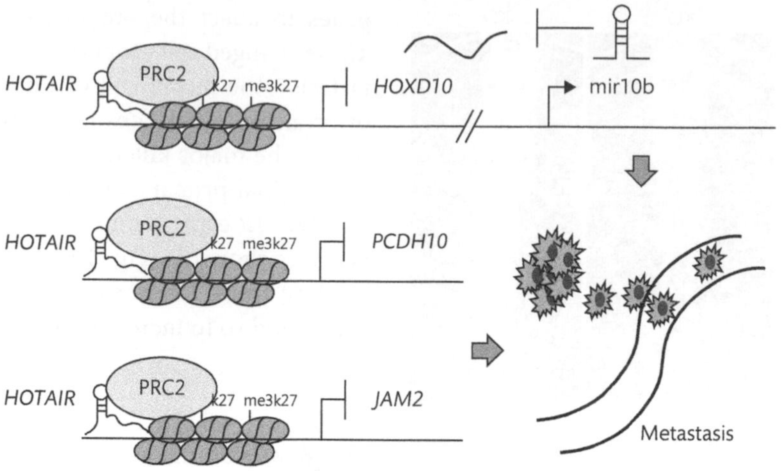

Figure 15.19 Increased HOTAIR expression in cancer epigenetically modifies expression of tumour suppressor genes. *HOTAIR* binds to the epigenetic writer protein PRC2. The function of PRC2 is to repress gene expression by writing the chromatin mark H3K27me3. *HOTAIR* targets PRC2 to tumour suppressors including *HOXD10*, *PCDH10*, and *JAM2*. This turns off these tumour suppressors, and in turn increases cancer cell metastasis. (From Wan and Chang[36].)

with the promoters of genes. Local transcription of long ncRNAs at enhancers provides a physical link to proteins assembling at promoters to help facilitate transcriptional initiation. These include interactions with a protein complex associated with RNA polymerase II called the mediator complex (see Fig. 15.20b).

As one example, an RNA is transcribed from an enhancer within the *HOXA* locus. This enhancer RNA is called *HOTTIP* and controls the *HOXA* genes, one of the *HOX* gene clusters that we discussed in the previous section. The position of *HOTTIP* in the *HOXA* gene cluster is shown in Fig. 15.17. As with other enhancers, even though the *HOTTIP* sequence is distant from *HOXA* genes, because of chromosome looping *HOTTIP* becomes physically located close to the promoters of the *HOXA* genes it regulates. Downregulation of *HOTTIP* also reduces expression of *HOXA* genes, and causes developmental abnormalities in chicken wing buds.

Key points

Transcriptional enhancers encode long ncRNAs. These enhancer RNAs may provide a physical bridge between enhancers and components of the transcriptional machinery assembled at the promoter that facilitate transcriptional initiation.

15.13 Antisense RNAs

Antisense RNAs are often lumped in with the larger group of ncRNAs, although not all antisense RNAs are non-coding.

Antisense RNAs are transcribed off the opposite strand to other genes (these other genes can be either protein-coding or non-protein-coding).[39,40] We have already discussed two antisense ncRNAs in this chapter: the *TSIX* gene involved in X chromosome silencing, and the *AIRN* ncRNA involved in silencing the *IGF2R* gene. In both these examples transcription of the antisense RNA acts to block transcription of the main gene. Here we are going to discuss other important examples of antisense RNAs from animals and plants.

In animals, expression of an antisense RNA called *ANRIL* is important in cancer.[7] The reason for this is that expression of *ANRIL* antisense RNA silences gene expression from two cell cycle regulator genes called *CDKN2A* and *CDKN2B*; unlike *ANRIL*, these two cell cycle regulator genes encode proteins (see Fig. 15.21). Both *CDKN2A* and *CDKN2B* are important tumour suppressors that can become inactivated by mutation or deletion in cancer. *ANRIL* epigenetically silences gene expression of these key cell cycle regulator genes by changing their DNA methylation and chromatin modification patterns.

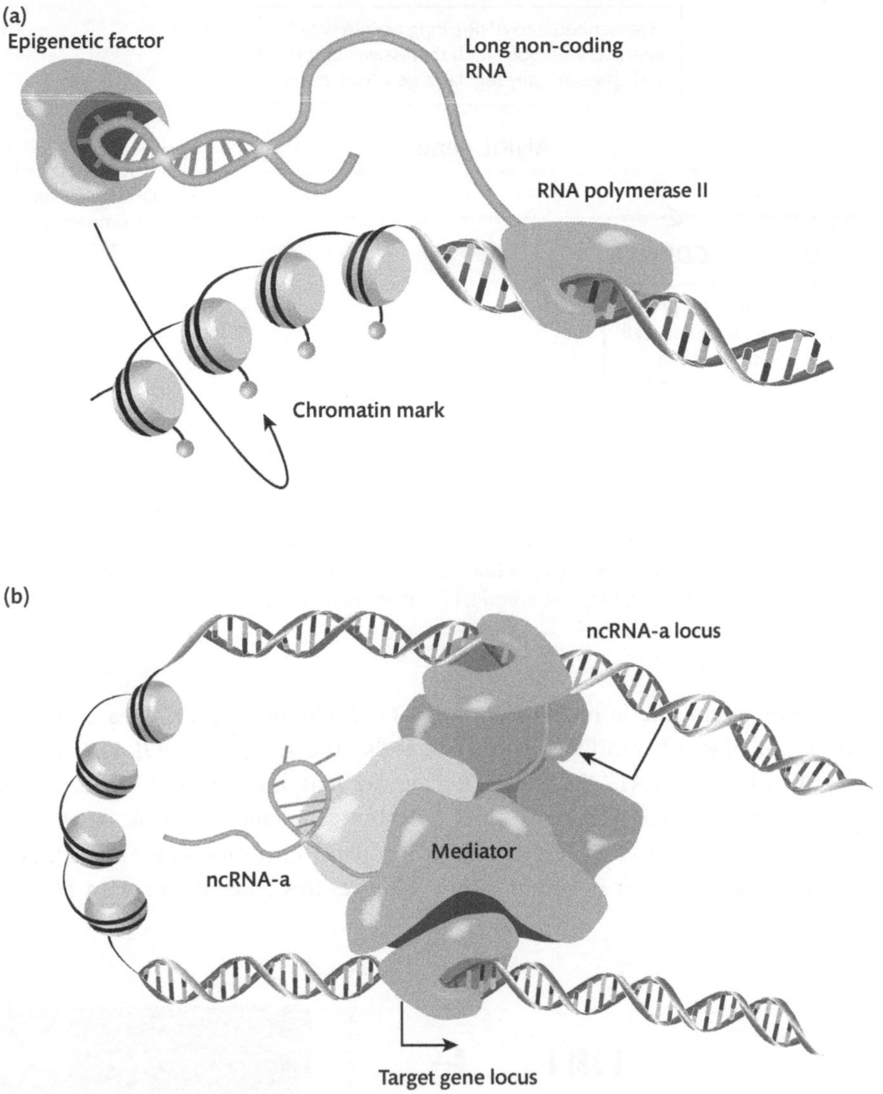

Figure 15.20 **Long RNAs encoded by enhancers can provide a physical bridge with promoters.** (a) RNA transcribed by RNA polymerase II can bind and locally recruit epigenetic modifiers proteins which then modify chromatin. (b) Long ncRNAs locally transcribed from enhancer regions can be bound by components associated with RNA polymerase II at the promoter. In this case, enhancer RNAs function as physical bridges between enhancers and promoters. (From Kung and Lee.[37])

ANRIL is implicated in several diseases. Increased levels of *ANRIL* expression in prostate cancer are thought to silence key tumour suppressor genes and help the disease to progress.[41] The *ANRIL* gene also has more general connections with disease, as polymorphisms in *ANRIL* are also associated with heart disease and diabetes.[42]

Another important role for antisense RNAs is in controlling gene expression patterns in plants that regulate flowering.[43–45] Flowering is induced in some plants after a cold spell, to tie in with spring after the cold of winter—a process called *vernalization*. In *Arabidopsis*, a plant that is widely used in molecular biology, such cold-induced flowering is controlled by epigenetic repression of a gene called *FLC* (abbreviation of *Flowering Control Locus*). Since the *FLC* gene encodes a protein transcription factor that is responsible for repressing genes encoding proteins involved in flowering, epigenetic repression of the *FLC* gene induces flowering.

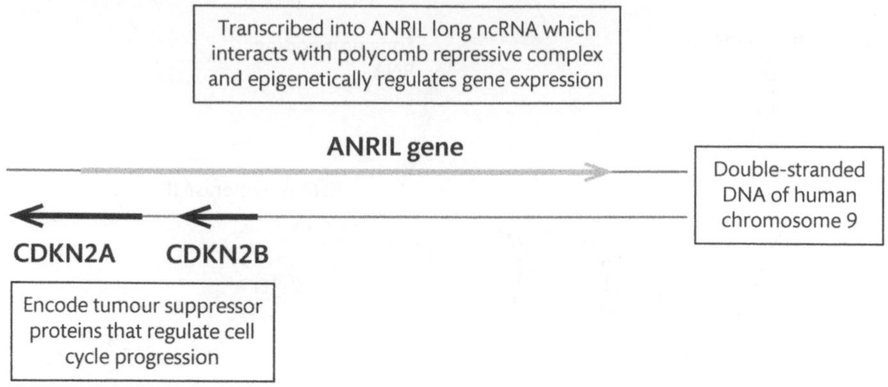

Figure 15.21 The *ANRIL* long ncRNA epigenetically controls cell cycle regulator protein gene expression. Chromosomal DNA is double-stranded, and either strand can contain genes. Sometimes both strands can contain genes, in which case one gene will be antisense to the other. The *ANRIL* ncRNA is transcribed from one strand of human chromosome 9; as a ncRNA, the *ANRIL* gene is transcribed but not translated. This chromosomal region also contains the genes that encode two important cell cycle regulators called CDKN2A and CDKN2B. Unlike *ANRIL*, both *CDKN2A* and *CDKN2B* are translated into proteins. Importantly, these two genes are on the opposite DNA strand to *ANRIL*, yet partially or totally overlap with *ANRIL* (the direction of these genes is indicated by the arrowheads). This chromosomal arrangement makes *ANRIL* RNA antisense to both these two protein-coding genes.

Antisense RNAs control epigenetic repression of the *FLC* gene. A cool spell of weather induces expression of a long ncRNA called *COOLAIR*.[46] *COOLAIR* is antisense to the *FLC* gene. *COOLAIR* transcripts are shown in the lower DNA strand in Fig. 15.22. Notice from this figure that a number of versions of *COOLAIR* antisense RNA are made with different 3′ ends. The longest *COOLAIR* transcripts cover the entire length of the *FLC* gene. *COOLAIR* antisense RNAs physically associate with the *FLC* region of the plant chromosome and initiate the epigenetic changes to silence the protein-coding *FLC* gene.

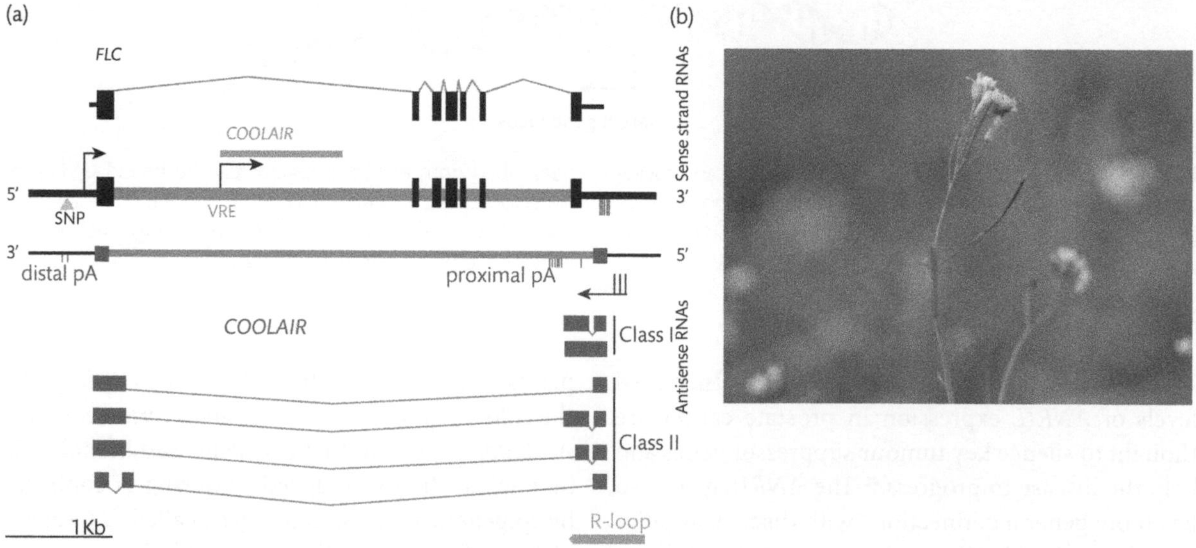

Figure 15.22 **Long ncRNAs control flowering time in plants.** (a) Structure of the *Flowering Control Locus (FLC)* from *Arabidopsis thaliana*. Both strands of DNA are shown, so both sense and antisense transcripts are included. *COOLAIR* is transcribed from the lower strand in an antisense direction to the *FLC* protein-coding gene and the *COLDAIR* long ncRNA gene. (b) Flowering *Arabidopsis* plant. (Part (a) is from Rataj and Simpson.[44] Part (b) was photographed by Kristian Peters and reproduced via a Creative Commons Attribution-Share Alike 3.0 Unported Licence.)

The *FLC* is even more complex. A second long ncRNA called *COLDAIR* is also made from this chromosomal region. Unlike *COOLAIR*, the *COLDAIR* long ncRNA is transcribed in the same sense as the *FLC* gene. Notice in Fig. 15.22 that the *COLDAIR* gene starts within intron 1 of the *FLC* gene, and is shorter than either the *FLC* gene or *COOLAIR*. The function of *COLDAIR* is to bind to the polycomb repressive complex (PRC2), and this further downregulates expression of the FLC transcription factor to induce flowering to take place.

Summary

This chapter has looked at how long ncRNAs are involved in epigenetic regulation of gene expression.

- DNA is packaged in eukaryotic cells, both to enable it to fit inside the volume of a cell and to provide a mechanism of gene regulation. DNA is wrapped around histone proteins to form chromatin. Highly compacted chromatin, called heterochromatin, is transcriptionally inert; it cannot be accessed by the transcription machinery. More loosely packed chromatin, euchromatin, is available to be transcribed.
- DNA is packaged into euchromatin and heterochromatin according to simple molecular tags added to DNA and histones by enzymes. These tags can be inherited on DNA replication and cell division and form the basis of epigenetics—information in the genome which is non-sequence-based but still very important in establishing patterns of gene expression in different cell types and tissues.
- A number of epigenetic modifying complexes are bound to long ncRNA molecules, including the polycomb repressive complex PRC2 that represses gene expression.
- Female mammals use a long ncRNA called *XIST* to inactivate one of their X chromosomes.
- Male flies use a dosage compensation complex containing two ncRNAs to epigenetically upregulate twofold expression from their single X chromosome so that it is equivalent to expression from XX female flies.
- Many imprinted genes occur in clusters containing both long ncRNAs and protein-coding genes. Expression of the long ncRNAs is linked to imprinting patterns of gene expression.
- Long ncRNAs are very important in development. These include the *HOTAIR* long ncRNA that is important for skeletal development, and the *COOLAIR* ncRNA that is important for regulating flowering time in plants.
- Some ncRNAs are very important in disease. In cancer cells, expression of the *HOTAIR* and *ANRIL* long ncRNAs can epigenetically change the properties of cancer cells so that they behave more aggressively.

Questions

15.1 What is chromatin and how can it be epigenetically modified?

15.2 Name an epigenetic regulator complex that is frequently associated physically with long ncRNAs.

15.3 How do XX females normalize X chromosome gene expression so it is equivalent to that of XY males?

15.4 How do insects solve the problem of dosage compensation between XY males and XX females, and how does this involve long ncRNAs?

15.5 What kinds of gene are imprinted in mammals and why?

15.6 Name two connections between imprinting and long non-coding RNAs.

15.7 Define what is meant by antisense RNA.

15.8 What is the name of the antisense RNA that controls flowering time in plants?

15.9 How does inappropriate expression of *HOTAIR* and *ANRIL* affect the properties of cancer cells?

References

1. **Mercer TR, Mattick JS.** Structure and function of long noncoding RNAs in epigenetic regulation. *Nat Struct Mol Biol* **20**, 300–7 (2013).
2. **Luger K, Mader AW, Richmond RK, et al.** Crystal structure of the nucleosome core particle at 2.8 Å resolution. *Nature* **389**, 251–60 (1997).

3. **Peters AH, Kubicek S, Mechtler A, et al.** Partitioning and plasticity of repressive histone methylation states in mammalian chromatin. *Mol Cell* **12**, 1577–89 (2003).
4. **Deaton AM, Bird A.** CpG islands and the regulation of transcription. *Genes Dev* **25**, 1010–22 (2011).
5. **Bernstein E, Allis CD.** RNA meets chromatin. *Genes Dev* **19**, 1635–55 (2005).
6. **Zhao J, Ohsumi TK, Kung JT, et al.** Genome-wide identification of polycomb-associated RNAs by RIP-seq. *Mol Cell* **40**, 939–53 (2010).
7. **Magistri M, Faghihi MA, St Laurent G, 3rd, Wahlestedt C.** Regulation of chromatin structure by long noncoding RNAs: focus on natural antisense transcripts. *Trends Genet* **28**, 389–96 (2012).
8. **Brockdorff N, Ashworth A, Kay GF, et al.** The product of the mouse *Xist* gene is a 15kb inactive X-specific transcript containing no conserved ORF and located in the nucleus. *Cell* **71**, 515–26 (1992).
9. **Brown CJ, Hendrich BD, Rupert JL, et al.** The human *XIST* gene: analysis of a 17kb inactive X-specific RNA that contains conserved repeats and is highly localized within the nucleus. *Cell* **71**, 527–42 (1992).
10. **Hall LL, Lawrence JB.** *XIST* RNA and architecture of the inactive X chromosome: implications for the repeat genome. *Cold Spring Harb Symp Quant Biol* **75**, 345–56 (2010).
11. **Hasegawa Y, Brockdorff N, Kawano S, et al.** The matrix protein hnRNP U is required for chromosomal localization of Xist RNA. *Dev Cell* **19**, 469–76 (2010).
12. **Jeon Y, Lee JT.** YY1 tethers Xist RNA to the inactive X nucleation center. *Cell* **146**, 119–33 (2011).
13. **Lee JT.** Epigenetic regulation by long noncoding RNAs. *Science* **338**, 1435–9 (2012).
14. **Zhao J, Sun BK, Erwin JA, et al.** Polycomb proteins targeted by a short repeat RNA to the mouse X chromosome. *Science* **322**, 750–6 (2008).
15. **Pinter SF, Sadreyev RI, Yildirim E, et al.** Spreading of X chromosome inactivation via a hierarchy of defined polycomb stations. *Genome Res* **22**, 1864–76 (2012).
16. **Grant J, Mahadevaiah SK, Khil P, et al.** Rsx is a metatherian RNA with Xist-like properties in X-chromosome inactivation. *Nature* **487**, 254–8 (2012).
17. **Meller VH, Wu KH, Roman G, et al.** roX1 RNA paints the X chromosome of male *Drosophila* and is regulated by the dosage compensation system. *Cell* **88**, 445–7 (1997).
18. **Ilik I, Akhtar A.** roX RNAs: non-coding regulators of the male X chromosome in flies. *RNA Biol* **6**, 113–21 (2009).
19. **Kageyama Y, Mengus G, Gilfillan G, et al.** Association and spreading of the *Drosophila* dosage compensation complex from a discrete roX1 chromatin entry site. *EMBO J* **20**, 2236–45 (2001).
20. **Ilik IA, Quinn JJ, Georgiev P, et al.** Tandem stem-loops in roX RNAs act together to mediate X chromosome dosage compensation in *Drosophila*. *Mol Cell* **51**, 156–73 (2013).
21. **Maenner S, Muller M, Frohlich J, et al.** ATP-dependent roX RNA remodeling by the helicase maleless enables specific association of MSL proteins. *Mol Cell* **51**, 174–84 (2013).
22. **Deng X, Meller VH.** Non-coding RNA in fly dosage compensation. *Trends Biochem Sci* **31**, 526–32 (2006).
23. **Wutz A.** Noncoding roX RNA remodeling triggers fly dosage compensation complex assembly. *Mol Cell* **51**, 131–2 (2013).
24. **Bartolomei MS, Zemel S, Tilghman SM.** Parental imprinting of the mouse H19 gene. *Nature* **351**, 153–5 (1991).
25. **Nordin M, Bergman D, Halje M, et al.** Epigenetic regulation of the Igf2/H19 gene cluster. *Cell Prolif* **47**, 189–99 (2014).
26. **Kallen AN, Zhou XB, Xu J, et al.** The imprinted H19 lncRNA antagonizes let-7 microRNAs. *Mol Cell* **52**, 101–12 (2013).
27. **Keniry A, Oxley D, Monnier P, et al.** The H19 lincRNA is a developmental reservoir of miR-675 that suppresses growth and Igf1r. *Nat Cell Biol* **14**, 659–65 (2012).
28. **Reik W, Murrell A.** Genomic imprinting. Silence across the border. *Nature* **405**, 408–9 (2000).
29. **Latos PA, Pauler FM, Koerner MV, et al.** AIRN transcriptional overlap, but not its lncRNA products, induces imprinted IGF2R silencing. *Science* **338**, 1469–72 (2012).
30. **Santoro F, Mayer D, Klement RM, et al.** Imprinted Igf2r silencing depends on continuous AIRN lncRNA expression and is not restricted to a developmental window. *Development* **140**, 1184–95 (2013).
31. **Wagner GP.** The developmental genetics of homology. *Nat Rev Genet* **8**, 473–9 (2007).
32. **Rinn JL, Kertesz M, Wang JK, et al.** Functional demarcation of active and silent chromatin domains in human HOX loci by noncoding RNAs. *Cell* **129**, 1311–23 (2007).
33. **Dasen JS.** Long noncoding RNAs in development: solidifying the Lncs to Hox gene regulation. *Cell Rep* **5**, 1–2 (2013).
34. **Tsai MC, Manor O, Wan Y, et al.** Long noncoding RNA as modular scaffold of histone modification complexes. *Science* **329**, 689–93 (2010).
35. **Li L, Liu B, Wapinski OL, et al.** Targeted disruption of Hotair leads to homeotic transformation and gene derepression. *Cell Rep* **5**, 3–12 (2013).
36. **Wan Y, Chang HY.** HOTAIR: flight of noncoding RNAs in cancer metastasis. *Cell Cycle* **9**, 3391–2 (2010).
37. **Kung JT, Lee JT.** RNA in the loop. *Dev Cell* **24**, 565–7 (2013).
38. **Lai F, Orom UA, Cesaroni M, et al.** Activating RNAs associate with Mediator to enhance chromatin architecture and transcription. *Nature* **494**, 497–501 (2013).
39. **Beiter T, Reich E, Williams RW, Simon P.** Antisense transcription: a critical look in both directions. *Cell Mol Life Sci* **66**, 94–112 (2009).

40. **Faghihi MA, Wahlestedt C.** Regulatory roles of natural antisense transcripts. *Nat Rev Mol Cell Biol* **10**, 637–43 (2009).
41. **Gutschner T, Diederichs S.** The hallmarks of cancer: a long non-coding RNA point of view. *RNA Biol* **9**, 703–19 (2012).
42. **Pasmant E, Sabbagh A, Vidaud M, Bieche I.** ANRIL, a long, noncoding RNA, is an unexpected major hotspot in GWAS. *FASEB J* **25**, 444–8 (2011).
43. **Heo JB, Sung S.** Vernalization-mediated epigenetic silencing by a long intronic noncoding RNA. *Science* **331**, 76–9 (2011).
44. **Rataj K, Simpson GG.** Message ends: RNA 3′ processing and flowering time control. *J Exp Bot* **65**, 353–63 (2014).
45. **Turck F, Coupland G.** Plant science. When vernalization makes sense. *Science* **331**, 36–7 (2011).
46. **Csorba T, Questa JI, Sun Q, Dean C.** Antisense COOLAIR mediates the coordinated switching of chromatin states at FLC during vernalization. *Proc Natl Acad Sci USA* **111**, 16160–5 (2014).

16 The short non-coding RNAs and gene silencing

Introduction

In the last chapter we discussed groups of long non-coding RNAs (**ncRNAs**) which have roles in regulating the transcriptional activity of eukaryotic chromosomes. The topic of this chapter is a group of very important RNAs with critical roles in regulating gene expression, but in this case the RNA molecules involved are much shorter, typically 22–30 nucleotides in length. Why should short RNA molecules have been co-opted into gene expression pathways? The reason is that they can hybridize very selectively to target RNA sequences, and so act as an efficient targeting mechanism for directing protein components to nucleic acids. These protein components can then carry out catalytic reactions that include the targeted destruction of RNA and the modification of chromatin.

16.1 Key concepts and common pathways

As a group, the short ncRNAs that we discuss in this chapter carry out three important and diverse roles in cells.

- Group 1: the siRNAs which generally target RNA for destruction.
- Group 2: the microRNAs which generally regulate protein translation from mRNAs.
- Group 3: short siRNAs which target chromatin for modification.

We use the word *generally* because there is significant overlap between these groups—for example, microRNAs can also promote mRNA degradation and epigenetic changes in the nucleus. These functional roles are shown in Fig. 16.1 and listed in Table 16.1. Notice here that analogous groups of short ncRNAs using similar protein components are important in animals, plants, and fungi; hence these short ncRNA systems must have evolved very early in the history of life before the last common ancestor of these three kingdoms.[1,2] The important new terms that we introduce in this chapter to describe these short ncRNAs and the processes they are involved in are included in the Glossary. Short ncRNAs in groups 1 and 2 are involved in **post-transcriptional gene silencing** (PTGS)—this form of gene silencing operates after transcripts have already been synthesized by transcription. The siRNAs work like an 'intracellular immune system' with the aim of incapacitating double-stranded (ds) RNAs that have invaded the cell (such as those produced by viruses). Other short PTGS ncRNAs usually target translational repression of endogenous mRNAs (the microRNA pathway). As we shall see in this chapter, the microRNAs are very important developmental regulators in animals and plants. Group 3 siRNAs are involved in **transcriptional gene silencing** (TGS). These work by directing epigenetic changes in the genome. Hence TGS prevents target genes from being transcribed into RNA in the first place.

Although the different small ncRNA pathways shown in Fig. 16.1 result in somewhat different

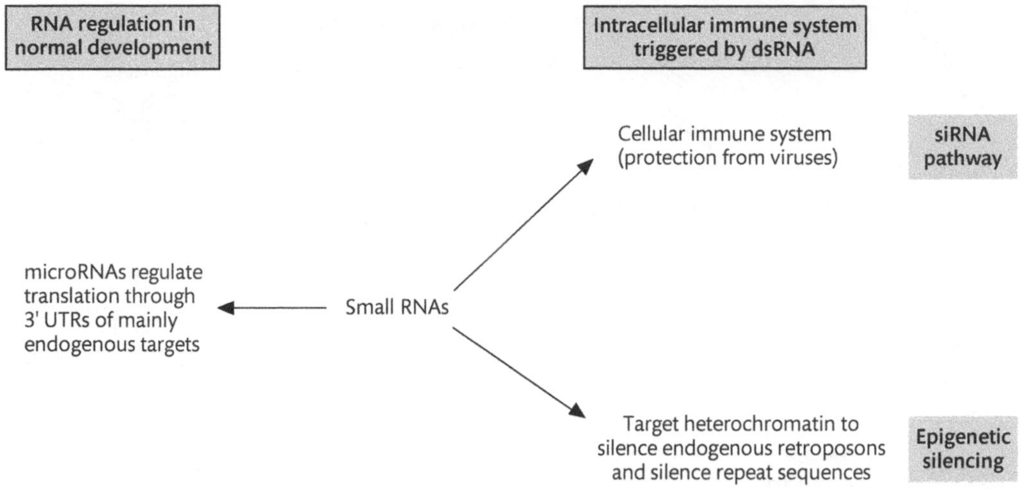

Figure 16.1 Three groups of short ncRNAs regulate distinct pathways within the cell. The siRNA pathway works as an intracellular immune system to destroy dsRNA (this is often viral in origin, although this pathway has been utilized by molecular biologists to inactivate specific mRNAs), the epigenetic silencing group of siRNAs target chromatin-modifying components to convert chromatin into transcriptionally inactive heterochromatin, and the microRNAs regulate target mRNAs during normal development.

Table 16.1 Summary of the properties and functions of the three groupings of different small ncRNAs*

Functional group of short ncRNAs	RNP complex	Trigger	Non-coding RNA and associated protein components	Role in cell
1. Destruction of dsRNA/mRNA (siRNA pathway)	RISC (RNA-induced silencing complex)	dsRNA	siRNAs, Argonaute, Dicer	mRNA cleavage (intracellular 'immune system') Monitoring dsRNA and targeting it for destruction
Destruction of RNA derived from transposable elements	piRNP	dsRNA	piRNAs, PIWI proteins	Cleavage of RNAs made by transposable elements
2. Regulation of endogenous mRNAs; microRNA (miRNA) pathway	miRNP complex	Base pair complementarity between miRNA and target mRNA	miRNAs, Argonaute, Dicer	mRNA translational repression—miRNAs regulate up to 90% of all human mRNAs
3. Directing formation of heterochromatin	RITS (RNA induced transcriptional silencing) complex	dsRNA	Argonaute, Dicer, RNA-dependent RNA polymerase	Directing epigenetic changes leading to transcriptional silencing
(Epigenetic regulation)	ScanRNAs SaRNAs/ microRNAs	dsRNA Complementarity to promoter sequences	ScRNAs, Argonaute, Dicer	Genome rearrangement in *Tetrahymena* Activation of transcription (RNAa) and transcriptional gene silencing (TGS)

*Small ncRNAs are first loaded onto ribonucleoprotein (RNP) complexes. These short ncRNAs then provide a targeting mechanism to direct them to their target nucleic acids. Once hybridized to complementary RNA targets, the complexes have biological effects including target RNA cleavage and translational repression.

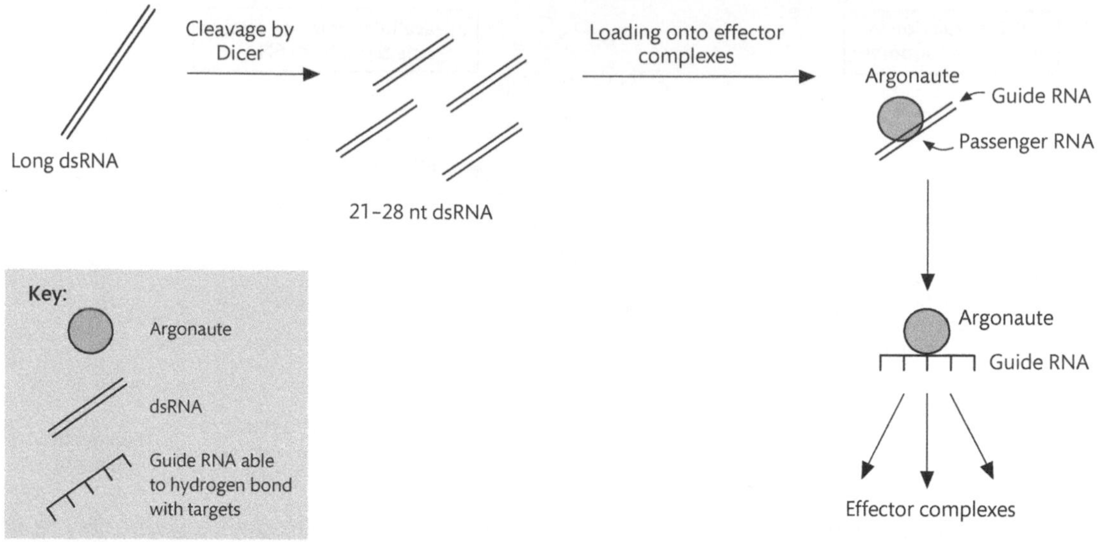

Figure 16.2 An overview of a typical pathway involving each of the groups of short ncRNAs. The Dicer and Argonaute proteins are common components in pathways which are regulated by short RNAs. Dicer proteins are cytoplasmic proteins which recognize long dsRNA precursors, chop them up into 25bp (base pair) dsRNAs, and help load these shorter dsRNAs onto Argonaute proteins. Argonaute binds to dsRNA, and then unwinds this dsRNA duplex, retaining one strand of RNA (the guide RNA) and discarding the other strand (the passenger RNA).

outcomes, they actually share some components and have similarities in the way they are triggered. We shall now look at the shared triggers and components, referring to Fig. 16.2.

16.1.1 Double-stranded RNA is targeted by Dicer

The key trigger for initiating these silencing pathways is the presence of dsRNA. While the initial trigger is dsRNA, in the next key step small RNAs are generated from dsRNA precursors by an enzyme called **Dicer**. Dicer is a cytoplasmic endonuclease (it cuts internally within RNA) which progressively cuts short fragments working from the 3′ end of longer dsRNA precursors.[3]

The atomic structure of Dicer has been solved, showing how this protein works both as a molecular ruler to measure and then as molecular scissors to cut dsRNA into 25bp fragments. The three-dimensional structure of Dicer is shown in Fig. 16.3. The Dicer protein is bound to a dsRNA helix, and the RNA helix has a 3′ overhang. Dicer releases a short dsRNA. Figure 16.3 includes a cartoon that illustrates how Dicer cuts RNA. Dicer is shown as an axe with a handle and a single head. The end of the axe stem is the PAZ domain which grips onto the dsRNA to be cut. The PAZ domain in Dicer binds to the two-nucleotide 3′ overhang in the substrate dsRNA, and the rest of the handle up to the axe head of Dicer is positively charged so that it holds on to the negatively charged dsRNA until it comes into contact with two closely juxtaposed RNAse (RNA cleavage) domains (the axe head). Metal ions in these RNAse domains catalyse a double cut in the substrate dsRNA, releasing a dsRNA fragment.

The length of the axe handle is very important because the distance between the PAZ domain and the RNAse domain determines the length of the dsRNA that is cut by Dicer. This distance is 65Å, which corresponds exactly to 25bp of dsRNA. Dicer does not just cut dsRNA once, but rather dices it up into small 25bp pieces. The two metal ions in the catalytic site are slightly offset so as to generate a new two-nucleotide overhang for a new Dicer protein to bind. This new Dicer protein will then chop off another 25-nucleotide dsRNA, making a new binding site for Dicer in the process, and so on until the entire dsRNA is chopped up. A mouse knockout of Dicer is embryonic lethal, confirming its importance in normal development.[4]

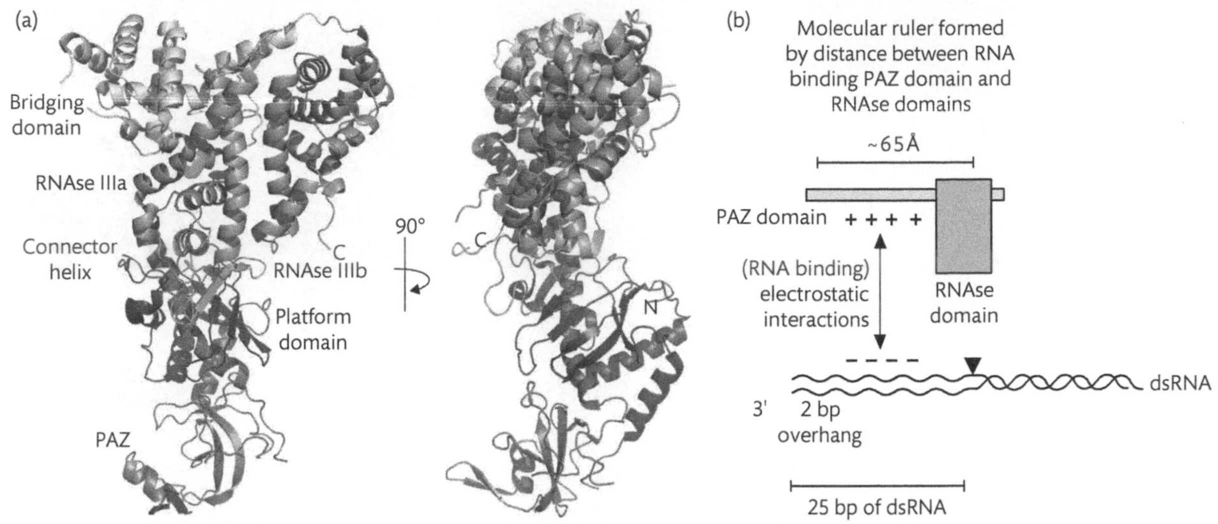

Figure 16.3 Functional organization and atomic structure of the Dicer endonuclease from the parasite *Giardia*. (a) Molecular model of the crystal structure of the Dicer protein bound to dsRNA. Notice that the dsRNA is shown as a double helix. At the end of this helix, the 3′ end has a two-nucleotide overhang and so is slightly longer than the 5′ end. Two views of the protein are shown, one with the RNA in front and the other a side view showing the flat positively charged surface of the Dicer protein which holds the RNA in place. RNA cutting (shown as white arrows) is catalysed by the metal ions in the active sites of the RNAse domains of Dicer. (b) Dicer protein acts like an axe to cut RNA into specific sizes. The PAZ domain binds to the 3′ end of dsRNA, and the RNAse domains behave like an axe head, cutting dsRNA 25 nucleotides upstream. By analogy with an axe, the length of the axe handle determines the length of dsRNA cut, and this is the distance between the PAZ domain and the RNAse domains. The cartoon of the Dicer protein and its associated dsRNA molecule shows the PAZ domain which binds to the 3′ overhang, the positive charges on the surface of Dicer which promote interaction with dsRNA, and the two RNAse domains which interact with each other to create two opposing cleavage sites in the dsRNA substrate. The molecular distance between the PAZ domain and the RNAse domains acts like a ruler to determine the length of RNA which is cut by Dicer. For a full colour version of Fig. 16.3(a), see Colour Illustration 17. (Part (a) based on Macrae et al.[3] with permission. Redrawn from the RCSB Protein Data Bank 2FFL by Alice Mumford, Oxford University Press.)

16.1.2 Argonaute proteins and slicer activity

After they have been generated by cleavage by Dicer, the short dsRNAs are transferred from Dicer proteins to a second set of RNA-binding proteins which belong to the **Argonaute** family. After a dsRNA is loaded onto an Argonaute protein, one RNA strand remains bound to the Argonaute (this is called the **guide RNA**) while the other RNA strand is discarded (this is called the **passenger RNA**). The guide RNA is so named because it is the sequence of this RNA which then can bind the Argonaute–RNA complex to target RNAs in the cell. Loading onto Argonaute proteins is an important step. When complexed with Argonaute, short RNAs form the functional complexes which carry out target RNA cleavage, the formation of heterochromatin, or translational repression in the cell.

Some Argonaute proteins can also act as molecular scissors to cut bound RNA. This RNA-cutting activity of Argonaute proteins is called the **slicer** activity. Argonaute proteins have also been studied at atomic resolution. The crystal structure of an Argonaute protein is shown in Fig. 16.4. Notice from this figure that Argonaute proteins are also RNA-binding proteins and have the same PAZ domain as Dicer which binds the projecting 3′ end of the bound dsRNA. A second domain, called the **MIDI domain**, binds the 5′ phosphate group at the other end of the 25-nucleotide dsRNA. In the C-terminus the **PIWI domain** is responsible for the cleavage activity. The PIWI domain contains an RNAse H fold responsible for the endonuclease activity. The PAZ domain assists with RNA binding; it recognizes, non-specifically, the 3′ end of siRNAs. Current models of slicer activity suggest that the guide strand of

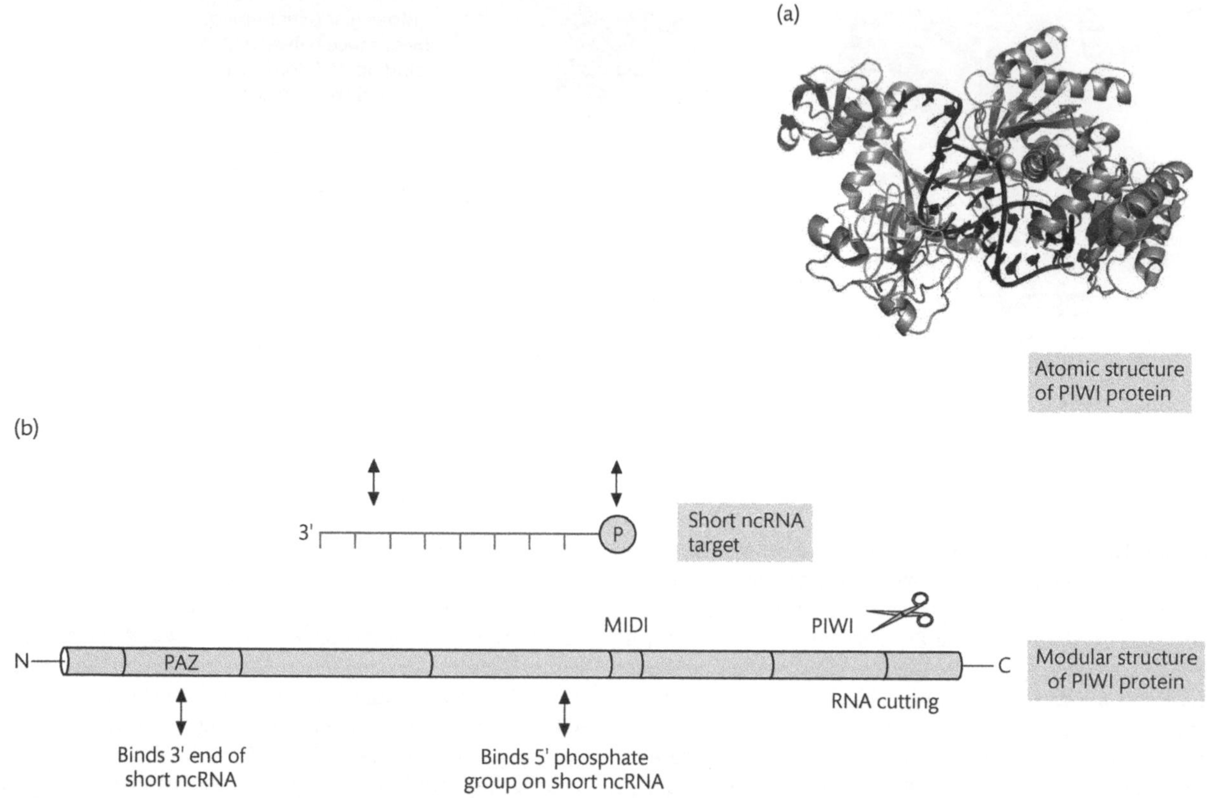

Figure 16.4 Argonaute family proteins. Once generated by Dicer, double-stranded short ncRNAs are loaded onto Argonaute proteins. One of these RNA strands (called the guide strand) remains bound to Argonaute, while the other RNA strand (called the passenger strand) is released. (a) Crystal structure of an Argonaute protein. (b) Cartoon of the PIWI protein showing the important domains and what they do. At atomic resolution a PIWI protein has two lobes. Binding of PIWI proteins to RNA is independent of the RNA sequence, and involves amino acids in either lobe of PIWI. The PAZ domain in one lobe of PIWI binds to the 3′ end of the short ncRNA by inserting into the phosphodiester backbone, and the MIDI domain in the other lobe binds to the phosphate group on the bound RNA. Many Argonaute proteins can act as molecular scissors since they contain a domain which can cut RNA. This is called slicer activity. (Part (a) is from Faradic et al.[1] with permission.)

an siRNA is bound at the 5′ end by the PIWI domain, and at the 3′ end by the PAZ domain. Binding to the target mRNA is initiated by the siRNA 'seed' region at the 5′ end of the siRNA guide strand.

Several Argonaute proteins have been identified. Interestingly, not all of them are necessarily cleavage competent, depending on the presence of specific amino acids in the catalytic site. The existence of non-catalytic Argonautes is explained when we consider the function of microRNAs later; microRNAs work like siRNAs but also repress translation. In the latter case, cleavage of target mRNAs is not required. Note that Argonaute-like proteins also exist in prokaryotes in which they are involved in a process called DNA interference, which prevents the propagation of foreign DNA.[5]

Key points

Two main types of silencing are mediated by short ncRNAs: post-transcriptional gene silencing (PTGS) and transcriptional gene silencing (TGS).

The triggers for the pathways involving each of these different short RNAs are dsRNA molecules which are chopped up by an enzyme called Dicer.

Once it has chopped up the dsRNAs, Dicer helps to load these shorter dsRNAs into further complexes that contain a protein called Argonaute which carries out either RNA silencing (PTGS) or heterochromatin formation (TGS). Argonaute remains bound to one strand of the short dsRNA—this is called the guide RNA, and it acts as an addressing system to find its intended targets through Watson-Crick base pairing.

16.2 Discovery and mechanism of RNA interference

16.2.1 Discovery of RNA interference

More than a decade ago a surprising observation was made during a research project on petunias. While trying to deepen the purple colour of these flowers, Rich Jorgensen and colleagues introduced a pigment-producing gene under the control of a powerful promoter. Instead of the expected deep purple colour, many of the flowers appeared variegated or even white. Jorgensen and colleagues named the observed phenomenon 'co-suppression', since the expression of both the introduced gene and the homologous endogenous gene was suppressed. First thought to be a quirk of petunias, co-suppression has since been found to occur in many species of plants.[6]

But what causes this gene silencing effect? Transgene-induced silencing in some plants appears to involve gene-specific methylation (TGS). In other cases nuclear run-on assays show that silencing occurs at the post-transcriptional level (PTGS). Nuclear run-on experiments are designed to identify genes that are being transcribed at any given time. Radioactive nucleotides are used to label new transcripts which can then be detected by hybridization to a target sequence. In PTGS the homologous transcript is made but then rapidly degrades in the cytoplasm. Once triggered, PTGS is mediated by a diffusible *trans*-acting molecule. From work done in invertebrates (nematodes and fruit flies), we now know that the *trans*-acting factor responsible for PTGS in plants is dsRNA. The process through which dsRNA is able to knock down (silence) genes is called *RNA interference* (RNAi). It turns out that RNAi is, in fact, an ancient process that occurs in both plants and animals.

The first evidence that dsRNA can cause gene silencing came from work in the nematode *Caenorhabditis elegans*.[7] Scientists were attempting to use antisense RNA to shut down expression of the *par-1* gene in order to assess its function (antisense RNA works by base pairing to a complementary sequence on mRNA, sterically blocking its translation). As expected, injection of the antisense RNA disrupted expression of *par-1*, but unexpectedly injection of the sense-strand control also disrupted *par-1* expression. This puzzle became even more interesting when *par-1* dsRNA was injected into *C. elegans*. The injection of dsRNA resulted in much more efficient silencing than injection of either the sense or the antisense strands alone. Indeed, the injection of just a few molecules of dsRNA per cell was sufficient to completely silence the expression of the target gene! Injection of dsRNA into the gut of the worms caused gene silencing not only throughout the worm, but also in its first-generation offspring. Quite surprisingly, soaking worms in dsRNA was sufficient to induce gene silencing (see Fig. 16.5). A Nobel Prize was awarded to Andrew Fire and Craig Mello for this pioneering work (Box 16.1).

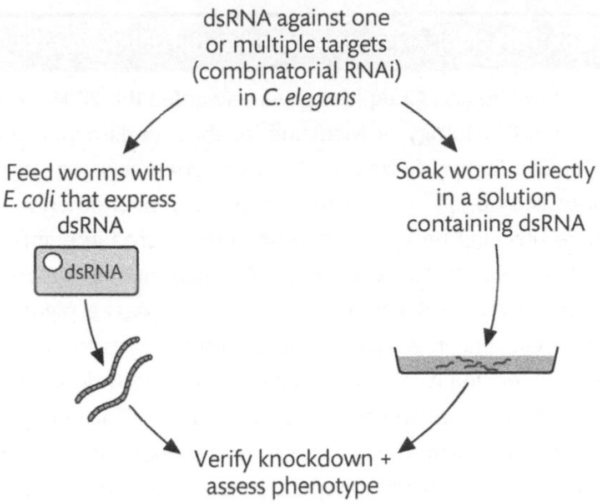

Figure 16.5 The ease of RNA interference in *C. elegans*. Gene expression can be knocked-down in *C. elegans* simply by feeding worms with bacteria that express a particular dsRNA. Silencing of the target gene can then occur in the entire worm. Multiple genes can be silenced at the same time by feeding them with bacteria that express several dsRNAs (this is described as combinatorial RNA interference). Alternatively, the worms can be soaked directly with solutions that contain siRNAs.

Both genetic and biochemical experiments clearly point to the fact that RNA interference as a mechanism has deep evolutionary roots. It has been proposed that RNA interference evolved as a defence mechanism against transposons (endogenously present in cells) and RNA viruses (which burden infected cells with a significant load of dsRNA). A classical example of RNA interference in action was described in 1928. Tobacco plants were infected with

Box 16.1 Nobel Prize awarded to the discoverers of RNA interference

Andrew Fire and Craig Mello were awarded the 2006 Nobel Prize in Physiology or Medicine for their fundamental contribution to the discovery of RNA interference (see http://nobelprize.org/). A bizarre unexplained phenomenon of gene silencing had previously been discovered in plants by Rich Jorgensen and colleagues. A transgene was incorporated into a plant genome in order to express a pigment-producing enzyme–but instead of promoting its expression, it was silencing it. This silencing effect was RNA-dependent, and could be observed in other organisms including the nematode worm *C. elegans*. Curiously, both the antisense and sense RNA could independently silence a homologous target; this was unexpected. Antisense RNA should hybridize to a corresponding mRNA and inhibit its translation, but not a sense RNA. Hence the phenomenon was not simply due to hybridization of the antisense RNA to mRNA, and the term RNA interference was coined by Craig Mello.

The salient features of the pioneering work described in their 1998 *Nature* paper are as follows. They looked at a couple of genes, most famously *Unc-22*. *Unc-22* encodes a myofilament protein, which when its expression is reduced causes a twitching phenotype in which the worms exhibit severe motion abnormalities. Injection of the sense or antisense *Unc-22* mRNA produced a very weak phenotype, whereas injection of dsRNA resulted in severe twitching. The effect was specific to the *Unc-22* mRNA. The dsRNA would only work when using processed mRNA sequence–promoter or intronic sequences did not work. This suggested that RNA interference was primarily post-transcriptional and occurred in the cytoplasm. They also determined that the targeted mRNA decreased substantially, indicating an active degradation process. Silencing was achieved sub-stoichiometrically, meaning that the process was probably catalytic (in contrast with the base pairing of antisense RNA or 'morpholino' oligonucleotides, often used to block translation, which is often effective but not catalytic). Most surprising of all, the silencing effect could spread from tissue to tissue and even to progeny.

It is important to note that many other key papers helped define RNA interference; but the work presented in this *Nature* paper is seen as a landmark in the discovery of RNA interference.

the tobacco ringspot virus. A striking observation was made; leaves were progressively less affected going up the plant, and the leaves at the top of the plant were completely normal, as shown in Fig. 16.6. The reason for this, unknown at the time, is RNA interference—a dsRNA-mediated defence against viruses by the plant. We now turn our attention to the mechanism of RNA interference, extending what we have learned so far.

16.2.2 Mechanism of RNA interference

How does the presence of dsRNA in a cell lead to gene knockdown? In one notable series of experiments, Phillip Zamore and colleagues found that dsRNA added to *Drosophila* embryo lysates was processed to shortened fragments of 21–25 nucleotides.[8] They also found that the homologous endogenous mRNA was cleaved only in the region corresponding to the introduced dsRNA and that cleavage occurred at intervals of 21–25 nucleotides. These were among the first insights into the mechanism of RNA interference.As discussed in Section 16.1, we now know that dsRNA is digested into 21–25-nucleotide dsRNAs; in the context of RNA interference these are known as **small interfering RNAs** (siRNAs). These siRNAs are produced by the enzyme Dicer in an ATP-dependent, processive manner. In the so-called effector step, the siRNA duplexes bind to a nuclease complex to form the **RNA-induced silencing complex** (RISC).[9,10] The RISC complex was first identified in *Drosophila* after fractionation of sequence-specific nuclease activities. (Thus many model organisms have contributed to the discovery and elucidation of RNA interference.)

Although Argonaute is an essential component of the RISC complex, found in all species, there are additional components. These include putative RNA helicases, which may assist in the unwinding of siRNAs, and dsRNA-binding proteins required for RISC assembly. It is important to stress that the RISC complex can work catalytically—a major strength of RNA interference. Thus a RISC complex is capable of multiple rounds of target mRNA recognition and cleavage or 'slicing' activity.[11,12] Once target mRNAs are sliced, they are no longer able to translate (after

Figure 16.6 Tobacco plant infected with the ringspot virus. This picture, originally taken in 1928, shows a tobacco plant infected with the ringspot virus. The symptoms of the infection are most noticeable at the bottom of the plant, whereas the top of the plant appears perfectly healthy. This is due to the activation of RNA interference which prevented the spread of the virus throughout the whole plant. (Reprinted from Baulcombe[6] with permission from Macmillan Publishers Ltd.)

all, only snipping the mRNA once in the ORF does the trick) and become prone to active mRNA degradation processes.

The RISC complex has no way of knowing which strand is the guide or passenger strand; but siRNAs can be designed in such a way as to favour a particular strand (see Box 16.2). The active RISC then targets the homologous transcript by base pairing interactions and cleaves the target mRNA about 12 nucleotides from the 3′ terminus of the siRNA through the slicer activity in the Argonaute proteins described in Section 16.1. The basic mechanism of RNA interference is summarized in Fig. 16.7.

Box 16.2 How to design an effective siRNA

The design of siRNAs has improved significantly, and is continuously improving. To a large extent, improvements in the design of siRNAs are based on empirical evidence. Research has shown that it is important that effective siRNAs carry two-nucleotide 3′ overhangs, that they have <50% GC content, and that they have a very stable duplex at the 5′ end of the guide strand. On the other hand, effective siRNAs tend to have a less stable duplex at the 3′ end of the guide strand where they are preferably AU rich. Lastly, a U is preferred at the cleavage site. All of these criteria facilitate the incorporation of single strands into the RISC complex (the RNA interference machinery has no way of knowing which is the guide strand and which the passenger strand).

It is important to ensure that the siRNA is specific to the target mRNA, otherwise other mRNAs may be affected (giving *off-target* effects). Even if an siRNA is ideal on paper, it still may not work. The reason is to do with mRNA secondary structure; mRNAs fold into complex structures and are packaged by proteins. Hence a particular target sequence in an mRNA may not be readily accessible.

There is an empirical way to determine effective siRNAs—microarray-based screening. In this technology, a large pool of siRNAs that span an entire mRNA are spotted onto a glass slide. The slide is then overlaid with a monolayer of, for example, mammalian cells. In essence the system allows multiple transfections to be performed simultaneously, and the knockdown is assessed using standard assays such as real-time PCR to assess the drop in mRNA levels. Western blotting or enzyme-linked immunosorbent assay (ELISA) must also be performed to verify that the protein level has decreased significantly—this is critical, especially when the protein has a long half-life.

Which part of the mRNA should be targeted with an siRNA? At first it was thought important to target the ORF. However, it is equally effective to target the UTRs. A special case is that of plants. In plants, RNA interference can also occur through modification of chromatin (transcriptional gene silencing). Therefore promoter sequences are commonly targeted by siRNAs in plants.

Key points

RNA interference was discovered through the observation of unusual, and at first unexplained, cases of gene silencing in plants and invertebrates. It is now known to be a widespread and ancient mechanism that probably evolved as a defence against viruses.

Key features of the mechanism are siRNAs, produced from dsRNA by Dicer, and the RISC complex with Argonaute proteins mediating the slicer (target cleavage) activity. Criteria that define what makes an effective siRNA have emerged.

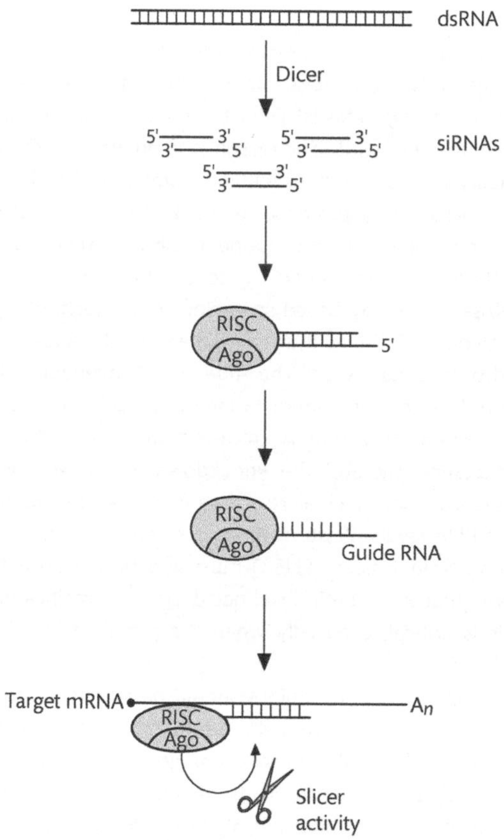

Figure 16.7 Basic mechanism of RNA interference. In RNA interference, a double-stranded molecule is recognized and chopped up by an enzyme called Dicer. The product of Dicer-mediated cleavage is a series of short double-stranded 25-nucleotide RNAs with 5′ overhangs known as siRNAs. The siRNAs are then incorporated into a ribonucleoprotein complex called RISC (the RNA-induced silencing complex). The siRNA is unwound, and the active RISC contains single-stranded 'guide' RNA which is able to hybridize to its intended target. Argonaute proteins (Ago) contain the 'slicer' activity responsible for the cleavage of target mRNA sequences.

16.3 The uses of RNA interference

The problem with expressing long dsRNA molecules in mammalian cells is that they elicit an unwanted interferon-like response which happens through the PKR (dsRNA-activated protein kinase) pathway. This pathway has evolved to warn the cell of viral infections and results in the shutdown of protein synthesis. Thus, for many years RNA interference was very successful in invertebrates but of no use to researchers working on mammalian cells. However, in 2001 there was a major breakthrough—Thomas Tuschl and colleagues realized that all you needed to do was to transfect mammalian cells with siRNAs.[13] dsRNAs less than 30 bases in length do not, in the main, activate the PKR pathway. This breakthrough has led to an explosion of RNA interference research, and mammalian gene knockdown by siRNAs is now a routine procedure in the laboratory.[14]

16.3.1 Limitations of RNA interference

RNA interference is achieved by specific siRNAs that are directed against target mRNAs. However, it is important to remember that RNA interference results in a gene knockdown and *not* knockout. siRNAs are not totally effective and some siRNAs are more efficient than others. There is useful information as to what makes an siRNA effective,[15,16] as discussed in Box 16.2. The structure of an mRNA at the site of the siRNA target undoubtedly influences the accessibility of the RISC complex. Transcription of the targeted genes is not suppressed; therefore the target mRNA is continuously being replenished. Pre-existing protein translated from target mRNAs may have a long half-life if it is a very stable protein; in this situation you can degrade the mRNA but the protein it encodes is still in the cell for a long time. Thus it is necessary to measure empirically the best conditions for knocking down each target gene of interest.

siRNAs are directly transfected into target cell cytoplasm, and this is relatively easy to achieve. siRNAs are commonly mixed with special liposome-based formulations in the same way as DNA is transfected into cells. However, despite being highly potent and catalytic, their effect is transient and repeat transfections are often required to maintain a gene knockdown over the long term.

siRNAs are readily available commercially, and increasingly less expensive to purchase from a number of biotechnology companies. Both strands may be ordered separately and hybridized in the laboratory, but more commonly a lyophilized duplex is purchased. Another method is to transcribe both strands of the siRNA in vitro in the laboratory, using DNA templates that include a prokaryotic promoter for in vitro transcription. This is a more economical method, but having to hybridize the strands is inconvenient and sensitive to ribonuclease contamination. It is also possible to purchase kits that contain the enzyme Dicer, so that long dsRNA molecules (generated from in vitro transcription) can be processed into siRNAs. Whatever the method, siRNAs are now readily available and routinely used.

16.3.2 Why are siRNAs so popular?

siRNAs enable researchers to study gene function in a matter of days. This is much better than having to generate a costly and time-consuming knockout. However, it is important to reiterate that RNA interference can only achieve gene knockdown, not knockout. This means that there is some residual expression of the target gene, which may or may not suit a particular experimental objective. The effectiveness of RNA interference has to be tested empirically for each target gene. In research, the gold standard is to achieve knockdown with at least two or three independent siRNAs to eliminate the possibility of off-target and non-specific effects.[17]

There is another way to achieve RNA interference which is to direct the synthesis of a single **short hairpin RNA** (shRNA) from a plasmid. shRNAs are processed into siRNAs and end up, like any siRNA, in the RISC complex. Plasmids are more difficult to transfect into cells (they need to get into the nucleus); but the advantage is that you can exploit viral delivery systems and generate stably transfected cells, and inducible promoters can be used to switch on the shRNA when required. Several vectors are now available; for example, tetracycline- or ecdysone-inducible expression vectors are available from biotechnology companies. The end-user can subclone an shRNA of choice into a suitable vector, or purchase ready-made shRNA clones from a number of companies and consortia. It is now possible to make transgenic plants and transgenic animals that stably or inducibly express shRNAs. The advantages of siRNA and shRNA approaches to gene knockdown are illustrated in Fig. 16.8.

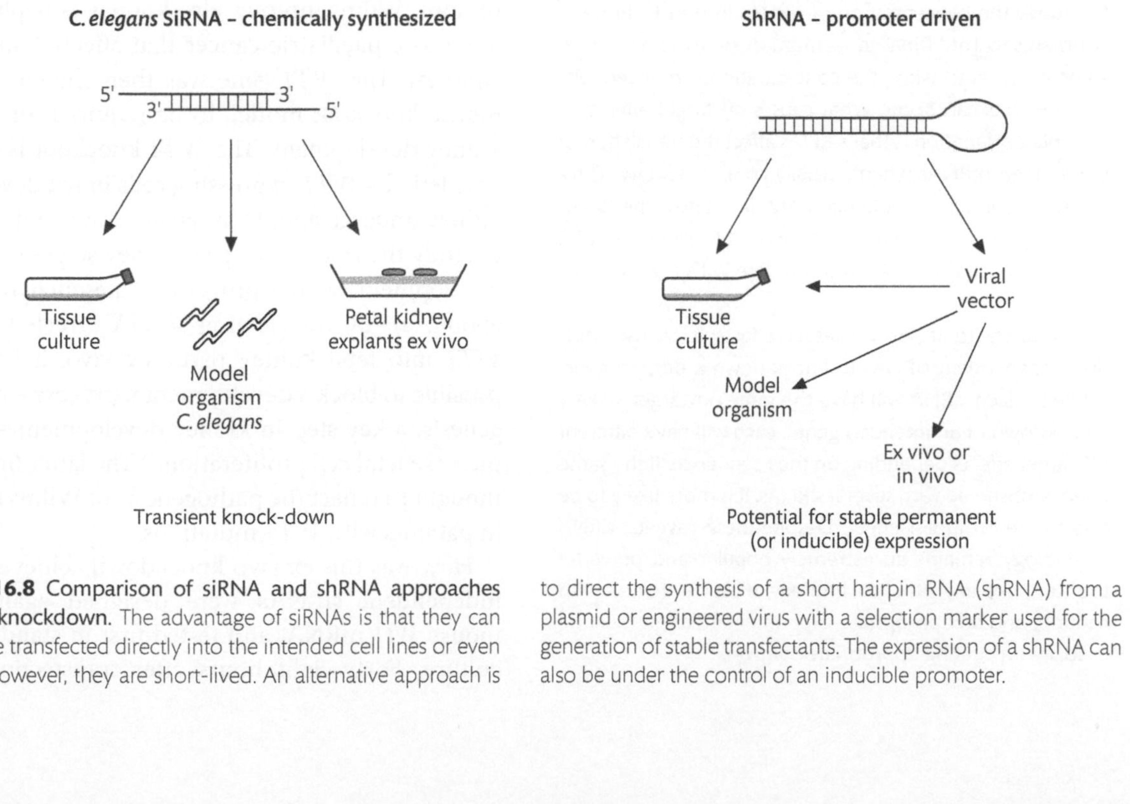

Figure 16.8 Comparison of siRNA and shRNA approaches to gene knockdown. The advantage of siRNAs is that they can simply be transfected directly into the intended cell lines or even tissues; however, they are short-lived. An alternative approach is to direct the synthesis of a short hairpin RNA (shRNA) from a plasmid or engineered virus with a selection marker used for the generation of stable transfectants. The expression of a shRNA can also be under the control of an inducible promoter.

16.3.3 Practical applications of RNA interference

We shall now describe some examples of the applications of RNA interference, but be aware that this is barely the tip of the iceberg. A survey of the RNA interference literature will immediately illustrate how much has been achieved with this technology[18–20] (bearing in mind that the technology does have its limitations, as described in Box 16.3).

One of the key model organisms associated with the discovery of RNA interference is the nematode worm *C. elegans*. The fact that you can simply feed *C. elegans* with bacteria that produce specific dsRNA and achieve knockdown inspired researchers to knock down its genes in a systematic manner. Thus a landmark *Nature* paper in the year 2000 described the knockdown of 90% of genes on *C. elegans*' chromosome I.[21] Functions were assigned to 13.9% of tested genes. It is now possible to interrogate a database (WormBase: http://www.wormbase.org/) for a phenotype associated with knockdown of a *C. elegans* gene of choice.

As well as systematically knocking down all of the genes on a given chromosome, researchers have targeted specific pathways or entire families of proteins. The SR proteins are a family of splice factors which we have discussed at length in earlier chapters. In a classic study, Javier Caceres and colleagues knocked down all the known SR protein genes in *C. elegans*.[22] Functional redundancy was observed in the sense that knocking down many individual SR proteins had no apparent phenotypic consequence. However, silencing more than one at a time (e.g. SRp20 (SRSF3) and SRp75 (SRSF4) together) was deleterious. Only the knockdown of ASF/SF2 (SRSF1) alone caused late embryonic lethality, indicating that some of its functions are both non-redundant and essential. The Wilms tumour suppressor gene *WT1* was discovered in 1990, when it was linked to an 11p13 deletion associated with WAGR syndrome, a congenital syndrome that includes Wilms tumour. Wilms tumour, also known as nephroblastoma, is a paediatric cancer that affects 1 in 10000 children. The *WT1* gene was then shown, using a mouse knockout model, to be required for normal kidney development. The *WT1* knockout is embryonic lethal—*WT1*-expressing cells in the developing kidney undergo apoptosis. Hence it was not possible to study the role of this gene at key stages of kidney development (as the progenitor mesenchymal cells apoptose). By transfecting siRNA directed against *WT1* into fetal kidney tissue ex vivo, it has been possible to block a developmental process—nephrogenesis, a key step in kidney development—and to increase fetal cell proliferation.[23] The latter finding is thought to reflect the pathogenesis of Wilms tumour in patients with *WT1* mutations.

How was this ex vivo knockdown achieved? Two independent siRNAs were designed against the mouse *Wt1* mRNA, and tested first in standard cell culture. Forty-eight hours after transfection, *Wt1*

Box 16.3 Limitations of RNA interference: off-target effects

Like any technology, it is important to note that RNA interference has its limitations. An obvious limitation is that it can only achieve knockdown, and where a small amount of residual gene product is present a clear phenotype may not always ensue. There is another potentially serious problem—off-target effects. Evidence suggests that siRNAs are not necessarily as specific as first thought.

When examining the expression of several genes at once using microarrays, researchers found siRNA-specific instead of target-specific effects. In many cases, a solution is to reduce the concentration of siRNAs in transfection experiments to 1nM (final concentration) or less; however, in some cases, even when the concentration is reduced, off-target effects can occur. What causes off-target effects? A possible explanation is that siRNAs affect the translation of unintended mRNAs when partially complementary. Thus, whereas perfect complementarity facilitates the slicer activity, partial complementarity may still allow the RISC complex to bind mRNAs and interfere with their normal translation.

A strategy to tackle off-target effects is to use multiple independent siRNAs to knock down a desired gene. Whereas each siRNA will have the same on-target activity (knockdown of an intended gene), each will have different off-target effects depending on their sequence. If the same effect is observed with several siRNAs, it is more likely to be due to the on-target effect. Despite these caveats, siRNA technology remains an extremely popular and powerful tool but, like any other technology, must be treated with caution, and an appropriate experimental design must be used, incorporating appropriate controls.

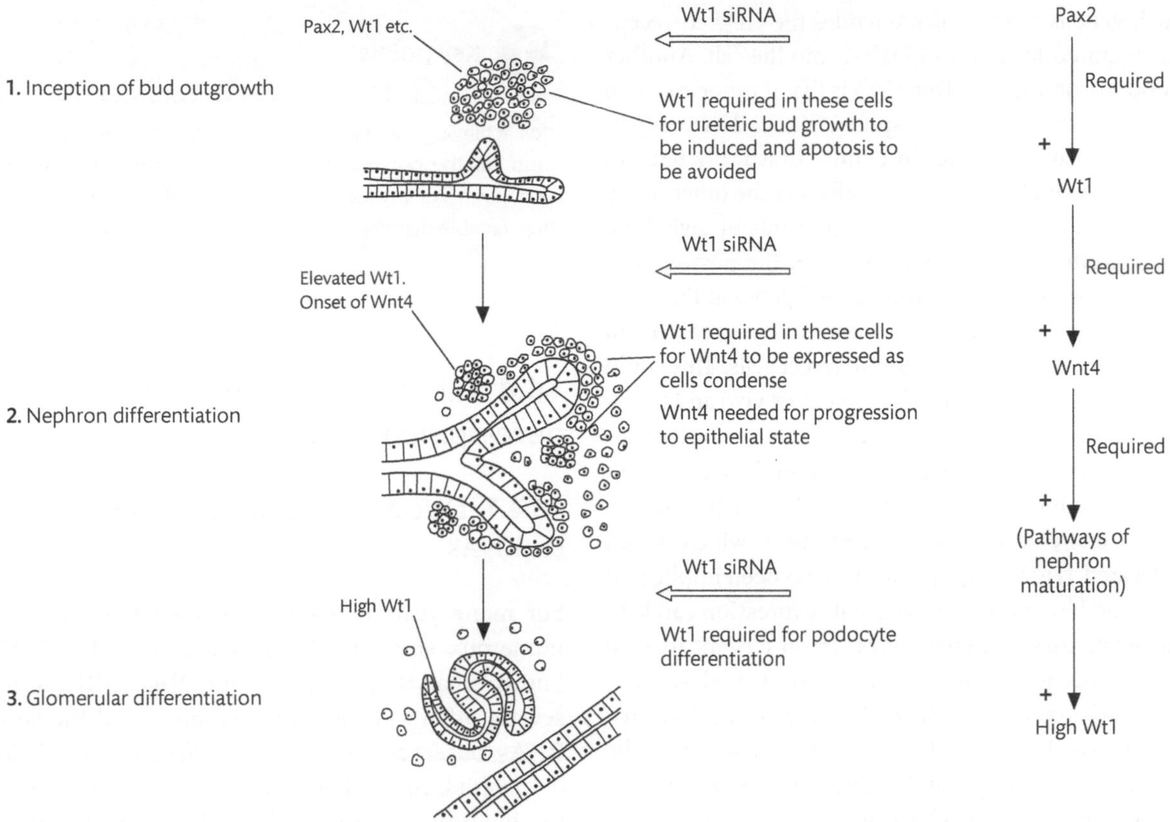

Figure 16.9 siRNAs directed against *WT1* block nephrogenesis. The human Wilms tumour suppressor gene *WT1* was discovered in 1990 and shown to be required for normal kidney development in vertebrates. A mouse *Wt1* knockout fails to develop kidneys at all because of fulminant apoptosis of mesenchymal cells. RNA interference has been used to knock down *Wt1* expression ex vivo in dissected mouse fetal kidney tissue. This has enabled researchers to show that *WT1* is required at distinct phases of kidney development including the formation of nephrons (nephrogenesis). (From Davies et al.[23] with permission from Oxford University Press.)

mRNA and protein levels had decreased substantially. Having validated the potency of the siRNAs in cell culture, they were then transfected into mouse fetal kidney explants. The siRNAs were able to diffuse across a tissue ten cells thick and achieve substantial knockdown of the *Wt1* gene. The formation of kidney 'nephrons' was scored and seen to be significantly reduced only by *Wt1*-specific siRNAs, as shown in Fig. 16.9.

An interesting consideration is whether or not RNA interference might be directed to knocking down specific splice isoforms. A good example is Bcl-x, a member of the Bcl-2 family of apoptosis modulators. There are two major alternative splice isoforms of Bcl-x which arise from an alternative 5′ splice site. One isoform results in a larger protein, Bcl-xL, and the other is Bcl-xS. Whereas Bcl-xL is anti-apoptotic, Bcl-xS has an antagonistic role. Thus, it may be possible to target Bcl-xL with an isoform-specific siRNA. This has been attempted in a cell culture model, and the result was an increased sensitization to agents that induce apoptosis.[24]

Another promising context in which to apply the power of siRNAs is the fight against viral infection; after all, RNA interference had originally been observed in infected tobacco plants. HIV-1 was the first human virus to be attacked by RNA interference.[25] However, the problem in targeting viruses with siRNAs is that they mutate rapidly and acquire resistance to siRNAs just as they acquire resistance to normal drugs. A possible answer is to target cellular genes that are required for the retrovirus to enter the cell and replicate. Cellular genes will not mutate at the same rate and therefore are more likely to be effective targets. One such gene is *Ccr5*, which encodes the CD4 co-receptor required for entry of HIV-1 into the cell. Another good reason for targeting *Ccr5* is that it is not essential for the normal immune response. Knockdown of *Ccr5* in human lymphocytes in culture drastically reduces the ability of HIV-1 to infect cells. On the other hand, HIV-1 may still find a way around this by switching away from a requirement for *Ccr5*; the answer may lie in knocking down multiple viral genes at the same time and delivering the RNA interference reagents in a highly specific manner to the target cells. This could be achieved by selecting CD4+ cells ex vivo and transducing them with a gene therapy viral vector that inserts a series of targeted shRNAs into the CD4+ cells. Treated cells could then be returned to the patient.

Fas is the so-called 'death receptor' which, when activated, triggers apoptosis. Fas has been implicated in several liver diseases where its expression can lead to severe loss of hepatocytes. In an elegant in vivo study, mice were injected intravenously with siRNAs directed against Fas. Both *Fas* mRNA and Fas protein levels decreased significantly as a result of the siRNA, suggesting that the siRNA was able to be absorbed from the bloodstream into the target cells in the liver. These cells were then shown to be resistant to apoptosis.[26] The volume of siRNA solution injected into the mice via the tail vein was substantial, meaning that a comparable treatment in humans is not technically feasible at this stage. However, the experiments demonstrated the principle that siRNA can potentially exert therapeutic effects in vivo.

There is no doubt that siRNAs hold promise in terms of clinical applications, but the key issue is how to target siRNA delivery in vivo effectively.[27,28] Several alternative siRNA delivery systems have been developed, including viral carriers, cationic peptides, liposomes, and lipid nanoparticles (LNPs). To illustrate the potential of siRNA therapeutics based on LNPs, Tabernero and colleagues prepared an LNP formulation of siRNAs that target *VEGF* (vascular endothelial growth factor) and *KSP* (kinesin spindle protein) mRNAs.[29] They considered both the pharmacokinetics and the clinical activity of the siRNAs. The siRNA drug was detected in tumour biopsies and they demonstrated cleavage of the targeted mRNAs in vivo. The LNP formulation was generally well tolerated, and complete regression of liver metastases was observed.

Key points

RNA interference is now a routine and popular method for achieving rapid gene knockdown in the laboratory; it also has the potential to be used as a therapeutic agent. However, like any technology, it has its limitations, including variable degrees of knockdown and off-target effects.

16.4 Discovery, biogenesis, and developmental roles of microRNAs

16.4.1 Discovery and main features of microRNAs

For many years it has been known that there are numerous small RNA molecules in cells.[2,30] The known small RNA species—snoRNAs, snRNAs, and several others—could not account for all the small RNAs observed. As RNA interference was being discovered, researchers also began to find out what the identity of these small RNAs might be. At the beginning of the twenty-first century a new class of small RNAs was identified and named **microRNAs** (miRNAs). MicroRNAs are now known to be widespread, existing in the plant and animal kingdoms, and numerous, being encoded by perhaps as many as 2–3% of all genes.[31] MicroRNAs are very important because they are involved in developmental and disease processes.

The first microRNA (associated with a phenotype) was described in the nematode *C. elegans*. In a screen for mutants with defects in the ability to govern the timing of the switching of cell fate in development, two genes, *lin*-4 and *let*-7, were identified. These genes are regulators of developmental timing events, and their own expression is temporally regulated. To their surprise, the researchers found that these genes encoded ncRNAs; these were in fact microRNAs. The Lin-4 microRNA and its target *Lin-14* mRNA are shown in Fig. 16.10. The *Drosophila* gene *bantam* also encodes a microRNA, which targets the 3′ UTR of *Hid* mRNA. Hid is a key activator of apoptosis, and its downregulation by *bantam* microRNA is a requirement in normal development. Both these *C. elegans* and *Drosophila* microRNAs behaved like

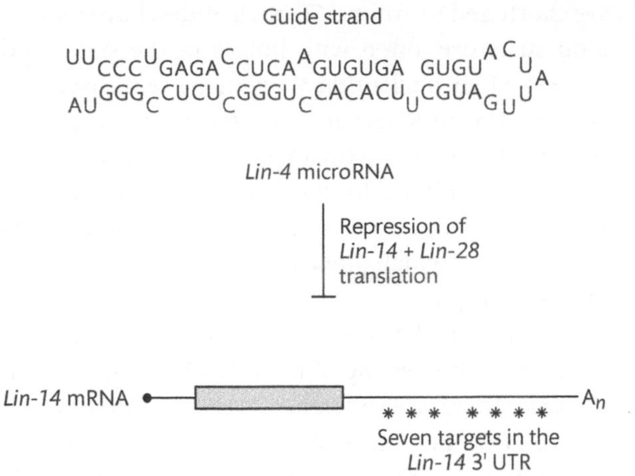

Figure 16.10 The Lin-4 microRNA and its target, the 3′ UTR of *Lin-14* mRNA. Lin-4 works during larval development to regulate the expression of the genes *Lin-14* and *Lin-28* involved in the regulation of development. Note the elegant simplicity of the Lin-4 microRNA. In length, it is effectively the equivalent of a guide strand of a siRNA. It has perfect base complementarity to a target in the 3′ UTR of the *Lin-14* and *Lin-28* mRNAs, illustrating the fact that a given microRNA can potentially target several mRNAs. Note that microRNAs are effective even if not all the bases are perfectly complementary; however, in such cases the effect is thought to be translational repression as opposed to destruction of mRNA.

any typical protein-coding gene that has a function in development. However, they were seen not to encode a protein but rather a short ncRNA molecule. Strikingly, their size was reminiscent of siRNAs, leading to the hypothesis that in fact microRNAs and siRNAs are part and parcel of the same process.[32,33] We now know that this is the case, and the distinction between microRNAs and siRNAs is blurred. As we saw in Section 16.1, there is considerable overlap in the cellular machinery that is associated with both siRNAs and microRNAs.

Several thousand microRNAs have been identified in eukaryotic cells in animals and plants, and in viruses. There are several ways of identifying microRNAs, as described in Box 16.4. A database of microRNAs is kept on the microRNA registry known as miRBase at the University of Manchester, UK (http://www.mirbase.org/). In total over 28 000 miRNAs have been identified to date (late 2014), and the number is growing rapidly.

MicroRNA genes are frequently found in clusters, which suggests that they might have arisen from duplication events. Several are **polycistronic**, which means that they are processed out of a single primary transcript. The genomic locations of microRNA genes vary. In humans, they are found in all chromosomes except the Y chromosome. In mammals, about 30% of microRNA genes are found in intergenic regions but, surprisingly, 70% are found within other larger transcription units. Within the larger transcription units they can be located within intronic and sometimes even within exonic sequences.[31] The full complement of microRNA genes in a genome has been described as the **microRNome (miRNome)**.

Box 16.4 Identification of microRNAs

How are microRNAs identified? Broadly speaking, there are three approaches.

The first is traditional genetics; the gene responsible for a phenotype is identified, cloned, and seen to encode an ncRNA.

The second approach is to use direct cloning of candidate microRNAs. Small RNAs of a size consistent with microRNAs (19–25 nucleotides) are isolated from a denaturing polyacrylamide gel, ligated to adaptor molecules at both the 5′ and 3′ ends, and amplified by PCR. Although this direct technique is very powerful, several of the clones are found to correspond to fragments of abundant RNAs such as rRNA or tRNA, but these can be easily identified and eliminated from the pool. Another drawback of the direct cloning method is that microRNAs that are not very abundant are more likely to be overlooked.

A third means of identifying microRNAs is to use a bioinformatics approach. MicroRNA genes have a number of recognizable features, such as sequences that when transcribed give rise to what looks like a typical microRNA precursor with an extensive hairpin structure. MicroRNAs also tend to exist in clusters; and if they show homology to other microRNA genes it makes them more likely to be bona fide microRNA genes.

As well as not being a fragment of another (larger) RNA, microRNAs need to satisfy a set of criteria, namely that they are derived from a longer precursor RNA which folds into a hairpin structure of typically 80 nucleotides in animals (of more variable length in plants). The expression of the mature microRNA and its precursor must be verified independently. In other words it must be shown to exist in multiple copies.

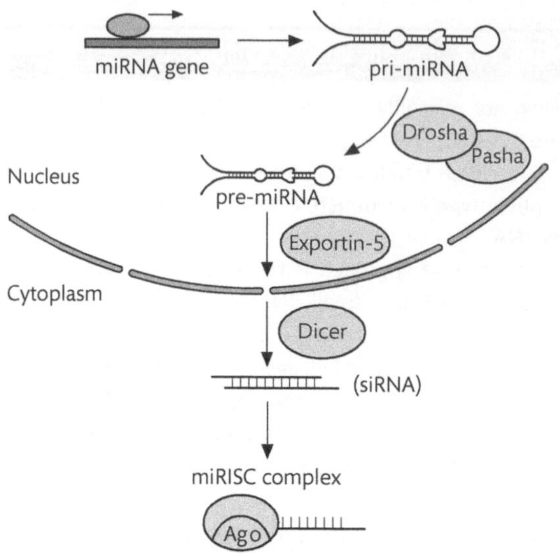

Figure 16.11 Biogenesis of microRNAs. The first step in microRNA (miRNA) biogenesis is the transcription of an miRNA gene into a transcript called the 'pri-miRNA'. The latter is trimmed by a nuclear protein called Drosha, a member of the RNAse III family, and by the dsRNA-binding protein Pasha, generating a stemloop structure known as a 'pre-miRNA' (not to be confused with a pri-miRNA). Pre-miRNAs are exported into the cytoplasm aided by a specific exportin, exportin-5. In the cytoplasm the pre-miRNA is recognized by Dicer and processed into the equivalent of an siRNA which is then incorporated into a miRISC complex.

16.4.2 Biogenesis of microRNAs and properties of miRISC complexes

The transcription of microRNA genes is mostly mediated by RNA polymerase II, which results in primary microRNA transcripts that are capped and polyadenylated. An important question is what regulates the expression of microRNA genes. Just like protein-coding genes, complex regulatory frameworks are likely to influence microRNA gene expression. As microRNAs are misexpressed in disease, their regulation is being investigated, but it is not yet well understood. Instead, at present more is known about their biogenesis[34,35] (Fig. 16.11). The primary transcript of a microRNA gene is known as a **pri-miRNA**. The pri-miRNAs are then essentially cropped by an enzyme called **Drosha**, part of a microprocessor complex. Like Dicer, Drosha is also a member of the RNAse-III-like family of enzymes. Drosha is assisted by **Pasha**, a dsRNA-binding protein. What arises from cropping is a **pre-miRNA** (not to be confused with a pri-miRNA)—pre-miRNAs are shortened to around 70 nucleotides, have a stem-loop structure, often with bulges in the stem, and possess a two-nucleotide 3′ overhang reminiscent of siRNAs. Pre-miRNAs are then specifically exported by a specialized exportin, exportin-5, in conjunction with Ran GTPase. In the cytoplasm, the enzyme Dicer processes the pre-miRNA into a 22-nucleotide intermediate which then assembles into a mature **miRISC** complex.

It transpires that much of the machinery responsible for the processing of microRNAs is shared with siRNAs. There are differences between siRNAs and microRNAs, but the distinction is at times somewhat blurred. The most important point about microRNAs is that they are encoded by nuclear genes. RISC complexes are known as siRISCs when they include siRNAs, and as miRISCs when they contain microRNA-derived sequences. The key similarity is that both siRNAs and microRNAs are unwound with a guide strand ending up in a RISC complex. RISC complexes, in general, can be subdivided into *cleaving* (with slicer activity) and *non-cleaving* (with no slicer activity, when the specific Argonaute does not cleave).

Therefore Argonaute proteins are not always enzymatically active. The consequence is that non-cleaving RISCs block translation, as opposed to cleaving target mRNAs. In a sense, they behave like antisense oligos. A further complication in understanding the action of siRISCs and miRISCs is that they can associate with a target mRNA even when they do not have perfect complementarity. This has repercussions in terms of off-target effects and increases their likelihood, but, by the same token, only RISC complexes that have active slicer and perfect complementarity will result in the cleavage of targets. Cleavage of targets releases RISCs for another round of cleavage (a catalytic process) and therefore is likely to be more potent. On the other hand, non-cleaving RISCs, or cleaving RISCs with imperfect complementarity, result in translational repression but the target mRNA is kept intact, which means that the mRNAs could in theory resume translation if the RISC complexes were removed, reversing the gene silencing effect. Therefore it should be borne in mind that the distinction between siRNAs and microRNAs is ambiguous, and that both can potentially direct mRNA degradation or translational repression.

16.4.3 MicroRNA target identification and association with P-bodies

A crucial question is how to identify bona fide mRNA targets of microRNAs. Current methods for target identification are rapidly evolving, but are still relatively limited.[36] By analogy with the mechanism of action of siRNAs, one of the most important criteria in a target is excellent complementarity between the 3′ end of the mRNA target and the 5′ end of the guide sequence of the siRNA or microRNA. The 'seeding' starts at position 2 in the guide sequence, up to the sixth or eighth nucleotide. Complementarity in the remainder of the microRNA can be looser, but only when the 5′ end is strongly complementary. On the other hand, effective microRNA targets can include a slightly weaker match at the 5′ end compensated by strong complementarity at the 3′ end. This indicates a degree of versatility in the definition of targets. In fact the definition of targets is further complicated by mRNA structure; as discussed in Box 16.2, a given siRNA (or microRNA) might be effective in theory, but only if a target mRNA sequence is sterically accessible. However, one could predict the existence of RNA helicases that assist microRNAs by unwinding potential target mRNAs.

Algorithms have been developed to search for potential mRNA targets, and these can be accessed directly on the internet. For example the program MiRanda has been used to identify mRNA targets in genomic sequences (http://www.microrna.org/). Whatever potential targets are identified, they need to be verified empirically. MicroRNA targets are also in the process of being collated in freely available collections, such as the microRNA registry at the Sanger Centre. In general, a given microRNA may have multiple targets because of the fact that partial complementarity can give rise to translational repression. Conversely, a given mRNA may be regulated by multiple microRNAs which in concert ensure that effective repression takes place. Undoubtedly, with microRNAs evolution has generated a complex machinery, and the challenge over the coming years is to define hundreds if not thousands of microRNA targets in several model organisms.

A further factor in microRNA action is a cytoplasmic structure called the **P-body**. P-bodies are cytoplasmic processing bodies which contain several proteins involved in both the degradation and translational repression of mRNA (described in Chapter 12). There is evidence that mRNAs that are targeted by microRNAs are physically adjacent to P-bodies.[37] Thus the P-body could be providing assistance to microRNAs with activities that help to remodel the mRNP particle, making it accessible to mRNA decay (decapping enzymes, endo- and exonucleases). In this scenario the miRISC would be tagging an mRNA for active degradation, despite not possessing a slicer-active Argonaute. A component of P-bodies is the RNA helicase p54/DHH1, independently identified as a major component of stored mRNP particles. P-bodies also contain the decapping enzyme Dcp1 and the nuclease Xrn1. RNA helicases would be instrumental in remodelling mRNPs that have been attached by a miRISC.

Alternatively, subsets of mRNAs may be sequestered with the aid of P-bodies and stored until physiological changes facilitate their release back into the pool of actively translating mRNAs. In this scenario, the microRNA is acting as a blocker of translation and the P-body would provide the framework for the storage of the translationally repressed mRNA.

16.4.4 Nuclear functions of microRNAs

At first microRNAs were thought to work exclusively in the cytoplasm, where they are known to cause mRNA translation repression and mRNA decay. However, it is now clear that most microRNAs are present in both the cytoplasm and the nucleus. Protein components of active miRISC complexes (including Argonaute proteins) are also detectable in the nucleus. Nuclear cytoplasmic shuttling of microRNAs involving CRM/exportin 1 and importin 8 has also been demonstrated. What then is the role of microRNAs in the nucleus? It is now clear that microRNAs also have roles in the nucleus, mainly in epigenetic regulation affecting transcription and alternative splicing.[38–40]

In the nucleus microRNAs can target promoters. They can help activate transcription; this class of microRNAs are described as **saRNAs** (small activating RNAs). saRNAs are involved in a process called **RNAa** (RNA activation of transcription). For example, miR-373 activates transcription of the *E-cadherin*

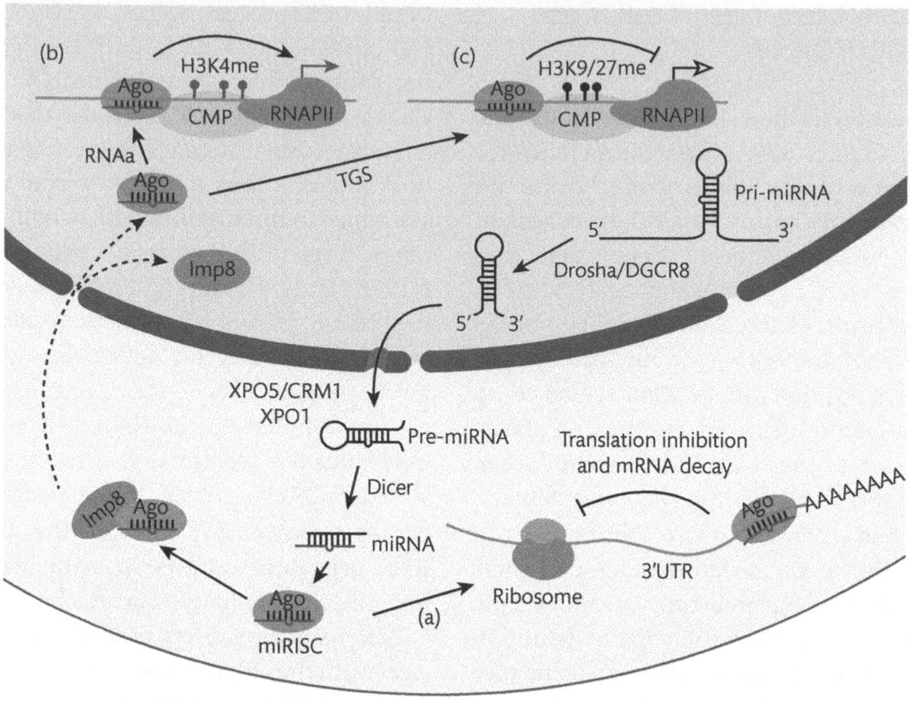

Figure 16.12 Effects of microRNAs on transcription. MicroRNA precursors are initially transcribed and trimmed in the nucleus and exported to the cytoplasm where they are further processed and associate with Argonaute proteins (Ago) in miRISC complexes. (a) The best-known function of microRNAs is in promoting translational repression and mRNA decay. However, microRNAs and associated Argonaute proteins are also imported into the nucleus with the aid of importin 8. In the nucleus they interact with target promoters. They can cause (b) transcriptional activation (RNAa) and (c) transcriptional gene silencing (TGS) by recruiting specific chromatin modifiers. (Adapted with permission from Huang and Li[38].)

gene; there are potential miR-373 target sites in its promoter. saRNAs complexed to Argonaute (Ago) can potentially bind to the promoter directly, or to promoter-derived nascent transcripts. SaRNAs are thought to recruit chromatin modifiers that increase H3K4 methylation (at lysine 4) which in turn helps activate transcription. On the other hand, the recruitment of a different set of chromatin modifiers can result in transcriptional gene silencing (TGS), for example by increasing H3K9/27 methylation. The involvement of microRNAs in the regulation of transcription through chromatin modifications is summarized in Fig. 16.12.

In Chapter 8 we discussed how changes to chromatin structure can have a profound effect on alternative splicing. If microRNAs can promote histone modifications, might they also affect alternative splicing? The answer is yes. Argonaute-1 is clearly involved in the regulation of alternative splicing; it controls the level of inclusion of the cassette exon 107 of the *SYNE2* gene and exon 33 of the *FN1* gene.[41] The effect of Argonaute-1 on alternative splicing is mediated by small ncRNAs that target intronic sequences; the interactions cause localized chromatin modifications which in turn affect alternative splicing.

16.4.5 MicroRNAs in development and disease

Several microRNAs are to be found in plants, where they have clear developmental roles.[42,43] In plants, microRNAs generally seem to exhibit much better matches to their target mRNAs, which may potentially facilitate target mRNA identification. MicroRNAs are clearly evolutionarily conserved in plants; several microRNA targets in non-flowering plants are homologous to those in the flowering plant model organism *Arabidopsis thaliana*. In fact there are clear homologies between microRNA families

Table 16.2 Examples of functionally characterized microRNAs and their mRNA targets in animals and plants*

Species	MicroRNA	Target mRNA
Caenorhabditis elegans	Lin-4	Lin-14, Lin-28
Caenorhabditis elegans	Let-7	Ras, Lin-41, Lin-57, Daf-12
Drosophila melanogaster	Bantam	Hid
Mus musculus	miR-1	Hand2, HDAC4
Mus musculus	miR-133	SRF
Homo sapiens	Let-7	Ras
Homo sapiens	miR-15	Bcl-2
Arabidopsis thaliana	miR-167	Auxin-responsive factors 6 and 8
Arabidopsis thaliana	miR-177	tRNA-Gly
Arabidopsis thaliana	miR-179	U1 snRNA

*Target mRNAs include those that encode proteins involved in development and cell proliferation; note that tRNAs and snRNAs are also targets.

all the way from mosses to *Arabidopsis*. MicroRNAs have been extensively studied in *Arabidopsis*, where they have been shown to control floral development and timing, leaf patterning and shape, vascular development, and fertility, among other things. Given their roles in plant development and physiology, the functions of microRNAs in plants are of interest to the biotechnology industry. The identification of microRNAs in several economically important crops is fast becoming an important issue.

MicroRNAs have also been implicated in normal vertebrate development. For example, during development of the mouse heart the activation of SRF (serum response factor) in myocytes results in expression of miR-1-1 and miR-1-2, which in turn repress the expression of the transcription factor Hand2 and the Notch ligand Delta, with profound consequences for the differentiation of progenitor cells. In a negative feedback loop, SRF also induces the expression of microRNAs miR-133a-1 and miR-133a-2 which actually inhibit the expression of SRF itself. Thus microRNAs are involved in developmental pathways including feedback loops.[44]

With hundreds and perhaps thousands of microRNAs in vertebrates, the potential for involvement in several developmental pathways is significant. It is also possible that microRNAs work by reinforcing the action of transcription factors and splice factors or other RNA-binding proteins on the expression of target genes in a coordinated fashion. Examples of microRNAs and their target mRNAs are listed in Table 16.2.

Several microRNAs have been implicated in cancer.[45] Like protein-coding genes, microRNAs can act as tumour suppressors or oncogenes. The reason for their involvement in cancer is that they regulate processes such as cell differentiation, proliferation, and apoptosis. Researchers have profiled the expression of several microRNAs in cancer patients and found significant differences between microRNAs in tumour cells and in the corresponding normal tissues. Thus the expression of the microRNA let-7 is reduced by more than 80% in 44% of lung cancer patients, and reduced levels are also associated with short post-operative survival.

Let-7 is a particularly interesting family of microRNAs. First identified in *C. elegans*, there are close homologues in humans. Let-7 microRNA regulates the expression of the *Ras* family of oncogenes, which are GTPases involved in cell signalling that are mutated in over 30% of all cancers. Both the Let-7 microRNA and its mRNA target (Ras) are strikingly conserved from *C. elegans* to humans. Over-expression of Let-7 results in reduced cell proliferation in a cell line model of lung cancer.

Thus, the following points can be made: microRNAs can clearly be conserved across evolution, just like protein-coding genes. MicroRNAs can potentially be viewed as therapeutic agents in that they might indicate suitable targets for the development of stable chemically modified siRNA-type drugs. Undoubtedly, future medical science will exploit nature's own RNA interference technology, the microRNAs.

Key points

MicroRNAs are short ncRNAs encoded by a significant proportion of genes. They can be discovered through standard genetic screens, direct cloning, or bioinformatics approaches.

MicroRNA precursors are trimmed down essentially into siRNA-like molecules that are incorporated into miRISCs. miRISCs work in two ways to achieve gene knockdown: they repress translation (non-cleaving RISC), or they cleave target mRNAs (cleaving RISC). MicroRNAs are also associated with mRNA decay and translational repression in P-bodies. They also work in the nucleus where they can induce epigenetic alterations and influence transcription and alternative splicing.

MicroRNAs have important roles in development and are also implicated in disease, including cancer.

16.5 Transcriptional silencing by non-coding RNAs in the centromere

We noted in Section 16.4.4 that microRNAs have nuclear functions in regulating gene expression. We now consider additional pathways used by short ncRNAs to regulate gene expression at the transcriptional level, namely TGS.[46,47] TGS works by inducing the formation of a transcriptionally inactive DNA called heterochromatin. Two particular types of chromosomal DNA are especially targeted for transcriptional inactivation: these are the **centromeres** (the parts of each eukaryotic chromosome which are important for chromosome movement during cell division) and mobile elements in chromosomes called transposable elements. The short ncRNAs involved in TGS also use Argonaute proteins and are triggered by dsRNA. However, one superficial difference is that some of the proteins involved in the complexes which form heterochromatin are slightly different from, although evolutionarily related to, Argonaute and have names such as PIWI and Aubergine.

Eukaryotic genomes are compartmentalized into chromosomes. Two regions of eukaryotic chromosomes are essential for their structure and function: the sequences at their ends called telomeres, and internal sequences called centromeres which are critical for chromosome segregation at cell division. Before cells divide, they replicate their chromosomes and then provide each daughter cell with a complete set through a process called mitosis. Centromeres are critical for mitosis. During mitosis a structure called the kinetochore assembles over each centromere and provides an attachment site for the microtubules which pull the duplicated chromosomes apart into daughter cells. Hence centromeres provide an anchor point to pull the chromosomes apart at cell division.

The chromatin at centromeres is highly condensed (heterochromatic) compared with adjacent non-centromeric chromatin, and this seems to be very important for centromere function—insertion of active genes into centromeres blocks their function. However, although centromeric DNA is condensed into heterochromatin, it is not totally transcriptionally silent, and is still transcribed at a low level. Centromeric DNA contains a number of repeated sequence elements which contain promoter elements pointing in either direction (towards each telomere). Thus these repeated elements are transcribed in both directions to give complementary RNAs which can hybridize through Watson–Crick base pairing to give dsRNAs (Fig. 16.13). These longer dsRNAs are then cut up into shorter dsRNAs which play a critical role in turning the centromeric chromatin into heterochromatin.

The role that short ncRNAs play in establishing centromeric heterochromatin is particularly well understood in yeast, where ncRNAs operate through a pathway called **RNA-induced transcriptional silencing** (RITS). An outline of the RITS pathway is shown in Fig. 16.14. The heterochromatin is shown as a zigzag line to indicate that it is highly condensed.

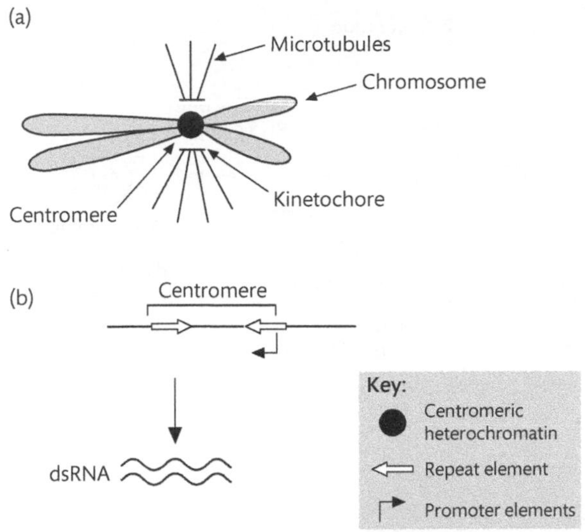

Figure 16.13 Centromeres are heterochromatic regions which generate low levels of dsRNA. Centromeres are a heterochromatic region within chromosomes which assemble a structure called the kinetochore at cell division. Because they are heterochromatic (containing more condensed chromatin), the centromeres are more compact and so appear thinner when the chromosomes are condensed during cell division. The kinetochore is the structure on which the microtubules assemble to pull the chromosomes apart during cell division. At the DNA level, centromeres contain several repeated sequences which contain promoter elements but are transcribed at a very low level. Since the promoters within these repeat elements point in both directions in the centromeric DNA, transcription from them results in RNAs being made from both strands of the DNA and so dsRNA is formed.

16.5.1 The three important steps in establishing RITS

The first step is the transcription of the centromeric sequences to generate dsRNA. Since it is within heterochromatin, centromeric DNA is not transcribed very much. However, the low level of observed transcription which does take place is, in fact, critically important for maintaining the heterochromatic state of the centromere. In this first step, an RNA polymerase II molecule makes a transcript from centromeric DNA.

Since promoters point in both directions within centromeric DNA, both strands are transcribed to give complementary RNAs. These complementary RNAs can hybridize with each other to give dsRNA. In yeast, single strands of centromeric RNA are also copied into double-stranded RNA by an **RNA-dependent RNA polymerase** (RDP) enzyme; these enzymes make an RNA strand from an RNA template. Therefore transcription from both strands of centromeric DNA and the RDP enzyme both produce dsRNA molecules. Note that dsRNA is the key trigger for the process of RITS.

Next, the dsRNA is exported from the nucleus into the cytoplasm. Once in the cytoplasm, the dsRNA is cut by the Dicer RNA cleaving enzyme to give a population of siRNA molecules. These siRNAs are then loaded onto a complex called the RITS complex. The RITS complex also contains an Argonaute protein and DNA methylating enzymes. The Argonaute protein remains attached to one RNA strand of the original double-stranded siRNA (the retained strand is called the guide strand) and this complex is imported into the nucleus. Thus the RITS complex is a chromatin-silencing complex with an address system provided by the short siRNA.

Once in the nucleus, the guide strand RNA of the RITS complex targets the whole complex exactly back to the centromere from which the guide RNA was first transcribed. This address system works because of **complementary base pairing** (see Chapter 2). Because the RNA component of the RITS complex was made first from dsRNAs transcribed from the centromeric sequence, the guide RNAs in the RITS complex will be complementary in sequence to nascent RNAs which are in the process of being newly transcribed from the same centromere.

16.5.2 The functions of RITS

Once targeted to the centromere by base pairing, the RITS complex has two important functions. The first of these is to further repress any transcriptional activity of the centromeres. Once targeted back to the general area of centromeric chromatin through tethering to nascent centromeric RNA, the protein components within the RITS complex methylate the DNA of chromatin and subsequently histone H3K9 to help maintain transcriptionally inactive heterochromatin.[48]

Secondly, since the Argonaute protein within the RITS complex is itself an RNA-cutting enzyme, it can cleave more nascent transcripts made from the centromere and so generate more cut dsRNAs to feed into more RITS complexes. Through these two mechanisms the low levels of transcription of centromeric heterochromatin act to reinforce transcriptional

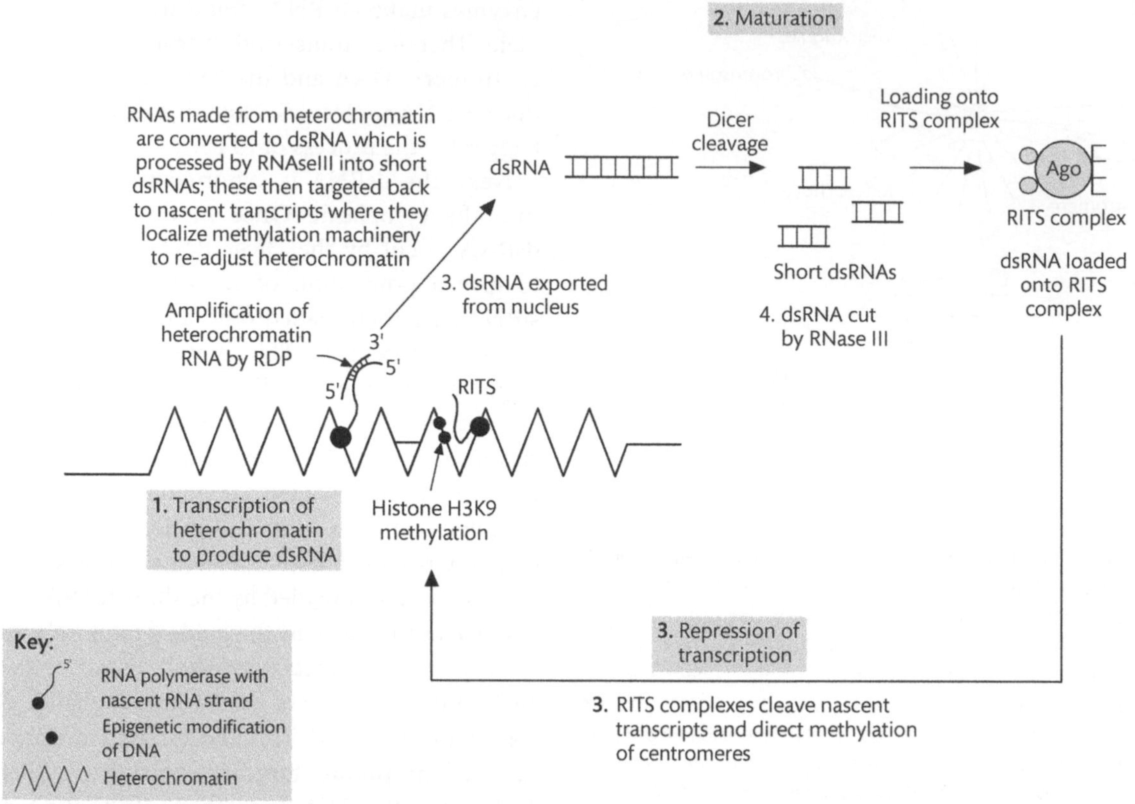
2. Maturation
RNAs made from heterochromatin are converted to dsRNA which is processed by RNAseIII into short dsRNAs; these then targeted back to nascent transcripts where they localize methylation machinery to re-adjust heterochromatin
dsRNA
Dicer cleavage
Loading onto RITS complex
Ago
RITS complex
Short dsRNAs
dsRNA loaded onto RITS complex
3. dsRNA exported from nucleus
4. dsRNA cut by RNase III
Amplification of heterochromatin RNA by RDP
3'
5'
5'
RITS
1. Transcription of heterochromatin to produce dsRNA
Histone H3K9 methylation
3. Repression of transcription
3. RITS complexes cleave nascent transcripts and direct methylation of centromeres
Key:
RNA polymerase with nascent RNA strand
Epigenetic modification of DNA
Heterochromatin

Figure 16.14 The low levels of dsRNAs made from centromeric DNA establish heterochromatin in fission yeast through a feedback mechanism. The RNA-induced transcriptional silencing (RITS) system relies on low levels of transcription taking place from centromeric heterochromatin. Since transcription takes place on both DNA strands, complementary RNAs are made which can hybridize with each other to give dsRNAs. The dsRNAs made by centromeric heterochromatin are cut by Dicer to give short siRNAs which are loaded onto the Argonaute protein. Argonaute selects one strand to keep hold of—this is the guide strand. The RNA guide strand of the Ago–RNA complex then acts as a very efficient homing device to target the RITS complex back to nascent transcripts at the centromere. Once localized at the centromere, the RITS complex can localize the methylation machinery to reinforce heterochromatinization of the centromere and prevent further transcription.

silencing of the centromeres. The repeat sequences of centromeres are important for establishing efficient RITS. Since each centromere has many copies of similar repeat sequences, the formation of a RITS complex will efficiently target the whole centromere from which it was made, rather than just the specific sequence which was first transcribed.

The RNA component of the RITS complex is sometimes called **rasiRNA** (RITS-complex-associated siRNA). Similar RITS complexes which target centromeric DNA targets for chromatin modification are found in yeast, *Arabidopsis thaliana* and *Drosophila melanogaster*. In each species rasiRNAs are similarly incorporated into RITS complexes, which in turn target DNA sequences for heterochromatinization.

Key points

Centromeric DNA is composed of repeated sequences that are transcriptionally 'quiet' as opposed to silent. This low level of transcription is important since it produces low levels of RNA from the centromeric repeats. These RNA molecules can then hybridize to form dsRNAs, which act to trigger a process called RITS which maintains transcriptional repression at the centromere.

The dsRNAs made from the centromere are chopped up by Dicer and then loaded onto complexes which contain Argonaute and proteins which convert DNA into heterochromatin (RITS complexes). The RNA component of the Argonaute complexes acts as an addressing system to target the RITS complexes back to the centromeres, where they maintain the heterochromatic state.

16.6 RNA-induced transcriptional silencing of transposons

In the final section of this chapter we turn our attention to the role of siRNAs in repressing the transcription of mobile elements called **transposons** or **transposable elements**.[49] Transposons are mobile genetic sequences which are able to spread around in the genome, but once inserted into chromosomes are replicated along with the rest of the chromosomal DNA. Transposable elements have been considered to be genetic parasites which live within the cell but do not make a direct contribution to normal cellular functions.

The number of transposable elements in the genome can be enormous. As an example, 50% of the genome in humans is derived from transposable elements (compared with 2% of the human genome comprising exons). While they are generally considered somewhat useless to the cells and organisms which contain them (and, indeed, bordering on harmful), transposable elements have some benefits; in evolutionary time some transposable elements have provided the raw material for new exons to develop, and some transposable elements provide important functions for their hosts. Most of the transposable elements within the human genome contain defects, meaning that they are no longer mobile.

Transposable elements are found in both eukaryotes and prokaryotes. The mechanisms by which transposable elements move are shown in Fig. 16.15. One class of transposable elements, called **retrotransposons**, move around the genome via RNA intermediates. Retrotransposons are related to the retroviruses which similarly use an RNA intermediate in their replication. The difference is that

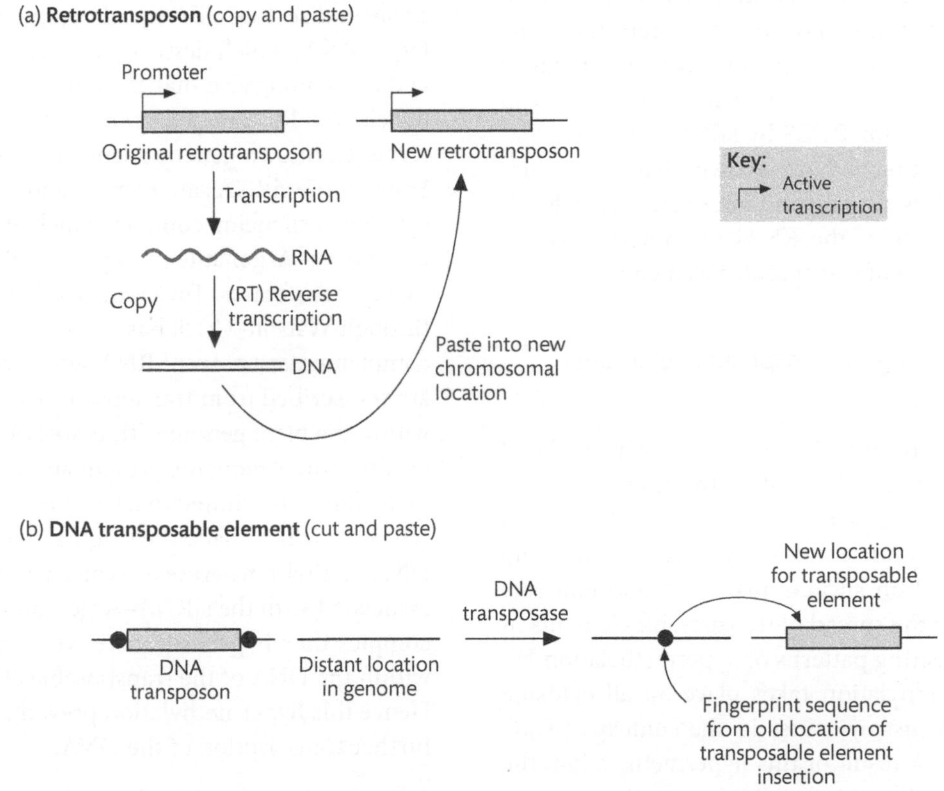

Figure 16.15 Two classes of transposable elements move around the genome in different ways. (a) Retrotransposons move via an RNA intermediate. These retrotransposons contain a promoter and are transcribed. The RNA product is converted to cDNA by the enzyme reverse transcriptase, and then this cDNA is copied into dsDNA before insertion back into the genome at a new position. (b) Rather than use an RNA intermediate, DNA transposons are recognized by enzymes called transposases which catalyse the excision of the DNA transposon and its re-integration elsewhere in the genome.

retrotransposons move around the genome *within* cells while retroviruses are infectious between cells. DNA transposable elements are also potentially mobile, but use enzymes called transposases to cut them out of the genome and reinsert them into new positions. As well as providing the cell with extra DNA to replicate, transposable elements can also cause mutations by inserting into important genes when they move in the genome. If insertions take place in important developmental genes they can result in genetic diseases, or if they take place in genes which control the cell cycle they may initiate cancer. Also, by hopping out of the genome, transposable elements can cause DNA breaks which need to be repaired to prevent genome instability. For these reasons it is very important for the cell to prevent the movement of transposable elements.

An important role for siRNAs in the eukaryotic cell is to prevent the movement of transposable elements. siRNAs direct the formation of heterochromatin which prevents the transcription which is required for the movement of retrotransposable elements. The use of an RNA intermediate in retrotransposons means that these RNAs can also be targeted for PTGS by siRNA-based mechanisms. Experiments have shown that nematodes (*Caenorhabditis elegans*) and mice containing defective components of the RNAi machinery have increased mobility of transposable elements.

16.6.1 Silencing of transposable elements in plants

In plants, as in animals, cytosines are methylated predominantly at CpGs, but methylation also takes place in other sequence contexts including CNG (where N is any nucleotide). An epigenetic silencing pathway based on siRNAs plays an important role in preventing the spread of transposable elements in plants by directing patterns of hypermethylation.[50,51] This hypermethylation takes place on all cytosine residues, not just cytosines in the context of CpG and CNG. As a result of this hypermethylation, the transposable elements containing methylated cytosines are unable to move efficiently around the genome because they cannot be transcribed. Figure 16.16 shows how the siRNA pathway directs epigenetic silencing of plant transposable elements (and other sequences which make dsRNAs). Let us go through each of the three steps in this process.

- The trigger for this silencing pathway is the appearance of dsRNA within plant cells. This can take place through the direct introduction of dsRNAs into plant cells by infectious RNA species (called viroids). Alternatively, dsRNA can be made between adjacent transposable elements by promoters facing towards each other which initiate transcription from both strands of DNA (in a similar way to the pairs of promoters which make dsRNA from the centromeres of *Schizosaccharomyces pombe*).
- After transcription, complementary RNAs can hybridize by Watson–Crick base pairing. The resulting dsRNA is then cleaved by the plant Dicer enzyme to generate shorter double-stranded siRNAs. The siRNAs produced in this way have two different fates: (1) some are recruited into a RISC complex, where they play a role in directing cleavage of complementary target RNAs which destroys the target RNA, and in so doing also makes more siRNAs from the cleaved RNA target; (2) other siRNAs are recruited into epigenetic silencing programmes.
- Some of the siRNAs are recruited into an epigenetic silencing complex which also contains an Argonaute protein and DNA methyltransferases. This complex then hybridizes through Watson–Crick base pairing with complementary nascent RNA sequences which are transcribed from transposable elements within the plant genome. Thus siRNA efficiently localizes the Argonaute protein and DNA methylation machinery back to the parent sequence which made the original dsRNA. The DNA methyltransferase enzymes which are associated with the siRNA–Argonaute protein complex then hypermethylate cytosine residues within the DNA of the transposable element. Hence this hypermethylation prevents any further transcription of the DNA.

An important feature about transposable elements is that they fall into families of elements with the same or similar sequences. Hence siRNAs made from one copy of a transposable element within a genome will act as an addressing system to target the

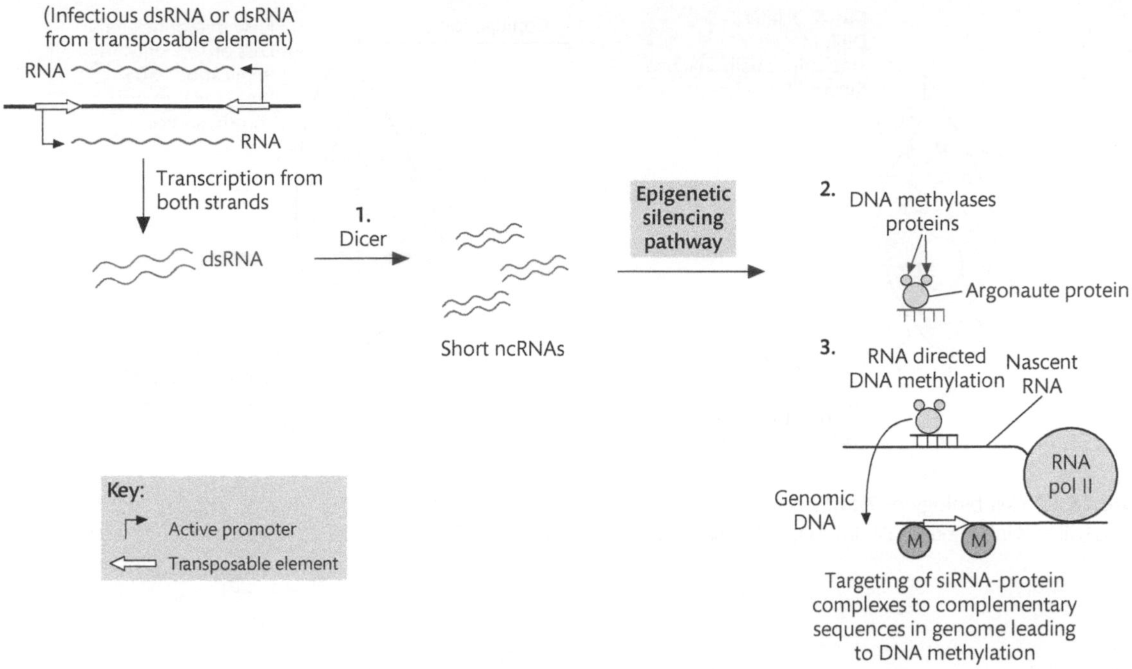

Figure 16.16 The siRNA pathway directs silencing of transposable elements in plants. The first step in this pathway is the generation of dsRNAs. These dsRNAs can either invade the plant cell directly or be generated by transcription of complementary DNA strands. In this figure dsRNA is shown being made by transcription occurring from two oppositely oriented transposable elements within the plant chromosome. Because both transposable elements contain promoters pointing towards each other, they will transcribe complementary RNAs which can hybridize to give dsRNA. Next the dsRNA is cleaved by Dicer to give short ncRNAs. These are loaded onto the Argonaute protein, where they can either join siRNA pathways to destroy the initiating transcript which started the pathway or join a similar Argonaute complex which directs epigenetic silencing of the originating transposable elements within the genome. The Argonaute protein will keep hold of one strand of this original dsRNA; this is called the guide strand. In the case of epigenetic silencing the guide RNA acts as a homing device to target a complex comprising the Argonaute protein and DNA methylases back to the part of the genome from which they were transcribed. Once back to the part of the genome from which it was transcribed, the guide RNA component of the complex will hybridize through complementary base pairing and the DNA methylases in the complex will direct DNA methylation.

transcriptional silencing of other transposable elements within the genome that have the same or a similar sequence. Hence retrotransposons in the plant genome are silenced by preventing their transcription through epigenetic modification of chromatin. We now discuss a very dramatic example of the role of short ncRNAs in getting rid of transposable elements. This example is from the fascinating single-celled organism *Tetrahymena thermophila*, and involves a special group of siRNAs called **scRNAs** (scan RNAs).

16.6.2 Silencing transposons in *Tetrahymena thermophila*

Perhaps the best, although most drastic, way to prevent the activity of transposable elements in a genome would be not just to silence them epigenetically but to actually throw them out. A process of this kind actually takes place in the single-celled ciliated protozoan *Tetrahymena thermophila*. *Tetrahymena* cells are somewhat unusual amongst eukaryotes because they contain two nuclei: a transcriptionally inactive diploid micronucleus, and a transcriptionally active polyploid macronucleus.

A diagram of a *Tetrahymena* cell containing two nuclei is shown in Fig. 16.17. The diploid micronucleus acts as the repository of genetic information which is passed on during sexual reproduction (this is called the germline), while the macronucleus has the job of producing all the RNA which is needed by the cell by acting as a template for transcription (the macronucleus is the somatic nucleus—it is not

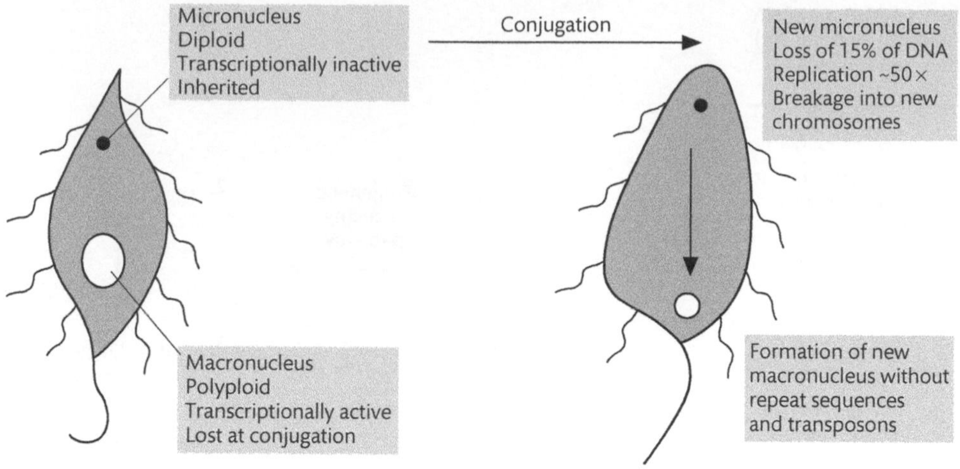

Figure 16.17 Cell biology of *Tetrahymena*. *Tetrahymena* is a single-celled ciliated organism which contains two nuclei: a micronucleus which is inherited at cell division, and a macronucleus which is transcribed into RNA and provides gene function.

passed on during reproduction). The micronucleus and macronucleus are quite different; the genome is massively amplified (polyploidy) in the macronucleus. Polyploidy of the macronucleus might be important for enabling the high levels of transcription required for the large cell size of *Tetrahymena*. Another difference is that the macronucleus does not contain any transposable elements.

Tetrahymena cells reproduce sexually under conditions of nutritional starvation. Unusually, sexual reproduction in *Tetrahymena* does not involve total cellular fusion of cells from opposite mating types (such as happens in the fusion of sperm and egg in metazoans, or the fusion of cells of different mating types in yeast). Instead, *Tetrahymena* cells partially fuse and exchange micronuclei; this process is called conjugation. As well as transfer of micronuclei during conjugation, the old macronucleus is destroyed—this necessitates the manufacture of a new macronucleus afterwards. The micronucleus acts as a template for the new macronucleus. However, the newly formed macronucleus is not a simple copy of the micronucleus. Instead, macronuclear formation requires (1) massive DNA replication to give polyploidy (multiple copies in the nucleus of each bit of DNA), (2) DNA breakage and telomere addition to the DNA fragments to give many new chromosomes, and (3) 'clean-up' of the new macronucleus by removing transposable elements.

Note that while transposable elements are removed from the macronucleus they still exist in the micronucleus, which is the germline of the *Tetrahymena* cell. Therefore removal of transposable elements from the macronucleus does not completely eradicate them, but only prevents them from interfering with transcription, while the transposable elements are maintained in the micronucleus. Since the micronucleus is generally not transcribed, transposable elements are relatively harmless there. *Tetrahymena* cells are thought to direct genome deletions of transposable elements in new macronuclei through a scanning mechanism which involves a set of small nuclear ncRNAs (scRNAs).[52] These scRNAs act as a scaffold for directing the large genome rearrangements that generate a new macronucleus without any transposable elements.

Let us now look at macronucleus formation as a three-step sequence. This sequence begins with a *Tetrahymena* cell containing a micronucleus, the old macronucleus, and a newly made macronucleus. In the first step in this process the micronucleus is completely transcribed. Although the micronucleus is generally transcriptionally silent, it is transcribed in both directions very early in conjugation, so at this time the whole *Tetrahymena* genome is copied and can form dsRNA by Watson–Crick base pairing. Since the whole genome is transcribed in both directions, dsRNA molecules will be made which contain RNA copies of everything present in the micronuclear genome—from genes to transposable elements. These dsRNAs are then cut by a Dicer enzyme to give short double-stranded scRNAs, which are about 28 nucleotides long. Once formed, scRNAs associate with an Argonaute protein. This Argonaute protein is also made by *Tetrahymena* cells, only early in conjugation.

scRNAs move to the old macronucleus before it is destroyed. Once in the macronucleus the scRNAs scan the macronuclear genome for complementary sequences that they can recognize by Watson–Crick base pairing. Those scRNAs that find a complementary DNA sequence are destroyed. This leaves an important subset of scRNAs remaining which cannot find a target. Transposable elements are missing from the old macronuclear genome. Therefore the scRNAs made from such transposable elements in the original micronucleus will not have targets present in the macronucleus and will survive.

The surviving scRNAs next move into the newly made unmodified macronucleus. Once in the new macronucleus they scan the macronuclear genome, and through Watson–Crick base pairing direct any complementary sequences (in this case they will be from transposable elements) to be converted into heterochromatin via DNA methylation and then deleted from the macronuclear genome. In this way transposable elements are removed from the new macronucleus. The mechanism of action of the scRNAs is summarized in Fig. 16.18.

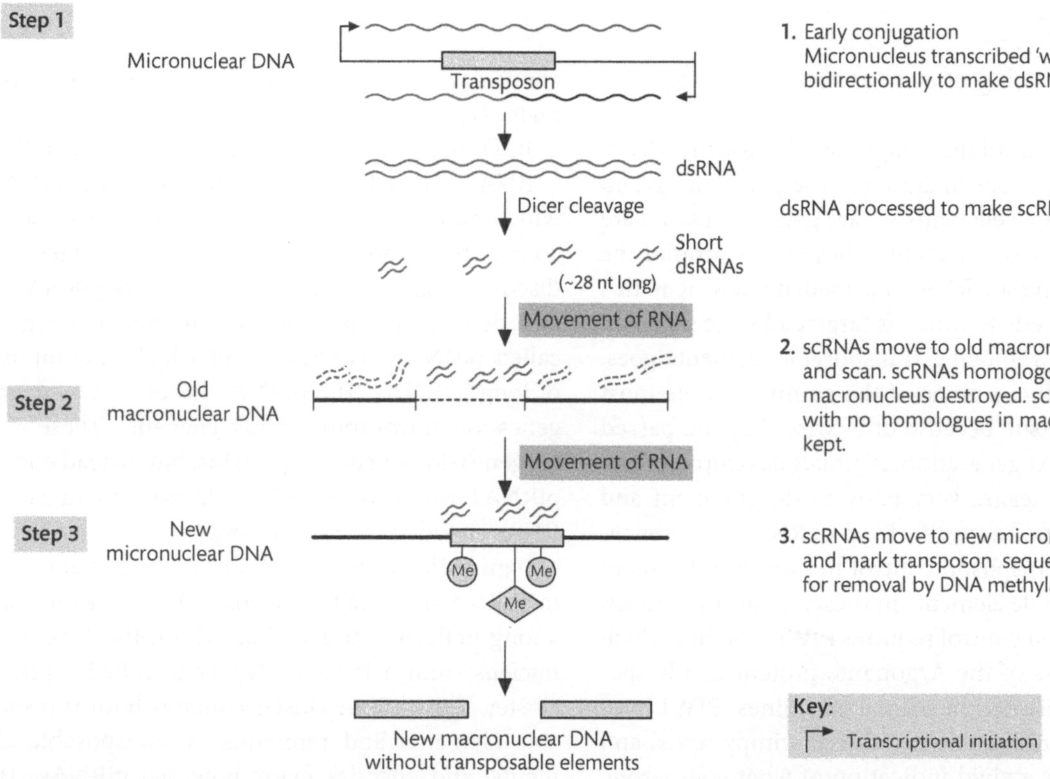

Figure 16.18 Mechanism of action of scRNAs. The scRNAs direct genome rearrangements in three stages. (1) The old micronucleus is transcribed in its entirety. This includes transposon sequences as well as genes. The DNA is transcribed on both strands, which can hybridize to give dsRNAs. These dsRNAs are cleaved by Dicer to generate short (approximately 28-nucleotides) scRNAs. (2) The scRNAs then 'scan' the old macronucleus to look for complementary sequences. Those scRNAs that find target RNA sequences are destroyed, while those that cannot find complementary sequences survive. These surviving scRNAs correspond in sequence to the transposable elements that are not present in the old macronucleus. (3) The surviving scRNAs then move into the new macronucleus where they direct machinery to remove complementary sequences, thus clearing the new macronucleus of transposable elements.

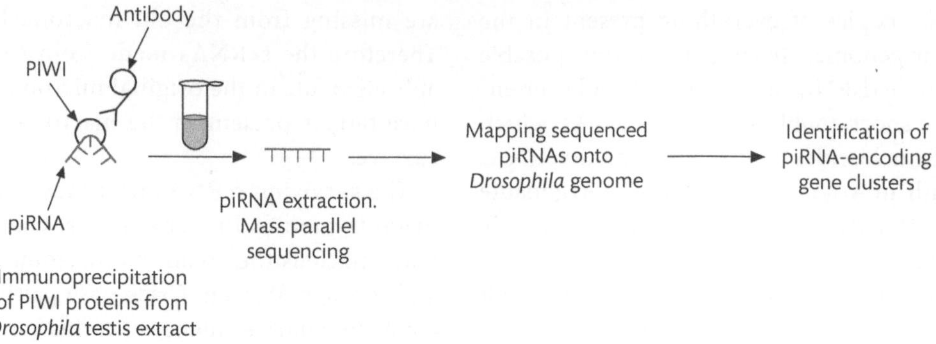

Figure 16.19 Identification of piRNA-encoding genes. piRNAs are encoded by long ncRNA genes. These genes were identified using antibodies directed against PIWI proteins, by immunoprecipitating piRNAs. Precipitated piRNAs were sequenced and the position of the genome which encoded them identified by sequence identity with the *Drosophila* genome. Most piRNAs were found to be encoded by a relatively small number of chromosomal regions which were already known to have roles in controlling transposable elements. Instead of encoding proteins, these chromosomal regions encode long ncRNAs which act as precursors for making piRNAs. Importantly, the regions of the *Drosophila* genome which encode piRNAs also act as hotspots for transposon integration. Hence the long non-coding piRNA precursor RNAs will also include RNA corresponding to transposable elements. Because transposons can insert in either orientation in the *Drosophila* genome (facing upstream or downstream), both sense and antisense piRNAs corresponding to transposable elements will usually be present in the long non-coding pre-piRNA.

16.6.3 Silencing of transposons in animals

In the final part of this chapter we discuss the silencing of transposons in animals. The particular group of transposable elements we are going to discuss are retrotransposable elements; these move around the genome using an RNA intermediate, and it is this RNA intermediate which is targeted by the siRNAs.

As far as removing transposable elements goes, preserving the genome of the germline is the most important issue because otherwise they are passed on to the next generation. Animals develop separate germ cell lineages very early in development and have developed a specific form of RNA interference-mediated transposon control to prevent movement of transposable elements in these.[53] This mechanism of transposon control requires **PIWI** proteins, which are members of the Argonaute protein family specifically expressed in animal germlines. PIWI is an acronym for 'P-element induced wimpy testis', and this name is a vivid indication of what goes wrong in mutants when the PIWI protein is missing. P-elements are transposable elements in *Drosophila* which are normally controlled by the PIWI protein.[54] However, when PIWI is inactivated by mutation, abnormal testis development occurs as a result of mass activation of P-element transposons—this abnormal development results in the wimpy testis phenotype.

In *Drosophila* a large and specialized class of short ncRNAs called PIWI interacting RNAs (**piRNAs**), which number in the tens of thousands, are used to control transposons. How these piRNA genes were discovered is shown in Fig. 16.19. These piRNAs are encoded by the genome in chromosomal regions called piRNA clusters, each of which is composed of long ncRNAs. The piRNA clusters act as master genes to control transposable elements. These master genes do not encode proteins, but instead encode piRNAs which direct RNA cleavage to inactivate RNAs encoded by transposons.

Figure 16.20 shows each of the three major steps in piRNA-mediated transposon inactivation. First, a long ncRNA is transcribed within the *Drosophila* nucleus from a long ncRNA gene called a piRNA cluster. This piRNA cluster contains both transposable elements and remnants of transposable elements, and encodes many potential piRNAs. This long ncRNA is the pre-piRNA.

The long pre-piRNA is then cut up to generate a number of 24–30-nucleotide long piRNAs, some of which are complementary to sense and some to antisense sequences of transposable elements. PiRNAs have a 2′-O-methyl modification on the 3′

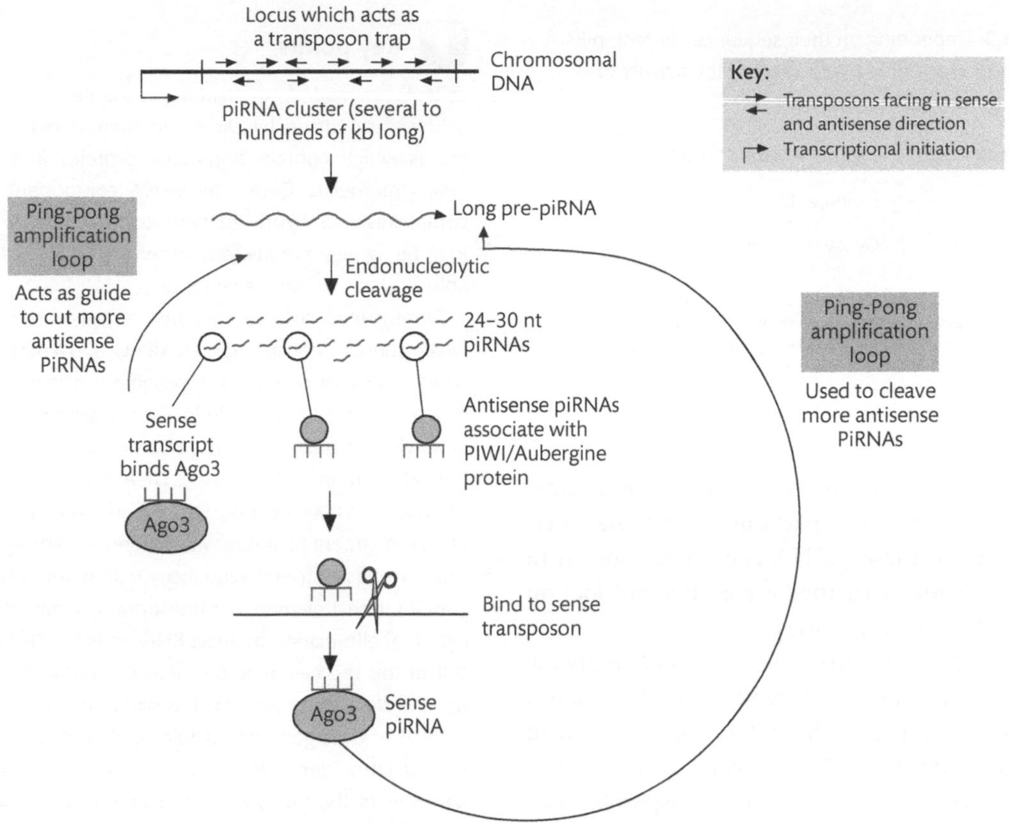

Figure 16.20 PIWI proteins are involved in a piRNA-directed pathway of RNA cleavage of transposon repressed RNAs.

end nucleotide and the 5′ end nucleotide is usually a U. The antisense piRNAs associate primarily with PIWI and a PIWI-like protein called Aubergine to form RNA–protein complexes. This incorporated antisense RNA now acts as the guide RNA, or address system, of the PIWI complex.

The guide RNA in these RNA–PIWI protein complexes then directs the complexes to bind through sequence complementarity to RNAs encoded by transposons. Because the antisense strand is loaded into the PIWI complex, this complex is targeted to hybridize with the sense strand of the transposons. Since PIWI and Aubergine are Argonaute family proteins, they can cut RNA internally (this is called RNA endonuclease activity, or slicing activity). Once sliced, the cut transposon-encoded RNAs are then bound by another Argonaute protein called Ago3. This Ago3-sense piRNA complex then directs the cleavage of more antisense piRNAs in the long pre-piRNA. This amplification of antisense piRNAs following cleavage of sense transposon sequences is called the 'ping-pong amplification loop'.

The piRNA clusters seem to be sites of highly efficient transposon insertion. Because of this, transposable elements moving around the genome are most likely to move into a piRNA cluster, i.e. these piRNA genes work as *transposon sinks*. Through making ncRNAs which work in the pathway described above, the piRNA clusters then act very efficiently to control movement of transposable elements in the *Drosophila* genome, i.e. to keep them at bay.

There are a number of transposable elements in the *Drosophila* genome. Different piRNA clusters in the *Drosophila* genome have become specialized to deal with different transposable elements depending on which sequences they contain. Each of these piRNA clusters can only produce piRNAs to regulate transposons which are present within them, and

Table 16.3 Depending on their sequence content, piRNA clusters are responsible for controlling the activity of different transposable elements*

piRNA cluster	Target retrotransposon
X-TAS locus	P elements
Flamenco locus	Gypsy retroposons

*Two different regions of the *Drosophila* genome (each region is called a locus) contain a piRNA cluster which encodes piRNAs which are complementary to and so target a different group of transposable elements

surprisingly different piRNA clusters have become specialized to look after particular transposable elements. Two of these piRNA clusters are shown in Table 16.3. Notice that they are at different loci on the *Drosophila* chromosomes.

PIWI proteins were first discovered in *Drosophila*. However, mice have analogous MIWI proteins, but these are not as well understood as those in *Drosophila*. Although PIWI proteins in *Drosophila* seem to target RNA encoded by transposable elements for destruction, the most recent data suggest that the equivalent mouse MIWI proteins might instead direct epigenetic silencing of transposable elements; this has been demonstrated in fetal germ cells.[55]

Key points

dsRNAs are made from plant transposable elements. These dsRNAs are cut up by Dicer and then loaded into complexes which contain Argonaute proteins and enzymes which methylate DNA. The siRNA component of these complexes can hybridize with complementary nascent RNA being newly made from other copies of the transposable element within the genome.

Tetrahymena cells contain two nuclei—a micronucleus which contains transposable elements, and a macronucleus which does not contain transposable elements and is the source of most RNAs made in *Tetrahymena* cells. The macronucleus is made using the micronucleus as a template, but rather than a direct copy, DNA rearrangements take place when a new micronucleus is made. The macronucleus DNA is fragmented and new telomeres are added onto the broken ends. Internal sequences within the DNA which contain repeat elements or transposable elements are targeted for elimination by short RNAs called scRNAs. As a result of this the new macronucleus is an amplified, broken up, and cleaned up version of its parent micronucleus.

PIWI is an Argonaute protein, and controls *Drosophila* transposable elements through an siRNA-directed PTGS pathway. In the fruit fly these mechanisms of transposon defence are based on RNA-directed cleavage of expressed transposon sequences rather than RNA-directed epigenetic regulation. The piRNAs act as a targeting mechanism to address Argonaute protein onto RNAs made from retrotransposons, leading to their destruction through slicing.

Table 16.4 List of RNA species described in this chapter and their principal characteristics

Type of RNA	Definition
ncRNA	Non-coding RNA (does not encode a protein)
siRNA	Small interfering RNA; 21-23-nucleotide dsRNA with two-nucleotide 3′ overhangs
microRNA (miRNA)	Small RNA, encoded by the genome, functionally akin to an siRNA
Guide RNA	The strand of a siRNA that is incorporated into the RISC complex
Passenger RNA	The strand of a siRNA that is discarded from the RISC complex
saRNA	Small activating RNA; targets promoters and activates transcription through chromatin modifications (RNAa)
rasiRNA	RITS complex-associated RNA; targets centromeric DNAs for chromatin modification
piRNA	PIWI-interacting RNA; used for controlling transposons
scRNA	Scan RNA; scaffold for directing the large genome rearrangements that generate a new macronucleus without any transposable elements

Summary

- Small ncRNAs are involved in gene silencing. Post-transcriptional gene silencing (PTGS) involves the cleavage of target mRNAs (RNA interference) or the repression of translation by microRNAs. Transcriptional gene silencing (TGS) is also mediated by short ncRNAs and works at the level of chromatin modification.
- RNA interference evolved as an ancient mechanism that allows eukaryotic cells to defend themselves against transposons and viruses. The key trigger for RNA interference is dsRNA, which is processed into shorter double-stranded siRNAs by an enzyme called Dicer.
- RNA interference is mediated by a complex called RISC (the RNA-induced silencing complex) which targets homologous mRNAs for degradation. Single strands of siRNAs are loaded into RISC complexes and then act to address these RISC complexes to other nucleic acids. RISC complexes also contain Argonaute proteins which can cut up RNA through a slicing activity.
- RNA interference is now widely used in research to knock down genes of interest, and is also being developed as a therapeutic agent.
- MicroRNAs are short ncRNAs encoded by the genome. They work in *trans* to silence genes by repressing translation and promoting mRNA degradation. MicroRNAs are also involved in TGS and in the activation of transcription (RNAa).
- MicroRNAs are involved in both normal development and disease processes.
- The mode of action and generation of microRNAs has a lot in common with the RNA interference machinery.
- siRNAs can also direct TGS by converting chromatin into heterochromatin. Short ncRNAs are implicated in transcriptional silencing at centromeres and in counteracting the deleterious consequences of transposons.
- A list of RNA species described in this chapter and their principal characteristics is provided in Table 16.4.

Questions

16.1 How do ncRNAs cause post-transcriptional and transcriptional gene silencing?

16.2 What is RNA interference and how did it evolve?

16.3 How does a RISC complex form and what does it do?

16.4 Why is RNA interference such a popular research tool?

16.5 What are microRNAs and how were they discovered?

16.6 Describe how microRNAs are involved in normal development.

16.7 Describe how microRNAs are involved in disease including cancer.

16.8 How do microRNAs regulate target gene expression in the cytoplasm and nucleus?

16.9 How do small ncRNAs cause transcriptional silencing in the centromere?

16.10 What is RITS and how is it established?

16.11 How can transposons be silenced by small ncRNAs in plants and animals?

References

1. **Faradic TA, Juranek SA, Tuschl T.** The growing catalog of small RNAs and their growing association with distinct Argonaute/PIWI family members. *Development* **135**, 1201–14 (2008).
2. **Kaikkonen MU, Lam MTY, Glass CK.** Non-coding RNAs as regulators of gene expression and epigenetics. *Cardiovascular Res* **90**, 430–40 (2011).
3. **Macrae IJ, Zhou K, Li F, Repic A, et al.** Structural basis for double-stranded RNA processing by Dicer. *Science* **311**, 195–8 (2006).
4. **Bernstein E, Kim SY, Carmell MA, et al.** Dicer is essential for mouse development. *Nat Genet* **35**, 215–17 (2003).
5. **Swarts DC, Makarova K, Wang Y, et al.** The evolutionary journey of Argonaute proteins. *Nat Struct Mol Biol* **21**, 743–53 (2014).
6. **Baulcombe D.** RNA silencing in plants. *Nature* **431**, 356–63 (2004).
7. **Fire A, Xu S, Montgomery MK, et al.** Potent and specific genetic interference by double-stranded RNA in *Caenorhabditis elegans*. *Nature* **391**, 806–11 (1998).
8. **Zamore PD, Tuschl T, Sharp PA, Bartel DP.** RNAi: double-stranded RNA directs the ATP-dependent

cleavage of mRNA at 21 to 23 nucleotide intervals. *Cell* **101**, 25–33 (2000).
9. **Meister G, Tuschl T.** Mechanisms of gene silencing by double-stranded RNA. *Nature* **431**, 343–9 (2004).
10. **Mello CC, Conte DC.** Revealing the world of RNA interference. *Nature* **431**, 338–42 (2004).
11. **Hammond SM.** Dicing and slicing: the core machinery of the RNA interference pathway. *FEBS Lett* **579**, 5822–9 (2005).
12. **Tolia NJ, Leemor JT.** Slicer and the Argonautes. *Nat Chem Biol* **3**, 36–43 (2006).
13. **Elbashir SM, Harborth J, Lendeckel W, et al.** Duplexes of 21-nucleotide RNAs mediate RNA interference in cultured mammalian cells. *Nature* **411**, 494–8 (2001).
14. **Mittal V.** Improving the efficiency of RNA interference in mammals. *Nat Rev Genet* **6**, 24–35 (2005).
15. **Sano M, Sierant M, Miyagishi M, et al.** Effect of asymmetric terminal structures of short RNA duplexes on the RNA interference activity and strand selection. *Nucleic Acids Res* **36**, 5812–21 (2008).
16. **Takasaki S.** Methods for selecting effective siRNA target sequences using a variety of statistical and analytical techniques. *Methods Mol Biol* **942**, 17–55 (2013).
17. **Jackson AL, Linsley PS.** Noise amidst the silence: off-target effects of siRNAs? *Trends Genet* **20**, 521–4 (2004).
18. **Cottrell TR, Doering TL.** Silence of the strands: RNA interference in eukaryotic pathogens. *Trends Microbiol* **11**, 37–43 (2003).
19. **Dorsett Y, Tuschl T.** SiRNAs: applications in functional genomics and potential as therapeutics. *Nat Rev Drug Discov* **3**, 318–19 (2004).
20. **Lieberman J, Song E, Lee SK, Shankar P.** Interfering with disease: opportunities and roadblocks to harnessing RNA interference. *Trends Mol Med* **9**, 397–403 (2003).
21. **Fraser AG, Kamath RS, Zipperlen P, et al.** Functional genomic analysis of *C. elegans* chromosome I by systematic RNA interference. *Nature* **408**, 325–30 (2000).
22. **Longman D, Johnstone IL, Caceres JF.** Functional characterisation of SR and SR-related genes in *Caenorhabditis elegans*. *EMBO J* **19**, 1625–37 (2000).
23. **Davies JA, Ladomery M, Hohenstein P, Michael L, Shafe A, Spraggon L, Hastie N.** Development of an siRNA-based method for repressing specific genes in renal organ culture and its use to show that the Wt1 tumour suppressor is required for nephron differentiation. *Hum Mol Genet* **13**, 235–46 (2004).
24. **Zhu H, Guo W, Zhang L, et al.** Enhancing TRAIL-induced apoptosis by Bcl-X(L) siRNA. *Cancer Biol Ther* **4**, 393–7 (2005).
25. **Rossi JJ.** RNAi as a treatment for HIV-1 infection. *BioTechniques* **40**, S25–9 (2006).
26. **Song E, Lee SK, Wang J, et al.** RNA interference targeting Fas protects mice from fulminant apoptosis. *Nat Med* **9**, 347–51 (2003).
27. **Borna H, Imani S, Iman M, Jamalkandi SA.** Therapeutic face of RNAi: in vivo challenges. *Expert Opin Biol Ther* **15**, 1–17 (2014).
28. **Kanasty R, Dorkin JR, Vegas A, Anderson D.** Delivery materials for siRNA therapeutics. *Nat Mater* **12**, 967–77 (2013).
29. **Tabernero J, Shapiro GI, LoRusso PM, et al.** First-in-humans trial of an RNA interference therapeutic targeting VEGF and KSP in cancer patients with liver involvement. *Cancer Discov* **3**, 406–17 (2013).
30. **Zamore PD, Haley B.** Ribo-gnome: the big world of small RNAs. *Science* **309**, 1519–24 (2005).
31. **Kim VN, Nam JW.** Genomics of microRNA. *Trends Genet* **22**, 165–73 (2006).
32. **Nakahara K, Carthew RW.** Expanding roles for miRNAs and siRNAs in cell regulation. *Curr Opin Cell Biol* **16**, 127–33 (2004).
33. **Tang G.** siRNA and miRNA: an insight into RISCs. *Trends Biochem Sci* **30**, 106–14 (2005).
34. **Murchison EP, Hannon GJ.** miRNAs on the move: miRNA biogenesis and the RNAi machinery. *Curr Opin Cell Biol* **16**, 223–9 (2004).
35. **Ha M, Kim VN.** Regulation of microRNA biogenesis. *Nat Rev Mol Cell Biol* **15**, 509–24 (2014).
36. **Ioshikhes, Roy S, Sen CK.** Algorithms for mapping of mRNA targets for microRNA. *DNA Cell Biol* **26**, 265–72 (2007).
37. **Shih-Peng C, Slack FJ.** MicroRNA-mediated silencing inside P-bodies. *RNA Biol* **3**, 97–100 (2006).
38. **Huang V, Li L-C.** MiRNA goes nuclear. *RNA Biol* **9**, 269–73 (2012).
39. **Roberts TC.** The microRNA biology of the mammalian nucleus. *Mol Ther Nuc Acids* **3**, e188 (2014).
40. **Li L-C.** Chromatin remodeling by the small RNA machinery in mammalian cells. *Epigenetics* **9**, 45–52 (2014).
41. **Alló M, Agirre E, Bessonov S, et al.** Argonaute-1 binds transcriptional enhancers and controls constitutive and alternative splicing in human cells. *Proc Natl Acad Sci USA* **111**, 15 622–9 (2014).
42. **Meyers BC, Souret FF, Lu C, Green PJ.** Sweating the small stuff: microRNA discovery in plants. *Curr Opin Biotech* **17**, 139–46 (2006).
43. **Carrington JC, Ambros V.** Role of microRNAs in plant and animal development. *Science* **301**, 336–8 (2003).
44. **Zhao Y, Shrivastava D.** A developmental view of microRNA function. *Trends Biochem Sci* **32**, 189–97 (2007).
45. **Hayes J, Peruzzi PP, Lawler S.** MicroRNAs in cancer: biomarkers, functions and therapy. *Trends Mol Med* **20**, 460–9 (2014).
46. **Allshire R.** Molecular biology. RNAi and heterochromatin—a hushed-up affair. *Science* **297**, 1818–19 (2002).
47. **White SA, Allshire RC.** RNAi-mediated chromatin silencing in fission yeast. *Curr Top Microbiol Immunol* **320**, 157–83 (2008).

48. **Volpe TA, Kidner C, Hall IM, et al.** Regulation of heterochromatic silencing and histone H3 lysine-9 methylation by RNAi. *Science* **297**, 1833–7 (2002).
49. **Girard A, Hannon GJ.** Conserved themes in small-RNA-mediated transposon control. *Trends Cell Biol* **18**, 136–48 (2008).
50. **Okamoto H, Hirochika H.** Silencing of transposable elements in plants. *Trends Plant Sci* **6**, 527–34 (2001).
51. **Bender J.** RNA-directed DNA methylation: getting a grip on mechanism. *Curr Biol* **22**, R400–1 (2012).
52. **Mochizuki K, Gorovsky MA.** Small RNAs in genome rearrangement in *Tetrahymena*. *Curr Opin Genet Dev* **14**, 181–7 (2004).
53. **Zamore PD.** RNA silencing: genomic defence with a slice of pi. *Nature* **446**, 864–5 (2007).
54. **O'Donnell KA, Boeke JD.** Mighty PIWIs defend the germline against genome intruders. *Cell* **129**, 37–44 (2007).
55. **Kuramochi-Miyagawa S, Watanabe T, Gotoh K, et al.** DNA methylation of retrotransposon genes is regulated by PIWI family members MILI and MIWI2 in murine fetal testes. *Genes Dev* **22**, 908–17 (2008).

17 RNA biology: future perspectives

Introduction

We hope that the preceding sixteen chapters have persuaded the reader that it is a very remarkable time in the field of RNA biology. There is a lot of information to take in—understanding how gene expression works in an integrated fashion is a real challenge. We now have to consider the complexities of chromatin modifications and epigenetics, the regulation of transcription, and co-transcriptional and post-transcriptional processes. It is also quite clear that these processes are integrated and connected; they are rarely compartmentalized and often share molecular components. Nonetheless, RNA biology now presents enormous opportunities for discovery. RNA biology allows us to study the evolution of life in a more complete way, to understand how genes and cells work, to improve our agriculture, and to search for new ways to diagnose and treat diseases.

In this final chapter we review some of the topical areas of RNA biology. We begin by describing the explosion of transcriptomics data and the opportunities and challenges that it brings. We consider the growing prominence of non-coding RNAs (ncRNAs)—one of the hottest current topics in the field. In the final section we describe the use of CRISPR, an RNA-guided genome editing system that will revolutionize both basic and applied research.

17.1 The emergence of transcriptomics

17.1.1 An explosion of transcriptomics data

Current biology has been transformed by massive amounts of data. Over 180 genomes have been sequenced (www.genomenewsnetwork.org). Several genomes are browsable on websites like the UCSC genome browser (www.genome.ucsc.edu) and ensembl (www.ensembl.org). One of the initial big surprises from the human genome was the comparatively small number of protein coding genes (20 364 protein coding genes in the December 2013 annotation) compared with other organisms that are apparently 'less complex' including *Caenorhabditis elegans* nematodes (20 447 protein coding genes annotated in the WS245 version of wormbase—more protein coding genes than humans!), *Drosophila melanogaster* fruit flies (13 397 protein coding genes, flybase release 5.46), and baker's yeast *Saccharomyces cerevisiae* (6692 genes, from the S288C genome).

Much to the excitement of RNA biologists **transcriptome** data (all of the RNAs that are transcribed) is now available. Transcriptome sequences are derived from copying RNA sequences into cDNA using reverse transcriptase, followed by sequencing. A key technological development is the advent of NGS or 'next-generation sequencing' (described in Box 17.1). This has made the analysis of complex transcriptomes a realistic prospect. It is now routine to sample a snapshot of the transcriptome. **RNA-Seq** (RNA sequence analysis or transcriptome profiling) reveals two layers of information: the identity of transcripts and their relative abundance. The analysis of transcriptomes shows that they are hugely complex. The advent of transcriptomic data has led to new, often unexpected discoveries.

Box 17.1 The advent of next-generation sequencing

DNA sequencing is the process of 'reading' the order of nucleotides. It started in the early 1970s when researchers were able to use DNA polymerase together with sequence specific primers. In 1977 Fred Sanger and colleagues, based at the MRC Laboratory of Molecular Biology in Cambridge UK, pioneered the use of chain-terminating inhibitors. It soon became a popular and widely used method. In brief, the method involves mixing dideoxy nucleotides (ddNTPs) with normal dNTPs. When a ddNTP is incorporated (at random) in a DNA strand, no further phosphodiester bonds can be created and the DNA strand cannot be extended further. The mixture of fragments can then be separated on a high resolution acrylamide gel and the sequence read directly. It used to be necessary to set up four separate reactions (with each ddNTP). Four samples were run on a gel for each sequence being read. Bands of a particular length meant that a particular ddNTP was incorporated at that position. The method was laborious and time consuming. It was improved with the use of fluorescently labelled ddNTPs. These could be combined in a single polymerase reaction, and the process of sequencing was facilitated by sequencer machines that could read the signal efficiently. With the aid of automation and capillary electrophoresis, by the end of the twentieth century it became possible to sequence entire genomes including the human genome. However, it was a very lengthy and expensive process and so it became vital to develop less expensive and high-throughput sequencing methods—enter 'next-generation sequencing' (NGS).

Thanks to NGS the cost of sequencing a human genome has gone down from 100 million dollars to a few thousand dollars (or less). Several NGS methods have been developed; two of the best known are pyrosequencing and illumina dye sequencing. Pyrosequencing follows the 'sequencing by synthesis' principle. It relies on the release of pyrophosphate each time a new nucleotide is incorporated. Illumina dye sequencing relies on the physical attachment of DNA fragments to primers on a slide. The DNA molecules are amplified to produce 'local colonies' whose sequence can then be read efficiently. 'Reversible terminator' bases are added, unincorporated nucleotides are removed, and a camera takes an image of the incorporated labelled nucleotides. The dye is then removed so that the next cycle can proceed. This type of sequencing means that it is possible to generate up to three billion reads per run. Crucially, the cost of sequencing a million bases is around $10, compared with over $2000 using the old Sanger chain termination method. New technologies are being actively developed and the costs are likely to come down even further. As well as allowing comprehensive analyses of transcriptomes, NGS is widely used to sequence genomes, epigenomes (the sum of all epigenetic changes to DNA sequence), and samples derived from chIP and RNA-chIP experiments (immunoprecipitations of DNA-protein and RNA-protein complexes).

Transcriptome-wide analyses show that gene number is not the only metric for a species. Almost every human protein coding gene produces multiple mRNA transcripts (with a plateau of 11 isoforms per gene), usually with two major mRNA isoforms per gene.[1] Human genes are very complex, complicating previous definitions of what a gene is.[2] Individual genetic loci containing protein coding genes produce multiple different kinds of RNA product, some of which are coding and some non-coding (see Figure 17.1).[3]

Although transcription was thought to be limited to known protein coding and ncRNA genes, transcriptome-wide analyses show that 70% of the human genome is transcribed.[1] Many thousands of genes for ncRNAs have been identified, largely within the regions between protein coding genes (see the long intervening RNA in Fig. 17.1, placed between two complex transcriptional loci). These 'intergenic' regions of the genome were once thought to be transcriptional deserts, but analysis at the whole-genome level has revealed that they often have epigenetic marks typical of protein coding genes. They can express transcripts that contain introns and exons but no open reading frames.[4]

Although they do not encode proteins, long ncRNAs can be very numerous and functionally important.[5,6] Given the caveat that genomes are not always completely annotated, some multicellular eukaryotes have more genes for ncRNAs than they have protein coding genes. For example, the human genome contains 20 364 protein coding genes, but there are 38 905 genes for ncRNAs in the 2013 release of the human genome (comprising short ncRNAs, long ncRNAs, and pseudogene derived transcripts). Hence, in humans there are almost as many genes for long ncRNA as there are for protein coding genes. Some single-celled organisms have a higher proportion of protein-coding genes to ncRNA genes than humans do (e.g. the genome of the budding yeast *Saccharomyces cerevisiae* contains 6692 annotated protein coding

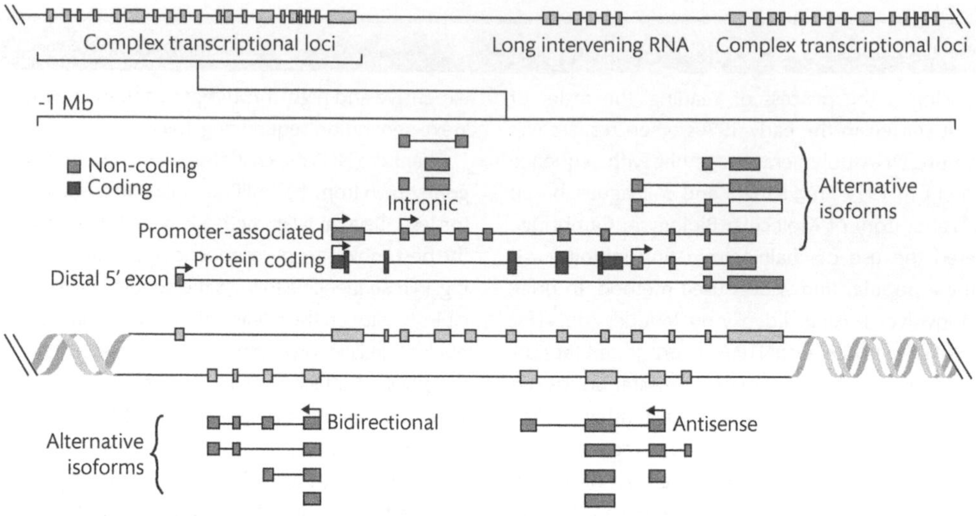

Figure 17.1 Complexity of human gene expression. Notice that a single locus on the chromosome can produce multiple different RNA products. In the intervals between these complex transcriptional loci which contain protein coding genes are the genes for long ncRNAs. (With permission from Mercer and Mattick.[3])

genes and 444 annotated ncRNA genes). Thus, ncRNAs have been proposed to have been pivotal in the development of complex eukaryotes and are of great evolutionary significance.

17.1.2 The power and challenges of RNA-Seq data

For many years microarrays were the method of choice when it came to observing how gene expression changes systematically in a particular experiment or sample. However, one of the major limitations of microarrays is that they do not necessarily represent the multitude of splice isoforms and ncRNAs that are present in the transcriptome. Companies that provide microarray services have incorporated alternative exons into their arrays. For example, the human exon 1.0 GeneChip array provided by Affymetrix contains probes that represent exon skipping, intron retention, mutually exclusive exons, alternative promoters, alternative polyadenylation sites, and alternative splice donor and acceptor sites—but, crucially, only when the difference in splice isoforms is over 25 base pairs. This is an issue because some alternative splice sites result in very small differences in transcripts. Moreover, the identity of all exons and splice isoforms in the genome is not necessarily known comprehensively. There are also technical problems to do with partial hybridization of probes to targets. On the other hand, RNA-Seq data obtained through NGS overcomes some of these limitations in that the transcriptome can be analysed directly without the need to decide what specific probes should be included on a microarray.

However, the use of NGS to examine alternative splicing does come with its own challenges.[7] One is *read mapping uncertainty*—it is not easy to reconstruct full-length transcripts. The other is the issue of *limited coverage*. Transcripts are expressed at vastly different levels and a considerable variation in size. Some regions may be particularly GC rich, making them harder to sequence. As a result it is easier to reach conclusions about genes that are more highly expressed. Because of these problems the precise quantification of expression levels of splice isoforms is not easy. One of the biggest challenges in using NGS is the ensuing bioinformatics analysis of the massive amount of data; in other words, it is very hard to make sense of it. Despite these caveats, NGS does offer some significant opportunities, including the ability to look at alternative splice sites that are in close proximity, meaning at distances of only 2–12 nuleotides.[8] These are known as **tandem alternative splice sites** (TASS), and they are the second most common class of alternative splicing in humans. As well as being able to detect TASS, NGS also facilitates the discovery of novel exons, novel ncRNAs, RNA editing events, fusion transcripts, and even novel genes that have not yet been described in a given genome.

The advent of NGS has undoubtedly brought a number of opportunities to the table in the context of biomedical research.[9] The study of cancer is a very good example. NGS reveals an astonishing heterogeneity in cancer transcriptomes, not only when comparing primary tumours with secondary tumours, but also when looking at different parts of a single tumour mass. Cancer cells are, by their very nature, genetically unstable and new mutations are affecting the genome all the time. This instability also impacts on the transcriptome; ratios of splice isoforms and levels of particular ncRNAs change all the time, perhaps even at a single cell level. NGS helps to illustrate one of the biggest challenges in understanding and treating cancer—its remarkable heterogeneity at a cellular and molecular level.

To illustrate the type of information that NGS can provide, let us briefly consider two examples taken from studies on breast and prostate cancer, both high incidence cancers. In breast cancer, NGS shows that an alternative splicing programme lies at the heart of a process called epithelial to mesenchymal transition (EMT). EMT is a necessary step in the process of metastasis. In prostate cancer, 121 unannotated ncRNAs and seven novel cancer-specific fusions have been discovered.

It is even possible to obtain RNA-Seq data from a single cell.[10] However, single-cell transcriptome analysis shows that cells that have an identical (or near-identical) genetic background (genome) may show significant differences in the transcriptome. Some of these single-cell differences could be due to a different landscape of epigenetic changes in each genome, or may simply be the consequence of the fact that stochastic (non-deterministic) processes influence the expression of several genes. The analysis of single-cell differences in transcriptomes illustrates the existence of yet another layer of complexity in gene expression.

17.2 The growing prominence of non-coding RNAs

The analysis of transcriptomes demonstrates a vast landscape of ncRNAs. What do all these ncRNAs do? In preceding parts of this book (particularly Chapters 15 and 16) we discussed some well-characterized examples, including how ncRNAs can act as a scaffold for epigenetic regulator proteins to assemble into active complexes and to target them. About a third of long ncRNAs are associated with epigenetic regulators, and so are candidates for controlling gene expression.[4]

17.2.1 Some pseudogenes are transcribed into ncRNAs

Pseudogenes are a big but still largely uncharacterized group of genes that express ncRNAs. Although pseudogenes do not themselves encode proteins, they are descended from protein coding genes that have 'degenerated' to lose their protein coding capacity. There are likely to be around 11 000 pseudogenes in the human genome.[6,11]

Because of their loss of protein coding capacity, pseudogenes have often been considered as 'junk' DNA, even though some are transcribed. This 'junk' description implies that pseudogenes are non-functional and decay over time. However, pseudogenes can have important functions. One example of an important pseudogene is the *XIST* ncRNA that we discussed in Chapter 15. *XIST* is the long ncRNA that controls X inactivation in female mammals. The probable evolutionary trajectory that created *XIST* involved a protein coding gene called the *Lnx3* gene. This *Lnx3* gene acquired mutations in mammals that incapacitated it in terms of protein coding capacity, thereby creating a pseudogene that evolved into the *XIST* long ncRNA that we find today.[12]

Pseudogenes are also important in diseases such as cancer. A pseudogene called *PTENB* controls the expression of the *PTEN* tumour suppressor gene.[13] The protein coding *PTEN* gene controls a signalling pathway to prevent cancer—in experimental animals, even a loss of 20% of *PTEN* expression is sufficient to shorten lifespan and increase cancer load.[14] Although *PTENP* is a pseudogene, *PTENP* is transcribed into a ncRNA that has a very similar 3′ untranslated region (3′ UTR) to *PTEN* mRNA. This similarity includes binding sites for the important microRNAs that bind to *PTEN* mRNA and down-regulate PTEN protein expression (see Fig. 17.2). Thus the *PTENP* pseudogene functions as a decoy

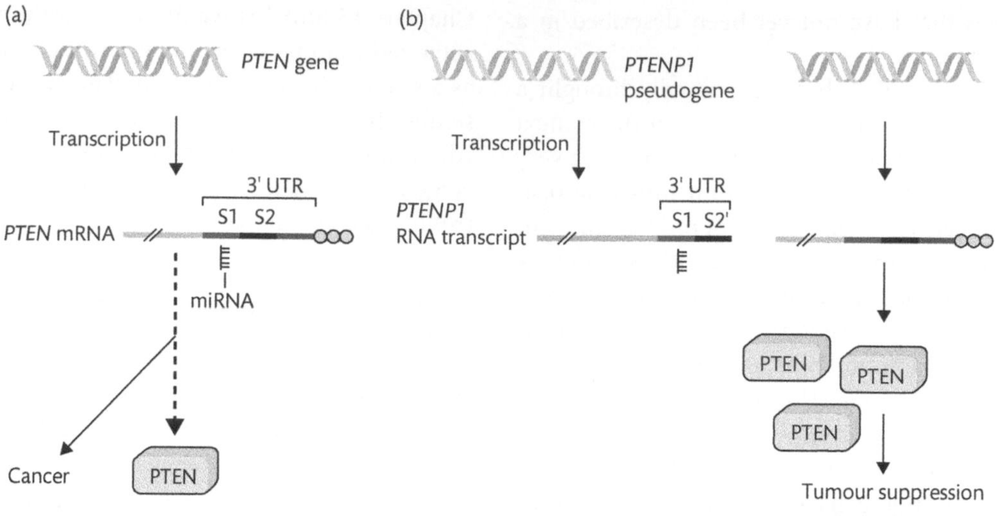

Figure 17.2 A pseudogene of the *PTEN* gene acts as a microRNA sponge. The *PTEN* gene is an important tumour suppressor. Levels of expression of *PTEN* mRNA are downregulated by miRNAs that bind to two regions of the *PTEN* mRNA 3′ UTR called S1 and S2. Loss of *PTEN* expression often causes cancer to develop. *PTENP1* is a transcribed pseudogene of *PTEN*, and importantly also contains binding sites for the repressive microRNAs. Expression of *PTENP* soaks up the microRNAs. This in turn results in increased expression of *PTEN* and helps to reduce cancer incidence. (With permission from Rigoutsos and Furnari.[14])

transcript, soaking up microRNAs that normally downregulate *PTEN* expression. This function of *PTENP* is so important that in some cancers the *PTENP* pseudogene, rather than *PTEN* itself, becomes lost through mutation.

17.2.2 Competing endogenous RNAs

The role of *PTENB* RNA, functioning as a decoy to control the expression of another RNA through having shared microRNA binding sites, has led to a current idea called the **ceRNA** (competing endogenous RNA) hypothesis. This hypothesis states that endogenous RNAs will frequently act as decoys to regulate other RNAs with common binding sites for miRNAs.[6,15] Another group of RNAs that might be important ceRNA decoys are the **circular RNAs** recently discovered at high levels in human cells.

Circular RNAs are long ncRNAs which, as their name suggests, are circular. Circular RNAs have been known about for a relatively long time. The first circular RNA to be discovered was found in the 1980s, encoded by a gene called *SRY* that controls male sex determination in mammals. Although a normal linear *SRY* RNA is made early in mouse development when sex is being determined, in adults a circular version of the *SRY* transcript is made instead.[16]

SRY was thought to be unusual in having a circular transcript, but recently thousands of circular RNAs have been identified in human cells by using bioinformatic programs to analyse RNA sequences.[17] Circular RNAs have unusual arrangements of exon junctions after sequencing—this is because they are formed by joining together exon junctions that are not normally joined together (how this happens is shown in Fig. 17.3).[18]

One function of circular RNAs is to act as sponges to soak up microRNAs that would otherwise bind to the linear transcripts. Therefore circular RNAs fit well into the ceRNA hypothesis. For example, one well-studied circular RNA that is highly expressed in the brains of humans and mice contains more than 70 microRNA binding sites.[19] By acting as a sponge, expression of this circular RNA (called ciRS-7) increases expression of the mRNAs normally targeted and repressed by these microRNAs.

Some genes produce circular transcripts as well as their better-known linear mRNA transcripts.[17] Many circular RNAs are produced at lower levels than linear mRNAs and are thought to be non-coding. But, because they do not have ends, circular RNAs are able to resist degradation by many nucleases within the cell (see Fig. 17.4 for the stability of circular

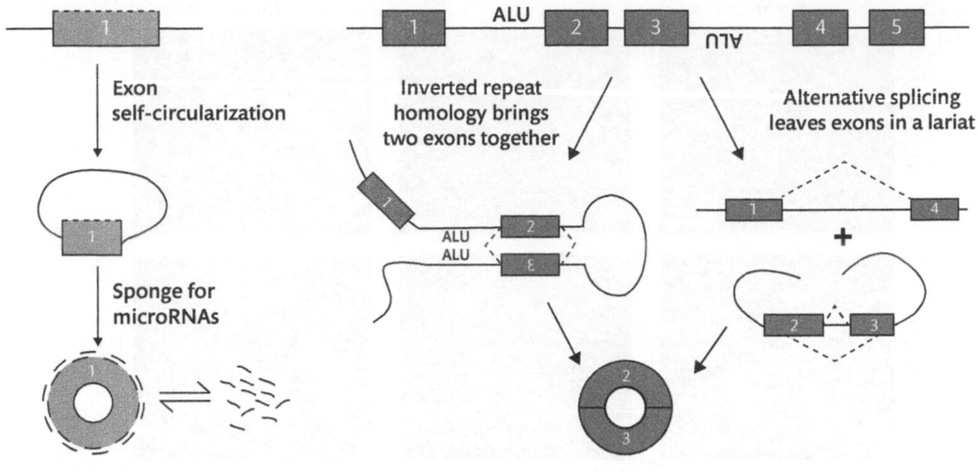

Figure 17.3 How circular RNAs are made. Circular RNAs are made by splicing pathways that utilize the same splice sites as used to make linear mRNA molecules (see Chapter 5), but in a way that makes circular RNAs rather than the normally linear mRNA molecules made from pre-mRNAs. In the first pathway (left-hand side), exon circularization results from an excised exon and its flanking introns; this RNA has both a 3′ splice site and a 5′ splice site that can splice together to make a circular RNA. In the second pathway (right-hand side) circular RNAs that contain several exons can be made. First, complementary sequences in flanking intron sequences (these can be provided, for example, by repeat sequences such as ALU sequences, one of which is inverted) can base pair to bring the splice sites together that should not normally be spliced. Splicing using these sites results in circular RNAs being made. Secondly, exon sequences can be released in intron lariats during normal splicing (see Chapter 5 for more details), and these exons can be spliced together to give circular RNAs. (Reproduced with permission from Valdmanis and Kay.[17])

RNAs to RNAses after their purification). Because of this, circular RNAs are very stable and so can be present at reasonably high levels in cells despite their low level of synthesis.

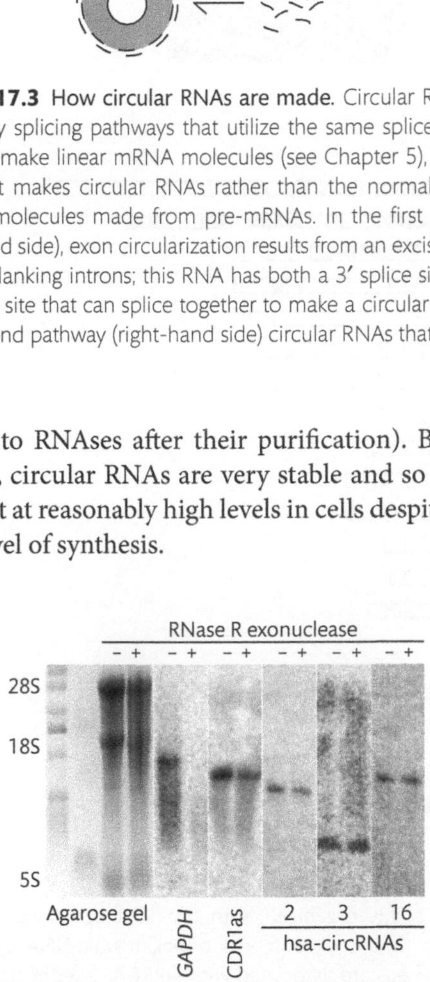

Figure 17.4 Circular RNAs are protected from digestion by exonuclease RNAse R. RNAse R digests from the ends of RNAs—it only digests linear RNA. In this experiment, RNA was electrophoresed on an agarose gel either after digestion with RNAse R or without treatment. The levels of particular RNAs were monitored in the RNA population using Northern blotting. Notice that the levels of the linear mRNA *GAPDH* go down after digestion, while the levels of the circular mRNAs (CDR1as and hsa-circRNAs) remain stable. (Reproduced with permission from Memczak et al.[18])

17.2.3 Long non-coding RNAs can be involved in nuclear organization

A final important new perspective we shall consider in this section is the role of long ncRNAs in nuclear organization. Several of the ncRNAs we have discussed in this book are highly localized within the nucleus. They include the X chromosome encoded *XIST* RNA that is initially transcribed from the non-coding X chromosome and then spreads out to coat the entire inactive X chromosome.[20] Individual chromosomes can have distinct territories within the nucleus. Another X chromosome encoded long ncRNA called *FIRRE*, which helps to organize the territories of chromosomes, including the X, within the nucleus, has been identified.[21] Although both are on the X chromosome, *FIRRE* and *XIST* are expressed from different parts of the X chromosome. After transcription *FIRRE* stays associated with its parent gene within nuclear foci. These nuclear foci can be seen in the image in Fig. 17.5. Notice there are two transcript foci in female cells, since both copies of the X chromosome (active and inactive) express *FIRRE*. *FIRRE*'s job seems to be to glue particular chromosome regions together within the nucleus.

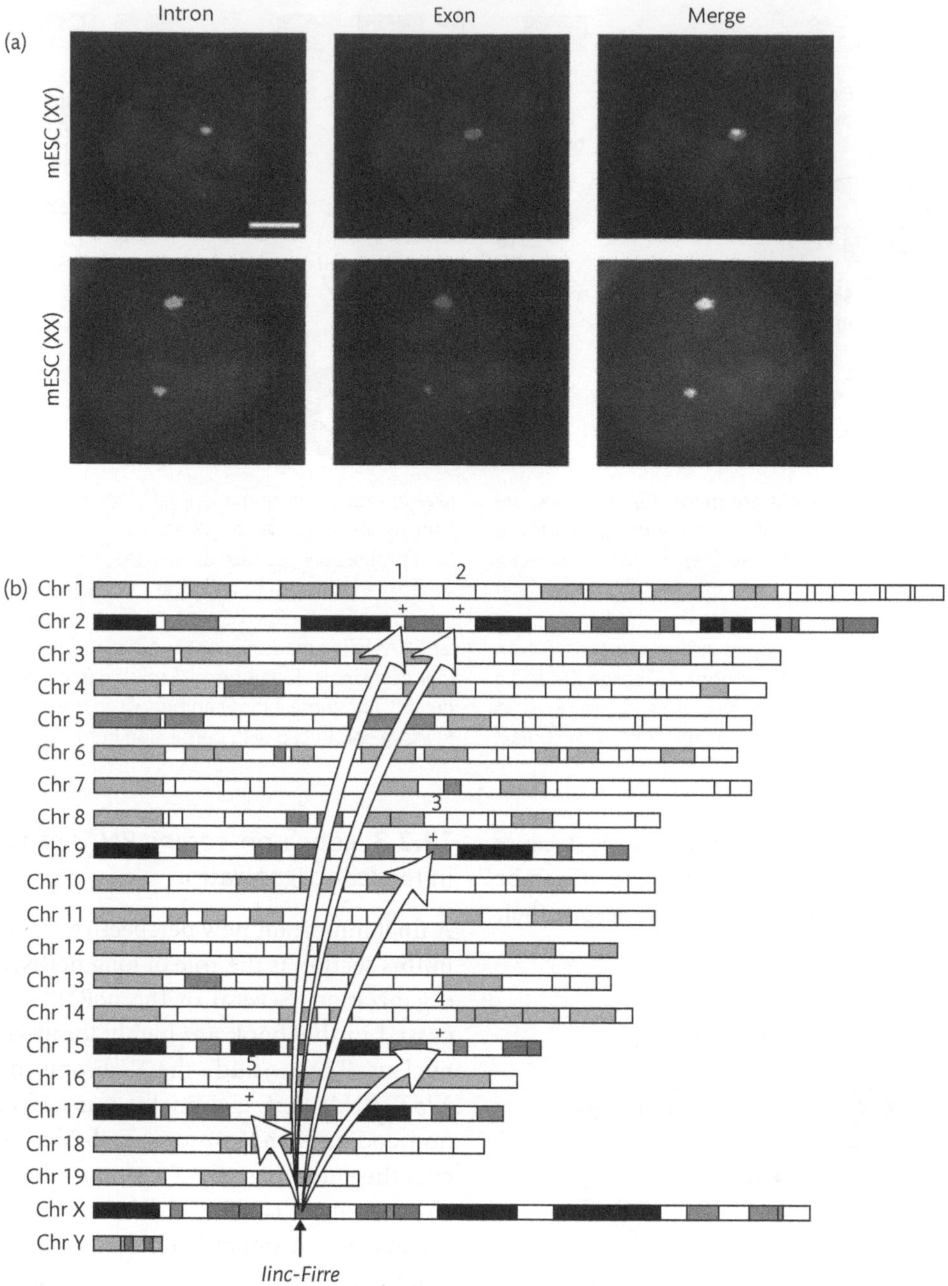

Figure 17.5 The *FIRRE* long ncRNA is expressed from the X chromosome and facilitates interactions between chromosomes. (a) The location of the *FIRRE* ncRNA was revealed by FISH using a probe against either intron or exon sequence—the intron probe shows the position of the gene, and the exon probe shows the position of the spliced long ncRNA transcribed from the gene. Two FISH signals are seen in female cells (XX), indicating that the *FIRRE* gene escapes X chromosome inactivation. These FISH signals indicate an overlap of intron and exon signals. (b) Individual chromosomes within the nucleus are typically highly organized, with distinct chromosomes within particular regions. Experiments to look at what DNA sequences in the nucleus are associated with FIRRE showed a clear binding to *FIRRE*. However, a major surprise was that there are also some other binding sites on four other chromosomes (2, 9, 15, and 17), and that these directly overlap within the nucleus. The function of *FIRRE* is to interact with these other chromosomal regions, to bring them into close connection with the X chromosome within the intact nucleus. (Reproduced with permission from Hacisuleyman et al.[21])

Before the advent of whole transcriptome analyses the existence and functions of many of these ncRNAs were totally unanticipated, and we predict that this area will continue to provide new surprises in the future.

17.3 RNA-guided genome editing

17.3.1 The CRISPR–Cas system

The CRISPR–Cas system is part of a prokaryotic adaptive immunity defence system that evolved to combat foreign DNA present in invading plasmids or bacteriophage. CRISPR stands for **c**lustered **r**egularly **i**nterspaced **s**hort **p**alindromic **r**epeats, and Cas stands for **C**RISPR-**as**sociated. The system is an RNA-directed endonuclease complex that in many ways echoes the process of RNA interference described in Chapter 16. At its core it consists of an array of repetitive sequences flanked by unique spacer sequences. The CRISPR arrays are transcribed and processed into individual **crRNAs**, each of which contains a spacer and part of a repeat. Cas proteins are endonucleases encoded by genes located next to the CRISPR arrays. We now look at the process in more detail.

There are three stages in the process: adaptation (sometimes referred to as acquisition), expression, and interference.[22] Adaptation involves the integration of short fragments of the foreign DNA into CRISPR loci. A bacteriophage-derived sequence approximately 30bp long is inserted at the leader side of a CRISPR locus. This is followed by the duplication of a repeat sequence, creating a new spacer–repeat unit. The selection of a *proto-spacer motif* from the invading DNA involves a short adjacent motif called the *proto-spacer adjacent motif* (PAM). This selection process is thought to involve two highly conserved proteins, Cas1 and Cas2. The adaptation process allows the system to be primed to react to a particular bacteriophage or plasmid. In the expression stage a primary transcript (called the pre-crRNA) is generated which is then cleaved into individual crRNAs. Note that there are variations in CRISPR–Cas systems across prokaryotes. These have been classified into three types: I, II, and III.[22] In type II a *trans*-encoded small RNA (tracrRNA) acts as a guide RNA for the processing of pre-crRNA by the enzyme RNAse III. In the interference stage, the foreign DNA is recognized thanks to complementarity to the proto-spacer sequence. The CRISPR–Cas process is summarized in Fig. 17.6.

17.3.2 Applications of CRISPR–Cas

Of the three CRISPR/Cas systems, type II is thought to be the simplest, mainly because only a single endonuclease, Cas9, is required for both pre-crRNA processing and target cleavage.[23] This raises the possibility that, as in the case of RNA interference, the system could be exploited relatively easily to target *any* sequence of interest not only in prokaryotes, but also in eukaryotic cells.[24] Moreover, instead of having to co-express a tracrRNA required for pre-crRNA processing, a synthetic guide RNA (gRNA) can simply be introduced (equivalent to a processed crRNA). This process works very well and has been successfully used to perform *genome editing* in yeast, plants, nematodes, fruit flies, and vertebrates, including human cells. The use of Cas9 in genome editing is summarized in Fig. 17.7.

What is especially exciting about this system is its flexibility; there are many potential applications.[25] It is possible to make deletions, for example taking out a specific cassette exon. Exogenous DNA can also be inserted at the sites where the genome is cut. Therefore genes can be disrupted, added, or even corrected by homologous recombination. This raises the possibility of using CRISPR therapeutically. Unlike RNA interference, the obvious advantage of CRISPR is that it potentially results in a permanent genetic fix, whereas in RNA interference the best you can achieve is long-term knockdown of a target gene.

However, as with any technology, there are technical issues that must be taken into account. The guide RNAs are relatively short and so there is a danger of targeting unwanted sites in the genome. The degree of specificity and safety of the system remain to be fully evaluated. When dealing with a diploid organism both copies of the target gene may need to be targeted—therefore the system needs to be very efficient. When performing gene editing in human cells in culture, the cells need to be highly transfectable, so that the various components (Cas9, the gRNA, and the DNA to be inserted) can be co-expressed. It is often necessary to introduce DNA that contains a selection marker to select for cells in which the modification has taken place; however, the selection marker may then need to be removed, which is a laborious process. After Cas9 has introduced double-strand breaks, the breaks are repaired by a non-homologous end-joining system which can

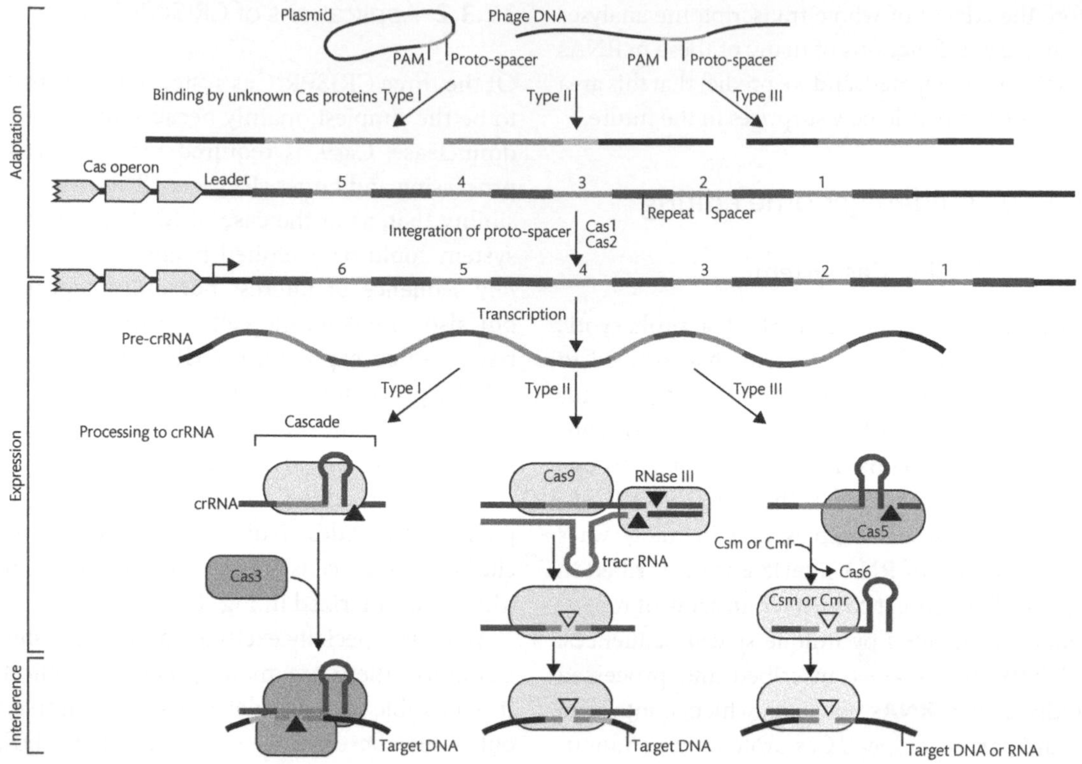

Figure 17.6 The three stages of the CRISPR–Cas adaptive immune response in prokaryotes. The process consists of adaptation, expression, and interference. There are three types of CRISPR–Cas systems, called I, II, and III. In adaptation in types I and II the selection of a proto-spacer derived from the invading DNA is dependent on the presence of a proto-spacer motif (PAM). The proteins Cas1 and Cas2, common to all three types, facilitate the integration of appropriate proto-spacers into the CRISPR loci. In the expression phase a pre-crRNA is transcribed and processed into individual crRNAs. Type II involves a *trans*-encoded small guide RNA (tracrRNA) that works with RNAse III to generate crRNAs. In the interference process the target DNA is recognized through hybridization with the complementary crRNA. Note that the three types of CRISPR–Cas involve different Cas endonucleases and that type III can target foreign RNA as well as DNA. (Reproduced wiith permission from Makarova et al.[22])

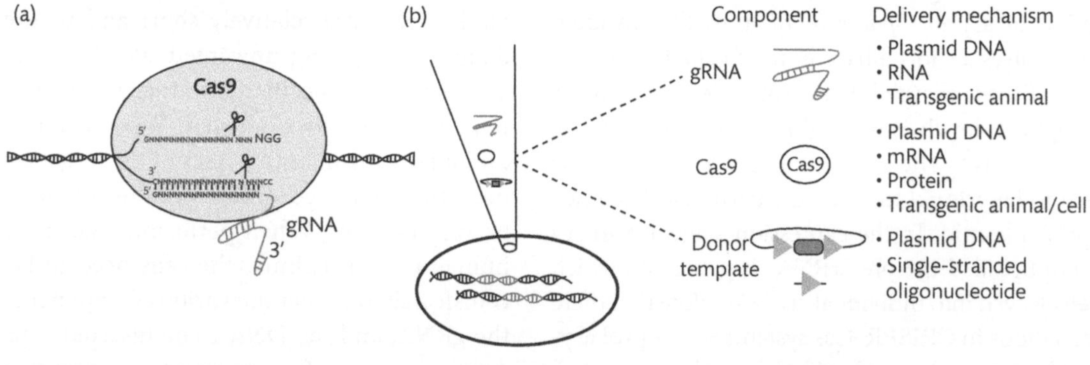

Figure 17.7 Overview of the CRISPR–Cas9 system. (a) A guide RNA (gRNA) hybridizes to its genomic target site (20bp long, enough to allow sufficient target sequence specificity). The PAM (protospacer adjacent motif) sequence in the complementary DNA strand is shown in bold. The PAM sequence is required for the Cas9 nuclease to cleave the target three bases upstream. (b) Components of the system can be delivered to donor cells in different ways to achieve genome editing. The gRNA can be delivered as an RNA, it can be transcribed from a plasmid, or it can be expressed in a transgenic organism. Cas9 can be introduced directly as an mRNA or protein, expressed from a plasmid, or similarly in a transgenic organism. When the aim is to introduce specific DNA sequences, a donor template can be included. The donor DNA can include a selection marker. (Reproduced with permission from Harrison et al.[25])

be error prone, often resulting in small insertions or deletions. It may be necessary to screen several genetically modified clones until the desired modification is achieved. Despite these necessary caveats, CRISPR technology is rapidly increasing in popularity because of its versatility and power. Applications include the genetic modification of crop plants,[26] systematic analysis of gene function in model organisms such as *Drosophila*[27] and zebrafish,[28] and the inhibition of specific microRNAs by cutting a microRNA gene at a specific site.[29] It is also possible to multiplex CRISPR. Box 17.2 summarizes how CRISPR has been used to model lung carcinogenesis in mice by targeting several genes at once.[30]

We end by describing an example of how CRISPR can be used to achieve very precise genome editing including germ-line transmission of targeted knock-ins.[31] In zebrafish the *albino* gene is required for the production of melanin pigment. *Alb* mutant larvae are pale. The *alb*^{b4} mutant is due to a G→T mutation in exon 6 that causes a premature stop codon. Researchers have used CRISPR–Cas9 to repair this mutation. They co-injected the CRISPR–Cas9 components with donor DNA that contains wild-type exon 6 sequence into one-cell stage embryos. They found that using circular donor DNA with flanking CRISPR target sites greatly increased the repair frequency from 1% to 46% in injected larvae. What is especially remarkable is the fact that the repaired allele could also be transmitted through the germ line. In effect, a single nucleotide change has been repaired, restoring the wild-type phenotype. The experiment is shown in Fig. 17.8.

Box 17.2 Use of CRISPR to model lung carcinogenesis in mice

The CRISPR-Cas system can be used in mice to model human disease including cancer. The power and versatility of the system is elegantly illustrated by the work of Platt and colleagues.[30] They first generated transgenic mice that express Cas9 either constitutively or in a Cre-dependent fashion. The Cre-Lox system allows tight control of expression in vivo by driving Cre expression with tissue-specific or inducible promoters. They then delivered gRNA to the brain, vasculature, immune, and lung cells using both viral and non-viral (particle-mediated) delivery methods. In lung tissue they generated loss-of-function mutations in the tumour suppressor genes *TP53* ('the guardian of the genome') and *Lkb1* (a protein kinase that regulates cell polarity). They also introduced a mutation in K-Ras (G12D) that is frequently encountered in lung adenocarcinoma (K-Ras encodes a G-protein involved in intracellular signal transduction often mutated in cancer). The result of these concurrent gene editing events was the generation of large tumours with a histopathology typical of adenocarcinoma.

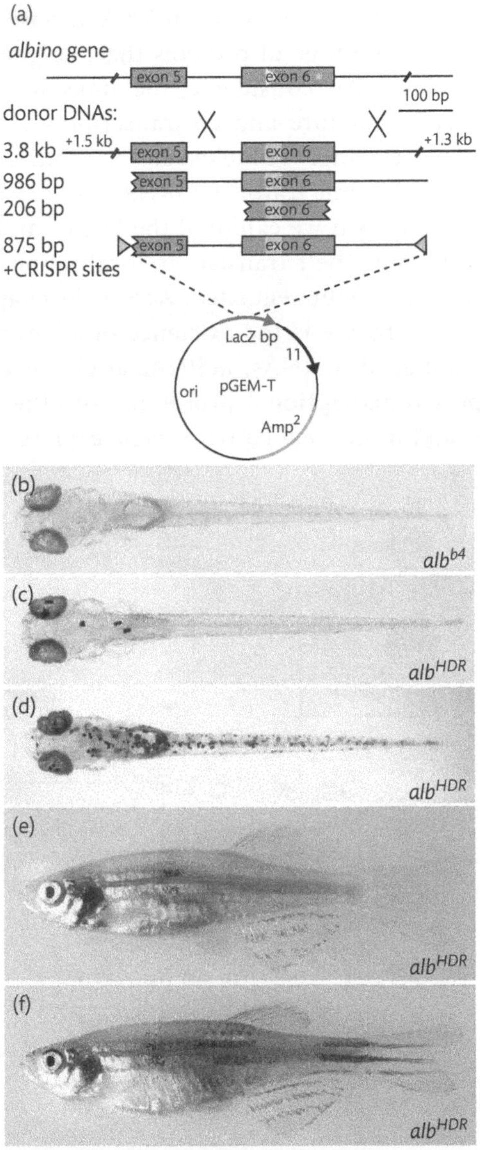

Figure 17.8 Homology-directed CRISPR–Cas9-mediated repair at the zebrafish *albino* locus. (a) The donor DNAs with the CRISPR target sites are shown. The donor DNA was cloned into a plasmid vector (pGEM-T) to increase the rate of homology-directed repair (HDR). Dorsal views of larvae five days after fertilization: (b) uninjected *alb*^{b4} control; HDR with (c) linear donor DNA and (d) circular donor DNA (more efficient); (e, f) two adult fish showing normal pigmented melanophores after successful HDR. For a full colour version of this figure, see Colour Illustration 18. (Reproduced with permission from Irion et al.[31])

17.4 Concluding remarks

In this textbook we have provided a comprehensive overview of the field of RNA biology. We started with the essential elements—an understanding of RNA structure, RNA-mediated catalysis, and RNA-binding proteins. We looked in some detail at how alternative pre-mRNA splicing can maximize the range of proteins that the genome can express. We considered the links between chromatin structure and co-transcriptional pre-mRNA processing. We looked at RNA editing and nucleocytoplasmic traffic of mRNAs and ncRNAs. In the cytoplasm we explored the fate of mRNAs including how their translation, localization, and degradation can be regulated. Across the chapters a common theme is the existence of a bewildering number of ncRNAs. ncRNAs are involved in the post-transcriptional processing of other ncRNAs and in the regulation of gene expression at multiple levels. You will have realized that all of the processes that we describe are strongly interconnected. We summarize this fact in Fig. 17.9, drawing attention to the central role of ncRNAs in all aspects of gene expression.

It is a very exciting time to be working in RNA biology, and much remains to be done. New technologies now allow us to study transcriptomes as well as genomes. The fact that in humans there are almost twice as many genes that produce ncRNAs as those that encode proteins is food for thought. It will take years to unlock the secrets of this incredible assortment of ncRNAs. The regulation of alternative splicing, of mRNA translation, and of localization and decay are still a long way from being comprehensively understood. The fact that RNA biology has come of age means that we can learn a lot more about how life evolved. But we can also exploit it to our advantage through biotechnology. So just as life probably evolved from RNA, RNA might also help make the modern world a better place.

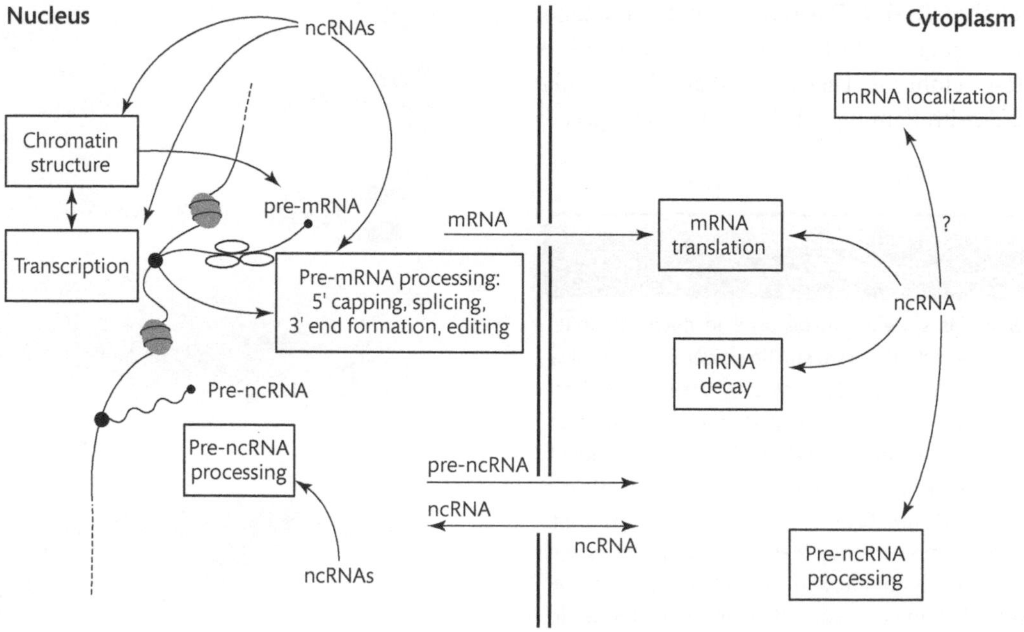

Figure 17.9 The interconnected nature of eukaryotic gene expression involves ncRNAs. In the nucleus chromatin structure influences both transcription and pre-mRNA processing. In turn chromatin structure and transcription are modified through ncRNAs (e.g. by the long ncRNA *XIST*). Co-transcriptional pre-mRNA processing is facilitated by ncRNAs (e.g. by snRNAs in splicing). Precursors of ncRNAs are also processed with the help of ncRNAs (e.g. snoRNAs aid pre-rRNA processing and scaRNAs aid snRNA maturation). The nucleocytoplasmic traffic of RNAs is a regulated process. Some ncRNA processing occurs in the cytoplasm; ncRNAs need to be re-imported into the nucleus. In the cytoplasm mRNA translation is catalysed in large part by rRNA. NcRNAs (notably microRNAs) are involved in the regulation of mRNA translation and decay. Note that ncRNAs are involved in essentially all processes. Similar principles (the interconnections and centrality of RNA) apply in prokaryotic cells.

Summary

- The advent of next-generation sequencing technologies has made it possible to sequence not only genomes but also transcriptomes. Transcriptomes are the total collection of RNAs transcribed from a genome. However, RNA-Seq (RNA sequence analysis) data is very complex and can be difficult to analyse bioinformatically.
- Transcriptome analysis makes it very obvious that there is an enormous amount of ncRNA. In humans there are almost twice as many ncRNA genes as protein coding genes. ncRNA is growing in prominence. Some pseudogenes transcribe ncRNAs that have functional roles. Long ncRNAs can be involved in nuclear organization, including interactions between chromosomes.
- The *competing endogenous RNA hypothesis* states that endogenous RNAs can act as decoys to regulate other RNAs, such as those with common binding sites for miRNAs. Examples of ceRNAs are the circular RNAs such as *SRY*, which is involved in male sex determination.
- The CRISPR–Cas system evolved as a prokaryotic adaptive immunity system aimed at counteracting plasmids or bacteriophage. At its core is an RNA-directed endonuclease complex. The system has been harnessed in order to perform sophisticated and precise genome editing in a wide range of species.
- RNA biology has come of age and there is an enormous amount of work left to be done. The function of a vast array of ncRNAs remains to be elucidated. Evidence shows that chromatin structure, transcription, and co-transcriptional and post-transcriptional processes are all interconnected, with ncRNA at the heart of most, if not all, processes.

Questions

17.1 Define what is meant by the terms transcriptome and transcriptomics.

17.2 Describe two types of next-generation sequencing techniques.

17.3 How does the proportion of ncRNAs versus protein coding genes vary in different species?

17.4 What is RNA-Seq data, what can it tell us, and what are the challenges involved in analysing it?

17.5 Might pseudogenes have important roles? If so, how are ncRNAs involved?

17.6 Explain the *competing endogenous RNA* hypothesis.

17.7 How are long ncRNAs involved in nuclear organization?

17.8 What are the three main types of CRISPR–Cas and how do they work?

17.9 How can the CRISPR–Cas system be exploited to perform genome editing? Are there any technical limitations?

References

1. **Djebali S, Davis CA, Merkel A, et al.** Landscape of transcription in human cells. *Nature* **489**, 101–8 (2012).
2. **Mudge JM, Frankish A, Harrow J.** Functional transcriptomics in the post-ENCODE era. *Genome Res* **23**, 1961–73 (2013).
3. **Mercer TR, Mattick JS.** Structure and function of long noncoding RNAs in epigenetic regulation. *Nat Struct Mol Biol* **20**, 300–7 (2013).
4. **Rinn JL, Chang HY.** Genome regulation by long noncoding RNAs. *Annu Rev Biochem* **81**, 145–66 (2012).
5. **Mattick JS.** Non-coding RNAs: the architects of eukaryotic complexity. *EMBO Rep* **2**, 986–91 (2001).
6. **Tay Y, Rinn J, Pandolfi PP.** The multilayered complexity of ceRNA crosstalk and competition. *Nature* **505**, 344–52 (2014).
7. **Capobianco E.** RNA-Seq data: a complexity journey. *Comput Struct Biotechnol J* **11**, 123–30 (2014).
8. **Szafranski K, Fritsch C, Schumann F, et al.** Physiological state co-regulates thousands of mammalian mRNA splicing events at tandem splice sites and alternative exons. *Nucleic Acids Res* **42**, 8895–904 (2014).
9. **Costa V, Aprile M, Esposito R, Ciccodicola A.** RNA-Seq and human complex diseases: recent accomplishments and future perspectives. *Eur J Hum Genet* **21**, 134–42 (2013).
10. **Saliba AE, Westermann AJ, Gorski SA, Vogel J.** Single-cell RNA-seq: advances and future challenges. *Nucleic Acids Res* **42**, 8845–60 (2014).
11. **ENCODE Project Consortium.** An integrated encyclopedia of DNA elements in the human genome. *Nature* **489**, 57–74 (2012).
12. **Duret L, Chureau C, Samain S, et al.** The Xist RNA gene evolved in eutherians by pseudogenization of a protein-coding gene. *Science* **312**, 1653–5 (2006).

13. **Poliseno L, Salmena L, Zhang J, et al.** A coding-independent function of gene and pseudogene mRNAs regulates tumour biology. *Nature* **465**, 1033–8 (2010).
14. **Rigoutsos I, Furnari F.** Gene-expression forum: decoy for microRNAs. *Nature* **465**, 1016–17 (2010).
15. **Tay Y, Kats L, Salmena L, et al.** Coding-independent regulation of the tumor suppressor PTEN by competing endogenous mRNAs. *Cell* **147**, 344–57 (2011).
16. **Capel B, Swain A, Nicolis S, et al.** Circular transcripts of the testis-determining gene Sry in adult mouse testis. *Cell* **73**, 1019–30 (1993).
17. **Valdmanis PN, Kay MA.** The expanding repertoire of circular RNAs. *Mol Ther* **21**, 1112–14 (2013).
18. **Memczak S, Jens M, Elefsinioti A, et al.** Circular RNAs are a large class of animal RNAs with regulatory potency. *Nature* **495**, 333–8 (2013).
19. **Hansen TB, Jensen TI, Clausen BH, et al.** Natural RNA circles function as efficient microRNA sponges. *Nature* **495**, 384–8 (2013).
20. **Rinn J, Guttman M.** RNA function. RNA and dynamic nuclear organization. *Science* **345**, 1240–1 (2014).
21. **Hacisuleyman E, Goff LA, Trapnell C, et al.** Topological organization of multichromosomal regions by the long intergenic noncoding RNA Firre. *Nat Struct Mol Biol* **21**, 198–206 (2014).
22. **Makarova KS, Haft DH, Barrangou R, et al.** Evolution and classification of the CRISPR-Cas systems. *Nat Rev Microbiol* **9**, 467–77 (2011).
23. **Sampson TR, Weiss DS.** Exploiting CRISPR/Cas systems for biotechnology. *Bioessays* **36**, 34–8 (2013).
24. **Jinek M, Chylinski K, Fonfara I, Hauer M.** A programmable dual-RNA-guided DNA endonuclease in adaptive bacterial immunity. *Science* **337**, 816–21 (2012).
25. **Harrison MM, Jenkins BV, O'Connor-Giles KM, Wildonger J.** A CRISPR view of development. *Genes Dev* **28**, 1859–72 (2014).
26. **Belhaj K, Chaparro-Garcia A, Kamoun S, et al.** Editing plant genomes with CRISPR/Cas9. *Curr Opin Biotechnol* **32**, 76–84 (2015).
27. **Housden BE, Lin S, Perrimon N.** Cas9-based genome editing in *Drosophila. Methods Enzymol* **546**, 415–39 (2014).
28. **Varshney GK, Burgess SM.** Mutagenesis and phenotyping resources in zebrafish for studying development and human disease. *Brief Funct Genomics* **13**, 82–94 (2014).
29. **Zhao Y, Dai Z, Liang Y, et al.** Sequence-specific inhibition of microRNA via CRISPR/CRISPRi system. *Sci Rep* **4**, 3943 (2014).
30. **Platt RJ, Chen S, Zhou Y, et al.** CRISPR-Cas9 knockin mice for genome editing and cancer modeling. *Cell* **159**, 440–55 (2014).
31. **Irion U, Krauss J, Nüsslein-Volhard C.** Precise and efficient genome editing in zebrafish using the CRISPR/Cas9 system. *Development* **141**, 4827–30 (2014).

GLOSSARY

A→I editing Conversion of adenosine to inosine, through modification of adenine base by deamination.

Acetylation Addition of an acetyl chemical group ($-COCH_3$).

ACF ApoBec complementing factor—a protein which forms part of the core complex which edits the *APOB* mRNA.

A complex Complex formed early in spliceosome assembly, composed of pre-mRNA, U1 snRNP, and U2 snRNP.

Activation energy The minimum energy which two molecules need to collide in order to react.

ADAR Adenosine deaminase acting on RNA—an enzyme which modifies nucleotides within RNA.

ADAT Adenosine deaminase acting on tRNA—an enzyme which modifies nucleotides within tRNA.

A-form helix Double-stranded nucleic acid helix formed between two RNA strands, or between a DNA and an RNA strand.

AID Activation-induced cytidine deaminase—a CDAR which works on DNA substrates to convert C to T.

AIDS Acquired immune deficiency syndrome.

Alternatively spliced exon An exon that is only spliced under particular circumstances by the spliceosome.

Alu element Repetitive sequence found in primate DNA, and a major target of A→I editing in humans because of regions of complementary base pairing in cognate RNAs. Named because it contains the restriction site recognized by the restriction enzyme *Alu*1.

Amino-acyl tRNA A tRNA that is charged with a specific amino acid.

Amino-acyl tRNA synthetase An enzyme that charges (i.e. loads) a tRNA with the appropriate amino acid.

Anticodon Triplet sequence in tRNA which forms hydrogen bonds with the codon of the mRNA.

Anti-sense oligonucleotides Short strings of DNA or RNA oligonucleotides complementary to mRNA targets.

APOB Apolipoprotein B—the mRNA encoding this protein is subject to C→U editing to produce a different form of the APOB protein in the intestine (APOB48) and the liver (APOB100).

ApoBec RNA-binding enzyme which carries out C→U RNA editing of the *APOB* transcript.

Aptamer Interacting RNA isolated through the SELEX procedure.

Archaea Prokaryotic organisms formerly regarded as primitive bacteria ('archaebacteria') but now known to be phylogenetically distinct from 'true' bacteria. They are genetically and metabolically different from other bacteria, and are descended from an ancient group of organisms that bridge the gap between bacteria and eukaryotes.

ARE AU-rich element.

Arginine fork The ability of a single arginine to use hydrogen bonding to interact with RNA.

Arginine-rich domain Arginine-rich region of a protein—generally part of an auxiliary RNA-binding domain.

Argonaute A family of proteins that are a critical component of the RNA-induced silencing complex (RISC) having two unique domains, PAZ (RNA binding) and PIWI (slicer activity), responsible for target RNA cleavage (slicing).

A site Site in the ribosome where the amino-acyl tRNA lands, charged with the next amino acid to be added to the polypeptide chain.

ATR Ataxia telangiectasia and Rad3-related protein.

Attacking group A chemical group which attacks and forms new covalent bonds with target atoms. Attacking groups are either nucleophiles (electron rich) or electrophiles (electron poor). *See also* Leaving group.

AU-rich element (ARE) Sequence element rich in As and Us often associated with mRNA instability.

Auxiliary RNA-binding domain A domain that helps other RNA-binding domains (e.g. an RRM) to bind stably to an RNA target sequence.

β-glycosidic bond Chemical bond between a pentose sugar and a base.

Balbiani rings Many copies of a chromosome bundled side by side—found primarily in the salivary glands of certain insects.

Base Heterocyclic molecule which is a component of nucleotides.

Base catalysis *See* General acid/base catalysis.

Base deamination Removal of an amino ($-NH_2$) group from a base. This is an important change since amino groups are frequent hydrogen bond donors.

Base modification Chemical change in a base which converts it into another base.

Base pairing Watson–Crick pairing of bases in RNA or DNA.

Base stacking Interactions between electrons in the aromatic rings of the bases, which contribute to the stability of RNA secondary structure.

B complex More mature splicing complex which contains all the U snRNPs.

B* complex Splicing complex which has released U1 and U4 snRNPs and started to reorganize before forming the first catalytically active splicing complex.

Becker muscular dystrophy (BMD) A less severe form of muscular dystrophy than DMD caused by missense mutation in the *dystrophin* gene.

B-form helix Double-stranded nucleic acid helix formed between two DNA strands.

Binding energy Free energy which is released when an enzyme binds to its substrates.

Bipartite processing element A set of two conserved sequences that help to define the 3′ end of histone mRNAs.

BMD Becker muscular dystrophy.

Box C/D snoRNA A class of snoRNAs generally involved in 2′-*O*-ribose methylation of pre-rRNA.

Box H/ACA snoRNA A class of snoRNAs generally involved in the pseudouridylation of pre-rRNA.

Brahmin Protein component of the SWI–SNF complex.

Branchpoint sequence A conserved sequence which is part of the 3′ splice site and is involved in the first catalytic step of splicing.

BRCA1 Breast cancer susceptibility gene.

Btub A structural component required to import coenzyme B_{12} into cells of the bacterium *Escherichia coli.*

C→U editing Chemical conversion of cytosine to uracil by deamination of a cytosine base.

C complex Catalytically active spliceosome.

Cajal body Nuclear site of snRNA transcription and snRNP assembly.

Cap-binding complex (CBC) The CBC is composed of two subunits, CBC80 and CBC20, and is loaded onto the 5′ cap structure found on all mRNAs. It binds to and stabilizes the TREX complex containing the RNA export adaptor REF.

Capping enzyme complex Enzymes required for the formation of the 5′ cap.

Carbohydrate A large group of organic compounds composed of hydrogen, carbon, and oxygen.

Cassette exon An exon that is either retained or wholly spliced out.

Catalyst A catalyst speeds up the rate of a chemical reaction, while itself remaining unaltered.

CBC Cap-binding complex.

CD44 Gene encoding an alternatively spliced panel of exons with prefix v1–v10. Exon v5 is regulated by the RNA-binding protein Sam68.

CDAR Cytosine deaminase acting on RNA.

cDNA Complementary DNA (copied from an RNA by reverse transcriptase).

CELF proteins A group of RNA-binding proteins related to CUG-BP.

Centromere The parts of each eukaryotic chromosome which are important for chromosome movement during cell division.

ceRNA Competing endogenous RNAs that act as decoys to regulate other RNAs with common binding sites for microRNAs.

Chromatin DNA within the eukaryotic nucleus which is packaged by wrapping around nucleosomes in the nucleus, and then coiled into a 30nm fibre.

Circular RNAs Long circular non-coding RNAs that behave as competing endogenous RNAs (ceRNAs) acting as decoys for microRNAs.

***Cis*-acting element** A sequence that influences the post-transcriptional regulation of an adjoining open reading frame.

***Cis*-splicing** The splicing together of exons within an RNA molecule.

Cleaving RISC A cleavage-competent RISC—contains endonuclease (slicer) activity.

CLIP assay In vivo crosslinking of protein–RNA complexes followed by immunoprecipitation—used to identify RNA targets.

Clk A CDC2-like kinase which adds phosphate groups to SR proteins in the nucleus to mobilize them from splicing speckles.

Closed-loop conformation A scenario in which the 5′ cap structure interacts with the poly(A) tail through the translation initiation factor eIF4G.

Coaxial stacking Stacking of two helices on top of each other so that they effectively form one long helix.

Codon Set of three nucleotides in an mRNA that specify a particular amino acid or translation stop.

Cold-shock domain (CSD) Bacteria express abundant small RNA-binding cold-shock proteins which correspond to the cold-shock domain in eukaryotic proteins.

Complementary base pairing The formation of Watson–Crick bonds between two strands that are said to be complementary.

Condensation reaction A reaction in which a small molecule like water is split off. The reaction between an –OH group and a phosphate group to form a phosphodiester bond is a condensation reaction.

Constitutive exon Exon with strong splice sites that is included in an mRNA by the spliceosome in every cell.

Coordinate bond Covalent bond in which both shared electrons are provided by one atom.

Co-transcriptional splicing Pre-mRNA splicing that generally occurs while transcription is still taking place.

Covalent bond Shared pair of electrons holding two atoms together.

CpG island Region of low methylation of CpG dinucleotides.

CpG methylation Methylation at a CpG site.

CpG site A region of DNA where a cytosine (C) nucleotide occurs next to a guanine (G) nucleotide in the linear sequence of bases along its length. 'CpG' is shorthand for '–C–phosphate–G–'.

CRISPR Clustered regularly interspaced short palindromic repeats—RNA-guided genome editing system.

CRM1 Chromosome region maintenance 1—a nuclear receptor protein (karyopherin) which exports rRNA and snRNA.

crRNA CRISPR arrays are transcribed and processed into individual crRNA, each of which includes a spacer and part of a repeat.

CSD Cold-shock domain.

CTD C-terminal domain.

C-terminal domain (CTD) The carboxy-terminal end of the enzyme RNA polymerase II, consisting of several heptad repeats. Required for the co-transcriptional recruitment of RNA processing factors.

CUG-BP CUG-binding protein—important in myotonic dystrophy.

Curly arrow In reaction diagrams indicates the movement of electrons as bonds break and form in chemical reactions.

Cytoplasmic polyadenylation Lengthening of the poly(A) tail at the 3′ end of mRNAs, generally associated with unmasking translationally repressed mRNAs.

DBP5 An RNA helicase which disassembles the mRNP export complex once it reaches the cytoplasm. DBP5 travels with the mRNP to the cytoplasm from the nucleus but is not active as a helicase until it reaches the cytoplasmic side of the NPC where it is activated by a protein called Gle1 along with its cofactor, the small molecule IP6.

DCC Dosage compensation complex.

Deadenylation The process of removing a poly(A) tail.

Deamination Removal of an amino group (e.g. from cytosine to give uracil).

Decapping The process of removing the 7-methyl guanosine cap.

Decoding centre The part of the ribosome required for reading the mRNA's genetic information.

Degradosome A complex of proteins that mediate mRNA degradation in bacteria.

Dense fibrillar centres (DFC) Nucleolar sites of pre-rRNA synthesis and early processing events.

Deprotonation *See* Protonate/deprotonate.

DFC Dense fibrillar centres.

Dicer RNAse III family enzyme that cleaves a dsRNA or an shRNA into 21–25nt siRNAs.

DMD Duchenne muscular dystrophy.

DMPK *Dystrophonia myotonica protein kinase*—gene which contains expanded triplet repeats in myotonic dystrophy.

DNA Deoxyribonucleic acid—in its Watson–Crick double-helix conformation, DNA is elegantly and easily replicated.

Dosage compensation complex (DCC) Protein–RNA complex which upregulates transcription from the single male X chromosome in fruit flies.

Drosha A nuclear RNAse III enzyme that specializes in miRNA biogenesis.

dsRBD RNA-binding domain that recognizes double-stranded RNA.

dsRNA Double-stranded RNA—RNA helices in which the bases are joined through hydrogen bonds according to Watson–Crick rules. Double-stranded RNA is a major target for both A→I editing in animals and uridine insertion/deletion editing in trypanosomes.

Duchenne muscular dystrophy (DMD) A more severe form of muscular dystrophy than BMD caused by insertion of a translational stop codon into the *dystrophin* gene.

dystrophin Gene encoding important muscle protein found in a complex which connects the cytoskeleton of muscle cells with the extracellular matrix.

E complex Early splicing complex, containing U1 snRNP bound to the 5′ splice site of pre-mRNA.

ECS Editing complementary sequence. Region of complementary RNA which is distant to the editing site but part of the same RNA molecule, and which base pairs with the RNA to be edited.

Editing Changing the nucleotide sequence of RNA. This can occur either through base modification of As or Us, or through the addition/deletion of Us.

EJC Exon junction complex.

Electrophilic catalyst Charged molecule (usually a metal ion in ribozymes) which stabilizes an electrical charge on a reaction intermediate.

Elongation factors Proteins that are involved in translation elongation.

EMSA *See* Gel retardation assay.

EMT Epithelial to mesenchymal transition.

Epigenetic code Information which controls whether chromatin is transcriptionally active or not, including DNA methylation and histone methylation and acetylation.

Epigenetics Information in the genome not directly carried by the nucleotide sequence, but inherited at cell division.

ESE Exonic splicing enhancer—bound by splice factors that activate splicing.

ESEFinder Online search tool used to find binding sites for splice factors in RNAs.

E site Exit site in the ribosome from which tRNA leaves once it has delivered its amino acid.

ESS Exonic splicing silencer.

ESSENCE Exon-specific splicing enhancers by chimaeric effectors.

EST Expressed sequence tag.

Euchromatin Corresponds to transcriptionally active DNA, and has a very 'open' structure which is readily accessible to transcription factors and RNA polymerase. As such, euchromatin includes the genes which are active in any given cell type.

Exon A stretch of RNA that is generally spliced into a mature mRNA.

Exon definition complex Early splicing complex which marks exons for splicing.

Exon junction complex (EJC) Complex of protein which remains bound to mRNA after splicing. The EJC is deposited 20–24 nucleotides upstream of the spliced exon junction, and has roles in RNA stability and mRNA export.

Exon skipping Alternative splicing process in which an exon is left out of an mRNA.

Exosome A protein complex with a core of nine subunits that form a ring-like structure which envelops the target RNA, promoting its degradation.

Exportin Nuclear receptor which mediates nuclear export through the nuclear pore. Exportins bind their cargo in the nucleus with RAN-GTP, and then release their partner proteins in the cytoplasm on releasing RAN-GDP.

Exportin-5 A specialized exportin that transports precursor miRNAs into the cytosol.

FCs Fibrillar centres.

FG repeat nucleoporin A group of nuclear pore proteins which have repeated phenylalanine (F) and glycine (G) amino acid residues.

Fibrillar centres (FCs) Sites within nucleoli that contain rRNA genes.

FISH Fluorescence in situ hybridization. Type of in situ hybridization in which the probe is labelled with a fluorescent group. *See* In situ hybridization.

Frameshift A frameshift is caused by a nucleotide insertion or deletion which is not a multiple of three. Since the genetic code is based on triplets, frameshifts introduce a change to the reading frame read by the translation machinery (this leads to the term frameshift).

Frameshift mutation A frameshift is caused by a nucleotide insertion or deletion which is not a multiple of three. Frameshifts introduce a change in the reading frame.

FRAP Fluorescence recovery after photobleaching—used in studying splice factor dynamics in living cells.

FTD-17 Frontotemporal dementia linked to chromosome 17.

Gel retardation assay Also known as electrophoretic mobility shift assay (EMSA). Binding of proteins to RNA causes the RNA to move more slowly through a gel—used to study protein–RNA interactions.

Gene expression The process whereby genes, encoded by DNA, express their inherited information usually in the form of proteins (but sometimes in the form of non-coding RNAs).

Gene expression factory A model in which all the molecular machines that carry out different steps of mRNA processing are physically linked together.

General acid A general acid is a proton (H^+) donor.

General acid/base catalysis A molecule other than water which catalytically speeds up a reaction by acting as a proton donor or acceptor. This activates nucleophilic oxygens for reactions, or stabilizes oxyanion leaving groups.

General base A general base is a proton (H^+) acceptor.

Gene regulation The processes through which genes are switched on and off and their expression regulated in a myriad of ways, both transcriptionally and post-transcriptionally.

Genome The genome is essentially the collection of all DNA in a given organism.

Genomics The study of genomes. Comparative genomics looks at the interrelationships between different genomes—their similarities and differences—and what this tells us about the evolutionary relationships between different organisms.

GFP Green fluorescent protein.

Gradient centrifugation The process of separating macromolecules and even organelles by very fast spinning, generating very strong centrifugal forces.

Granular component (GC) Part of the nucleolus involved in late RNA processing and ribosome formation.

Group I and II introns Self-splicing parasitic elements.

GTP Guanosine 5′-triphosphate.

GTPase Enzyme which converts GTP to GDP. A notable example is RAN, a key player in nuclear RNA export.

Guide RNAs Short RNAs which act as templates for U insertion and U deletion editing in mitochondrial transcripts from trypanosomes. Also used to describe the RNA strand associated with RNA-induced silencing complexes. In trypanosomes the guide RNAs are transcribed from both the maxicircles and minicircles of the mitochondrial genome.

H19 Non-coding RNA made from an imprinted gene cluster on human chromosome 11. *H19* does not direct epigenetic changes, but its transcription seems to be important in directing an enhancer away from activating the *IGF2* gene on the maternal copy of chromosome 11.

Hairpin stemloop Intramolecular helix in which a strand of RNA folds back on itself and hydrogen bonds to form a stem, leaving an unpaired sequence (the loop).

Hammerhead ribozyme Cleaves the product of rolling-circle replication of plant viruses such as TRSV (tobacco ringspot virus).

H complex Non-specific complex which forms on pre-mRNA in nuclear extracts, and contains hnRNP proteins.

HDAC Histone deacetylase. Removes acetyl groups from histones.

Helical junctions Joining regions in RNA structures that link different helices together.

Helicase *See* RNA helicase.

Heptad repeat A repeat sequence of YSPTSPS present in the C-terminal domain of RNA polymerase II.

Heterochromatin Tightly packed chromatin containing transcriptionally inactive DNA.

Heterocyclic Ring-shaped molecules in purines and pyrimidine bases which contain other atoms in addition to carbon.

HGPS Hutchinson–Gifford progeria syndrome.

Histone code Modifications to histones which affect transcriptional activity. These modifications are frequently methylation and acetylation.

Histone deacetylase (HDAC) Removes acetyl groups from histones.

Histone methyltransferase (HMT) Adds methyl groups to histones.

Histones Small basic proteins which are the building blocks of nucleosomes. Histones H2A, H2B, H3, and H4 are assembled into nucleosomes (there are two copies of each histone in a nucleosome), and adjacent nucleosomes are linked together by a fifth histone H1.

HIV Human immune deficiency virus.

hnRNA Heterogeneous nuclear RNA—full-length nuclear pre-mRNA.

hnRNPs Proteins that package nuclear pre-mRNA.

Homeodomain A 60 amino acid helix-turn-helix DNA-binding domain present in several transcription factors. Also thought to bind RNA.

Homeotic genes Any of a group of genes that control the pattern of body formation during early embryonic development.

HOTAIR Non-coding RNA made from the *HOXC* cluster of human genes. *HOTAIR* binds to the polycomb group of epigenetic regulators, and represses another cluster of *HOX* genes called *HOXD*.

***HOX* gene** A homeotic gene which controls body plans in animals.

HP1 Heterochromatin protein 1. HP1 proteins are found in animals, plants, and fungi, and repress transcription. HP1 proteins contain a chromodomain which can bind to methylated histones. Animal HP1 binds to H3 methylated at lysine 9, and causes widespread heterochromatinization across the genome.

Hydrogen bond Interaction between the partial positive charge of a hydrogen atom bonded to an electronegative atom like oxygen or nitrogen.

ICR Imprinting control region.

Immunoprecipitation Isolation of complexes through the specific interaction between an antibody and a component of the complex.

Importin Nuclear receptor which mediates nuclear import through the nuclear pore. Importins release these partner proteins in the nucleus on binding RAN-GTP, but assemble in the cytoplasm in the absence of RAN binding.

Importin β Nuclear import receptor for snRNPs to which it binds via the import receptor snurportin.

Imprinting control region (ICR) Region in imprinted gene clusters which controls the imprinting pattern. ICRs are often promoters for long ncRNAs.

Initiation factors Proteins that are involved in translation initiation.

Initiator tRNA A special type of tRNA that delivers *N*-formyl methionine as the very first amino acid.

In-line configuration Alignment of the incoming oxygen atom, the phosphorus atom under attack, and the leaving oxygen atom.

In-line probing Method for probing RNA secondary structure.

Inosine Modified base which hydrogen bonds with cytidine.

Interchromatin granule cluster Storage compartment for splicing located next to perichromatin fibrils.

Internal loop Structures that form when two strands of a helix have an unequal number of unpaired bases.

In situ hybridization A technique whereby an anti-sense labelled nucleic acid probe is hybridized through Watson–Crick base pairing to determine the intracellular or tissue distribution of a particular sequence (e.g. a localized mRNA). *See* FISH.

Intron A stretch of RNA that is generally spliced out from a pre-mRNA.

Intron definition The formation of the early splicing complex which marks introns for removal by splicing.

Intron retention Retention of an intron in an mRNA.

In vitro translation The use of cell extracts to translate specific mRNAs in the laboratory.

IRE Iron response element—involved in the post-transcriptional regulation of mRNAs associated with iron metabolism.

IRES Internal ribosome entry site—required for the initiation of translation downstream of the first effective AUG start codon.

ISE Intronic splice enhancer—bound by splice factors that activate splicing.

ISS Intronic splice silencer—bound by splice factors that repress splicing.

Karyopherin A group of nuclear transport receptors which fall into two groups. Importins bind to proteins with an NLS and move them into the nucleus. Exportins bind to proteins with an NES and move them out of the nucleus.

KH domain K-homology domain—an RNA-binding domain first identified in the protein hnRNP K.

Kinetic coupling model Model explaining the connection between transcription elongation speeds and selection of alternatively spliced exons in which the elongation speed controls the production rate of competing exons.

Kinetoplast Mitochondrial genome of trypanosomes, which is composed of maxicircles and minicircles of DNA.

Kissing loop complexes RNA structures formed by hydrogen bonding between single-stranded regions of loops.

Knockdown (silencing) Drastic reduction in the expression of a gene due to RNA interference targeting an mRNA. Not to be confused with a knockout where no gene product is made.

Knockout The manipulation of a gene such that it is not expressed at all (e.g. via its deletion).

Lariat intermediate Loop-shaped intermediate formed by the splicing reaction.

Leaving group Molecule or atom which is released in a substitution reaction.

Leptomycin B Drug which binds to the nuclear export receptor CRM1 and blocks nuclear export of rRNA, snRNA, and some unspliced HIV pre-mRNAs.

Lewis acid Molecule or functional group which can interact with unpaired electrons.

Ligand Substance able to form a complex with a biological molecule.

LMNA Lamin A.

Major spliceosome The most abundant spliceosome in a cell, which uses U2 snRNP and the other major snRNPs.

MAP kinase Mitogen-activated protein kinase.

Masked message An mRNA that is translationally repressed long term.

Maternal mRNA An mRNA that is expressed in oocytes.

Mdx mouse strain Mouse containing a DMD mutation.

Messenger RNA (mRNA) A transcript that encodes a polypeptide.

Metazoan Multicellular animal, with a body made up of differentiated cells arranged in tissues and organs.

Methylation Addition of a methyl group ($-CH_3$).

Methyl cytosine DNA-binding proteins Contain a methyl cytosine-binding domain (MBD). Bind to methylated cytosine, and complex with HDACs to form heterochromatin.

7-Methyl guanosine cap (m^7G) The very 5′ end of an mRNA which is modified with an extra guanosine joined in an unusual 5′–5′ bond.

microRNA (miRNA) A short non-coding RNA expressed from a microRNA gene. MicroRNAs are involved in gene silencing.

microRNome (miRNome) The collection of all microRNAs expressed in a given genome.

MIDI domain Part of an Argonaute protein, this domain binds to the 5′ phosphate of the guide RNA.

Miller spread Electron microscopy technique devised in the 1960s that allows the direct visualization of chromatin and nascent transcripts.

Minigene Shorted version of a gene which might contain particular regions of interest. For example, minigenes often contain particular exons and introns from genomic DNA, and often include shortened versions of introns.

Minor spliceosome Less abundant spliceosome which uses U12 snRNP and other minor snRNPs.

miRISC A RISC complex that has incorporated a microRNA.

miRNA *See* microRNA.

Missense mutation Mutation which changes the amino acid encoded.

MMTV Mouse mammary tumour virus.

Molecular motors Molecular machines that mediate the intracellular transport of vesicles, organelles, and macromolecules.

Morpholino A modified oligonucleotide in which the normal phosphodiester bond in the backbone is replaced by a non-ionic phosphorodiamidate linkage.

MPMV Mason–Pfizer monkey virus.

mRNA decay mRNA degradation—the process whereby mRNAs are actively degraded.

mRNA stability A parameter that reflects the half-life of a mRNA.

mRNA surveillance Active elimination of faulty mRNAs that encode dysfunctional and potentially toxic proteins.

mRNP Messenger ribonucleoprotein; mRNA coated with proteins.

MSF Macrophage stimulating factor.

Muscleblind (MBLN) RNA-binding protein important in myotonic dystrophy.

Myotonic dystrophy (DM) Multi-systemic disease caused by splicing defects in different tissues.

Nascent transcript An RNA copy of a DNA sequence in the act of being produced by RNA polymerase.

ncRNA *See* non-coding RNA.

NES Nuclear export sequence.

Neurofibrillary plaque Insoluble aggregate in brains of patients with neurodegeneration. Plaques contain the tau protein.

***N*-formyl methionine** The first amino acid in a polypeptide chain; deposited by the initiator tRNA.

NGS Next-generation sequencing—facilitates rapid and large-scale sequence analysis of genomes and transcriptomes.

NLS Nuclear localization sequence.

No-go decay (NGD) Degradation of mRNAs that stall during the process of translation elongation.

Non-cleaving RISC A RISC that lacks slicer activity; it only blocks the translation of targeted mRNAs.

Non-coding RNA (ncRNA) An RNA that does not encode for a polypeptide, i.e. all RNAs other than mRNA.

Nonsense-mediated decay (NMD) A process that removes mRNAs that carry premature termination codons (PTCs).

Nonsense mutation A mutation which adds a translational stop codon.

NOR Nucleolar organizing region.

Northwestern blot A technique used to study protein–RNA interactions in which proteins are first transferred to a membrane and then incubated with a riboprobe (RNA ligand probe).

NOVA Neuro-oncological ventral antigen—an RNA-binding protein important in paraneoplastic diseases.

NPC Nuclear pore complex.

nt nucleotide.

NTF2 Nuclear import receptor for RAN-GDP—needed to recycle export components back into the nucleus.

Nuclear export adaptor Protein which binds to RNA and to nuclear export receptors.

Nuclear export receptor Protein which enables mRNP particles to exit through the nuclear pore.

Nuclear export sequence (NES) Short peptide sequence in a protein which mediates its nuclear export.

Nuclear localization sequence (NLS) Short peptide sequence in a protein which mediates its nuclear import.

Nuclear pore complex (NPC) Aperture in the nuclear envelope through which all nucleocytoplasmic traffic takes place.

Nuclear speckles *See* Splicing factor compartments.

Nucleolar organizing region (NOR) Clusters of ribosomal RNA genes around which nucleoli form.

Nucleolus Prominent and multifunctional nuclear organelle associated with rRNA transcription and processing.

Nucleophile Molecules or ions with lone pairs of electrons which can form new chemical bonds.

Nucleoporin Protein component of the nuclear pore complex.

Nucleoside Sugar joined only to a base. Four bases are used in RNA: adenine, cytosine, guanine, and uracil.

Nucleosome A length of chromosomal DNA coiled around a set of core histone proteins.

Nucleotide Sugar connected to one or more phosphate groups and a base. Four nucleotides are used in RNA: adenosine, cytidine, guanosine, and uridine.

OB-fold Oligosaccharide/oligonucleotide binding fold reminiscent of the cold-shock domain.

Oligo(dT) Oligodeoxynucleothymidine—generally used to hybridize to poly(A) tails.

Oligonucleotide Short chain of nucleotides (oligos).

OMIM Online Mendelian Inheritance in Man.

Oncogene A gene which, when mutated or expressed at high levels, helps turn a normal cell into a tumour cell.

Open reading frame (ORF) A string of codons in an mRNA that encode a polypeptide sequence from start to stop codon.

ORF Open reading frame.

Orphan snoRNAs snoRNAs whose function and targets are not known.

Oxyanion Negatively charged oxygen ions, denoted O^-. When attached to RNA these are more potent nucleophiles than -OH groups.

Pan editing Extensive RNA editing which adds multiple uridines to some trypanosome mitochondrial transcripts.

Paraneoplastic disorder Disorder/symptom associated with cancer.

PARN Polyadenylated ribonuclease.

Pasha A double-stranded RNA-binding protein that works with Drosha in the nucleus to initiate miRNA biogenesis.

Passenger RNA The strand of an siRNA that is not associated with the RNA-induced silencing complex.

PAZ domain Named after the proteins Piwi, Argonaute, and Zwille. Found in Dicer proteins and involved in RNA-binding and protein–protein interactions.

P-body Cytoplasmic body in which mRNA decay and mRNA translation repression occur.

Peptidyl transferase centre The part of the ribosome where amino acids are added to the nascent polypeptide.

Perichromatin fibrils Splicing factors localized on a nascent transcript in the process of being spliced.

PHAX A nuclear export receptor for snRNA.

Phosphodiester backbone The alternating phosphodiester bonds and sugars in chains of RNA and DNA.

Phosphodiester bond A single phosphorus atom joined to two oxygen atoms through ester bonds. These are the covalent bonds which hold RNA and DNA nucleotides together.

Phosphorothioate A modified oligonucleotide in which an oxygen in the backbone phosphate is replaced by a sulphur atom to increase its half-life.

Pictogram Pictorial representation of a consensus sequence showing the frequency of each different nucleotide in each position.

Pioneer round of translation The first time an mRNA is translated many mRNA-bound complexes, such as the EJC, are removed.

piRNA PIWI-interacting RNAs used for controlling transposons.

PIWI domain Named after the P-element induced wimpy testis phenotype in *Drosophila*—found in Argonaute proteins and involved in RNA cleavage.

Plasmodesmata Microscopic channels that traverse the cell walls of plants connecting two protoplasts (cytoplasms).

Polyadenylate polymerase The enzyme responsible for adding a sequence of As at the 3′ end of an mRNA.

Poly(A) tail Up to 200 A nucleotides added at the 3′ end of most mRNAs. The poly(A) tail is added at the site of cleavage that occurs between the AAUAAA and G/U-element polyadenylation signals.

Polycistronic mRNA An mRNA from which the translation of more than one protein can be achieved through independent open reading frames.

Polycomb group (PcG) proteins PcG proteins contain a chromodomain which binds to H3 methylated at lysine 27. PcG proteins perform point repression of specific targets within genome (notably *HOX* genes) and maintain repression of the inactive X chromosome in female mammals. Some PcG proteins can bind to RNA, so they can be targeted to chromatin in this way.

Polynucleotide A long chain of nucleotides.

Polyribosome Several ribosomes loaded onto an actively translating mRNA.

Polysome *See* Polyribosome.

Polytene chromosomes Giant chromosomes that consist of many copies of carefully lined up replicated copies of chromosomes. They are found in some insect larvae, notably in the salivary gland of *Drosophila*.

POMA Paraneoplastic opsoclonus myoclonus ataxia.

Post-transcriptional gene silencing (PTGS) A form of gene silencing that operates after transcription has occurred. It includes RNA interference.

Post-transcriptional process A gene expression or gene regulation process that occurs once transcription is complete.

PP1 Protein phosphatase 1—regulates the activity of some RNA splicing regulators through dephosphorylation.

PPR Pentatricorepeat RNA-binding domain consisting of 35 amino acids that form two antiparallel alpha helices—abundant in mitochondria and chloroplasts of terrestrial plants.

Premature termination codon (PTC) A stop codon that occurs too early, resulting in premature termination of translation.

pre-miRNA The precursor of a microRNA.

pre-mRNA The precursor of a mRNA which is unspliced.

pre-rRNA The precursor of a ribosomal RNA.

pre-tRNA The precursor of a transfer RNA.

Primary structure For nucleic acids this is the linear sequence of nucleotides (nucleic acid sequence) in the molecule.

pri-miRNA The primary transcript of a microRNA gene, first processed by the enzyme Drosha with the assistance of Pasha.

Progerin Abnormal form of lamin A protein without 50 amino acids at the C-terminus which are important for protein processing and localization.

Proteome Complete repertoire of proteins expressed in a cell, tissue, or organism.

Protonate/deprotonate Add/remove a hydrogen ion (H^+, also called a proton).

Provirus Viral genome inserted into and maintained as part of the host genome.

PRP Pre-RNA processing protein.

PRP8 Evolutionarily conserved spliceosomal protein which forms part of the active site of the spliceosome.

Pseudoexon Intronic sequence with flanking splice sites which appears very similar to an authentic exon, but is rarely spliced.

Pseudogene A gene that is descended from a protein coding gene that has degenerated, losing protein coding capacity.

Pseudoknot Structures that form from base pairing of RNA sequences in loops.

PSI P-element somatic inhibitor.

P site Peptidyl-tRNA site in the ribosome, occupied by the most recent amino acid to be added to the growing polypeptide.

PTC Premature termination codon.

PTGS Post-transcriptional gene silencing.

PUF domain Pumilio and FBF homology RNA-binding domain consisting of eight repeats of a three alpha-helical bundle of 36 amino acids.

Quasi-RRM (qRRM) A domain with loose similarity to the RRM with discrepancies in the canonical RNP1 and RNP2 motifs.

RAN RAS-related nuclear protein. Small GTPase which binds GTP and can hydrolyse it to GDP, but does not efficiently release GDP. RAN can be in the GTP-bound form (cytoplasm) or the GDP-bound form (nucleus).

RANBP1 Protein which assists in the release of RAN–GDP from karyopherins.

RANBP2 Protein which assists in the release of RAN–GDP from karyopherins.

RAN-GAP Cytoplasmic protein which activates the GTPase activity of RAN, thereby converting RAN–GTP to RAN–GDP.

RAS A small GTPase which acts as a molecular switch in the cell (from 'rat sarcoma', where the gene was first discovered).

RAS-related nuclear protein *See* RAN.

rasiRNA RITS complex-associated siRNA. Targets centromeric DNAs for chromatin modification.

Rate-determining step The slowest step in a reaction. This also requires the highest activation energy to progress.

RCC1 Regulator of chromatin condensation 1—nuclear protein which converts RAN-GDP to RAN-GTP by switching the GDP for a GTP.

Recruitment model Model explaining the connection between transcription complexes assembled on different promoters and alternative splicing. According to this model, different transcription complexes recruit different splicing factors.

REF RNA export factor—an RNA export adaptor.

Release factors Proteins that are involved in translation termination.

Retinitis pigmentosa (RP) A retinal disease.

Retrotransposons Transposons that move around the genome via RNA intermediates.

REV RNA export adaptor which binds to an RNA sequence in incompletely spliced HIV pre-mRNAs called the RRE and to the nuclear export adaptor CRM1.

RGG boxes Auxiliary RNA-binding domains consisting of a glycine-rich stretch interspersed with arginine.

Ribocyte Simple replicating microbes in which fundamental biochemical processes were entirely dependent on RNA.

Ribonucleases Enzymes that digest RNA—subdivided into exoribonucleases (that attack free ends) and endoribonucleases (that attack RNA internally).

Ribonucleolytic ribozymes Ribozymes which speed up RNA cleavage reactions.

Ribonucleoprotein *See* RNP.

Riboprobe An RNA molecule used as a ligand in the study of RNA–protein interactions; usually radiolabelled.

Ribose zipper RNA tertiary structure which brings two strands of RNA into close proximity, usually a region of single-stranded RNA and an RNA helix or helix-containing structure.

Ribosome Macromolecule composed of ribosomal RNA (rRNA) and ribosomal proteins that carries out mRNA translation—subdivided into large and small ribosomal subunits.

Riboswitch Naturally occurring RNA aptamers which can bind target molecules but can also switch between different conformations depending on whether they are bound to their ligand or not.

Ribozyme Ribonucleic acid enzyme—an enzyme made out of RNA.

RISC RNA-induced silencing complex. Can cleave a target mRNA or repress its translation. Contains Argonaute proteins.

RITS RNA-induced transcriptional silencing—RNA-induced silencing complex that directs chromatin remodelling in transcriptional gene silencing.

R looping RNA–DNA hybridization method used to discover introns.

RNA (ribonucleic acid) Equally as important as DNA—it is less stable but more versatile, to the extent that RNA molecules can even catalyse their own replication. The RNA world hypothesis suggests that RNA preceded DNA as the repository of genetic information.

RNAa RNA-mediated activation of transcription.

RNA-binding domain Part of a protein that has the ability to bind RNA.

RNA chaperone A protein that facilitates RNA processing or translation generally by keeping RNA in a single-stranded conformation.

RNA-dependent RNA polymerase (RDP) Enzymes that produce an RNA strand from an RNA template.

RNA editing Alteration of the sequence of an RNA through direct chemical modification of RNA bases.

RNA era *See* RNA world hypothesis.

RNA export complex RNP complex containing RNA assembled with RNA export adaptors and RNA export receptors.

RNA helicase Enzyme which reorganizes RNA–RNA and RNA–protein complexes, usually using NTP hydrolysis.

RNA-induced silencing complex (RISC) A complex of proteins that incorporates the guide RNA from an siRNA. A RISC binds to target mRNAs and results in their cleavage or translational repression through the Argonaute proteins.

RNA-induced transcriptional silencing (RITS) A form of transcriptional gene silencing that is mediated by short non-coding RNAs.

RNA interference A form of post-transcriptional gene silencing dependent on siRNAs.

RNA ligase An enzyme that catalyses the conversion of linear RNA to a circular form by the transfer of the 5′-phosphate to the 3′-hydroxyl terminus.

RNA nuclear export adaptor A group of proteins which binds to RNA and to nuclear export receptors. Different non-coding RNAs are bound by different nuclear export adaptors.

RNA nuclear export receptor Proteins which enable RNP to leave through the nuclear pore. The group of proteins which act as nuclear export receptors for non-coding RNAs are called karyopherins.

RNA polymerase Enzyme which uses a DNA template to make an RNA copy.

RNAse *See* Ribonuclease.

RNA-Seq RNA sequence analysis for transcriptome profiling.

RNA surveillance A process that exists to check the correct processing of RNA precursors.

RNA world hypothesis The hypothesis according to which in primitive life RNA was solely responsible for carrying and replicating genetic information. This time is called the *RNA era*.

RNA zipcode A sequence element, often in the 3′ UTR, that directs mRNA localization.

RNAP RNA polymerase.

RNP Ribonucleoprotein—a complex of RNA and protein(s).

RON Transmembrane receptor tyrosine kinase which controls cellular behaviour.

Rox Long non-coding RNAs which help to upregulate gene expression on the single X chromosome in the male fruit fly.

RP Retinitis pigmentosa.

RRE REV response element. Structured RNA sequence which is bound by REV nuclear export adaptor.

RRM RNA recognition motif—a protein domain which interacts generally with ssRNA in a sequence-specific fashion.

rRNA Ribosomal RNA.

RT-PCR Reverse-transcription polymerase chain reaction.

saRNA Small activating RNA involved in the activation of transcription.

S_N2 mechanism Chemical reaction in which an attacking nucleophile forms a new bond at the same time as the leaving group which it replaces departs.

scaRNA Small Cajal-body-associated RNA, generally involved in snRNA processing.

scaRNPs Small Cajal RNPs.

Scissile phosphodiester bond Phosphodiester bond which is going to be cut—this is just a normal phosphodiester bond.

scRNA Scan RNAs. Associated with the silencing of transposons in *Tetrahymena*.

Secondary structure Two-dimensional folding of RNA into a set of motifs, including helices, hairpin stemloops, and pseudoknots.

SELEX Systematic evolution of ligands by exponential enrichment—a technique used to isolate interacting RNAs through an iterative process of selection and enrichment.

Self-splicing intron An intron able to splice itself out of a precursor RNA without the help of any proteins.

Sex lethal protein (Sxl) RNA-binding protein which controls fruit fly sex determination.

Shine-Dalgarno sequence Ribosome binding site in bacterial mRNA, just upstream of AUG, at which translation initiates.

Short hairpin RNA (shRNA) A stemloop structure that is processed into an siRNA.

Signal recognition particle (SRP) Targets a nascent peptide to the endoplasmic reticulum (ER).

Silencing Refers to a drastic reduction in expression of a gene (knockdown).

siRNA Small interfering RNA that becomes incorporated in the RNA-induced silencing complex in the context of RNA interference.

Skipped exon *See* Cassette exon.

Slicer Endonuclease activity within a RISC complex mediated by Argonaute proteins.

SMA Spinal muscular atrophy.

Small interfering RNA *See* siRNA.

SMN Survival of motor neurons, mutated in SMA.

Sm proteins Group of core proteins associated with snRNAs.

snoRNA Small nucleolar RNA.

snoRNP Small nucleolar ribonucleoprotein mainly involved in ribosomal RNA processing.

snRNA Small nuclear RNA.

snRNP Small nuclear ribonucleoprotein mainly involved in pre-mRNA splicing (pronounced 'snurp').

Snurportin Nuclear import receptor for snRNPs.

Spliced leader Exon spliced onto the 5′ end of trypanosome mRNAs.

Spliceosome RNA–protein complex which catalyses the removal of introns from eukaryotic pre-mRNAs.

Spliceosome cycle Series of assembly and disassembly steps to build up and then dismantle a catalytically active spliceosome.

Splice site Conserved sequence at intron–exon junctions critical for splicing, and cut and joined in the splicing reaction.

Splice site commitment The point at which the splicing machinery is committed to initiate the splicing reaction.

Splicing The process whereby exons are joined together precisely after two *trans*-esterification reactions.

Splicing code Embedded target sites for RNA-binding proteins within pre-mRNAs which control RNA processing.

Splicing factor compartments Sites of storage of splice factors in the nucleus—also known as nuclear speckles.

Splicing speckle Storage sites for splicing components and often located adjacent to sites of active gene expression. Splicing speckles also contain nuclear export adaptors.

SR domain Serine–arginine-rich domain—commonly found in splice factors together with RRMs.

SR protein Serine–arginine-rich protein—generally a splicing factor.

SR protein kinase (SRPK) An enzyme that adds phosphate groups to SR proteins, modulating their intracellular localization.

SRSF1 An SR protein splicing regulator protein. Previously known as ASF/SF2.

SSU processome U3 snoRNP, required for the formation of the 18S rRNA in the small subunit of the ribosome.

Stemloop *See* Hairpin stemloop.

Stochastic Random or probabilistic, but with a degree of direction.

Stress granules (SGs) Dense cytosolic aggregates that contain RNA and protein. They appear during cellular stress (including oxidative stress, heat shock, osmotic shock, and viral infection) and are associated with translation repression and mRNA decay.

Substitution reaction Chemical reaction in which an atom or group of atoms is replaced (substituted) by another atom or group of atoms.

Sugar Water-soluble carbohydrate.

Supraspliceosome Fully formed spliceosome added to pre-mRNA.

SWI/SNF Chromatin-modifying complex first described in yeast—stands for 'mating type switch/sucrose non fermentable medium'.

Synonymous codon Groups of similar codons which encode the same amino acid but just differ at the most 3′ position which is called the wobble base.

Synonymous mutation A mutation which does not change the amino acid encoded.

TAP Tip-associated protein, a nuclear export receptor.

TASS Tandem alternative splice sites that are in close proximity (2–12 nucleotides apart).

Tau Microtubule-associated protein which has an important role in neurodegeneration.

Telomerase RNP complex responsible for the formation of telomeres.

Telomere A chromatin structure that protects chromosomes from degradation, fusion, and undesirable recombination.

Terminator stem Stemloop structure which causes transcriptional termination in bacteria when followed by a run of U residues.

Tertiary structure Folding of RNA molecules into complex globular structures.

THO complex Contains a group of proteins involved in mRNP packaging and transcriptional elongation. Associated with the spliceosome in metazoans and with RNA polymerase II in yeast.

TMG Trimethyl guanosine.

TOES Targeted oligonucleotide enhancers of splicing.

TPA 12-*O*-tetradecanoylphorbol-13-acetate.

Tra2 β Transformer 2 beta. This is an RNA-binding protein which functions as a splicing regulator.

tracrRNA *Trans*-encoded small guide RNA that works with RNAse III to generate crRNAs.

***Trans*-acting factor** A protein or other factor that binds to a particular sequence in order to regulate gene expression.

Transcription The process of copying a DNA template into a corresponding RNA.

Transcriptional gene silencing (TGS) A form of gene silencing that operates at the chromatin level, effectively preventing transcription.

Transcription factor A protein involved in basal or core or regulated transcription.

Transcriptome Complete repertoire of transcripts expressed in cells, tissues, or organisms.

Transcriptomics The analysis of the complete repertoire of transcripts expressed in cells, tissues, or organisms.

Transfer RNA (tRNA) The molecular adaptor which associates a 'codon' in the mRNA with a specific amino acid.

Transition state A combination of reacting chemicals when they are at the top of the activation energy barrier between substrates and products.

Translation The process of reading codons in mRNA into a polypeptide sequence. Subdivided into initiation, elongation, and termination.

Translational attenuation A form of translation regulation in which a particular mRNA conformation reduces the amount of protein produced, often in the context of a feedback loop.

Translocation The process whereby the ribosome moves on to the next mRNA codon.

Transport granules RNP complexes that contain localized mRNAs in the act of being transported.

Transposable element Mobile genetic element in the genome such as the Alu sequence element.

Transposon *See* Transposable element.

***Trans*-splicing** Splicing together of physically separate RNA molecules.

TREX complex Transcription/export complex, comprising the RNA export adaptor REF, the RNA helicase UAP56, and the THO complex. In yeast the TREX complex loads REF and UAP56 onto nascent transcripts.

Trigonal bipyramid Shape of the transition state found in RNA self-cleavage reactions.

Trithorax group (TrxG) proteins TrxG proteins epigenetically activate chromatin by trimethylating H3K4 and then bind to this trimethylated histone. Some TrxG proteins can bind to RNA, so they can be targeted to chromatin in this way.

tRNA *See* Transfer RNA.

TSIX Long antisense RNA which is made from the X chromosome, and is complementary in sequence to *XIST*. *TSIX* is expressed from the active X chromosome.

U2AF U2 auxiliary factor—early component of spliceosome to bind to pre-mRNA.

UAP56 RNA-binding RNA helicase which recruits the RNA export adaptor REF to mRNPs and is also a component of the spliceosome.

Ultrabithorax (Ubx) *HOX* gene which controls thoracic development in fruit flies.

Unmasking mRNA The process of activating the translation of mRNAs whose translation has been repressed long term.

UTR Untranslated region, found at the 5′ and 3′ ends of mRNAs.

VEGF Vascular endothelial growth factor.

Wobble base Codons are made up of three nucleotides which encode either specific amino acids or stop codons. The wobble base is the most 3′ nucleotide in the codon and is frequently modified by A→I editing.

WT1 Wilms tumour 1.

XIC X inactivation centre in mammals, which contains three important non-coding RNAs including *XIST*.

XIST Long non-coding RNA which is made from the inactive X chromosome only. *XIST* RNA coats the inactive female X chromosome and targets it for epigenetic inactivation.

XITE Enhancer for the *TSIX* gene which produces long non-coding RNAs as part of this function.

Zinc finger A protein domain characterized by a series of four cysteines and histidines that coordinate a zinc ion. Able to bind both DNA and RNA.

Zipcodes *See* RNA zipcodes.

ZNF9 *Zinc finger 9*—a gene which contains expanded triplet repeats in myotonic dystrophy.

Zone of silencing Region of a pre-mRNA which is bound by splicing repressor proteins and is invisible to the spliceosome.

INDEX

A

A site (acceptor site), ribosome 231
A–A base pair, hydrogen bonding 7
Acetabularia, mRNA localization 217–20
acetylation 337
ACF (ApoBec complementing factor) 283
acids, general acid 37
actin mRNA, localization 212, 215
activation-induced cytidine deaminase (AID) 286
adaptive immunity, C to U RNA editing 286
ADARs (adenine deaminases acting on RNA) 70, 270–4
  ADAR2 protein, autoregulation of own splicing 278
  editing complementary sites (ECSs) 275
  mouse knockouts 70
  selective A to I RNA editing 274–5
ADATs (adenosine deaminases acting on tRNA) 280
adenomatous polyposis coli (APC) 213
adenosine
  coaxial stacking 26
  editing to inosine 270–4, 279–80
  polyadenylation 193
  structure 269
adenovirus genome 85
adenovirus hexon RNA 85
affinity antibodies 286
ageing, Hutchinson–Gilford progeria (HGPS) 141–2
AIRN (antisense IGF2R RNA) 348
allosteric enzymes 32
alpha-amanitin *(table)* 168, 250
alternative polyadenylation 172
alternative splicing 111–37
  alternative exons 105, 111–12
  defects and disease
  exon skipping 112
  increase of coding capacity 111
  insulin receptor 259
  intron retention 112
  isoforms 111
  microarray services 392
  pathways 111
  regulation by signal transduction pathways 126
  regulators 119, 121
  RT–PCR 113
  splice sites 153
  see also exons
Alu elements 274
Amanita fungi 250
amino acids, structure 227
amino-acyl tRNA synthetases 228
anaemia, Diamond-Blackfan anaemia 297
animal models, CaMKIIα mRNA 215
animals, silencing of transposons 384
ANRIL
  antisense RNA 352
  expression 353
antibiotics
  effects on translation in bacteria *(table)* 235
  and stop codons 261
antibody responses
  adaptive immunity 286
  affinity antibodies 286
  innate immunity 284
anticodon wobble, tRNA 229
antisense IGF2R RNA (AIRN), long ncRNA 348
antisense oligonucleotides 213, 245–6
  gene therapy 245
  phosphorothioates 246
antisense RNA 352–4
  CDKN2A and CDKN2B 352
  'ping-pong amplification loop' 385
APC (adenomatous polyposis coli) 213
APOB mRNA
  intestine 282
  multicomponent editing complex 283
APOB protein 281–4
ApoBec proteins 284
  AID (activation-induced cytidine deaminase) 286
  ApoBec1 (apolipoprotein B editing enzyme, catalytic polypeptide-like) 284–6
  hypermutation of invading retroviruses 286
  infectivity of retroviruses 285
  isoforms 284–6
  role in innate immunity 284–6
apolipoprotein B (APOB) 281–4
apoptosis, Fas 154, 370
aptamers, SELEX 78
Arabidopsis
  Flowering Control Locus (FLC) 353–4
  gene numbers 111
  microRNAs and their mRNA targets 375
  microRNAs and their mRNA targets *(table)* 375
Archaea
  base pairing, phylogenetic conservation 22
  KH (K-homology) domain 65–6
  RNA polymerase 168
  snoRNA-like molecules 300
AREs, AU-rich elements *(table)* 257
arginine 227
arginine fork 75
arginine-rich domains 73–4
Argonaute proteins 360–2
  activity 359, 372
  alternative splicing 374
  RITS complex 378
  RNA-binding proteins 74, 364
  slicer activity 361–2
ASF/SF2 65, 145, 155–6, 158, 179, 312, 368
  see also SRSF1 (new nomenclature)
aspartate 227
ataxia telangiectasia 140
AU-rich element (ARE) 255–7
Aubergine, RNA–protein complexes 385
autoimmune diseases 94, 134
auxiliary domains 60

B

Bacillus stearothermophilus, enzyme BstCCA 308
bacterial cell
  antibiotics and their effects on translation *(table)* 235
  btuB mRNA 31–2, 34
  mRNA degradation 264
  organization 186–7
Balbiani rings 176, 196
base, general 37
base pairing
  non-Watson–Crick 16–18, 26
  phylogenetic conservation 22
bases
  catalysis 46
  general bases 37
  hydrogen bonding 9
  (table) 7
  triplex 25
Becker muscular dystrophy (BMD) 158, 160
beta-barrel structure, CspA 66–7
(G)beta-galactosidase synthesis 223
beta-glycosidic bond 7
bicoid 70
bioinformatics, SELEX experiments 54–6
bipartite processing element 181
BRCA1 (breast cancer 1) gene 144–6
  defects in splicing 144, 146
  exon skipping 144
  mutation of exonic splicing enhancer 144
breast cancer, epithelial to mesenchymal transition (EMT) 393
bromodomain 336
Brr2 99
btuB mRNA 31–2, 34
  E. coli 31–2, 34
  structure 32

C

c-fos 65, 69
c-myc 215, 241, 250, 256, 304
C-terminal domain (CTD) 172–3, 184
  phosphorylation 167, 173, 184
  RNA polymerase II 172–3
  transcriptional surveillance 172
Caenorhabditis elegans
  ADARs (adenine deaminases acting on RNA) 278
  epigenetic silencing 363
  gene knockdown 368
  genes 111
  microRNAs 370
    mRNA translation regulation 238–9
  microRNAs and their mRNA targets *(table)* 375
  protein LIN-28 68
  RNA interference (RNAi) 363–7
  trans-splicing 106
Cajal bodies 295, 300, 323
  snRNP complexes 324
Cajal body-associated RNAs *see* scaRNAs
CaMKIIα mRNA, mouse models 215
cancer (cells)
  BRCA1, exonic splicing enhancers (ESEs) 144
  and contact inhibition 155
  growth factor VEGF-A 153
  hallmarks 152
  metastasis, epithelial to mesenchymal transition (EMT) 393
  six differences from normal cells 152
  spliceosomal components 151
  splicing changes 153–4
  splicing factor levels 154
  splicing factor SRSF1 156
cap-binding complex (CBC) 170
cap-binding proteins, eIF4E 237
Caper proteins 130
capping enzyme complex (CEC) 169
cardiac troponin T 133
cargo mRNA 188
cargo snRNP 326
Cas (CRISPR-associated) proteins, genome editing 397
casein 250
cassette exon 113, 397
cat mRNA 235
catalytic RNAs 36–63
  base catalysis 37
  deprotonation 46
  group I introns 42
  group II self-splicing introns 43
  in-line configuration 46, 47
  metal ions, active catalyst role 23, 26
  nucleophile 306
  ribozymes 36
  RNA polymerases, and RNA ligases 53–4
  RNA world hypothesis 36, 52–3
  see also ribozymes; RNAse P
CCA 304
Ccr5 370
CDKN2A and CDKN2B, antisense RNAs 352
cDNA (complementary DNA) 285
  transcriptional gene silencing (TGS) 376
cell cycle, oncogenes, and tumour suppressor genes 155–7
centromeres 377
ceRNA (competing endogenous RNA) hypothesis 394
Chironomus, Balbiani rings 196
chloride channel 133
chromatin
  co-transcriptional splicing 177
  euchromatin 334
  functions 308, 334
  gene expression 334, 335
  heterochromatin 334, 337, 376
  perichromatin fibrils (PFs) 176
  splicing and transcription 173–8
  transcriptionally active/inactive 337
chromosome maintenance (CRM1) 317
chromosomes
  maternal/paternal, differential gene expression 345
  polytene 176
  pronuclear transplantation experiments 346
  X and Y 338–9
    dosage compensation mechanisms 343
chylomicrons 282
circular RNAs 394
  increased expression of mRNAs normally targeted and repressed 394
  stability 395
cis-acting elements 259
  (table) 257
cis-splicing 104–6
citric acid cycle 153
classical protein import pathway 326
CLIP assay 78–9
CLIP genome-wide map of NOVA RNA-binding sites 80
co-transcriptional pre-mRNA processing 166–82
co-transcriptional splicing 175
coaxial stacking 24–6
codons
  premature termination codons (PTCs) 118, 260
  stop codons 118, 232, 260
  synonymous codons 224, 281
coenzyme B12, and riboswitches 31
cold shock-induced proteins 61, 66
  beta-barrel structure 66–7

CspA in *E. coli* 234–5
cold-shock domain (CSD) 66–9, 234
Colour illustration section 415–22
competing endogenous RNA (ceRNA) hypothesis 394
complementary base pairing 377
complementary DNAs (full-length cDNAs) 285
congenital contracture syndrome 203
consensus structure 22
COOLAIR 354
COXII, COXIII 289
CpG
  hypomethylation 337
  methylation of DNA 336
CPSF multiprotein complex 170
CRISPR
  homologous recombination 397
  modelling lung carcinogenesis in mice 399
  RNA-guided genome editing 397–9
  vs RNA interference 397
CRISPR–Cas system, adaptive immune response in prokaryotes 398
crRNAs 397
cryptic genes 287
CspA, *E. coli* 234
CspB, beta-barrel structure 67
CstF (cleavage stimulation factor) 170
CTD phosphorylation 167, 173, 184
cytoplasmic polyadenylation element (CPE) 243
cytoplasmic processing bodies *see* P-bodies
cytoplasmic proteins, RAN-GTPase activating protein (RAN-GAP) 316
cytosine to uracil (C to U) RNA editing 8
  in innate immunity 284

D

Dcp (decapping) bodies 252
DEAD box helicases 99
deamination 8
decoys, RNAs with common binding sites for miRNAs 394
degradosome 264
denaturing gel electrophoresis 100–1, 371
dendritic spine formation, fragile X syndrome (FXS) 214
dense fibrillar centres (DFC) 302
deoxyribose 4
  structure 6
Diamond–Blackfan anaemia 297
Dicer enzyme 70, 360–1
  dsRNA 359–60
  functional organization and atomic structure 361
  short dsRNAs 361–2
  small interfering RNAs (siRNAs) 364
dideoxy nucleotides (ddNTPs) 391
DNA
  nucleosomes 334–5
  storage 334–5
  transposable elements 379
DNA acetylation 336
DNA double helix 11–16
  A and B forms 4, 16, 17
  B form 16, 17
  RNA/DNA differences, summary 13
DNA interference 362
DNA methylation
  regions of reduced CpG methylation 337
  transcriptional silencing 336–7
DNA methyltransferases 380
DNA nucleotides, structure 6
dosage compensation complex (DCC) 343–4
dosage compensation mechanisms, *Drosophila* and mammals 344–5
double helix
  A and B forms 4, 16, 17
  base stacking 15
  RNA/DNA differences, summary 13
  structure 11–16
double-stranded RNA binding proteins (dsRBPs) 69–70
double-stranded RNAs (dsRNAs) 361–2
Down syndrome cell adhesion molecule (DSCAM) 114
Drosha enzyme 372
Drosophila
  body plan, *HOX* genes 349–50
  dosage compensation complex (DCC) 343
  Doublesex gene 123
  DSCAM gene 114–16
  gene expression, cf mammals 343
  gene numbers 111
  long ncRNAs 349
  male and female differentiation 344
  microRNAs and their mRNA targets *(table)* 375
  mRNA localization 205–7, 210
  P element transposon 384
  piRNAs 384
  roX1, roX2 343
  selective A to I RNA editing 274–8
  sex differences 124
  sex-lethal (Sxl) 125
  Staufen dsRBDs 61, 70, 208, 211
  Tap gene 196
  Ubx, active and inactive 349
  without ADAR genes 274–5
DSCAM 114
dsRBD domain 69, 70
dsRNA-activated protein kinase (PKR) 69, 127, 236, 366
dsRNAs 376, 378
  epigenetic silencing 235–6
  gene silencing 363
  targeted by Dicer 360
Duchenne muscular dystrophy (DMD) 158, 160
dwarfism 148
dystrophin 158–61
  ORF 160
  size 86, 158
  splicing 108, 175

E

E site (exit site), ribosome 231
early (E) complex 95
  3′ splice site 94
early onset myopathy 228
Ebola virus, VP35 70–1
editing complementary sites (ECSs) 275
editosomes 291
  molecular components 292
eIF4E-binding protein (4E-BP) 237
electron microscopy, Miller spread 297
electrophoretic mobility shift assay (EMSA) 76
elongation factors (EFs) 231
endonucleases, Dicer 70, 359–60
enhancers 351–2
  transcriptional enhancers 88
epigenetic code 336
epigenetic regulation of gene expression 334–58
  ncRNAs 334–58
epigenetic silencing 359
epithelial to mesenchymal transition (EMT) 157, 393
ERKs *see* MAP kinase
Escherichia coli
  CspA 234
  rpoH heat shock protein 28
ESE finder program 145
ESEs *see* exonic splicing enhancers
ESSENCE *see* exon specific splicing enhancers by chimaeric effectors

Eukaryotes, base pairing, phylogenetic conservation 22
eukaryotic cell 186–7
eukaryotic genes
  modification of rRNA 297
  mRNA translation regulation 245–6
  ribosomal proteins 296
  see also gene expression
evolution, RNA world hypothesis 52–3
exon complementary sequences (ECSs) 279
exon definition complex 103, 105
exon definition *(table)* 105
exon junction arrays 115
exon junction complex (EJC) 207, 260–3
exon skipping 112, 143, 161
  BRCA1 (breast cancer 1) gene 144
exon specific splicing enhancers by chimaeric effectors (ESSENCE) 314
exon–intron junctions *see* splice sites
exonic splicing enhancers (ESEs) 117, 162
  BRCA1 (Breast Cancer 1) gene 144
  ESE finder program 145
  SR proteins 120
exonic splicing silencers (ESSs) 117
exons 104, 112–13
  alternative exon size 112
  alternatively spliced exons 104
  cassette exon 113
  constitutive exons 104
  exon skipping 112
  nuclear localization sequence (NLS) 205, 325–6
  poison 133
  recognition by splicing machinery 115–17, 120
  splicing efficiency 121
  splicing enhancers in *Doublesex* gene 123
  vs intron definition rules 105
  see also splicing
exosome 255–6
export adaptors
  nuclear export sequence (NES) 320
  REF export adaptor 189
exportin-1 and importin-8 373
eye, structure 148

F

Fas
  'death receptor' 154, 370
  and Fas protein 154, 370
ferritin mRNA 237
FG-repeat nucleoporins 188, 199
fibrillar centres (FC) 302
filter retention assay 77
FIRRE 395–6
Flowering Control Locus (FLC) 353–4
fluorescence recovery after photobleaching (FRAP) 178
fluorescence in situ hybridization (FISH) 205
fragile X syndrome (FXS) 75, 214
fucoids, mRNA localization 217–20
future RNA research 390–401

G

'G quartet', consensus (DWGG) 75
G–U base pair, hydrogen bonding 7
gel electrophoresis 100–1, 146
  native gels 100–1, 113, 146
  polyacrylamide gel 100, 146
  SDS-PAGE 77, 183
gel retardation assay 76
gene expression
  chromatin 334–7
  differential, maternal/paternal chromosomes 345
  epigenetic regulation 334
  evolution, splicing code 138–42
  genetic imprinting 345–7
  introns 87
  short ncRNAs 358–98
  terminology 3
gene expression factory 178
gene number paradox 111
gene regulation, history 223
gene silencing
  and short ncRNAs 358–98
  see also epigenetic silencing
gene therapy 162, 245, 370
  mutated dystrophin gene 159
gene-specific methylation (TGS) 363
general acid/base catalysis 37
genetic code, synonymous codons 224–5, 281
genetic imprinting, IGF2R 345–7
genome browsers 390
genome editing, CRISPR 397–9
germ cells, masked messages 240–3
Giardia, Dicer enzyme 361
Gle1 199
glutamate 227
glycine-rich RNA-binding proteins (GRP), evolutionary conservation 64
gradient centrifugation 60
green fluorescent protein (GFP) 192–3, 195, 208
gRNAs *see* guide RNAs
gRNAs (guide RNAs) 289, 298, 361
growth factor, VEGF-A 153
GTPases 315
guanine nucleotide exchange factor (GEF) 314
guide RNAs 289, 298, 361

H

H complex 90, 95, 103
H19 ncRNA 347
hairpin ribozyme 39
hairpin stemloop 18
hammerhead ribozyme 39, 48–50
  nucleotides G12 and G8 49
HDAC inhibitors 313
helical junction 18, 19
helix-turn-helix motifs 75–6
hepatitis B virus (HBV) 50
hepatitis delta virus (HDV), small ribozyme 50–1
heptad repeat 171
heterochromatin 334, 337
  transcriptional gene silencing (TGS) 376
heterocyclic molecules 7
heterogeneous ribonucleoproteins *see* hnRNPs
hexon protein 85
histidine, as acid or base 227
histone acetyl transferases (HATS) 336–7
histone code 337
histone deacetylases (HDACs) 313, 336
histone downstream element (HDE) 181–2
histone mRNA
  3′ end formation 181–3
  bipartite processing element 181–3
histone proteins 334, 336
  base pairing interactions 182
  post-translational modification 177
HIV
  ApoBec proteins 285
  CD4 and *Ccr5* 370
  life cycle 285, 329
  pre-mRNA splicing control 162
  provirus 162
  Rev protein 328
  reverse transcriptase, and RNA aptamers 30–1
  RNA genome, secondary structure 14
  RNA processing 163
  splicing as a route to therapy 172–3
  SRSF1 163
  vif (virion infectivity factor) 295
hnRNAs 60, 104
  see also pre-mRNA

hnRNPs 60, 62, 64, 103, 118
A1 protein 126
effects of poison exons 118–19
effects of stress 128
gradient centrifugation 60
H complex 95
KH (K-homology) domain 64–6
naming 118
phosphorylation 118, 127
structure 75
homeodomain 75–6
homology-directed repair (HDR) 399
Hoogsteen bonds 26
HOTAIR 349–51
expression in cancer, tumour suppressor genes 352
long ncRNA 349–51
HOX genes, long ncRNAs 349–51
HOX transcription factor 349
hTR, RNA component of telomerase 295, 300
human genetic diseases
affinity antibodies 286
mutations of mRNA export *(table)* 203
splicing as a route to therapy 162
see also specific diseases
human genome
microRNAs and their mRNA targets *(table)* 375
mitochondrial genome 308
ncRNAs 391
RNPs 257
transposable elements 379
HuR 60
Hutchinson–Gilford progeria (HGPS) 141–2
hydrogen bonding 9, 15
hypomethylation of CpG 337

I

IGF2R
genetic imprinting 345–7
repression of gene expression 348
IL-3 (interleukin 3), expression and stability 251
illumina dye sequencing 391
immune responses 284
immunoprecipitation 78
importin beta 326
imprinting 345–7
imprinting control region (ICR) 347, 349
in situ hybridization 205
in vitro analysis, pre-mRNA splicing 100
in-line configuration 46
catalytic RNAs 46
infectious diseases
affinity antibodies 286
splicing as a route to therapy 162
initiation factors (IFs) 229
'initiator' element 167
innate immunity, C to U RNA editing 284
inosine structure 269
insulin receptor, preproinsulin 259
insulin transcription 260
interchromatin granule clusters (IGCs) 178
interference, RNAi 193, 195, 363–7
internal loops 18
intestine, *APOB* mRNA 282
introns 86–33
affect transcription 88–9
contain genes 88
definition 104
different 'AT–AC' rule 101
in evolution 89–92
evolution of new proteins 89
gene expression 87
Group I
function 40–1
metal ions 42
splicing 41–2
structure 25
Group II 44
splicing 43–5, 97
increase of eukaryotic gene expression levels 87, 88–90
insertion 86
mutations 139
R-looping 85
removal 86
retention 112
S. cerevisiae 86, 107
spliceosomal introns 86–9
splicing enhancers (ISEs) 117
splicing silencers (ISSs) 117
U12-dependent 101
IPTG (isopropyl-beta-*D*-thiogalactopyranoside) 223
iron metabolism, translational regulation 237
iron response elements (IREs) 237–8

K

karyopherins 314, 317
CRM1 317, 320
exportin-t and exportin-5 318, 321–2
HEAT domain 318
interaction with RAN–GTP 319
nuclear export *(table)* 318
karyotype image 339
KH (K-homology) domain 61, 65
kinetic coupling model 127
kinetochore 377
kinetoplast 287
maxicircles and minicircles 287
kissing loop complex 18, 19
knockdown of genes
ASF/SF2 (SRSF1) 368
assessing 365
by dsRNA 364
by siRNAs 366–7
Ccr5 in human lymphocytes 370
database (WormBase) 368
ex vivo 368
miRSCs 376
morpholinos 246
RNA interference 363, 366, 369, 397
RNBPs 78
RPS19 298
siRNA and shRNA approaches 367
targeting by antisense oligonucleotides 245
knockins 399
knockout, defined 245, 367
Knox genes 219
Kozak consensus sequence 230

L

lac operon mRNA synthesis 223
lamin A protein 141
'lampbrush' chromosomes 68
lariat intermediates 43–4, 176
leaky scanning 230
leptomycin B 320
let-7 370, 375
lin-4 microRNA and its target 370, 371
liver, *APOB* mRNA 282
LMNA 141–2
Lnx3, evolved into *XIST* 393
long ncRNAs *see* ncRNAs, long ncRNAs
LSm protein 327
lung cancer 152
lung carcinogenesis, mice 399
lysine 227
riboswitch, Thermatoga 235

M

M-fold program 23
macro RNAs 334–58
magnesium ions
A, B and C 306–7
active site of RNAse P 306
binding to RNA 23, 26

mammals
  dosage compensation mechanisms 338
  marsupials and placentals 342
  pronuclear transplantation experiments 346
MAPK pathway 259
masked messages
  translation of mRNA 240–3
  unmasking 242–3
maskin 243
Mason Pfizer monkey virus (MPMV) 202
MasonMPMV virus, cell RNA export machinery 202
maternal mRNAs 241
Mdx mouse, gene therapy 162
meayamycins 159
mesenchymal transition (EMT) 157
messenger ribonucleoproteins *see* mRNPs
messenger RNA *see* mRNA
metal ions
  binding to RNA 23, 26
  direct/indirect interactions 24
  Group I introns 42
  magnesium ions, A, B and C 306–7
  ribozymes 36
  specific catalytic roles in ribozymes 37
metallothionein 1 (*MT1*) mRNA 216
metazoan cell *see* eukaryote cell
7-methyl guanosine cap 169
microarray services, alternative exons 392
microarrays 115
microcephalic osteodysplastic primordial dwarfism 148
microfilament mRNA, localization 212
microRNome (miRNome) 371
MIDI domain 361–2
Miller spread electron microscopy 297
miRISC complexes 372
miRNAs/microRNAs 21, 215, 253, 359, 370
  biogenesis 372–3
  in development and disease 374–6
  editing, trypanosomes 287–91
  genome subcellular organization in trypanosomes 288
  identification 371
  mRNA translation regulation 238–9
  nuclear export receptor 321
  nuclear functions 373–4
  polycistronic microRNAs 371
  precursors 376
  pri-miRNA transcript 372
  registry (miRBase) 371
  target identification and association with P-bodies 373
  transcription by RNA polymerase II 167
miRNP complex 396
miRSCs, knockdown of genes 376
missense mutations 160, 226
mitochondrial cytochrome c oxidase subunit II (COXII) 289
mitochondrial genome, human 308
mitochondrial mRNA (mt-mRNA) 309
mitochondrial tRNA (mt-tRNA) 308–9
molecular mimicry 232
molecular motors 208
morpholinos 246
motor neuron disease 203
mouse models
  CRISPR, modelling lung carcinogenesis 399
  GluR2 RNA editing 274–7
  Mdx gene therapy 161–2
  microRNAs and their mRNA targets *(table)* 375
  MIWI proteins 386
  muscleblind gene knockouts 133
  nude mice tumours 156
mRNA
  cleavage and polyadenylation at 3′ end 170–1, 181–3
  closed loop conformation 254
  discovery 223
  formation of ends 170
  half-life 250
  no-go decay 263–4
  nuclear export 187
  polyadenylation 192–3
  protein classes *(table)* 188
  secondary structure 21
  structure 170–1
  surveillance 252
  translational activation 244
  translocation 193
  transport from polyribosome to P-bodies 253
  wobble base 270
mRNA degradation 250–64
  AU-rich element (ARE) 255–7
  bacteria 264
  cleavage (breaking bond) 254
  deadenylation (removing poly(A) tail) 254
  decapping (removal of cap at 5′ end) 254
  in eukaryotes 254
  exosome 255–6
  multiple targets 258
  nonsense-mediated 260–2
  pioneer round of translation 263
  plants 265
  sites and mechanisms 252–9
  surveillance complex 172, 252
mRNA editing
  fruitfly (*Drosophila*) 272–4
  mouse 274–7
  trypanosomes 290–2
mRNA export
  classes of protein *(table)* 188
  mechanism 188–96
mRNA isoforms
  RT–PCR 113
  see also alternative splicing
mRNA localization 205–19
  classical examples in development 207–8
  cytoskeletal motors 207
  examples in differentiated somatic cells 212–14
  exon junction complex 260–3
  metastatic potential 213
  perinuclear 215
  in plants 217–20
  zipcodes 207
mRNA nuclear export 186–203
  step 1 188–93
  step 2 193–4
  step 3 194–9
  step 4 199–200
  step 5 200–3
mRNA stability 250–64
  cis-elements 257
mRNA structure 170–1
mRNA surveillance 252
mRNA transcription
  by RNA polymerase, capping 169
  random nuclear diffusion 193
mRNA translation 222–49
  blocking with antisense oligonucleotides 245
  coupling with other posttranscriptional processes 239–40
  defined 222
  elongation and termination 231–2
  eukaryote cell 245–6
  initiation 229–30
  masked messages 240–3
  prokaryote cell 231–2

regulation 234–40
release factors (RFs) 231–2
translocation 193
in vitro translation 244–5
mRNA transport, from polyribosome to P-bodies 253
mRNPs 60–81
cap-binding complex (CBC) 170
complexes (mRNA packaged by proteins) 207
exiting 200
localization factors 207–9
masked messages 240–3
movement through NPCs 197
RNA recognition motif (RRM) 62–3
RNA-binding domains 61
see also hnRNPs
MS2:MBP method, protein-RNA complexes 78
MSF (macrophage stimulating factor) 157
MT1 (metallothionein 1) mRNA 216
muscular dystrophy (BMD and DMD) 141, 158
mutations
missense mutations 160, 226
nonsense mutations 226
point mutations 139
in splicing, leading to disease 138–82
myelodysplastic syndrome 151
myopathy, early onset and SEPN1 228
myotonic dystrophy
DMPK mRNA export 202–3
exon skipping 161
SRSF1 155

N

N-formyl methionine 229
nascent transcripts 60
native gel electrophoresis 100–1
ncRNAs (long ncRNAs) 338–9, 341, 349
cancer prognosis 351
circular RNAs 394
control of flowering time in plants 354–5
H19 transcription 347
imprinting 345–7
nuclear organization 395
ncRNAs (non-coding RNAs) 295–305, 314–40
compartment-specific transport complexes 314–12
definition 295
epigenetic regulatory 337
evolutionary significance 292
functions 393
imprinting 346–7
nuclear export 315, 316
RAN as location guide 315
short *(table)* 358–9
transcription 347
X-inactive specific transcript 338, 341
XITE, XIST 338–42
nervous system development 138
neural splicing regulator, NOVA 1 and NOVA 79–81, 118, 134–5
neurofibromatosis 140
neurogranin mRNA 215
neurons
mRNA localization 208, 213–14
P-bodies 215
stress granules 215
survival of motor neuron (SMN) protein 215
transport granules 215
next generation sequencing (NGS) 391–2
NIH 3T3 cells
assay 155
transformation 157
NMIA (N-methylisatoic anhydride) 21
no-go mRNA decay 263–4
Nobel Prize laureates in RNA biology 2
non-coding RNAs *see* ncRNAs
non-stop decay (NSD) 263
non-Watson–Crick base pairing 16–17
hydrogen bond interactions 26
phylogenetic conservation 22
nonsense-mediated decay (NMD) 260–2
mechanism 262–4
pioneer round of translation 263
Northern blot technique 101
Northwestern blot technique 77
Notch signalling pathway, Numb 152
NOVA 1 and NOVA 2 79–81, 118, 134–5
NTF2 323
nuclear export
exportin-t and exportin-5 318
mRNA 187
ncRNAs 315
(table) 318
nuclear export adaptors 188, 189, 201
nuclear export receptors 194–6
nuclear export sequence (NES) 205, 320
nuclear import complex 323
nuclear import receptor protein (NFT2) 323
nuclear localization sequence (NLS) 205, 325–6
nuclear phosphorylation 180
nuclear pore complexes (NPCs), movement of mRNP 197–9
nuclear pore proteins 188
nuclear proteins 180
Nuclear Protein Databse 180
stochastic model of organization 180
nuclear speckles (SFCs) 178–9
nucleic acids
four important properties 15
primary structure 12–13
secondary structure 13
nucleocytoplasmic traffic of mRNA 186–200
exon junction complex (EJC) 207, 260–3
random diffusion 193
nucleocytoplasmic traffic of non-coding RNA 314–20
nucleolus
dense fibrillar components (DFC) 302
fibrillar centres (FC) 302
granular componen (GC) 302
nucleolar-organizing region (NOR) 302
proteomic analysis 303
structure and function 295, 302–4
nucleophiles
2′-OH group of RNA 44, 45–6
Group II introns 44
GTP 42
nucleoporins 188
FG-repeat 188, 199
nucleosides 4
nucleosomes 334–5
nucleotides
addition or deletion in RNA editing 268–9
backbone 8
bonds 8
changes via editing (A to I) 269
structure 6

O

OB-fold (oligosaccharide/oligonucleotide-binding fold) 67
oligo(dT) chromatography 76
oligonucleotides 8, 246
see also antisense oligonucleotides

oligosaccharide/oligonucleotide binding (OB) fold 67
oncogenes
  assay 155
  microRNAs 375
  Ras 375
  and tumour suppressor genes 155–7, 300
  see also proto-oncogenes
oocytes, masked messages 241
open reading frame (ORF) 31, 226, 268
  created from frameshifted transcripts 287–8
  frameshift mutations 226
  for GFP 195
  pan editing 287
  protein size 88
oskar 70
oskar mRNA 210
osteogenesis imperfecta type I 203

P

P element transposon 384
P elements 275
P site (peptidyl-RNA site), ribosome 231
P-bodies (processing bodies) 215, 242, 252, 286, 373
  Dcp (decapping bodies) 252
paraneoplastic disorders 134
parasitic DNA elements 98
Pasha, dsRNA-binding protein 372
passenger RNA 361
PAZ RNA-binding domain 74
pentatricopeptide repeats (PPRs) 75
peptide bond formation 224–5
perichromatin fibrils (Pfs)
  interchromatin granule clusters (ICGCs) 178
  sites of nascent transcripts 178
petunia, co-suppression 363
phage display 73
phosphodiester bonds 8, 9, 15, 53–4
  hammerhead ribozymes 39
phosphorothioates 246
'ping-pong amplification loop' 385
piRNAs 384, 396
PIWI interacting RNAs (piRNAs) 384, 396
PIWI proteins 384–5
  atomic structure 371
  piRNA-directed pathway 385
  RNA-binding domain 73–4, 361–2
PKC 259
PKR (dsRNA-activated protein kinase) pathway 69, 127, 366
plants
  co-suppression 363
  glycine-rich RNA binding protein (GRP) family 64
  microRNAs and their mRNA targets 375
  mRNA degradation 264
  mRNA localization 217–20
  regulation of flowering 353
  root hairs, mRNA localization 218
  silencing of transposons 380
  siRNA pathway 381
plasmids 367, 398
plasmodesmata 219
'poison' exons 133
polar bonds 6
poly(A) tail 170, 230
polyacrylamide gel 100, 146
polyadenylation
  alternative 172
  cytoplasmic 243
  fast 171
polycistronic mRNAs 86, 299
polycomb group (PcG) 350
polycomb repressive complex 2 (PRC2) 336, 338
polynucleotides 8
polytene chromosomes 176
post-transcriptional gene silencing (PTGS) 358, 363
post-transcriptional processes 60
pre-microRNA 372
pre-mRNA 86
  radiolabelling 100
pre-mRNA processing 166–85
  CTD 172–3
  ends of mRNA 86
  histone mRNA 3′ end formation 181–3
  RNA polymerase(s) 166–9
  spatial organization 178
  splicing and transcription 173–8
  transcription 297
pre-mRNA splicing 90–2
  by spliceosome 104–36
  defects leading to disease 138–92
  evolved from parasitic DNA elements 98
  see also alternative splicing
premature ageing 141
premature termination codons (PTCs) 118, 260
preproinsulin mRNA 259–60
PRMT proteins (protein arginine methyl transferases) 75
progerin 142
prokaryote cell
  mRNA degradation 264
  mRNA regulation 234–5
  mRNA translation 166
  organization 186–7
proline 227
protein import pathway 326
protein nucleic acid (PNA) oligonucleotide 314
protein phosphorylation 126–7, 127, 131
protein–RNA interactions 76–81
  CLIP assay 78–9
  MS2:MBP method 78
  yeast three-hybrid approach 78
protein-coding genes
  and RNA polymerase II 167
  TATA box 167
proteome 3, 111
proto-oncogenes
  c-fos 65, 69
  c-myc 215, 241, 250, 256, 304
  defined 157
  mRNA localization 215
  RON 157
protospacer adjacent motif (PAM) 397
provirus 328
PRP28 99
PRP8 149
pseudogenes 393
pseudoknot 18–19
  viral translation 228
PTEN gene, acts as microRNA sponge 394
PTENP 393
PUF (Pumilio and FBF homology) domains 75
purines 7
pyrimidines 7
pyrosequencing 391
pyruvate kinase M 153

Q

quasi-RRMSs (qRRMs) 65

R

R-looping 85
radiolabelling, pre-mRNA 100
RAN system 315–16
RAN–GTPase activating protein (RAN-GAP) 316
RANBP1 and RANBP2 316
RAS-related nuclear protein *see* RAN
rasiRNAs 378, 396
RCC1 316
  guanine nucleotide exchange factor (GEF) 314
read mapping uncertainty 392
recruitment model 130

REF export adaptor 188
REF (RNA export factor) 189
REF–GFP 201
release factors (eRFs in eukaryotes) 232
release factors (RFs in prokaryotes) 232
retinitis pigmentosa (RP) 148–51
retinoic acid-binding protein CRABPI 216
retrotransposons 379
retroviruses 328–30
  ApoBec proteins 286
  life cycle 285, 329
Rev protein 328–30
Rev-Response Element (RRE) 75
reverse transcription-polymerase chain reaction (RT-PCR)29 55, 113–14, 146, 162, 175, 268, 342
'reversible terminator' bases 391
RGG boxes 75
ribonucleases *see* RNAses
ribonucleolytic enzymes *see* ribozymes
ribonucleoprotein (RNP) 60
  see also mRNPs
riboprobes 76
ribose 4, 6
ribose zippers 26
ribosomal proteins 222
  RPL32 122
ribosomal RNA *see* rRNA
ribosome
  A site (acceptor site) 231
  decoding centre 223
  E site (exit site) 231
  formation, eukaryotes 296
  internal ribosome entry site (IRES) 230
  P site (peptidyl-RNA site) 231
  peptidyl transferase centre 223
  polyribosome (aka polysome) 229
  structure 222–4
  subunit protein in Diamond–Blackfan anaemia 297
ribosome binding site (RBS) 28, 32
riboswitches 31–3, 235–6
  binding to coenzyme B12 31
  similarity to allosteric enzymes 31–2
  structure 236
ribozymes 36–63, 96
  deprotonation 46
  Group I/II 39
  hammerhead 39, 48–50
  HDV 50–1
  naturally occurring (table) 40
  ribonucleolytic 48
  small/large *(table)* 40
  summary 57
  synthetic 193
  true catalysts? *(table)* 52
  in viruses 40
RISC complexes 364
RITS complex 359
  rasiRNAs (RITS-complex-associated siRNA) 378, 396
RNA
  7-methyl guanosine cap 169
  bases accept/release protons 37
  bulges 18
  chaperones 67, 234, 242
  export adaptors 189
  functional classes *table* 40
  functional repertoires 12
  histones 242
  nuclear export sequence (NES) 320
  nucleocytoplasmic traffic 186–200
  polycistronic 371
  snoRNAs 295, 298–9
  snRNAs 323
  surveillance 297
  terminology 3
  vs DNA *(tables)* 5, 28
  World hypothesis 52–3
  zipcodes 206
  see also RNA structures
RNA editing 70, 268–95
  A to I RNA editing 270–4
  adaptive immunity 286
  ADAR proteins 70, 270–4, 278
  APOB mRNA 282
  ApoBec1 284–6
  biological functions 274–5
  C to U RNA editing 282–90
  innate immunity 284
  nucleotide addition or deletion 268–9
  pan editing 270–4, 287
  through base modification 268
  uridine insertion and deletion 287, 291
RNA endonuclease 292
RNA export complexes, nuclear shuttle 321
RNA export factor (REF) 189
RNA export receptor 316
RNA helicases 67, 74, 99
  DEAD box 99
  DPB5 199
RNA helices
  environment 23
  positively charged molecules 23
RNA interference (RNAi) 195, 363–7
  co-suppression 363
  discovery 364
  limitations 366, 368
  mechanism 364–6
  off-target effects 368
  practical applications 368–70
  short hairpin RNA (shRNA) 367
  uses 366
  vs CRISPR 397
  see also siRNAs
RNA ligases
  and RNA polymerases, catalysis 53–4
  SELEX 54–6
RNA polymerase I, transcription of pre-mRNA 166–70
RNA polymerase II
  C-terminal domain 172–3
  CTD phosphorylation 167, 172–3
  elongation times 129
  mRNA processing 170–2, 186–200
  phosphorylation 169
  splicing speed 128
  transcription of protein-coding genes, miRNAs and snRNAs 167–8
RNA polymerase III, transcription of tRNA 167
RNA polymerase(s) 167–70
  experimental formation 54–5
  phosphodiester bonds 53–4
  and RNA ligases 53–4
  summary *(table)* 168
RNA recognition motif (RRM) 62–5, 257
  amino acid consensus sequences 62
  quasi-RRM 64–5
  RRM-containing proteins 64–6
RNA species, principal characteristics 386
RNA splicing defects, and disease 138–92
RNA structures 4–33
  aptamers 31, 78
  binding to target molecules 28–9
  coaxially stacked helices 24
  complex folded 30–3
  double helix (A form) 16
  folding patterns 21
  hydrogen bonding 24
  loops 18
  motifs 17
  nucleotides, structure 5–6
  secondary 11, 22–5, 28, 40, 42, 67, 74
  secondary structure motifs 18–19
  stabilized by metal ions 26
  tertiary 11, 24–5
  tertiary *(table)* 24, 25
  thermosensors 27
  tRNA 305

RNA-binding assays 77–8
RNA-binding domains (RBDs) 60–6
arginine-rich domains 74
cold shock domains 66
dsRBDs 69, 70
homeodomains 75
KH domains 61, 65
known domains *(table)* 61
PAZ 73, 74
pentatricopeptide repeats 74
PIWI 73–4
PUF domains 75
RNA helicases 67, 74, 99, 199
zinc-finger domains 71–3
RNA-binding proteins (RBPs) 60–84, 123
argonaute family 74
coldshock domains 69
RPL30 121
RRM-containing proteins 64–6
trans-acting factors 259
RNA-chIP experiments (immunoprecipitations) 391
RNA-containing macromolecules, roles in RNA biogenesis 295–6
RNA-dependent RNA polymerase (RDP) enzyme 377
RNA-guided genome editing 397–9
RNA-induced silencing complex (RISC) 364, 366
RNA-induced transcriptional silencing (RITS) 376
rasiRNAs (RITS-complex-associated siRNA) 378, 396
transposons 379
RNA-Seq (RNA sequence analysis) 390, 392–5
next generation sequencing (NGS) 391–2
RNAse A, acid–base catalysis 37–8
RNAse III family of ribonucleases 70
RNAse P 306
active site, magnesium ions 306
operates as a ribozyme 307
shape, bacterial species 22
structure 5, 20
true catalyst 52
RNAseq 113, 115
RNA–PIWI protein complexes 385
RNA–RNA base pairing, spliceosome 93
RNP complexes 359
rolling circle replication 48
RON, autophosphorylation 157
RON receptor tyrosine kinase 157
root hairs, mRNA localization 218
RPL30 121
RRM *see* RNA recognition motif (RRM)
rRNA 222
non-coding RNA in biogenesis 295
principal modifications 298
synthesis steps, pre-rRNA processing 297
transcription by RNA polymerase I 166–70
RT–PCR 29, 55, 113–14, 146, 162, 175, 268, 342
mRNA isoforms 113

S

S-adenosylmethionine (SAM) riboswitch 33
Saccharomyces cerevisiae
commitment complex 176
genome 391–2
intron-containing RNAs 86
Mer1, Mer2 133
Mer1-regulated alternative splicing 124
mRNA localization 210
saRNAs (small activating RNAs) 373
scan RNAs (scRNAs) 381–3
scaRNAs (small Cajal RNAs) 295–6, 300, 324
role in RNA biogenesis 296, 299
scaRNPs (small Cajal RNPs) 324
Schizosaccharomyces pombe, dsRNA 380
SDS–PAGE 77
SECIS RNA 228
Seckel syndrome 146
selenoproteins 228
SELEX experiments 30–3
bioinformatics 54–6
RNA aptamers 30–1, 78
RNA polymerase(s) 54–7
SEPN1 228
serine/threonine kinases (PKRs) 69, 127, 236, 366
sex chromosomes
dosage compensation 338
inactivation of X chromosome 338
see also X chromosomes
SFRS10, Tra2β 312
SHAPE (selective 2′-hydroxyl acylation analysed by primer extension) 21
Shine–Dalgarno sequence 12, 230, 309
short hairpin RNA (shRNA) 367
short ncRNAs 358–98
short siRNAs 60, 358
signal recognition particle (SRP) 205, 217, 233
in translation 233
signal transduction pathways 65
regulation of alternative splicing 126
silencers *see* splicing silencers
silencing of transposons
animals 384
plants 381
Tetrahymena thermophila 381–3
siRNA pathway, eukaryotes 380, 381
siRNAs 358–60, 364, 365
delivery systems 370
design 367
directed against WT1, blocking nephrogenesis 369
popularity 367
slicer activity 361
Sm proteins 94
small Cajal body-associated RNAs *see* scaRNAs
small interfering RNAs *see* siRNAs
small ncRNAs *(table)* 359
small nuclear mRNAs *see* snRNAs
small nuclear ribonucleoproteins *see* snRNPs
small nucleolar mRNAs *see* snoRNAs
SMN complex 309
SMN proteins 214
and snRNP assembly 309
SMN1, SMN2 309, 310–11
activator model 312
and mRNA localization 214
silencer model 312
snoRNAs (small nucleolar RNAs) 295, 301
age/origin 300
Box C/D (SNORDs) 298
Box H/ACA snoRNAs 299
families 298
orphan snoRNAs 299
role in RNA biogenesis 296
snoRNPs 297
snRNAs (small nuclear RNAs) 318
catalytically active spliceosome 97
consensus sequence 31
non-coding RNA involved in its biogenesis 295
nuclear export receptor CRM1 320
nuclear transport 315–16
secondary stem and loop structures 93
transcription by RNA polymerase II 167
snRNPs 94–6, 314, 323
cargo snRNP 323

reimport to nucleus 326
Sm proteins 94
U2, U5, U6 95
U4/U6 324
snurportin (SNUPN) 326
spinal muscular atrophy (SMA) 151, 214
SMN proteins 309
SMN1/SMN2 and mRNA localization 214
splice factor SRSF1 163, 239
splice site commitment 176
splice sites 40, 90–7
3′ and 5′ 95
5′ and 3′ splice site sequences 90–2
alternative splicing 153
branchpoint sequence 92
commitment 176
conserved 91–2
degenerate splice site sequences 92
lariat intermediate 91
pictogram 92
polypyrimidine tract 92
tandem alternative (TASS) 392
spliced leader (SL) RNA 106–7
spliceosomal assembly and reaction intermediates 100
spliceosomal introns 85–9, 147
alternative splicing 111–12
spliceosomal proteins, retinitis pigmentosa (RP) 148–51
spliceosomal splicing 84–100
spliceosomal U snRNAs 326
spliceosome 86, 93–100
mechanism 96–8
mutations 147
spliceosome assembly 94–9, 176
A complex 96
B complex 96
C complex 96
DEAD box RNA helicases 99
E complex 95
exon definition 103–5
intron definition 104
spliceosome cycle 94–7
stepwise addition of proteins 98–9
U1–U6 93–8
spliceosome complexes
complex H 101
U2 and U12-dependent assembly pathways 103
visualized on native gels 101
spliceosome cycle 94–7
in vitro extracts 100
spliceosomes
major/minor coexist in eukaryotes 102
U2 and U12-dependent assembly pathways 103
spliceostatin 158, 159
splicing 40
alternative polyadenylation 172
and chromatin 173–8
cis-splicing 84
co-transcriptional splicing 175
coupling with transcription 173
evolution from parasitic DNA elements 98
link with transcription 90–7
models 174
mutations leading to disease 138–92
nuclear export of mRNA 191–3
splicing activator proteins (splicing enhancers) 116
splicing co-regulators
Caper 130
recruitment model 130
splicing code 105, 117–18
binding sites for splicing activator proteins (splicing enhancers) 116
discovery 117
enhancers and silencers 140, 303
implications 138
mutations 138–42, 163–4
predicted online 118–19
splicing defects, and disease 138–65
splicing enhancers 116, 117, 303
splicing factor compartments (SFCs) 178–80
splicing factors 180
ASF/SF2 (SRSF1) 156, 163
CHECK PAGE enrichment, subnuclear sites *(table)* 179
enrichment in cell nucleus 179
levels in cancer cells 154
oncogenicity 155
splicing proteins, U2AF65 64, 93, 95, 96, 130
splicing repressors 118–19, 135
hnRNPs 64, 118, 126
PTB 61, 64
R-looping 84–5
see also hnRNPs
splicing silencers 116–17, 140
ESSs and ISSs 117
PTGS and TGS 358, 363
splicing therapy 158
SR proteins 64, 117, 118, 188–90
C. elegans 368
as RNA export adaptors 190
splice factor SRSF1 163, 239
see also ASF/SF2; SRSF1
SROOGLE 120
SRSF1 155
HIV 163
SRSF1 *formerly known as* ASF/SF2 368
ssRNA regions 18
SSU processome 297
Staufen 61, 70, 208, 211
stemloop binding protein (SLBP) 181–3
steroid hormones, Caper proteins 130
steroid receptor superfamily 71
stop codons 232
premature termination codons (PTCs) 118, 260
stress granules (SGs) 242, 252–3
translation initiation factor eIF2α 253
sugars 4–6
surveillance, transcriptional 172
survival of motor neuron *see* SMN protein
Svedberg unit (S) 60
'symbol (prime) 5
symplekin 170
Systematic Evolution of Ligands by EXponential enrichment *see* SELEX

T

tandem alternative splice sites (TASS) 392
TAP tip-associated protein 194–6
targeted oligonucleotide enhancers of splicing (TOES) 313
TATA box 167
telomerase 300–1
RNA component (hTR) 295
telomeres 300
terminal uridylyl transferase (TUTase) 292
terminator bases 391
tertiary structures of RNA 11, 24
Tetrahymena thermophila 381–3
macronucleus formation 387
ribozyme 39–40
scRNAs 381–3
tetraloop-tetraloop receptors 25, 26
TFIIH, transcription initiation complex 173
TFIIIA, zinc-finger domain 72
Thermatoga
L-box 235
lysine riboswitch 235–6
thermosensors, RNA structures 27–8

THO proteins 189
thymine, alternative to uracil 8
TIAR 259, 260
tobacco ringspot virus 365
Tra2β, SFRS10 312
TRAMP, and RNA surveillance 306
trans-acting factors, RNA-binding proteins 259
trans-encoded small RNA (tracrRNA) 397
trans-splicing 106–8
transcription 166–9
  core transcription factors 169
  coupling with splicing and chromatin 173–8
  introns, time lag 129
  kinetic coupling model 127
  link with splicing 173–8
  recruitment model 130
  regulation 169
transcription factors
  Early Growth Response family 72
  TFIIIA 72
transcription initiation complex, TFIIH 173
transcription and RNA export (TREX complex) 188–90
transcriptional activation 195, 244
transcriptional elongation 88, 108
  speeds 129
transcriptional enhancers 88
  long ncRNAs 351–2
transcriptional gene silencing (TGS) 358, 376
transcriptome 111
transcriptomics 115
  data 390
transfer RNA *see* tRNA
transferrin receptor (TfR) 237
transformation, NIH 3T3 cells 157
transgene-induced silencing (TGS) 363
translation initiation factor eIF2α, phosphorylation 253
translation of mRNA *see* mRNA translation
translational attenuation 234–5
translocation, defined 231–2
translocon 233
transport granules 215
transposon repressed RNAs, cleavage 385
transposons/transposable elements 379–86
  silencing in plants 380
  see also P elements
TREX complex 188–90
tristetraprolin (TTP) 259
tRNA 222, 228–30
  A to I RNA editing 279–90
  amino-acyl tRNA synthetases 228
  anticodon loop 228
  anticodon wobble 229
  nuclear export receptor 321
  pre-tRNA processing 304–5
  RNA surveillance 305
  structure 229
  transcription by RNA polymerase III 167
tRNA precursors (pre-tRNA) 304–5
trypanosome mitochondrial RNA editing 287–91
  COXII transcript 289
  creating ORFs from frameshifted transcripts 106, 287–91
  discovery 289
  nuclear-encoded proteins as therapeutic targets 291–2
trypanosomes
  genome 288
  kinetoplast 287
  RNA-binding proteins 61
  spliced leader (SL) RNA 106
  trans-splicing 106–8
TSIX 341
  X chromosome silencing 352
tubulin binding domain 70
tumour suppressor genes 155–7
  PTEN 393–4
tunnel vision 149
tyrosine kinases, RON receptor 157

U

U1 snRNP 96, 161–2, 173, 323
U12-dependent introns 101
  pictograms 102
U2 snRNA 189, 300
U2AF65 64, 93, 95, 96, 130
U3 snoRNA 296–7, 302
U4 snRNP 96, 323
U5 snRNP 97, 98, 102–3, 108, 327
U6 snRNP 304, 327
UAP56 189, 192, 262
UBF (upstream binding factor), transcription by RNA polymerase I 167
Ubx, active and inactive 349
UCSC genome browser 113, 390
UHG gene, snoRNAs 88–9
uridine insertion and deletion 287, 291
UTRs (untranslated regions) 31, 259
UV crosslinking assay 77

V

vascular plants *see* plants
'Venice model', mRNA transport 208–9
vernalization, *Arabidopsis* 353
vimentin 216
vitamin B12, and riboswitches 31

W

WAGR syndrome 368
Warburg effect 153
Watson–Crick base pairing 15, 16, 380
Western blot technique 77, 101
Wilms tumour suppressor gene, WT1 73, 155, 368–9
wobble base 270
World hypothesis of RNA 52–3
WT1, nephrogenesis 73, 155, 368–9

X

X chromosome
  inactivation 338–40
  placental and marsupial mammals 342
X-inactivation centre (Xic) 338–40
  ncRNAs and XIST 341
X-inactive specific transcript *see XIST*
Xenopus
  masked messages 241
  morpholinos 246
  mRNA localization 207, 211
XIST 338–42
  coating inactive X chromosome 395
  long ncRNA, Repeat A 338, 340–1
  pseudogene 393
XITE (X inactivation intergenic transcription element) 342

Y

Y-box proteins 67–8, 75
  conserved CSD 234
  function 242
yeast *see Saccharomyces cerevisiae*
yeast three-hybrid approach 78
YSPTSPS heptad repeat 172

Z

zebrafish, CRISPR, modelling albino locus repair 399
zinc-finger motifs 71–3
  artificial RNA-binding 73
ZRANB2 family of proteins 73

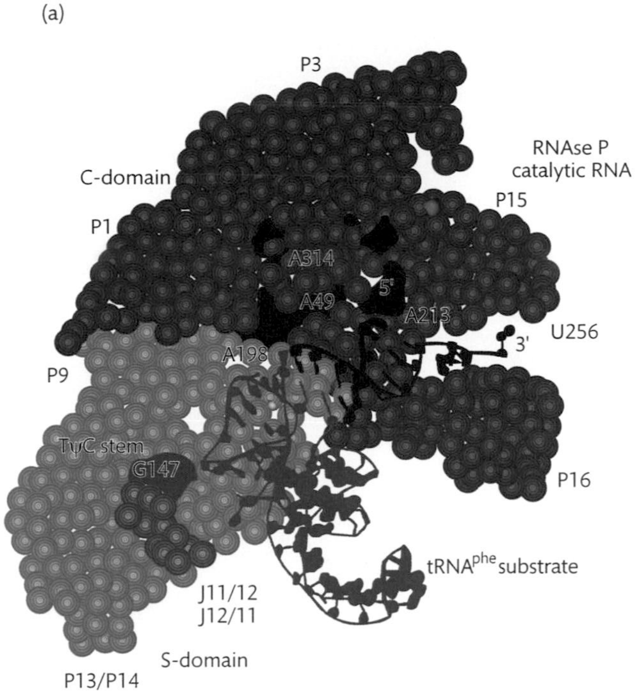

RNAse P attached to substrate tRNA

Figure 1.2(a) (CI 1)

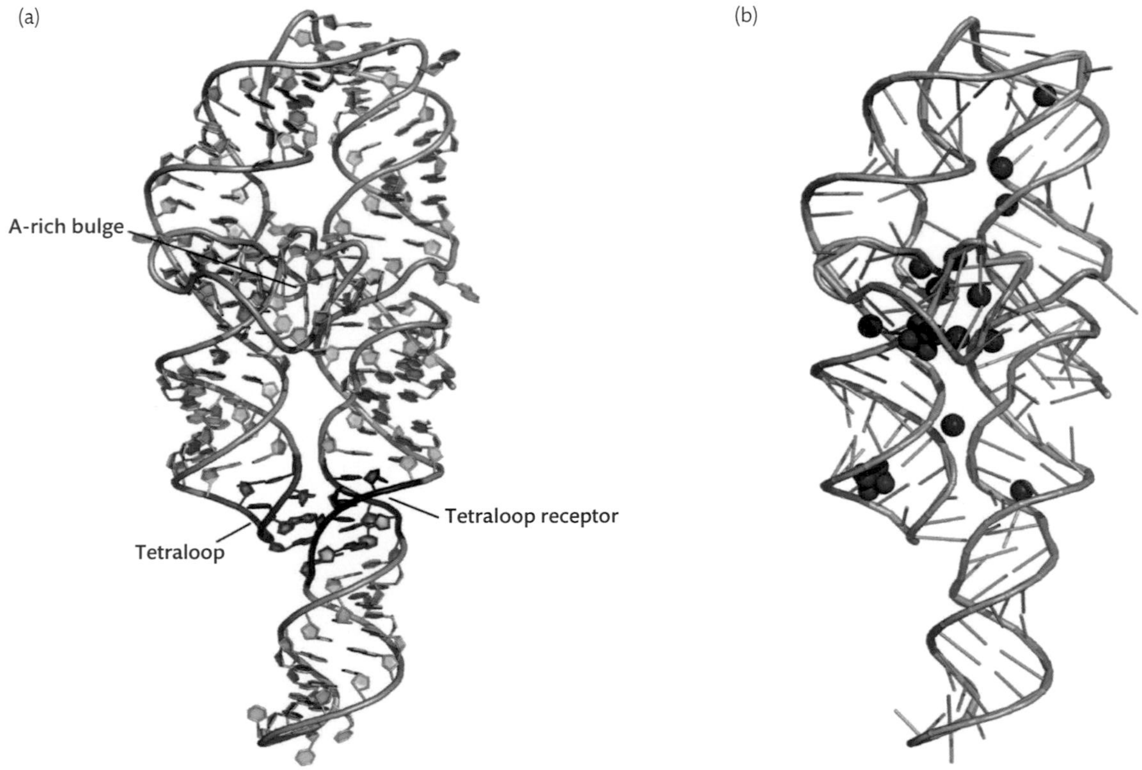

Figure 2.14 (a) and (b) (CI 2)

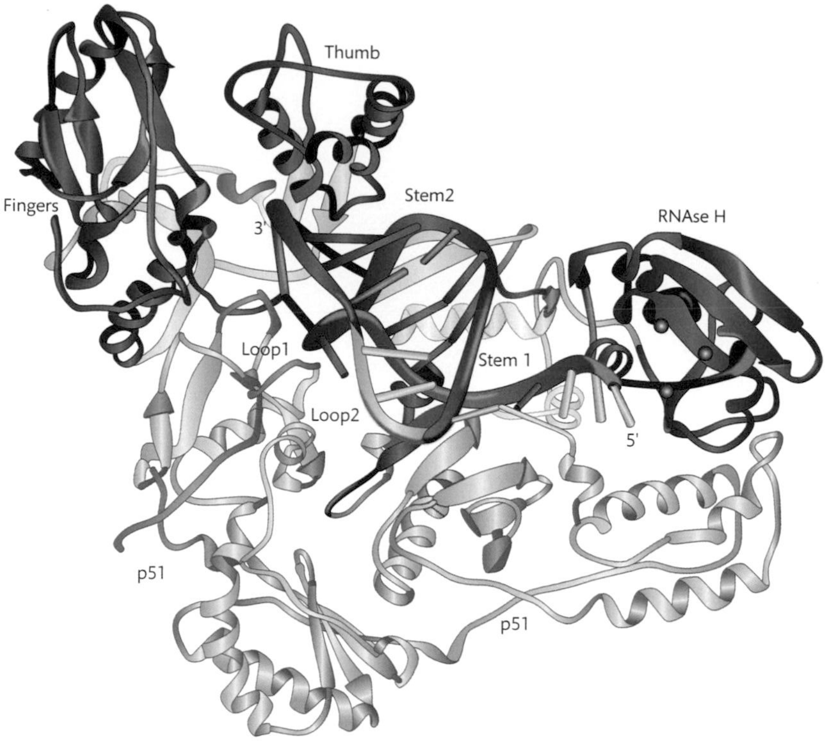

Figure 2.20 (CI 3)

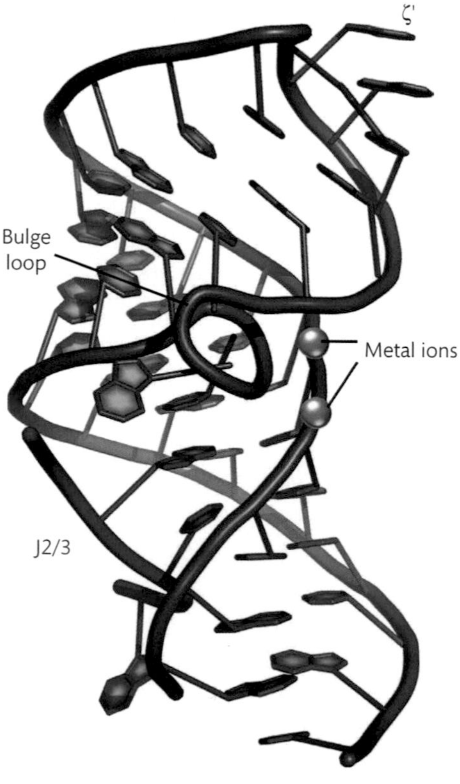

Figure 3.9 (CI 4)

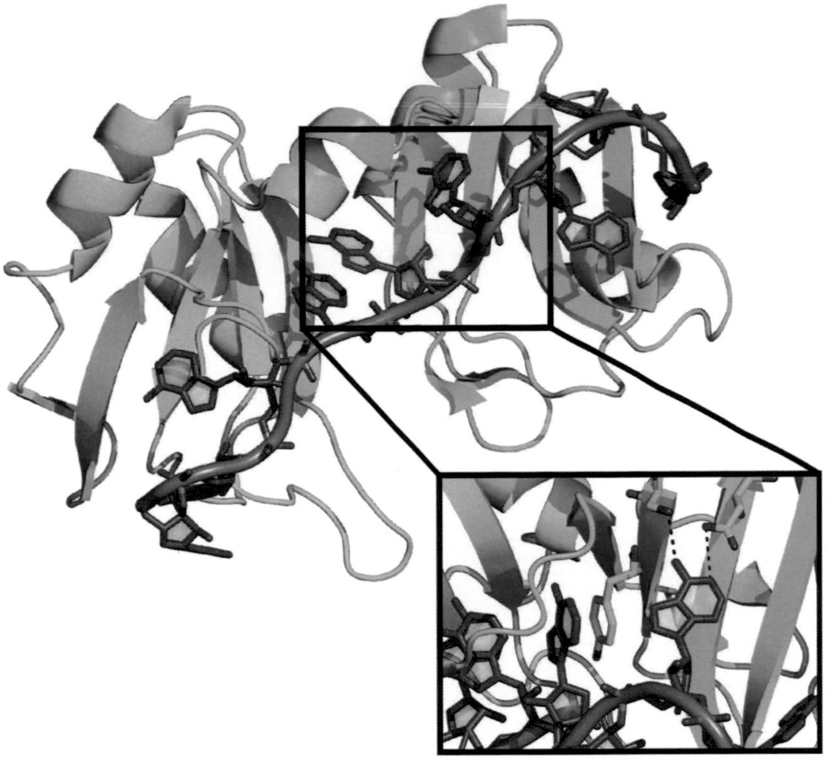

Figure 4.1 (CI 5)

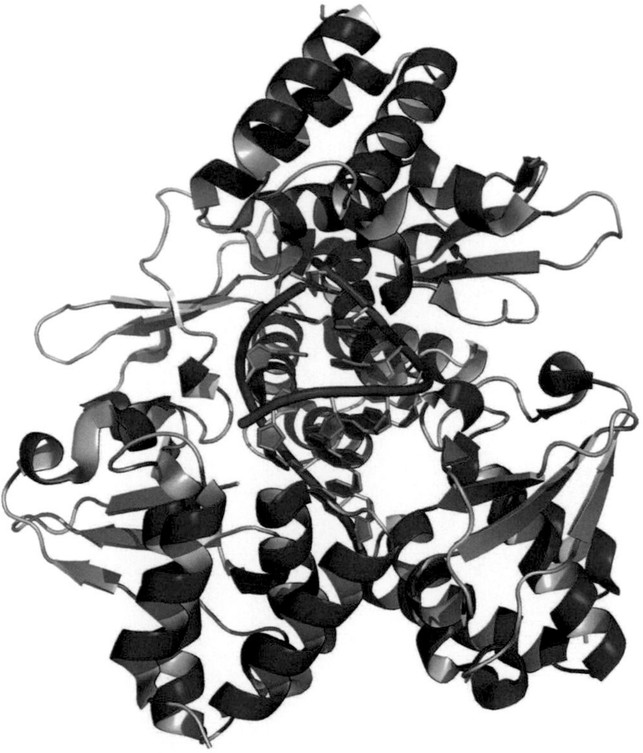

Figure 4.7 (CI 6)

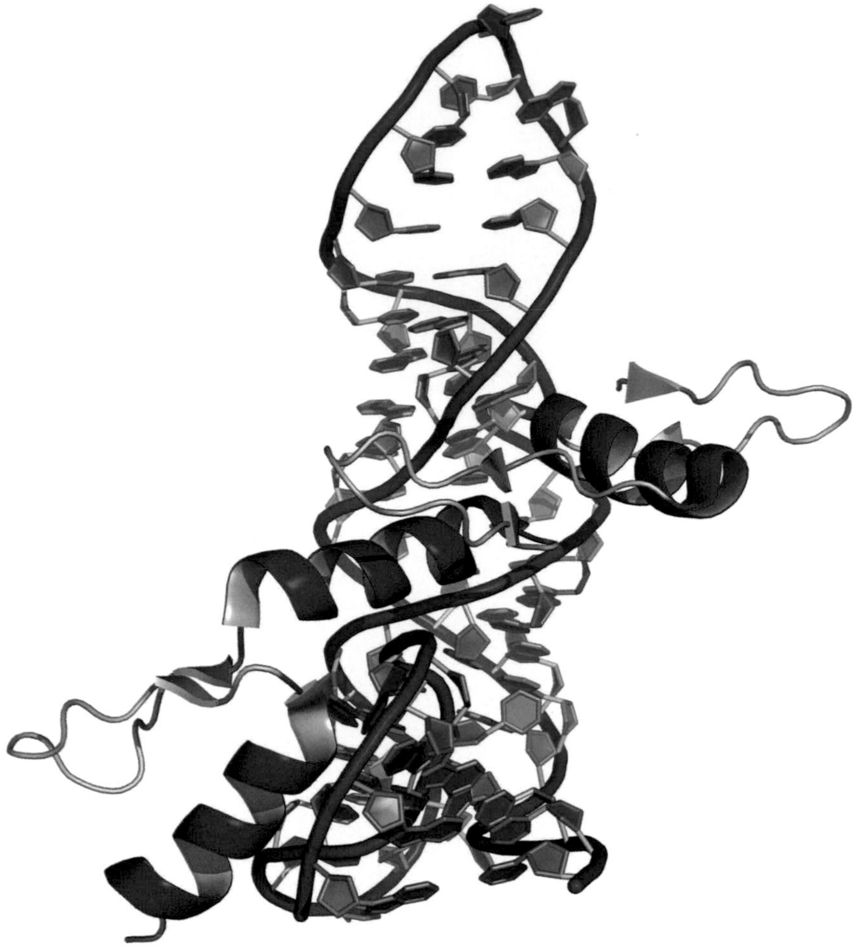

Figure 4.8 (CI 7)

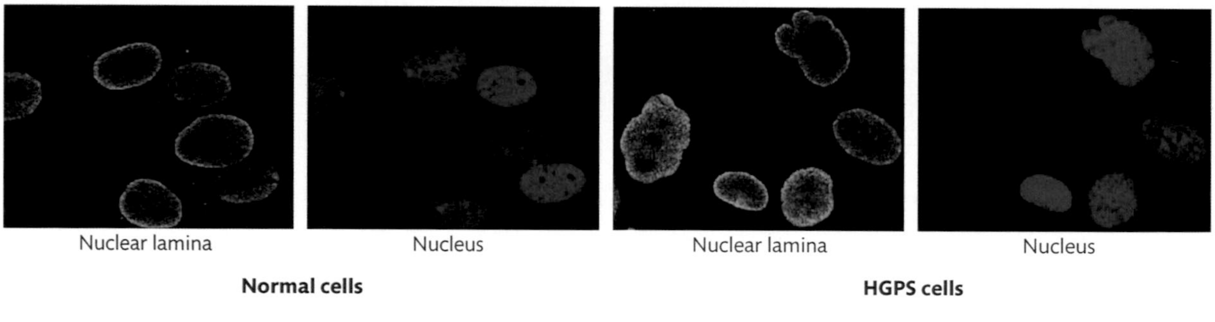

Figure 7.5(b) (CI 8)

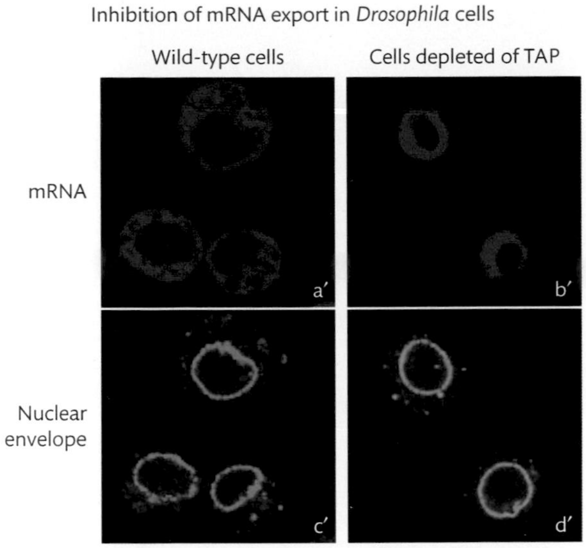

Figure 9.12 (CI 9)

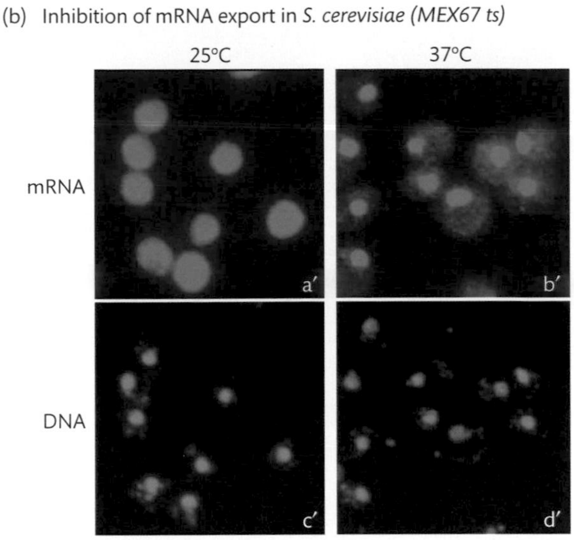

Figure 9.13b (CI 10)

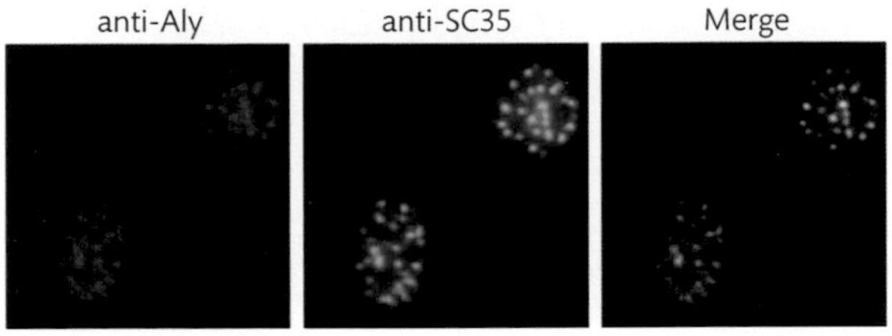

Figure 9.18 (CI 11)

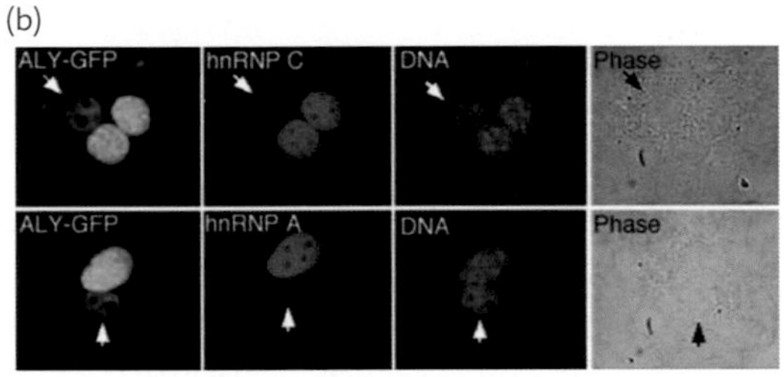

Figure 9.19b (CI 12)

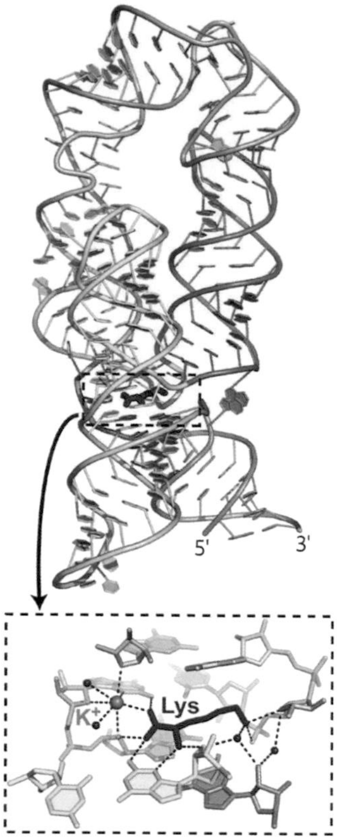

Figure 11.11 (CI 13)

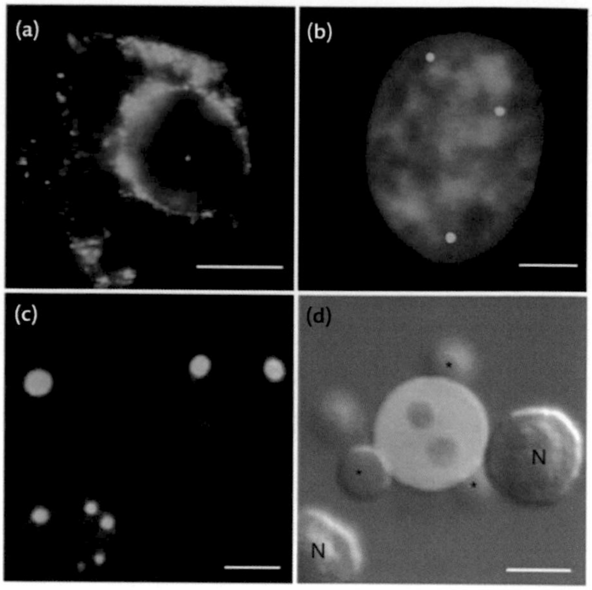

Figure 14.25 (CI 14)

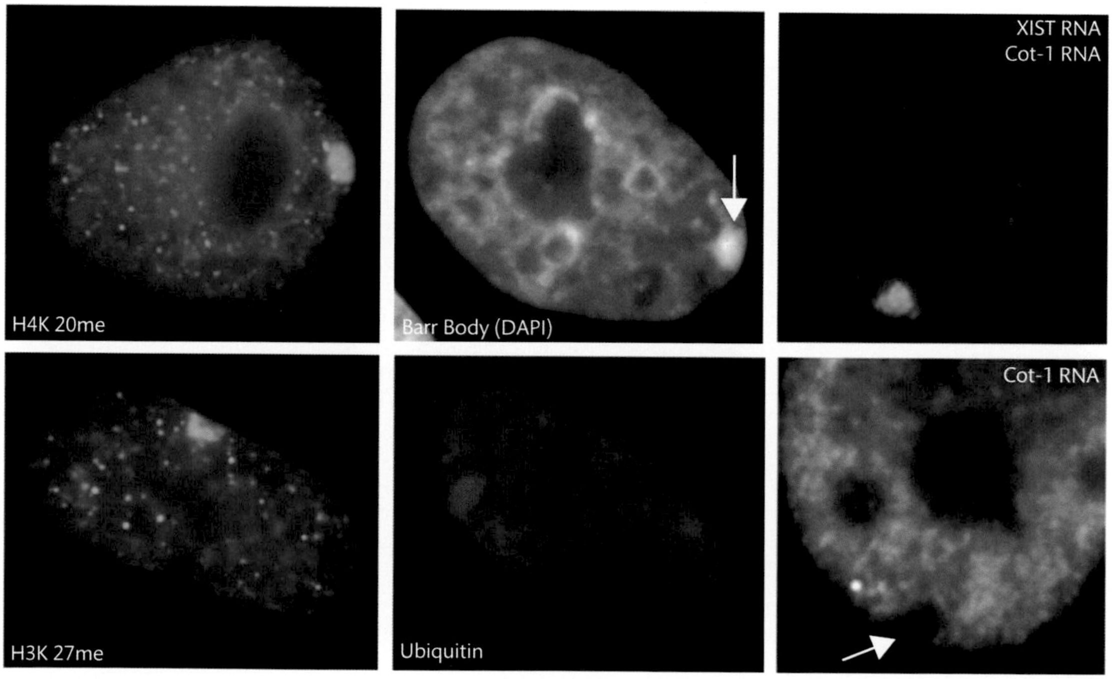

Figure 15.7 (CI 15)

Figure 15.11 (CI 16)

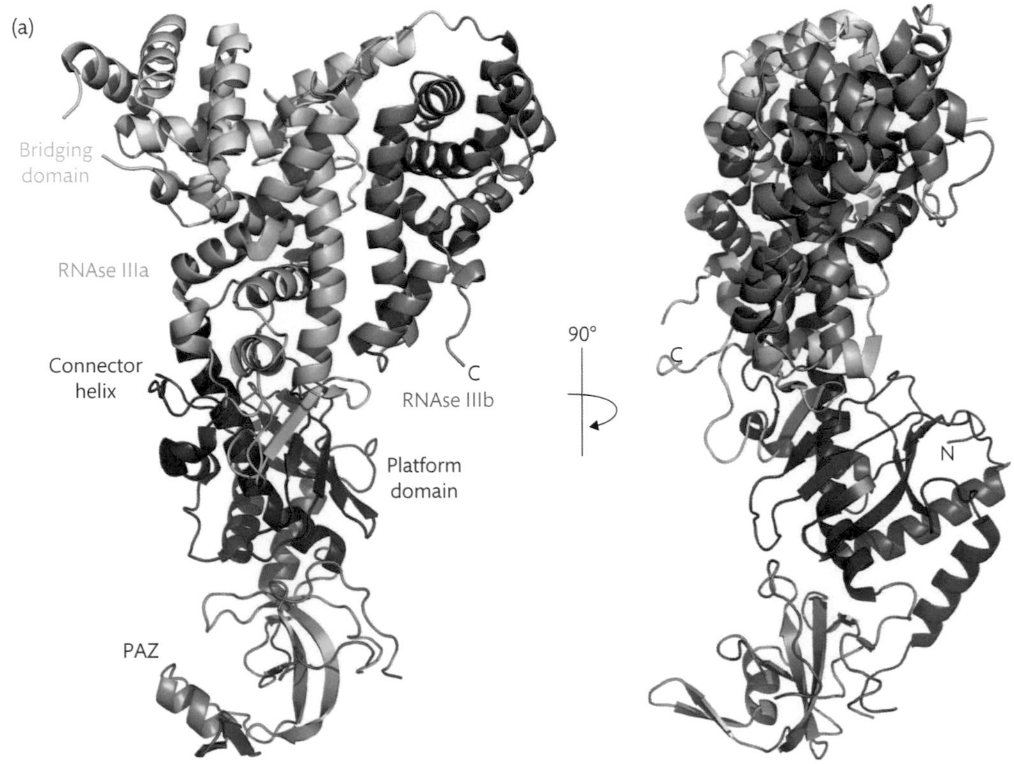

Figure 16.3a (CI 17)

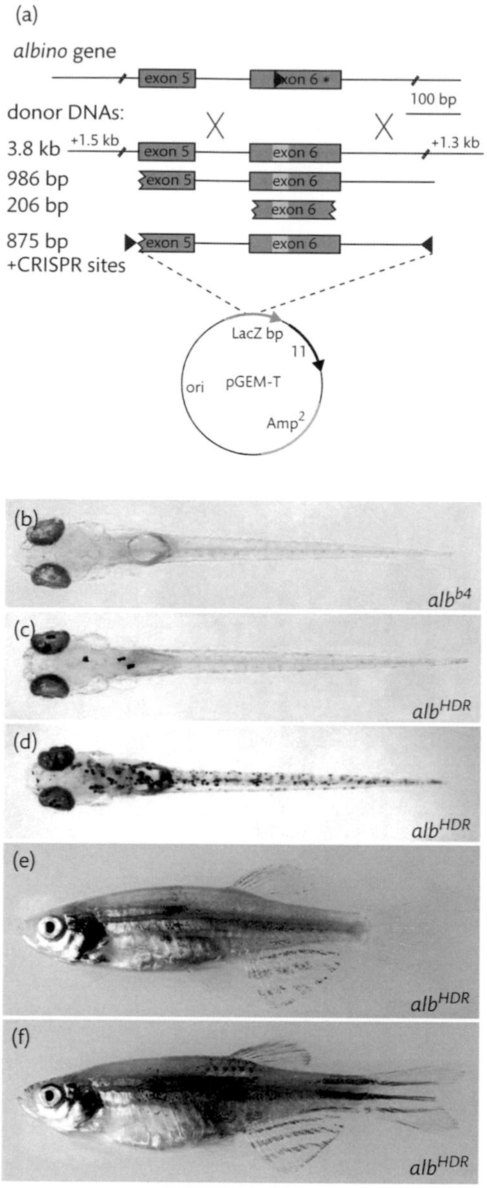

Figure 17.8 (CI 18)